Reliability of RoHS-Compliant 2D and 3D IC Interconnects

Reliability of RoHS-Compliant 2D and 3D IC Interconnects

John H. Lau, Ph.D.

McGraw Hill

New York Chicago San Francisco
Lisbon London Madrid Mexico City
Milan New Delhi San Juan
Seoul Singapore Sydney Toronto

The McGraw-Hill Companies

Cataloging-in-Publication Data is on file with the Library of Congress

1 2 3 4 5 6 7 8 9 0 CTP/CTP 1 6 5 4 3 2 1 0

ISBN 978-0-07-175379-1
MHID 0-07-175379-6

Sponsoring Editor
Michael Penn

Editorial Supervisor
Stephen M. Smith

Production Supervisor
Richard C. Ruzycka

Acquisitions Coordinator
Michael Mulcahy

Project Manager
Sapna Rastogi, Glyph International

Copy Editor
Lisa McCoy

Proofreader
Shivani Arora, Glyph International

Indexer
Robert Swanson

Art Director, Cover
Jeff Weeks

Composition
Glyph International

This book is printed on acid-free paper.

About the Author

John H. Lau, Ph.D., has been an ITRI Fellow of the Industrial Technology Research Institute (ITRI in Taiwan) since January 2010. Prior to that, he was a visiting professor at HKUST (Hong Kong University of Science & Technology) for 1 year; the Director of the Microsystems, Modules, and Components (MMC) Laboratory at the Institute of Microelectronics (IME in Singapore) for 2 years; and a Senior Scientist/MTS at HP/Agilent in California for more than 25 years.

Dr. Lau's professional competences are design, analysis, materials, process, manufacturing, qualification, reliability, testing, and thermal management of electronic, optoelectronic, LED, and MEMS components and systems (which include devices, substrates, packages, and PCBs), with emphases on RoHS-compliant products, SMT, flip-chip WLP, TSV, and other enabling technologies for 3-D IC integration and SiP.

With more than 35 years of R&D and manufacturing experience, he has authored or co-authored more than 300 peer-reviewed technical publications, more than 20 issued and pending U.S. patents, and more than 100 book chapters, and has given more than 260 lectures/workshops/keynotes worldwide. He has authored or co-authored 16 textbooks on 3-D MEMS packaging, reliability of 3-D IC integrations, advanced packaging, BGA, CSP, TAB, flip-chip WLP, high-density interconnects, COB, DCA, SMT, and lead-free materials, soldering, manufacturing, and reliability.

Dr. Lau earned a Ph.D. degree in theoretical and applied mechanics from the University of Illinois at Urbana-Champaign, an MASc degree in structural engineering from the University of British Columbia, a second M.S. degree in engineering physics from the University of Wisconsin at Madison, and a third M.S. degree in management science from Fairleigh Dickinson University. He also has a BE in civil engineering from National Taiwan University.

He was a member of the Editorial Boards of *ASME Transactions; Journal of Electronic Packaging; IEEE Transactions on Component, Packaging, and Manufacturing Technology; Circuit World; Soldering and Surface Mount Technology;* and others. He was the General Chair and Program Chair of the IEEE ECTC (Electronic Components and Technology Conference) from 1990 to 1995 and the IEMTS (International Electronic Manufacturing Technology Symposium) from 1987 to 1992, Solder Mechanics Symposium Organizer of the ASME Winter Annual Meeting (1987–2002), and 3-D IC Integration Symposium Organizer of the ASME IMECE 2010. Also, he was the Publications Chair of the IEEE ECTC from 1995 to 2006. He served on the Board of Governors of IEEE/CPMT and has been one of its distinguished lecturers every year for the past 10 years.

Dr. Lau has received many awards from ASME, IEEE, SME, and other societies. He is an elected ASME Fellow and has been an IEEE Fellow since 1994.

Contents

Foreword

What good is a new design, material, process, or technology if a product made by any one or combination of them is not reliable? Thus, reliability assessments of a product's new design, material, process, and/or technology is very important in deciding when to ship the product and how to define its warranty.

The major trend in the electronics and optoelectronics industry today is to make products more personal by making them smarter, lighter, smaller, thinner, shorter, and faster, while at the same time making them more user-friendly, functional, powerful, reliable, robust, innovative, and creative, and less expensive. As the trend toward miniature and compact products continues, the introduction of cool products that are more user-friendly and contain a wider variety of functions and use less power will stimulate growth in the market. One of the key technologies that is expected to help to make these novel product design goals possible is three-dimensional (3-D) integration. For these integrations, the solder bump is the electrical and mechanical "glue" of 3-D integrated circuit (IC) packaging interconnects (e.g., the ordinary solder bumps) and of 3-D IC integration interconnects (e.g., the micro-solder bumps). Thus, the reliability of solder interconnects is extremely important and absolutely necessary for the development of 3-D IC packaging and integration technologies.

RoHS (restriction of the use of certain hazardous substances) has been a law in many countries. Currently, it bans lead (Pb), mercury (Hg), cadmium (Cd), hexavalent chromium (Cr^{6+}), two types of brominated flame retardants, polybrominated biphenyls (PBBs), and polybrominated diphenyl ethers (PBDEs). Since 2000, the electronics industry has spent billions of dollars on compliance with the law. More than 95% of the effort has been directed at eliminating Pb, for example, creating lead-free materials, developing soldering processes, performing characterization measurements and reliability assessments, etc. In this book, only Pb-free solders for electronics and optoelectronics products are considered.

Unfortunately, for most of the practicing engineers and managers, as well as scientists and researchers, lead-free solder interconnect reliability (compounded with the latest 3-D integration technologies) in electronics and optoelectronics packaging and interconnect systems is not well understood. Thus, there is an urgent need, both in industry and research institutes, to create a comprehensive book on the current state of knowledge of design for reliability, reliability testing, and failure analysis of lead-free solder interconnects. This book is written so that readers can quickly learn about the basics of problem-solving methods and understand the trade-offs inherent in making system-level decisions.

To meet these needs, John H. Lau at Electronics and Optoelectronics Laboratories has collected a great deal of useful information from the latest technical publications and produced *Reliability of RoHS-Compliant 2D and 3D IC Interconnects*. This is an excellent book for industry and research institutes as well as universities. It is equally appropriate as an introduction to lead-free solder interconnect reliability for those just entering the field and as an up-to-date reference for those already engaged in interconnect design and process developments for 3-D IC packaging and 3-D IC integration.

This book has 16 chapters that cover the whole spectrum of lead-free solder interconnect reliability, from the fundamentals to the latest developments in this field. Chapter 1 briefly discusses the RoHS-compliant IC semiconductor and packaging technologies. Chapters 2 and 3 provide fundamental knowledge and necessary tools on reliability engineering. Chapters 4 and 5 present the lead-free interconnect reliability of two conventional electronics packages. The lead-free interconnect reliability of two conventional optoelectronic packages is presented in Chaps. 6 and 7. Chapters 8 and 9 provide the interconnect reliability of two new lead-free solders that are of great interest to the industry. Chapters 10 through 13 present the latest results on the reliability of 3-D IC integration, which includes chip-to-wafer bonding, wafer-to-wafer bonding, through-silicon-via (TSV), active and passive interposers, thermal management, and electromigration of microbumps. The last three chapters provide the effects of some important factors (e.g., dwell time and ramp time, high-speed strain rate, and void) on lead-free solder interconnect reliability.

Will lead-free solders provide the characteristics necessary to allow the world to depend on them in the future? This book may not be able to answer this question; however, it will help all participants in the electronics and optoelectronics assembly world better understand what needs to be done in order to answer this question, plan for the future, and make it happen.

Dr. Ian Yi-Jen Chan, Vice President
Director of Electronics and Optoelectronics Laboratories
Industrial Technology Research Institute
Hsinchu, Taiwan

Preface

Almost 30 years ago, when the author was a newcomer in the electronics industry, he believed that, mechanically, solders were very poor at resisting thermal fatigue and should be avoided as joining materials for interconnections. Since then, after assembling many electronic chips and optoelectronic devices on substrates, electronic and optoelectronic packages on printed circuit boards, and three-dimensional (3-D) integrated circuit (IC) integration systems, the author has completely changed his mind about solders as joining materials and greatly appreciates the "convenience" solders provide. Solders have given us remarkable flexibility to interconnect electronics and optoelectronics components. The unique properties of solders have facilitated assembly choices that have fueled creative advances in semiconductor assembly developments, for example, 3-D IC packaging and 3-D IC integration. For these technologies, solder is the electrical and mechanical "glue," and, thus, solder interconnect reliability is one of the most critical issues in the development of these technologies. In this book, the solder interconnect (joint) reliability of various high-density conventional packaging, 3-D IC packaging, and 3-D IC integration assemblies will be presented. Also, in order to comply with RoHS (restriction of the use of certain hazardous substances) regulations, a necessity in this competitive and heavily regulated world market, only RoHS-compliant materials, for example, lead-free solders, will be considered in this book.

This book is unique and special. The materials covered are not only for electronic manufacturing services (EMS) on the second-level interconnects, but also for packaging assembly on the first-level interconnects and for the semiconductor back-end on the 3-D IC integration interconnects. Thus, this book is extremely useful and absolutely necessary for the complete food chain of electronics and optoelectronics products.

There are seven major subjects in this book, namely: (1) RoHS-compliant semiconductor and packaging technologies (Chap. 1); (2) fundamentals of reliability engineering (Chaps. 2 and 3); (3) reliability of conventional electronics packaging assemblies

(Chaps. 4 and 5); (4) reliability of conventional optoelectronic packaging assemblies (Chaps. 6 and 7); (5) reliability of special lead-free solder assemblies (Chaps. 8 and 9); (6) reliability of 3-D IC integration assemblies (Chaps. 10 through 13); and (7) important factors affecting interconnect reliability (Chaps. 14 through 16).

Chapter 1 briefly discusses the RoHS-compliant conventional IC packaging, 3-D IC packaging, 3-D IC integration, and 3-D Si integration technologies.

Chapter 2 presents the basics of reliability engineering, which consist of: (1) design for reliability (DFR), (2) reliability testing and data analysis, and (3) failure analysis. Several important notes on failure criteria are discussed in Chap. 3.

Chapter 4 provides the DFR, reliability testing and data analysis, and failure analysis of a high pin-count ceramic column grid array (CCGA) package with Sn3Ag0.5Cu (SAC) solder paste. The reliability of an SAC plastic ball grid array (PBGA) package on a lead-free printed circuit board (PCB), with and without underfills, is presented in Chap. 5.

Chapter 6 presents the reliability of a light-emitting diode (LED) display lead-free assembly as well as the recommendation for 3-D LED and IC integration. Chapter 7 discusses the reliability of a set of parallel vertical-cavity surface-emitting lasers (VCSELs) lead-free–soldered on a silicon substrate, as well as the reliability of a VCSEL on a PCB with polymer waveguides.

Chapter 8 presents the reliability of a low-temperature solder (SnBiAg) for a surface mount technology (SMT) assembly of various conventional packages on a PCB. The reliability of a new lead-free solder (SACX) is discussed in Chap. 9, where X is a very low level of doping added to SAC.

The next four chapters are specifically for 3-D IC integration. Chapter 10 presents the chip-to-wafer (C2W) bonding and lead-free interconnect reliability, and Chap. 11 presents the wafer-to-wafer (W2W) bonding and lead-free interconnect reliability. In addition, the materials and processes are provided and discussed. Chap. 12 presents the through-silicon-via (TSV) interposer interconnect reliability due to both local and global thermal expansion mismatches and a new 3-D IC integration SiP with a TSV interposer. Finally, the electromigration of lead-free microbumps for 3-D IC integration is presented in Chap. 13.

The next three chapters examine some important factors affecting the reliability of lead-free interconnects discussed in Chaps. 4 through 13. Chapter 14 presents the effects of dwell time and ramp rate on SAC thermal cycling reliability test results, Chap. 15 presents the effects of high strain rate (impact) on SAC solder joints, and Chap. 16 discusses the effects of voids on solder interconnect reliability.

For whom is this book intended? Undoubtedly it will be of great value to all those faced with the challenging problems created by the ever-increasing interest in lead-free interconnect reliability of conventional IC packaging, 3-D IC packaging and 3-D IC integration. For three groups of specialists in particular—(1) those who are active or intend to become active in research and development of lead-free solder interconnects; (2) those who have encountered practical lead-free interconnect reliability problems and wish to understand and learn more methods for solving such problems; and (3) those who have to choose reliable, creative, high-performance, robust, and cost-effective 3-D IC packaging and 3-D IC integration techniques for their RoHS-compliant products—this book will hopefully provide a fundamental base that will aid in stimulating further research and development on mechanical, optical, electrical, and thermal designs, materials, processes, manufacturing, testing, reliability, and more sound applications of 2-D and 3-D interconnects in RoHS-compliant products. This book also can be used as a text for college and graduate students who have the potential to become our future leaders, scientists, and engineers in the electronics and optoelectronics industry.

The organizations that learn how to design and manufacture reliable interconnects in their conventional IC packaging, 3-D IC packaging, and 3-D IC integration systems have the potential to make major advances in the electronics and optoelectronics industry, and to gain great benefits in cost, performance, functionality, quality, size, and weight. It is my hope that the information presented in this book may assist in removing roadblocks, avoiding unnecessary false starts, and accelerating design, materials, and process development of these packaging (especially 3-D IC integration) technologies. I am strongly against the notion that packaging is the bottleneck of advanced applications. Rather, I would like to consider this as a golden opportunity to make a major contribution to the industry by developing innovative, high-performance, cost-effective, and reliable RoHS-compliant interconnects for electronics and optoelectronic products.

John H. Lau, Ph.D.

Acknowledgments

Development and preparation of *Reliability of RoHS-Compliant 2D and 3D IC Interconnects* was facilitated by the efforts of a number of dedicated people. I would like to thank them all, with special mention of Michael McCabe, Michael Mulcahy, Richard Ruzycka, and Stephen Smith of McGraw-Hill, Sapna Rastogi and Shivani Arora of Glyph International, and Lisa McCoy for their unswerving support and advocacy. My special thanks go to Michael Penn and Steve Chapman, who made my dream of this book come true by approving and effectively sponsoring the project, patiently listening to my excuses for delay, and solving many problems that arose during the book's preparation. It has been a great pleasure and fruitful experience to work with all of them in transferring my messy manuscripts into a very attractive printed book.

The material in this book clearly has been derived from many sources, including individuals, companies, and organizations, and I have attempted to acknowledge by citations in the appropriate parts of the book the assistance that I have been given. It would be quite impossible for me to express my thanks to everyone concerned for their cooperation in producing this book, but I would like to extend due gratitude.

Also, I would like to thank several professional societies and publishers for permitting me to reproduce some of their illustrations and information in this book, including the American Society of Mechanical Engineers (ASME) conference proceedings (e.g., *International Intersociety Electronic Packaging Conference*) and transactions (e.g., *Journal of Electronic Packaging*), the Institute of Electrical and Electronic Engineers (IEEE) conference proceedings (e.g., *Electronic Components and Technology Conference* and *Electronics Packaging and Technology Conference*) and transactions (e.g., *Advanced Packaging, Components and Packaging Technologies,* and *Manufacturing Technology*), the International Microelectronics and Packaging Society (IMAPS) conference proceedings (e.g., *International Symposium on Microelectronics*) and transactions (e.g., *International Journal of Microcircuits & Electronic Packaging*), *IBM Journal of Research and Development, Intel Technology Journal, Electronic Packaging & Production, Advanced Packaging, Circuits*

Assembly, Surface Mount Technology, Connection Technology, Solid State Technology, Circuit World, Microelectronics International, and *Soldering and Surface Mount Technology.*

I would like to thank my former employers, the Hong Kong University of Science & Technology (HKUST), the Institute of Microelectronics (IME), Agilent, and HP, for providing me with excellent working environments that have nurtured me as a human being, fulfilled my need for job satisfaction, and enhanced my professional reputation. Also, I would like to thank Dr. Don Rice (HP), Dr. Steve Erasmus (Agilent), Professor Dim-Lee Kwong (IME), and Professor Ricky Lee (HKUST) for their kindness and friendship while I was at their organizations. Furthermore, I would like to thank Dr. Ian Yi-Jen Chan, VP and Director of Electronics and Optoelectronics Laboratories (EOL) for his trust, respect, and support of my work at the Industrial Technology Research Institute (ITRI) in Taiwan. Finally, I would like to thank the following colleagues for their stimulating discussions and significant contributions to this book (the arrangement of the names reflects the order in which our collaborations appear in the chapters): Dr. Zhang Xiaowu, Dr. Tang Gongyue, Professor Ricky Lee, Dr. Walter Dauksher, Dr. Paul Vianco, Dr. Dongkai Shangguan, Dr. Dennis Lau, Dr. Joe Chu, Dr. Fubin Song, Mr. Nick Hoo, Mr. Joe Smetana, Mr. David Love, Mr. Ed Ott, Mr. Todd Castello, Dr. Jeffery Lo, Mr. Woai-Sheng Chow, Mr. Eng-Leong Soon, Dr. Jimmy Lam, Professor Matthew Yuen, Professor Philip Chan, Dr. Yida Zou, Mr. Sergio Camerlo, Dr. Lim Teck Guan, Mr. Khong Chee Houe, Professor Dim-Lee Kwong, Mr. Jerry Gleason, Dr. Gregory Henshall, Dr. Valeska Schroeder, Professor John Pang, Dr. Ning-Cheng Lee, Dr. Luhua Xu, Dr. Kok Ee Tan, Dr. Danniel Rooney, Mr. Edward Richard, Mr. Po Tse, Dr. Sang Liu, Dr. Zhi Wei Song, Dr. David Geiger, Ms. Sharon Lim Pei-Siang, Mr. Vempati Srinivasa Rao, Dr. Vaidyanathan Kripesh, Mr. Juan Milla, Mr. Andy Fenner, Dr. Daquan Yu, Dr. Aibin Yu, Mr. Navas Khan, Dr. Li Ling Yan, Professor Chengkuo Lee, Dr. Won Kyoung Chou, Dr. Seung Wock Yoon, Ms. Cheryl Selvanayagam, Dr. Y. S. Chan, Mr. Chai Tai Chong, Mr. Shiguo Liu, Mr. Charles Vath, Mr. John Doricko, Ms. Germaine Yen, Dr. Wen Hui Zhu, Dr. Stephen Pan, Dr. Don Rice, Dr. Steve Erasmus, Dr. Chi Jiaa, Ms. Sin-Yee Goh, and Ms. Gek-Yeng Tay. Definitely, I would like to thank my eminent colleagues (the enumeration of whom would not be practical here) at ITRI, HKUST, IME, Agilent, EPS, HP, Sandia National Lab, and throughout the electronics industry for their useful help, strong support, and stimulating discussions. Working and socializing with them has been a privilege and an adventure. I learned a lot about life and advanced integrated circuit (IC) packaging and 3-D IC interconnection technologies from them.

Last, I would like to thank my daughter Judy and my wife Teresa for their love, consideration, and patience by allowing me to work peacefully on this book. Their simple belief that I am making my small contribution to the electronics industry was a strong motivation for me. Thinking that Judy has earned her Ph.D. degree in physics from Princeton University and is working for Stanford University and that Teresa and I are in good health, I want to thank God for His generous blessings.

John H. Lau, Palo Alto, CA

Reliability of RoHS-Compliant 2D and 3D IC Interconnects

CHAPTER 1

Introduction to RoHS-Compliant Semiconductor and Packaging Technologies

1.1 Introduction

The RoHS-compliant semiconductor and 3-D integration technologies are briefly presented and discussed in this chapter. Three-dimensional integration consists of 3-D integrated circuit (IC) packaging, 3-D IC integration, and 3-D Si integration. They are different—in general, the TSV (through-silicon-via) separates the 3-D IC packaging and 3-D IC/Si integrations, that is, the latter two use TSV, but 3-D IC packaging does not. TSV is the heart of 3-D IC/Si integration and is a more-than-26-year-old technology. Even the TSV (for electrical feed-through) was invented by William Shockley in 1962 (the patent was filed on October 23, 1958), but it was not originally designed for 3-D integration. In this chapter, TSV (with a new concept that every chip can have two surfaces with circuits) is the focus. State-of-the-art key differences, trends of these three technologies, and a 3-D integration roadmap are presented.

1.2 Electronics Industry

Thousands of industries are present in the past four great Western transformations—the Renaissance, the Reformation, the Industrial Revolution, and the Computer Age. The electronics industry, also called the electronics equipment industry, is the largest and most important industry. Since 1996, the electronics industry has been the largest industry in the world (~$1.5 trillion by 2010). It is the most dynamic, fascinating, and important area of research, design, material, process, manufacturing, assembly, tests and inspections, quality, reliability, workmanship, etc.

There are many electronics products, such as the categories of electrical and electronic equipment (EEE) covered by the EU (European Union) WEEE (The Waste Electrical and Electronic Equipment): (1) large household appliances; (2) small household appliances; (3) IT and telecommunications equipment; (4) consumer equipment; (5) lighting equipment; (6) electrical and electronic tools; (7) toys, leisure, and sports equipment; (8) medical devices; (9) monitoring and control instruments; and (10) automatic dispensers. Beside the electronics products covered by the WEEE, there are many others, for example, defense equipment.

Most electronics products are made of electronics, a branch of science and technology that makes use of the controlled motion of electrons through different media, for example, silicon, gallium arsenide, silicon carbide, germanium, and copper. The ability to control electron flow is usually applied to information handling or device control. Most electronics devices use semiconductor devices to perform the electron control. In order for an electronics product to work, the semiconductors must be packaged and connected on a printed circuit rigid/flexible board (PCB) in certain fashions to create an electronic circuit with the intended function of the electronics system.

1.3 IC Semiconductor Technology

The global semiconductor industry is dominated by the United States, Korea, Japan, and the EU (i.e., Intel, Samsung, Toshiba, Texas Instruments, STMicroelectronics, Renesas, Sony, Qualcomm, Hynix, and Infineon) and could be a $260 billion market by 2010. A semiconductor is a material that has an electrical conductivity between that of a conductor and an insulator. Devices made from semiconductor materials, such as transistors and diodes, are the foundation of modern electronics.

The invention of the transistor by Bardeen, Brattain, and Shockley (Fig. 1.1) in 1947, which earned them the 1956 Nobel Prize in Physics, foreshadowed the development of generations of computers, smart phones and unknown products yet to come. Many consider the transistor to be the greatest invention of the electronics industry because it is the key element in practically all electronics products. Today, several companies still produce billions of discrete (individually packaged) transistors every year.

The invention of the integrated circuit (IC), that is, the integration of two or more of the tiny transistors into a small chip, by Jack Kilby of Texas Instruments in 1958 (Fig. 1.2, US Patent 2,981,877, "Semiconductor Device and Lead Structure," filed in July 1959, which earned

A replica of the transistor invented by Bardeen, Brattain, and Shockley at Bell Labs

Bell Labs junction transistor

FIGURE 1.1 Bell Labs transistor.

him the 2000 Nobel Prize in Physics) and six months later by Robert Noyce of Fairchild Semiconductor (Fig. 1.3, US Patent 3,138,743, "Miniaturized Electronic Circuits," filed in February 1959, who didn't share the Nobel Prize with Jack Kilby because Robert Noyce passed away in 1990) excited the development of integrations such as small-scale integration (SSI), medium-scale integration (MSI), large-scale integration (LSI), very large-scale integration (VLSI), ultra large-scale integration (ULSI), and wafer-scale integration (WSI). Kilby's chip was made of germanium, and Noyce's chip was made of silicon. Today, most of the transistors (>90%) produced are in silicon IC chips.

The proposal of doubling the number of transistors on an IC chip (for a minimum cost) every 24 months by Gordon Moore in 1965

Figure 1.2 Texas Instrument's integrated circuit by Jack Kilby.

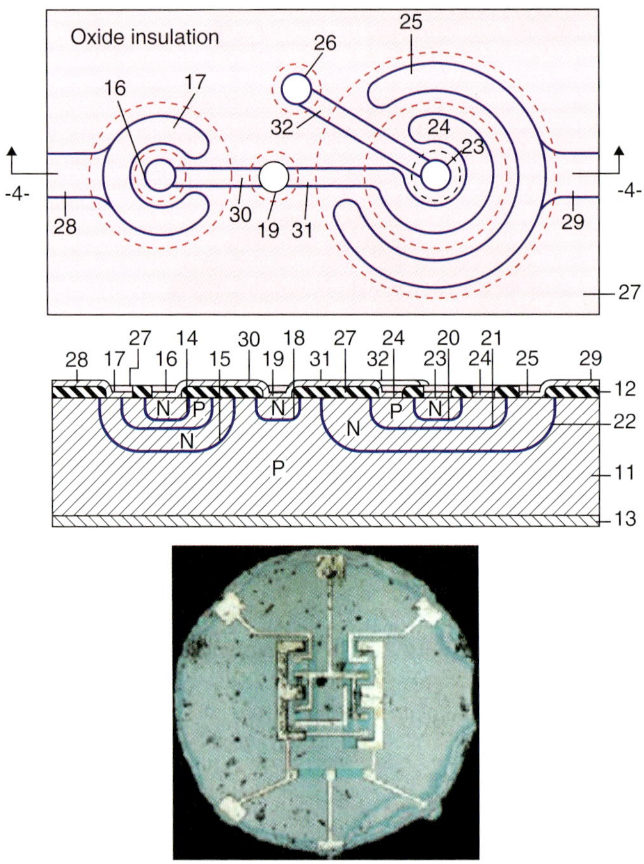

Figure 1.3 Fairchild's integrated circuit by Robert Noyce.

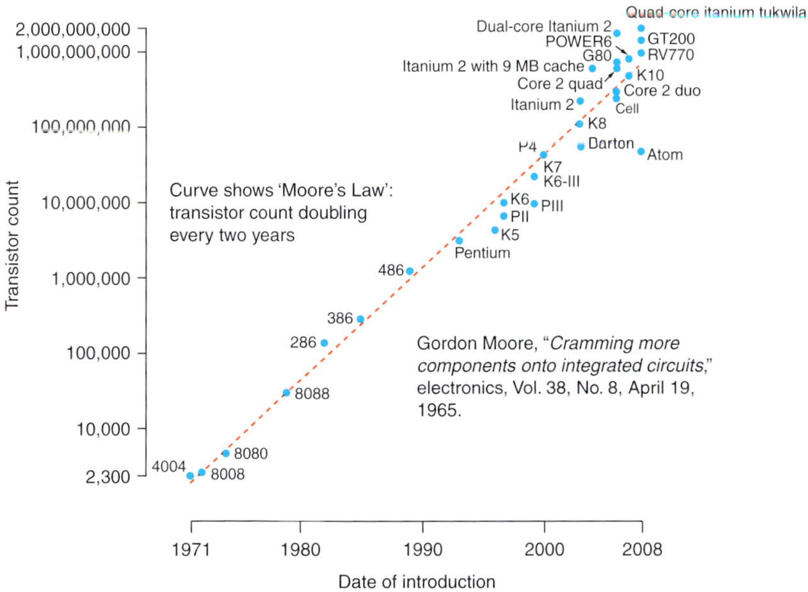

FIGURE 1.4 Moore's law.

(also called Moore's law) has been the most powerful driver for the development of the microelectronics industry in the past 44 years (Fig. 1.4). [It is noted that the dynamic random access memory (DRAM) has been doing better than the Moore's law, i.e., the number of transistors on a DRAM chip doubles approximately every 18 months.] This law emphasizes lithography scaling and integration (in 2-D) of all functions on a single chip, perhaps through system-on-chip (SoC). Today, the feature size of 32 nm is in volume production, and more than billions of transistors have been fabricated on many 2-D IC chips.

1.4 3-D IC Packaging Technology

1.4.1 Conventional IC Packaging Technology

The semiconductor IC chips are usually packaged in components such as the plastic or ceramic quad flat pack (QFP), plastic or ceramic ball grid array (BGA), plastic or ceramic column grid array (CCGA), bump chip carrier (BCC), thin outline package (TOP), and wafer level chip scale package (WLCSP) [1–15]. These chips can be placed either face up (with wire bonding) or face down (with solder bumped flip chip), either on a substrate (either organic or ceramic) or on a lead frame (Fig. 1.5). These components are then assembled on the PCB [1–15].

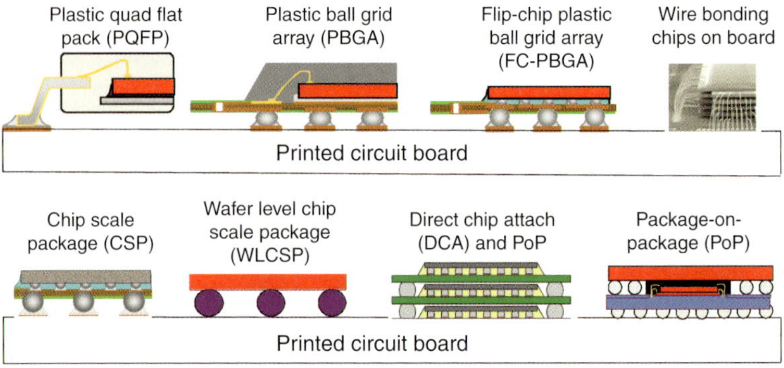

FIGURE 1.5 Various conventional packaging technologies.

In some applications, the semiconductor IC chips can be directly attached on the PCB with either wire bonds (chip-on-board) or solder bumps (direct chip attached), as shown in Fig. 1.5.

1.4.2 3-D IC Packaging Technology

Three-dimensional packaging consists of two or more conventional components (packages) stacked in the vertical direction. The most common one is stacking up the memory chips with wire bonds and die attachments, and is called 3-D memory stacking. It can be seen from Fig. 1.5 that 8 bare memory chips and from Fig. 1.6 that 20 bare memory chips are stacked with wire bonds and die attachments in the vertical direction. The other is shown in Figs. 1.5 and 1.7, and is called package-on-package (PoP). It can be seen that a flip chip is

FIGURE 1.6 Akita Elpida's 20 stacked chips.

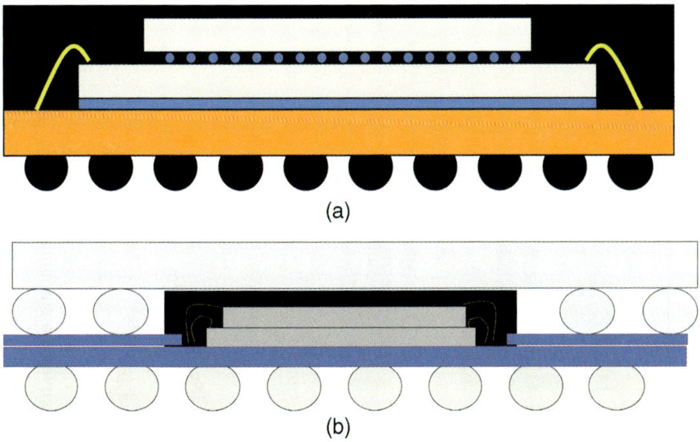

Figure 1.7 Package-on-package (PoP).

solder-bumped on the upper surface of a chip, which is wire-bonded on a substrate with a die attached (Fig. 1.7*a*). Also, a flip chip with solder bumps is attached to a substrate, which is supporting two stacked chips with wire bonds (Fig. 1.7*b*). All these 3-D memory stacks and PoP are mature technologies and in high volume production, as shown in the maturity status of 3-D technology [16] (Fig. 1.8) and the 3-D integration roadmap (Fig. 1.9) [17].

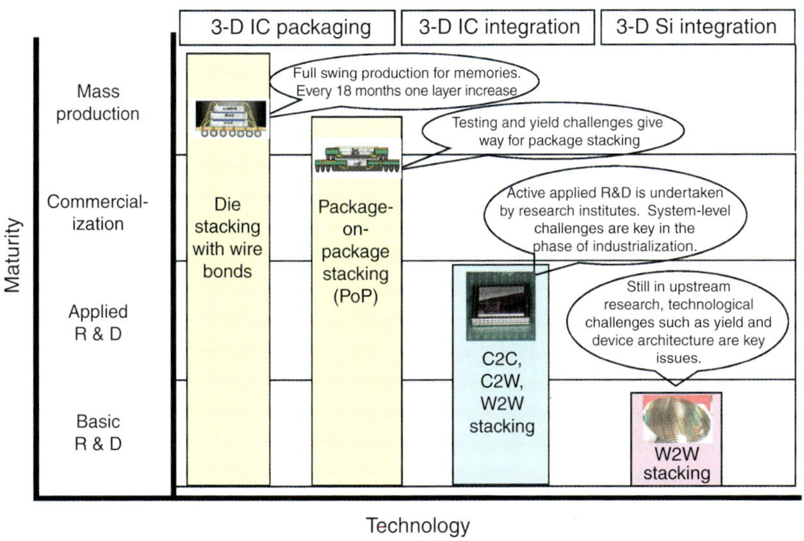

Figure 1.8 Maturity of 3-D technologies.

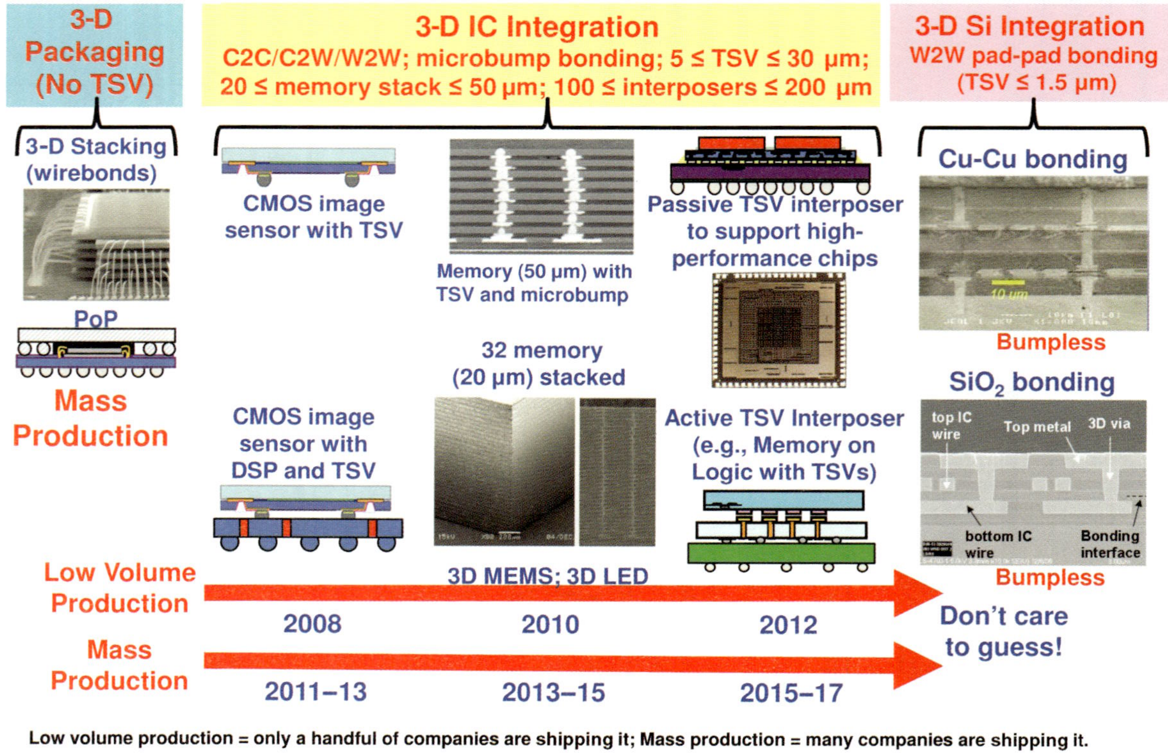

FIGURE 1.9 Three-dimension technologies road map.

1.5 3-D IC Integration Technology

As shown in Fig. 1.10, both 3-D IC integration and 3-D Si integration are more than Moore. They are different, and 3-D Si integration will be discussed later.

1.5.1 FEOL, BEOL, and MEOL

FEOL (front end of line) is usually performed in semiconductor fabrication plants (commonly called fabs) to pattern the active devices, for example, transistors. The process is from a bare wafer to passivation, which covers everything except the bonding pads for the next level of interconnects, that is, BEOL.

BEOL (back end of line) is usually performed in packaging assembly and test houses, and it involves everything after passivation, for example, under bump metallurgy (UBM), wire bonding, metallization, wafer bumping, back grinding, dicing, and testing.

It should be pointed out that this definition is from a packaging point of view. For the semiconductor people, FEOL is defined as the first portion of IC fabrication where the individual devices such as the transistors and resistors are patterned. This process is from a bare wafer to (but not including) the deposition of metal layers. BEOL is defined as the fabrication in which the active devices are interconnected with wiring on the wafer. This process starts from the first layer of metal to bonding pads with passivation. It also includes contacts, insulator, and dicing the wafer into individual IC chips.

MEOL (mid end of line) is new and performed by combining some of the FEOL and BEOL technologies into a 3-D IC integration technology,

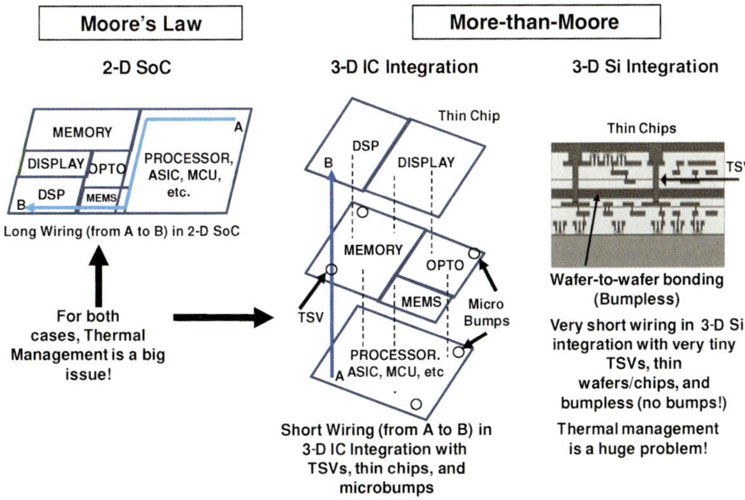

FIGURE 1.10 Moore's law (SoC) vs. more-than-Moore (3-D IC/Si) integrations.

which involves, for example, TSV, microbumps, thin-wafer handling, metallization, UBM, wafer bumping, back grinding, dicing, and testing. Thus, 3-D IC integrations must be executed in the fabs and packaging assembly and test houses. However, since the fabs' equipment and semiconductor personals are too expensive to make the 3-D IC integration and final test right before the PCB assembly (which are not their core competence and major business), eventually (when there are volumes), the packaging assembly and test houses will do it all.

1.5.2 More Than Moore

Apart from SoC (Moore's law), integration of all these functions can be achieved through either 3-D IC integration or 3-D Si integration, which are some of the *more-than-Moore* technologies. Based on the silicon platform technology, anything that involves the integration of electronics, photonics, mechanics, chemistry, heat, magnetics, biology, etc., for functionality and system performance when interacting with people and the environment is known as more-than-Moore, for example, 3-D MEMS (microelectromechanical systems) [18] and 3-D light-emitting diode (LED) [19].

1.5.3 3-D IC Integration Technology

Three-dimensional IC integration is the future key technology for electronic products that must meet the following requirements:

(1) Form factor: Smaller is better, for example, personal digital assistants (PDAs), cameras, mobiles phones [complementary metal oxide semiconductors (COMS) image sensors]

(2) Performance: Faster is better, for example, shorter local and global interconnect lengths

(3) Cost: Cheaper is better, for example, system improvements without the high cost of scaling

(4) Reliability: Low risk is better, for example, it still works after drops

The Holy Grail of 3-D IC integration (e.g., heterogeneous integration) is shown in Fig. 1.10, where some of the chips, for example, microdisplay, MEMS, opto, memory, microprocessor, optoelectronic, multiple outputs DC–DC converter, application-specific IC (ASIC), microcontroller unit (MCU), digital signal processor (DSP), microbattery, and analog-to-digital (A/D) mixed signals are combined and stacked in three dimensions. The simplest examples are shown in Figs. 1.9 and 1.11*a*. It can be seen that eight memory chips, each 50 μm thick, are stacking up with microbumps and communicating with TSVs. Even with eight chips, however, their total thickness is still less than the thickness of an ordinary wafer. Figure 1.11*b* shows 32 chips (each 20 μm thick) stacked by Samsung. In general, the electrical performance of 3-D IC integration is better than that of 2-D SoC (Fig. 1.10). However, there are many critical issues that need to be considered.

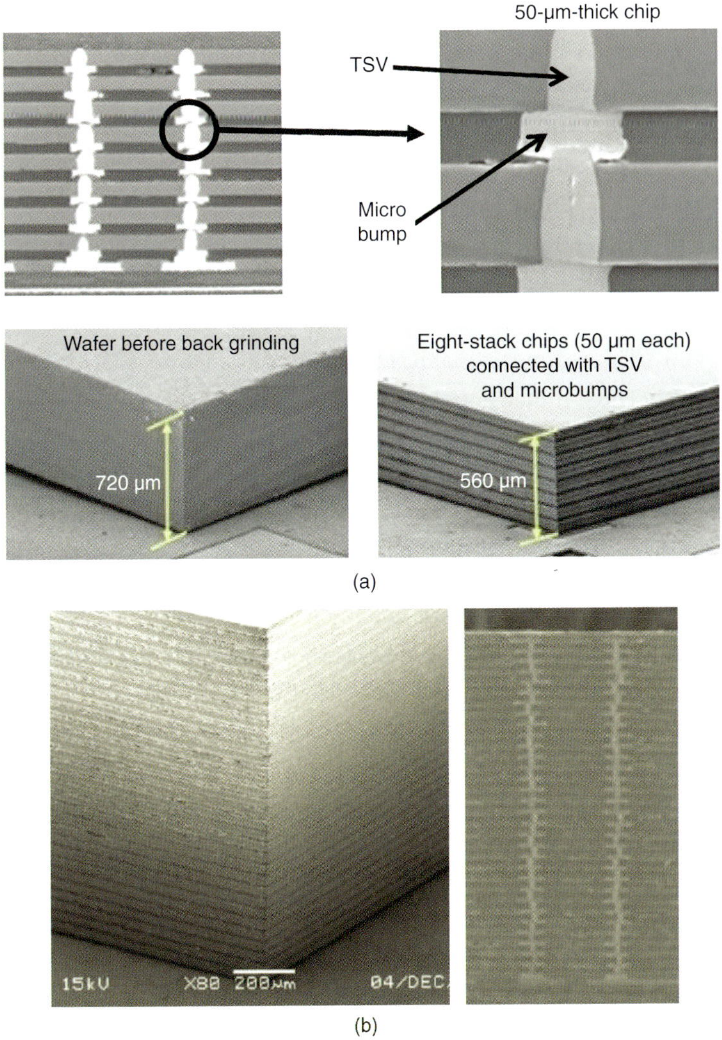

FIGURE 1.11 (a) Samsung's eight stacked chip with TSV and microbumps; (b) Samsung' 32 stacked chips each with 20 μm thick.

1.5.4 Critical Issues of 3-D IC Integration

Just as with many other new technologies, 3-D IC integration still faces many critical issues. In the development of 3-D IC integration, at least the following must be noted and understood [16–223]:

(1) Known good dies (KGDs) are required.

(2) Design guidelines and software are not available.

(3) Test methods and equipment are lacking.

(4) TSV with redistribution layers (RDL) usually are required.

(5) Microbumps usually are required.
(6) Wafer thinning and thin-wafer handling during processing are necessary.
(7) High equipment accuracy is necessary for alignments.
(8) Fast chips must be mixed with slow chips.
(9) Large chips must be mixed with small chips.
(10) Three-dimensional IC stacking inspection methodology is needed.
(11) Three-dimensional IC stacking expertise is lacking.
(12) Three-dimensional IC stacking infrastructure is lacking.
(13) Three-dimensional IC stacking standards are lacking.
(14) Thermal issues: The heat flux generated by stacked multifunctional chips in miniature packages is extremely high.
(15) Thermal issues: 3-D circuits increase total power generated per unit surface area.
(16) Thermal issues: Chips in the 3-D stack may be overheated if proper and adequate cooling is not provided.
(17) Thermal issues: The space between the 3-D stack may be too small for cooling channels (i.e., no gap for fluid flow).
(18) Thermal issues: Thin chips may create extreme conditions for on-chip hot spots.

It can be seen that the key enabling technologies for 3-D IC integration are, among others: (1) TSV, with or without RDL, (2) thin-wafer strengthening and handling, (3) microbumping and assembly, and (4) thermal managements. They will be briefly discussed in the following sections.

It should be pointed out that electronic design automation (EDA) is the soul of 3-D IC integration. EDA is a category of software tools for designing electronic systems and the focuses of the software are design, simulation, analysis and verification, manufacturing preparation, and testing. The key players of EDA are Synopsys, Cadence, Mentor Graphics, Zuken, Agilent, etc. These vendors are not seriously, full-force, and systematically working on the EDA because the ecosystem (e.g., standard and infrastructure) of 3-D IC integration does not exist. Thus, an ecosystem for 3-D IC integration is desperately needed, now! Without it, we would not have a soul (EDA) for 3-D IC integration.

1.5.5 TSV, With or Without RDL

The most important key enabling technology for 3-D IC integration is TSV, with or without RDL. It was invented by William Shockley in 1958, as shown in Fig. 1.12. It provides advanced vertical (electrical

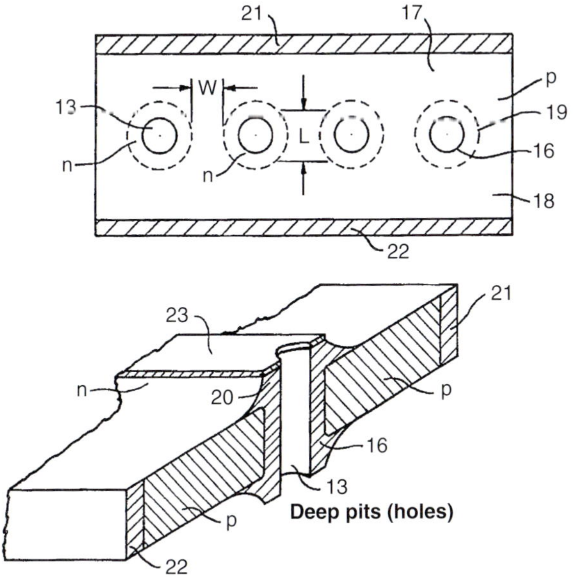

FIGURE 1.12 TSV invented by William Shockley, U.S. Patent #3,044,909, filed on October 23, 1958 and granted on July 17, 1962.

feed-through) interconnects and system-in-package (SiP) solutions, such as C2C (chip-to-chip), C2W (chip-to-wafer), and W2W (wafer-to-wafer) stacking; wafer-level packaging and redistribution; interposer packaging; and the shortest electrical path (vertical electrical feed-through) and lowest signal loss between two sides of a silicon chip or interposer.

Unlike the conventional packaging and 3-D packaging (BEOL), which put, for example, the wire bonds, solder bumps, gold stud bumps, or anisotropic conductive adhesives on the bonding pads of the passivation wafer made by the FEOL, the first thing that the 3-D IC integration (MEOL) does is "dig" holes (TSVs) through the passivation and between the devices/circuits of the wafer. This is a novel approach that opens up the doors for many new and useful applications. In addition to those shown in Figs. 1.11 and 1.12, Fig. 1.13 shows that a MEMS device is integrated with an ASIC chip into a 3-D structure [18]. Figure 1.14 shows that LED devices can be mounted on an active chip [19], for example, LED driver, ASIC, sensor, and processer. Figure 1.15 shows that a memory is using the CPU as its active interposer [20, 21]. All of these are shown in the 3-D IC integration roadmap (Fig. 1.9), and their advantages are: (1) better performance, (2) smaller form factor, (3) smaller footprint, (4) lighter weight, and (5) potentially lower cost.

It should be emphasized that a few years ago, once the (active) front side of a chip is flipped on a substrate or PCB, nothing can be done on its back side except attaching a heat spreader/sink. With the

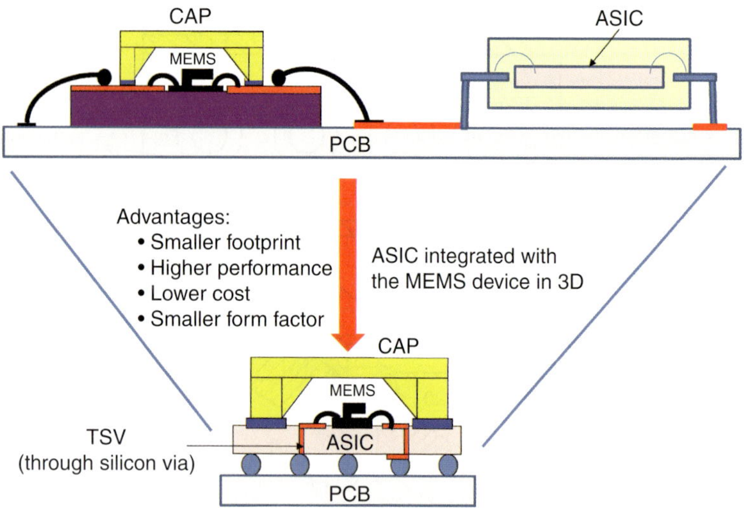

Figure 1.13 3-D MEMS IC (ASIC) integration. For large ASICs, a substrate/underfill is needed.

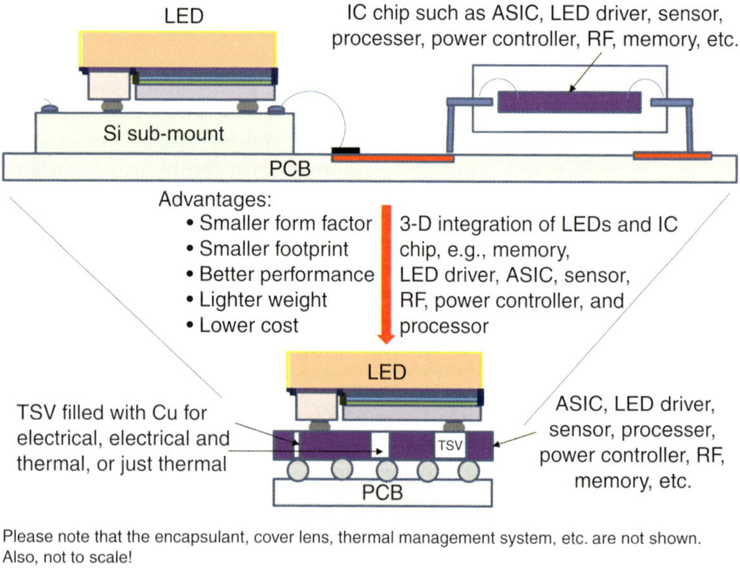

Figure 1.14 3-D LED IC (e.g., ASIC and LED driver) integration. For large LED driver or ASICs, a substrate/underfill is needed.

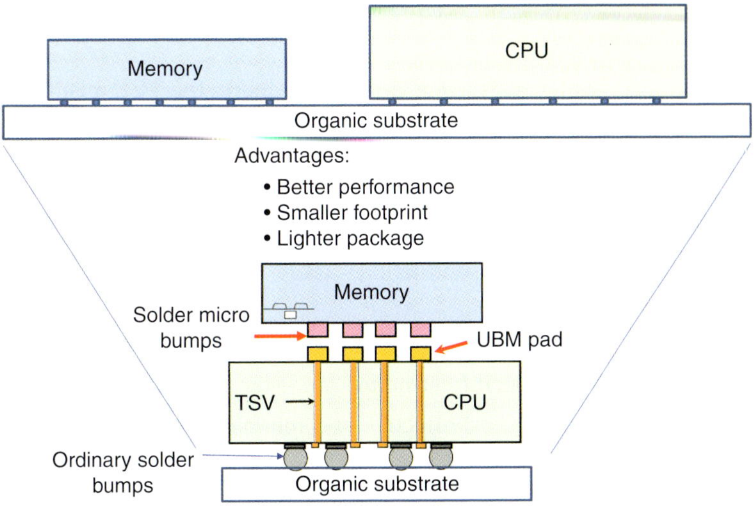

FIGURE 1.15 3-D IC (memory and CPU) integration. Underfills are needed between the memory chip and TSV CPU and the TSV CPU and organic substrate.

TSVs: (1) there is no distinction between the front side and back side of any chip anymore; (2) any chip can have two sides (top and bottom) with circuits; and (3) just use your imagination, and the sky is the limit.

As with many other new technologies, TSVs still face many critical issues. In the development of TSVs, the following must be noted and understood [16–223]:

(1) Today, the only TSV 3-D IC volume product is the CMOS image sensor. (Straightly speaking, it is not a 3-D IC integration.)

(2) The cost of TSV is higher than that of wire bonding.

(3) TSV design software is lacking.

(4) TSV design guidelines are not commonly available.

(5) Copper filling helps on thermal problems, but increases TCE (thermal coefficient of expansion).

(6) Void-free copper filling usually takes a long time to produce (lower throughput).

(7) The TSV cost for poor-yield IC wafers is high because many TSVs are wasted on the bad die.

(8) There is a high cost for low TSV wafer manufacturing yield, especially for high-cost die.

(9) TSV wafer yields are high (>99.99%).

(10) Single-point touch-up on the TSV wafer is difficult.

(11) TSVs with high aspect ratios are difficult to manufacture at high yield.

(12) TSV wafer warpage is a problem due to the TCE mismatch between the silicon and copper.

(13) Thin TSV wafer handling is necessary during most of the processes.

(14) Test methodology and software for TSV are lacking.

(15) High-volume production tools are lacking and/or expensive.

(16) TSV inspection methodology and software are lacking.

(17) TSV expertise, infrastructure, and standards are lacking.

(18) What are the cost-effective and reliable TSVs, and for what IC devices?

(19) How large is the TSV market?

(20) What is the life cycle of TSV?

Today the only volume product with TSV is Toshiba's CMOS image sensors. In order to reduce the footprint and enhance the performance of the product, they converted the wire-bonding technology to flip-chip technology, as shown in Fig. 1.16 [22]. (It should be noted that the system houses would kill for reducing the product size, weight, and cost, no matter how little reductions there are. From this point of view, the designs shown in Figs. 1.11 through 1.15 are very attractive.) In order to let the current flow from the PCB, through the chip, to the sensor, they "dig" holes on the chip, as shown in Fig. 1.17. Toshiba doesn't call the holes TSV, but rather through-chip-via (TCV) [22] because they used a PCB fabrication process.

The process of their novel low-cost TCV technology is shown in Fig. 1.18 [22]. It can be seen that the major steps are: (1) the first via drilling from the back of the silicon wafer with alignment to the back of the Al pads (with Ni bumps) by using a YAG laser (with the energy of the laser pulse below 0.05 mJ so the Ni will not be ablated or melted—see Fig. 1.19); (2) fill the via by dielectric film lamination (with epoxy resin and silica filler); (3) the second via (dielectric film) drilling on the epoxy resin with alignment to the center of the first via using the same YAG laser; (4) copper pattern plating with subtractive or semi additive PCB electroplating processes (Fig. 1.19); and (5) solder resist coating and patterning. Thus, based on the ordinary PCB fabrication process, the TCV technology has been helping Toshiba shipping their CMOS image sensors at a very low cost. The so-called "via-last" TSV manufacturing process using the DRIE (deep reactive ion etching), CMP (chemical and mechanical polishing), etc., is not needed [22].

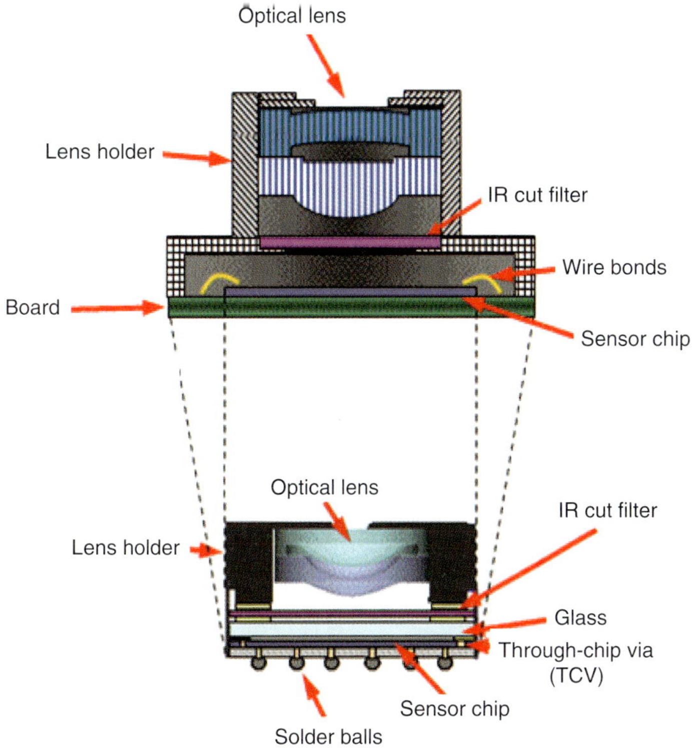

FIGURE 1.16 Toshiba's CMOS image sensor (converted from wire bond to flip chip).

1.5.6 Manufacturing Process for TSV 3-D IC Integration

In addition to the PCB fabrication-based TCV technology used by Toshiba for a niche application (e.g., very simple circuit, very low number of vias, very large pitch, and very large via size), there are other ways to make the TSV with or without RDL for 3-D IC integration (with a large number, small size, and/or fine-pitch of vias). The one would be used by the industry the most is shown in Fig. 1.20. There are five key steps to make the TSV:

1. Via formation by either DRIE or laser drilling
2. SiO_2 deposition by either thermal oxidation for passive interposers or PECVD (plasma enhanced chemical vapor deposition)
3. Barrier and seed layer deposition by physical vapor deposition (PVD) or electrografting (eG)
4. Cu plating or W (tungsten) sputtering to fill the vias
5. CMP of Cu plating residues (overburden)

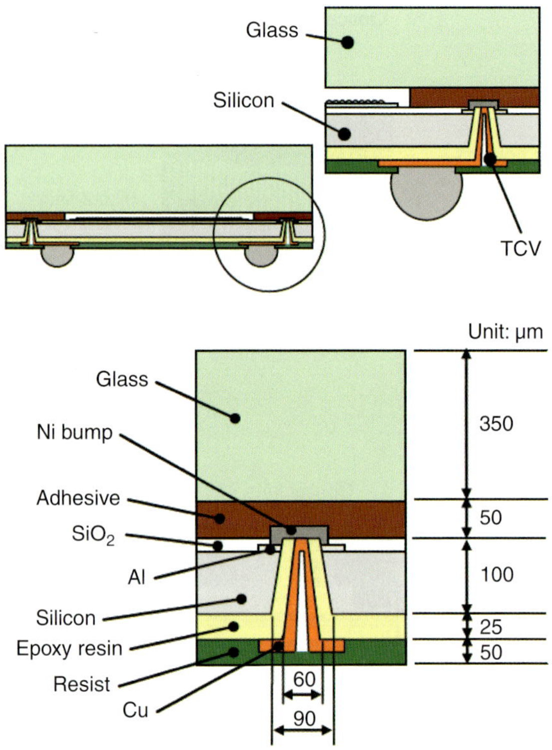

FIGURE 1.17 Toshiba's TCV (through-chip via).

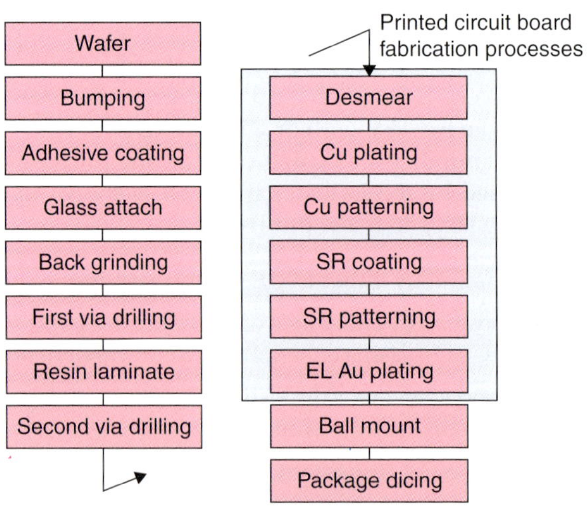

FIGURE 1.18 Toshiba's low-cost PCB process.

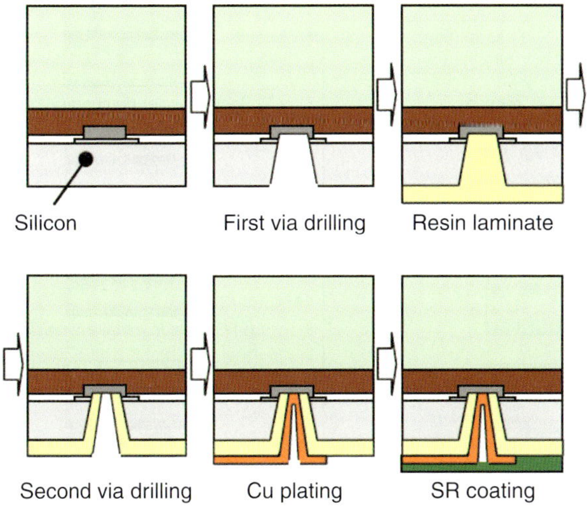

FIGURE 1.18 (*Continued*)

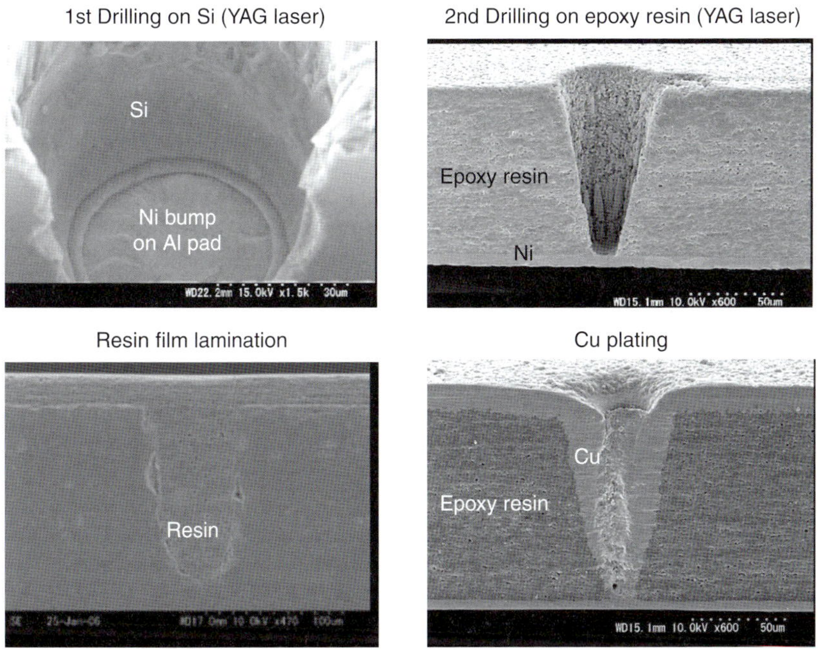

FIGURE 1.19 Results of Toshiba's TCV technology.

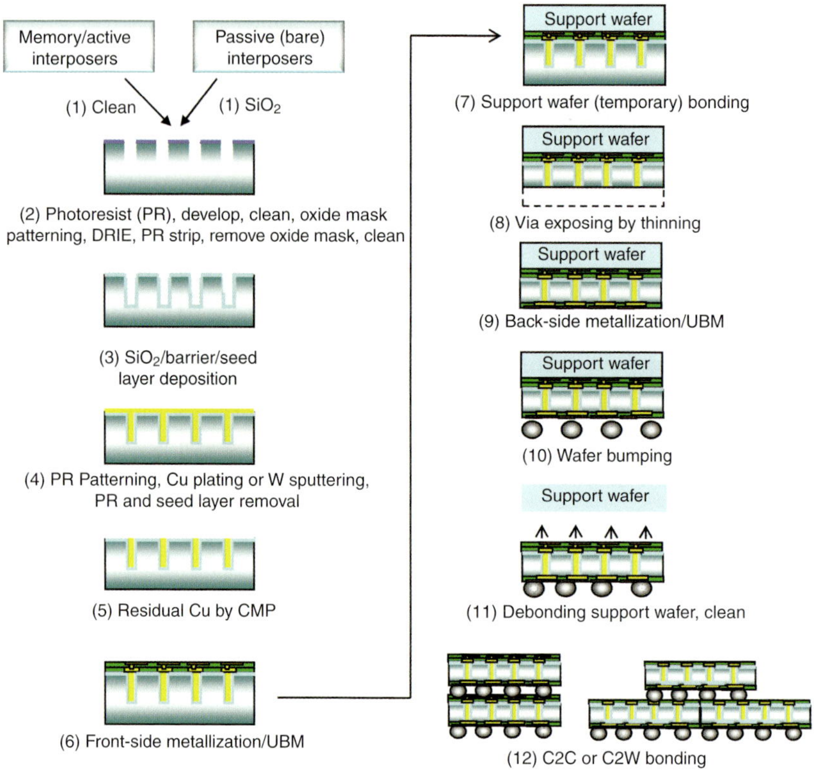

Figure 1.20 The most likely manufacturing process for memory stacked and active/passive interposers.

It should be noted that for bare wafers (e.g., passive interposers), an SiO_2 layer is needed before the photoresist. For active memory chips and active interposers (e.g., ASIC and processor) because of their passivation layer, only cleaning is needed. Next, the front-side metallization/UBM can be made on the blind TSV wafer, which is usually thicker than 300 μm.

In most of the 3-D IC integration applications, the thickness of the passive/active interposers ranges from 100 μm to 200 μm and that of the stacking memory chips ranges from 20 μm to 50 μm. Thus, most of the TSVs fabricated are blind vias and a support wafer (carrier) is needed for the remaining processes.

As soon as the support wafer is temporarily bonded to the blind TSV wafer, it can be thinned down to expose the TSV. Then it is followed by, for example, the back-side metallization/UBM and wafer bumping. Next, the support wafer is debonded from the TSV wafer. Finally, the TSV with RDL wafer is ready for either W2W bonding (recommended only for very high chip yield wafers or of one of them

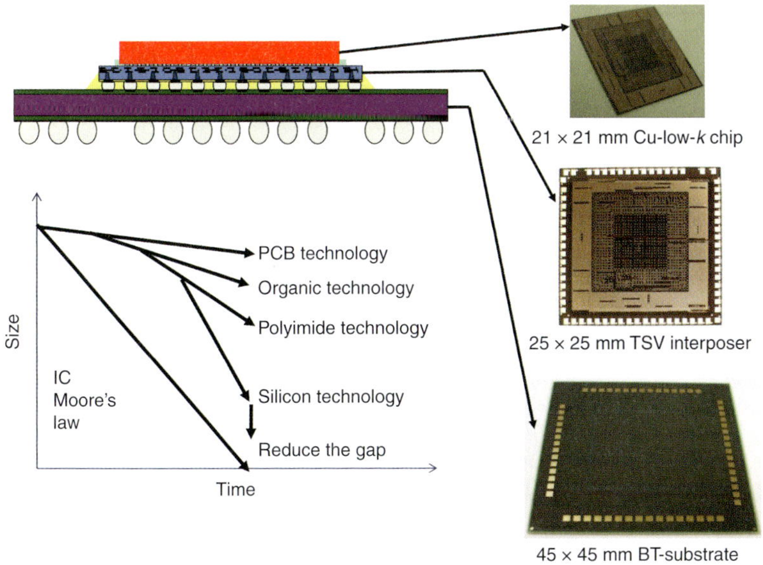

Figure 1.21 Various substrate technologies to support a high-performance chip

is not an active wafer, for example, the cap wafer for MEMS applications) or diced into individual chips for C2C or C2W bonding.

Today, all the steps on the left side of Fig. 1.20 are under-controlled by the industry and there is room for improvement. However, all the steps on the right side of Fig. 1.20 require more research and development (R & D) effort because of the very thin TSV wafer, which is required by 3-D IC integration applications.

Figure 1.21 shows a high performance chip that cannot be supported by the organic bismaleimide triazene (BT) substrate but that is supported by a passive TSV interposer (Si-substrate), with two redistribution layers on top and one at the bottom [23]. A simple BT substrate is used to fan-out all the I/Os, power, and ground connections to the PCB. It should be pointed out that the advance of semiconductor technology (because of Moore's law) is many times faster than that of the PCB, BT, and polyimide technologies; thus, the silicon is used as the substrate to reduce the cap. Figure 1.22 shows the TSV with RDL wafer and the chip with its cross-sections clearly showing the RDLs of the interposer [23].

The TSV size should be as small as possible (≤30 μm) because (1) there is less thermal expansion mismatch between the Si (TEC = 2.5×10^{-6}/°C) and Cu (TEC = 18.5×10^{-6}/°C), (2) less Cu plating, (3) higher throughput, and (4) more space for routing. Five to 10 μm vias are not uncommon and for some very high-performance applications, the via diameter could go down to one micron.

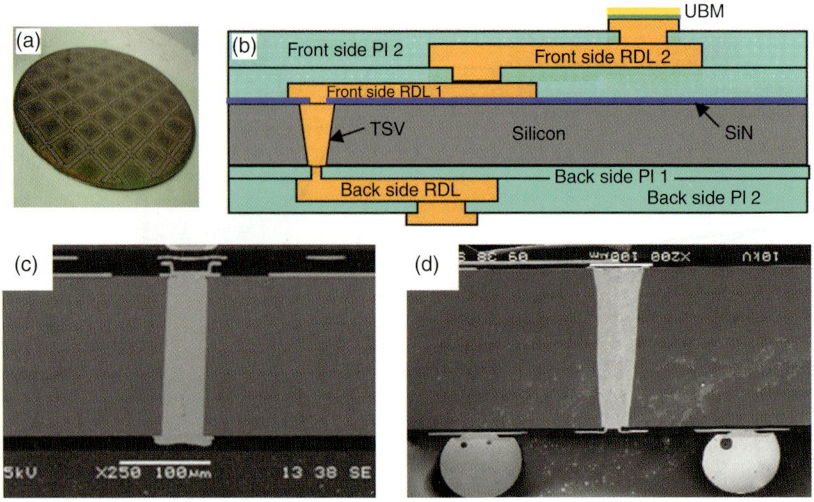

FIGURE 1.22 Si interposer with two RDLs on top and one RDL at the bottom (IME).

1.5.7 How to Improve TSV Manufacturing Yield?

As mentioned earlier, there are five important steps in making the TSV. In order to improve the TSV manufacturing yield and enhance the quality, at least the following must be noted [223]:

Via Forming by DRIE

What we want:

(a) Sidewall roughness: As smooth as possible

(b) Via straightness: Vertical or slightly taped is preferred

(c) Via uniformity: <± 5% variations across the whole wafer

(d) Via forming speed: Reasonable

What we should do:

(a) The etch rate (R_t) (= etch cycle time/passivation cycle time) (Fig. 1.23) should be balanced. (SF_6 will be used in the etch cycle, whereas C_4F_8 is used in the passivation cycle.) If the etch cycle time is too long and the passivation cycle time is too short, the via will not be straight and the sidewall will be rough. If the etch cycle time is too short and the passivation cycle time is too long, then the etch rate will be slow. $R_t = 7s/5s$ is a good starting point.

(b) The most common mask processes are photoresist and photoresist+SiO_2. Photoresist is simple, but the undercut could be larger. Photoresist+SiO_2 is better, but it needs more processing steps.

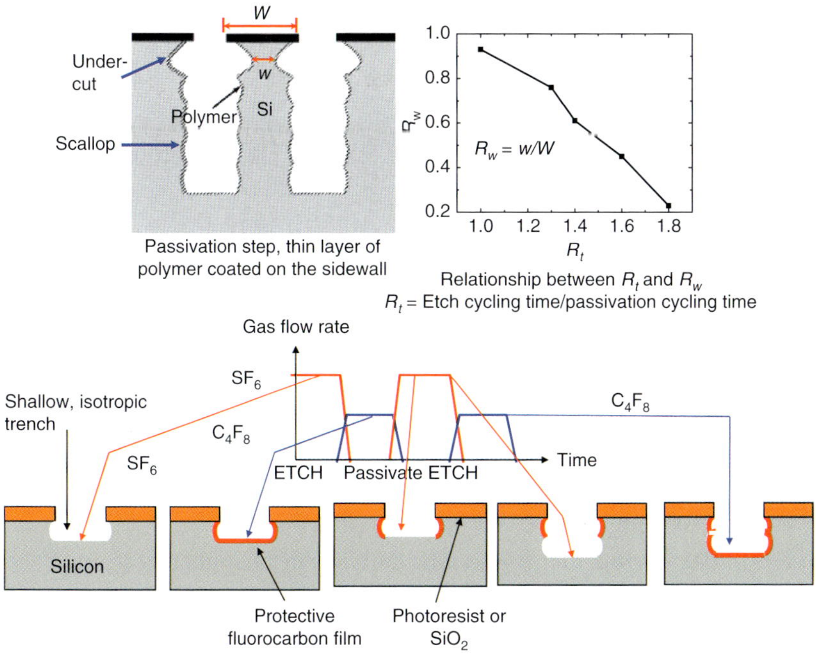

Figure 1.23 TSV by deep reactive ion etch (DRIE), which goes through many alternating passivation and etching cycles (SF_6 is for etch and C_4F_8 is for passivation).

(c) Higher etch power will increase etch rate. However, if etch power is too high, the mask layer will be etched away.

Dielectric Deposition by Thermal Oxidation and PECVD

What we want:

(a) Conformity: The via sidewall (from top to bottom) is covered by SiO_2 (~1 μm)

(b) Adhesion: The adhesion between Si and SiO_2 is good

(c) Uniformity: < ±5% variations across the whole wafer

(d) Sidewall roughness: < 100 nm

What we should do:

(a) Thermal oxidation can achieve a uniform isolation layer, but its operating temperature is high (~1050°C). and thus it is suitable for the via-first process or passive interposers.

(b) PECVD of SiO_2 is operated at a temperature between 250°C and 400°C, and it is not easy to cover the sidewall of high aspect ratio vias uniformly. Gas flow [SiH_4, N_2O, or $Si(OC_2H_5)_4$]

need to be controlled. In addition, lower temperatures (≤350°C) should be used because 400°C is too close to the allowable temperature of most IC devices.

Barrier/Adhesion/Seed Metal Layer by PVD and Electrografting

What we want:

(a) Conformity: 100% sidewall coverage

(b) Dimensions: ~100 nm of Ti (for adhesion) and ~1 μm of Cu (for seed metal layer)

(c) Uniformity: < ±5% variations across the whole wafer

What we should do:

(a) PVD sputtering power is important. Usually, a higher power can increase adhesion and provide better sidewall coverage

(b) Before sputtering, precleaning the wafer in the sputtering chamber (using Ar plasma to bomb the wafer surface) is important.

Wet deposition of the insulation and barrier layers inside high-aspect-ratio TSV can also be achieved by eG, which transients the chemical bond initiated by small electric current followed by electroless chemical propagation. This could replace the dry vapor deposition with less expensive wet processing using low-cost electroplating tools.

Via Filling with Cu and W

What we want:

(a) Uniformity: < ±5% variation across the whole wafer

(b) Void free: No voids in the Cu filling

(c) Plating speed: Reasonable

(d) Overburden: <20 μm of Cu on wafer surface

What we should do:

(a) The plating current is the most important parameter. A smaller current (e.g., 0.1 A) will help achieve better via filling, especially for high-aspect-ratio vias, but the plating speed is slower.

(b) Plating solution additives/components (suppressor, leveler, and accelerator) need to be controlled. The suppressors are typically polyalkylene glycol-type polymers, which are used to improve the morphology and profile of the via filling. The levelers are typically alkane surfactants containing sulfonic acids and amine or amide functionalities, which are also used to improve the morphology. The accelerators are typically sulfur derivatives of propane sulfonic acid, which are used to brighten the plated Cu color, that is, to prevent Cu from oxidation.

(c) Besides plating current and chemistry, pretreatment (wetting + DI rinsing + preabsorbing accelerator) is also a key contributor to void-free filling.

(d) Plating uniformity can also be improved by agitating the solution and rotating the wafer during the plating.

(e) Plating waveform such as the multistep pulse-reverse (damascene) plating process is recommended.

For 3-D Si integration, since the via is very small (~1 μm), tungsten (W) is usually used and deposited by CVD (using WF_6 and H_2 as gas). Normally, the flow rate and volume of WF_6 and H_2 need to be matched and the temperature needs to be controlled at about 450°C. For 3-D IC integration, since the via is much larger than that of 3-D Si integration, if W is used, sputtering would do it.

CMP

What we want:

(a) Uniformity: Wafer total thickness variation (TTV) should be as small as possible ($< \pm 5\%$)

(b) Residual stress: No microcracks on the wafer

What we should do:

(a) Slurry flow rate and down force should be optimized to achieve high removal rate, better uniformity, and less stress. A 180 cc/min slurry flow rate and a 330 g/cm^2 down pressure are good starting points.

(b) The head (clockwise) speed should be one to two times the table (counterclockwise) speed.

(c) To reduce the warpage and stress, 500 Å of SiN is deposited on the Cu and then annealed at 200°C for 80 min under N_2.

1.5.8 Thin-Wafer Handling

In order to have high-performance, low-profile, and lightweight products with 3-D IC integration technology, the thickness of the chips/wafers is usually very thin. Making the wafer thin is not a big problem. Most of the back-grinding machines (e.g., Disco) can do the job and grind the wafers to as thin as 5 μm. However, handling thin wafers through all the semiconductor fabrication and packaging assembly processes is difficult. Usually, the thin-device wafer is temporarily bonded on a support wafer. Then it goes through all the semiconductor fabrication processes, such as metallization, passivation, UBM, and the packaging processes, such as back grinding and solder bumping. After all these are done, removing the thin wafer from the support wafer poses another big challenge. This section briefly discusses a few methods for handling thin wafers.

The new 3M+SUSS wafer support system [24] enables conventional back-grinding equipment to produce wafers with a final thickness as low as 20 μm. The key to the 3M system is its ability to provide a rigid, uniform support surface to minimize stress on the wafer as the silicon is removed, resulting in less cracking and chipping. The system includes both the equipment (by SUSS) and consumables (3M UV-Curable Liquid Adhesive LC-2201; glass support plate, which typically can be reused many times; and 3M light-to-heat conversion solution) necessary for mounting, demounting, and removing adhesive from the wafer.

In the 3M+SUSS system, a glass plate is used to support the wafer through the back-grinding process. An ultraviolet (UV)-curable liquid adhesive is used as the bonding agent between the device wafer and the glass plate (support wafer). After the back-grinding process the thinned device wafer is transferred onto a dicing tape, and the support glass is removed by laser debonding of the adhesive-glass interface using a light-to-heat conversion (LTHC) layer. The adhesive then can be removed from the wafer, leaving behind fewer residues than seen with typical back-grinding tapes. This system also works for other semiconductor and packaging processes as long as the thermal-expansion mismatch between the device wafer and the glass plate is within the allowable tolerance.

EVG+Brewer Science have developed a solution that enables temporary bonding of a device wafer to a right carrier substrate (support wafer) and allows not only thinning but also a full range of subsequent processes, including high-temperature deposition, etching, lithography, dielectric application and curing, plating, and chemical cleaning [25].

Temporary Bonding

The front side of the device wafer and the support wafer are coated with the WaferBond, an HT Brewer Science adhesive material, by using a coating chamber that includes spin and spray coat capability. Both wafers then are transferred to a bond chamber, where they are carefully centered and vacuum-bonded at elevated temperatures. Once the device wafer is temporarily bonded to the support wafer, it is ready for back-side processing, including back-grinding, etching, metallization, TSV formation, etc.

Debonding

During the debonding process, the thin wafer is debonded, first via a thermally activated slide lift-off approach from the support wafer,

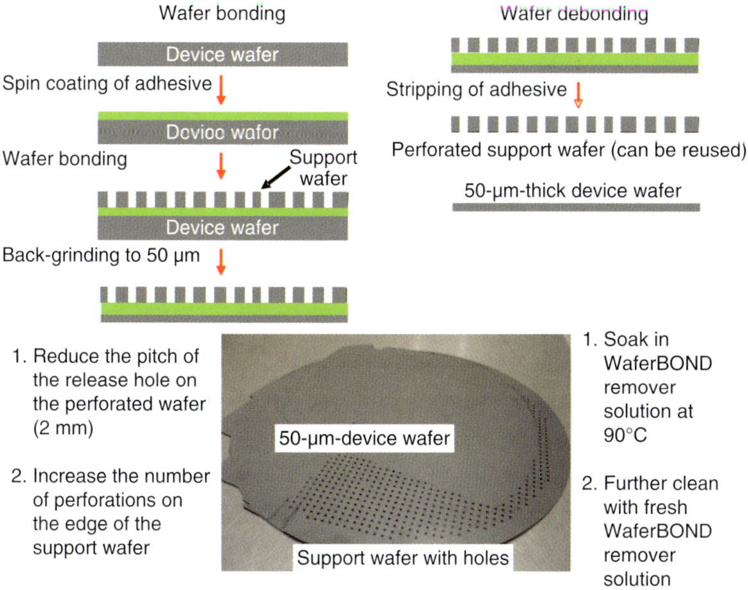

FIGURE 1.24 A simple wafer-supporting method (went through SiO_2 process).

cleaned in a single-wafer cleaning chamber to remove the remaining adhesive residues, and then transferred to the appropriate output format, such as a film-frame carrier, a dedicated wafer cassette, or a coin-stack packing canister. The support wafer is also cleaned and then can be reused immediately for another debonding process.

Figure 1.24 shows a simple and low-cost support-wafer method for thin-wafer (50 μm) handling [108]. In order to reduce the residual stress and become crack-free during debonding, one must reduce the pitch of the release holes (1 mm in diameter) on the perforated wafer to 2 mm, and increase the number of perforations on the edge of the support wafer. These two optimized debonding methods allow more chemical solution to uniformly penetrate into the bonding adhesive and eventually debond the wafers successfully.

1.5.9 Low-Cost, Lead-Free Solder Microbumps

As mentioned earlier, the thickness of chips for 3-D IC integration application is thin, and thus the conventional flip-chip solder bumps (~100 μm) cannot be used for connecting the thin chips. Instead, very tiny bumps (<25 μm) are needed. Figure 1.25 shows the lead-free Sn bump (on a high-performance memory chip) and Ni-Au UBM (on a logic chip) formation and their assembly. It can be seen that the lead-free solder joints are on an 8-μm pad, 15-μm pitch, and 16-μm standoff height [20].

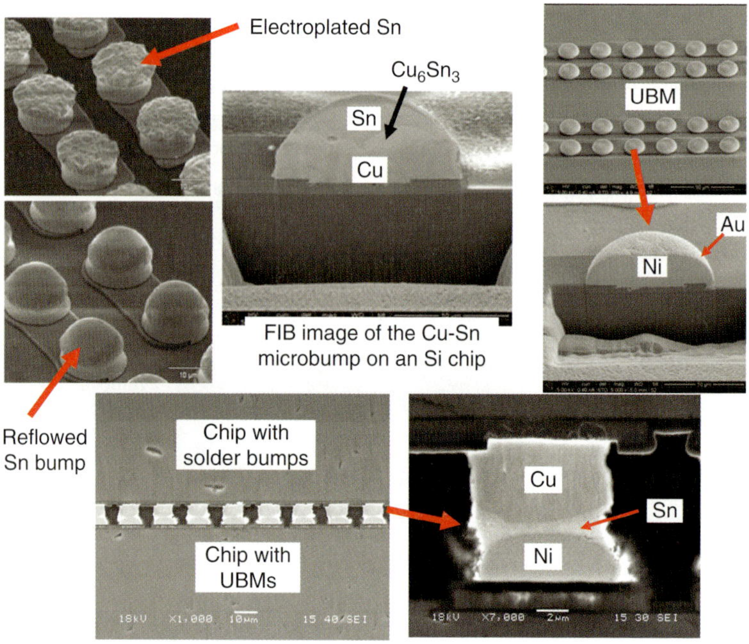

FIGURE 1.25 C2W assembly with microbumps on 8 μm pads and 15 μm pitch.

Underfill is usually needed between two chips (one with the Cu-filled TSVs and the other one without) because of the thermal expansion mismatch [7, 21]. (The equivalent TCE of a Cu-filled TSV silicon chip could be as high as $10 \times 10^{-6}/°C$.) Due to the tiny gap between the chips, underfills with small filler sizes (maximum ≤3 μm and average ≤0.5 μm) are recommended.

1.5.10 Thermal Management of 3-D IC Integration

Thermal management of 3-D IC integration system is very important, and low-cost and effective thermal management design guidelines and solutions are desperately needed for the widespread use of 3-D IC integrations. Even with the most advanced software and high-speed hardware, it is impossible to model all the TSVs in a 3-D IC integration system. In this section, based on the theory of heat transfer and computational fluid dynamics (CVD), the equivalent thermal conductivity of a TSV interposer/chip with various TSV diameters, pitches, and aspect ratios are developed. These equivalent thermal conductivities are then used (with the TSV chip/interposer as a block) to perform all the simulations [26].

Figure 1.26 schematically shows a high-performance packaging system. The high-performance chip is connected to the TSV interposer (or chip) through microbumps, and the bottom of the TSV

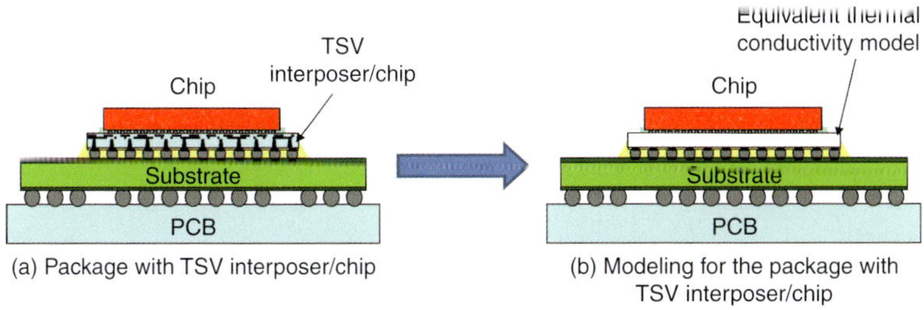

FIGURE 1.26 Equivalent thermal conductivity model TSV interposer.

interposer/chip is connected to the organic substrate by the ordinary solder bumps. Due to the large thermal expansion mismatch between the silicon interposer/chip and the organic substrate, underfill is needed so the solder joints are reliable. Underfill may also be necessary between the high-performance chip and the copper-filled TSV chip (or interposer) since, depending on the number (pitch) of vias and the via diameter, these two chips may create a very large thermal expansion mismatch and the microbump solder joints may not be able to survive.

As mentioned earlier, the equivalent thermal conductivity of the TSV interposer/chip is determined through detailed 3-D CFD analyses. The TSV chip/interposer is schematically shown in the upper-right corner of Fig. 1.27. It can be seen that, unlike the thermal conductivity of the ordinary silicon chip, which is isotropic, because of the

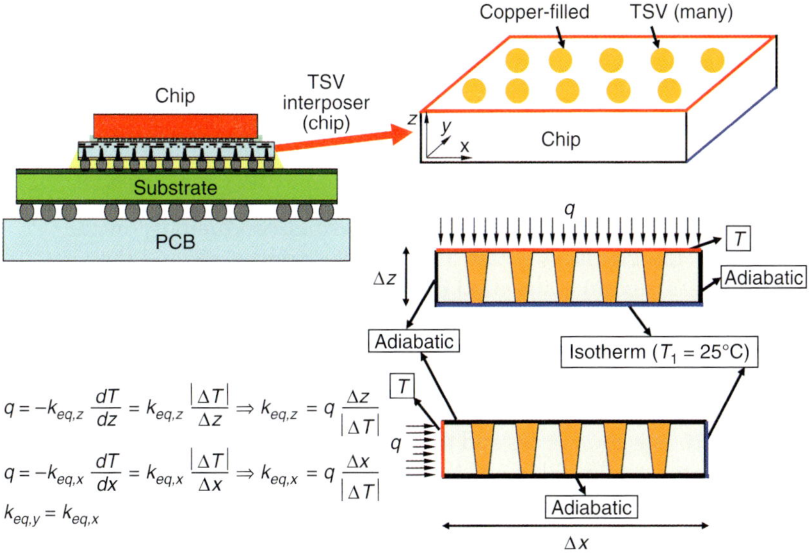

FIGURE 1.27 Equivalent thermal conductive models for computational fluid dynamic.

<table>
<tr><th>Component</th><th>Interposer</th><th>TSV</th><th colspan="4">TSV filler</th></tr>
<tr><td>Material</td><td>Si</td><td>Cu</td><td>Polymer with filler particles</td><td>Polymer with filler particles</td><td>Al</td><td>Cu</td></tr>
<tr><td>Thermal conductivity (W/m°C)</td><td>150</td><td>390</td><td>4.0</td><td>25</td><td>237</td><td>390</td></tr>
<tr><td>Dimension detail (mm)</td><td>1.4 × 1.4 × 0.3</td><td>Varied</td><td colspan="4">Copper plating thickness for partial filled via (5~25 μm)</td></tr>
<tr><td>TSV pitch (mm)</td><td colspan="6">Ranging from 0.15 to 0.6</td></tr>
</table>

TABLE 1.1 Geometry and Material of the Structure

copper-filled TSVs, the thermal conductivity of the TSV silicon chip is anisotropic, that is, the thermal conductivity in the *xy*-planar directions ($k_{eq,x} = k_{eq,y}$) is not equal to that in the *z*-normal direction ($k_{eq,z}$).

The approaches for extracting ($k_{eq,z}$) and ($k_{eq,x} = k_{eq,y}$) are shown, respectively, in the middle and bottom of the right side of Fig. 1.27. First, construct the geometry of the copper-filled TSV chip with various diameters, pitches, and aspect ratios. Then, input the thermal material properties (Table 1.1) of all the elements in the model. Finally, apply the kinetic and kinematic boundary conditions, and calculate the temperature distributions. The equivalent thermal conductivity can be obtained with the equations shown:

$$q = -k_{eq,z}\frac{dT}{dz} = k_{eq,z}\frac{|\Delta T|}{\Delta z} \Rightarrow k_{eq,z} = q\frac{\Delta z}{|\Delta T|}$$

$$q = -k_{eq,x}\frac{dT}{dx} = k_{eq,x}\frac{|\Delta T|}{\Delta x} \Rightarrow k_{eq,x} = q\frac{\Delta x}{|\Delta T|}$$

$$k_{eq,y} = k_{eq,x}$$

For example, to extract the equivalent thermal conductivity in the *z*-direction, the geometry of the TSV chip is constructed, and then a uniform heat flux (q) is imposed on the top surface of the TSV chip, and the bottom surface is set as the isotherm boundary (25°C), while the four surrounding boundaries are set as adiabatic boundaries. By using the Flowtherm software, the average temperature on the top surface of the TSV chip can be calculated and consequently ($k_{eq,z}$) can be obtained. The geometries and thermal properties used for detailed modeling are listed in Table 1.1. As shown in [26], the equivalent thermal conductivity in the *x*-, *y*-, and *z*-directions is:

$$k_{eq,z} = 150 + 188\, D^2/P^2$$

$$k_{eq,x} = k_{eq,y} = 150 + 105\, D^2/P^2$$

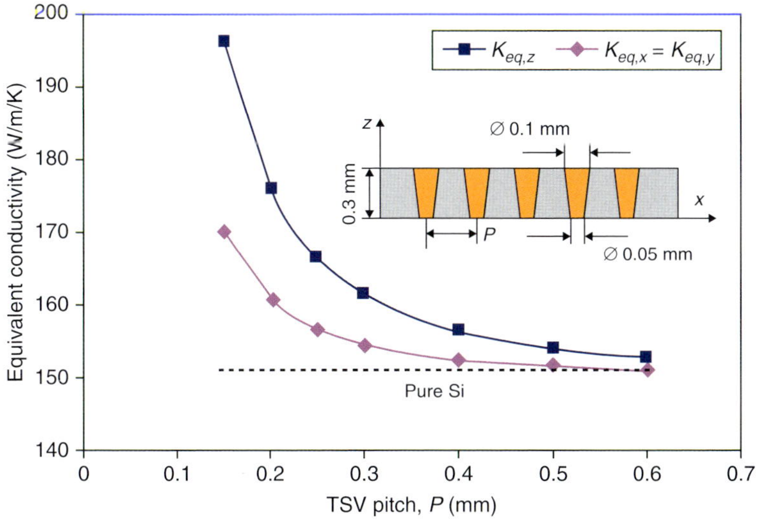

FIGURE 1.28 Thermal conductivity of copper-filled TSV chips/interposers with various pitches.

where P is the pitch, $D = (D_1 + D_2)/2$ is the diameter, and D_1 and D_2 are the diameters of a tapered TSV.

Figure 1.28 shows the pitch effect of the Cu-filled TSV on the equivalent thermal conductivity of the TSV interposer/chip with a thickness of 300 μm and an average diameter of 75 μm. It can be seen that: (1) the equivalent thermal conductivity of the TSV interposer/chip increases when the TSV pitch decreases (i.e., more vias for a given chip size); (2) the equivalent thermal conductivity is larger than that of the pure silicon material [the thermal conductivity of silicon is 150 W/(m·°C) and that of copper is 390 W/(m·°C)]; (3) the equivalent thermal conductivity in the z-direction ($k_{eq,z}$) is larger than that in the x- and y-directions ($k_{eq,x}$ and $k_{eq,y}$); and (4) the equivalent thermal conductivity is more sensitive to the smaller TSV pitches, that is, there are lots of Cu-filled vias.

Figures 1.29 and 1.30 show the equivalent thermal conductivity of the TSV interposer/chip with different filler materials (a) in the normal direction and (b) in the planar direction. The wafer is assumed to be 0.3 mm thick, the copper plating thickness is 10 μm, and the aspect ratio is 4. It can be seen that the overall trend is that the equivalent thermal conductivity increases with the thermal conductivity of the filler materials. However, the filler material has negligible effect on the equivalent thermal conductivity for TSV with larger pitches (e.g., pitch equals 0.6 mm). Furthermore, when the thermal conductivity of filler materials (e.g., 4 W/m°K and 25 W/m°K) is much less than that of the pure silicon interposer, the filler material has a much greater effect on the equivalent thermal conductivity in the x- and y-directions

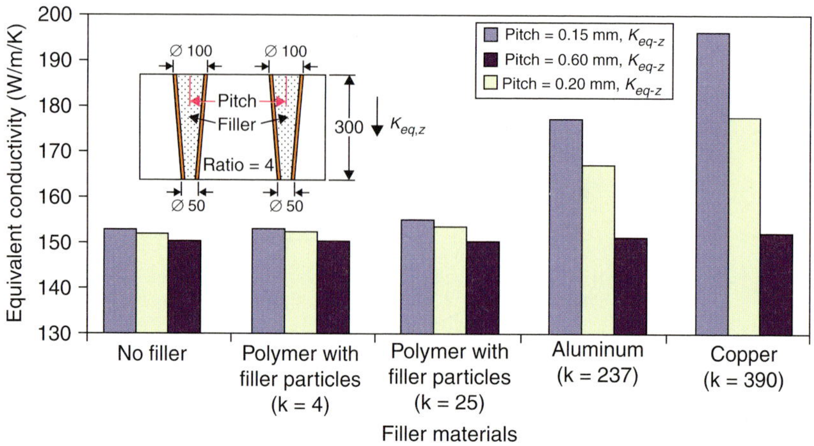

Figure 1.29 Equivalent thermal conductivity of TSV interposers/chips with different fillers (*Z*-direction).

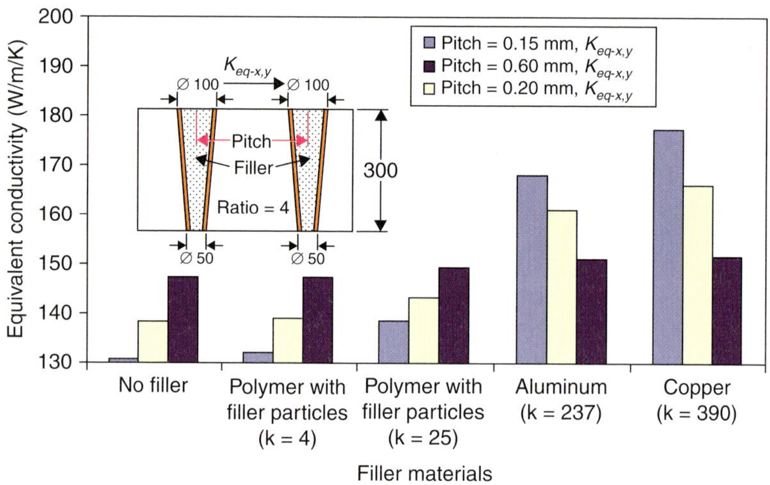

Figure 1.30 Equivalent thermal conductivity of TSV interposer/chip with different fillers (*X*- and *Y*-directions).

than that in the z-direction. This is due to the fact that the thermal path in the x- or y-direction is partially isolated by the low thermally conductive material, while the thermal path goes smoothly through the silicon in the z-direction.

Figure 1.31 [180] shows the schematic for the compact modeling of the 3-D IC integration with the TSV chip/interposer, which is modeled as a block (without TSV as shown in Fig. 1.26*b*) with the determined

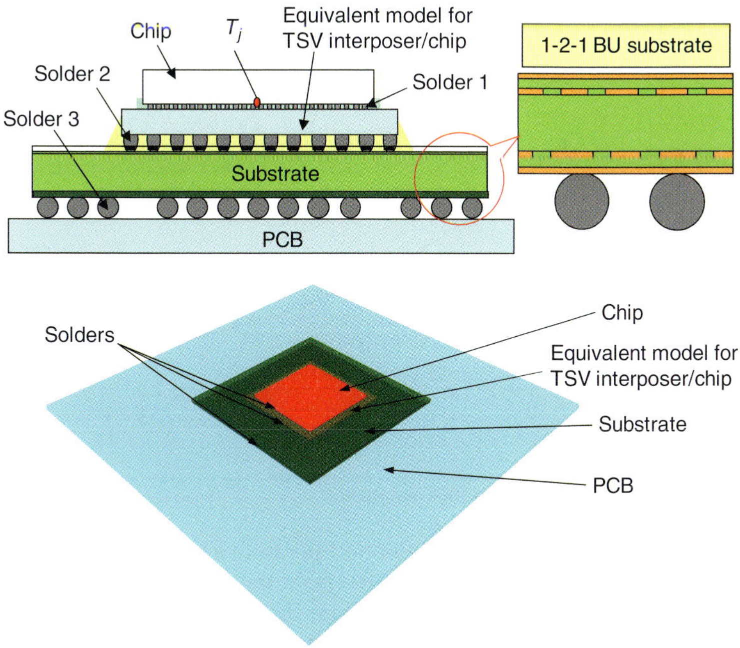

FIGURE 1.31 A simple model for a high performance chip on a TSV interposer and then on a BT-substrate.

equivalent thermal conductivities. This model consists of a large chip, a block (for the TSV interposer/chip), a 1-2-1 BU (build-up) substrate, a PCB, and solder joints. Their dimensions and thermal properties are listed in Table 1.2.

The junction to ambient thermal resistance of a package is often used as an index to describe the thermal performance of the package—the higher the thermal resistance, the poorer the thermal performance of the package. The junction to ambient thermal resistance of the package (R_{ja}) can be expressed as $R_{ja} = (T_j - T_a)/P$ where T_j and T_a

	Chip	TSV	Bumps	Underfill	PCB
Material	Si (TSV)	Cu	SnAg	Polymer	FR4
k (W/m/°C)	Empirical equation	390	57	0.5	// 0.8 ⊥ 0.3
Dimension (mm)	5 × 5	Ø 0.05	Ø 0.20 Height = 0.15	5 × 5 × 0.15	76 × 114 × 1.6
Power (W)	0.2 W	N.A	N.A	N.A	N.A

TABLE 1.2 Material Properties, Dimension, and Power of the Structure

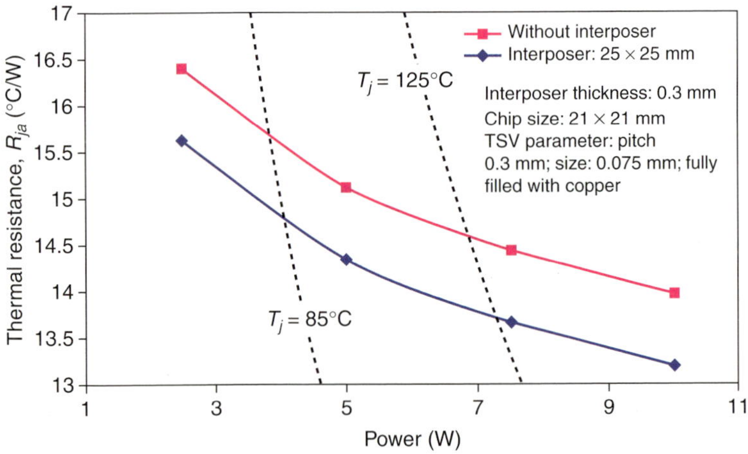

FIGURE 1.32 Thermal resistance vs. power.

respectively represent the junction temperature and ambient temperature, and P is total power dissipated from the chip. Figure 1.32 [180] shows the junction to ambient thermal resistance of packages with different chip power dissipations. It can be seen that: (1) the thermal resistance reduces as the chip power increases (this is because high chip power induces high package temperature; as such, high heat transfer coefficient is induced and more heat is removed from the package); and (2) the thermal resistance of the TSV package is lower than that of the package without TSV. (This is because of the TSV interposer/chip's spreading effect.) The dashed lines in the Fig. 1.32 represent the correlations between the thermal resistance and chip power when the junction temperature equals 85°C and 125°C, respectively.

Figure 1.33 [180] shows the effect of the TSV interpose/chip area on the package thermal performance. It can be seen that the thermal resistance (R_{ja}) decreases about 14% when the TSV interposer size increases from 21 × 21 mm to 45 × 45 mm. This indicates that package thermal performance can be improved by increasing the TSV interposer/chip size.

Figure 1.34 [180] shows the effect of the chip size for different size ratios of the TSV interposer to the chip on the thermal performance of the package. It can be seen that for the same size ratio of chip to TSV interposer, the smaller sizes of chips induce higher thermal resistance of the package.

1.5.11 3-D IC Packaging versus 3-D IC Integration

The biggest difference between 3-D IC integration and 3-D IC packaging is that 3-D IC integration utilizes the TSV to connect chips in the vertical direction, while the 3-D IC packaging doesn't. The advantages

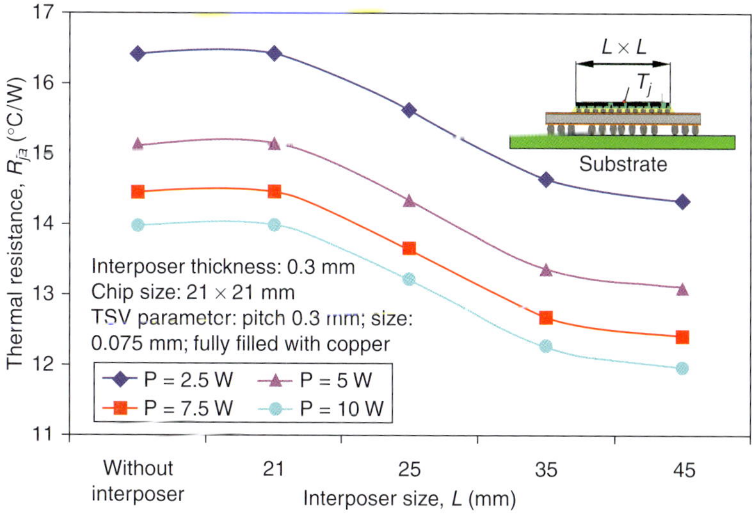

FIGURE 1.33 Thermal resistance vs. interposer size.

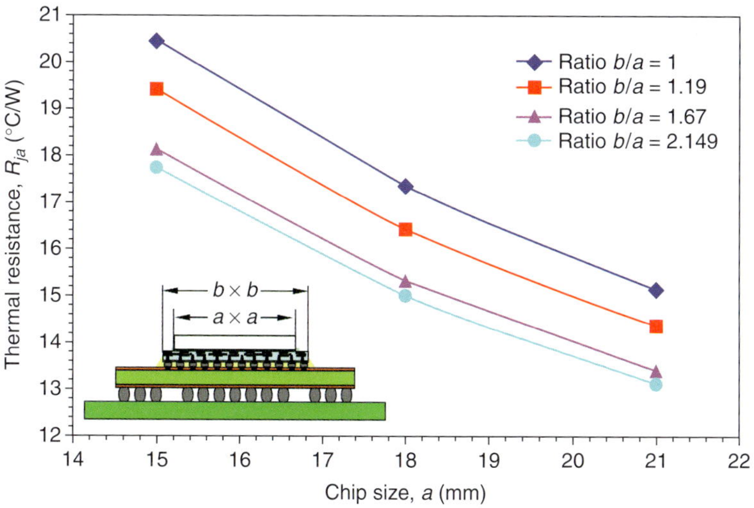

FIGURE 1.34 Thermal resistance vs. chip size.

of the 3-D IC integration over the 3-D IC packaging are: (1) smaller footprint, (2) smaller form factor, (3) less weight, (4) higher performance, (5) less power, and (6) potentially lower cost, which are the key reasons that system houses would kill for. The challenges (opportunities) facing 3-D IC integration, among others, are: (1) high-yield TSV fabrication, (2) thin chip and wafer strengthening and handling, (3) microbumping and assembly, (4) thermal management, (5) inspection and testing,

(6) cost reductions, and (7) reliability. Of course, designs and characterizations are always important.

1.6 3-D Si Integration Technology

1.6.1 3-D IC Integration versus 3-D Si Integration

As mentioned earlier, 3-D Si integration is one of the more-than-Moore concepts, and is a very old idea [27, 28], which consists of two or more layers of active electronic components (in wafer form) that are integrated vertically into a single circuit. It was trigged by the advance of the silicon-on-insulator (SOI) technology first reported by Gat and his colleagues more than 30 years ago [29], when people thought Moore's law could be hitting the wall by the 1990s. (Of course, it didn't.)

The biggest difference between the 3-D IC integration and 3-D Si integration is that 3-D Si integration doesn't use bumps such as solder, gold, or conductive adhesive to bond/connect two wafers together, that is, it is bumpless. It should be emphasized that there is no (or an infinitesimal) gap between wafers, and thermal management could be a big problem. Also, W2W is the only way to perform the bonding operation. Thus, unlike 3-D IC integration, which utilizes C2C, C2W, or W2W bonding methods, yield is a big issue (e.g., some bad chips are forced to bond on the good chips). Furthermore, the TSV size for 3-D Si integration is much smaller (~1 μm and eventually will be in submicro scales), as shown in Fig. 1.10, and thus its TSV manufacturing yield loss is higher. Finally, the requirements of the bonding conditions, such as the surface cleanness, surface flatness, and the class of cleanroom (which are heavily affecting the yield) for 3-D Si integration are much tougher than those for 3-D IC integration. On the other hand, the advantages of 3-D Si integration over the 3-D IC integration are: (1) better electrical performance, (2) less power, (3) lower profile, (4) less weight, and (5) potentially lower cost. There are two different W2W bonding methods for 3-D Si integration—namely, Cu-to-Cu bonding and oxide-to-oxide bonding, as shown in Fig. 1.10. Again, the thermal management of 3-D Si integration is much tougher than that of 3-D IC integration.

1.6.2 W2W (Cu-to-Cu) Bonding

Figure 1.35*a* schematically shows the IBM/RPI bonding structure of two device layers bonded in a face-to-face approach, and Fig. 1.35*b* shows one with the face-to-back approach [30–32]. A typical Cu-to-Cu interconnect is shown in Fig. 1.36, which shows a high-quality bonding interface. Before bonding, the Cu interconnects (pads) are fabricated with the standard BEOL damascene process, followed by the oxide CMP processed (oxide touch-up) to recess the oxide level to 40 nm lower than the Cu surface. The bonding temperature is ramped up to

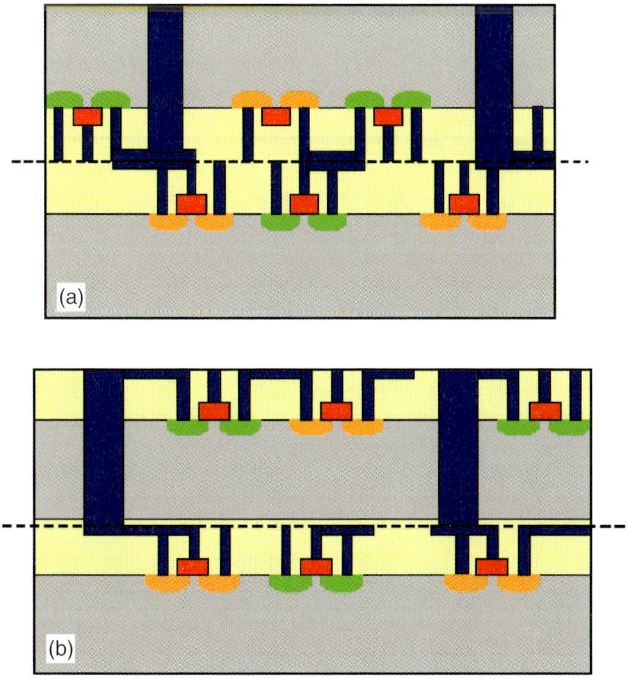

Figure 1.35 (a) Face-to-face W2W (Cu-to-Cu) bonding and (b) face-to-back W2W (Cu-to-Cu) bonding.

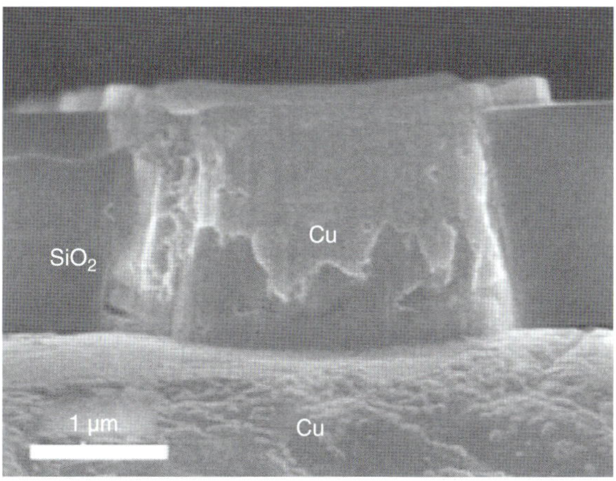

Figure 1.36 IBM's Cu-to-Cu (W2W) bonding structure.

400°C. Excellent electrical connectivity and good thermal reliability have been achieved [30–32].

Figure 1.37 schematically shows the NIMS/AIST/Toshiba/University of Tokyo bonding structure of two device layers bonded at

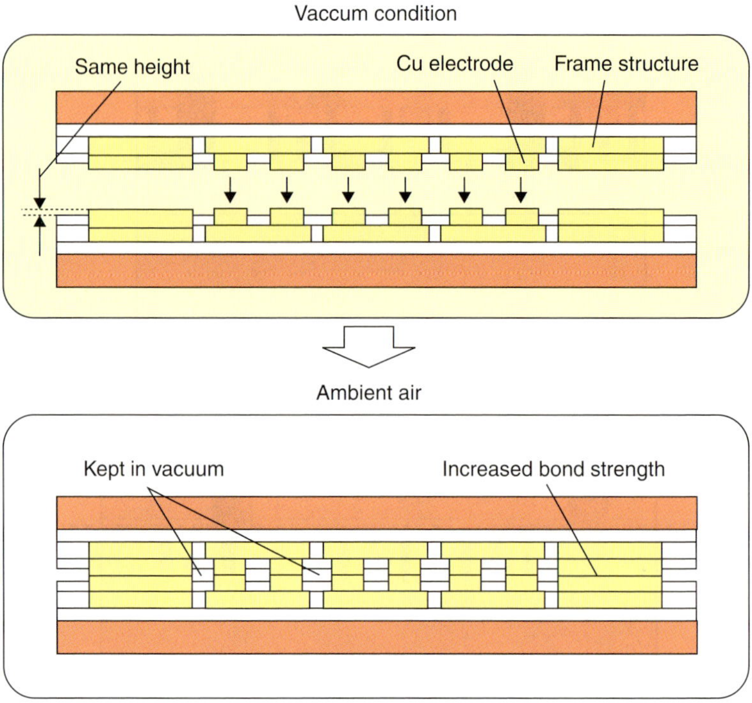

FIGURE 1.37 Face-to-face Cu-to-Cu (W2W) bonding at room temperature.

room temperature [33–37]. A typical cross-sectional scanning electron microscope (SEM) image of the interface between the bumpless electrodes (pads) after a high-temperature storage test is shown in Fig. 1.38. It can be seen that a tight adhesion between surfaces is maintained even after exposed to 150°C for 1000 hours [33–37].

1.6.3 W2W (SiO_2-to-SiO_2) Bonding

Figure 1.39 schematically shows MIT's oxide-to-oxide bonding structure of three device layers bonded at 275°C [38–44]. It can be seen that: (a) two completed circuit wafers (Tier 1 and Tier 2) are planarized, aligned, and bonded face-to-face together; (b) a wet etch of the handle silicon exposes the buried oxide (BOX) of the upper wafer; (c) 3-D vias are patterned and etched through the BOX and deposited oxides expose metal contacts in both layers; (d) a Ti/TiN liner and 1 μm of tungsten (W) are deposited to fill the 3-D vias (with the larger diameter equal to 1.5 μm) and electrically connect the two layers; and (e) a third layer can then be bonded with its face to the BOX (back) of Tier (layer) 2 and 3-D vias are formed. Figure 1.40 shows a typical cross-section of the three-layer 3-D (ring oscillator) structure. It can be seen that: (1) the layers are bonded and interconnected with W-plugs,

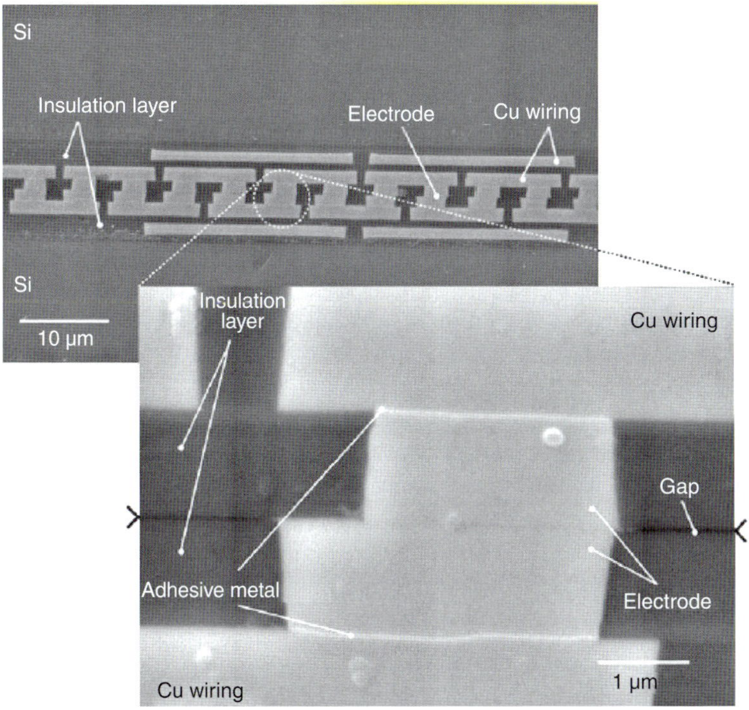

FIGURE 1.38 Cu-to-Cu (W2W) room temperature (bumpless) bonding

(2) the conventional interlevel connections are in the bottom two layers, and (3) the 3-D vias are located in the isolation (field) region between transistors. A few functional 3-D structures/circuits have been created and demonstrated [38–44].

Figure 1.41 schematically shows Leti/Freescale/STMicroelectronics' dielectric-to-dielectric bonding structure of two device layers bonded at < 400°C [45–47]. It can be seen that: (a) first, a metal level is formed on a 200 mm bulk wafer and SOI wafer; next, these wafers are bonded face to face, and then the bulk silicon of the SOI wafer is removed down to the BOX layer; (b) the interstrata vias (ISVs) are formed, which make contact from upper strata to lower strata; and (c) a metal layer is formed at the top of the back side of the SOI wafer. The typical cross-section of an ISV is shown in Fig. 1.42 [45–47], and it can be seen that this ISV (~1.5 μm) makes good contact with both the top and bottom metal layers, and is void-free [45–47].

1.6.4 Future Outlook of 3-D Si Integration

Unlike the 3-D IC integration, which will be in volume production in the next few years, a lot of work still needs to be done to achieve 3-D Si integration. Besides thin-wafer handling, vias formation, and thermal

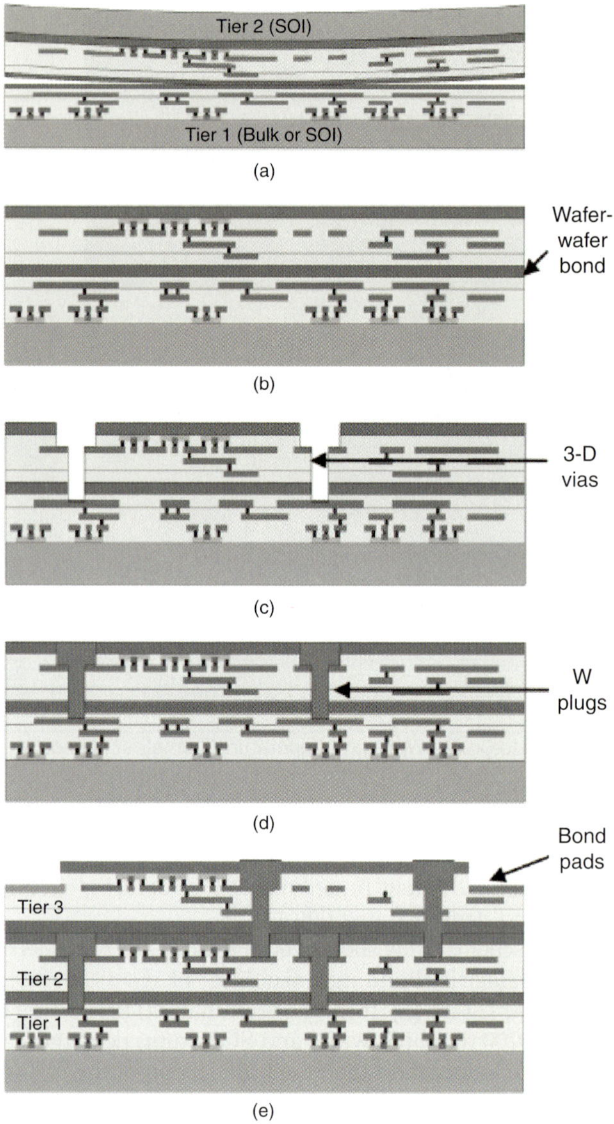

FIGURE 1.39 MIT's 3-D Si integration: (a) Two completed circuit wafers are planarized, aligned, and bonded face-to-face, (b) the handle silicon is removed, (c) 3-D vias are etched through the deposited BOX and the field oxides, (d) tungsten plugs are formed to connect circuits in both tiers, and (e) after tier 3 is transferred, bond pads are etched through the BOX for testing and packaging.

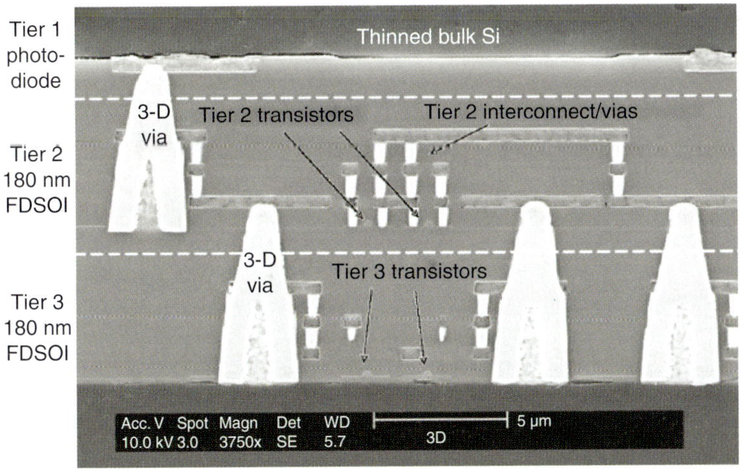

FIGURE 1.40 SEM image of MIT's SiO_2-to-SiO_2 (W2W) bonding, which shows that the tiers are bonded and interconnected with tungsten plugs. Note that the vias are located in the isolation (field) region between transistors.

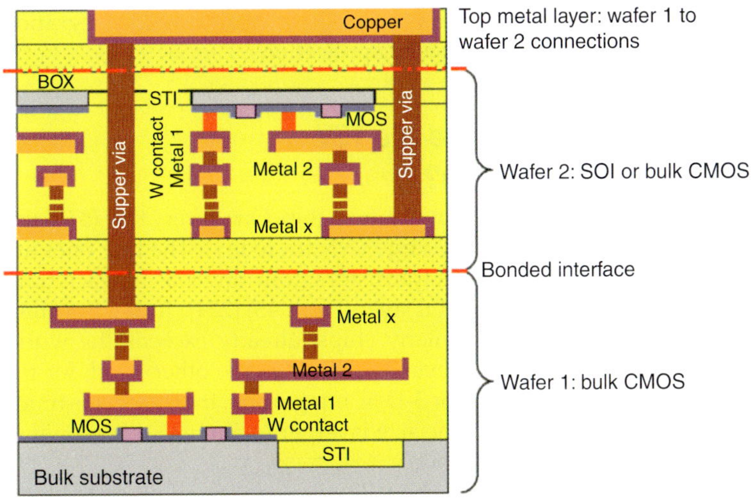

FIGURE 1.41 Leti's 3D Si integration SiO_2-to-SiO_2 bonding.

management, more research and development efforts should be placed on, for example, cost reduction, design and process parameter optimization, bonding environment, W2W bonding alignment, wafer distortion, wafer bow (warpage), inspection and testing, contact performance, contact integrality, contact reliability, and manufacturing yield issues. Also, how to systematically and reliably package the 3-D Si integration to the next level of interconnect is a new challenge.

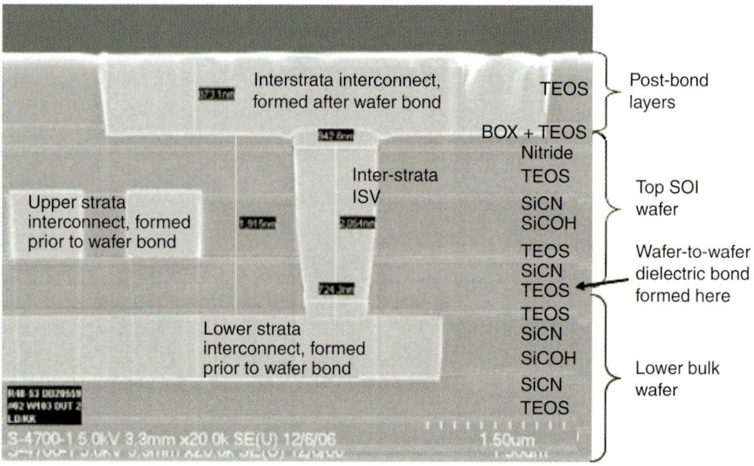

FIGURE 1.42 Leti's inter-strata connection, SiO_2-to-SiO_2 (W2W) bonding.

After all, the 3-D Si integration must be attached to the PCB or flexible board in order for a product to function. In the next 10 years, it would be hard-press for high volume products with the 3-D Si integration technology, except for very niche applications.

1.7 Who Is Going to Make the TSV for 3-D IC Integration?

There are at least two different vias on a chip. One is very tiny (≤ 0.1 μm today), and the other is very large (≥ 5 μm today). The tiny ones are connected to devices such as transistors (four tiny vias for each transistor) and build the first metal layer (Fig. 1.43) [224]. Today, the number of these tiny vias, for many chips, already exceeds the world population, which is more than 7 billion. On the other hand, for the large vias that we call TSV for 3-D IC integration, the number is much less (< 100,000 today) and the size is larger (a 5-μm TSV is shown in Fig. 1.43 to show their contrast).

Based on what happened about 20 years ago to the low-cost, solder-bump, flip-chip, and wafer-level packaging technologies [4], it will not be a surprise to see that, at the early stage, semiconductor foundries would like to do TSV with or without RDLs and wafer bumping, but when the volumes pick up, the packaging assembly and test houses will do it all because they are more flexible. Also, their core competence and major business are to build packages for the chips from the wafers given (made and tested) by the semiconductor foundries and to perform the final packaging tests. Then they ship only the good ones to EMS (electronics manufacturing services), who perform the PCB assembly, in-circuit test, system (also called final or functional) test,

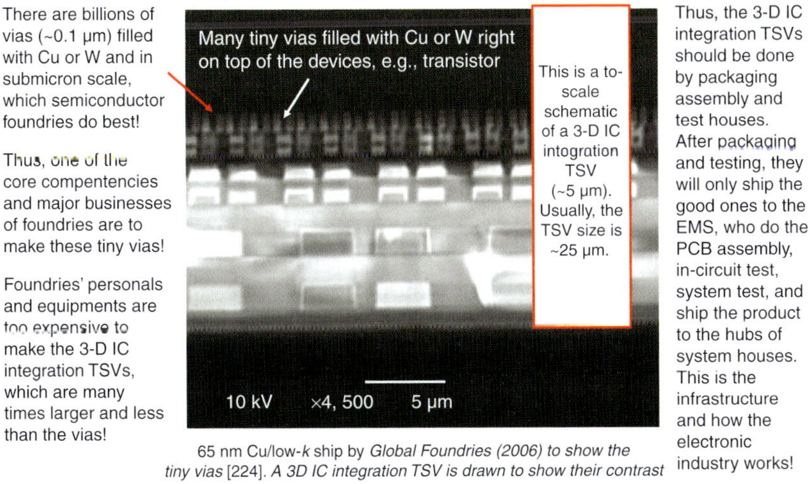

FIGURE 1.43 Who is going to make the 3D IC integration TSVs?

and ship the product to the system houses' hubs for distribution. Some system houses prefer to have their EMS ship them the in-circuit tested, good PCBs, perform the final test in-house, and then ship the product. This is the infrastructure (as shown in Fig. 1.44) and how the electronics industry works. Of course, for certain special reasons or niche applications, there are always a few exceptions.

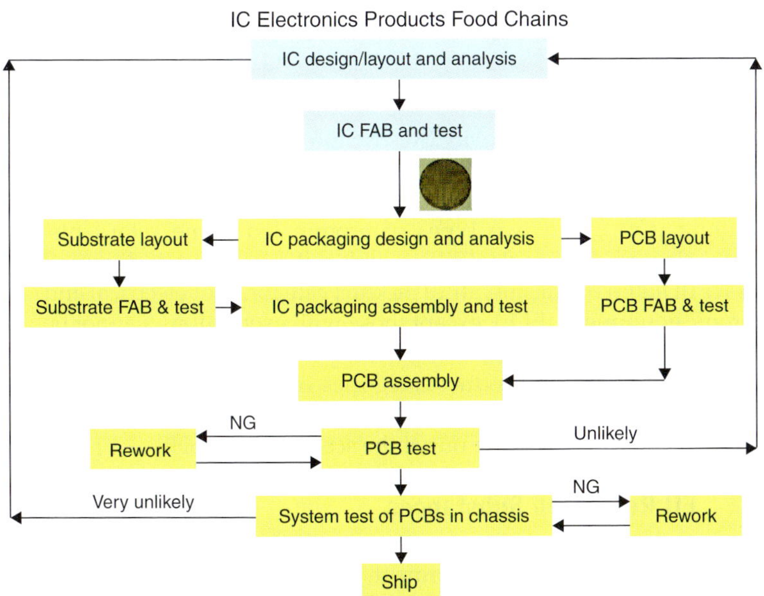

FIGURE 1.44 Food chain of electronics products.

1.8 Notes on TSV Manufacturing Costs

IC chip yield (Y_T) plays the most important role in TSV manufacturing cost. If Y_T is low for a particular IC device, then it is not cost-effective to use the TSV technology because it makes the good die too expensive, unless it is compensated for by density, performance, weight, and form factor.

TSV manufacturing yield (Y_{TSV}) plays the second most important role in TSV manufacturing cost. Since this is the first post-wafer processing after the IC semiconductor fabs, the packaging assembly and test houses should strive to make $Y_{TSV} > 99.99\%$. Otherwise, the subsequent steps will be expensive, with wasted material and processes on the damaged TSV die.

Wafer-bumping yield ($Y_B > 99.99\%$) plays the third most important role in TSV manufacturing cost. The wafer-bumping/packaging assembly and test houses should strive to make $Y_{TSV}Y_B > 99.99\%$ to minimize the hidden cost [222, 223], since they cannot afford to damage good die that already have TSVs.

Based on these three considerations, the dates in the 3-D IC integration roadmap (Fig. 1.9) most likely will be postponed because of the tough requirements of TSV manufacturing yield ($Y_{TSV} > 99.99\%$) and the very high hidden costs [222, 223]. Meantime, passive interposers may help the electronic industry ship some 3-D IC integration products. Please see Chapter 12, Sec. 12.6 of this book.

1.9 EU RoHS Update

1.9.1 EU RoHS

Since July 1, 2006, the RoHS (Restriction of the Use of Certain Hazardous Substances) directive has been implemented in EU countries. It bans lead (Pb), mercury (Hg), cadmium (Cd), hexavalent chromium (Cr^{6+}), and two types of brominated flame retardants: polybrominated biphenyls (PBBs) and polybrominated diphenyl ethers (PBDEs). That means, all 10 product categories of electrical and electronic equipment (EEE), except those with exemptions, cannot be put on the market in the EU if they contain those six banned materials. For a list of the 10 product categories of EEE and exemptions, please consult EU's RoHS websites. It should be noted that currently category 8 and category 9 are out of the scope of EU RoHS.

1.9.2 EU Restricted Substances

Recently, a draft proposal for a recast of the RoHS directive has been published by the European Parliament, who proposed to increase the list of restricted substances (such as PVC, chlorinated plasticizers, organohalogens, flame retardants, and the phthalates BBP, DBP, and DEHP) and include all electrical products. Also, the EU Commission proposed to change the definition of homogeneous materials, which would now align itself more along the lines of China's RoHS. The definition states that a homogeneous material is one that consists of only one material throughout, a combination of multiple materials that cannot be mechanically disjointed into different materials, or finally, a surface coating.

1.9.3 EU Product Category

The European Parliament proposed to add an eleventh product category to the existing 10. The scope of the new category would simply be "other EEE not covered by any of the categories 1–10." In addition, the current exclusion of large-scale stationary industrial tools (LSIT) has been deleted, so all EEE, including manufacturing production line equipment, would be in scope. All these would come into scope in July 2014. Furthermore, the controversial "equipment that is part of another type of equipment that does not fall in scope and can only fulfill its function if it is part of that equipment," is proposed to change to "part of stationary installations or transport equipment that is not electrical or electronic equipment." This would bring, for example, the car radio, into scope.

1.9.4 EU Exemptions

With regard to exemptions, the European Parliament proposed that the expiry period be "up to" four years and not the current four years. The EU Commission will decide within six months of an exemption expiring whether or not it will be renewed. Grace periods will be allowed, but for no more than 18 months after the exemption expires. The exemption of the spare parts will be limited to 42 months after the amended directive enters into force. The substances mentioned above will also be restricted in spare parts.

All these proposals will be debated well into 2010, but could have a significant impact on the industry, at the very least leading to more substance data collection on products falling within scope and several new restricted substances. Eventually, RoHS would cover just about every EEE!

1.10 China RoHS Update

1.10.1 China RoHS

China's RoHS was promulgated on February 28, 2006, by seven different organizations, namely Ministry of Information Industry (MII); National Development and Reform Commission (NDRC); Ministry of Commerce; Administration of Customs; Administration of Industry and Commerce; Administration of Quality Supervision, Inspection, and Quarantine (AQSIQ); and State Environmental Protection Administration (SEPA). There are eight chapters with 41 articles in the China

RoHS and it bans the same six materials as the EU RoHS and undefined others. The implementation date of the China RoHS was March 1, 2007. China's RoHS requires the formulation of "standards" and a "product catalogue." This product catalogue will be issued in batches and gradually expanded.

China's RoHS uses a "two-step" method to deal with the banned materials. The first step has been enforced since March 1, 2007, where the electronic information products (EIP) put into the Chinese market and all the EIP produced in China have to be marked according to the "Marking for control of pollution caused by electronic information products" standard. It should be noted that EIP include just about every electronic product, even printed circuit boards, components, and materials! The good news is that at this stage, even the maximum concentrate values (MCV) of the banned substances in all the homogenous materials of your products exceed the "Requirements for concentration limits for certain hazardous substances in electronic information products" standard. Thus, you still can put your products into the Chinese market and produce your products in China.

The second step is serious! If your product is listed in the China RoHS product catalogue, before you can even put your product in the Chinese market, the MCV of the banned substances in all the homogenous materials of your products must meet the required standards and then must pass the China Compulsory Certification (CCC).

1.10.2 China Standards

Nine standards accompany the Chinese RoHS that pertain to marking for control of pollution caused by electronic information products; requirements for concentration limits for certain hazardous substances in electronic information products; testing methods for regulated substances in electronic information products; lead-free solder-chemical compositions and forms; test method for lead-free solders; lead-free solder powders standard; lead-free solder pastes standard; lead-free solder fluxes standard; and general guidelines for the environmental friendly use period of electronic information products.

1.10.3 China's Product Catalogue

The selection criteria for the product catalogue will be based on the feasibility of economics and maturity of technologies to make the products, as well as most of the electronic enterprises that can use the replaceable (substitute) products. Also, it should be noted that all the products in the product catalogue are green products, which meet the requirements for concentration limits for all the banned hazardous substances and pass the CCC.

The first draft of the product catalogue was published on October 9, 2009, allowing for a one-month consultation period. The restrictions will come into force 10 months after the adoption of the legislation, unless changes are made. The first batches of products are principally telephones (such as mobile phones, networked handsets, and land-line telephones) and all types of printers that attach to a computer. The restricted substances are the same as in the EU RoHS (excluding deca-BDE), as are the MCV of homogeneous material, but China's RoHS specifically refers to coatings and very small components. Also, they allow some exemptions that are in the EU RoHS: 10 for telephones and 12 for printers.

1.10.4 China Compulsory Certificate (CCC)

In order to pass the CCC, the product must be tested and analyzed. The testing can only be carried out in China via authorized Chinese laboratories. Unfortunately, the precise policies and procedures of CCC for the China RoHS have not established by the Chinese government yet. MII and the Certification and Accreditation Administration of China (CNCA) of AQSIQ are working very hard on this area. CNCA will soon announce the policies/procedures and recommend a few organizations, such as the System Certification Center of China Electronics Standardization Institute (CESI) and China Quality Certification Centre (CQC) to oversee the CCC for the China RoHS.

1.11 Impact of RoHS on the Electronics Industry

Since 2000, the electronics industry has spent billions of dollars to comply with the RoHS directive. More than 95 percent of the effort has been directed at eliminating lead (Pb), that is, creating lead-free materials, developing soldering processes, performing characterization measurements and reliability assessments, etc. [224–610].

The eutectic SnPb solder (63wt%Sn and 37wt%Pb), with a melting point of 183°C, has been the predominant choice for assembling electronics products for decades due to its outstanding solderability and reliability. Because of the RoHS directive, Pb is banned in most electronics products made by the conventional IC packaging, 3-D IC packaging, and 3-D IC integration technologies as shown in Figs. 1.5 and 1.9.

Not long ago, there were more than 100 lead-free solder alloys to choose from, as shown in Table 3.1 of [4]. However, because of the very tight regulatory deadlines, the industry was "rushed" to select SnAgCu solder alloys and was "in a great hurry" to build the standards and infrastructure around the Sn-(1–4 wt%)Ag-(0.5–0.7 wt%)Cu lead-free solder family. The melting point of these solders is around 217°C, which is 34°C higher than that of the SnPb solder. Thus, during assembly, the ICs, packages and PCBs will be subjected to higher soldering temperatures, and their cost, performance, and reliability are of great concern to manufacturers [224–610]. Higher energy consumption is another concern.

Component	Chip	Solder 1	Interposer	Solder 2	Substrate	Solder 3	PCB
Material	Si	SnAg	Si+TSV (Cu)	SnAg	Buildup	SnAg	FR4
Thermal Conductivity (W/m/C)	150	57	Keq	57	// 100 ⊥ 0.5	57	// 0.8 ⊥ 0.3
Dimension (mm)	21 × 21 × 0.75	Pitch: 0.15 Height: 0.08 Aver D: 0.08	Variable	Pitch: 0.5 Height: 0.1 Aver D: 0.1	1-2-1 (45 × 45 × 1)	Pitch: 1.0 Height: 0.6 Aver D: 0.6	101 × 114 × 1.6

Table 1.3 Dimensional and Thermal Conductivity of the Structure

Due to a lack of experimental and simulation data, one of the critical issues of lead-free soldering is solder-joint reliability, especially for high-reliable (>10 years of life) products. As discussed earlier, and as shown in Figs. 1.5 and 1.9, that solder is the electrical and mechanical "glue" of electronic assemblies. For semiconductor electronics, solder materials are used for the 3-D IC integration interconnects (e.g., microbumps), first-level interconnects (e.g., solder bumps), and second-level interconnects (e.g., solder balls). For opto-electronics, LED, and MEMS, solder materials are used for sealing, die attachment, solder bumps, and/or solder balls. Thus, the reliability of lead-free (RoHS-compliant) solder joints (interconnects) is under scrutiny [16–610] and the focus of this book.

1.12 References

1. Lau, J. H., Lee, C., Premachandran, C., and Yu, A., *Advanced MEMS Packaging*, McGraw-Hill, New York, 2010.
2. Lau, J. H., C. P. Wong, N. C. Lee, and R. Lee, *Electronics Manufacturing with Lead-Free, Halogen-Free, and Adhesive Materials*, McGraw-Hill, New York, 2003.
3. Lau, J. H., and R. Lee, *Microvias for Low Cost, High Density Interconnects*, McGraw-Hill, New York, 2001.
4. Lau, J. H., *Low Cost Flip Chip Technologies for DCA, WLCSP, and PBGA Assemblies*, McGraw-Hill, New York, February 2000.
5. Lau, J. H., and R. Lee, *Chip Scale Package: Design, Materials, Process, Reliability, and Applications*, McGraw-Hill, New York, February 1999.
6. Lau, J. H., C. P. Wong, J. Prince, and W. Nakayama, *Electronic Packaging: Design, Materials, Process, and Reliability*, McGraw-Hill, New York, February 1998.
7. Lau, J. H., and Y. Pao, *Solder Joint Reliability of BGA, CSP, Flip Chip, and Fine Pitch SMT Assemblies*, McGraw-Hill, New York, August 1997.
8. Lau, J. H., *Flip Chip Technologies*, McGraw-Hill, New York, 1996.
9. Lau, J. H., *Ball Grid Array Technology*, McGraw-Hill, New York, 1995.
10. Lau, J. H., *Chip On Board Technologies for Multichip Modules*, Van Nostrand Reinhold, New York, March 1994.
11. Lau, J. H., *Handbook of Fine Pitch Surface Mount Technology*, Van Nostrand Reinhold, New York, 1993, 1994.
12. Frear, D., Morgan, H., Burchett, S. and Lau, J. H., *The mechanics of Solder Alloy Interconnects*, Van Nostrand Reinhold, New York, 1994.
13. Lau, J. H., *Thermal Stress and Strain in Microelectronics Packaging*, Van Nostrand Reinhold, New York, 1993.
14. Lau, J. H., *Handbook of Tape Automated Bonding*, Van Nostrand Reinhold, New York, November 1992.
15. Lau, J. H., *Solder Joint Reliability: Theory and Applications*, Van Nostrand Reinhold, New York, April 1991.
16. Lau, J. H. "Critical Issues of 3-D IC Integrations." In *Proceedings of IMAPS International Symposium on Microelectronics*, San Jose, CA, November 2009, pp. 585–592. Also, to be published in IMAPS Transactions, *Journal of Microelectronics and Electronic Packaging*.
17. Lau, J. H. "Key Enabling Technologies for 3-D IC Integrations," *Professional Development Course given at IEEE Electronic & Components Technology Conference*, San Diego, CA, May 2009.
18. Lau, J. H. "Design and Process of 3-D MEMS Packaging." In *Proceedings of IMAPS International Symposium on Microelectronics*, San Jose, CA, November 2009, pp. 1–9. Also, to be published in IMAPS Transactions, *Journal of Microelectronics and Electronic Packaging*.

19. Lau, J. H., Lee, R., Yuen, M., and Chan, P. "3-D LED and IC Wafer Level Packaging," *Journal of Microelectronics International,* Vol. 27, No. 2, 2010, pp. 98–105.
20. Yu, A., Lau, J. H., Ho, S., Kumar, A., Wai, Y., Yu, D., Jong, M., Kripesh, V., Pinjala, D., and Kwong, D. "Study of 15-µm-pitch solder microbumps for 3-D IC integration." In *IEEE Proceedings of Electronic Components and Technology Conference,* San Diego, CA, May 2009, pp. 6–10.
21. Yu, A., Lau, J. H., Ho, S., Kumar, A., Yin, H., Ching, J., Kripesh, V., Pinjala, D., Chen, S., Chan, C., Chao, C., Chiu, C., Huang, M., and Chen, C. "Three dimensional interconnects with high aspect ratio TSVs and fine pitch solder microbumps." In *IEEE Proceedings of Electronic Components and Technology Conference,* San Diego, CA, May 2009, pp. 350–354. Also, to be published in *IEEE Transactions on Advance Packaging*.
22. Sekiguchi, M., Numata, H., Sato, N., Shirakawa, T., Matsuo, M., Yoshikawa, H., Yanagida, M., Nakayoshi, H., and Takahashi, K. "Novel Low Cost Integration of Through Chip Interconnection and Application to CMOS Image Sensor." In *IEEE Proceedings of Electronic Components and Technology Conference,* San Diego, CA, May 2006, pp. 1367–1374.
23. Zhang, X., Chai, T., Lau, J. H., C. Selvanayagam, K. Biswas, S. Liu, D. Pinjala, G. Tang, Y. Ong, S. Vempati, E. Wai, H. Li, B. Liao, N. Ranganathan, V. Kripesh, J. Sun, J. Doricko, and C. Vath. "Development of Through Silicon Via (TSV) Interposer Technology for Large Die (21 × 21 mm) Fine-pitch Cu/Low-*k* FCBGA Package." In *IEEE Proceedings of Electronic Components and Technology Conference,* San Diego, CA, May 2009, pp. 305–312. Also, to be published in *IEEE Transactions in Advanced Packaging*.
24. www.3m.com/electronicmaterial.
25. Pargfrieder, S., Kettner, P., Privett, M., and Ting, J. "Temporary Bonding and Debonding Enabling TSV Formation and 3-D Integration for Ultra-thin Wafers." In *IEEE Proceedings of Electronics Packaging Technology Conference,* Singapore, December 2008, pp. 1301–1305.
26. Lau, J. H., and Tang, G. Y. "Thermal Management of 3-D IC Integration with TSV (Through Silicon Via)." In *IEEE Proceedings of Electronic Components & Technology Conference,* San Diego, CA, May 25–29, 2009, pp. 635–640.
27. Akasaka, Y. "Three-dimensional IC Trends." In *Proceedings of the IEEE,* Vol. 74, No. 12, December 1986, pp. 1703–1714.
28. Akasaka, Y., and Nishimura, T. "Concept and Basic Technologies for 3-D IC Structure." In *IEEE Proceedings of International Electron Devices Meetings,* Vol. 32, 1986, pp. 488–491.
29. Gat, A., L. Gerzberg, J. Gibbons, T. Mages, J. Peng, and J. Hong "CW Laser of Polyerystalline Silicon: Crystalline Structure and Electrical Properties," *Applied Physics Letter,* Vol. 33, Issue 8, October 1978, pp. 775–778.
30. Chen, K., S. Lee, P. Andry, C. Tsang, A. Topop, Y. Lin, Y., J. Lu, A. Young, M., Ieong, and W. Haensch, W. "Structure, Design and Process Control for Cu Bonded Interconnects in 3-D Integrated Circuits." In *IEEE Proceedings of International Electron Devices Meeting,* (IEDM 2006), San Francisco, CA, December 11–13, 2006, pp. 367–370.
31. Liu, F., Yu, R., Young, A., Doyle, J., Wang, X., Shi, L., Chen, K., Li, X., Dipaola, D., Brown, D., Ryan, C., Hagan, J., Wong, K., Lu, M., Gu, X., Klymko, N., Perfecto, E., Merryman, A., and Kelly K., Purushothaman, S., Koester, S., Wisnieff, R., and Haensch, W. "A 300- Wafer-Level Three-Dimensional Integration Scheme Using Tungsten Through-Silicon Via and Hybrid Cu-Adhesive Bonding." In *IEEE Proceedings of IEDM,* December 2008, pp. 1–4.
32. Yu, R., Liu, F., Polastre, R., Chen, K., Liu, X., Shi, L., Perfecto, E., Klymko, N., Chace, M., Shaw, T., Dimilia, D., Kinser, E., Young, A., Purushothaman, S., Koester, S., and Haensch W. "Reliability of a 300-mm-compatible 3DI Technology Base on Hybrid Cu-adhesive Wafer Bonding." In *Proceedings of Symposium on VLSI Technology Digest of Technical Papers,* 2009, pp. 170–171.
33. Shigetou, A. Itoh, T., Sawada, K., and Suga, T. "Bumpless Interconnect of 6-um pitch Cu Electrodes at Room Temperature." In *IEEE Proceedings of Electronic & Components Technology Conference,* Lake Buena Vista, FL, May 27–30, 2008, pp. 1405–1409.

34. Shigetou, A. Itoh, T., Matsuo, M., Hayasaka, N., Okumura, K., and Suga, T. "Bumpless Interconnect Through Ultrafine Cu Electrodes by Mans of Surface-Activated Bonding (SAB) Method," *IEEE Transaction on Advanced Packaging*, Vol. 29, No. 2, May 2006, pp. 218–226.
35. Wang, C., and Suga, T. "A Novel Moire Fringe Assisted Method for Nanoprecision Alignment in Wafer Bonding." In *IEEE Proceedings of Electronic Components & Technology Conference*, San Diego, CA, May 25–29, 2009, pp. 872–878.
36. Wang, C., and Suga, T. "Moire Method for Nanoprecision Wafer-to-Wafer Alignment: Theory, Simulation and Application." In *IEEE Proceedings of International Conference on Electronic Packaging Technology & High Density Packaging*, August 2009, pp. 219–224.
37. Higurashi, E., Chino, D., Suga, T., and Sawada, R. "Au-Au Surface-Activated Bonding and its Application to Optical Microsensors with 3-D Structure." In *IEEE Journal of Selected Topic in Quantum Electronics*, Vol. 15, No. 5, September/October 2009, pp. 1500–1505.
38. Burns, J., Aull, B., Keast, C., Chen, C., Chen, C. Keast, C., Knecht, J., Suntharalingam, V., Warner, K., Wyatt, P., and Yost, D. "A Wafer-Scale 3-D Circuit Integration Technology." In *IEEE Transactions on Electron Devices*, Vol. 53, No. 10, October 2006, pp. 2507–2516.
39. Chen, C., Warner, K., Yost, D., Knecht, J., Suntharalingam, V., Chen, C., Burns, J., and Keast, C. "Sealing Three-Dimensional SOI Integrated-Circuit Technology." In *IEEE Proceedings of International SOI Conference*, 2007, pp. 87–88.
40. Chen, C., Chen, C., Yost, D., Knecht, J., Wyatt, P., Burns, J., Warner, K., Gouker, P., Healey, P., Wheeler, B., and Keast, C. "Three-dimensional Integration of Silicon-on-Insulator RF Amplifier," *Electronics Letters*, Vol. 44, No. 12, June 2008, pp. 1–2.
41. Chen, C., Chen, C., Yost, D., Knecht, J., Wyatt, P., Burns, J., Warner, K., Gouker, P., Healey, P., Wheeler, B., and Keast, C. "Wafer-Scale 3-D Integration of Silicon-on-Insulator RF Amplifiers." In *IEEE Proceedings of Silicon Monolithic Integrated Circuit in RF Systems*, 2009, pp. 1–4.
42. Chen, C., Chen, C., Wyatt, P., Gouker, P., Burns, J., Knecht, J., Yost, D., Healey, P., and Keast, C. "Effects of Through-BOX Vias on SOI MOSFETs." In *IEEE Proceedings of VLSI Technology, Systems and Applications*, 2008, pp. 1–2.
43. Chen, C., Chen, C., Burns, J., Yost, D., Warner, K., Knecht, J., Shibles, D., and Keast, C. "Thermal Effects of Three Dimensional Integrated Circuit Stacks." In *IEEE Proceedings of International SOI Conference*, 2007, pp. 91–92.
44. Aull, B., Burns, J., Chen, C., Felton, B., Hanson, H., Keast, C., Knecht, J., Loomis, A., Renzi, M., Soares, A., Suntharalingam, V., Warner, K., Wolfson, D., Yost, D., and Young, D. "Laser Radar Imager Based on 3-D Integration of Geiger-Mode Avalanche Photodiodes with Two SOI Timing Circuit Layers." In *IEEE Proceedings of International Solid-State Circuits Conference*, 2006, pp. 16.9–16.12.
45. Chatterjee, R., Fayolle, M., Leduc, P., Pozder, S., Jones, B., Acosta, E., Charlet, B., Enot, T., Heitzmann, M., Zussy, M., Roman, A., Louveau, O., Maitreqean, S., Louis, D., Kernevez, N., Sillon, N., Passemard, G., Pol, V., Mathew, V., Garcia, S., Sparks, T., and Huang, Z. "Three Dimensional Chip Stacking using a Wafer-to-Wfer Integration." In *IEEE Proceedings of IITC*, 2007, pp. 81–83.
46. Ledus, P., Crecy, F., Fayolle, M., Fayolle, M., Charlet, B., Enot, T., Zussy, M., Jones, B., Barbe, J., Kernevez, N., Sillon, N., Maitreqean, S., Louis, D., and Passemard, G. "Challenges for 3-D IC Integration: Bonding Quality and Thermal Management." In *IEEE Proceedings of IITC*, 2007, pp. 210–212.
47. Poupon, G., Sillon, N., Henry, D., Gillot, C., Mathewson, A., Cioccio, L., Charlet, B., Leduc, P., Vinet, M., and Batude, P. "System on Wafer: A New Silicon Concept in Sip." In *Proceedings of the IEEE*, Vol. 97, No. 1, January 2009, pp. 60–69.
48. Henry, D., Baillin, X., Lapras, V., Sillon, N., Bunne, B., Hernandez, C., and Vigier-Blanc, E. "High Aspect Ratio Vias First for Advanced Packaging." In *IEEE Proceedings of Electronics Packaging Technology Conference*, Singapore, December 10–12, 2007, pp. 215–221.

49. Iker, F., Tezcan, D. S., Teixeira, R. C., Soussan, P., Moor, P., Beyne, E., and Baert, K. "3-D Embedding and Interconnection of Ultra Thin (< 20 μm) Silicon Dies." In *IEEE Proceedings of Electronics Packaging Technology Conference*, Singapore, December 10–12, 2007, pp. 222–226.
50. Mathewson, A., Brun, J., Franiatte, R., Nowodzinski, A., Sillon, N., and Bonvalot, B. "Reliability Aspects of Microinsert Based Interconnection Technologies." In *IEEE Proceedings of Electronics Packaging Technology Conference*, Singapore, December 10–12, 2007, pp. 226–232.
51. Lee, W. S. V., Khan, N., Ebin, L., Yoon, S. W., and Kripesh, V. "Cu via Exposure by Backgrinding for TSV Applications." In *IEEE Proceedings of Electronics Packaging Technology Conference*, Singapore, December 10–12, 2007, pp. 874–881.
52. Beyne, E. "Solving Mechnical and Economical Barriers to the Adoption of Through-Si-Via 3-D Integration Technologies." In *IEEE Proceedings of Electronics Packaging Technology Conference*, Singapore, December 9–12, 2008, pp. 29–34.
53. Henry, D., Charbonnier, J., Chausse, Pascal., and Jacquet, F. "Through Silicon Vias Technology for Cmos Image Sensors Packaging: Presentation of Technology and Electrical Results." In *IEEE Proceedings of Electronics Packaging Technology Conference*, Singapore, December 9–12, 2008, pp. 35–44.
54. Jouve, A., Fowler, S., Privett, M., and Puligadda, R. "Facilitating Ultrathin Wafer Handling for TSV Processing." In *IEEE Proceedings of Electronics Packaging Technology Conference*, Singapore, December 9–12, 2008, pp. 45–50.
55. Jacquet, F., Henry, D., Charbonnier, J., Bouzaida, Nicole, Sillon, N., Tsai, Colin., Balut, C., and Sraine, J. "Low Cost Lithography Solution for Advanced Packaging and Application to through Silicon via Process." In *IEEE Proceedings of Electronics Packaging Technology Conference*, Singapore, December 9–12, 2008, pp. 51–57.
56. Deshpande, A., Natarajan, V., and Methani, J. "Pareto-Optimal Die Orientations for 3-D Stacking of Identical Dies." In *IEEE Proceedings of Electronics Packaging Technology Conference*, Singapore, December 9–12, 2008, pp. 193–199.
57. Zhao, L., Liao, H., Miao, M., and Jin, Y. "Design and Analysis of an I-shaped TSV Structure for 3-D SiP." In *IEEE Proceedings of Electronics Packaging Technology Conference*, Singapore, December 9–12, 2008, pp. 200–205.
58. Curran, B., Ndip I., Guttovski, S,. and Reichl, H. "Managing Losses in through Silicon vias with Different Return Current Path Configurations." In *IEEE Proceedings of Electronics Packaging Technology Conference*, Singapore, December 9–12, 2008, pp. 206–211.
59. Beica, R., Siblerud, P., Sharbono, C., and Bernt, M. "Advanced Metallization for 3-D Integration." In *IEEE Proceedings of Electronics Packaging Technology Conference*, Singapore, December 9–12, 2008, pp. 212–218.
60. Saadaoui, M., Wien, W., Zeijl, H., Schellevis, H., Laros, M., and Sarro L. "New Front to Back-side 3-D Interconnects Based High Aspect Ratio Through Silicon Vias." In *IEEE Proceedings of Electronics Packaging Technology Conference*, Singapore, December 9–12, 2008, 219–223.
61. Jiang, T., and Luo, S. "3-D Integration—Present and Future." In *IEEE Proceedings of Electronics Packaging Technology Conference*, Singapore, December 9–12, 2008, 373–378.
62. Zhang, X., Premachandran, C. S., Chong, S., Wai, L. C., Sheng, W., Lee, V., Chai, T. C., Vaidyanathan, K., and Lau, J. H. "Development of a 65-nm Cu/low-k Stack Die FBGA Package for SiP Applications." In *IEEE Proceedings of Electronics Packaging Technology Conference*, Singapore, December 9–12, 2008, 379–386.
63. Yu, A., Kumar, A., Ho, S. W., Yin, H. W., Lau, J. H., Khong, C. H., Siang, P., Lim, S., and Zhang, X. "Development of Fine Pitch Solder Microbumps for 3-D Chip Stacking." In *IEEE Proceedings of Electronics Packaging Technology Conference*, Singapore, December 9–12, 2008, 387–392.
64. Watanabe, N., Mori, T., and Asano, T. "Low-Temperature 3-D Chip-Stacking using Compliant Bump." In *IEEE Proceedings of Electronics Packaging Technology Conference*, Singapore, December 9–12, 2008, 393–398.

65. Ang, X., Lin, A. T., Wei, J., Chen, Z., and Wong, C. C. "Low Temperature Copper-Copper Thermocompression Bonding." In *IEEE Proceedings of Electronics Packaging Technology Conference*, Singapore, December 9–12, 2008, 399–404.
66. Huang, M. S., Li, D. W. P., Suthiwongsunthorn, N., and Chelvam, P. M. "Temporary Bonding of Wafer to Carrier for 3-D-Wafer Level Packaging." In *IEEE Proceedings of Electronics Packaging Technology Conference*, Singapore, December 9–12, 2008, 405–411.
67. Pargfrieder, S., Kettner, P., and Privett, M. "Temporary Bonding and DeBonding Enabling TSV Formation and 3-D Integration for Ultra-thin Wafers." In *IEEE Proceedings of Electronics Packaging Technology Conference*, Singapore, December 9–12, 2008, 859–866.
68. Vempati, S. R., Ho, D. S. W., Lee, V. W. S., Li, H. Y., Liao, E., Nagarajan, R., and Chai, T. C. "TSV Interposer Fabrication for 3-D IC Packaging." In *IEEE Proceedings of Electronics Packaging Technology Conference*, Singapore, December 9–11, 2009, pp. 431–437.
69. Roth, H., He, Z., and Mayer, T. "Inspection of Through Silicon Vias (TSV) and other Interconnections in IC Packages by Computed Tomography." In *IEEE Proceedings of Electronics Packaging Technology Conference*, Singapore, December 9–11, 2009, pp. 438–441.
70. Miao, M., Zhu, Y., Ji, M., Ma, S., Sun, X., and Jin, Y. "Bottom-up Filling of Through Silicon Via (TSV) with Parylene as Sidewall Protection Layer." In *IEEE Proceedings of Electronics Packaging Technology Conference*, Singapore, December 9–11, 2009, pp. 442–446.
71. Ong, Y. Y., Chai, T. C., Yu, D., Thew, M. L., Myo, E., Wai, L. C., and Jong, M. C. "Assembly and Reliability of Micro-bumped chips with Through-silicon Vias (TSV) Interposer." In *IEEE Proceedings of Electronics Packaging Technology Conference*, Singapore, December 9–11, 2009, pp. 452–458.
72. Pares, G., Minoret, S., Lugand, J., Huet, S., Lappas, V., Anciant, R., and Henry, D. "Mid-process through Silicon vias Technology using Tungsten Metallization: Process Optimization and Electrical Results." In *IEEE Proceedings of Electronics Packaging Technology Conference*, Singapore, December 9–11, 2009, pp. 772–777.
73. Hoe, Y. Y. G., Tang, G. Y., Pinjala, D., Chai, T. C., Lau, J. H., Zhang, X., and Kpipesh, V. "Effect of TSV Interposer on the Thermal Performance of FCBGA Package." In *IEEE Proceedings of Electronics Packaging Technology Conference*, Singapore, December 9–11, 2009, pp. 778–786.
74. Chausse, P., Boouchoucha, M., Henry, D., Sillon, N., and Chapelon, L. "Polymer Filling of Medium Density Through Silicon Via for 3-D-packaging." In *IEEE Proceedings of Electronics Packaging Technology Conference*, Singapore, December 9–11, 2009, pp. 790–794.
75. Wolf, M. J., Zoschke, K., Klumpp, A., Wieland, R., Klein, M., Nebrich, L., and Heinig, A. "3-D Integration of Image Sensor SiP using TSV Silicon Interposer." In *IEEE Proceedings of Electronics Packaging Technology Conference*, Singapore, December 9–11, 2009, pp. 795–800.
76. Pan, S. P., Zhang, X., and Pinjala, D. "Prediction of Hotspots on 3-D Packages due to Joule Heating in Through Silicon Vias (TSV)." In *IEEE Proceedings of Electronics Packaging Technology Conference*, Singapore, December 9–11, 2009, pp. 801–806.
77. Wai, L. C., Zhang, X., Chai, T. C., Srinivas, V. R., Ho, D., Pinjala, D., Myo, E. P. P., Jong, M. C., Lim, S., Ong, J., Thew, S., Chen, K., and Shekar, V. N. "Assembly of Large Dies Fine Pitch Cu/low-k FCBGA Package with Through Silicon Via (TSV) Interposer." In *IEEE Proceedings of Electronics Packaging Technology Conference*, Singapore, December 9–11, 2009, pp. 44–48.
78. Ho, S. W., Su, N., Lim, L. S., Ong, S. C., Lee, W. S., and Rao, V. S. "Development of Thin Film Dielectric Embedded 3-D Stacked Package." In *IEEE Proceedings of Electronics Packaging Technology Conference*, Singapore, December 9–11, 2009, pp. 185–190.
79. Kim, Y., Kang, S. G., and Kim, E. K. "Assessment of Thinned Si Wafer Warpage in 3-D Stacked Wafers." In *IEEE Proceedings of Electronics Packaging Technology Conference*, Singapore, December 9–11, 2009, pp. 964–967.

80. Bauer, C. E., and Neuhaus, H. J. "3-D Device Integration." In *IEEE Proceedings of Electronics Packaging Technology Conference*, Singapore, December 9–11, 2009, pp. 427–430.
81. Zeijl, H. W. V., and Sarro, P. M. "Alignment and Overlay Characterization for 3-D Integration and Advanced Packaging." In *IEEE Proceedings of Electronics Packaging Technology Conference*, Singapore, December 9–11, 2009, pp. 447–451.
82. Lau, J. H., Lim, Y., Lim, T., Tang, G., Khong, C., Zhang, X., Ramana, P., Zhang, J., Tani, C., Chandrappan, J., Chai, J., Li, J., Tangdiongga, G., and Kwong, D. "Design and analysis of 3-D stacked optoelectronics on optical printed circuit boards." In *Proceedings* of *SPIE, Photonics Packaging, Integration, and Interconnects VIII*, San Jose, CA, January 19–24, 2008, Vol. 6899, pp. 07.1–07.20.
83. Khong, C. H., Yu, A., Zhang, X., Kripesh, V., Pinjala, D., Kwong, D. L., Chen, S., Chan, C. F., Chao, C. C., Chiu, C. H., Huang, C. M., and Chen, C. "Sub-modeling Technique for Thermo-Mechanical Simulation of Solder Microbumps Assembly in 3-D Chip Stacking." In *IEEE Proceedings of Electronics Packaging Technology Conference*, Singapore, December 9–11, 2009, pp. 591–595.
84. Selvanayagam, C. S., Zhang, X., Rajoo, R., and Pinjala, D. "Modelling Stress in Silicon with TSVs and its Effect on Mobility." In *IEEE Proceedings of Electronics Packaging Technology Conference*, Singapore, December 9–11, 2009, pp. 612–618.
85. Kettner, P., Burggraf, J., and Kim, B. "Thin Wafer Handling and Processing-Results Achieved and Upcoming Tasks in the Field of 3-D and TSV." In *IEEE Proceedings of Electronics Packaging Technology Conference*, Singapore, December 9–11, 2009, pp. 787–789.
86. Yoon, S. W., Bahr, A., Baraton, X., Marimuthu, P. C., and Carson, F. "3-D eWLB (embedded wafer level BGA) Technology for 3-D-Packaging/3-D-SiP (Systems-in-Package) Applications." In *IEEE Proceedings of Electronics Packaging Technology Conference*, Singapore, December 9–11, 2009, pp. 915–919.
87. Tanaka, N., Sato, T., Yamaji, Y., Morifuji, T., Umemoto, M., and Takahashi, K. "Mechanical Effects of Copper Through-Vias in a 3-D Die-Stacked Module." In *IEEE Proceedings of Electronic Components & Technology Conference*, San Diego, May 28–31, 2002, pp. 168–176.
88. Tomisaka, M., Yonemura, H., Hoshino, M., Takahashi, K., Okamura, T., Sun, J. J., and Kondo, K. "Electroplating Cu Fillings for Through-Vias for Three-Dimensional Chip Stacking." In *IEEE Proceedings of Electronic Components & Technology Conference*, San Diego, May 28–31, 2002, pp. 399–403.
89. Tanaka, N., Yamaji, Y., Sato, T., and Takahashi, K. "Guidelines for Structural and Material-System Design of a Highly Reliable 3-D Die-Stacked Module with Copper Through-vias." In *IEEE Proceedings of Electronic Components & Technology Conference*, New Orleans, May 27–30, 2003, pp. 597–602.
90. Premachandran, C. S., Nagarajan, R., Yu, C., Zhang, X., and Choong, C. S. "A Novel Electrically Conductive Wafer Through Hole Filled Vias Interconnect for 3-D MEMS Packaging." In *IEEE Proceedings of Electronic Components & Technology Conference*, New Orleans, May 27–30, 2003, pp. 627–630.
91. Spiesshoefer, S. "IC Stacking Technology using Fine Pitch, Nanoscale through Silicon Vias." In *IEEE Proceedings of Electronic Components & Technology Conference*, New Orleans, May 27–30, 2003, pp. 631–633.
92. Rasmussen, F. E., Heschel, M., and Hansen, O. "Batch Fabrication of Through-Wafer Vias in CMOS Wafers for 3-D Packaging Applications." In *IEEE Proceedings of Electronic Components & Technology Conference*, New Orleans, May 27–30, 2003, pp. 634–639.
93. Klumpp, A., Merkel, R., Wieland, R., and Ramm, P. "Chip-to-Wafer Stacking Technology for 3-D System Integration." In *IEEE Proceedings of Electronic Components & Technology Conference*, New Orleans, May 27–30, 2003, pp. 1080–1083.
94. Tanida, K., Umemoto, M., Tomita, Y., Tago, M., Nemoto, Y., Ando, T., and Takahashi, K. "Ultra-High-Density 3-D Chip Stacking Technology." In *IEEE Proceedings of Electronic Components & Technology Conference*, New Orleans, May 27–30, 2003, pp. 1084–1089.
95. Miettinen, J., Tanskanen, J., and Ristolainen, E. O. "Stacked 3-D MCP with Plastic Ball Vertical Interconnections." In *IEEE Proceedings of Electronic Components & Technology Conference*, New Orleans, May 27–30, 2003, pp. 1101–1105.

96. Spiesshoefer, S., Schaper, L., Burkett, S., Vangara, G., Rahman, Z., and Arunasalam, P. "Z-Axis Interconnects Using Fine Pitch, Nanoscale Through-Silicon Vias: Process Development." In *IEEE Proceedings of Electronic Components & Technology Conference*, New Orleans, May 27–30, 2003, pp. 466–471.
97. Zhang, X., and Tee, T. Y. "Advanced Warpage Prediction Methodology for Matrix Stacked Die BGA During Assembly Processes." In *IEEE Proceedings of Electronic Components & Technology Conference*, Las Vegas, June 1–4, 2004, pp. 593–600.
98. Takahashi, K., Taguchi, Y., Tomisaka, M., Yonemura, H., Hoshino, M., Ueno, M., Egawa, Y, Nemoto, Y., Yamaji, Y., Terao, H., and Kondo, K. "Process Integration of 3-D Chip Stack with Vertical Interconnection." In *IEEE Proceedings of Electronic Components & Technology Conference*, Las Vegas, June 1–4, 2004, pp. 601–609.
99. Miettinen, J., Mantysalo, M., Kaija, K., and Ristolainen, E. "System Design Issues for 3-D System-in-Package (Sip)." In *IEEE Proceedings of Electronic Components & Technology Conference*, Las Vegas, June 1–4, 2004, pp. 610–615.
100. Umemoto, M., Tanida, K., Nemoto, Y., Hoshino, M., Kojima, K., Shirai, Y., and Takahashi, K. "High-Performance Vertical Interconnection for High-Density 3-D Chip Stacking Package." In *IEEE Proceedings of Electronic Components & Technology Conference*, Las Vegas, June 1–4, 2004, pp. 616–623.
101. Kikuchi, K., Segawa, S., Jung, E., Nemoto, Y., Umemoto, M., Nakagawa, H., Tokoro, K., and Aoyagi, M. "Ultra Wide Bandwidth Performance of High-Density Wiring Interposer for 3-D Packaging." In *IEEE Proceedings of Electronic Components & Technology Conference*, Las Vegas, June 1–4, 2004, pp. 630–634.
102. Barton, J., Majeed, B., Dwane, K., Delaney, K., Bellis, S., Rodgers, K., and Mathuna, S. C. O. "Development and Characterization of Ultra Thin Autonomous Modules for Ambient System Applications Using 3-D Packaging Techniques." In *IEEE Proceedings of Electronic Components & Technology Conference*, Las Vegas, June 1–4, 2004, pp. 635–641.
103. Banerjee, S. R., and Drayton, R. F. "Wafer Level Interconnects for 3-D Packaging." In *IEEE Proceedings of Electronic Components & Technology Conference*, Las Vegas, June 1–4, 2004, pp. 1513–1518.
104. Takahashi, K., Umemoto, M., Tanaka, N., Tanida, K., Nemoto, Y., Tomita, Y., Tage, M., and Bonkohara, M. "Ultra-High-Density Interconnection Technology of Three-Dimensional Packaging." *Microelectronics Reliability* 43:1267–1279, 2003.
105. Lee, S. W. R., Hon, R., Zhang, S. X. D., and Wong, C. K. "3-D Stacked Flip Chip Packaging with Through Silicon Vias and Copper Plating or Conductive Adhesive Filling." In *IEEE Proceedings of Electronic Components & Technology Conference*, Lake Buena Vista, FL, May 31-June 3, 2005, pp. 795–801.
106. Pargfrieder, S., Kettner, P., Privett, M., and Ting, J. "Temporary Bonding and Debonding Enabling TSV Formation and 3-D Integration for Ultra-thin Wafers." In *IEEE Proceedings of Electronics Packaging Technology Conference*, Singapore, December 2008, pp. 1301–1305.
107. Patel, C. S., Tsang, C. K., Schuster, C., Doany, F. E., Nyikal, H., Baks, C. W., Budd, U., Paivikki, L., Andry, P. S., Canaperi, D. F., Edelstein, D. C., Horton, R., and Knickerbocker, J. "Silicon Carrier with Deep Through-Vias, Fine Pitch Wiring and Through Cavity for Parallel Optical Transceiver ." In *IEEE Proceedings of Electronic Components & Technology Conference*, Lake Buena Vista, FL, May 31-June 3, 2005, pp. 1318–1324.
108. Zhang, X., Kumar, A., Zhang, Q., Ong, Y., Ho, D., Khong, C., Kripesh, V., Lau, J., H., Kwong, D., Sundaram, V., and Tummula, R. "Application of Piezoresistive Stress Sensors in Ultra Thin Device Handling and Characterization." *Journal of Sensors & Actuators: A. Physical*, Vol. 156, Nov. 2009, pp. 1–7.
109. Mcmahon, J. J. Lu, J. Q., and Gutmann, R. J. "Wafer Bonding of Damascene-Patterned Metal/Adhesive Redistribution Layers for Via-First Three-Dimensional (3-D) Interconnect." In *IEEE Proceedings of Electronic Components & Technology Conference*, Lake Buena Vista, FL, May 31-June 3, 2005, pp. 331–336.
110. Lo, W., Chen, Y., Ko, J., Kuo, T., Chien, C., Chen, Y., Chen, W., Leu, F., and Hu, H. "Development and Characterization of Low Cost Ultrathin 3-D Interconnect." In *IEEE Proceedings of Electronic Components & Technology Conference*, Lake Buena Vista, FL, May 31-June 3, 2005, pp. 337–342.

111. Ranganathan, N., Balasubramanian, N., and Hwee, S. C. "High Aspect Ratio Through-Wafer Interconnect for Three Dimensional Integrated Circuits." In *IEEE Proceedings of Electronic Components & Technology Conference*, Lake Buena Vista, FL, May 31-June 3, 2005, pp. 343–348.
112. Tanaka, N., and Yoshimura, Y. "Ultra-Thin 3-D Stacked SIP Formed using Room-Temperature Bonding between Stacked Chips." In *IEEE Proceedings of Electronic Components & Technology Conference*, Lake Buena Vista, FL, May 31-June 3, 2005, pp. 795–801.
113. Premachandran, C. S., Ranaganathan, N., Mohanraj, S., Choong, C. S., and Lyer, M. K. "A Vertical Wafer Level Packaging using Through Hole Filled Via Interconnects by Lift Off Polymer Method for MEMS and 3-D Stacking Applications." In *IEEE Proceedings of Electronic Components & Technology Conference*, Lake Buena Vista, FL, May 31-June 3, 2005, pp. 1094–1098.
114. Nagarajan, R., Ebin, L., Balasubramanian, N., Lee, D., Seng, S. C., and Prasad, K. "Development of a Novel Deep Silicon Tapered Via Etch Process for Through-Silicon Interconnection in 3-D Integrated Systems." In *IEEE Proceedings of Electronic Components & Technology Conference*, San Diego, CA, May 30-June 2, 2006, pp. 383–387.
115. Henry, D., Belhachemi, D., Souriau, J. C., Brunet-Manquat, C., Puget, C., Ponthenier, G., Vallejo, J. L., Lecouvey, C., and Sillon, N. "Low Electrical Resistance Silicon Through Vias: Technology and Characterization." In *IEEE Proceedings of Electronic Components & Technology Conference*, San Diego, CA, May 30-June 2, 2006, pp. 1360–1366.
116. Yoon, S. W., Witarsa, D., Lim, S. Y. L., Ganesh, V., Chai, T. C., Navas, K. O., Kripesh, V., and Viswanath, A. G. K. "Reliability Studies of a Through Via Silicon Stacked Module for 3-D Microsystem Packaging." In *IEEE Proceedings of Electronic Components & Technology Conference*, San Diego, CA, May 30-June 2, 2006, pp. 1449–1453.
117. Andry, P. S., Tsang, C., Sprogis, E., Patel, C., Wright, S. L., Webb, B. C., Buchwalter, L. P., Manzer, D., Horton, R., and Knickerbocker, J. "A CMOS-Compatible Process for Fabricating Electrical Through-Vias in Silicon." In *IEEE Proceedings of Electronic Components & Technology Conference*, San Diego, CA, May 30-June 2, 2006, pp. 831–837.
118. Kim, B., Sharbono, C., Ritzdorf, T., and Schmauch, D. "Factors Affecting Copper Filling Process within High Aspect Ratio Deep Vias for 3-D Chip Stacking." In *IEEE Proceedings of Electronic Components & Technology Conference*, San Diego, CA, May 30-June 2, 2006, pp. 838–843.
119. Greisen, C. C., Shiv, L., Weichel, S., and Hesschel, M. "Processing and Performance of Broadband Integrated Resistor Structures on Non-Planar Topologies in Hermetic Silicon Enclosure with vertical Micro Vias." In *IEEE Proceedings of Electronic Components & Technology Conference*, San Diego, CA, May 30-June 2, 2006, pp. 1449–1453.
120. Newman, M. W., Muthukumar, S., Schuelein, M., Dambrauskas, T., Dunaway, P., Jordan, J. M., Kulkami, S., Linde, C., Opheim, T., Stingel, R., and Topic, L. A. "Fabrication and Electrical Characterization of 3-D Vertical Interconnects." In *IEEE Proceedings of Electronic Components & Technology Conference*, San Diego, CA, May 30-June 2, 2006, pp. 394–398.
121. Bower, C. A., Malta, D., and Temple, D. "High Density, Vertical Interconnects for 3-D Integration of Silicon Integrated Circuits." In *IEEE Proceedings of Electronic Components & Technology Conference*, San Diego, CA, May 30-June 2, 2006, pp. 399–403.
122. Stoukatch, S., Winters, C., Beyne, E., Raedt, W. D., and Hoof, C., V. "3-D-SIP Integration for Autonomous Sensor Nodes." In *IEEE Proceedings of Electronic Components & Technology Conference*, San Diego, CA, May 30-June 2, 2006, pp. 404–408.
123. SKhan, N., Yoon, S. W., Ganesh, V. P., Lim, S., and Vaidyanathan, K. "Development of 3-D Stack Package using Silicon Interposer for High Power Application." In *IEEE Proceedings of Electronic Components & Technology Conference*, San Diego, CA, May 30-June 2, 2006, pp. 756–760.

124. Tanaka, N., and Yoshimura, Y. "Low Cost Through-Hoe Electrode Interconnection for 3-D-Sip using Room-Temperature Bonding." In *IEEE Proceedings of Electronic Components & Technology Conference*, San Diego, CA, May 30-June 2, 2006, pp. 814–818.
125. Dietrich, L., Topper, M., Ehrmann, O., and Reichi, H. "Conformance of ECD Wafer Bumping to Future Demands on CSP, 3-D Integration, and MEMS." In *IEEE Proceedings of Electronic Components & Technology Conference*, San Diego, CA, May 30-June 2, 2006, pp. 1050–1059.
126. Periasamy, G. V., Lim, S., Yin, H., Yoon, S. W., and Kripesh, V. "Assembly Technology Development for 3-D Silicon Stacked Module for Handheld Products." In *IEEE Proceedings of Electronic Components & Technology Conference*, San Diego, CA, May 30-June 2, 2006, pp. 404–408.
127. Wong, E., Minz, J., and Lim, S. K. "Effective Thermal Via and Decoupling Capacitor Insertion for 3-D System-On-Package." In *IEEE Proceedings of Electronic Components & Technology Conference*, San Diego, CA, May 30-June 2, 2006, pp. 1795–1801.
128. Jang, D. M., Ryu, C., Cho, B. H., Kim, J., Lee, W. J., Yu, J., Lee, K. Y., and Oh, T. S. "Development and Evaluation of 3-D SiP with Vertically Interconnected Through Silicon Vias (TSV)." In *IEEE Proceedings of Electronic Components & Technology Conference*, Reno, NV, May 29-June 1, 2007, pp. 847–852.
129. Xie, B., Shi, X., and Fan, X. "Sensitivity Investigation of Substrate Thickness and Reflow Profile on Wafer Level Film Failures in 3-D Chip Scale Packages by Finite Element Modeling." In *IEEE Proceedings of Electronic Components & Technology Conference*, Reno, NV, May 29-June 1, 2007, pp. 242–248.
130. Okoro, C., Eneman, G., Gonzalez, M., Vandevelde, B., Swinnen, B., Stoukatch, S., Beyne, E., and Vandepitte, D. "Analysis of the Induced Stresses in Silicon During Thermocompression Cu-Cu Bonding of Cu-Through-Vias in 3-D-SIC Architecture." In *IEEE Proceedings of Electronic Components & Technology Conference*, Reno, NV, May 29-June 1, 2007, pp. 249–255.
131. Sakuma, K., Andry, P. S., Dang, B., Tsang, C. K., Patel, C., Wright, S. L., Webb, B., Kang, S. K., Polastre, R., Horton, R., Knickerbocker, J. U., Maria, J., and Sprogis, E. "3-D Chip Stacking Technology with Low-Volume Lead-Free Interconnections." In *IEEE Proceedings of Electronic Components & Technology Conference*, Reno, NV, May 29-June 1, 2007, pp. 627–632.
132. Barkey, D., Ruff, A., Trezza, J., Keigler, A., Liu, Z., and Wu, B. "Reaction Engineering of Through-Chip Via Filling for Wafer-Level 3-D Packaging." In *IEEE Proceedings of Electronic Components & Technology Conference*, Reno, NV, May 29-June 1, 2007, pp. 638–642.
133. Tezcan, D. S., Pham, N., Majeed, B., Moor, P. D., Ruythooren, W., and Baert, K. "Sloped Through Wafer Vias for 3-D Wafer Level Packaging." In *IEEE Proceedings of Electronic Components & Technology Conference*, Reno, NV, May 29-June 1, 2007, pp. 643–647.
134. Wang, W. C., Lee, F., Weng, G. L., Tai, W., Ju, M., Chuang, R., and Fang, W. "Platform of 3-D Package Integration." In *IEEE Proceedings of Electronic Components & Technology Conference*, Reno, NV, May 29-June 1, 2007, pp. 743–747.
135. Kurita, Y., Matsui, S., Takahashi, N., Soejima, K., Komuro, M., Itou, M., Kakegawa, C., Kawano, M., Egawa, Y., Saeki, Y., Kikuchi, H., Kato, O., Yanagisawa, A., Mitsuhashi, T., Ishino, M., Shibata, K., Uchiyama, S., Yamada, J., and Ikeda, H. "A 3-D Stacked Memory Integrated on a Logic Device Using SMAFTI Technology." In *IEEE Proceedings of Electronic Components & Technology Conference*, Reno, NV, May 29-June 1, 2007, pp. 821–829.
136. Wagner, M., and Zou, J. "Method for Creating Low Cost 3-D MEMS Chemical Sensors by Using Through Wafer Vias & Wafer Bonding." In *IEEE Proceedings of Electronic Components & Technology Conference*, Reno, NV, May 29-June 1, 2007, pp. 1119–1126.
137. Lee, H., Choi, Y. S., Song, E., Choi, K., Cho, T., and Kang, S. "Power Delivery Network Design for 3-D SIP Integrated over Silicon Interposer Platform." In *IEEE Proceedings of Electronic Components & Technology Conference*, Reno, NV, May 29-June 1, 2007, pp. 1193–1198.

138. Schaper, L., Burkett, S., Gordon, M., Cai, L., Liu, Y., Jampana, G., and Abhulimen, I. U. "Integrated System Development for 3-D VLSI." In *IEEE Proceedings of Electronic Components & Technology Conference*, Reno, NV, May 29-June 1, 2007, pp. 853–857.
139. Worwag, W., and Dory, T. "Copper Via Plating in Three-Dimensional Interconnects." In *IEEE Proceedings of Electronic Components & Technology Conference*, Reno, NV, May 29-June 1, 2007, pp. 842–846.
140. Fukushima, T., Yamada, Y., Kikuchi, H., Tanaka, T., and Koyanagi, M. "Self-Assembly Process for Chip-to-Wafer Three-Dimensional Integration." In *IEEE Proceedings of Electronic Components & Technology Conference*, Reno, NV, May 29-June 1, 2007, pp. 836–841.
141. Dixit, P., Lin, N., Miao, J., Wong, W. K., and Teo, K. C. "Concept and Analytical Analysis of Silicon Micro/Nanopillars Based 3-D Stacked Microchannel Heat Sink for Advanced Heat Dissipation Applications." In *IEEE Proceedings of Electronic Components & Technology Conference*, Reno, NV, May 29-June 1, 2007, pp. 1149–1154.
142. Shen, L. C., Chien, C. W., Jaung, J. Y., Hung, Y. P., Lo, W. C., Hsu, C. K., and Lee, Y. C., Cheng, H. C., and Lin, C. T. "A Clamped Through Silicon Via (TSV) Interconnection for Stacked Chip Bonding Using Metal Cap on Pad and Metal Column Forming in Via." In *IEEE Proceedings of Electronic Components & Technology Conference*, Lake Buena Vista, FL, May 27–30, 2008, pp. 544–549.
143. Khan, N., Rao, V. S., Lim, S., We, H. S., Lee, V., Zhang, X. W., Rui, Y., Ebin, L., Ranganathan, Chai, T. C., Kripesh, V., and Lau, J. "Development of 3-D Silicon Module with TSV for System in Packaging." In *IEEE Proceedings of Electronic Components & Technology Conference*, Lake Buena Vista, FL, May 27–30, 2008, pp. 550–555.
144. Knickerbocker, J. U., Andry, P. S., Buchwalter, L. P., Deutsch, A., Horton, R. R., Jenkins, K. A., Kwark,Y. H., McVicker, G., Patel, C. S., Polastre, R. J., Schuster, C., Sharma, A., Sri-Jayantha, S. M., Surovic, C. W.,Tsang, C. K., Webb, B. C., Wright, S. L., McKnight, S. R., Sprogis, E. J., and Dang, B. "Development of next-generation system-on-package (SOP) technology based on silicon carriers with fine-pitch interconnection." *IBM J. Res. Dev.* 49:725–754, 2005.
145. Wolf, M. J., Wunderle, B., Jürgensen, N., Engelmann, G., Ehrmann, O., and Michel, B., Dretschkow, T., Uhlig, A., and Reichl, H. "High Aspect Ratio TSV Copper Filling with Different Seed Layers." In *IEEE Proceedings of Electronic Components & Technology Conference*, Lake Buena Vista, FL, May 27–30, 2008, pp. 563–570.
146. Kumagai, K., Yoneda, Y., Izumino, H., Shimojo, H., Mabuchi, Y., Sunohara, M., Kurihara, T., and Higashi, M. "A Silicon Interposer BGA Package with Cu-Filled TSV and Multi-Layer Cu-Plating Interconnect." In *IEEE Proceedings of Electronic Components & Technology Conference*, Lake Buena Vista, FL, May 27–30, 2008, pp. 571–576.
147. Sunohara, M., Tokunaga, T., Kurihara, T., and Higashi, M. "Silicon Interposer with TSVs (Through Silicon Vias) and Fine Multilayer Wiring." In *IEEE Proceedings of Electronic Components & Technology Conference*, Lake Buena Vista, FL, May 27–30, 2008, pp. 847–852.
148. Selvanayagam, C. S., Lau, J. H., Zhang, X., Seah, S. K. W., Kripesh, V., and Chai, T. C. "Nonlinear Thermal Stress/Strain Analyses of Copper Filled TSV (Through Silicon Via) and Their Flip-Chip Microbumps." In *IEEE Proceedings of Electronic Components & Technology Conference*, Lake Buena Vista, FL, May 27–30, 2008, pp. 1073–1081.
149. Hu, G., Kalyanam, H., Krishnamoorthy, S., and Polka, L. "Package technology to address the memory bandwidth challenge for tera-scale computing." *InTEL Technol. J.* 11, pp. 197–206, 2007.
150. Ho, S. W., Yoon, S. W., Kripesh, V., Lau, J. H., Zhou, Q., and Pasad, K. "High RF Performance TSV Silicon Carrier for High Frequency Application." In *IEEE Proceedings of Electronic Components & Technology Conference*, Lake Buena Vista, FL, May 27–30, 2008, pp. 1946–1952.
151. Henry, D., Jacquet, F., Neyret, M., Baillin, X., Enot, T., Lapras, V., Brunet-Manquat, C., Charbonnier, J., Aventurier, B., and Sillon, N. "Through Silicon

Vias Technology for CMOS Image Sensors Packaging." In *IEEE Proceedings of Electronic Components & Technology Conference*, Lake Buena Vista, FL, May 27–30, 2008, pp. 556–562.

152. Beica, R., Sharbono, C., and Ritzdorf, T. "Through Silicon Via Copper Electrodeposition for 3-D Integration." In *IEEE Proceedings of Electronic Components & Technology Conference*, Lake Buena Vista, FL, May 27–30, 2008, pp. 577–583.
153. Ramm, P., Wolf, M. J., Klumpp, A., Wieland, R., Wunderle, B., Michel, B. and Reichl, H. "Through Silicon Via Technology – Processes and Reliability for Wafer-Level 3-D System Integration." In *IEEE Proceedings of Electronic Components & Technology Conference*, Lake Buena Vista, FL, May 27–30, 2008, pp. 841–846.
154. Chen, K. S., Ayon, A. A., Zhang, X., and Spearing, S. M. "Effect of process parameters on the surface morphology and mechanical performance of silicon surfaces after deep reactive ion etching (DRIE)." *J. Microelectromech. Syst.* 11:264–275, 2002.
155. Kuo, T. Y., Chang, S. M., Shih, Y. C., Chiang, C. W., Hsu, C. K., Lee, C. K., Lin, C. T., Chen, Y. H., and Lo, W. C. "Reliability Tests for a Three Dimensional Chip Stacking Structure withThrough Silicon Via Connections and Low Cost." In *IEEE Proceedings of Electronic Components & Technology Conference*, Lake Buena Vista, FL, May 27–30, 2008, pp. 853–858.
156. Majeed, B., Pham, N. P., Tezcan, D. S., and Beyne, E. "Parylene N as a Dielectric Material for Through Silicon Vias." In *IEEE Proceedings of Electronic Components & Technology Conference*, Lake Buena Vista, FL, May 27–30, 2008, pp. 1556–1561.
157. Andry, P. S., Tsang, C. K., Webb, B. C., Sprogis, E. J., Wright, S. L., Bang, B., and Manzer, D. G. "Fabrication and characterization of robust through-silicon vias for silicon-carrier applications." *IBM Journal of Research and Development* 52:571–581, 2008.
158. Ranganathan, N., Ebin, L., Linn, L., Lee, W. S. V., Navas, O. K., Kripesh, V., and Balasubramanian, N. "Integration of High Aspect Ratio Tapered Silicon Via for Through-Silicon Interconnection." In *IEEE Proceedings of Electronic Components & Technology Conference*, Lake Buena Vista, FL, May 27–30, 2008, pp. 859–865.
159. Wright, S. L., Andry, P. S., Sprogis, E., Dang, B., and Polastre, R. J. "Reliability Testing of Through-Silicon Vias for High-Current 3-D Applications." In *IEEE Proceedings of Electronic Components & Technology Conference*, Lake Buena Vista, FL, May 27–30, 2008, pp. 879–883.
160. Tian, J., and Bartek, M. "Simultaneous Through-Silicon Via and Large Cavity Formation Using Deep Reactive Ion Etching and Aluminum Etch-Stop Layer." In *IEEE Proceedings of Electronic Components & Technology Conference*, Lake Buena Vista, FL, May 27–30, 2008, pp. 1787–1792.
161. King, C. R. Jr., Sekar, D., Bakir, M. S., Pikarsky, J., Meindl, J. M., and Dang, B. "3-D Stacking of Chips with Electrical and Microfluidic I/O Interconnects." In *IEEE Proceedings of Electronic Components & Technology Conference*, Lake Buena Vista, FL, May 27–30, 2008, pp. 1–7.
162. Huang, Z., Chatterjee, R., Justison, P., Hernandez, R., Pozder, S., Jain, A., Acosta, E., Gajewski, D. A., Mathew, V., and Jones, R. E. "Electromigration of Cu-Sn-Cu Micropads in 3-D Interconnect." In *IEEE Proceedings of Electronic Components & Technology Conference*, Lake Buena Vista, FL, May 27–30, 2008, pp. 12–17.
163. White, G., Dalmia, S., Carastro, L., Russell, C., Sundaram, V., and Swaminathan, M. "A Novel Methodology (Low Temperature Laminated Organics) for 3-D Integration Using Multilayer Organics." In *IEEE Proceedings of Electronic Components & Technology Conference*, Lake Buena Vista, FL, May 27–30, 2008, pp. 34–37.
164. Premachandran, C. S., Lau, J., Xie, L., Khairyanto, A., Chen, K., Ei, M. P. P., Chew, M., and Choi, W. K. "A Novel, Wafer-Level Stacking Method for Low-Chip Yield and Non-Uniform, Chip-Size Wafers for MEMS and 3-D SIP Applications." In *IEEE Proceedings of Electronic Components & Technology Conference*, Lake Buena Vista, FL, May 27–30, 2008, pp. 314–318.
165. Rida, A., Yang, L., Chaisilwattana, N., Travis, S., Bhattacharya, S., and Tentzeris, M. M. "3-D Packaging Architecture Using Paper as a Dielectric Medium."

In *IEEE Proceedings of Electronic Components & Technology Conference*, Lake Buena Vista, FL, May 27–30, 2008, pp. 371–373.

166. Knickerbocker, J. U., Andry, P. S., Dang, B., Horton, R. R., Patel, C. S., Polastre, R. J., Sakuma, K., Sprogis, E. S., Tsang, C. K., Webb, B. C., and Wright, S. L. "3-D Silicon Integration." In *IEEE Proceedings of Electronic Components & Technology Conference*, Lake Buena Vista, FL, May 27–30, 2008, pp. 538–543.
167. Lühn, O., Radisic, A., Vereecken, P. M., Swinnen, B., Ruythooren, W., Van Hoof, C., and Celis, J. P. "Reducing the Electrodeposition Time for Filling Microvias with Copper for 3-D Technology." In *IEEE Proceedings of Electronic Components & Technology Conference*, Lake Buena Vista, FL, May 27–30, 2008, pp. 866–870.
168. McMahon, J. J., Chan, E., Lee, S. H., Gutmann, R. J., and Lu, J. Q. "Bonding Interfaces in Wafer-Level Metal/Adhesive Bonded 3-D Integration." In *IEEE Proceedings of Electronic Components & Technology Conference*, Lake Buena Vista, FL, May 27–30, 2008, pp. 871–878.
169. Houston, P., Baldwin, D. F., Li, Z., Stout, G., and Tessier, T. "A 3-D-WLCSP Package Technology: Processing and Reliability Characterization." In *IEEE Proceedings of Electronic Components & Technology Conference*, Lake Buena Vista, FL, May 27–30, 2008, pp. 936–943.
170. Sekhar, V. N., Kumar, A., Chai, T. C., Lee, W. S. V., Wang, X. L. S., Zhang, X., Premchandran, C. S., Kripesh, V., Lau, J. H., and Shen, L. "Effect of Wafer Back Grinding on the Mechanical Behavior of Multilayered Low-k for 3-D Stack Packaging Applications." In *IEEE Proceedings of Electronic Components & Technology Conference*, Lake Buena Vista, FL, May 27–30, 2008, pp. 1517–1524.
171. Toh, C. H., Metha, G., Tan, H. H., and Ong, W. P. L., Die Attach Adhesives for 3-D Same-Sized Dies Stacked Packages." In *IEEE Proceedings of Electronic Components & Technology Conference*, Lake Buena Vista, FL, May 27–30, 2008, pp. 1538–1543.
172. Peterson, D., Chen, C. F., and Karulkar, P. C. "Characterization of Drop Impact Survivability of a 3-D CSP Stack Module." In *IEEE Proceedings of Electronic Components & Technology Conference*, Lake Buena Vista, FL, May 27–30, 2008, pp. 1648–1653.
173. Das, R. N., Lauffer, J. M., Poliks, M. D., Egitto, F. D., and Markovich, V. R. "Nano-Micro Filled Conductive Adhesive-Based 3-D Micro Arrays for Z-Axis Interconnections." In *IEEE Proceedings of Electronic Components & Technology Conference*, Lake Buena Vista, FL, May 27–30, 2008, pp. 1970–1977.
174. Zhang, W., Matin, A., Beyne, E., and Ruythooren, W. "Optimizing Au and In Micro-Bumping for 3-D Chip Stacking." In *IEEE Proceedings of Electronic Components & Technology Conference*, Lake Buena Vista, FL, May 27–30, 2008, pp. 1984–1989.
175. Posada, G., Carchon, G., de Raedt, W., and Nauwelaers, B. "Thin-Film MCM-D Technology with Through-Substrate Vias for the Integration of 3-D SiP Modules." In *IEEE Proceedings of Electronic Components & Technology Conference*, Lake Buena Vista, FL, May 27–30, 2008, pp. 2060–2066.
176. Iker, F., Soussan, P., Beyne, E., and Baert, K. "Technology Platform for 3-D Stacking of Thinned Embedded Dies." In *IEEE Proceedings of Electronic Components & Technology Conference*, Lake Buena Vista, FL, May 27–30, 2008, pp. 8–11.
177. Sunohara, M., Shiraishi, A., Taguchi, Y., Murayama, K., Higashi, M., Shimizu, M., Matsumura, K., Ikeda, A., Kuroki, Y., and Tsurushima, T. "Development of Silicon Module with TSVs and Global Wiring (L/S=0.8/0.8 μm)." In *IEEE Proceedings of Electronic Components & Technology Conference*, San Diego, CA, May 25–29, 2009, pp. 25–31.
178. Töpper, M., Baumgartner, T., Klein, M., Fritzsch, T., Roeder, J., Lutz, M., von Suchodoletz, M., Oppermann, H., and Reichl, H. "Low Cost Wafer-Level 3-D Integration without TSV." In *IEEE Proceedings of Electronic Components & Technology Conference*, San Diego, CA, May 25–29, 2009, pp. 339–344.
179. Liu, X., Chen, Q., Dixit, P., Chatterjee, R., Tummala, R. R., and Sitaraman, S. K. "Failure Mechanisms and Optimum Design for Electroplated Copper

Through-Silicon Vias (TSV)." In *IEEE Proceedings of Electronic Components & Technology Conference*, San Diego, CA, May 25–29, 2009, pp. 624–629.

180. Tang, G., O. Navas, D. Pinjala, J. H. Lau, A. Yu, and V. Kripesh "Integrated Liquid Cooling Systems for 3-D Stacked TSV Modules", In *IEEE Transactions in Components and Packaging Technologies*, Vol. 33, Issues 1, 2010, pp. 184–195.
181. Charbonnier, J., Cheramy, S., Henry, D., Astier, A., Brun, J., Sillon, N., Jouve, A., Fowler, S., Privett, M., Puligadda, R., Burggraf, J., and Pargfrieder, S. "Integration of a Temporary Carrier in a TSV Process Flow." In *IEEE Proceedings of Electronic Components & Technology Conference*, San Diego, CA, May 25–29, 2009, pp. 865–871.
182. Rieske, R., Landgraf, R., and Wolter, K. J. "Novel Method for Crystal Defect Analysis of Laser Drilled TSVs." In *IEEE Proceedings of Electronic Components & Technology Conference*, San Diego, CA, May 25–29, 2009, pp. 1139–1146.
183. Raynal, F., Zahraoui, S., Frederich, N., Gonzalez, J., Couturier, B., Truzzi, C., and Lerner, S. "Electrografted Seed Layers for Metallization of Deep TSV Structures." In *IEEE Proceedings of Electronic Components & Technology Conference*, San Diego, CA, May 25–29, 2009, pp. 1147–1152.
184. Khan, N., Li, H. Y., Tan, S. P., Ho, S. W., Su, N., Wai, Y. H., Kripesh, V., Pinjala, Lau, J., and Chuan, T. K. "3-D Packaging with Through Silicon Via (TSV) for Electrical and Fluidic Interconnections." In *IEEE Proceedings of Electronic Components & Technology Conference*, San Diego, CA, May 25–29, 2009, pp. 1153–1158.
185. Chang, H. H., Shih, Y. C., Hsiao, Z. C., Chiang, C. W., Chen, Y. H., and Chiang, K. N. "3-D Stacked Chip Technology Using Bottom-Up Electroplated TSVs." In *IEEE Proceedings of Electronic Components & Technology Conference*, San Diego, CA, May 25–29, 2009, pp. 1177–1184.
186. Chen, Z., Song, X., and Liu, S. "Thermo-Mechanical Characterization of Copper-Filled and Polymer-Filled TSVs Considering Nonlinear Material Behaviors." In *IEEE Proceedings of Electronic Components & Technology Conference*, San Diego, CA, May 25–29, 2009, pp. 1374–1380.
187. Limansyah, I., Pechlaner, A., Weber, W., Wolf, M. J., Klumpp, A., Zoschke, K., Wieland, R., Klein, M., Oppermann, H., Nebrich, L., Reichl, H., and Heinig, A. "3-D Image Sensor SiP with TSV Silicon Interposer." In *IEEE Proceedings of Electronic Components & Technology Conference*, San Diego, CA, May 25–29, 2009, pp. 1430–1436.
188. Itabashi, T. "Integrated Materials Enabling TSV/3-D-TSV." In *IEEE Proceedings of Electronic Components & Technology Conference*, San Diego, CA, May 25–29, 2009, pp. 1759–1763.
189. Laviron, C., LaPras, V., Henry, D., Moreau, S., Anciant, R., Brunet-Manquat, C., Sillon, N., Dunne, B., Galbiati, P., and Toia, F. "Via First Approach Optimization for Through Silicon Via Applications." In *IEEE Proceedings of Electronic Components & Technology Conference*, San Diego, CA, May 25–29, 2009, pp. 14–19.
190. Lu, K. H., Zhang, X., Ryu, S. K., Im, J., Huang, R., and Ho, P. S. "Thermo-Mechanical Reliability of 3-D ICs Containing Through Silicon Vias." In *IEEE Proceedings of Electronic Components & Technology Conference*, San Diego, CA, May 25–29, 2009, pp. 630–634.
191. Curran, B., Ndip, I., Guttovski, S., and Reichl, H. "The Impacts of Dimensions and Return Current Path Geometry on Coupling in Single Ended Through Silicon Vias." In *IEEE Proceedings of Electronic Components & Technology Conference*, San Diego, CA, May 25–29, 2009, pp. 1092–1097.
192. Tezcan, D. S., Duval, F., Philipsen, H., Luhn, O., Soussan, P., and Swinnen, B. "Scalable Through Silicon Via with Polymer Deep Trench Isolation for 3-D Wafer Level Packaging." In *IEEE Proceedings of Electronic Components & Technology Conference*, San Diego, CA, May 25–29, 2009, pp. 1159–1164.
193. Jacquet, F., Henry, D., Charbonnier, J., Bouzaida, N., Sillon, N., and Raine, J. S. "Cost Effective Dry Lithography Solution for Through Silicon Via Technology." In *IEEE Proceedings of Electronic Components & Technology Conference*, San Diego, CA, May 25–29, 2009, pp. 1170–1176.

194. Malta, D., Gregory, C., Temple, D., Wang, C., Richardson, T., and Zhang, Y. "Optimization of Chemistry and Process Parameters for Void-Free Copper Electroplating of High Aspect Ratio Through-Silicon Vias for 3-D Integration." In *IEEE Proceedings of Electronic Components & Technology Conference*, San Diego, CA, May 25–29, 2009, pp. 1301–1306.

195. Nilsson, P., Ljunggren, A., Thorslund, R., Hagström, M., and Lindskog, V. "Novel Through-Silicon Via Technique for 2d/3d SiP and Interposer in Low-Resistance Applications." In *IEEE Proceedings of Electronic Components & Technology Conference*, San Diego, CA, May 25–29, 2009, pp. 1796–1801.

196. Dang, B., Wright, S. L., Andry, P., Sprogis, E., Tsang, C., Polastre, R., Knickerbocker, J., and Ketkar, S. "3-D Chip Stack with Integrated Decoupling Capacitors." In *IEEE Proceedings of Electronic Components & Technology Conference*, San Diego, CA, May 25–29, 2009, pp. 1–5.

197. Jourdain, A., Soussan, P., Swinnen, B., and Beyne, E. "Electrically Yielding Collective Hybrid Bonding for 3-D Stacking of Ics." In *IEEE Proceedings of Electronic Components & Technology Conference*, San Diego, CA, May 25–29, 2009, pp. 11–13.

198. Lee, Y. J., and Lim, S. K. "Routing Optimization of Multi-modal Interconnects In 3-D ICs." In *IEEE Proceedings of Electronic Components & Technology Conference*, San Diego, CA, May 25–29, 2009, pp. 32–39.

199. Choi, W. K., Premachandran, C. S., Chiew, O. S., Xie, L., Liao, E., Khairyanto, A. B. R., Sheng, K. W. C., Thaw, P. P., and Lau, J. H. "Development of Novel Intermetallic Joints using Thin Film Indium Based Solder by Low Temperature Bonding Technology for 3-D IC Stacking." In *IEEE Proceedings of Electronic Components & Technology Conference*, San Diego, CA, May 25–29, 2009, pp. 333–338.

200. Agarwal, R., Zhang, W., Limaye, P., and Ruythooren, W. "High Density Cu-Sn TLP Bonding for 3-D Integration." In *IEEE Proceedings of Electronic Components & Technology Conference*, San Diego, CA, May 25–29, 2009, pp. 345–349.

201. Lannon, J. Jr., Gregory, C., Lueck, M., Huffman, A., and Temple, D. "High Density Cu-Cu Interconnect Bonding for 3-D Integration." In *IEEE Proceedings of Electronic Components & Technology Conference*, San Diego, CA, May 25–29, 2009, pp. 355–359.

202. Kawashita, M., Yoshimura, Y., Tanaka, N., Shimokawa, H., Kinoshita, N., Uematsu, T., Fujisawa, M., Naito, T., and Akazawa, T. "Characterization of MOS Transistor after Through-Hole Electrode Fabrication and 3-D-Assembly by Mechanical Caulking." In *IEEE Proceedings of Electronic Components & Technology Conference*, San Diego, CA, May 25–29, 2009, pp. 360–364.

203. Nagai, M., Tamari, Y., Saito, N., Kuriyama, F., Fukunaga, A., Owatari, A., and Shimoyama, M. "Electroplating Copper Filling for 3-D Packaging." In *IEEE Proceedings of Electronic Components & Technology Conference*, San Diego, CA, May 25–29, 2009, pp. 648–653.

204. Li, Z., Evans, J. L., Houston, P. N., Baldwin, D. F., Stout, E. A., and Tessier, T. G. "Design, Processing and Reliability Characterizations of a 3-D-WLCSP Packaged Component." In *IEEE Proceedings of Electronic Components & Technology Conference*, San Diego, CA, May 25–29, 2009, pp. 972–979.

205. Vempati, S. R., Su, N., Khong, C. H., Lim, Y. Y., Kripesh, V., Lau, J. H., Liew, B. P., Au, K. Y., Tanary, S., Fenner, A., Erich, R., and Milla, J. "Development of 3-D Silicon Die Stacked Package Using Flip Chip Technology with Micro Bump Interconnects." In *IEEE Proceedings of Electronic Components & Technology Conference*, San Diego, CA, May 25–29, 2009, pp. 980–987.

206. Lu, H. C., Wu, G. M., Pan, C., and Chou, Y. T. "Coupling Coefficient Improvement for Inductor Coupled Vertical Interconnect in 3-D IC Die Stacking." In *IEEE Proceedings of Electronic Components & Technology Conference*, San Diego, CA, May 25–29, 2009, pp. 1207–1212.

207. Healy, M. B., and Lim, S. K. "A Study of Stacking Limit and Scaling in 3-D ICs: An Interconnect Perspective." In *IEEE Proceedings of Electronic Components & Technology Conference*, San Diego, CA, May 25–29, 2009, pp. 1213–1220.

208. Iker, F., Teixeira, R. C., Ruythooren, W., and Funaya, T. "Diamond Bit Cutting as Alternative to Polymer Patterning for 3-D Interconnections Technologies."

In *IEEE Proceedings of Electronic Components & Technology Conference*, San Diego, CA, May 25–29, 2009, pp. 1284–1288.

209. Kumar, A., Sekhar, V. N., Lim, S., Keng, C., Sharma, G., Vempati, S. R., Kripesh, V., Lau, J. H., and Kwong, D. L. and Xia, D. "Wafer Level Embedding Technology for 3-D Wafer Level Embedded Package." In *IEEE Proceedings of Electronic Components & Technology Conference*, San Diego, CA, May 25–29, 2009, pp. 1289–1296.
210. Massad, J. E., Bauer, T. M., and Shinde, S. L. "Thermomechanical Modeling of Back-End-of-the-Line 3-D Interconnects." In *IEEE Proceedings of Electronic Components & Technology Conference*, San Diego, CA, May 25–29, 2009, pp. 1361–1367.
211. Bearda, T., Travaly, Y., Wostyn, K., Halder, S., Swinnen, B., Mölders, T., Vorghese, I., and Cheng, P. "Post-dicing Particle Control for 3-D Stacked IC Integration Flows." In *IEEE Proceedings of Electronic Components & Technology Conference*, San Diego, CA, May 25–29, 2009, pp. 1513–1516.
212. Brun, J., Boutry, H., Franiatte, R., Hilt, T., and Sillon, N. "3-D Via Belt Technology." In *IEEE Proceedings of Electronic Components & Technology Conference*, San Diego, CA, May 25–29, 2009, pp. 1670–1675.
213. Ahr, A., Yun, H., Balut, C., and Huffman, A. "Microbump Creation System for 3-D Advanced Packaging Applications." In *IEEE Proceedings of Electronic Components & Technology Conference*, San Diego, CA, May 25–29, 2009, pp. 1788–1791.
214. Taya, A., Iwata, Y., Satoh, R., Murata, H., Morinaga, E., Kudo, K., and Okamoto, K. "Research on a Method to Derive an Ideal 3-D SiP." In *IEEE Proceedings of Electronic Components & Technology Conference*, San Diego, CA, May 25–29, 2009, pp. 2075–2080.
215. Sakuma, K., Nagai, N., Mizuno, J., and Shoji, S. "Vacuum Ultraviolet (VUV) Surface Treatment Process for Flip Chip and 3-D Interconnections." In *IEEE Proceedings of Electronic Components & Technology Conference*, San Diego, CA, May 25–29, 2009, pp. 641–647.
216. Lau, J. H., Yue, T. G., Hoe, G. Y. Y., Zhang, X. W., Chong, C. T., Damaruganath, P., and Vaidyanathan, K. "Effects of TSV (Through Silicon Via) Interposer/Chip on the Thermal Performances of 3-D IC Packaging." *ASME Paper: IPACK2009–89380*, July 19–23, 2009, San Francisco, CA, pp. 1–8.
217. Ladani, L. J., and Rodriguez, O. "Thermo-Mechanical Reliability of Through Silicon Vias (TSVS) and Solder Interconnects in 3-Dimensional Integrated Circuits." *ASME Paper: IPACK2009–89056*, July 19–23, 2009, San Francisco, CA, pp. 1–6.
218. Huang, Y. H., Lee, H. H., Hwang, S. J., and Huang, D. Y. "Copper Plating Process for Through Silicon Via With High Aspect Ratio in Advanced Packaging." *ASME Paper: IPACK2009–89163*, July 19–23, 2009, San Francisco, CA, pp. 1–6.
219. Watanabe, N., Kawashita, M., Yoshimura, Y., Tanaka, N., and Asano, T. "High-Density Room-Temperature 3-D Chip-Stacking Using Mechanical Caulking With Compliant Bump and Through-Hole-Electrode." *ASME Paper: IPACK2009–89274*, July 19–23, 2009, San Francisco, CA, pp. 1–6.
220. Jain, A., Alam, S., Pozder, S., and Jones, R. E. "Thermal-Electrical Co-Optimization of Block-Level Floorplanning in 3-D Integrated Circuits." *ASME Paper: IPACK2009–89017*, July 19–23, 2009, San Francisco, CA, pp. 1–7.
221. Ladani, L. J. "Effect of Design Parameters on Thermo-Mechanical Stresses in 3-D ICS." *ASME Paper: IPACK2009–89083*, July 19–23, 2009, San Francisco, CA, pp. 1–5.
222. Lau, J. H. "State-of-the-art and Trends in 3-D Integration," *Chip Scale Review*, April 2010, pp. 22–28.
223. Lau, J. H. "TSV Manufacturing Yields and Hidden Costs for 3-D IC Integration." In *IEEE Proceedings of Electronic & Components Technology Conference*, Las Vegas, June 2–4, 2010, pp. 1031–1041.
224. Ong, Y., S. Ho, K. Vaidyanathan, V. Sekhar, M. Jong, S. Long, V. Lee, C. Leong, V. Rao, J. Ong, X. Ong, X. Zhang, Y. Seung, J. H. Lau, Y. Lim, D. Yeo, K. Chan, Z.

Yanfeng, J. Tan, and D. Sohn. "Design, assembly and reliability of large die and fine-pitch Cu/low-k flip chip package", *Journal of Microelectronics Reliability*, Vol. 50, 2010, pp. 986–994.

225. Kang, S. K., Horkans, J., Andricacos, P. C., Carruthers, R. A., Cotte, J., Datta, M., Gruber, P., Harper, J. M. E., Kwietniak, K., Sambucetti, C., Shi, L., Brouillette, G., and Danovitch, D. "Pb-Free Solder Alloys for Flip Chip Applications." In *IEEE Proceedings of Electronic & Components Technology Conference*, San Diego, June 2–5, 1999, pp. 283–288.
226. Detert, M., Herzog, T., Wolter, K. J., and Zerna, T. "Screen Printing, Placement and Joining Technologies with Lead-Free Joining Materials." In *IEEE Proceedings of Electronic & Components Technology Conference*, Las Vegas, May 21–24, 2000, pp. 125–128.
227. Gaffney, K., Poarch, J., and Delaney, D. "Lead-Free Flip Chip Process Development." In *IEEE Proceedings of Electronic & Components Technology Conference*, Las Vegas, May 21–24, 2000, pp. 129–133.
228. Ye, L., Lai, Z., Liu, J., and Tholen, A. "Microstructural Coarsening of Lead-Free Solder Joints During Thermal Cycling." In *IEEE Proceedings of Electronic & Components Technology Conference*, Las Vegas, May 21–24, 2000, pp. 134–137.
229. Baynham, G., and Baldwin. D. "Flip Chip Processing of Lead-Free Solders and Halogen-Free High Density Microvia Substrates." In *IEEE Proceedings of Electronic & Components Technology Conference*, Las Vegas, May 21–24, 2000, pp. 1101–1105.
230. Korhonen, T. M., Su, P., Hong, S. J., Korhonen, M. A., and Li, C. Y. "Under Bump Metallizations for Lead Free Solders." In *IEEE Proceedings of Electronic & Components Technology Conference*, Las Vegas, May 21–24, 2000, pp. 1106–1110.
231. Zama, S., Baldwin, D. F., Hikami, T., and Murata, H. "Flip Chip Interconnect Systems Using Wire Stud Bumps and Lead Free Solder." In *IEEE Proceedings of Electronic & Components Technology Conference*, Las Vegas, May 21–24, 2000, pp. 1111–1117.
232. Levis, K., and Mawer, A. "Assembly and Solder Joint Reliability of Plastic Ball Grid Array with Lead-Free Versus Lead-Tin Interconnect." In *IEEE Proceedings of Electronic & Components Technology Conference*, Las Vegas, May 21–24, 2000, pp. 1198–1204.
233. Seelig, K., and Suraski, D. "The Status of Lead-Free Solder Alloys." In *IEEE Proceedings of Electronic & Components Technology Conference*, Las Vegas, May 21–24, 2000, pp. 1405–1409.
234. Whitten, G. "Lead-Free Solder Implementation for Automotive Electronics." In *IEEE Proceedings of Electronic & Components Technology Conference*, Las Vegas, May 21–24, 2000, pp. 1410–1415.
235. Snowdon, G., Tanner, C., and Thompson, J. "Lead-Free Soldering Electronic Interconnect: Current Status and Future Developments." In *IEEE Proceedings of Electronic & Components Technology Conference*, Las Vegas, May 21–24, 2000, pp. 1416–1419.
236. Butterfield, A., Visintainer, V., and Goudarzi, V. "Lead-Free Solder Paste Flux Evaluation and Implementation in Personal Communication Devices." In *IEEE Proceedings of Electronic & Components Technology Conference*, Las Vegas, May 21–24, 2000, pp. 1420–1425.
237. Nguty, T. A., Budiman, S., Rajkumar, D., Solomon, R., Ekere, N. N., and Currie, M. A. "Understanding the Process Window for Printing Lead-Free Solder Pastes." In *IEEE Proceedings of Electronic & Components Technology Conference*, Las Vegas, May 21–24, 2000, pp. 1426–1435.
238. Takeda, S., and Masuko, T. "Novel Die Attach Films Having High Reliability Performance for Lead-Free Solder and CSP." In *IEEE Proceedings of Electronic & Components Technology Conference*, Las Vegas, May 21–24, 2000, pp. 1616–1622.
239. Li, H., and Wong, C. "Development of New No-Flow Underfill Materials for Both Eutetic Solder and a High Temperature Melting Lead-Free Solder." In *IEEE Proceedings of Electronic & Components Technology Conference*, Orlando, May 29-June 1, 2001, pp. 298–303.

240. Kang, S. K., Shih, D. Y., Fogel, K., Lauro, P., Yim, M. J., Advocate, G., Griffin, M., Goldsmith, C., Henderson, D. W., Gosselin, T., King, D., Konrad, J., Sarkhel, A., and Puttlitz, K. J. "Interfacial Reaction Studies on Lead(Pb)-Free Solder Alloys." In *IEEE Proceedings of Electronic & Components Technology Conference*, Orlando, May 29-June 1, 2001, pp. 448–454.
241. Lin, K. J., Silva, A., Frear, D., Guo, Y., Jang, J. W., Li, L., Mitchell, D., Yeung, B., and Zhang, C. "Characterization of Lead-Free Solders and Under Bump Metallurgies for Flip-Chip Package." In *IEEE Proceedings of Electronic & Components Technology Conference*, Orlando, May 29-June 1, 2001, pp. 455–462.
242. Zhang, C., Lin, J. K., and Li, L. "Thermal Fatigue Properties of Lead-free Solders on Cu and Ni-P Under Bump Metallurgies." In *IEEE Proceedings of Electronic & Components Technology Conference*, Orlando, May 29-June 1, 2001, pp. 463–470.
243. Salam, B., Ekere, N. N., and Rajkumar, D. "Study of the Interface Mircrostructure of Sn-Ag-Cu Lead-Free Solders and the Effect of Solder Volume on Intermetallic Layer Formation." In *IEEE Proceedings of Electronic & Components Technology Conference*, Orlando, May 29-June 1, 2001, pp. 471–477.
244. Kripesh, V., Teo, P. S., Chong, C. T., Vishwanadam, G., Mui, Y. C. "Development of a Lead Free Chip Scale Package for Wireless Applications." In *IEEE Proceedings of Electronic & Components Technology Conference*, Orlando, May 29-June 1, 2001, pp. 665–670.
245. Taguchi, T., Kato, R., Akita, S., Okuno, A., Suzuki, H., and Okuno, T. "Lead-Free Interfacial Structures and Their Relationship to Au Plating." In *IEEE Proceedings of Electronic & Components Technology Conference*, Orlando, May 29-June 1, 2001, pp. 675–680.
246. Zribi, A., Zavalij, L., Borgesen, P., Primavera, A., Westby, G., Cotts, E. J. "The Kinetics of Formation of Ternary Intermetallic Alloys in Pb-Sn and Cu-Ag-Sn Pb-Free Electronic Joints." In *IEEE Proceedings of Electronic & Components Technology Conference*, Orlando, May 29-June 1, 2001, pp. 687–692.
247. Zeng, K., Vuorinen, V. and Kivilahti, J. "Intermetallic Reactions between Lead-Free SnAgCu Solder and Ni(P)/Au Surface Finish on PWBs." In *IEEE Proceedings of Electronic & Components Technology Conference*, Orlando, May 29-June 1, 2001, pp. 693–699.
248. Wiese, S., Schubert, A., Walter, H., Dudek, R., Feustel, F., Meusel, E., and Michel, B. "Constitutive Behavior of Lead-Free Solders vs. Lead-Containing Solders Experiments on Bulk Speciments and Flip-Chip Joints." In *IEEE Proceedings of Electronic & Components Technology Conference*, Orlando, May 29-June 1, 2001, pp. 890–902.
249. Farooq, M., Ray, S., Sarkhel, A., and Goldsmith, C. "Evaluation of Lead(Pb)-Free Ceramic Ball Grid Array (CBGA): Wettability, Microstructure and Reliability." In *IEEE Proceedings of Electronic & Components Technology Conference*, Orlando, May 29-June 1, 2001, pp. 978–986.
250. Baynham, G., Baldwin, D. F., Boustedt, K., and Wennerholm, C. "Flip Chip with Lead-Free Solders on Halogen-Free Microvia Substrates." In *IEEE Proceedings of Electronic & Components Technology Conference*, Orlando, May 29-June 1, 2001, pp. 1135–1139.
251. Siewert, T. A., Smith, D. R., Liu, S., and Madeni, J. C. "Database on Lead-Free Solders." In *IEEE Proceedings of Electronic & Components Technology Conference*, Orlando, May 29-June 1, 2001, pp. 1312–1314.
252. Sattiraju, S. V., Dang, B., Johnson, R. W., Li, Y., Smith, J. S., and Bozack, M. J. "Wetting Characteristics of Lead-Free Solder Pastes and Pb-Free PWB Finishes." In *IEEE Proceedings of Electronic & Components Technology Conference*, Orlando, May 29-June 1, 2001, pp. 1135–1139.
253. Chung, C. K., Aspandiar, R., Leong, K. F., and Tay, C. S. "The Interactions of Lead (Pb) in Lead Free Solder (Sn/Ag/Cu) System." In *IEEE Proceedings of Electronic & Components Technology Conference*, San Diego, May 28–31, 2002, pp. 168–175.
254. Chee, C. K., Chin, Y. T., Sterrett, T., He, Y., Sow, H. P., Manepali, R., and Chandran, D. "Lead-free Compatible Underfill Materials For Flip Chip

Applications." In *IEEE Proceedings of Electronic & Components Technology Conference*, San Diego, May 28–31, 2002, pp. 417–424.
255. Joo, D., and Yu, J. "Effects of Microstructure on the Creep Properties of the Lead-free Sn-3.5 Ag-Cu Solders." In *IEEE Proceedings of Electronic & Components Technology Conference*, San Diego, May 28–31, 2002, pp. 1221–1225.
256. Grossmann, G., Nicoletti, G., and Solèr, U. "Results of Comparative Reliability Tests on Lead-free Solder Alloys." In *IEEE Proceedings of Electronic & Components Technology Conference*, San Diego, May 28–31, 2002, pp. 1232–1237.
257. Lee, J. C. B., Yao, Y. L., Chiang, F. Y., Zheng, P. J., Liao, C. C., and Chou, Y. S. "Characterization Study of Lead-free Sn-Cu Plated Packages." In *IEEE Proceedings of Electronic & Components Technology Conference*, San Diego, May 28–31, 2002, pp. 1238–1245.
258. Schubert, A., Dudek, R., Walter, H., Jung, E., Gollhardt, A., Michel, B., and Reichl, H. "Reliability Assessment of Flip-Chip Assemblies with Lead-free Solder Joints." In *IEEE Proceedings of Electronic & Components Technology Conference*, San Diego, May 28–31, 2002, pp. 1246–1255.
259. Arra, M., Xie, D., and Shangguan, D. "Performance of Lead-Free Solder Joints Under Dynamic Mechanical Loading." In *IEEE Proceedings of Electronic & Components Technology Conference*, San Diego, May 28–31, 2002, pp. 1256–1262
260. Balkan, H., Patterson, D., Burgess, G., Carlson, C., Elenius, P., Johnson, M., Rooney, B., Sanchez, J., Stepniak, D., and Wood, J. "Flip-Chip Reliability: Comparative Characterization of Lead Free (Sn/Ag/Cu) and 63Sn/Pb Eutectic Solder." In IEEE Proceedings of Electronic & Components Technology Conference, San Diego, May 28–31, 2002, pp. 1263–1269.
261. Guo, Y., Lin, J. K., and Silva, A. "Reliability Evaluations of Chip Interconnect in Lead-Free Solder Systems." In *IEEE Proceedings of Electronic & Components Technology Conference*, San Diego, May 28–31, 2002, pp. 1275–1280.
262. Patwardhan, V., Kelkar, N., and Nguyen, L. "Lead-Free Wafer Level-Chip Scale Package: Assembly and Reliability." In *IEEE Proceedings of Electronic & Components Technology Conference*, San Diego, May 28–31, 2002, pp. 1355–1358.
263. Lee, N., and Bixenman, M. "Flux technology for Lead-Free Alloys & Its Impact on Cleaning." In *IEEE Proceedings of Electronic & Components Technology Conference*, San Diego, May 28–31, 2002, pp. 1484–1490.
264. Horsley, R., Ekere, N., and Salam, B. "Effect of Lead-free BGA Rework on Joint Microstructure and Intermetallic Compound Formation." In *IEEE Proceedings of Electronic & Components Technology Conference*, San Diego, May 28–31, 2002, pp. 1497–1501.
265. Sundelin, J. J., Nurmi, S. T., Lepistö, T. K., and Ristolainen, E. O. "Pull Testing of Lead-free QFP Solder Joints." In *IEEE Proceedings of Electronic & Components Technology Conference*, San Diego, May 28–31, 2002, pp. 1622–1626.
266. Kang, S. K., Choi, W. K., Shih, D. Y., Lauro, P., Henderson, D. W., Gosselin, T., and Leonard, D. N. "Interfacial Reactions, Microstructure and Mechanical Properties of Pb-Free Solder Joints in PBGA Laminates." In *IEEE Proceedings of Electronic & Components Technology Conference*, San Diego, May 28–31, 2002, pp. 146–153.
267. Kujala, A., Reinikainen, T., and Ren, W. "Transition to Pb-free Manufacturing Using Land Grid Array Packaging Technology." In *IEEE Proceedings of Electronic & Components Technology Conference*, San Diego, May 28–31, 2002, pp. 359–364.
268. Choi, W., Lee, Y., and Tu, K. "Structure and Kinetics of Sn Whisker Growth on Pb-free Solder Finish." In *IEEE Proceedings of Electronic & Components Technology Conference*, San Diego, May 28–31, 2002, pp. 628–633.
269. Tu, K., and Zeng, K. "Reliability Issues of Pb-free Solder Joints in Electronic Packaging Technology." In *IEEE Proceedings of Electronic & Components Technology Conference*, San Diego, May 28–31, 2002, pp. 1194–1200.
270. Gan, H., and Tu, K. "Effect of Electromigration on Intermetallic Compound Formation in Pb-free Solder-Cu Interfaces." In *IEEE Proceedings of Electronic & Components Technology Conference*, San Diego, May 28–31, 2002, pp. 1206–1212.

271. Jang, S. Y., Wolf, J., Kwon, W. S., and Paik, K. W. "UBM (Under Bump Metallization) Study for Pb-Free Electroplating Bumping: Interface Reaction and Electromigration." In *IEEE Proceedings of Electronic & Components Technology Conference*, San Diego, May 28–31, 2002, pp. 1213–1220.
272. Bradley, E. "Lead-Free Solder Assembly: Impact and Opportunity." In *IEEE Proceedings of Electronic & Components Technology Conference*, New Orleans, May 27–30, 2003, pp. 41–46.
273. Ku, A., Shapiro, A., and Schoenung, J. M. "Lead-Free Solders: Issues of Toxicity, Availability and Impacts of Extraction." In *IEEE Proceedings of Electronic & Components Technology Conference*, New Orleans, May 27–30, 2003, pp. 47–53.
274. Lau, J., Dauksher, W., and Vianco, P. "Acceleration Models, Constitutive Equations, and Reliability of Lead-Free Solders and Joints." In *IEEE Proceedings of Electronic & Components Technology Conference*, New Orleans, May 27–30, 2003, pp. 229–236.
275. Amagai, M., Toyoda, Y., and Tajima, T. "High Solder Joint Reliability with Lead Free Solders." In *IEEE Proceedings of Electronic & Components Technology Conference*, New Orleans, May 27–30, 2003, pp. 317–322.
276. Yau, E. W. C., Gong, F. F., Hong, B. F .W., and Chan, P. C. H. "A Reliability Comparison of Lead-Free and Eutectic Sloder for Stencil Printing based Flip-Chip Applications." In *IEEE Proceedings of Electronic & Components Technology Conference*, New Orleans, May 27–30, 2003, pp. 323–330.
277. Theuss, H., Kilger, T., and Ort, T. "Solder Joint Reliability of Lead-Free Solder Balls Assembled with SnPb Solder Paste." In *IEEE Proceedings of Electronic & Components Technology Conference*, New Orleans, May 27–30, 2003, pp. 331–337.
278. Pang, J. H. L., Xiong, B. S., Neo, C. C., Zhang, X. R., and Low, T. H. "Bulk Solder and Solder Joint Properties for Lead Free 95.5Sn-3.8Ag-0.7Cu Solder Alloy." In *IEEE Proceedings of Electronic & Components Technology Conference*, New Orleans, May 27–30, 2003, pp. 673–679.
279. Jang, J. W., Silva, A. D., Lin, J. K., and Frear, D. "Mechanical Tensile Fracture Behaviors of Solid-State-Annealed Eutectic SnPb and Lead-Free Solder Flip Chip Bumps." In *IEEE Proceedings of Electronic & Components Technology Conference*, New Orleans, May 27–30, 2003, pp. 680–684.
280. Teutsch, T., Blankenhorn, R. G., and Zakel, E. "Lead-Free Solder Bumping Process for High Temperature Automotive Application." In *IEEE Proceedings of Electronic & Components Technology Conference*, New Orleans, May 27–30, 2003, pp. 1468–1471.
281. Zhang, Q., Dasgupta, A., and Haswell, P. "Viscoplastic Constitutive Properties and Energy-Partitioning Model of Lead-Free Sn3.9Ag0.6Cu Solder Alloy." In *IEEE Proceedings of Electronic & Components Technology Conference*, New Orleans, May 27–30, 2003, pp. 1862–1868.
282. Hua, F. "Pb-Free Solder Challenges in Electronic Packaging and Assembly." In *IEEE Proceedings of Electronic & Components Technology Conference*, New Orleans, May 27–30, 2003, pp. 58–63.
283. Kim, N. S., Jung, S. Y., and Oh, S. Y. "Reliability of the Pb-Free Solder Bump on Electroless Ni Under Bump Metallurgies." In *IEEE Proceedings of Electronic & Components Technology Conference*, New Orleans, May 27–30, 2003, pp. 77–82.
284. Lin, K. L., Chen K., Hsu, H. M., and Shi, C. L. "Improvement in the Properties of Sn-Zn Eutectic Based Pb-free Solder." In *IEEE Proceedings of Electronic & Components Technology Conference*, New Orleans, May 27–30, 2003, pp. 658–663.
285. Ezawa, H., Miyata, M., Set, M., and Honma, S. "Pb-Free Bumping by Alloying Electroplated Metal Stacks." In *IEEE Proceedings of Electronic & Components Technology Conference*, New Orleans, May 27–30, 2003, pp. 664–667.
286. Farooq, M., Goldmann, L., Martin, G., Goldsmith, C., and Bergeron, C. "Thermo-Mechanical Fatigue Reliability of Pb-Free Ceramic Ball Grid Arrays: Experimental Data and Lifetime Prediction Modeling." In *IEEE Proceedings of*

Electronic & Components Technology Conference, New Orleans, May 27–30, 2003, pp. 827–833.

287. Kim, Y. B, Noguchi, H., and Amagai, M. "Vibration Fatigue Reliability of BGA-IC Package with Pb-free Solder and Pb-Sn Solder." In IEEE Proceedings of Electronic & Components Technology Conference, New Orleans, May 27–30, 2003, pp. 891–897.
288. Choi, W., Kang, S. K., Sohn, Y. C., and Shih, D. Y. "Study of IMC Morphologies and Phase Characteristics Affected by the Reactions of Ni and Cu Metallurgies with Pb-Free Solder Joints." In *IEEE Proceedings of Electronic & Components Technology Conference*, New Orleans, May 27–30, 2003, pp. 1190–1196.
289. Chang, S., Uang, R. H., Liou, D. C., Hu, T. H., Chen, K. C., Chen, Y. F., and Chen, Y. H. "Investigation of Electroplating Ni UBM for Pb-Free Solders." In *IEEE Proceedings of Electronic & Components Technology Conference*, New Orleans, May 27–30, 2003, pp. 1209–1214.
290. Chen, L., and Li, Y. "Effects of Antimony on the Growth of Intermetallic Compounds in Sn-Ag-Cu Pb-free Solder Joints." In *IEEE Proceedings of Electronic & Components Technology Conference*, New Orleans, May 27–30, 2003, pp. 1235–1242.
291. Jackson, G., et.al. "Establishing Control Factors of Intermetallic Formation within Pb-free Solder Bumps at Flip Chip Geometries." In *IEEE Proceedings of Electronic & Components Technology Conference*, New Orleans, May 27–30, 2003, pp. 1472-1479.
292. Sohn, Y. C., Yu, J., Kang, S. K., Shih, D. Y., and Lee, T. Y. "Study of Spalling Behavior of Intermetallic Compounds During the Reaction Between Electroless Ni-P Metallization and Lead-free Solders." In *IEEE Proceedings of Electronic & Components Technology Conference*, Las Vegas, June 1–4, 2004, pp. 76–81.
293. Park, S., Joshi, R., and Goldmann, L. "Reliability of Lead-Free Copper Columns in Comparison with Tin-Lead Solder Column Interconnects." In *IEEE Proceedings of Electronic & Components Technology Conference*, Las Vegas, June 1–4, 2004, pp. 82–89.
294. Ebersberger, B., Bauer, R., and Alexa, L. "Qualification of SnAg Solder Bumps for Lead-Free Flip Chip Applications." In *IEEE Proceedings of Electronic & Components Technology Conference*, Las Vegas, June 1–4, 2004, pp. 683–691.
295. Amagai, M., Toyoda, Y., Ohnishi, T., and Akita, S. "High Drop Test Reliability: Lead-free Solders." In *IEEE Proceedings of Electronic & Components Technology Conference*, Las Vegas, June 1–4, 2004, pp. 1304–1309.
296. Osenbach, J. W, Shook, R. L., Vaccaro, B. T., and Amin, P. A. "Lead Free Packaging and Sn-Whiskers." In *IEEE Proceedings of Electronic & Components Technology Conference*, Las Vegas, June 1–4, 2004, pp. 1314–1324.
297. John, H., Xiong, B., and Low, T. "Creep and Fatigue Characterization of Lead Free 95.5Sn-3.8Ag-0.7-Cu Solder." In *IEEE Proceedings of Electronic & Components Technology Conference*, Las Vegas, June 1–4, 2004, pp. 1333–1337.
298. Tunga, K., Kacker, K., Pucha, R. V., and Sitaraman, S. K. "Accelerated Thermal Cycling: Is It Different for Lead-free Solder?" In *IEEE Proceedings of Electronic & Components Technology Conference*, Las Vegas, June 1–4, 2004, pp. 1579–1585.
299. Moon, K., Li, Y., Xu, J., and Wong, C. "Lead-free Solder Interconnect by Variable Frequency Microwave (VFM)." In *IEEE Proceedings of Electronic & Components Technology Conference*, Las Vegas, June 1–4, 2004, pp. 1579–1585.
300. Paik, K. W., Jeon, Y. D., and Cho, M. G. "Interfacial Reactions and Bump Reliability of Various Pb-Free Solder Bumps on Electroless Ni-P UBMs." In *IEEE Proceedings of Electronic & Components Technology Conference*, Las Vegas, June 1–4, 2004, pp. 675–682.
301. Dunford, S., Canumalla, S and Viswanadham, P. "Intermetallic Morphology and Damage Evolution Under Thermomechanical Fatigue of Lead(Pb)-Free Solder Interconnections." In *IEEE Proceedings of Electronic & Components Technology Conference*, Las Vegas, June 1–4, 2004, pp. 726–736.

302. Rinne, G. "Electromigration in SnPb and Pb-Free Solder Bumps." In *IEEE Proceedings of Electronic & Components Technology Conference*, Las Vegas, June 1–4, 2004, pp. 974–978.
303. Lin, K. L., Liu, P. C., and Song, J. M. "Wetting Interaction between Pb-free Sn-Zn Series Solders and Cu, Ag Substrates." In *IEEE Proceedings of Electronic & Components Technology Conference*, Las Vegas, June 1–4, 2004, pp. 1579–1585.
304. Hayashi, E., Baba, S., Kimura, M., Idaka, S., Maeda, A., and Satoh, M. "Realization of Pb-Free FC-BGA Technology on Low-k Device." In *IEEE Proceedings of Electronic & Components Technology Conference*, Lake Buena Vista, FL, May 31-June 3, 2005, pp. 9–13.
305. Dai, S. X., Pan, N., Castro, A., Culler, J., Hussain, M., Lewis, R., and Michalka, T. "High I/O Glass Ceramic Package Pb-Free BGA Interconnect Reliability." In *IEEE Proceedings of Electronic & Components Technology Conference*, Lake Buena Vista, FL, May 31-June 3, 2005, pp. 23–29.
306. Sohn, Y. C., Yu, J., Kang, S. K., Shih, D. Y., and Lee, T. Y. "Effect of Intermetallics Spalling on the Mechanical Behavior of Electroless Ni(P)/Pb-Free Solder Interconnection." In *IEEE Proceedings of Electronic & Components Technology Conference*, Lake Buena Vista, FL, May 31-June 3, 2005, pp. 83–88.
307. Yoon, S. W., Kripesh, V., Li, H. Y., Iyer, M. K., Viswanath, A. "150 μm Pitch Pb-Free Flipchip Packaging with Cu/Low-k Interconnects." In *IEEE Proceedings of Electronic & Components Technology Conference*, Lake Buena Vista, FL, May 31-June 3, 2005, pp. 100–106.
308. Morganelli, P., Shah, J., Wheelock, B., Mohan, V., and Laffey, M. "Partial Pre-Applied Underfill for Assembly and Reliability of Pb-Free CSPs." In *IEEE Proceedings of Electronic & Components Technology Conference*, Lake Buena Vista, FL, May 31-June 3, 2005, pp. 223–227.
309. Kay, R. W., Gourcuff, E., Desmulliez, M. P. Y., Jackson, G. J., Steen, H. A. H., Liu, C., and Conway, P. P. "Stencil Printing Technology for Wafer Level Bumping at sub-100-Micron Pitch Using Pb-Free Alloys." In *IEEE Proceedings of Electronic & Components Technology Conference*, Lake Buena Vista, FL, May 31-June 3, 2005, pp. 848–854.
310. Setty, K., Subbarayan, G., and Nguyen, L. "Powercycling Reliability, Failure Analysis and Acceleration Factors of Pb-Free Solder Joints." In *IEEE Proceedings of Electronic & Components Technology Conference*, Lake Buena Vista, FL, May 31-June 3, 2005, pp. 907–915.
311. Lal, A., Bradley, E., and Sharda, J. "Effect of Reflow Profiles on the Board Level Drop Reliability of Pb-Free (SnAgCu) BGA Assemblies." In *IEEE Proceedings of Electronic & Components Technology Conference*, Lake Buena Vista, FL, May 31-June 3, 2005, pp. 945–953.
312. Gleason, J., Roubaud, P., Reynolds, C., Lyjak, K., Bath, J., Chu, Q., and Kelly, M. "Pb-free Assembly, Rework, and Reliability Analysis of IPC Class 2 Assemblies." In *IEEE Proceedings of Electronic & Components Technology Conference*, Lake Buena Vista, FL, May 31-June 3, 2005, pp. 959–969.
313. Yoon, S., Chen, Z., Osterman, M., Han, B., and Dasgupta, A. "Effect of Stress Relaxation on Board Level Reliability of Sn Based Pb-Free Solders." In *IEEE Proceedings of Electronic & Components Technology Conference*, Lake Buena Vista, FL, May 31-June 3, 2005, pp. 1210–1214.
314. Su, P., Uehling, T., Wontor, D., Ding, M., and Ho, P. S. "An Evaluation of Electromigration Performance of SnPb and Pb-free Flip Chip Solder Joints." In *IEEE Proceedings of Electronic & Components Technology Conference*, Lake Buena Vista, FL, May 31-June 3, 2005, pp. 1431–1436.
315. Kanagavel, S., Painaik, M., Hurley, J. M., and Dhoble, A. "Improving Solder Joint Reliability in Pb-Free Flip Chips with Advanced Underfill Adhesives." In *IEEE Proceedings of Electronic & Components Technology Conference*, Lake Buena Vista, FL, May 31-June 3, 2005, pp. 1475–1479.
316. Burlacu, D., Kivilahti, J., and Nguyen, L. "Effects of the Solder Oxide Layer on High-Frequency Signal Propagation in Pb-Free Interconnections." In *IEEE Proceedings of Electronic & Components Technology Conference*, Lake Buena Vista, FL, May 31-June 3, 2005, pp. 1874–1879.

317. Kiumi, R., Takeda, S., Yoshioka, J., Kuriyama, F., and Saito, N. "Composition Control for Lead-Free Alloy Electroplating on Flip Chip Bumping." In *IEEE Proceedings of Electronic & Components Technology Conference*, Lake Buena Vista, FL, May 31-June 3, 2005, pp. 120–126.
318. Hu, G., and Tay, A. A. O. "On the Relative Contribution of Temperature, Moisture and Vapor Pressure to Delamination in a Plastic IC Package during Lead-Free Solder Reflow." In *IEEE Proceedings of Electronic & Components Technology Conference*, Lake Buena Vista, FL, May 31-June 3, 2005, pp. 172–178.
319. Ou, S., Xu, Y., Tu, K. N., Alam, M. O., and Chen, Y. C. "Micro-Impact Test on Lead-Free BGA Balls on Au/Electrolytic Ni/Cu Bond Pad." In *IEEE Proceedings of Electronic & Components Technology Conference*, Lake Buena Vista, FL, May 31-June 3, 2005, pp. 467–471.
320. Chong, D. Y. R., Ng, K., Tan, J. Y. N., Low, P. T. H., Pang, J. H. L., Che, F. X., Xiong, B. S., and Xu, L. H. "Drop Impact Reliability Testing for Lead-Free and Leaded Soldered IC Packages." In *IEEE Proceedings of Electronic & Components Technology Conference*, Lake Buena Vista, FL, May 31-June 3, 2005, pp. 622–629.
321. Lin, K. L., and Chiu, Y. T. "The Bonding of Sn-Zn-Ag-Al-Ga Lead-Free Solder Balls on Cu/Ni-P/Au BGA Substrate." In *IEEE Proceedings of Electronic & Components Technology Conference*, Lake Buena Vista, FL, May 31-June 3, 2005, pp. 692–695.
322. Kang, S. C., Kim, C., Muncy, J., Schmidt, M., Lee, S., and Baldwin, D. "Wetting Dynamics of Eutectic and Lead-Free Solders with Various Reflow Conditions and Surface Finish Metallizations." In *IEEE Proceedings of Electronic & Components Technology Conference*, Lake Buena Vista, FL, May 31-June 3, 2005, pp. 832–842.
323. Doraiswami, R., and Tummala, R. "Nano-Composite Lead-Free Interconnect and Reliability." In *IEEE Proceedings of Electronic & Components Technology Conference*, Lake Buena Vista, FL, May 31-June 3, 2005, pp. 871–873.
324. Fan, X., Rasier, G., and Vasudevan, V. S. "Effects of Dwell Time and Ramp Rate on Lead-Free Solder Joints in FCBGA Packages." In *IEEE Proceedings of Electronic & Components Technology Conference*, Lake Buena Vista, FL, May 31-June 3, 2005, pp. 901–906.
325. Liu, Y., Geiger, D. A., and Shangguan, D. "Determination of Components Candidacy for Second Side Reflow with Lead-Free Solder." In *IEEE Proceedings of Electronic & Components Technology Conference*, Lake Buena Vista, FL, May 31-June 3, 2005, pp. 970–976.
326. Clech, J. P. "An Extension of the Omega Method to Primary and Tertiary Creep of Lead-Free Solders." In *IEEE Proceedings of Electronic & Components Technology Conference*, Lake Buena Vista, FL, May 31-June 3, 2005, pp. 1261–1271.
327. Ng, H. S., Tee, T. Y., Goh, K. Y., and Luan, J. E. "Absolute and Relative Fatigue Life Prediction Methodology for Virtual Qualification and Design Enhancement of Lead-free BGA." In *IEEE Proceedings of Electronic & Components Technology Conference*, Lake Buena Vista, FL, May 31-June 3, 2005, pp. 1282–1291.
328. Lau, J., and Dauksher, W. "Effects of Ramp-Time on the Thermal-Fatigue Life of SnAgCu Lead-Free Solder Joints." In *IEEE Proceedings of Electronic & Components Technology Conference*, Lake Buena Vista, FL, May 31-June 3, 2005, pp. 1292–1298.
329. Lall, P., Islam, N., and Suhling, J. "Prognostication and Health Monitoring of Leaded and Lead-Free Electronic and MEMs Packages in Harsh Environments." In *IEEE Proceedings of Electronic & Components Technology Conference*, Lake Buena Vista, FL, May 31-June 3, 2005, pp. 1305–1313.
330. Ebersberger, B., Bauer, R., and Alexa, L. "Reliability of Lead-Free SnAg Solder Bumps: Influence of Electromigration and Temperature." In *IEEE Proceedings of Electronic & Components Technology Conference*, Lake Buena Vista, FL, May 31-June 3, 2005, pp. 1407–1415.
331. Li, Y., Moon, K. S., and Wong, C. P. "Development of Conductive Adhesives with Novel Corrosion Inhibitors for Stabilizing Contact Resistance on Non-Noble Lead-Free Finishes." In *IEEE Proceedings of Electronic &

Components Technology Conference, Lake Buena Vista, FL, May 31-June 3, 2005, pp. 1462–1467.

332. Ghosh, K., McCabe, M., and Tian, G. "Advances in Flip Chip Underfill Technology for Lead-Free Component Packaging." In *IEEE Proceedings of Electronic & Components Technology Conference*, Lake Buena Vista, FL, May 31-June 3, 2005, pp. 1480–1485.
333. Reiff, D., and Bradley, E. "A Novel Mechanical Shock Test Method to Evaluate Lead-free BGA Solder Joint Reliability." In *IEEE Proceedings of Electronic & Components Technology Conference*, Lake Buena Vista, FL, May 31-June 3, 2005, pp. 1519–1525.
334. Lall, P., Singh, N., Suhling, J., Strickland, M., and Blanche, J. "Decision-Support Models for Thermo-Mechanical Reliability of Leadfree Flip-Chip Electronics in Extreme Environments." In *IEEE Proceedings of Electronic & Components Technology Conference*, Lake Buena Vista, FL, May 31-June 3, 2005, pp. 127–136.
335. Tsai, I., Tai, L. J., Yen, S. F., Chuang, T. H., Wu, E., Lo, R., and Ku, T. "Identification of Mechanical Properties of Intermetallic Compounds on Lead Free Solder." In *IEEE Proceedings of Electronic & Components Technology Conference*, Lake Buena Vista, FL, May 31-June 3, 2005, pp. 687–691.
336. Darveaux, R. "Shear Deformation of Lead Free Solder Joints." In *IEEE Proceedings of Electronic & Components Technology Conference*, Lake Buena Vista, FL, May 31-June 3, 2005, pp. 882–893.
337. Che, F. X., Pang, J. H. L., Xiong, B. S., Xu, L., and Low, T. H. "Lead Free Solder Joint Reliability Characterization for PBGA, PQFP and TSSOP Assemblies." In *IEEE Proceedings of Electronic & Components Technology Conference*, Lake Buena Vista, FL, May 31-June 3, 2005, pp. 916–921.
338. Dong, H., Fan, L., Moon, K. S., Wong, C. P. and Baskes, M. I. "Molecular Dynamics Simulation of Lead Free Solder for Low Temperature Reflow Applications." In *IEEE Proceedings of Electronic & Components Technology Conference*, Lake Buena Vista, FL, May 31-June 3, 2005, pp. 983–987.
339. Nicholls, L., Darveaux, R., Hansen, B., Carey, C., Aoki, T., Akimoto, T., and Chang, J. "Fine Pitch (150μm) Pb-Free Flip-Chip Bumping and Packaging." In *IEEE Proceedings of Electronic & Components Technology Conference*, San Diego, CA, May 30-June 2, 2006, pp. 131–138.
340. Ray, S., Kiyono, S. S., Waite, K., and Nicholls, L. "Qualification of Low-K 90nm Technology Die with Pb-Free Bumps on a Build-up Laminate Package (PBGA) with Pb-Free Assembly Processes." In *IEEE Proceedings of Electronic & Components Technology Conference*, San Diego, CA, May 30-June 2, 2006, pp. 139–144.
341. Tsai, J. Y., Sundaram, V., Wiedenman, B., Sun, Y., Wong, C. P., and Tummala, R. R. "A Novel 20–100μm Pitch IC-to-Package Interconnect and Assembly Process for Pb-Free Solder, Copper or Gold Stud Bumps." In *IEEE Proceedings of Electronic & Components Technology Conference*, San Diego, CA, May 30-June 2, 2006, pp. 263–268.
342. Burlacu, D., Kivilahti, J., and Nguyen, L. "Influence of Solder Microstructure and Oxide Layers on High Frequency Electrical Losses of WL-CSP Pb-Free Interconnections." In *IEEE Proceedings of Electronic & Components Technology Conference*, San Diego, CA, May 30-June 2, 2006, pp. 577–583.
343. Chae, S. H., Zhang, X., Chao, H. L., Lu, K. H., Ho, P. S., Ding, M., Su, P., Uehling, T., and Ramanathan, L. N. "Electromigration Lifetime Statistics for Pb-Free Solder Joints with Cu and Ni UBM in Plastic Flip-Chip Packages." In *IEEE Proceedings of Electronic & Components Technology Conference*, San Diego, CA, May 30-June 2, 2006, pp. 650–656.
344. Nah, J. W., Suh, J. O., Tu, K. N., Yoon, S. W., Chong, C. T., Kripesh, V., Su, B. R., and Chen, C. "Electromigration in Pb-Free Solder Bumps with Cu Column as Flip-Chip Joints." In *IEEE Proceedings of Electronic & Components Technology Conference*, San Diego, CA, May 30-June 2, 2006, pp. 657–662.
345. Lin, K. L., and Kuo, S. M. "The Electromigration and Thermomigration Behaviors of Pb-Free, Flip-Chip Sn-3Ag-0.5Cu Solder Bumps." In *IEEE Proceedings of Electronic & Components Technology Conference*, San Diego, CA, May 30-June 2, 2006, pp. 667–672.

346. Xie, D., Rooney, D. T., Geiger, D., and Shangguan, D. "A Comparative Study on Drop Test Performance of Fine Pitch BGA Assemblies using Pb-Free and Tin-Lead Solders." In *IEEE Proceedings of Electronic & Components Technology Conference*, San Diego, CA, May 30-June 2, 2006, pp. 865–874.
347. Osterman, M., Dasgupta, A., and Han, B. "A Strain Range-Based Model for Life Assessment of Pb-Free SAC Solder Interconnects." In *IEEE Proceedings of Electronic & Components Technology Conference*, San Diego, CA, May 30-June 2, 2006, pp. 884–890.
348. Cavasin, D. "Yellow Solder: An Assessment of the Quality and Reliability of Pb-Free Lead Finishes and Solder Ball Alloys Exhibiting Excessive Sn Oxidation." In *IEEE Proceedings of Electronic & Components Technology Conference*, San Diego, CA, May 30-June 2, 2006, pp. 1060–1065.
349. Cavasin, D. "The Impact of Interfacial Void Formation on Pb-Free Sn-4.0Ag-0.5Cu BGA Solder Joint Integrity." In *IEEE Proceedings of Electronic & Components Technology Conference*, San Diego, CA, May 30-June 2, 2006, pp. 1191–1195.
350. Gan, H., Wright, S. L., Polastre, R., Buchwalter, L. P., Horton, R., Andry, P. S., Patel, C., Tsang, C., Knickerbocker, J., Sprogis, E., Kang, S. K., Lee, K. W. and Pavlova, A. "Pb-Free Micro-Joints (50μm Pitch) for the Next Generation Micro-Systems: The Fabrication, Assembly and Characterization." In *IEEE Proceedings of Electronic & Components Technology Conference*, San Diego, CA, May 30-June 2, 2006, pp. 1210–1215.
351. Zhao, X. J., and Caers, J. F. J. M. "Evaluation of Soldering and Reliability Performance of NiPdAu Finished PPF with Pb-Free Solder Alloy 95.5Sn4.0Ag0.5Cu." In *IEEE Proceedings of Electronic & Components Technology Conference*, San Diego, CA, May 30-June 2, 2006, pp. 1442–1448.
352. Bieler, T. R., Jiang, H., Lehman, L. P., Kirkpatrick, T., and Cotts, E. J. "Influence of Sn Grain Size and Orientation on the Thermomechanical Response and Reliability of Pb-Free Solder Joints." In *IEEE Proceedings of Electronic & Components Technology Conference*, San Diego, CA, May 30-June 2, 2006, pp. 1462–1467.
353. Paquet, M. C., Gaynes, M., Duchesne, E., Questad, D., Bélanger, L., and Sylvestre, J. "Underfill Selection Strategy for Pb-Free, Low-K and Fine Pitch Organic Flip-Chip Applications." In *IEEE Proceedings of Electronic & Components Technology Conference*, San Diego, CA, May 30-June 2, 2006, pp. 1595–1603.
354. Chong, S. C., Chai, T. C., Lim, S., Hnin, W. Y., Tan, Y. M., and Cheng, C. K. "Assembly Challenges of High Density Large Fine Pitch Lead-Free, Flip-Chip Package." In *IEEE Proceedings of Electronic & Components Technology Conference*, San Diego, CA, May 30-June 2, 2006, pp. 10–15.
355. Kumar, K. M., Kripesh, V., and Tay, A. A. O. "Sn-Ag-Cu Lead-Free Composite Solders for Ultra-Fine-Pitch Wafer-Level Packaging." In *IEEE Proceedings of Electronic & Components Technology Conference*, San Diego, CA, May 30-June 2, 2006, pp. 237–243.
356. Gao, F., and Takemoto, T. "The Evolution of Plastic Deformation of Sn3.5Ag-Based Lead-Free Solders." In *IEEE Proceedings of Electronic & Components Technology Conference*, San Diego, CA, May 30-June 2, 2006, pp. 269–274.
357. Xu, L., and Pang, J. H. L. "Effect of Intermetallic and Kirkendall Voids Growth on Board Level Drop Reliability for SnAgCu Lead-Free BGA Solder Joint." In *IEEE Proceedings of Electronic & Components Technology Conference*, San Diego, CA, May 30-June 2, 2006, pp. 275–282.
358. Yin, W., and Lee, N. C. "Reworkable No-Flow Underfilling for Both Tin-Lead and Lead-Free Reflow for CSP Assembled under Air." In *IEEE Proceedings of Electronic & Components Technology Conference*, San Diego, CA, May 30-June 2, 2006, pp. 502–509.
359. Lall, P., Hande, M., Bhat, C., Suhling, J., Islam, N., and Lee, J. "Feature Extraction and Damage-Precursors for Prognostication of Lead-Free Electronics." In *IEEE Proceedings of Electronic & Components Technology Conference*, San Diego, CA, May 30-June 2, 2006, pp. 718–727.
360. Ma, H., Suhling, J. C., Lall, P., and Bozack, M. J. "Reliability of the Aging Lead-Free Solder Joint." In *IEEE Proceedings of Electronic & Components Technology Conference*, San Diego, CA, May 30-June 2, 2006, pp. 848–864.

361. Song, F., and Lee, S. W. R. "Corrosion of Sn-Ag-Cu Lead-Free Solders and the Corresponding Effects on Board Level Solder Joint Reliability." In *IEEE Proceedings of Electronic & Components Technology Conference*, San Diego, CA, May 30-June 2, 2006, pp. 891–898.
362. Darveaux, R., Reichman, C., and Islam, N. "Interface Failure in Lead-Free Solder Joints." In *IEEE Proceedings of Electronic & Components Technology Conference*, San Diego, CA, May 30-June 2, 2006, pp. 906–917.
363. Lau, J., and Dauksher, W. "Thermal Stress Analysis of a Flip-Chip Parallel VCSEL (Vertical-Cavity Surface-Emitting Lasers) Package with Low-Temperature Lead-Free (48Sn-52In) Solder Joints." In *IEEE Proceedings of Electronic & Components Technology Conference*, San Diego, CA, May 30-June 2, 2006, pp. 1009–1017.
364. Ren, F., Zhang, X., Nah, J. W., Tu, K. N., Xu, L., and Pang, J. H. L. "In-Situ Study of the Effect of Electromigration on Strain Evolution and Mechanical Property Change in Lead-Free Solder Joints." In *IEEE Proceedings of Electronic & Components Technology Conference*, San Diego, CA, May 30-June 2, 2006, pp. 1160–1163.
365. Amagai, M. "A Study of Nano Particles in SnAg-Based, Lead-Free Solders for Intermetallic Compounds and Drop Test Performance." In *IEEE Proceedings of Electronic & Components Technology Conference*, San Diego, CA, May 30-June 2, 2006, pp. 1170–1190.
366. Song, F. and Lee, S. W. R. "Investigation of IMC Thickness Effect on the Lead-Free Solder Ball Attachment Strength: Comparison between Ball Shear Test and Cold Bump Pull Test Results." In *IEEE Proceedings of Electronic & Components Technology Conference*, San Diego, CA, May 30-June 2, 2006, pp. 1196–1203.
367. Li, Y., Moon, K. S., and Wong, C. P. "Novel Lead-Free Nano-Scale, Non-Conductive Adhesive (NCA) for Ultra-Fine Pitch Interconnect Applications." In *IEEE Proceedings of Electronic & Components Technology Conference*, San Diego, CA, May 30-June 2, 2006, pp. 1239–1245.
368. Lau, J., Lee, S. W. R., Song, F., Lau, D., and Shangguan, D. "Isothermal Fatigue Tests of Plastic Ball Grid Array (PBGA) SnAgCu Lead-Free Solder Joints at 60°C." In *IEEE Proceedings of Electronic & Components Technology Conference*, San Diego, CA, May 30-June 2, 2006, pp. 1476–1483.
369. Perez, M. G., O'Keefe, M. J., Colfax, R., Murry, D., Smith, J., Kleine, D. W., and Amick, P. "Vibration Testing of Repaired Lead-Tin/Lead-Free Solder Joints." In *IEEE Proceedings of Electronic & Components Technology Conference*, San Diego, CA, May 30-June 2, 2006, pp. 1493–1498.
370. Lal, A., and Bradley, E. "Relationship of Tensile Interfacial Strength to Lead-Free BGA Impact Performance." In *IEEE Proceedings of Electronic & Components Technology Conference*, San Diego, CA, May 30-June 2, 2006, pp. 1628–1633.
371. Li, Y., Xiao, F., Moon, K. S., and Wong, C. P. "A Novel Environmentally Friendly and Biocompatible Curing Agent for Lead-Free Electronics." In *IEEE Proceedings of Electronic & Components Technology Conference*, San Diego, CA, May 30-June 2, 2006, pp. 1639–1644.
372. Ghaffarian, R., Ramkumar, S. M., and Varanasi, A. "Thermal Cycle/Aging Reliability of Lead-Free 0201 Assemblies." In *IEEE Proceedings of Electronic & Components Technology Conference*, San Diego, CA, May 30-June 2, 2006, pp. 1940–1956.
373. Ray, S., Muncy, J., McLaughlin, P. V., and Nicholls, L. "Qualification of Low-k 65nm Technology Die with Pb-Free Bumps on a 2–2-2 Laminate Package (PBGA) with Pb-Free Assembly Processes." In *IEEE Proceedings of Electronic & Components Technology Conference*, Reno, NV, May 29-June 1, 2007, pp. 1–7.
374. Dang, B., Wright, S. L., Andry, P. S., Tsang, C. K., Patel, C., Polastre, R., Horton, R., Sakuma, K., Webb, B. C., Sharma, A., Knickerbocker, J. U., Sprogis, E., and Zhang, G. "Assembly, Characterization, and Reworkability of Pb-Free Ultra Fine Pitch C4s for System-on-Package." In *IEEE Proceedings of Electronic & Components Technology Conference*, Reno, NV, May 29-June 1, 2007, pp. 42–48.
375. Park, S., Tang, Z., and Chung, S. "Temperature Effect of Interfacial Fracture Toughness on Underfill for Pb-Free Flip Chip Packages." In *IEEE Proceedings*

of *Electronic & Components Technology Conference*, Reno, NV, May 29-June 1, 2007, pp. 105–109.

376. Vasudevan, V., Fan, X., Liu, T., and Young, D. "Slow Cycle Fatigue Creep Performance of Pb-Free (LF) Solders." In *IEEE Proceedings of Electronic & Components Technology Conference*, Reno, NV, May 29-June 1, 2007, pp. 116–123.
377. Zhang R. "Assembly and Application Specific Methodology to Determine Accelerated Temperature Cycling Requirements for Pb-Free Solder Joint Reliability." In *IEEE Proceedings of Electronic & Components Technology Conference*, Reno, NV, May 29-June 1, 2007, pp. 124–128.
378. Cavasin, D., Anani, M., Rittman, G., and Casto, J. "Board Level Characterization of Pb-Free Sn-3.5Ag Versus Eutectic Pb-Sn BGA Solder Joint Reliability." In *IEEE Proceedings of Electronic & Components Technology Conference*, Reno, NV, May 29-June 1, 2007, pp. 129–135.
379. Syed, A., Scanlon, J., Kim, T. S., Cha, S. W., and Ryu, C. G. "Effect of Pb-Free Alloy Composition on Drop/Impact Reliability of 0.4, 0.5 & 0.8mm Pitch Chip Scale Packages with NiAu Pad Finish." In *IEEE Proceedings of Electronic & Components Technology Conference*, Reno, NV, May 29-June 1, 2007, pp. 951–956.
380. Ramakrishna, G., Nandagopal, B., Dick, T., Burton, A., Higdon, R., Anvari, M., Hu, M., Teng, S., and Xue, J. "Comprehensive Assembly and Reliability Study of 0201's for High Reliability Applications for Utilizing both a Pb-Free and Sn/Pb Assembly Process." In *IEEE Proceedings of Electronic & Components Technology Conference*, Reno, NV, May 29-June 1, 2007, pp. 1313–1319.
381. Chae, S. H., Chao, B., Zhang, X., Im, J., and Ho, P. S. "Investigation of Intermetallic Compound Growth Enhanced by Electromigration in Pb-Free Solder Joints." In *IEEE Proceedings of Electronic & Components Technology Conference*, Reno, NV, May 29-June 1, 2007, pp. 1442–1449.
382. Lin, K. L., and Lin, G. P. "The Electromigration Investigation on the Newly Developed Pb-Free Sn-Zn-Ag-Al-Ga Solder Ball Interconnect." In *IEEE Proceedings of Electronic & Components Technology Conference*, Reno, NV, May 29-June 1, 2007, pp. 1467–1472.
383. Kwon, Y. M., and Paik, K. W. "Electromigration of Pb-Free Solder Flip-Chip Using Electroless Ni-P/Au UBM." In *IEEE Proceedings of Electronic & Components Technology Conference*, Reno, NV, May 29-June 1, 2007, pp. 1472–1477.
384. Zhao, X. J., Caers, J. F. J. M., Vries, J. W. C., Wong, E. H., and Ranjoo, R. "A Component Level Test Method for Evaluating the Resistance of Pb-Free BGA Solder Joints to Brittle Fracture under Shock Impact." In *IEEE Proceedings of Electronic & Components Technology Conference*, Reno, NV, May 29-June 1, 2007, pp. 1522–1529.
385. Kang, S. K., Cho, M. G., Lauro, P., and Shih, D. Y. "Critical Factors Affecting the Undercooling of Pb-Free, Flip-Chip Solder Bumps and In-Situ Observation of Solidification Process." In *IEEE Proceedings of Electronic & Components Technology Conference*, Reno, NV, May 29-June 1, 2007, pp. 1597–1603.
386. Kim, D., Suh, D., Millard, T., Kim, H., Kumar, C., Zhu, M., and Xu, Y. "Evaluation of High Compliant Low Ag Solder Alloys on OSP as a Drop Solution for the 2nd Level Pb-Free Interconnection." In *IEEE Proceedings of Electronic & Components Technology Conference*, Reno, NV, May 29-June 1, 2007, pp. 1614–1619.
387. Choi, J. Y., Kang, I. S., Park, B. J., Park, Y. M., and Jung, G. J. "Study on Inter-Metallic Compound (IMC) at Interface of Pb-Free Solder and Nickel UBM for Mobile Application." In *IEEE Proceedings of Electronic & Components Technology Conference*, Reno, NV, May 29-June 1, 2007, pp. 1802–1808.
388. Ruhmer, K., Laine, E., O'Donnell, K., Kostetsky, J., Hauck, K., Manessis, D., Ostmann, A., Toepper, M., and Juergensen, N. "Alternative UBM for Lead-Free Solder Bumping Using C4NP." In *IEEE Proceedings of Electronic & Components Technology Conference*, Reno, NV, May 29-June 1, 2007, pp. 15–21.
389. Jiang, H., Moon, K. S., Wong, C. P., and Hua, F. "Thermal Properties of Tin/Silver Alloy Nanoparticles for Low Temperature Lead-Free Interconnect Technology." In *IEEE Proceedings of Electronic & Components Technology Conference*, Reno, NV, May 29-June 1, 2007, pp. 54–58.

390. Ramkumar, S. M., and Srihari, K. "Influence of PCB Surface Finish and Thermal and Temperature/Humidity Aging on the Performance of a Novel Anisotropic Conductive Adhesive for Lead-Free Surface Mount Assembly." In *IEEE Proceedings of Electronic & Components Technology Conference*, Reno, NV, May 29-June 1, 2007, pp. 65–73.
391. Kim, I., and Lee, S. B. "Fatigue Life Evaluation of Lead-Free Solder under Thermal and Mechanical Loads." In *IEEE Proceedings of Electronic & Components Technology Conference*, Reno, NV, May 29-June 1, 2007, pp. 95–104.
392. Pei, M., and Qu, J. "Effect of Rare Earth Elements on Lead-Free Solder Microstructure Evolution." In *IEEE Proceedings of Electronic & Components Technology Conference*, Reno, NV, May 29-June 1, 2007, pp. 198–204.
393. Gao, F., Nishikawa, H., and Takemoto, T. "Nano-Scale Mechanical Responses of Sn-Ag Based Lead-Free Solders." In *IEEE Proceedings of Electronic & Components Technology Conference*, Reno, NV, May 29-June 1, 2007, pp. 205–210.
394. Che, F. X., Zhu, W. H., Sun, W., Sun, A., Wang, C. K., Tan, H. B., and Pang, J. H. L. "Comprehensive Modeling of Stress-Strain Behavior for Lead-Free Solder Joints under Board-Level Drop Impact Loading Condition." In *IEEE Proceedings of Electronic & Components Technology Conference*, Reno, NV, May 29-June 1, 2007, pp. 528–535.
395. Johnson, M. E., Lu, H. Y., Lawhead, D., Tessier, T., Scott, D., and Curtis, A. "Using High Speed Shear and Cold Ball Pull to Characterize Lead-Free Solder Alloys and Predict Board Level Drop Test Performance." In *IEEE Proceedings of Electronic & Components Technology Conference*, Reno, NV, May 29-June 1, 2007, pp. 536–542.
396. Sakuma, K., Andry, P. S., Dang, B., Tsang, C. K., Patel, C., Wright, S. L., Webb, B., Kang, S. K., Polastre, R., Horton, R., Knickerbocker, J. U., Maria, J., and Sprogis, E. "3-D Chip Stacking Technology with Low-Volume Lead-Free Interconnections." In *IEEE Proceedings of Electronic & Components Technology Conference*, Reno, NV, May 29-June 1, 2007, pp. 627–632.
397. Ma, H., Suhling, J. C., Zhang, Y., Lall, P., and Bozack, M. J. "The Influence of Elevated Temperature Aging on Reliability of Lead-Free Solder Joints." In *IEEE Proceedings of Electronic & Components Technology Conference*, Reno, NV, May 29-June 1, 2007, pp. 653–668.
398. Pandher, R. S., Lewis, B. G., Vangaveti, R., and Singh, B. "Drop Shock Reliability of Lead-Free Alloys—Effect of Micro-Additives." In *IEEE Proceedings of Electronic & Components Technology Conference*, Reno, NV, May 29-June 1, 2007, pp. 669–676.
399. Kobayashi, T., Tanaka, M., Kariya, Y., Sasaki, T., and Tatsumi, K. "Effect of Ni Addition on Bending Properties of Sn-Ag-Cu Lead-Free Solder Joints." In *IEEE Proceedings of Electronic & Components Technology Conference*, Reno, NV, May 29-June 1, 2007, pp. 684–688.
400. Darveaux, R., and Reichman, C. "Mechanical Properties of Lead-Free Solders." In *IEEE Proceedings of Electronic & Components Technology Conference*, Reno, NV, May 29-June 1, 2007, pp. 695–706.
401. Ishikawa, S., Uchiyama, T., Hashino, E., Kohno, T., Tanaka, M., and Tatsumi, K. "Lead-Free Solder Micro-Ball Bumps for the Next Generation of Flip Chip Interconnection: Micro-Ball Materials, Bump Formation Process and Reliability." In *IEEE Proceedings of Electronic & Components Technology Conference*, Reno, NV, May 29-June 1, 2007, pp. 872–877.
402. Uchida, M., Ito, H., Yabui, K., Nishiuchi, H., Togasaki, T., Higuchi, K., and Ezawa, H. "Low-Stress Interconnection for Flip Chip BGA Employing Lead-Free Solder Bump." In *IEEE Proceedings of Electronic & Components Technology Conference*, Reno, NV, May 29-June 1, 2007, pp. 885–891.
403. Kim, H., Zhang, M., Kumar, C. M., Suh, D., Liu, P., Kim, D., Xie, M., and Wang, Z. "Improved Drop Reliability Performance with Lead-Free Solders of Low Ag Content and Their Failure Modes." In *IEEE Proceedings of Electronic & Components Technology Conference*, Reno, NV, May 29-June 1, 2007, pp. 962–967.
404. Laine, E., Perfecto, E., Campbell, B., Wood, J., Busby, J., Garant, J., and Guerin, L. "C4NP Technology for Lead-Free Solder Bumping." In *IEEE Proceedings of*

Electronic & Components Technology Conference, Reno, NV, May 29-June 1, 2007, pp. 1320–1325.
405. Lavoie, H., Paquet, M. C., Sylvestre, J., Ouimet, S., Duchesne, E., Barbeau, S., Gauvin, M., and Oberson, V. "From Leaded to Lead-Free Assembly and New Packaging Technology Challenges." In *IEEE Proceedings of Electronic & Components Technology Conference*, Reno, NV, May 29-June 1, 2007, pp. 1333–1339.
406. Nguyen, J., Geiger, D., Rooney, D., and Shangguan, D. "Solder Joint Characteristics and Reliability of Lead-Free Area Array Packages Assembled under Various Tin-Lead Soldering Process Conditions." In *IEEE Proceedings of Electronic & Components Technology Conference*, Reno, NV, May 29-June 1, 2007, pp. 1340–1349.
407. Li, Y., Yim, M. J., and Wong, C. P. "Novel Monolayer-Enhanced Non-Conductive Film (NCF) for Ultra-Fine Pitch High Performance Interconnect in Lead-Free Electronics." In *IEEE Proceedings of Electronic & Components Technology Conference*, Reno, NV, May 29-June 1, 2007, pp. 1911–1915.
408. Osenbach, J., Bachman, M., Weachock, R., Delucca, J., Baiocchi, F., Crouthamel, D., and Amin, A. "Thermal Degradation of Sputtered Al/Ni(V)/Cu-UBM in Pb-Free Flip Chip Solder Joints Connected to Substrate with Different Surface Finishes." In *IEEE Proceedings of Electronic & Components Technology Conference*, Lake Buena Vista, FL, May 27–30, 2008, pp. 123–130.
409. Chae, S. H., Im, J., Ho, P. S., and Uehling, T. "Effects of UBM Thickness, Contact Trace Structure and Solder Joint Scaling on Electromigration Reliability of Pb-Free Solder Joints." In *IEEE Proceedings of Electronic & Components Technology Conference*, Lake Buena Vista, FL, May 27–30, 2008, pp. 354–359.
410. Lu, M., Lauro, P., Shih, D. Y., Polastre, R., Goldsmith, C., Henderson, D. W., Zhang, H., and Cho, M. G. "Comparison of Electromigration Performance for Pb-Free Solders and Surface Finishes with Ni UBM." In *IEEE Proceedings of Electronic & Components Technology Conference*, Lake Buena Vista, FL, May 27–30, 2008, pp. 360–365.
411. Kang, S. K., Shih, D. Y., Cho, M. G., Seo, S. K., and Lee, H. M. "Controlling the Interfacial Reactions in Pb-Free Interconnections by Adding Minor Alloying Elements to Sn-Rich Solders." In *IEEE Proceedings of Electronic & Components Technology Conference*, Lake Buena Vista, FL, May 27–30, 2008, pp. 478–484.
412. Ajmera, A., Ramkumar, S. M., and Liu, T. L. "Finite Element Modeling and Experimental Validation of Conventional and High Speed Shear Testing in Pb-Free Environment." In *IEEE Proceedings of Electronic & Components Technology Conference*, Lake Buena Vista, FL, May 27–30, 2008, pp. 1021–1030.
413. Caers, J. F. J. M., Zhao, X. J., Wong, E. H., Seah, S. K. W., Selvanayagam, C. S., van Driel, W. D., Owens, N., Leoni, M., Tan, L. C., Eu, P. L., Lai, Y. S., and Yeh, C. L. "A Study of Crack Propagation in Pb-Free Solder Joints under Drop Impact." In *IEEE Proceedings of Electronic & Components Technology Conference*, Lake Buena Vista, FL, May 27–30, 2008, pp. 1166–1172.
414. Nah, J. W., Buchwalter, S. L., Gruber, P. A., Shih, D. Y., and Furman, B. K. "C4NP Cu-Cored Pb-Free Flip Chip Interconnections." In *IEEE Proceedings of Electronic & Components Technology Conference*, Lake Buena Vista, FL, May 27–30, 2008, pp. 1351–1356.
415. Dang, B., Shih, D. Y., Buchwalter, S., Tsang, C., Patel, C., Knickerbocker, J., Gruber, P., Knickerbocker, S., Garant, J., Semkow, K., Ruhmer, K., and Hughlett, E. "50μm Pitch Pb-Free Microbumps by C4NP Technology." In *IEEE Proceedings of Electronic & Components Technology Conference*, Lake Buena Vista, FL, May 27–30, 2008, pp. 1505–1510.
416. Pandher, R., and Healey, R. "Reliability of Pb-Free Solder Alloys in Demanding BGA and CSP Applications." In *IEEE Proceedings of Electronic & Components Technology Conference*, Lake Buena Vista, FL, May 27–30, 2008, pp. 2018–2023.
417. Lee, J. H., Lim, G. T., Park, Y. B., Yang, S. T., Suh, M. S., Chung, Q. H., and Byun, K. Y. "Size Effect on Electromigration Reliability of Pb-Free, Flip-Chip Solder Bump." In *IEEE Proceedings of Electronic & Components Technology Conference*, Lake Buena Vista, FL, May 27–30, 2008, pp. 2030–2034.

418. Tsai, T. F., Chang, J. Y., Chien, C. W., Juang, J. Y., and Shen, L. C. "Ternary Lead-Free SnAgCu Micro-Bumps for Ultra-Fine Pitch Chip-to-Chip Interconnection." In *IEEE Proceedings of Electronic & Components Technology Conference*, Lake Buena Vista, FL, May 27–30, 2008, pp. 54–58.
419. Ebersberger, B., and Lee, C. "Cu Pillar Bumps as a Lead-Free Drop-in Replacement for Solder-Bumped, Flip-Chip Interconnects." In *IEEE Proceedings of Electronic & Components Technology Conference*, Lake Buena Vista, FL, May 27–30, 2008, pp. 59–66.
420. Rooney, D. T., Geiger, D., Shangguan, D., and Lau, J. "Metallurgical Analysis and Hot Storage Testing of Lead-Free Solder Interconnects: SAC versus SACC." In *IEEE Proceedings of Electronic & Components Technology Conference*, Lake Buena Vista, FL, May 27–30, 2008, pp. 89–98.
421. Vasudevan, V., and Fan, X. "An Acceleration Model for Lead-Free (SAC) Solder Joint Reliability under Thermal Cycling." In *IEEE Proceedings of Electronic & Components Technology Conference*, Lake Buena Vista, FL, May 27–30, 2008, pp. 139–145.
422. Ramkumar, S. M., and Srihari, K. "Influence of Process Parameters on Component Assembly and Drop Test Performance Using a Novel Anisotropic Conductive Adhesive for Lead-Free Surface Mount Assembly." In *IEEE Proceedings of Electronic & Components Technology Conference*, Lake Buena Vista, FL, May 27–30, 2008, pp. 225–233.
423. Gao, F., Qu, J., Nishikawa, H., and Takemoto, T. "Temperature Dependence of Mechanical Properties of Individual Phases in Sn-3.0Ag-0.5Cu Lead-Free Solder Alloy." In *IEEE Proceedings of Electronic & Components Technology Conference*, Lake Buena Vista, FL, May 27–30, 2008, pp. 466–471.
424. Xie, D., Shangguan, D., Chin, C., Ang, K., and Lau, D. "A New Method to Evaluate BGA Pad Cratering in Lead-Free Soldering." In *IEEE Proceedings of Electronic & Components Technology Conference*, Lake Buena Vista, FL, May 27–30, 2008, pp. 893–898.
425. Lall, P., Bhat, C., Hande, M., More, V., Vaidya, R., Suhling, J., Pandher, R., and Goebel, K. "Interrogation of System State for Damage Assessment in Lead-Free Electronics Subjected to Thermo-Mechanical Loads." In *IEEE Proceedings of Electronic & Components Technology Conference*, Lake Buena Vista, FL, May 27–30, 2008, pp. 918–929.
426. Farris, A., Pan, J., Liddicoat, A., Toleno, B. J., Maslyk, D., Shangguan, D., Bath, J., Willie, D., and Geiger, D. A. "Drop Test Reliability of Lead-Free Chip Scale Packages." In *IEEE Proceedings of Electronic & Components Technology Conference*, Lake Buena Vista, FL, May 27–30, 2008, pp. 1173–1180.
427. Jiang, H., Moon, K. S., and Wong, C. P. "Tin/Silver/Copper Alloy Nanoparticle Pastes for Low-Temperature, Lead-Free Interconnect Applications." In *IEEE Proceedings of Electronic & Components Technology Conference*, Lake Buena Vista, FL, May 27–30, 2008, pp. 1400–1404.
428. Klein, M., Hutter, M., Oppermann, H., Fritzsch, T., Engelmann, G., Dietrich, L., Wolf, J., Brämer, B., Dudek, R., and Reichl, H. "Development and Evaluation of Lead-Free Reflow Soldering Techniques for the Flip Chip Bonding of Large GaAs Pixel Detectors on Si Readout Chips." In *IEEE Proceedings of Electronic & Components Technology Conference*, Lake Buena Vista, FL, May 27–30, 2008, pp. 1893–1899.
429. Quintero, P., Oberc, T., and McCluskey, P. "Reliability Assessment of High Temperature Lead-Free Device Attach Technologies." In *IEEE Proceedings of Electronic & Components Technology Conference*, Lake Buena Vista, FL, May 27–30, 2008, pp. 2131–2138.
430. Krause, M., Petzold, M., Müller, M., Wiese, S., and Wolter, K. J. "Scaling Effects on Grain Size and Texture of Lead Free Interconnects—Investigations by Electron Backscatter Diffraction and Nanoindentation." In *IEEE Proceedings of Electronic & Components Technology Conference*, Lake Buena Vista, FL, May 27–30, 2008, pp. 75–81.
431. Lall, P., Iyengar, D., Shantaram, S., Panchagade, D., Suhling, J., and Pandher, R. "Design Envelopes and Optical Feature Extraction Techniques for Survivability

of SnAg Leadfree Packaging Architectures under Shock and Vibration." In *IEEE Proceedings of Electronic & Components Technology Conference*, Lake Buena Vista, FL, May 27–30, 2008, pp. 1036–1047.
432. Anson, S. J., Slezak, J. G., and Srihari, K. "Investigation of Enhanced Solder Wetting in 63Sn/37Pb and Sn-Ag-Cu Lead Free Alloy." In *IEEE Proceedings of Electronic & Components Technology Conference*, Lake Buena Vista, FL, May 27–30, 2008, pp. 1737–1744.
433. Zhao, X. J., Caers, J. F. J. M., Wong, E. H., Seah, S. K. W., van Driel, W. D., Owen, N., and Lai, Y. S. "Frequency Dependent S-N Curves for Predicting Drop Impact Robustness of Pb-Free Solder Interconnects." In *IEEE Proceedings of Electronic & Components Technology Conference*, San Diego, CA, May 25–29, 2009, pp. 93–99.
434. Xie, D., Geiger, D., and Gektin, V. "Reliability Study of High-end Pb-Free CBGA Solder Joint under Various Thermal Cycling Test Conditions." In *IEEE Proceedings of Electronic & Components Technology Conference*, San Diego, CA, May 25–29, 2009, pp. 109–116.
435. Chen, H. Y., and Chen, C. "In-Situ Observation of the Failure Induced by Thermomigration of Interstitial Cu in Pb-Free, Flip-Chip Solder Joints." In *IEEE Proceedings of Electronic & Components Technology Conference*, San Diego, CA, May 25–29, 2009, pp. 319–324.
436. Coyle, R., Popowich, R., Read, P., Fleming, D., Reid, M., Ryan, C., Collins, M., Punch, J., and Chatterji, I. "The Influence of the Pb-Free Solder Alloy Composition and Processing Parameters on Thermal Fatigue Performance of a Ceramic Chip Resistor." In *IEEE Proceedings of Electronic & Components Technology Conference*, San Diego, CA, May 25–29, 2009, pp. 423–430.
437. Logterman, M., and Gopalakrishnan, L. "A Product Feasibility Study of Assembling Pb-Free BGAs in a Eutectic Sn/Pb Process." In *IEEE Proceedings of Electronic & Components Technology Conference*, San Diego, CA, May 25–29, 2009, pp. 742–751.
438. Su, P., Li, L., Lai, Y. S., Chiu, Y. T., and Kao, C. L. "A Comparison Study of Electromigration Performance of Pb-Free, Flip-Chip Solder Bumps." In *IEEE Proceedings of Electronic & Components Technology Conference*, San Diego, CA, May 25–29, 2009, pp. 903–908.
439. Nicholls, L., Darveaux, R., Syed, A., Loo, S., Tee, T. Y., Wassick, T. A., and Batchelor, B. "Comparative Electromigration Performance of Pb-Free Flip Chip Joints with Varying Board Surface Condition." In *IEEE Proceedings of Electronic & Components Technology Conference*, San Diego, CA, May 25–29, 2009, pp. 914–921.
440. Lu, M., Shih, D. Y., Lauro, P., Kang, S., Goldsmith, C., and Seo, S. K. "The Effects of Ag, Cu Compositions and Zn Doping on the Electromigration Performance of Pb-Free Solders." In *IEEE Proceedings of Electronic & Components Technology Conference*, San Diego, CA, May 25–29, 2009, pp. 922–929.
441. Gadepalli, H., Dhanasekaran, R., Ramkumar, S. M., Jensen, T., and Briggs, E. "Influence of Reflow Profile and Pb-Free Solder Paste in Minimizing Voids for Quad Flat Pack No-Lead (QFN) Assembly." In *IEEE Proceedings of Electronic & Components Technology Conference*, San Diego, CA, May 25–29, 2009, pp. 2016–2024.
442. Ma, H., Lee, T. K., Kim, D. H., Liu, K. C., Kim, S. H., and Park, H. G. "Isothermal Aging Effects on the Dynamic Performance of Lead-Free Solder Joints." In *IEEE Proceedings of Electronic & Components Technology Conference*, San Diego, CA, May 25–29, 2009, pp. 390–397.
443. Kwon, D., Azarian, M. H., and Pecht, M. "Detection of Solder Joint Failure Precursors on Tin-Lead and Lead-Free Assemblies Using RF Impedance Analysis." In *IEEE Proceedings of Electronic & Components Technology Conference*, San Diego, CA, May 25–29, 2009, pp. 663–667.
444. Ahmad, M., Xie, W., Liu, K. C., Xue, J., and Towne, D. "Parametric Acceleration Transforms for Lead-Free Solder Joint Reliability under Thermal Cycling Conditions." In *IEEE Proceedings of Electronic & Components Technology Conference*, San Diego, CA, May 25–29, 2009, pp. 682–691.

445. Liu, W., and Lee, N. C., Porras, A., Ding, M., Gallagher, A., Huang, A., Chen, S., and Lee, J. C. B. "Achieving High Reliability Low Cost Lead-Free SAC Solder Joints Via Mn or Ce Doping." In *IEEE Proceedings of Electronic & Components Technology Conference*, San Diego, CA, May 25–29, 2009, pp. 994–1007.
446. Suganuma, K., Kim, K. S., Kim, S. S., Kim, D. S., Kang, M., and Kim, S. J. "Joining Characteristics of Various High Temperature Lead-Free Interconnection Materials." In *IEEE Proceedings of Electronic & Components Technology Conference*, San Diego, CA, May 25–29, 2009, pp. 1764–1768.
447. Sawada, Y., Sato, M., Abe, T., Tokunaga, M., Baba, S., and Hatanaka, Y. "Low Thermal Stress Flip-Chip Package for Ultra Low-k Die and Lead-Free Bumps." In *IEEE Proceedings of Electronic & Components Technology Conference*, San Diego, CA, May 25–29, 2009, pp. 1775–1780.
448. Meier, K., Wolter, K. J., Wiese, S., and Roellig, M. "Mechanical Characterization of Lead-Free Solder Alloys under High Strain Rate Loads." In *IEEE Proceedings of Electronic & Components Technology Conference*, San Diego, CA, May 25–29, 2009, pp. 1781–1787.
449. Ji, Q., Zhao, R., Huang, Q., and Zhu, P. "Underfills for Lead-Free and Low-K Flip Chip Packages." In *IEEE Proceedings of Electronic & Components Technology Conference*, San Diego, CA, May 25–29, 2009, pp. 1830–1835.
450. Bang, W. H., Chen, L. S., Kim, C. U., Lee, T. K., and Liu, K. C. "Rate Dependence of Bending Fatigue Failure Characteristics of Lead-Free Solder Joint." In *IEEE Proceedings of Electronic & Components Technology Conference*, San Diego, CA, May 25–29, 2009, pp. 2070–2074.
451. Lall, P., Shirgaokar, A., and Arunachalam, D. "Principal Component Analysis Based Development of Norris-Landzberg Acceleration Factors and Goldmann Constants for Leadfree Electronics." In *IEEE Proceedings of Electronic & Components Technology Conference*, San Diego, CA, May 25–29, 2009, pp. 251–261.
452. Venkatadri, V., Xing, Y., Cotts, E., Srihari, K., Yin, L., and Borgesen, P. "Accelerating the Effects of Aging on the Reliability of Lead Free Solder Joints in a Quantitative Fashion." In *IEEE Proceedings of Electronic & Components Technology Conference*, San Diego, CA, May 25–29, 2009, pp. 398–405.
453. Lall, P., Shantaram, S., Angral, A., and Kulkarni, M. "Explicit Submodeling and Digital Image Correlation Based Life-Prediction of Leadfree Electronics under Shock-Impact." In *IEEE Proceedings of Electronic & Components Technology Conference*, San Diego, CA, May 25–29, 2009, pp. 542–555.
454. Lall, P., More, V., Vaidya, R., and Goebel, K. "Prognostication of Latent Damage and Residual Life in Leadfree Electronics Subjected to Multiple Thermal-Environments." In *IEEE Proceedings of Electronic & Components Technology Conference*, San Diego, CA, May 25–29, 2009, pp. 1381–1392.
455. Nowottnick, M., Berek, H., Bell, H Herwig, H and Moschallski, A. "Condensation Heating Process for Lead-Free Soldering." In *SMTA Proceedings of International Surface Mount Technology Conference*, Rosemont, September 22–26, 2002, (SMT3–1).
456. Jo, H., Nieman, B. E., and Lee, N. C. "Voiding of Lead-Free Soldering at Microvia." In *SMTA Proceedings of International Surface Mount Technology Conference*, Rosemont, September 22–26, 2002, (SMT3–3).
457. Klenke, B. "Lead-Free Selective Soldering: The Wave of the Future." In *SMTA Proceedings of International Surface Mount Technology Conference*, Rosemont, September 22–26, 2002, (SMT4–2).
458. Chen, W. T., Tsai, R. Y., and Kao, C. R. "Effect of Copper Concentration on the Solid-State Aging Reactions Between Tin-Copper Lead-Free Solders and Nickel." In *SMTA Proceedings of International Surface Mount Technology Conference*, Rosemont, September 22–26, 2002, (SOL1–2).
459. Ramkumar, S. M., Ganeshan, V., and Thenalur, K. "Reliability and Paste Process Optimization of Eutectic and Lead-Free for Mixed Packaing." In *SMTA Proceedings of International Surface Mount Technology Conference*, Rosemont, September 22–26, 2002, (SOL2–1).

460. Hunt, C., and Lea, D. "Evaluation of the Comparative Solderability of Lead-Free Solders in Nitrogen-Part II." In *SMTA Proceedings of International Surface Mount Technology Conference*, Rosemont, September 22–26, 2002, (SOL2–2).
461. Herzog, T., Rudolph, S., and Wolter, K. J. "Process Capability, Wetting Behavior and Temperature Dependent Shear Strength of Alternative Lead Free Solder Joints." In *SMTA Proceedings of International Surface Mount Technology Conference*, Rosemont, September 22–26, 2002, (SOL2–3).
462. Albrecht, H. J., Wilke, K., and Brodie, D. S. "Board Level Reliability of Lead-Free Soldered Interconnections." In *SMTA Proceedings of International Surface Mount Technology Conference*, Rosemont, September 22–26, 2002, (SOL3–2).
463. Kelly, M., Colnago, D., and Sirtori, V. "Component Temperature Study on Tin-Lead and Lead-Free Assembilies." In *SMTA Proceedings of International Surface Mount Technology Conference*, Rosemont, September 22–26, 2002, (SOL3–3).
464. Seelig, K., and Suraski, D. "A Study of Lead-Contamination in Lead-Free Electronics Assembly and Its Impact on Reliability." In *SMTA Proceedings of International Surface Mount Technology Conference*, Rosemont, September 22–26, 2002, (SOL4–3).
465. Zribi, A., Kinyanjui, R. K., Varano, W., and Cotts, E. J. "Inhibiting Growth of Au-Ni-Sn Ternary Compounds at the Interfaces of Pb-Based and Pb-Free Solder Joints." In *SMTA Proceedings of International Surface Mount Technology Conference*, Rosemont, September 22–26, 2002, (SOL1–1).
466. Chromik, R. R., Vinci, R. P., Allen, S. L. and Notis, M. R. "Investigation of the Mechanical Properties of Pb-Free Solder Joints by Nanoindentation." In *SMTA Proceedings of International Surface Mount Technology Conference*, Rosemont, September 22–26, 2002, (SOL1–3).
467. Henshall, G., Roubaud, P., Chew, G., Prasad, S., Carson, F., Keeffe, E., and Bulwith R. "Impact of Component Terminal Finish on the Reliability of Pb-Free Solder Joints." In *SMTA Proceedings of International Surface Mount Technology Conference*, Rosemont, September 22–26, 2002, (SOL4–1).
468. Dalderup, M., Barbini, D., and Diepstraten, G. "Selective Soldering with Lead-Free Alloys." In *SMTA Proceedings of International Surface Mount Technology Conference*, Chicago, September 21–25, 2003, pp. 31–34.
469. Syed, M., Trotter, V., and Hoult, A. "Automated Selective Soldering of Lead-Tin and Lead-Free Solders-A Comparative Study Using Diode Laser and Xenon Lamp." In *SMTA Proceedings of International Surface Mount Technology Conference*, Chicago, September 21–25, 2003, pp. 41–46.
470. Wood, P. "Rework of BGA/CSP Components Using Lead Free Solders." In *SMTA Proceedings of International Surface Mount Technology Conference*, Chicago, September 21–25, 2003, pp. 64–69.
471. Shea, C., and Howell, K. "Lead-Free Wave Soldering-Tighter Process Windows Need Tighter Process Controls." In *SMTA Proceedings of International Surface Mount Technology Conference*, Chicago, September 21–25, 2003, pp. 494–500.
472. Janet, E. "Evaluation of Lead-Free Solder Bonds and Effects of the Lead-Free Process on Devices Using Acoustic Micro Imaging." In *SMTA Proceedings of International Surface Mount Technology Conference*, Chicago, September 21–25, 2003, pp. 560–565.
473. Handwerker, C., Siewert, T., Bath, Jasbir., Benedetto, E., Bradley, E., Gedney, R., Sohn, J., and Snugovsky, P. "NEMI Lead-Free Assembly Project: Comparison Between PbSn and SnAgCu Reliability and Microstructures." In *SMTA Proceedings of International Surface Mount Technology Conference*, Chicago, September 21–25, 2003, pp. 664–669.
474. Brian, E. "Joint Group on Pollution Prevention Lead-Free Solder Project." In *SMTA Proceedings of International Surface Mount Technology Conference*, Chicago, September 21–25, 2003, pp. 679–685.
475. Handwerker, C., Walter, H., and Lantzsch, M. "Alternative Lead Free Ball Materials with Improved Interface Properties." In *SMTA Proceedings of International Surface Mount Technology Conference*, Chicago, September 21–25, 2003, pp. 686–692.

476. Shina, S., Harriman, L., MacFadden, T., Anderson, R., Pasquito, H., Wilkish, G., Abbott, D., Pinsky, D., Quealy, M., Kistler, M., and Walters, K. "A Comparative Analysis of Lead Free Materials and Processes Using Design of Experiments Techniques." In *SMTA Proceedings of International Surface Mount Technology Conference*, Chicago, September 21–25, 2003, pp. 705–714.
477. Astrom, A. "The Effect of Nitrogen Reflow Soldering in a Lead-Free Process." In *SMTA Proceedings of International Surface Mount Technology Conference*, Chicago, September 21–25, 2003, pp. 715–721.
478. Qi, Y., CGhorbani, H. R. and Spelt, J. K. "Accelerated Thermal Cycling of Tin-Lead and Lead-Free Solder Joints." In *SMTA Proceedings of International Surface Mount Technology Conference*, Chicago, September 21–25, 2003, pp. 735–740.
479. Dusek, M., and Hunt, C. "Measurement of Materials Properties of Lead-Free Solders for Modeling Requirements." In *SMTA Proceedings of International Surface Mount Technology Conference*, Chicago, September 21–25, 2003, pp. 753–758.
480. Mattsson, F., Geiger, D. A., Shangguan, D., and Castello, T. "PCB Design and Assembly Process Study of 01005 Size Passive Components Using Lead-Free Solder." In *SMTA Proceedings of International Surface Mount Technology Conference*, Chicago, September 26–30, 2004, pp. 99–105.
481. Tosun, U. "Lead-Free (No-Clean)-Unheated Cleaning?" In *SMTA Proceedings of International Surface Mount Technology Conference*, Chicago, September 26–30, 2004, pp. 152–156.
482. Monette, F. "Material Control for Lead-Free Manufacturing." In *SMTA Proceedings of International Surface Mount Technology Conference*, Chicago, September 26–30, 2004, pp. 178–183.
483. Huang, B., and Lee, N. "Effect of Lead-Free Alloy Composition on Tombstoning." In *SMTA Proceedings of International Surface Mount Technology Conference*, Chicago, September 26–30, 2004, pp. 219–225.
484. Suraski, D., and Seeling, K. "Controlling Copper Build Up in Automatic Soldering Equipment Using Lead-Free Solder." In *SMTA Proceedings of International Surface Mount Technology Conference*, Chicago, September 26–30, 2004, pp. 236–238.
485. Bruno, F., and Grano, F. "Effect of the Reflow Profile on BGA Voiding in a Lead Free Process." In *SMTA Proceedings of International Surface Mount Technology Conference*, Chicago, September 26–30, 2004, pp. 291–296.
486. Privett, M., Toleno, J. B., Poole, N., and Steen , H. "Lead Contamination Risks in Lead Free Solder joints for High Temperature Applications." In *SMTA Proceedings of International Surface Mount Technology Conference*, Chicago, September 26–30, 2004, pp. 388–393.
487. Geiger, D., Castello, T., and Shangguan, D. "X-Ray Inspection of Area Array Packages Using Tin-Lead and Lead-Free Solders." In *SMTA Proceedings of International Surface Mount Technology Conference*, Chicago, September 26–30, 2004, pp. 553–555.
488. Pollock, A. "Inspection of Lead-Free Solder Joints on Printed Circuit Assemblies Using Automated Inspection Techniques." In *SMTA Proceedings of International Surface Mount Technology Conference*, Chicago, September 26–30, 2004, pp. 560–568.
489. Farrell, R., Beck, S., and Gamick R. "Lead Free Process Transition-Phase II: Paste Performance Assessment by Surface Finish and Mechanical Strength Correlation." In *SMTA Proceedings of International Surface Mount Technology Conference*, Chicago, September 26–30, 2004, pp. 574–592.
490. Lau, J., Chu, J., Chen, W., Khaw, R., Cheng, M., and Lee, J. "3-D X-Ray Tests and Analyses of Lead-Free PBGA Package Solder Joints." In *SMTA Proceedings of International Surface Mount Technology Conference*, Chicago, September 26–30, 2004, pp. 593–605.
491. Havener, M., and Farrell, R. "Lead-Free Solder Paste Evaluation." In *SMTA Proceedings of International Surface Mount Technology Conference*, Chicago, September 26–30, 2004, pp. 740–746.

492. Paulraj, P., and Armendariz, N. "Surface Mount Challenges with Lead-Free Reflow." In *SMTA Proceedings of International Surface Mount Technology Conference*, Chicago, September 26–30, 2004, pp. 751–756.
493. Herzog, T., and Wolter, K. "Investigations of the Microstructure of Lead-Free Solder Joints and Reliability Behavior by the Example of Chip Resistance Components." In *SMTA Proceedings of International Surface Mount Technology Conference*, Chicago, September 26–30, 2004, pp. 769–775.
494. Clech, J. "An Obstacle-Controlled Creep Model for Sn-Pb and Sn-Based Lead-Free Solders." In *SMTA Proceedings of International Surface Mount Technology Conference*, Chicago, September 26–30, 2004, pp. 776–802.
495. Dasgupta, A., Huang, B., and Lee, N. C. "Effect of Lead-Free Solder Composition on Voiding at Microvia." In *SMTA Proceedings of International Surface Mount Technology Conference*, Chicago, September 26–30, 2004, pp. 841–846.
496. Sinha, A., and Kuczynski, J. "Mechanical and Thermal Characterization of Lead-Free Solder." In *SMTA Proceedings of International Surface Mount Technology Conference*, Chicago, September 26–30, 2004, pp. 769–775.
497. Lau, J., and Liu, K. "Update of Global Trends in Lead-free Soldering." In *SMTA Proceedings of International Surface Mount Technology Conference*, Chicago, September 26–30, 2004, pp. 810–832.
498. Gowda, A., and Srihari, K. "Process and Reliability Issues with Lead-Free CSP Rework." In *SMTA Proceedings of International Surface Mount Technology Conference*, Chicago, September 25–29, 2005, pp. 99–105.
499. Bath, J., Chu, Q., and Ghalib, N. "iNemI Lead-Free Micro-BgA, CBGA, Mictor Connector and Through-Hole DIP Rework Evaluations." In *SMTA Proceedings of International Surface Mount Technology Conference*, Chicago, September 25–29, 2005, pp. 144–151.
500. James, W., and Zarrow, P. "Material Logistics for Lead-Free." In *SMTA Proceedings of International Surface Mount Technology Conference*, Chicago, September 25–29, 2005, pp. 183–186.
501. Liu, Y., Manning, W., Huang, B., and Lee, N. C. "A Kinetic Approach of Profiling for Voiding Control at Lead-Free Reflow Soldering." In *SMTA Proceedings of International Surface Mount Technology Conference*, Chicago, September 25–29, 2005, pp. 197–205.
502. Havener, M. "Lead-Free Wave Solder Flux Evaluation." In *SMTA Proceedings of International Surface Mount Technology Conference*, Chicago, September 25–29, 2005, pp. 229–233.
503. Rooney, D., Gullo, L, Xie, D., and Castello, T. T. "Evaluation of Reliability and Metallurgical Integrity of Wire Bonds and Lead Free Solder Joints on Flex Circuit Modules." In *SMTA Proceedings of International Surface Mount Technology Conference*, Chicago, September 25–29, 2005, pp. 410–417.
504. Zhao, J., Mackessy, D., and Jackson, J. "A study of the Failure Mechanisms in Lead-Free and Eutectic Tin-Lead Solder Bumps for Flip Chip Assembly." In *SMTA Proceedings of International Surface Mount Technology Conference*, Chicago, September 25–29, 2005, pp. 470–477.
505. Sweatman, K., Suenaga, S., and Nishimura, T. "Intermetallic Growth in Lead-Free Solders on PCB Substrates." In *SMTA Proceedings of International Surface Mount Technology Conference*, Chicago, September 25–29, 2005, pp. 544–547.
506. Geiger, D., Liu, Y., and Shangguan, D. "Evaluation of OSP Surface Finishes for Lead-Free Soldering." In *SMTA Proceedings of International Surface Mount Technology Conference*, Chicago, September 25–29, 2005, pp. 563–567.
507. Sampathkumar, M., Rajesnayagham, S., Ramkumar, S. M., and Anson, S. J. "Investigation of the Performance of SAC and SACBi Lead-Free Solder Alloys with OSP and Immersion Silver PCB Finishes." In *SMTA Proceedings of International Surface Mount Technology Conference*, Chicago, September 25–29, 2005, pp. 568–575.
508. Albrecht, H. "Lead-Free Solder Joints and Failure Mechanism on Different Finishes." In *SMTA Proceedings of International Surface Mount Technology Conference*, Chicago, September 25–29, 2005, pp. 584–590.

509. Dusek, M., and Hung, C. "Effects of Microstructure, PCB Finish and Processing on Lead-Free Reliability." In *SMTA Proceedings of International Surface Mount Technology Conference*, Chicago, September 25–29, 2005, pp. 591–600.
510. Bemard, D., and Willis, B. "Measurement Variation in the Fill of Intrusively Reflowed Lead Free Solder Joints with Different Board Finishes by X-Ray Inspection." In *SMTA Proceedings of International Surface Mount Technology Conference*, Chicago, September 25–29, 2005, pp. 697–703.
511. Ashmore, C. "Mass Imaging of Lead-Free Materials-What Impact Does Stencil Technology Have?." In *SMTA Proceedings of International Surface Mount Technology Conference*, Chicago, September 25–29, 2005, pp. 734–737.
512. Bradford, J., Felty, J., and Russell, B. "JCAA/JG-PP Lead-Free Solder Project: Combined Environments Test." In *SMTA Proceedings of International Surface Mount Technology Conference*, Chicago, September 25–29, 2005, pp. 824–840.
513. Whiteman, L. "Salt Atmosphere, Temperature Humidity, and Mechanical Shock Environmental Stress Testing Results of the JG-PP/JCAA Lead Free Soldering Program." In *SMTA Proceedings of International Surface Mount Technology Conference*, Chicago, September 25–29, 2005, pp. 841–845.
514. Bath, J., Sethuraman, S., Zhou, X., Willie, D., and Hyland, K. "Reliability Evaluation of Lead-Free SnAgCu PBGA676 Components Using Tin-Lead and Lead-Free SnAgCu Solder Paste." In *SMTA Proceedings of International Surface Mount Technology Conference*, Chicago, September 25–29, 2005, pp. 891–901.
515. Clech, J. "Acceleration Factors and Thermal Cycling Test Efficiency for Lead-Free Sn-Ag-Cu Assemblies." In *SMTA Proceedings of International Surface Mount Technology Conference*, Chicago, September 25–29, 2005, pp. 197–205.
516. Chung, C. K., Mustapha, F., Hua, F., and Aspandiar, R. "An Assessment of Lead Free Solder (Sn3.7Ag0.8Cu) Wettability." In *IEEE Proceedings of Electronics Packaging Technology Conference*, Singapore, December 10–12, 2002, pp. 1–5.
517. Anand, A., and Mui, Y. "Lead-free Solder Evaluation for Ball Attache Process." In *IEEE Proceedings of Electronics Packaging Technology Conference*, Singapore, December 10–12, 2002, pp. 6–10.
518. Beelen-Hendrikx, C. C. M., Klerk, J., and Water, J. T. "Lead-Free Semiconductor Packaging." In *IEEE Proceedings of Electronics Packaging Technology Conference*, Singapore, December 10–12, 2002, pp. 11–19.
519. Wu, J., and Pecht, M. "Fretting Corrosion Studies for Lead-Free Alloy Plated Contacts." In *IEEE Proceedings of Electronics Packaging Technology Conference*, Singapore, December 10–12, 2002, pp. 20–24.
520. Egli, A., Zhang, W., and Schwager, F. "New Approaches to Whisker Free Tin Deposits." In *IEEE Proceedings of Electronics Packaging Technology Conference*, Singapore, December 10–12, 2003, pp. 55–58.
521. Sriyarunya, A., and Schetty, R. "Lead-Free Plating for Semiconductor Devices Qualification & Implementation." In *IEEE Proceedings of Electronics Packaging Technology Conference*, Singapore, December 10–12, 2003, pp. 59–64.
522. Lim, P. S., Chua, K. L., Lee, C., and Teo, P. S. "High Density Fine Pitch Pb-Free Flip-Chip Interconnect Assembly." In *IEEE Proceedings of Electronics Packaging Technology Conference*, Singapore, December 10–12, 2003, pp. 65–71.
523. Rinne, G. "Emerging Issues in the Physics and Control of Electromigration in SnPb and Pb-Free Solder Bumps." In *IEEE Proceedings of Electronics Packaging Technology Conference*, Singapore, December 10–12, 2003, pp. 72–76.
524. See, B. "Effects of Gold on the Properties of Tin-Antimony Solder in lip-Chip-Pin-Grid-Array (FCGPA) Packages." In *IEEE Proceedings of Electronics Packaging Technology Conference*, Singapore, December 10–12, 2003, pp. 59–64.
525. Lo, W. C., Chen, Y. C., Chen, Y. F., Fu, H. C., Chang, S. M., Chang, L., Shen, H. T. and Chen, K. C. "Reliability and Characterization of Dielectric Layers Used for Leaded and Lead-Free Solder Application." In *IEEE Proceedings of Electronics Packaging Technology Conference*, Singapore, December 10–12, 2003, pp. 574–577.
526. Lim, H., Kwan, H., and Shi, X. "Intermetallic Growth Study on Lead-Free Solder Joint Under Thermal Cycling and Isothermal Aging." In *IEEE Proceedings of Electronics Packaging Technology Conference*, Singapore, December 10–12, 2003, pp. 59–64.

527. Limaye, P., Maurissen, W., Lambrinou, K., Duflos, F., Vandevelde, B., Allaert, B., Hillaert, J., Vandepitte. D., and Verlinden, B. "Low-Temperature Embrittlement of Lead-Free Solders in Joint Level Impact Testing." In *IEEE Proceedings of Electronics Packaging Technology Conference*, Singapore, December 10–12, 2007, pp. 140–151.
528. Müller, W., Weinberg, K., and Böhme, T. "A Higher Gradient Theory of Multiphase Solid Mixtures and its Application to Lead-Free Solder Micro-Morphology." In *IEEE Proceedings of Electronics Packaging Technology Conference*, Singapore, December 10–12, 2003, (B4.2).
529. Lu, S., Wang, B., and Chen, J. "Evolutions in Microstructure and Property of Sn-Ag-Cu Lead-Free Solders with Trace Addition of Magnesium." In *IEEE Proceedings of Electronics Packaging Technology Conference*, Singapore, December 10–12, 2003, (C4.2).
530. John., and Yang, Z. "Performance Evaluation for Electrical Plate Lead-Free Solder Bumping on Sapphire Wafers." In *IEEE Proceedings of Electronics Packaging Technology Conference*, Singapore, December 10–12, 2003, (C4.3).
531. Bergmann, R., Tang, P. T., Hansen, H. N., and Moller, P. "In-Situ Investigation of Lead-Free Solder Alloy Formation Using a Hot-Plate Microscope." In IEEE Proceedings of Electronics Packaging Technology Conference, Singapore, December 10–12, 2007, pp. 252–256.
532. Ruhmer, K., Hughlett, E., Ishizuka, M., Kojima, T., Asaka, T., Dang, B., Buchwalter, S and Shih, D. Y. "Lead Free Micro Bumping—Cost & Yield Challenges." In *IEEE Proceedings of Electronics Packaging Technology Conference*, Singapore, December 10–12, 2007, pp. 331–337.
533. Song, F. B, Lee, S. W. R., Newman, K., Reynolds, H., Clark, S., and Sykes, Bob. "Comparison of Joint Strength and Fracture Energy of Lead-Free Solder Balls in High Speed Ball Shear/Pull Tests and their Correlation with Board Level Drop Test." In *IEEE Proceedings of Electronics Packaging Technology Conference*, Singapore, December 10–12, 2007, pp. pp. 450–458.
534. Yang, Y., Lim, Y. J. Kumar, A., Lee, T. K. and Chen, Z. "Interface Reactions and Shear Strength of Lead-Free Sn-3.5Ag Solder with Ni-W-P Metallization." In *IEEE Proceedings of Electronics Packaging Technology Conference*, Singapore, December 10–12, 2007, pp.527–532.
535. Chotchakornpant, S., Tippayamontri, J and Kongsubto, P. "Behavior of Lead Free Solder Joint after Thermal Treatment." In *IEEE Proceedings of Electronics Packaging Technology Conference*, Singapore, December 10–12, 2007, pp. 541–545.
536. Lau, J., Liu, S., Shangguan, D Song, Z. W. and Geiger, D. "Process Development and Solder Joint Reliability of a New Lead-Free Solder: Sn3wt%Ag0.5wt%Cu0.019wt%Ce (SACC)." In *IEEE Proceedings of Electronics Packaging Technology Conference*, Singapore, December 10–12, 2007, pp. 546–552.
537. Dusek, M., and Hunt, C. "Low Cycle Isothermal Fatigue Properties of Lead-Free Solders." In *IEEE Proceedings of Electronics Packaging Technology Conference*, Singapore, December 10–12, 2007, pp. 760–763.
538. Pun, K., Eu, P. L., Islam, M,. and Cui C. Q. "Effect of Ni Layer Thickness on Intermetallic Formation and Mechanical Strength of Sn-Ag-Cu Solder Joint." In *IEEE Proceedings of Electronics Packaging Technology Conference*, Singapore, December 9–12, 2008, pp. 487–493.
539. Herkommer, D., Reid, M., and Punch, J. "In-Situ Optical Creep Observation and Constitutive Modelling of Joint-Scale SAC Solder Shear Samples." In *IEEE Proceedings of Electronics Packaging Technology Conference*, Singapore, December 9–12, 2008, pp. 506–515.
540. Tan, K., Xu, L,. Lye, H., Pang, J., Lau, J., and Zhang. X. "Creep Properties of Sn3wt%Ag0.5wt%Cu0.019wt%Ce (SACC) Lead-Free Solder." In *IEEE Proceedings of Electronics Packaging Technology Conference*, Singapore, December 9–12, 2008, pp. 521–526.
541. Yamashita, M., Hidaka, N., and Shohji, I. "The Effect of Ag, Ni and Ge Elements in Lead-free Sn Base Solder Alloy." In *IEEE Proceedings of Electronics Packaging Technology Conference*, Singapore, December 9–12, 2008, pp. 582–587.

542. Eu, P. H., Ding, M., Wong, T. L., Amin, N., Ahmad, I., and Mok, Y. L. "A Study of SnAgNiCo vs. Sn3.8Ag0.7Cu C5 Lead Free Solder Alloy on Mechanical Strength of BGA Solder Joint." In *IEEE Proceedings of Electronics Packaging Technology Conference*, Singapore, December 9–12, 2008, pp. 588–594.
543. Han, Y., Jing, H., Ling, M., Nai, S., Tan, C. M., Wei, J., Xu, L., and Zhang, S. "A New Creep Model for SnAgCu Lead-Free Composite Solders: Incorporating Back Stress." In *IEEE Proceedings of Electronics Packaging Technology Conference*, Singapore, December 9–12, 2008, pp. 689–695.
544. Lee, N. "Future Lead-Free Solder Alloys and Fluxes—Meeting Challenges of Miniaturization." In *IEEE Proceedings of Electronics Packaging Technology Conference*, Singapore, December 9–12, 2008, pp. 864–872.
545. Eu, P., Wong, T. L., Amin, N., and Ahmad, I. "FCPBGA with SOP Pad Finishing: A Study of Lead-Free Solder Ball Attach Improvement." In *IEEE Proceedings of Electronics Packaging Technology Conference*, Singapore, December 9–12, 2008, pp. 1040–1045.
546. Zhang, L., Wang, D., and Tong W. "A Numerical Simulation Study of Wetting and Bridging of SnAgCu Solder Liquid." In *IEEE Proceedings of Electronics Packaging Technology Conference*, Singapore, December 9–12, 2008, pp. 1347–1350.
547. Xu, H., Li, M., Kim, J. M. and Kim, D. W. "Lead-Free Soldering Technique by using Medium-Frequency Electromagnetic Field." In *IEEE Proceedings of Electronics Packaging Technology Conference*, Singapore, December 9–12, 2008, pp. 1351–1357.
548. Song, J. M., Lin, L. W., Lee, N. C. and Lai, Y. S, and Chiu, Y. T. "Metallurgical Perspective on Alloying Modification of Sn-Ag-Cu Solders." In *IEEE Proceedings of Electronics Packaging Technology Conference*, Singapore, December 9–12, 2008, pp. 1358–1363.
549. Ibitayo, D., Everhart, L., Morgenstern, M., Geil, B., and Mitchell, J. "Reliability Assessment of Lead-Free Universal Solder for Direct Bonding in Power Electronics Packaging." In *IEEE Proceedings of Electronics Packaging Technology Conference*, Singapore, December 9–12, 2008, pp. 1364–1368.
550. Huang, Y., Lin, K. L., and Liu, D. "Effect of Impact Rate on the Strength of Lead-free Solder Joints." In *IEEE Proceedings of Electronics Packaging Technology Conference*, Singapore, December 9–12, 2008, pp. 1369–1374.
551. Xu, J., Hu, Q., He, H., Zhang, F., and Zhao, Z. "Study of Sn-Bi-Cu Lead-free Solder." In *IEEE Proceedings of Electronics Packaging Technology Conference*, Singapore, December 9–12, 2008, pp. 1375–1380.
552. Feng, W., Wang, C., and Morinaga, M. "Quantum Computational Approach to Design Sn-Based Pb-Free Solder Alloys in Microelectronic Mounting and Packaging." *ASME Paper: IPACK2001–15872*, July 8–13, 2001, Kauai, HI, pp. 1–8.
553. Chalco, P., and Blackshear, E. "Reliability Issues of BGA Packages Attached With Lead-Free Solder." *ASME Paper: IPACK2001–15820*, July 8–13, 2001, Kauai, HI, pp. 1–8.
554. McCluskey, P. "Fatigue and Intermetallic Formation in Lead Free Solder Die Attach." *ASME Paper: IPACK2001–15823*, July 8–13, 2001, Kauai, HI, pp. 1–7.
555. Nurmi, S. T., Limnell, T. J., Hinkka, L. M., Ristolainen, E. O., and Lepistö, T. K. "Improving the Reliability of PB-Free BGA Solder Joints by Using the Second Reflow." *ASME Paper: IPACK2001–15781*, July 8–13, 2001, Kauai, HI, pp. 1–5.
556. Wu, J. C. L., Lee, C. B., Chou, Y. S., Lee, S. C., Ding, Y. C., and Zheng, P. Z. "Development of Lead-Free Package and Its Reliability." *ASME Paper: IPACK2001–15708*, July 8–13, 2001, Kauai, HI, pp. 1–10.
557. Amano, H., and Yu, Q. "Effect of Interfacial Factors on Fatigue Lifetime of Lead-Free Die-Attach Solder Joint." *ASME Paper: IPACK2001–15721*, July 8–13, 2001, Kauai, HI, pp. 1–8.
558. Yu, Q., Kim, D. S., Shiratori, M., Kakino, M., and Aoki, H. "The Effect of Intermetallic Compound on Thermal Fatigue Reliability of Lead-Free Solder Joints." *ASME Paper: IPACK2001–15751*, July 8–13, 2001, Kauai, HI, pp. 1–8.
559. Gruber, P. A., Jalloh, M. A., and Takacs, M. A. "Injection Molded Solder IMS for Fluxless Lead-Free Solder Bumping of Decoupling Capacitor Package." *ASME Paper: IPACK2001–15813*, July 8–13, 2001, Kauai, HI, pp. 1–6.

560. Lee, J. C. B., Wu, S., Chou, H. L., and Lai, Y. S. "Development of Lead-Free Flip Chip Package and Its Reliability." *ASME Paper: IPACK2003–35046*, July 6–11, 2003, Maui, HI, pp. 1–7.
561. Liu, L., Andersson, C., Liu, J., and Chan, Y. C. "Thermodynamic Assessment of Sn-Co-Cu Lead-Free Solder Alloy." *ASME Paper: IPACK2003–35126*, July 6–11, 2003, Maui, HI, pp. 1–5.
562. Liu, C., Conway, P., Li, D., and Hendriksen, M. "Analysis of the Micro-Mechanical Properties in Aged Lead-Free, Fine Pitch Flip Chip Joints." *ASME Paper: IPACK2003–35130*, July 6–11, 2003, Maui, HI, pp. 1–7.
563. Nurmi, S. T., Sundelin, J. J., Ristolainen, E. O., and Lepistö, T. K. "The Effect of Multiple Reflow Times on Lead-Free Solder Joint Microstructure." *ASME Paper: IPACK2003–35150*, July 6–11, 2003, Maui, HI, pp. 1–10.
564. Dutta, I., Park, C., and Choi, S. "Creep and Microstructural Evolution in Lead-Free Microelectronic Solder Joints." *ASME Paper: IPACK2003–35209*, July 6–11, 2003, Maui, HI, pp. 1–6.
565. Kim, D. S., Yu, Q., Shibutani, T., and Shiratori, M. "Nonlinear Behavior Study on Effect of Hardening Rule of Lead Free Solder Joint." *ASME Paper: IPACK2003–35250*, July 6–11, 2003, Maui, HI, pp. 1–7.
566. Lu, H., and Bailey, C. "Modelling the Performance of Lead-Free Solder Interconnects for Copper Bumped Flip-Chip Devices." *ASME Paper: IPACK2003–35356*, July 6–11, 2003, Maui, HI, pp. 1–6.
567. Zhang, Q., Dasgupta, A., and Haswell, P. "Creep and High-Temperature Isothermal Fatigue of Pb-Free Solders." *ASME Paper: IPACK2003–35361*, July 6–11, 2003, Maui, HI, pp. 1–6.
568. Gadepalli, H., Dhanasekaran, R., Ramkumar, S. M., Jensen, T., and Briggs, E. "Characterizing and Minimizing Voids in Quad Flat No-Lead (QFN) Device Assembly Using Pb-Free Solder Alloys." *ASME Paper: IPACK2009–89304*, July 19–23, 2009, San Francisco, CA, pp. 1–9.
569. Xie, W., Liu, K. C., and Brillhart, M. "Pb-Free Thin Small Outline Package (TSOP) Board Level Reliability Study." *ASME Paper: IPACK2009–89183*, July 19–23, 2009, San Francisco, CA, pp. 1–6.
570. Huang, C. Y., Chien, T. M., Ku, C. L., Hsieh, H. C., and Huang, H. H. "Reliability Assessment for Printed Circuit Board in Lead-Free Process." *ASME Paper: IPACK2009–89253*, July 19–23, 2009, San Francisco, CA, pp. 1–7.
571. Gao, F., Jing, J., Liang, F. Z., Williams, R. L., and Qu, J. "Plastic Strain Distribution as a Precursor for Transition From Ductile to Brittle Failure in Lead-Free Solder Joints." *ASME Paper: IPACK2009–89098*, July 19–23, 2009, San Francisco, CA, pp. 1–7.
572. Lee, S. W. R., Yang, C., and Chan, Y. S. "Assessment of Relative Thermal Fatigue Life of SAC Lead-Free and Tin-Lead Solders With Custom-Made BGA Assemblies Creating Various Stress Ranges." *ASME Paper: IPACK2009–89260*, July 19–23, 2009, San Francisco, CA, pp. 1–5.
573. Pang, J. H. L., and Tan, K. E. "Strain-Rate Effects on Constitutive Behavior of Sn3.8-Ag0.7-Cu Lead-Free Solder." *ASME Paper: IPACK2009–89278*, July 19–23, 2009, San Francisco, CA, pp. 1–7.
574. Quintero-Aguiló, P. O., Cáceres-Valencia, P., and Valentín, R. "Combinatorial Studies for the Minimization of Tin Whiskers Growth in Lead Free Electronics." *ASME Paper: IPACK2009–89029*, July 19–23, 2009, San Francisco, CA, pp. 1–6.
575. Hua, F., Garner, C. M., Song, H. G., and Morris, J. W. "Creep Behavior of Pb-Free Solders." *ASME Paper: IMECE2001/EPP-24728* , November 11–16, 2001, New York, pp. 1–11.
576. Sharma, P., Dasgupta, A., Ganti, S., and Loman, J. "Prediction of Rate-Independent Constitutive Behavior of PB-Free Solders Based on First Principles." *ASME Paper: IMECE2002–39254*, November 17–22, 2002, New Orleans, LA, pp. 1–10.
577. Zhang, Q., Dasgupta, A., and Haswell, P. "Cyclic Mechanical Durability of Sn-3.9ag-0.6cu and Sn-3.5ag Lead-Free Solder Alloys." *ASME Paper: IMECE2002–39250*, November 17–22, 2002, New Orleans, LA, pp. 1–6.

578. Gopakumar, S., Poyyapakkam, V., Blass, D., Borgesen, P., and Srihari, K. "Lead-Free Flip Chip Assembly." *ASME Paper: IMECE2002–39251*, November 17–22, 2002, New Orleans, LA, pp. 1–5.
579. Yunus, M., Venkateswaran, M., and Borgesen, P. "Flip Chip Process Development and Relaibility Evaluation With Lead-Free Solder Alloy." *ASME Paper: IMECE2002–39253*, November 17–22, 2002, New Orleans, LA, pp. 1–7.
580. Rafanelli, A. J. "The Lead-Free Interconnection Movement: A Pro-Active Approach to Understand the Effects on COTS Usage in the Aerospace and Defense Industries." *ASME Paper: IMECE2002–39497*, November 17–22, 2002, New Orleans, LA, pp. 1–11.
581. Yu, Q., Kim, D., Jin, J., Takahashi, Y., and Shiratori, M. "Fatigue Strength Evaluation for Sn-Zn-Bi Lead-Free Solder Joints." *ASME Paper: IMECE2002–39686*, November 17–22, 2002, New Orleans, LA, pp. 1–7.
582. Venkateswaran, M., Borgesen, P., and Srihari, K. "Evaluation of a Conductive Adhesive Based Approach to Lead Free Flip Chip Assembly." *ASME Paper: IMECE2002–39252*, November 17–22, 2002, New Orleans, LA, pp. 1–5.
583. Nakatsuka, T., Serizawa, K., Saga, T., Shimokawa, H., and Nishimura, A. "Reliability of Pb-free Solder Joints of Surface-mounted LSI Packages after Flow-soldering." In *IMAPS Proceedings of International Symposium on Microelectronics*, Boston, MA, September 20–22, 2000, pp. 308–313.
584. Fukaya, M., and Higuchi, C. "Pb-Free Resistor for Pb-Free LTCC System." In *IMAPS Proceedings of International Symposium on Microelectronics*, Boston, MA, September 20–22, 2000, pp. 636–641.
585. Sakai, T., Amita, H., Shibuya, Y., Kurozumi, T., Nagasaki, S., and Shoji, T. "Reliability of Solder Paste and Solder Pm-Coating Technology by Lead-Free Sn-Zn Base Solder Alloy." In *IMAPS Proceedings of International Symposium on Microelectronics*, Boston, MA, September 20–22, 2000, pp. 293–302.
586. Lee, N. C. "Lead-Free Soldering and Low Alpha Solders for Wafer Level Interconnects." In *IMAPS Proceedings of International Symposium on Microelectronics*, Boston, MA, September 20–22, 2000, pp. 541–550.
587. Seelig, K., and Suraski, D. "Lead-Free Soldering: An Overview and Comparison of the Viable Solutions." In *IMAPS Proceedings of International Symposium on Microelectronics*, Boston, MA, September 20–22, 2000, pp. 558–563.
588. Rae, A., Lasky, R., Belmonte, J., McLenaghan, J., and Gosselin, D. "Developing and Benchmarking Lead-Free Processes." In *IMAPS Proceedings of International Symposium on Microelectronics*, Boston, MA, September 20–22, 2000, pp. 605–610.
589. Solberg, V. "Lead-Free Soldering for CSP: The Impact of Higher Temperature SMT Assembly Processing." In *IMAPS Proceedings of International Symposium on Microelectronics*, Boston, MA, September 20–22, 2000, pp. 611–616.
590. Lau, J. H., and Pan, S. H. "Creep Behaviors of Flip Chip on Board with 96.5Sn-3.5Ag and 100In Lead-Free Solder Joints." In *IMAPS Proceedings of International Symposium on Microelectronics*, Boston, MA, September 20–22, 2000, pp. 866–873.
591. Tanaka, Y., Takahashi, J., and Kawashima, K. "Lead Free Soldering Technology for Mobile Equipment." In *IMAPS Proceedings of International Symposium on Microelectronics*, Boston, MA, September 20–22, 2000, pp. 314–319.
592. Jung, E., Kloeser, J., Kallmayer, C., Coskina, P., Aschenbrenner, R., and Reichl, H. "Lead Free Solders for Area Array Packaging." In *IMAPS Proceedings of International Symposium on Microelectronics*, Boston, MA, September 20–22, 2000, pp. 551–557.
593. Prasad, S., Carson, F., Kim, G. S., Lee, J. S., Jeong, T. S., and Kim, Y. S. "Reliability of Lead Free BGA Packages." In *IMAPS Proceedings of International Symposium on Microelectronics*, Boston, MA, September 20–22, 2000, pp. 722–726.
594. Feingold, A. H., Wahlers, R. L., and Stein, S. J. "Lead Free Dielectric Tape System for High Frequency Applications." In *IMAPS Proceedings of International Symposium on Microelectronics*, Baltimore, MD, October 9–11, 2001, pp. 133–137.

595. Vandevelde, B., Chandrasekhar, A., Driessens, E., Beyne, E., van Puymbroeck, J., and Heerman, M. "The PSGA, A Lead-Free CSP for High Performance & High Reliable Packaging." In *IMAPS Proceedings of International Symposium on Microelectronics*, Baltimore, MD, October 9–11, 2001, pp. 260–265.
596. Suzuki, M., Matsuoka, H., Kono, E., and Igarashi, M. "Sn-Zn Lead-Free Solder Applied to Notebook Personal Computer." In *IMAPS Proceedings of International Symposium on Microelectronics*, Baltimore, MD, October 9–11, 2001, pp. 298–303.
597. Ogawa, T., Erbil, A., and Bhattacharya, S. K. "Lead-Free High K Dielectrics for Integral Capacitor Using MOCVD." In *IMAPS Proceedings of International Symposium on Microelectronics*, Baltimore, MD, October 9–11, 2001, pp. 458–463.
598. Suganuma, K., Kim, K. S., and Huh, S. H. "Selection of Sn-Ag-Cu Lead-Free Alloys." In *IMAPS Proceedings of International Symposium on Microelectronics*, Baltimore, MD, October 9–11, 2001, pp. 529–534.
599. Nishina, T., and Okamoto, K. "Solder Paste with Polymerizing Flux for Lead-Free Solder Alloy." In *IMAPS Proceedings of International Symposium on Microelectronics*, Baltimore, MD, October 9–11, 2001, pp. 535–540.
600. Nomura, S., Miyazaki, M., Oki, K., Yoshida, A., Ogata, S., and Takei, T. "Improvement of Wave Soldering Equipment for Lead-Free Solder." In *IMAPS Proceedings of International Symposium on Microelectronics*, Baltimore, MD, October 9–11, 2001, pp. 541–546.
601. Jonnalagadda, K., Bai, T., and Olson, B. "Thermal and Bending Fatigue of PBGA Assemblies with Lead-Free Solder Pastes." In *IMAPS Proceedings of International Symposium on Microelectronics*, Baltimore, MD, October 9–11, 2001, pp. 553–558.
602. Moriwaka, Y., Takahashi, S., Kohinata, M., and Uchiyama, N. "Ultra Low Alpha Emission Lead Free Solder for Flip Chip Bumps." In *IMAPS Proceedings of International Symposium on Microelectronics*, Baltimore, MD, October 9–11, 2001, pp. 559–564.
603. Popelar, S., Strandjord, A., and Niemet, B. "A Compatibility Evaluation of Lead-Based and Lead-Free Solder Alloys in Conjunction with Electroless Nickel/Immersion Gold Flip Chip UBM." In *IMAPS Proceedings of International Symposium on Microelectronics*, Baltimore, MD, October 9–11, 2001, pp. 575–580.
604. Karim, Z. S., Martin, J., and Schetty, R. "Lead-Free Solder Bump Technologies for Flip-Chip Packaging Applications." In *IMAPS Proceedings of International Symposium on Microelectronics*, Baltimore, MD, October 9–11, 2001, pp. 581–587.
605. Prasad, S., Carson, F., Zahn, B., Lee, T. K., and Lee, H. T. "Thermally Enhanced PBGA: Package Characterization, Thermal Performance and Reliability at High Temperatures (Pb-free)." In *IMAPS Proceedings of International Symposium on Microelectronics*, Denver, CO, September 4–6, 2002, pp. 1–7.
606. Danielsson, H. "Packaging Technologies for Automotive Electronics in the Lead-Free Era." In *IMAPS Proceedings of International Symposium on Microelectronics*, Denver, CO, September 4–6, 2002, pp. 1–6.
607. Hormadaly, J. "New Lead-Free Thick Film Resistors." In *IMAPS Proceedings of International Symposium on Microelectronics*, Denver, CO, September 4–6, 2002, pp. 1–5.
608. Lugwig, R., Lee, N. C., Fan, C., and Zhang, Y. "Evaluation of Two Novel Lead-Free Surface Finishes." In *IMAPS Proceedings of International Symposium on Microelectronics*, Denver, CO, September 4–6, 2002, pp. 1–7.
609. Kim, S. H., Kim, J. Y., and Yu, J. "Investigation of Electroplated Ni and Ni-Cu Alloy UBM (Under Bump Metallurgy) with Lead-Free Solders for Flip Chip Packages." In *IMAPS Proceedings of International Symposium on Microelectronics*, Denver, CO, September 4–6, 2002, pp. 1–5.
610. Chandran, D. P. K., Chee, C. K., He, Y., Sterret, T., Sow, H. P., Rudge, A. V., and Abdullah, A. S. "Thermal Characterization of High Temperature Reflow Compatible Epoxy Molding Compound Used in Lead-Free Packaging." In *IMAPS Proceedings of International Symposium on Microelectronics*, Denver, CO, September 4–6, 2002, pp. 1–6.

CHAPTER 2

Reliability Engineering of Lead-Free Interconnects

2.1 Introduction

Many people in the electronic industry like to use the word "reliability"—especially the people in packaging (e.g., solder-bump reliability of the first-level interconnects), printed circuit board (PCB) assembly (e.g., solder-joint reliability of the second-level interconnects), and now 3-D IC integration (e.g., microbump reliability of the stacked-chip interconnects). Unfortunately, more than 90% of these people don't even know what reliability means. In this chapter, reliability basics and reliability engineering [which includes design for reliability (DFR), reliability testing and data analysis, and failure analysis] of lead-free interconnects will be discussed. Also, the material properties of SnAgCu and other lead-free solders are presented [1–17].

2.2 Reliability Basics

In this section, some fundamental concepts and practical methods of reliability are presented and discussed without getting too deep in mathematics.

2.2.1 Definition of Reliability

In this book, reliability of an interconnect (e.g., solder bump, solder joint, or microbump) of a particular package in an electronic product is defined as the *probability* that the interconnect will perform its intended function for a specified period under a given operating condition without failure [17, 18]. (The key word is probability, i.e., reliability is probability, and its answers are between 0 and 1.) Numerically speaking, reliability is the percent of survivors; that is, $R(x) = 1 - F(x)$, where $R(x)$ is the reliability (survival) function and $F(x)$ is the cumulative distribution function (CDF). Life distribution is a theoretical population model used to describe the lifetime of an interconnect, and is defined as the CDF, that is, $F(\mathrm{x})$ for the interconnect population. Thus, the one and only way to determine the interconnect reliability is by reliability testing to determine the $F(x)$.

2.2.2 Objective of Reliability Tests

The objective of reliability tests is to obtain failures (the more, the better) and to best fit the failure data to determine the parameters of the CDF of a chosen probability distribution (e.g., Weibull). The number of items (i.e., sample size) to be tested should be such that the final data are statistically significant. The reliability test time is unknown, but usually takes a while (e.g., a few months for thermal cycling tests). It should be noted and emphasized that as soon as the life distribution $F(x)$ of the lead-free interconnects is estimated by reliability testing, the reliability $R(x)$, failure rate, cumulative failure rate, average failure rate, mean time to failure, etc., of the lead-free interconnects are readily determined [19, 20, 21].

Most reliability tests are accelerated tests, with increased intensity of exposure to aggressive environmental conditions and realistic sample sizes and test times. Thus, acceleration models are needed to map (transfer) the failure probability, reliability function, failure rate, and mean time to failure from a test condition to an operating condition. In establishing the acceleration models for lead-free interconnects, their surrounding materials (e.g., solder, molding plastic, ceramic, copper, fiber-reinforced glass epoxy, and silicon), loadings (e.g., stress, strain, temperature, humidity, current density, and voltage), and failure mechanisms and modes (e.g., overload, fatigue, corrosion, and electromigration) must be considered.

2.2.3 Life Distributions

There are a few life distributions for interconnect reliability, for example, Weibull, lognormal, and exponential, and the most used one is Weibull. The three-parameter (x_o, θ, β) Weibull cumulative distribution function is [22–30]:

$$F(x) = 1 - \exp\left[-\left(\frac{x - x_0}{\theta - x_0}\right)^{\beta}\right] \tag{2.1}$$

where x is the random variable (e.g., life); x_o is the expected minimum value of x (it is also referred to as the location parameter); θ is the characteristic value at which $F(x = \theta) = 63.2\%$, or the scale parameter, which could be used to represent the quality of a product; and β is the Weibull slope, or the shape parameter, which is a measure of the uniformity of a product (the larger the slope, the more uniform is the product). It is frequently reasonable to assume that the expected minimum value of life (x_o) of the population, that is, the lower bound of life of the population is equal to zero; then we have the two-parameter (θ, β) Weibull distribution.

It should be pointed out that Weibull distribution has been used frequently for life phenomena, for example, situations where the percent occurrences (say, failure rates) may decrease, increase, or remain constant with an increase in the characteristic measured, for parts at debug, wear out, and chance failure stages of a product's life. Examples are the life of electronic components, fatigue life, corrosion life, wear-out life, etc.

2.2.4 Reliability, Failure Rate, and MTTF

The two-parameter (θ, β) Weibull cumulative distribution function, $F(x)$, is:

$$F(x) = 1 - \exp\left[-\left(\frac{x}{\theta}\right)^{\beta}\right] \tag{2.2}$$

The Weibull probability density function (PDF), $f(x)$, is:

$$f(x) = \frac{dF(x)}{dx} = \left(\frac{\beta}{x}\right)\left(\frac{x}{\theta}\right)^{\beta} \exp\left[-\left(\frac{x}{\theta}\right)^{\beta}\right] \tag{2.3}$$

The reliability function, $R(x)$, is:

$$R(x) = 1 - F(x) = \exp\left[-\left(\frac{x}{\theta}\right)^{\beta}\right] \tag{2.4}$$

The failure rate, $h(x)$, is:

$$h(x) = \frac{f(x)}{R(x)} = \left(\frac{\beta}{\theta}\right)\left(\frac{x}{\theta}\right)^{\beta-1} \tag{2.5}$$

The cumulative failure rate, $H(x)$, is:

$$H(x) = -\ln R(x) \tag{2.6}$$

The average failure rate between time x_1 and x_2, $AFR(x_1, x_2)$, is:

$$AFR(x_1, x_2) = \frac{\ln R(x_1) - \ln R(x_2)}{x_2 - x_1} \tag{2.7}$$

The mean-time-to-failure, $MTTF$ (or mean life), is:

$$MTTF = \int_0^{\infty} x f(x)\,dx = \theta\Gamma\left(1 + \frac{1}{\beta}\right) \tag{2.8}$$

where Γ is the gamma function. It should be pointed out that some special distributions could be degenerated from the Weibull distribution, for example, when $0 < \beta < 1$ (the PDF exponentially decreases from infinity); $\beta = 1$ (exponential distribution); $\beta > 1$ (the PDF rises to peak and then decreases); $\beta = 2$ (Rayleigh distribution); $3 < \beta < 4$ (normal bell-shape appearance); $\beta = 10$ (similar to Tape I extreme value distribution).

Figure 2.1 shows a typical Weibull probability plot from the thermal cycling reliability test result (–25°C–125°C, one cycle per hour) of a 256-pin plastic ball grid array (PBGA) package SnAgCu(SAC) lead-free solder joints. It can be seen that after 6000 cycles there were 16 failures and the test was stopped. By best-fitting these data to the two-parameter Weibull life distribution, we have $\theta = 5466$ cycles and $\beta = 2.8$.

For instruction purposes, let's assume that the operating condition is exactly the same as the test condition (usually it is not and thus acceleration models are needed); then we have the life distribution $F(x)$, reliability function $R(x)$, failure rate $h(x)$, and $MTTF$ as the following:

$$F(x) = 1 - \exp(-x/5466)^{2.8}$$

$$R(x) = \exp(-x/5466)^{2.8}$$

$$h(x) = (2.8/5466)(x/5466)^{1.8}$$

$$MTTF = 5466\Gamma(1 + 1/2.8)$$

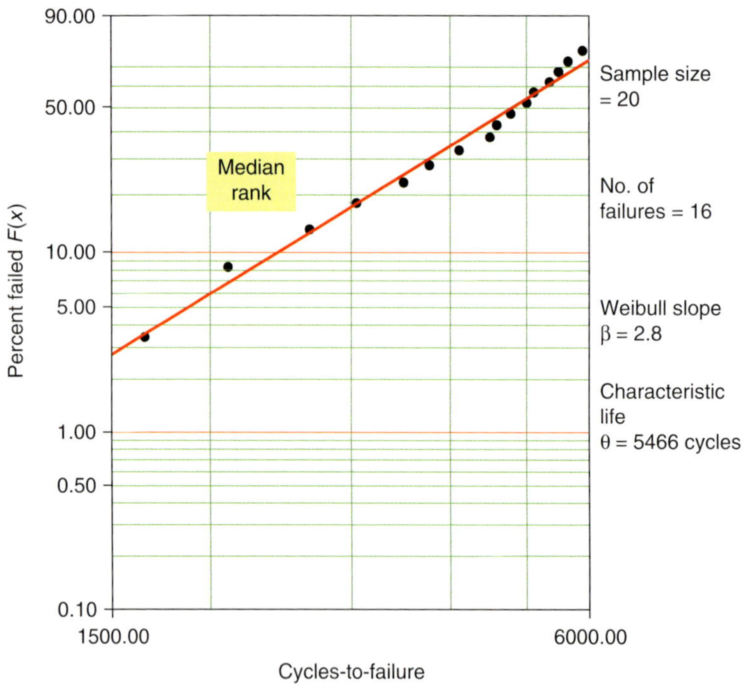

FIGURE 2.1 Thermal cycling test results of lead-free 256-pin PBGA (–25°C–125°C, one cycle per hour, failure criterion = 50% resistance increase).

With these equations, we can answer many questions that are important when deciding to ship a product or not and can define the product's warranty with this PBGA.

What is the probability that the 256-pin lead-free solder joints will fail (a) by 1000 cycles, (b) by 4000 cycles, or (c) between 1000 and 4000 cycles? and (d) if we use 2000 units of them, how many do we expect to fail in the first 1000 cycles? The answers are:

(a) $F(1000) = 1 - \exp(-1000/5466)^{2.8} = 0.01$

(b) $F(4000) = 0.34$

(c) $F(4000) - F(1000) = 0.33$

(d) $2000 \times 0.01 = 20$ units

What is the probability that the 256-pin lead-free solder joints will survive (a) by 1000 cycles, (b) by 4000 cycles, and (c) between 1000 and 4000 cycles?

(a) $R(1000) = 1 - F(1000) = 1 - 0.01 = 0.99$

(b) $R(4000) = \exp(-4000/5466)^{2.8} = 0.66$

(c) $1 - [F(4000) - F(1000)] = 0.67$

What is the failure rate of the 256-pin PBGA lead-free solder joints at 1000 cycles?

$$h(1000) = 24081 \times 10^{-9} \text{ per cycle} = 24{,}081 \text{ FITs (ppm/k)}$$

What is the MTTF of the 256-pin PBGA lead-free solder joints?

$$\text{MTTF} = 5466\Gamma(1 + 1/2.8) = 5466\Gamma(1.357) = 5466 \times 0.89 = 4865 \text{ cycles}$$

2.2.5 Simple System Reliability

The probability that k independent identical units, each with reliability $R(x)$, all survive past x hours is:

$$R_s(x) = [R(x)]^k \tag{2.9}$$

The probability that by time x at least one of k independent identical units will fail is given by:

$$F_s(x) = 1 - [R(x)]^k = 1 - [1 - F(x)]^k \tag{2.10}$$

A tragic example: the *Challenger* Space Shuttle disaster! The reliability of a single O-ring (the failing component) had been estimated to be 0.99, that is, in 99 out of 100 cases, the O-ring would survive. However, there were eight of them in the Shuttle and all had to work properly. Thus, the system reliability is $[0.99]^8 = 0.92$, that is, in 92 out of 100 cases, the O-ring would survive. Therefore, the potential (probability) for the O-ring's failure was 0.08, that is, in 1 out of approximately 12 cases. Sadly, the *Challenger* mishap occurred on its twelfth Shuttle launch in 1986.

Another example: A computer contains four identical 256-PBGAs, all operating independently and each critical to the operation of the computer. If each 256-PBGA has a reliability estimated at 0.90, what is the reliability of the computer? It is $[0.90]^4 = 0.66$, that is, only in 66 out of 100 cases, the computer will survive.

2.2.6 Ranking

In Fig. 2.1, the Weibull plot is with the median rank. Rankings are values assigned to items in a sample to determine their relative (failure) occurrence in a population. The median of these values (numbers) is then used as a true rank of the failure out of the group tested. This is known as median, or 50% rank, which can be approximated by [23–25]:

$$\text{Median or 50\% rank} = (j - 0.3)/(n + 0.4) \tag{2.11}$$

where j is the failure order number and n is the sample size. For example, if there are five samples, then $n = 5$. The median (50%) rank is 12.94% for $j = 1$, 31.47% for $j = 2$, 50% for $j = 3$, 68.53% for $j = 4$, and 87.06% for $j = 5$.

In order to make sure the estimated parameters (e.g., the characteristic life) are within the intervals for a given confidence level, rankings other than the median are necessary. For other required ranking (G), the percent rank (z) can be determined by (z is the percent rank of the jth value in n samples) [23–25]:

$$\begin{aligned} &1-(1-z)^n - nz(1-z)^{n-1} - \frac{n(n-1)}{2!}z^2(1-z)^{n-2} - \cdots \\ &- \frac{n(n-1)\cdots(n-j+1)}{(j-1)!}z^{j-1}(1-z)^{n-j+1} = G \end{aligned} \tag{2.12}$$

For example, given the sample size $n = 10$ and the required ranking $G = 95\%$, the percent failure rank (z) is 25.89% for $j = 1$, 39.42% for $j = 2$, 50.69% for $j = 3$, 60.76% for $j = 4$, 69.65% for $j = 5$, 77.76% for $j = 6$, 85% for $j = 7$, 91.27% for $j = 8$, 96.32% for $j = 9$, and 99.49% for $j = 10$. The 95% failure rank of the lowest of ten measurements is 25.89%. This means that in 95% of the cases, the lowest of ten would represent as much as 25.89% of the population. Only in 5% of cases would the lowest of ten represent more than 25.89% of the population.

2.2.7 True Weibull Slopes

Confidence is defined as the probability that a given interval determined from the test data will contain the population parameters, such as (using Weibull as an example) the mean/characteristic life and Weibull slope. Confidence applies to the test itself and reliability applies to the product!

To predict the true Weibull slope (β_t) of the population, it is necessary to estimate the Weibull slope error in relation to the number of failures and the required confidence level. The error (E) in the Weibull slope depends on the number of failures (N) and the required confidence level (C), which can be obtained from the following equation [23–25]:

$$\frac{1}{\sqrt{2\pi}}\int_{-\infty}^{E\sqrt{2N}} e^{-\frac{t^2}{2}}dt = \frac{(1+C)}{2} \tag{2.13}$$

The integral in Eq. (2.13) cannot be carried out in closed form and is approximated by the following:

$$\frac{1}{\sqrt{2\pi}}\int_{-\infty}^{E\sqrt{2N}} e^{-\frac{t^2}{2}}\,dt = 1 - Z(x)(b_1t + b_2t^2 + b_3t^3 + b_4t^4 + b_5t^5) + \varepsilon(x) \tag{2.14}$$

where

$$Z(x) = \frac{1}{\sqrt{2\pi}}e^{-\frac{x^2}{2}}$$

$$t = \frac{1}{1+px}$$

$$x = E\sqrt{2N}$$

$$\begin{aligned} p &= .2316419 \\ b_1 &= .31938153 \\ b_2 &= -.356563782 \\ b_3 &= 1.781477937 \\ b_4 &= -1.821255978 \\ b_5 &= 1.330274429 \\ |\varepsilon(x)| &< 7.5 \times 10^{-8} \end{aligned}$$

Thus, for a given C (the confidence level) and N (the number of failures), the E (error) can be searched from the previous equations. For engineering practice convenience, Eq. (2.13) has been plotted into a design chart (Fig. 2.2) for various values of C and N. For example, for a given $C = 0.9$ (that means in 90 out of 100 cases), the Weibull slope error of the example shown in Fig. 2.1 (with $N = 16$ failures) is $E = 30\%$. Thus, the true Weibull slope β_t of the 256-PBGA SnAgCu solder joints will happen in the intervals of $(2.8 - 0.3 \times 2.8) \le \beta_t \le (2.8 + 0.3 \times 2.8)$ or $1.96 \le \beta_t \le 3.64$.

2.2.8 True Characteristic Life

For a given C (the confidence level), in order to predict the population (true) characteristic value (θ_t) from the sample test data, it is necessary to determine the CDFs (life distributions) with certain percent ranks for the required confidence interval from Eq. (2.12). The lower percent ranks and upper percent ranks for $C = 95\%$ are, respectively, 2.5% and 97.5%; for $C = 90\%$, they are

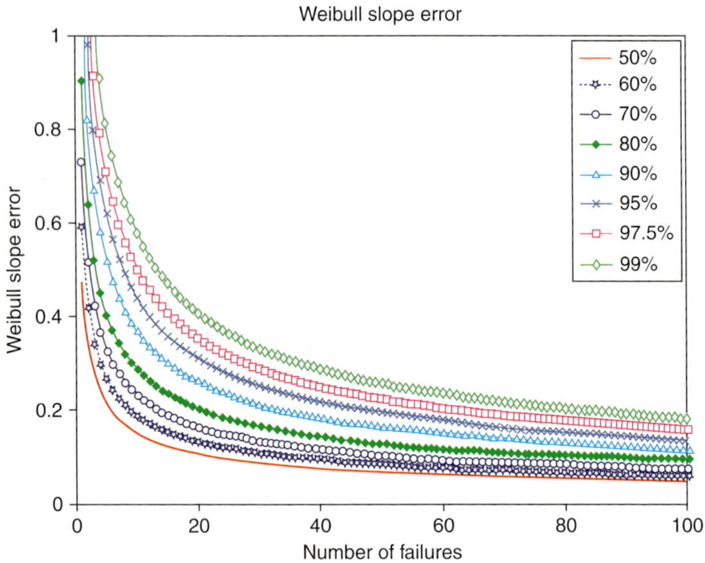

FIGURE 2.2 Weibull slope error for different confidences (shown by symbols) and number of failures.

5% and 95%; for $C = 80\%$, they are 10% and 90%; for $C = 70\%$, they are 15% and 85%; and for $C = 60\%$, they are 20% and 80%.

For example, for a given $C = 0.9$, the true characteristic life (63.2% failed), θ_t, of the 265-pin PBGA SnAgCu solder joints will happen (by determining the life distributions for the 95% rank and 5% rank) within the intervals $4480 \leq \theta_t \leq 6020$ cycles, as shown in Fig. 2.3.

2.2.9 True Mean Life

The mean life, that is, *MTTF*, from test samples can be determined by Eq. (2.8) with $\theta = 5466$ and $\beta = 2.8$, and is equal to $[MTTF = 5466\Gamma(1 + 1/2.8) = 5466\Gamma(1.357) = 5466 \times 0.89]$ 4865 cycles, which occurs at 51% failed [by Eq. (2.2) with $x = 4865$, $\theta = 5466$, and $\beta = 2.8$]. The population true mean life, μ, of the 265-pin PBGA SnAgCu solder joints will happen within the intervals $3822 \leq \mu \leq 4865$ cycles. The values of the intervals have been determined by rewriting Eq. (2.2) and with the following forms:

$$x = \theta\left(-\ln[1 - F(x)]^{\frac{1}{\beta}}\right)$$

$$\mu_{\text{lower}} = \theta_{95\%\text{rank}}\left(-\ln(1 - 0.51)^{\frac{1}{\beta_{95\%\text{rank}}}}\right) = 3822$$

$$\mu_{\text{upper}} = \theta_{5\%\text{rank}}\left(-\ln(1 - 0.51)^{\frac{1}{\beta_{5\%\text{rank}}}}\right) = 4865$$

where $\theta_{5\%\text{rank}} = 6020$, $\beta_{5\%\text{rank}} = 4.2$, and $\theta_{95\%\text{rank}} = 4480$, $\beta_{95\%\text{rank}} = 2$.

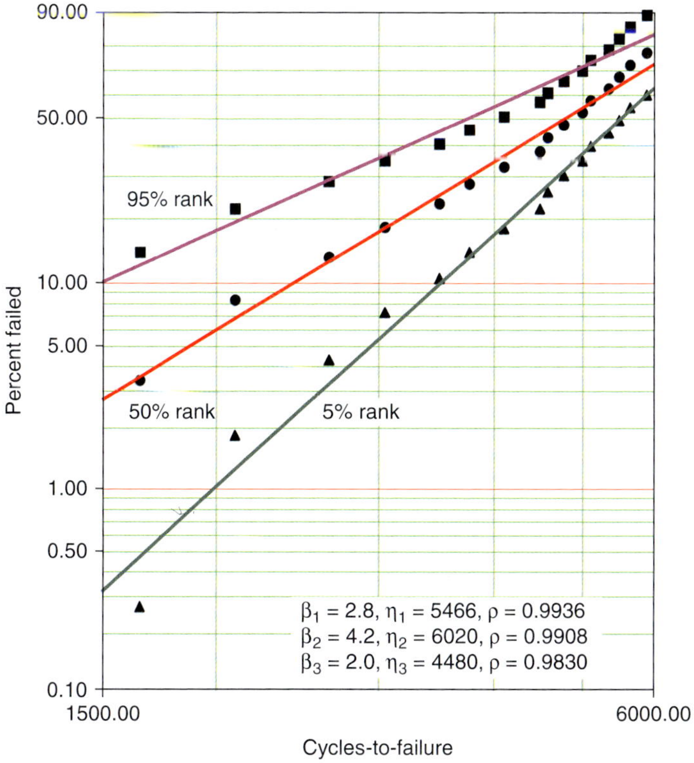

FIGURE 2.3 A 256-pin PBGA with 90% confidence (–25°C–125°C, one cycle per hour, failure criterion = 50% resistance increase).

2.2.10 Comparison of the MTTF of Two Different Sets of Samples

In many situations, the qualities and uniformities of two products are compared based on the knowledge of the limited test data. One of the difficult tasks in life testing is to draw conclusions about a population from a small sample size. It is even more difficult to compare the populations of two products based on the knowledge of their limited test data.

If one product is found to be superior to another, how confident, P (this is different from the confidence, C, discussed earlier), can it be that the same is true of their populations? Herein, a simple approach is used to determine whether one product is better than the other, without inquiring what the actual differences are [23–25]:

$$P = \frac{1}{1 + \dfrac{\log 1/q}{\log 1/(1-q)}} \tag{2.15}$$

$$\text{where}\quad q = 1 - \frac{1}{\left[1+\left(\frac{t+4.05}{6.12}\right)^5\right]^{40/7}}$$

$$t = \frac{\sqrt{1+\sqrt{T}}\,(\rho-1)}{\rho\Omega_2+\Omega_1}$$

$$\rho = \frac{M_2}{M_1}$$

$$T = (r_1-1)(r_2-1)$$

$$\Omega_1 = \sqrt{\frac{\Gamma(1+2/\beta_1)}{\Gamma^2(1+1/\beta_1)}-1}$$

$$\Omega_2 = \sqrt{\frac{\Gamma(1+2/\beta_2)}{\Gamma^2(1+1/\beta_2)}-1}$$

and M_1 is the mean life of sample 1 (S1), M_2 is the mean life of sample 2 (S2), r_1 is the number of failures in S1, r_2 is the number of failures in S2, β_1 is the Weibull slope of S1, and β_2 is the Weibull slope of S2. It can be seen that this confidence P is different from the confidence C (a given value.) The use of these equations will be shown in Chapter 4.

2.2.11 Accelerated Models

In reliability tests, the ideal situation is to have the test conditions as close as possible to the use (operating) conditions. However, because of time-to-market and cost savings, this is almost impossible. The practical way is to run accelerated tests with increased intensity, and realistic sample sizes and test times. (Thus, most reliability tests are accelerated tests.) In this case, we construct the acceleration factors to map (transfer) the failure probability, reliability function, and failure rate from a test condition to the service operating (use) condition. Unfortunately, this is the price we have to pay. Acceleration models are needed for determining the acceleration factors [1].

As mentioned earlier, the best-fitting Weibull life distribution of a set of lead-free interconnects is:

$$F_T(x_T) = 1 - \exp[-(x_T/\theta_T)^{\beta_T}] \tag{2.16}$$

where F_T, x_T, θ_T, and β_T are the life distribution, time-to-failure, characteristic life, and shape parameter (or Weibull slope), respectively, at test condition. Let's consider the following acceleration model (transformation or mapping):

$$x_0 = \alpha x_T^m \tag{2.17}$$

where x_0 is the time-to-failure at operating condition, α is the acceleration factor, and m is a larger-than-zero real number, for example, 8.88. Then, we have:

$$F_0(x_0) = 1 - \exp[-(x_0/\theta_0)^{\beta_0}] \tag{2.18}$$

$$R_0(x_0) = \exp[-(x_0/\theta_0)^{\beta_0}] \tag{2.19}$$

$$h_0(x_0) = \frac{\beta_0}{\theta_0}\left(\frac{x_0}{\theta_0}\right)^{\beta_0 - 1} \tag{2.20}$$

$$\beta_0 = \frac{\beta_T}{m} \tag{2.21}$$

$$\theta_0 = \alpha\theta_T^m \tag{2.22}$$

where F_0, R_0, h_0, x_0, θ_0, and β_0 are the life distribution, reliability, failure rate, time-to-failure, characteristic life, and Weibull slope, respectively, at operating condition.

When $m = 1$, that is, linear acceleration,

$$x_0 = \alpha x_T \tag{2.23}$$

then we have:

$$\beta_0 = \beta_T \tag{2.24}$$

$$\theta_0 = \alpha\theta_T \tag{2.25}$$

$$F_0(x_0) = 1 - \exp[-(x_0/\alpha\theta_T)^{\beta_T}] \tag{2.26}$$

$$h_0(x_0) = \frac{\beta_T}{\alpha\theta_T}\left(\frac{x_0}{\alpha\theta_T}\right)^{\beta_T - 1} \tag{2.27}$$

It can be seen that for a linear acceleration ($m = 1$), the Weibull CDF plots of the operating condition and testing condition should have the same Weibull slope (i.e., the shape parameter remains the same), and the characteristic life is different by a factor of the acceleration factor. It should be emphasized that the same Weibull slope is not an assumption but just came out naturally for linear acceleration. Thus, if two or more different test cells (with different temperature conditions) of the same lead-free interconnects yield very different Weibull slopes, then the linear acceleration model is not good (i.e., $m \neq 1$), or the Weibull distribution is not fitting with the test data, or both. Usually, the Weibull distribution adequately represents fatigue failures; thus, other acceleration models such as more terms and/or higher orders may need to be determined.

When $m = 2$ (quadratic acceleration, $x_0 = \alpha x_T^2$), then we have ($\beta_0 = \beta_T/2$ and $\theta_0 = \alpha\theta_T^2$):

$$F_0(x_0) = 1 - \exp[-(x_0/\alpha\theta_T^2)^{\beta_T/2}] \tag{2.28}$$

$$R_0(x_0) = \exp[-(x_0/\alpha\theta_T^2)^{\beta_T/2}] \tag{2.29}$$

$$h_0(x_0) = \frac{\beta_T}{2\alpha\theta_T^2}\left(\frac{x_0}{\alpha\theta_T^2}\right)^{\frac{\beta_T}{2}-1} \tag{2.30}$$

In this case, the Weibull slope of the operating condition is half that of the test condition, while the characteristic life of the operating condition is $\alpha\theta_T$ times that of the test condition.

Finally, let's consider $m = 1/2$ (square-root acceleration, $x_0 = \alpha x_T^{1/2}$); then we have ($\beta_0 = 2\beta_T$ and $\theta_0 = \alpha\sqrt{\theta_T}$):

$$F_0(x_0) = 1 - \exp\left[-(x_0/\alpha\sqrt{\theta_T})^{2\beta_T}\right] \tag{2.31}$$

$$R_0(x_0) = 1 - \exp\left[-(x_0/\alpha\sqrt{\theta_T})^{2\beta_T}\right] \tag{2.32}$$

$$h_0(x_0) = \frac{2\beta_T}{\alpha\sqrt{\theta_T}}\left(\frac{x_0}{\alpha\sqrt{\theta_T}}\right)^{2\beta_T-1} \tag{2.33}$$

It can be seen that, in this case, the Weibull slope at the operating condition is two times that at test condition, while the characteristic life at operating condition is equal to the acceleration factor times the square root of the characteristic life at test condition. Similar results can be generated for other values of m, that is, other accelerations.

Figure 2.4 shows the Weibull plot of a set of lead-free test data with $\beta_T = \beta_1 = 4$ and $\theta_T = h_1 = 3000$. Assuming $\alpha = 2$, then the acceleration models for $m = 1$, 1.5, and 0.5 are given in Fig. 2.4. It can be seen that when $m = 1$, the Weibull plot is parallel to the tested one.

In most solder joint reliability tests of IC packages on PCBs, the luxury of running multiple thermal cycling conditions is not available (due to the test duration, chamber occupation, and man power). Thus, acceleration models are required to predict the life distribution and failure rate of the solder joints under anticipated service conditions. In most applications pertaining to SnPb solder joints, the linear acceleration has been assumed to be valid.

In the case of lead-free solder joints, not enough test data are available representing a variety of IC packages tested and temperature cycling conditions to establish a suitable acceleration model. However, prior empirical studies have reported that lead-free solder joints exhibited large variation in Weibull slopes for thermal cycling

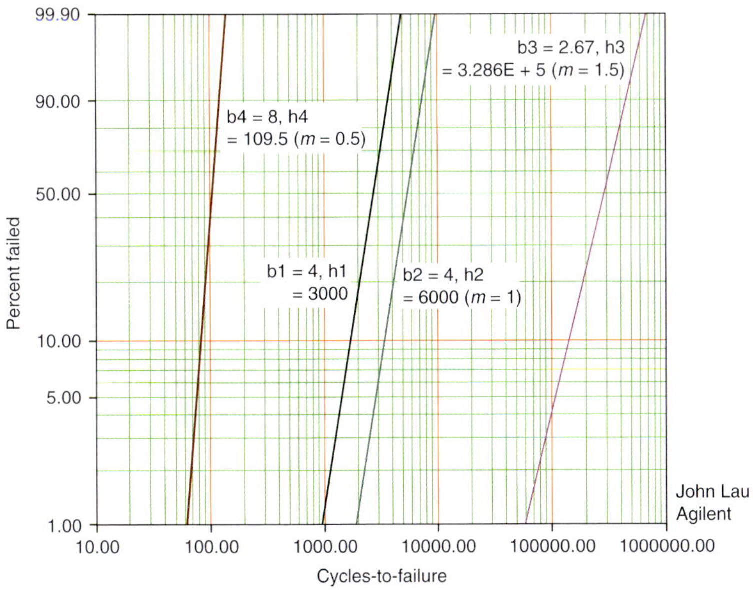

FIGURE 2.4 Life Distributions (β = b1 = 4 and θ = h1 = 3000 cycles) with various acceleration models ($x_0 = \alpha x_t^m$, assuming $\alpha = 2$, then $x_0 = 2x_t^m$ and let $m = 1, 1.5, 0.5$).

tests with different temperature conditions (i.e., $m \neq 1$). Thus, the study reported upon here should be helpful for selecting the appropriate acceleration model(s).

2.2.12 Acceleration Factors

How do you determine α (the acceleration factor)? The answer is that you need an acceleration model, which has been discussed in the previous section (2.2.11). How do you choose an acceleration model? The answer is cycling tests and failure mechanisms! For SnPb solder joints, the following acceleration factors have been frequently used [31–33]:

$$\alpha = \left(\frac{f_0}{f_T}\right)^q \left(\frac{\Delta T_T}{\Delta T_0}\right)^c \frac{\Phi_0}{\Phi_T} \tag{2.34}$$

$$\alpha = \left(\frac{f_0}{f_T}\right)^q \left(\frac{\Delta T_T}{\Delta T_0}\right)^c \exp[1414(1/T_0 - 1/T_T)] \tag{2.35}$$

$$\alpha = \left(\frac{f_0}{f_T}\right)^q \left(\frac{\Delta \gamma_T}{\Delta \gamma_0}\right)^c \exp[1414(1/T_0 - 1/T_T)] \tag{2.36}$$

In these equations, $T_T, f_T, \Delta T_T, \Delta\gamma_T$ and $T_0, f_0, \Delta T_0, \Delta\gamma_0$ are the temperatures (in degrees Kelvin), the temperature cycling frequency, the temperature range (in degree Celsius), and the shear strain range, respectively, at testing condition and at operating condition. Also, ϕ_T and ϕ_o are the isothermal fatigue test lifetimes at the peak temperature (in degrees Celsius) attained during the testing cycle and operating cycle, respectively. For SnPb solders, $q = 1/3$ and $-c = 1.9 \sim 2.0$ have been used. A critical discussion of these equations has been given by Solomon [33].

Do these acceleration factors apply to the lead-free solder joints? The answer is: We don't know! More cycling tests and a better understanding of the failure mechanisms of the lead-free solder joints are necessary in order to select a reliable acceleration model (most likely, it is not linear) with the correct acceleration factor that contains the most important parameters of the lead-free solders (e.g., activation energy and creep deformation for thermal loading) and of the test conditions (e.g., cycling frequency and temperature range).

2.3 Reliability Engineering of Lead-Free Interconnects

Figure 2.5 shows the concept of reliability engineering of interconnects [17, 18]. It consists of three major tasks, namely, design for reliability (DFR), reliability testing and data analysis, and failure analysis (FA).

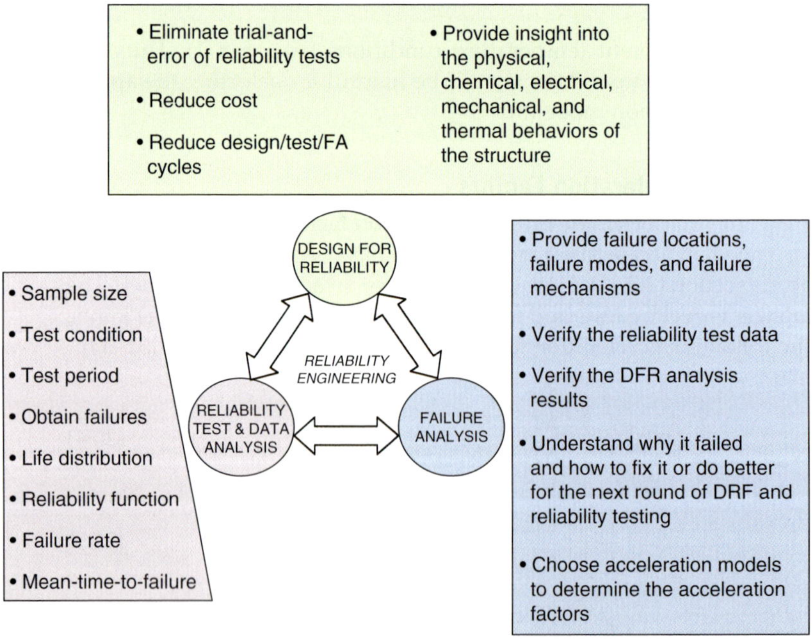

FIGURE 2.5 Reliability engineering.

Usually, the procedure starts with a design of the interconnects of a particular package with, for example, the given chip size, the solder alloys, the through-silicon-via (TSV), the molding compound, and the corresponding PCB, and demonstrates that the design is electrically, thermally, and mechanically sound. As an example, the DFR activity is often performed with a finite-element simulation using the material properties of all the structural elements and the imposed boundary conditions. The next step in the process is for a certain number of samples of the sound or reliable design to be built and tested under certain conditions for a certain period. The test data (failures) then are analyzed and fitted into a life-distribution (or reliability) designation for the interconnects. FA then should be done on the failed samples to find out the root cause and understand the reason for their failure. In this chapter, DFR, reliability testing, and failure analysis will be briefly mentioned and demonstrated throughout this book.

2.4 Reliability Testing and Data Analysis

The common reliability tests used for 3-D IC integration, packaging, and PCB assemblies are as follows.

- Temperature cycling tests (e.g., –40°C–125°C)
- Power cycling tests (depends on devices)
- Functional cycling tests (depends on devices)
- High-/low-temperature storage tests (120°C/–20°C)
- Biased 85/85 tests (e.g., 85°C/85 %RH, 1.8 V)
- High-voltage extended-life tests (e.g., 100°C, 1.8 V)
- Pressure-cook tests (autoclave tests) (e.g., 121°C, 2 atm)
- Salt-atmosphere tests (MIL-STD-883D)
- Moisture-sensitivity tests (e.g., IPC/JEDEC-020C)
- Shock (drop) tests (e.g., 1.3- to 1.5-mm drops)
- Vibration tests (e.g., random and harmonic vibrations)
- Mechanical bending, twisting, and shearing tests (e.g., IPC/JEDEC-9702)
- Electromigration tests (e.g., see Chap. 13 of this book)
- Others

Some of these tests will be used and discussed throughout this book. During the tests, the data can be statistically analyzed using the equations and methods shown earlier in this chapter, which are demonstrated in a few chapters of this book.

2.5 Failure Analysis

During and/or after the test, FA should be done on the failed samples to find out the root cause, understand the reason for their failure, and propose solutions to fix/improve the design/test. This information is useful for:

1. Providing the failure locations, failure modes, and failure mechanisms
2. Verifying the reliability-test data
3. Verifying the DFR simulation results
4. Understanding why the sample failed and how to fix it (or to do better)
5. The next round of DFR and reliability testing
6. Choosing acceleration models to determine the acceleration factors because most of the reliability tests are naturally "accelerated"

2.5.1 Failure Analysis Methods

The common methods for interconnect FA are:

- Visual inspection
- X-ray inspection
- A-, B-, and C-mode scanning acoustic microscopy (SAM)
- Dye and pry
- Cross-sectioning
- High-power microscopy
- Scanning electron microscopy (SEM)
- Focused ion beam (FIB) and SEM
- X-ray diffraction (XRD)
- Transmission electron microscopy (TEM)
- Energy-dispersive x-ray spectrometry (EDX)
- Tomographic acoustic microimaging (TAMI)
- Others

Some of these methods will be used and discussed throughout the remaining of this book. Among these methods, the one most often used is the x-ray inspection, since most of the interconnects discussed in this book are area-array and 3-D IC integration, and x-ray inspection will be presented herein with more details.

2.5.2 3-D X-ray Inspections and Analyses of Area Array Lead-Free Interconnects

There are at least two different nondestructive interconnect inspection methods, namely, visual inspection and x-ray inspection. Visual inspection is the simplest, lowest-cost, and frequently used method. However, since the wetting of the SnAgCu alloys is not as good as that of the SnPb solder alloys, the lead-free solder joints are not as shiny as the SnPb solder joints and the wetting angles are somewhat different between these two sets of solder joints [34]. New visual inspection specifications for lead-free solder joints are needed.

Due to the limitations of the human eye, visual inspection can only observe the surface portions of the solder joints. The solder joints underneath the chips, package substrates, PCBs, etc., can only be observed through x-ray inspection. In general, the x-ray-addressable defects are (1) opens, (2) shorts, (3) voids, (4) misalignments, (5) insufficient solders, and (6) excess solders, although this can vary, depending on the x-ray machine being used. Since there is no lead (Pb) in the lead-free (mostly tin) solder joints, the reflections of the x-ray from the lead-free solder joints are different from the tin-lead solder joints.

Two of the most frequently asked questions are: Can we use the existing machine to do x-ray inspections of lead-free solder joints, and what do we have to change to make it work? The objective of this section is to answer these questions. Emphasis is placed on determining the difference (shift) of the x-ray measurement parameters (e.g., solder diameter and solder thickness) between the lead-free and tin-lead (Sn37wt%Pb) solder joints. Also, the threshold adjustment values for the lead-free solder thickness are recommended and discussed.

Test Vehicles

The test vehicles (boards) consist of the 1156-pin PBGA with Sn4Ag0.5Cu solder balls, and the PCB (printed circuit board) with an organic solderability preservative (OSP) surface finish. The leaf-free solder paste used for the current surface mount assembly is Sn3Ag0.5Cu.

Test Board Assembly

The standard 1156-pin PBGA package is made with the Sn37Pb solder balls. On the other hand, the lead-free 1156-pin PBGA is made with the Sn4Ag0.5Cu solder balls (Fig. 2.6). The diameter of these solder balls is the same: 0.63 + 0.07/–0.13 mm. The ball pitch for both cases is 1 mm. The bismaleimide triazene (BT) substrate thickness for both cases is the same, and is equal to 0.56+/–0.06 mm.

The PCB (eight layers) is made of FR-4 fiberglass laminate materials and is 280 × 200 × 2.4 mm. The surface finish of the copper pad is OSP. The assembly process of the test boards is robust and with regular

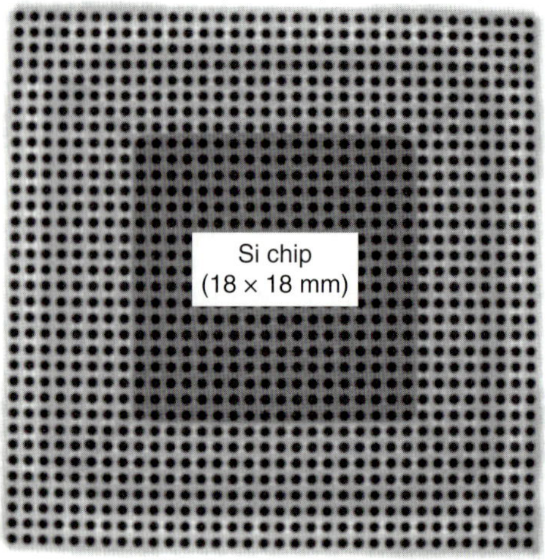

FIGURE 2.6 X-ray image of the 1156-pin PBGA (35 × 35 mm) package with either SnPb or SnAgCu solder balls.

equipments. For solder paste printing, the stainless steel stencil is manufactured with a laser-cutting process followed by electro polishing. The stencil thickness is 127 μm and the aperture is 0.5-mm round, which gives an area ratio (of the aperture opening area to the aperture wall area) of 1.0.

The lead-free solder paste used is a no-clean solder paste with Sn3Ag0.5Cu composition and 88.5wt% metal content. The powder (particle) size is Type 3 (i.e., more than 80wt% of the nominal diameter is between 20 μm and 45 μm, less than 1wt% of the nominal diameter is larger than 45 μm, no more than 10wt% of the nominal diameter is less than 20 μm, and none of the powder size is larger than 50 μm). The tin-lead solder paste used is a no-clean solder paste with Sn37Pb alloys and 90wt% metal content. The powder size is also Type 3. The solder pastes are printed with the following parameters: print speed = 30 mm/s, squeegee pressure = 8 kg, and separation speed = 1.2 mm/s. The PBGAs are picked and placed using a production fine-pitch placer.

A force-convection reflow oven with ten heating zones is used for reflowing the tin-lead/lead-free solder balls and solder pastes in air. The reflow temperature profiles on PCB are: (1) for the SnPb reflow process, the low limit and high limit for the time above reflow –183°C are, respectively, 30 s and 90 s, and for the peak temperature are, respectively, 205°C and 235°C; and (2) for the SnAgCu reflow process, the low limit and high limit for the time above reflow –220°C are, respectively, 30 s and 60 s, and for the peak temperature are, respectively, 230°C and 245°C. The test boards and sample size are shown in Table 2.1. It can be seen that there are (1) four SnPb assemblies, that is, SnPb solder paste and SnPb solder balls; (2) four lead-free assemblies, that is, lead-free solder paste and lead-free solder balls; and (3) four

Board	Solder Paste	SnPb PBGA	SAC PBGA
1	Sn37Pb	X	X
2	Sn37Pb	X	X
3	Sn37Pb	X	X
4	Sn37Pb	X	X
5	Sn3Ag0.5Cu		X
6	Sn3Ag0.5Cu		X
7	Sn3Ag0.5Cu		X
8	Sn3Ag0.5Cu		X

TABLE 2.1 Test Boards

mixed assemblies, that is, Sn4Ag0.5Cu solder balls with Sn37Pb solder paste.

5DX Inspections

One of the highest-power and most popular x-ray machines available today is the 5DX. This machine can do automatic x-ray inspections, 3-D laminography x-ray inspections, hidden solder-joint inspections, and mirror ball grid array inspections. The parameters that the 5DX used to address those six defects mentioned earlier are diameter, thickness, flattening, and circulation. Among these parameters, diameter and thickness play the most important roles.

The basic principles and theories for determining the solder diameter and thickness can be found in the manuals of the 5DX, and they will not be repeated here. However, simply stated: The diameter is calculated from either the circle profile method (i.e., calculating the diameter using the circle profile and averaging the distances between the pairs of located edges, as shown in the Fig. 2.7*a*) or the contour method (i.e., calculating the diameter using the contour method finds the center of the circle, calculates the average radius, and doubles it), as shown in the Fig. 2.7*b*.

After the diameter is determined, the gray level of each pixel in the circle bases on the diameter is measured and averaged, resulting in the thickness measurement for the solder joint. The darker the image, the higher in thickness the measurement is! In this study, emphasis is placed on determining the difference in solder diameter and solder thickness between the lead-free and tin-lead solder joints. Also, the changes (threshold adjustment values) for the lead-free solder thickness, depending on the slice location, are recommended. Figure 2.7 shows the slice locations, namely, pad slice (solder near the pad of the PCB) and ball slice (at the middle of the solder joint) for the present study.

The steps in using the 5DX inspection machine are quite simple: (1) Prepare the computer-aided design (CAD) files. (2) Use a test

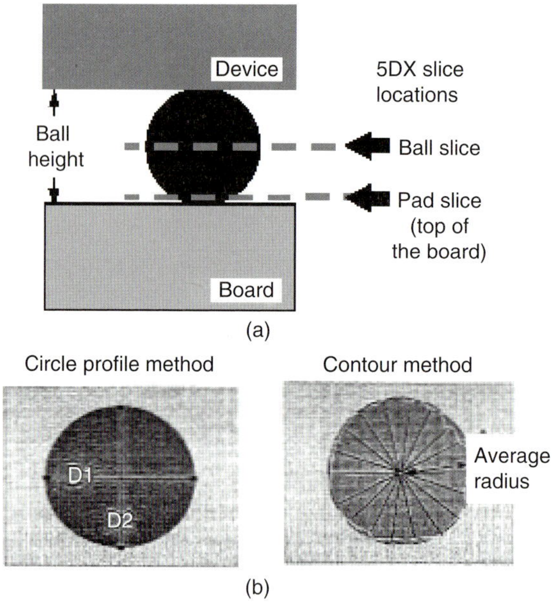

FIGURE 2.7 5DX measurement locations: pad slice and ball slice. Solder diameter determination by circle profile method and contour method.

development workstation (TDW) to define the kind of components, their location, pitch, etc. (3) Put the test boards in the machine. (4) Fine-tune the TDW program and pay special attention to (a) alignment point; (b) surface mapping; (c) verified images; (d) the gray level; and (e) threshold parameters for all failures (defects), such as open, short, void, misalignment, and insufficient or excess solders. (5) Run the machine for the test boards. (6) Analyze and report the defects. (7) Make conclusions and recommendations.

Figures 2.8 and 2.9 show the 5DX images at the pad-slide location and ball-slide location for the SnPb, SnAgCu, and mixed (SnAgCu solder ball and SnPb solder paste) assemblies, respectively. It can be seen that: (1) for all the cases, as expected, the diameter of the ball slide is larger than that of the pad slide; and (2) for both pad slice and ball slice, the images of the SnPb assemblies are darker and more concentrate (focus) than those of the SnAgCu assemblies.

5DX Test Results and Analyses

The solder diameter and solder thickness measurements of the 1156-pin PBGA package at the ball-slide and pad-slide locations of the SnPb, SnAgCu, and mixed assemblies are shown in Figs. 2.10 through 2.21. A total of 27,744 measurements have been taken by the 5DX. Figures 2.10 through 2.13 compare the measurement results between

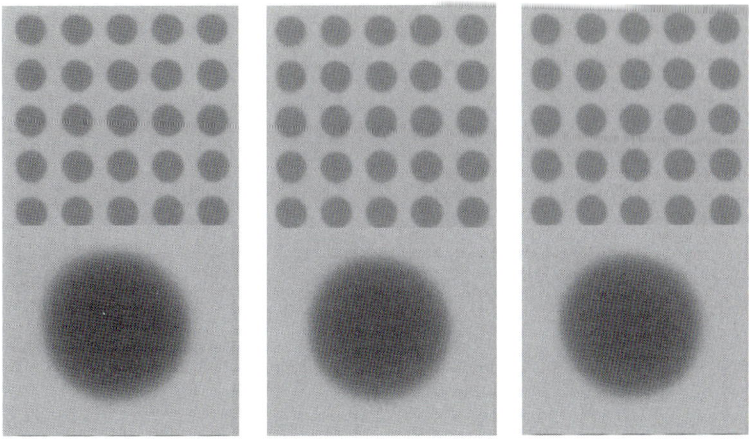

SnPb ball and SnPb paste LF ball and SnPb paste LF ball and LF paste

FIGURE 2.8 5DX images (ball slice) for SnPb, SnAgCu, and mixed assemblies.

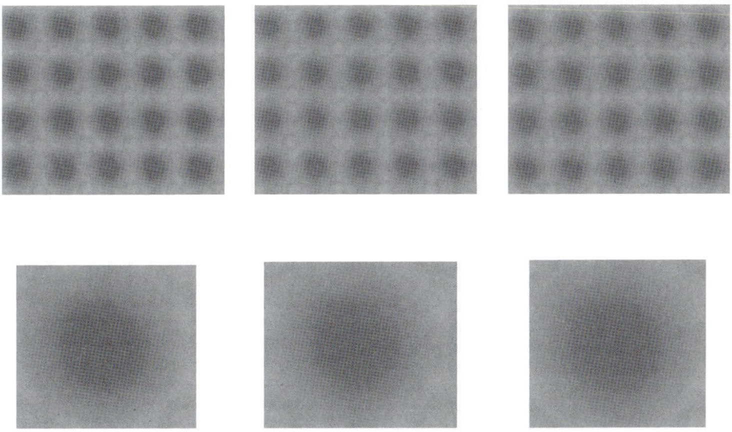

SnPb ball and SnPb paste LF ball and SnPb paste LF ball and LF paste

FIGURE 2.9 5DX images (pad slice) for SnPb, SnAgCu, and mixed assemblies.

the SnPb assemblies and the mixed (SnAgCu solder ball and SnPb solder paste) assemblies. It can be seen that: (1) for both cases, as expected, the diameter of solder balls at the center (ball slice) is larger than that near the pad slice; (2) for both cases, as expected, the solder thickness near the pad slice of the solder joints is less than that at the center of the solder joints; (3) the solder diameter of SnPb solder balls with SnPb paste assembly is larger than that of SnAgCu solder

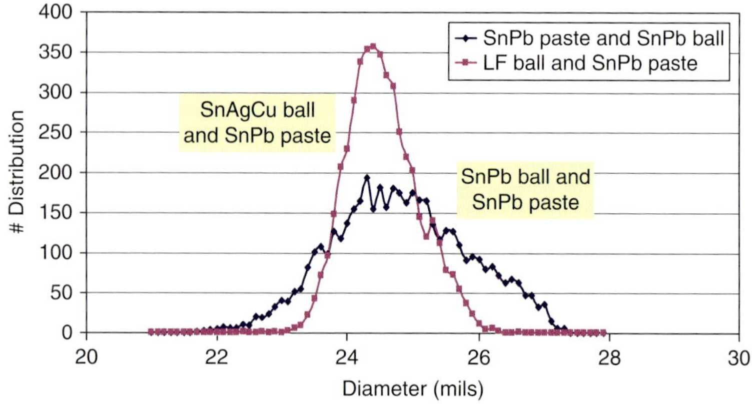

FIGURE 2.10 Diameter measurement (ball slice) (SnPb ball and SnPb paste vs. SnAgCu ball and SnPb paste).

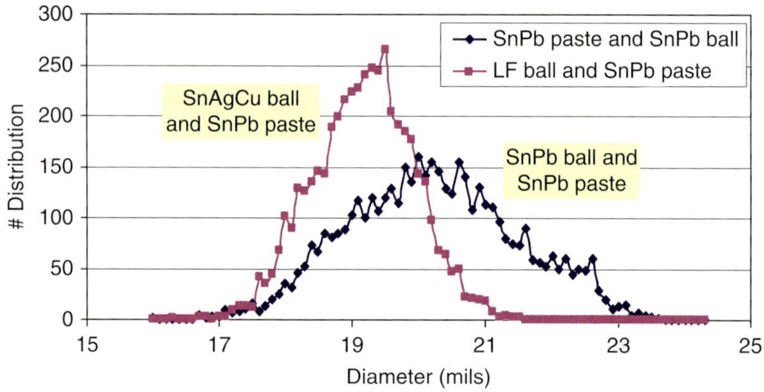

FIGURE 2.11 Diameter measurement (pad slice) (SnPb ball and SnPb paste vs. SnAgCu ball and SnPb paste).

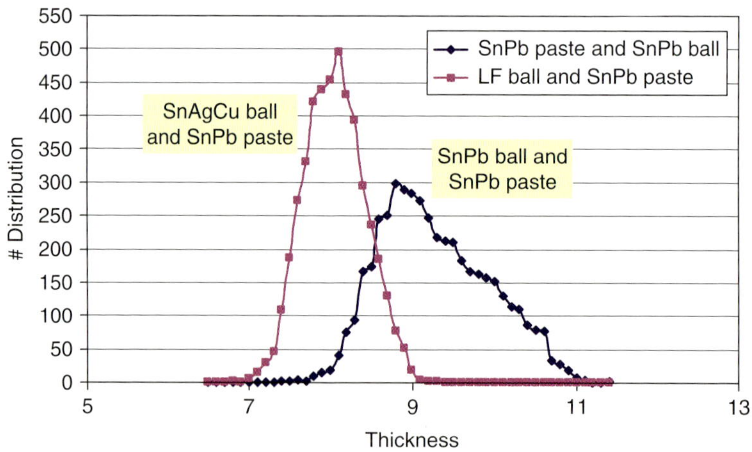

FIGURE 2.12 Thickness measurement (ball slice) (SnPb ball and SnPb paste vs. SnAgCu ball and SnPb paste).

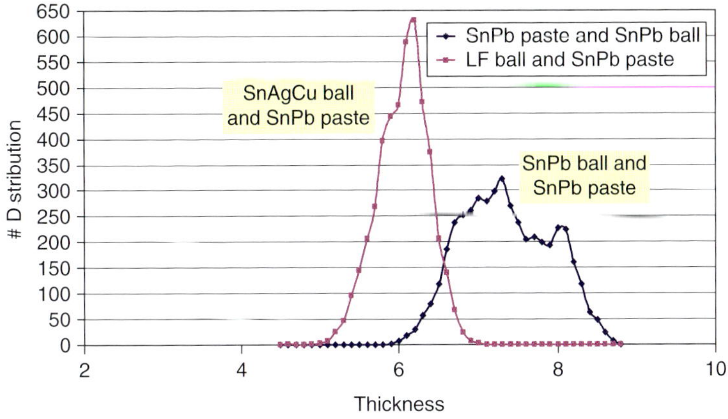

FIGURE 2.13 Thickness measurement (pad slice) (SnPb ball and SnPb paste vs. SnAgCu ball and SnPb paste).

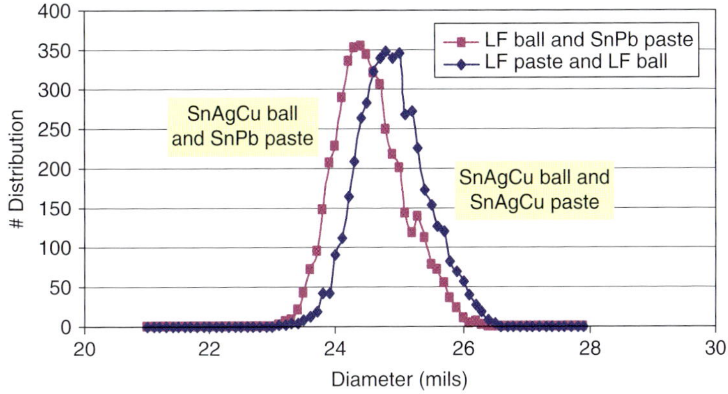

FIGURE 2.14 Diameter measurement (ball slice) (SnAgCu ball and SnPb paste vs. SnAgCu ball and SnAgCu paste).

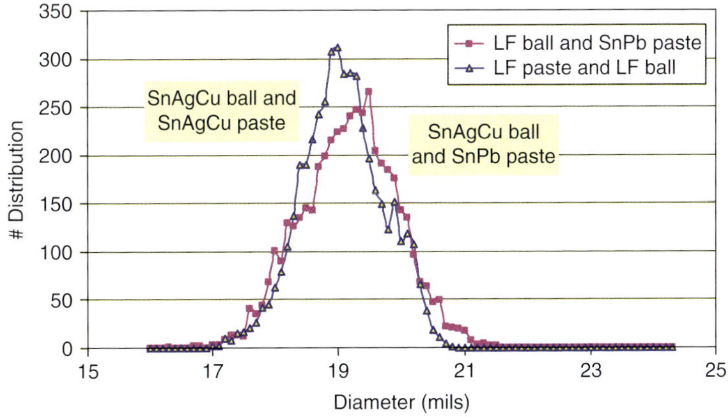

FIGURE 2.15 Diameter measurement (pad slice) (SnAgCu ball and SnPb paste vs. SnAgCu ball and SnAgCu paste).

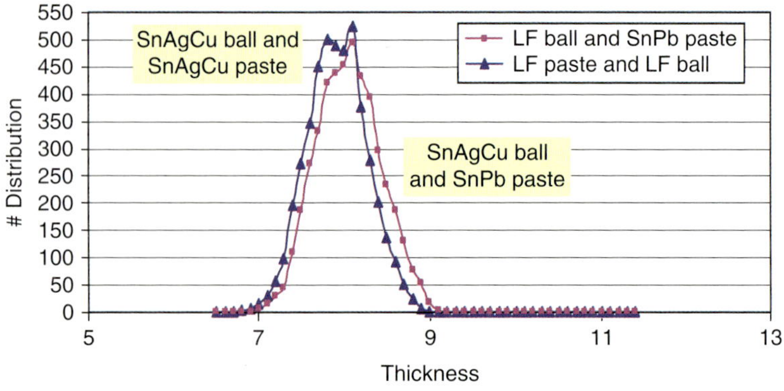

FIGURE 2.16 Thickness measurement (ball slice) (SnAgCu ball and SnPb paste vs. SnAgCu ball and SnAgCu paste).

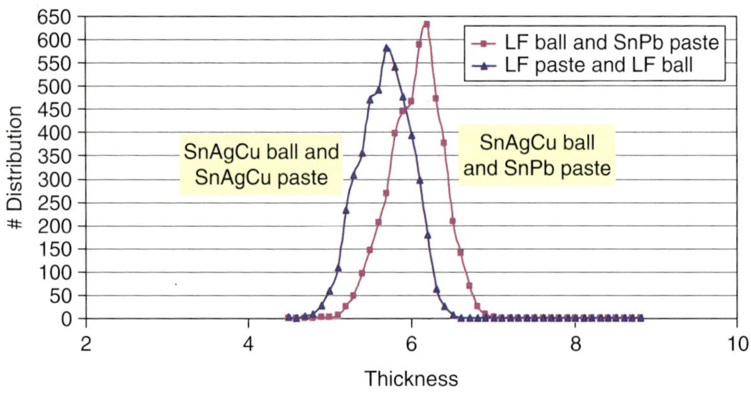

FIGURE 2.17 Thickness measurement (pad slice) (SnAgCu ball and SnPb paste vs. SnAgCu ball and SnAgCu paste).

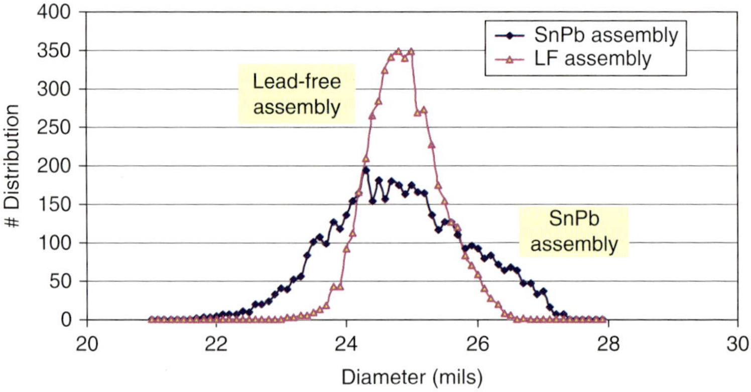

FIGURE 2.18 Diameter measurement (ball slice) (SnPb assembly vs. lead-free assembly).

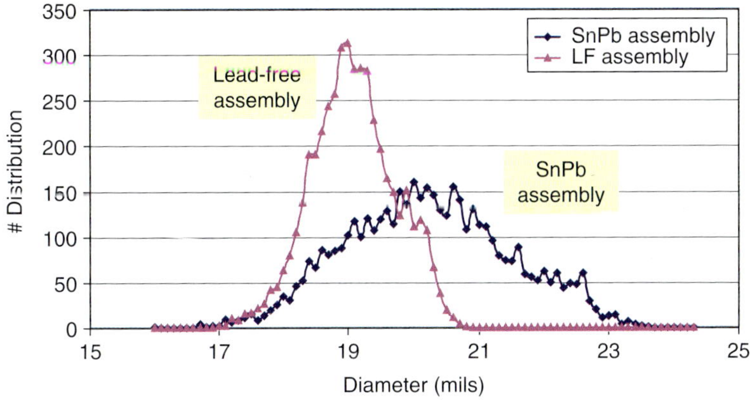

FIGURE 2.19 Diameter measurement (pad slice) (SnPb assembly vs. lead-free assembly).

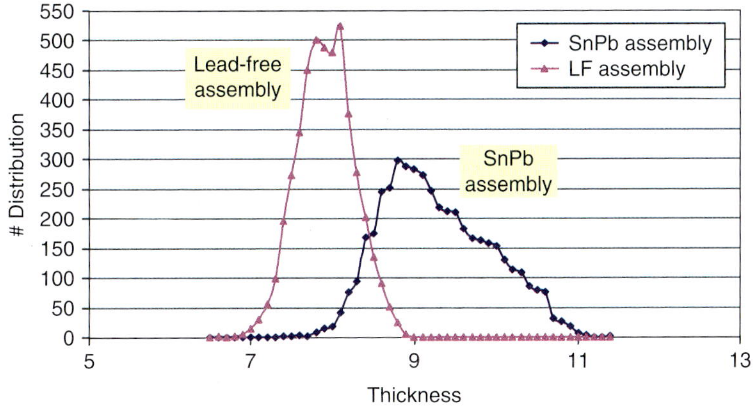

FIGURE 2.20 Thickness measurement (ball slice) (SnPb assembly vs. lead-free assembly).

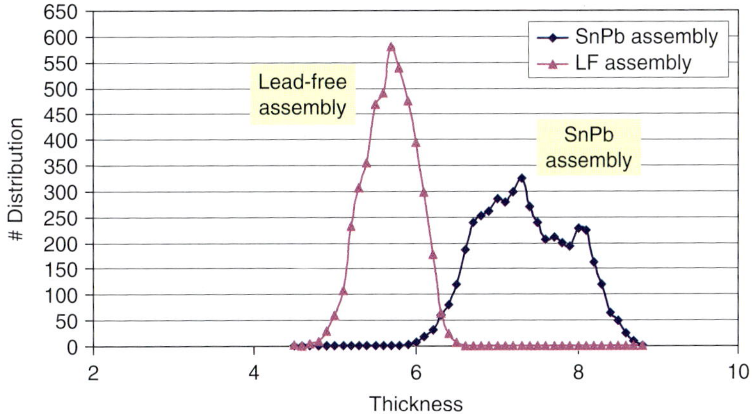

FIGURE 2.21 Thickness measurement (pad slice) (SnPb assembly vs. lead-free assembly).

balls with SnPb solder paste assembly, especially near the pad area. This is due to the lower reflow temperatures of the SnPb assembly process, that is, the temperatures are not high enough to fully reflow the lead-free solder alloys; and (4) at both locations, the solder thickness of the SnAgCu solder ball with SnPb solder paste assembly is smaller than that of the SnPb assembly. The solder thickness at the peak of the spectrum is shifted by 15% near the pad location and by 8% at the center of the solder joint.

Figures 2.14 through 2.17 compare the measurement results between the SnAgCu assembly with the mixed assembly (SnAgCu solder ball and SnPb solder paste). It can be seen that at both (pad slice and center slice) locations, the solder diameter and solder thickness of these assemblies are similar (no significant difference). The small amount of SnPb solder paste slightly increases the solder thickness. Figures 2.18 through 2.21 compare the measurement results between the SnAgCu assembly with the SnPb assembly. It can be seen that the diameter of the SnPb solder joints is larger than that of the SnAgCu solder joints, especially near the pad area. The solder thickness of the SnAgCu solder joints is less than that of the SnPb solder joints (22% near the pad slice and 12% at the center of the solder joints).

Conclusions and Recommendations

(1) In all cases, the diameter of solder joints at the ball slice (center) is larger than that near the pad slice.

(2) In all cases, the solder thickness at the center of the solder joints is larger than that near the copper pad of the solder joints.

(3) The diameter of mixed assembly (SnAgCu solder ball with SnPb solder paste) is smaller than that of the SnPb assembly (SnPb solder ball and paste).

(4) The solder thickness of the mixed assembly is smaller than that of the SnPb assembly.

(5) To re-adjust the threshold from SnPb solder joints to mixed solder joints depending on the slice location, the solder thickness values for the mixed assembly should be 15% (near the pad location) and 8% (at the center) lower than those of the SnPb solder joints.

(6) The diameter and solder thickness of the mixed assembly are similar (no significant difference) to those of the SnAgCu assembly (SnAgCu solder ball with SnAgCu solder paste).

(7) The solder diameter of the SnAgCu assembly is less than that of the SnPb assembly, especially near the copper pad area.

(8) The solder thickness of the SnAgCu assembly is less than that of the SnPb assembly.

(9) To re-adjust the threshold from SnPb assembly to SnAgCu assembly, the solder thickness values for the SnAgCu solder joints should be 22% (near the pad slice) and 12% (near the center) lower than those of the SnPb solder joints.

2.6 Design for Reliability (DFR)

It should be noted and emphasized that the reliability obtained from design of reliability (DFR) is not the same as the reliability (probability) obtained from reliability tests. For example, the life (number of cycles to failure) of an interconnect from reliability tests depends on the percent failed (or survived). Statistically, there are an infinite number of possible lives; for example, the characteristic life corresponds to 63.2% failed (or 36.8% survived), and the life corresponding to other percentages of failure will have other values. On the other hand, the best we can do from DFR is to predict a (one) life of the interconnect for a given boundary condition. As another example, the failure rate (i.e., the number of interconnects that failed in the first year, second, third, etc.) can be determined from the reliability test. However, the DFR does not even know what failure rate means. Finally, the reliability test result depends on sample size and the required confidence level. Again, DFR doesn't know any of these.

2.6.1 Advantages of DFR

In the past three decades, with the advancement of computers, the DFR of interconnects has become as important as the reliability test. Even though it cannot predict interconnect reliability, failure rate, MTTF, etc., DFR is usually performed by a mathematical modeling (e.g., through finite element methods) based on the geometry and material properties of the structure, the boundary condition, and laws of engineering and physics in order to:

1. Eliminate trial-and-error reliability tests
2. Reduce cost
3. Reduce design/test/FA cycles
4. Reduce time to market
5. Provide insight into the physical, chemical, electrical, mechanical, and thermal behaviors of solder joints (e.g., the maximum stress/strain locations) to help test board designs to capture the most likely (solder-joint) failure locations and to help FA find these failure sites

2.6.2 A Note on DFR

With the advances in computers and the rapid development of commercially available finite element codes, boundary-value problems with coupling/nonlinearity/anisotropicity material properties and time-temperature–dependent effects can be solved. However, it should be pointed out that no boundary-value problem can completely describe the complex world around us. Every boundary-value problem aims at a certain class of phenomena, formulates its essential features, and disregards what is of minor importance. The finite element method is merely a technique for obtaining approximate solutions to boundary-value problems. Thus, the basic assumptions, governing equations, and special features of the boundary-value problem should be kept in mind while constructing and analyzing a finite element model of the problem. Verify the finite element results by actual tests, if possible. If this is not possible, then verify the analysis procedures and the accuracy of the finite element codes by solving similar types of problems that have experimental results or closed-form solutions. In fact, this last step is vital and actually should be done first [35].

2.6.3 Material Properties of Lead-Free Solder Interconnects

In order to perform DFR and fully understand the engineering physics of solder-joint reliability, the material properties of solders must be known (through tests) and understood. Some important material properties of solders are:

- Phase-transition temperatures (i.e., solidus, liquidus, plastic, and eutectic)
- Melting point
- Boiling point
- Softening point
- Hardness
- Density
- Dispensability
- Glass-transition temperature
- Coefficient of thermal expansion (CTE)
- Vapor pressure
- Surface tension
- Storage modulus
- Loss modulus
- Viscosity
- Miscibility

- Electrical conductivity
- Thermal conductivity
- Stress-strain curves
- Shear strength
- Tensile strength
- Ultimate strength
- Elongation
- Yield strength
- Young's modulus
- Poisson's ratio
- Creep curves
- Stress-relaxation curves
- Creep rates
- Creep rupture
- Fracture toughness
- Viscoelasticity
- Viscoplasticity
- Fatigue strength
- Isothermal fatigue curves
- Fatigue crack-growth curves
- Moisture absorption
- Others

In order to fully understand the lead-free interconnects, all these material properties have to be determined by tests and measurements. Do we have all these material properties of lead-free solder today? No, we don't, and thus lots of work needs to be done! In this chapter and the remainder of this book, the solderability, creep curves, creep rates, isothermal fatigue, fatigue crack-growth, Young's modulus, Poisson ratio, thermal conductivity, stress-strain curves, shear strength, tensile strength, ultimate strength, elongation, yield strength, and stress-relaxation curves will be used and discussed.

2.6.4 Effects of Halogen on Solderability and Reliability of SAC Solder Pastes

It is well known that SAC lead-free solder alloys have poorer solderability than SnPb solder alloys because of their metallic characteristics and much higher soldering temperatures. Hence, most of the lead-free solder pastes have higher halogen and activator contents than the tin-lead solder pastes to improve their solderability. However, higher halogen contents reduce the reliability of lead-free solder joints. This section investigates the solderability and reliability of solder pastes with different halogen contents. By choosing the proper halogen content, lead-free solder pastes can be achieved that have good solderability and better reliability.

Activators play important roles in fluxing and soldering processes. Generally speaking, activators can be divided into two groups: halogens and organic acids. Halogens are usually more powerful and reduce oxidation during the soldering process. Unfortunately, the key factors that affect the solder joint reliability, such as surface insulation resistance (SIR) and electron migration resistance (EMR), decrease as the halogen contents increase. This is due to halogen's higher activities, which make the flux residue unstable.

Experiments

Solder Paste Samples

Five solder paste samples with different halogen contents (0.01%, 0.05%, 0.10%, 0.20%, and 0.30%), as shown in Table 2.2, are tested in this study. The alloy composition of all solder pastes is Sn-3.0wt%Ag-0.5wt%Cu with 11.0wt% flux contents. While activator content in flux varies, the solvent content is slightly reduced as the activator is increased. The rest of the compounds in the solder paste flux are the same for all samples.

Spread Test (JIS Z3197 8.3.1.1)

All solder pastes are printed on a scrubbed and oxidized copper plate with dimensions of 30 × 30 × 0.3 mm by a stencil with one round hole of 6.5 mm in diameter and 0.2 mm in thickness. This copper plate is heated by a hot plate at 270°C (~50°C above the liquidus temperature of the SnAgCu solder alloy) for 30 s and cooled, the highest thickness of solder is measured, and the spread coefficient is calculated by the following equation:

$$S_R = (D - H)/D \times 100\%$$

Sample	Halogen Content
1	0.01%
2	0.05%
3	0.10%
4	0.20%
5	0.30%

Table 2.2 Samples with Various Halogen Contents

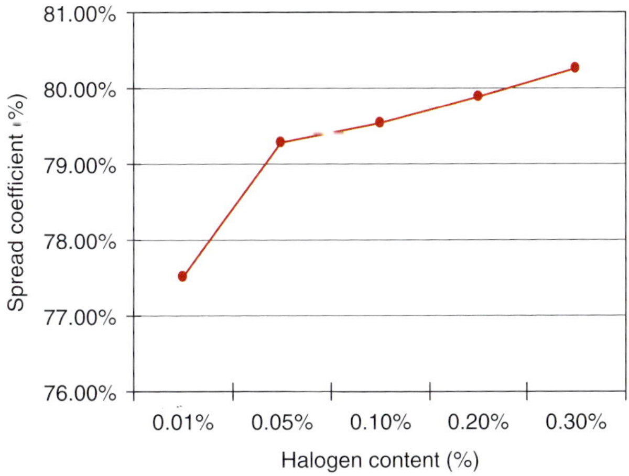

FIGURE 2.22 Spread coefficients with various halogen contents.

where S_R = spread coefficient of solder paste (%)
H = highest thickness of solder (mm)
D = diameter of the solder if it forms a sphere (mm)
$D = 1.24\ V^{(1/3)}$
V = volume of the solder (mass/density)

Five specimens are tested for each solder paste sample, and the average spread coefficient is calculated, which is shown in Fig. 2.22. (The larger the spread coefficient, the better the solderability.) It can be seen that: (1) The spread coefficient is a function of the halogen content; the higher the halogen contents, the larger the spread coefficients, that is, the better the solderability. (2) The spread coefficient increases dramatically when the halogen contents are increasing from 0.01% to 0.05%. (3) The rate of increase of the spread coefficient is basically constant when the halogen content is larger than 0.05%.

Solder Ball Test (IPC TM-650 2.4.43)

All solder pastes are printed on a ceramic plate with dimensions of 75 × 25 × 0.8 mm by a stencil with three round holes of 6.5 mm in diameter and 0.1 mm in thickness. This is heated by a hot plate at 245°C (~25°C above the liquidus temperature of the SnAgCu solder alloy) for 30 s and cooled. Then the number of solder balls is counted and recorded under a microscope with 30× magnification.

Three specimens are tested for each solder paste sample, and the results are shown in Fig. 2.23. (The fewer solder ball numbers, the better solderability. Usually, when the solder ball number is less than 3, it is accepted.) It can be seen that: (1) The solder ball number is a function of the halogen content; the higher the halogen contents, the smaller the solder ball number. (2) The solder ball number reduces

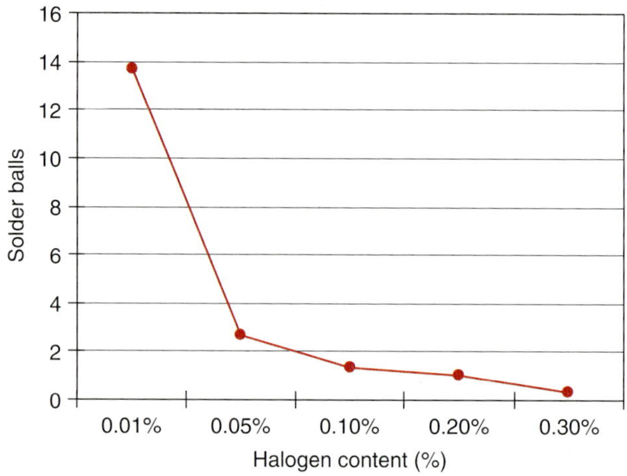

FIGURE 2.23 Solder balls with various halogen contents.

dramatically when the halogen contents are increasing from 0.01% to 0.05%. (3) There could be solderability problems when the halogen contents are less than 0.05%.

Corrosion Test (IPC TM-650 2.6.15)

All solder pastes are placed on an acid-washed copper plate with dimensions of 50 × 50 × 0.5 mm. The copper plates are heated by a hot plate at 270°C (~50 °C above the liquidus temperature of the SnAgCu solder alloy) for 60 s and cooled. The test coupons are observed carefully at 20× magnifications for any discoloration, and the results are recorded. Put all the coupons into a chamber with 40°C and 93% relative humidity for 240 h (10 days), observe the coupons, and record the results. For all the samples and the controls, there is no discoloration found, that is, no corrosion occurred on the coupons.

Surface Insulation Resistance (IPC TM-650 2.6.3.3)

All solder pastes are printed on the pattern B (12.5 mil line/space) of IPC-B-25 coupons and reflowed. In addition to the two control coupons (precleaned and unprocessed), all the samples are tested under 85°C/85%RH and 50V DC bias voltage, and their resistances are measured with a 100-V DC voltage at 24, 96, and 168 hours. (When the measured resistance of a sample is lower than 10^9 Ω, the sample is determined to have failed. The test will be repeated if the control resistance is lower than 10^9 Ω.)

All specimens are inspected under a 10× to 30× microscope using backlighting within 24 h of completion of the test. If the dendrite spans 25% or more of the original spacing, the sample is determined to have failed. Three specimens are tested for each solder paste sample, and the average resistance is calculated, which is shown in Fig. 2.24.

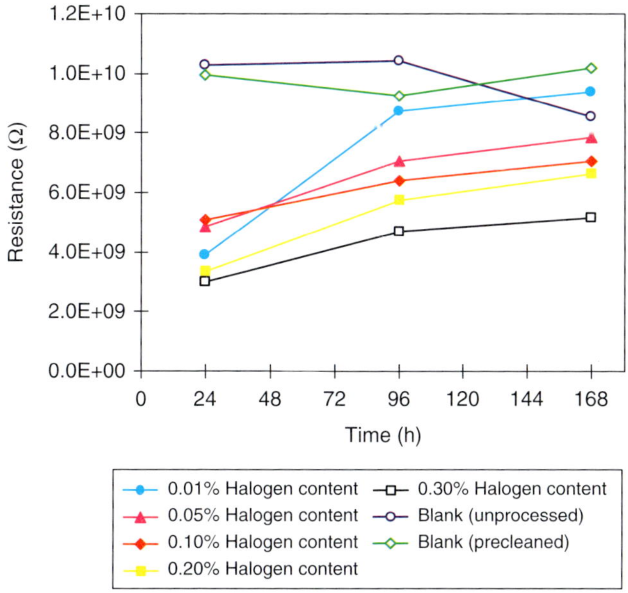

FIGURE 2.24 Surface insulation resistances with various halogen contents.

It can be seen that: (1) The surface insulation resistance is a function of the halogen content; in general, the higher the halogen contents, the lower the surface insulation resistance. (2) For all the samples, in general, the surface insulation resistance is a function of time; the longer the times, the larger the surface insulation resistance. (3) For all the samples, their surface insulation resistance is larger than 10^9 Ω. Also, no dendrite is found on the comb pattern.

Electrochemical Migration Resistance Test (IPC TM-650 2.6.14.1)

All solder pastes are printed on the pattern B (12.5 mil line/space) of IPC-B-25 coupons and reflowed. In addition to the one control coupon (precleaned), all the samples are tested under 65°C/85%RH, and their resistances are measured with a 100-V DC voltage at 96 hours. Then, under the same condition, a 10-V DC bias voltage is applied to the samples, and their resistances are measured with a 100-V DC voltage at 596 hours. (When the measured resistance of a sample is lower than 10^9 Ω, then the sample is determined to have failed. The test will be repeated if the control resistance is lower than 10^9 Ω.)

All specimens are inspected under a 10× to 30× microscope using backlighting within 24 hours of completion of the testing. If the dendrite spans 25% or more of the original spacing, the sample is determined to have failed. Three specimens are tested for each solder paste sample, and the average resistance is calculated, which is shown in Fig. 2.25. It can be seen that: (1) The electrochemical migration resistance

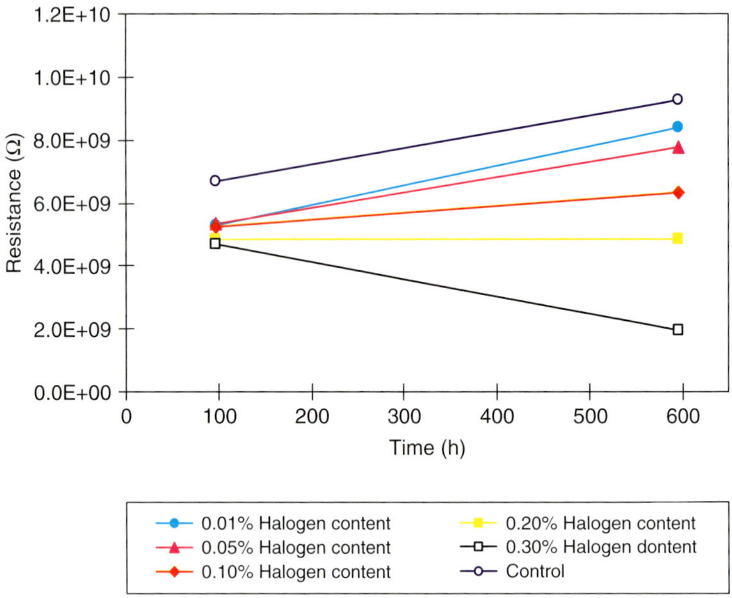

FIGURE 2.25 Electrochemical migration resistances with various halogen contents.

is a function of halogen content; the higher the halogen contents, the lower the electrochemical migration resistances. (2) The electrochemical migration resistance is a function of time; in general, the longer the times, the higher the electrochemical migration resistance, especially for lower halogen contents, for example, 0.01% and 0.05%. (3) For all the samples, no dendrite spanning is found on the comb pattern.

In the first 96 h of the tests, the electrochemical migration resistance of all the samples is above 4.0×10^9 Ω. After the bias voltage is applied for 500 h, Sample group no. 5 (with 0.30% halogen contents) shows electrochemical migration resistance degradation. The measured resistance of 2 out of 12 Sample group no. 5 is below 10^9 Ω. Although the resistance of those two Sample group no. 5 is below 10^9 Ω and is determined to have failed, no dendrite spanning is found on the comb pattern.

Summary and Recommendations

Important results are summarized in Table 2.3 and are as follows:

(1) Higher halogen contents in lead-free solder pastes lead to better solderability, since the spread coefficient is higher and the solder ball number is fewer.

(2) For halogen contents between 0.01% and 0.05%, there is a dramatic change in spread coefficient and solder ball number.

Sample	1	2	3	4	5
Halogen content	0.01%	0.05%	0.10%	0.20%	0.30%
Spread coefficient	77.52%	79.28%	79.53%	79.89%	80.26%
Solder ball	13.7	2.7	1.3	1.0	0.3
Wetting test	Pass	Pass	Pass	Pass	Pass
Copper plate corrosion	Pass	Pass	Pass	Pass	Pass
SIR	Pass	Pass	Pass	Pass	Pass
ECM	Pass	Pass	Pass	Pass	Fail

TABLE 2.3 Effects of Halogen Contents on Lead-free SnAgCu Alloy Solderability and Reliability

In order to ensure good solderability, the halogen content should be no less than 0.05%.

(3) Higher halogen contents in lead-free solder pastes lead to poor reliability, since the surface insulation resistance and the electrochemical migration resistance are lower.

(4) All halogen contents considered passed the surface insulation resistance tests. However, when the halogen contents are larger than 0.2%, the electrochemical migration resistance decreases with time and the value is less than 10^9 Ω, which is not acceptable for reliability.

(5) For the trade-off between solderability and reliability, it is recommended to use the halogen contents between 0.05% and 0.2%.

(6) Since different types of activators yield different solderability and reliability results, halogen-free types of flux should be studied.

2.6.5 Creep, Young Modulus, CTE, and Poisson's Ratio of SnAgCu and Other Solders

Useful constitutive equations of SnAgCu have been given in [1–17] and Table 2.4. Schubert et al., based on 108 data points from the literature (Fig. 2.26) and their own measurements, presented a new creep constitutive equation [7]. At the same time, Lau, Dauksher, and Vianco, based on 32 data points measured by Sandia (Fig. 2.27), presented another creep constitutive equation [1, 2]. A new set of creep constitutive equations was proposed by Lau and Dauksher [3], which is based on the average values of [1, 2] and [7].

Alloy No.	Alloy Compositions	$E = E_0 + E_1^* T(C)$ (GPa)		$\alpha = \alpha_0 + \alpha_1^* T(C)$ (ppm/°C)		ν	C_1 (s^{-1}) or (s^{-1} MPa^{-C_3}) for Norton
		E_0	E_1	α_0	α_1		
1	Sn3.9Ag0.6Cu	53	–0.08	21.301	0.017	0.34	441000
2	Sn3.8Ag0.7Cu Sn3.5Ag0.75Cu Sn3.5Ag0.5Cu Castin™	45.056	–0.0574	20		0.36	277984
3	Sn(3.5–3.9)Ag (0.5–0.8)Cu	49	–0.07	21.301	0.017	0.35	500000
4	Sn3.8Ag 0.7Cu						~0.5
5	Sn3.8Ag 1Cu						0.0026
6	Sn3.8Ag 0.7Cu			14.9	0.018		
7	Sn4Ag 0.5Cu	41.75	–0.075				2×10^{-21}
8	Sn3.9Ag 0.6Cu	18.596	–0.0206				143.41
9	Sn3.8Ag 0.7Cu	56.571	–0.25				
10	Sn4Ag 0.5Cu						2.474×10^{-14}
11	Sn3Ag 0.5Cu						
12	Sn3Ag 0.5Cu	54					
13	Sn3.9Ag 0.6Cu					0.34	
14	SnPb	34.4	–0.152	24.5			926[508-T(K)]/T(K)
15	Sn	44.3		21.743	0.0139	0.33	

1. Lau, J., W. Dauksher, and P. Vianco, "Acceleration Models, Constitutive Equations and Reliability of Lead-Free Solders and Joints," *IEEE Electronic Components and Technology Conference Proceedings*, New Orleans, Louisiana, June 2003, pp. 229–236.
2. Viance, P., J. Rejent, and A. Kilgo, "Creep Behavior of the Ternary 95.5Sn-3.9Ag-0.6Cu Solder: Part I—As-Cast Condition," *Journal of Electronics Material*, Vol. 33, 2004, pp. 1389–1400.
3. Lau, J. H., and W. Dauksher, "Creep Constitutive Equations of Sn(3.5-3.9)wt%Ag(0.5-0.8)wt%Cu Lead-Free Solder Joints," *MicroMaterails and Nanomaterials*, Berlin, Germany, Editor: B. Michel, 2004, pp. 54–62.
4. Zhang, Q., A. Dasgupta, and P. Haswell, "Viscoplastic Constitutive Properties and Energy-Partitioning Model of Lead-Free Sn3.9Ag0.6Cu Solder Alloy," *IEEE Electronic Components and Technology Conference Proceedings*, New Orleans, Louisiana, June 2003, pp. 1862–1868.
5. Pang, J. H. L., B. S. Xiong, C. C. Neo, X. R. Zhang, and T. H. Low, "Bulk Solder and Solder Joint Properties for Lead Free 95.5Sn-3.8Ag-0.7Cu Solder Alloy," *IEEE Electronic Components and Technology Conference Proceedings*, New Orleans, Louisiana, June 2003, pp. 673–679.
6. Pao, Y. H., R. Govila, and E. Jih, "An Experimental and Modeling Study of Thermal Cyclic Behavior of Sn-Cu and Sn-Pb Joints," *Proceedings of Materials Science Society Symposium*, Vol. 323, 1994, pp. 153–158.
7. Schubert, A., R. Dudek, H. Walter, E. Jung, A. Gollhardt, B. Michel, and H. Reichl, 2002, "Reliability Assessment of Flip-Chip Assemblies with Lead-Free Solder Alloys," *IEEE Electronic Components and Technology Conference Proceedings*, San Diego, CA, May 2003, pp. 1246–1255.
8. Schubert, A., R. Dudek, E. Auerswald, A. Gollhardt, B. Michel, and H. Reichl, "Fatigue Life Models for SnAgCu and SnPb Solder Joints Evaluated by Experiments and Simulation," *IEEE Electronic Components and Technology Conference Proceedings*, New Orleans, Louisiana, June 2003, pp. 603–610.

TABLE 2.4 SnAgCu Material Properties

C_2 (MPa^{-1})	C_3	C_4(eV)	Lab	Comments
.005	4.2	0.400	Sandia	Refs. 1, 2, bulk sample, compression, strain rate = 4.2×10^{-5}, and TMA, –25°C–160°C
0.02447	6.41	0.56	Fraunhofer & Dresden U.	Ref. 8, various specimens types, fit to numerous researchers results, –55°C–150°C
0.01	5	0.5	Agilent	Ref. 3, average of values for solder alloys in 1 and 2
0.1878	3.07	0.545	Fraunhofer & Dresden U.	Ref. 7, bulk, tension, C_1 obtained by inspection, 23°C–150°C
0.185	3	0.401	Fraunhofer & Dresden U.	Ref. 9, bulk, tension, 20°C–180°C
			Fraunhofer & Dresden U.	Ref. 9, bulk, TMA, 20°C–150°C, calculated at 100°C and 150°C
1	18	0.861	Fraunhofer & Dresden U.	Ref. 10, flip-chip bumps, shear, 5°C and 50°C
0.10854	3.7884	0.652	CALCE	Ref. 4, thin disk, shear, 25°C, 75°C, 125°C
			Nanyang Tech. U.	Ref. 5, bulk-dog bone, tension, bulk, 25°C, 75°C, 125°C, strain rate = 5.6×10^{-3} /s
–	8.36	0.062	Ford	Ref. 6, shear, 40°C–140°C (Norton)
6.6 to 10.7		0.776 to 0.983	U.C. Berkley & Intel	Refs. 11, 12, BGA-like, 60°C–130°C, anomalous creep behavior
			Nagaoka U.	Ref. 13, bulk–ASTM E606, tension, strain rate = 4×10^{-2} /s, 20°C
			NIST & Sandia	Ref. 14
1/(37.8-.07*T(K)	3.3	0.548		Ref. 15
			Textbook Sn	Refs. 16, 17

9. Wiese, S., A. Schubert, H. Walter, R. Dudek, F. Feustal, E. Meuesel, and B. Michel, "Constitutive Behavior of Lead-Free Solders vs. Lead-Containing Solders—Experiments on Bulk Specimens and Flip-Chip Joints," *IEEE Electronic Components and Technology Conference Proceedings*, Orlando, FL, pp. 890–902.
10. Wiese, S., S. Rzepka, and E. Meusel, "Time Dependent Plastic Behavior of Solders and Its Effect on FEM Simulations for Electronic Packages," *Proceedings of the 8th International Symposium on Advanced Packaging Materials*, 2002, pp. 101–111.
11. Song, H. J., J. W. Morris, and F. Hua, "The Creep Properties of Lead-Free Solder Joints," *Journal of Materials*, June 2002, pp. 30–32.
12. Hua, F., "Pb-Free Solder Challenges in Electronic Packaging and Assembly," *IEEE Electronic Components and Technology Conference Proceedings*, New Orleans, Louisiana, June 2003, pp. 58–63.
13. Kanchanomai, C., Y. Miyashita, and Y. Mutoh, "Low Cycle Fatigue Behaviors of Sn-Ag, Sn-Ag-Cu and Sn-Ag-Cu-Bi Lead-Free Solders," *Journal of Electronic Materials*, Vol. 31, No. 5, pp. 456–465.
14. Siewert, T., S. Liu, D. Smith and J. Madeni, 2002, "Properties of Lead-Free Solders, Release 4.0," NIST and Colorado School of Mines.
15. Lau, J. H., *Low Cost Flip Chip Technologies for DCA, WLCSP, and PBGA Assemblies*, McGraw-Hill, New York, 2000.
16. Touloukian, Y. S., and C. Y. Ho, *Thermophysical Properties of Matter*, The TPRC Data Series, Plenum, New York, 1970.
17. Lau, J. H., and Y. Pao, *Solder Joint Reliability of BGA, CSP, Flip Chip, and Fine Pitch SMT Assemblies*, McGraw-Hill, New York, 1997.

TABLE 2.4 SnAgCu Material Properties (*Continued*)

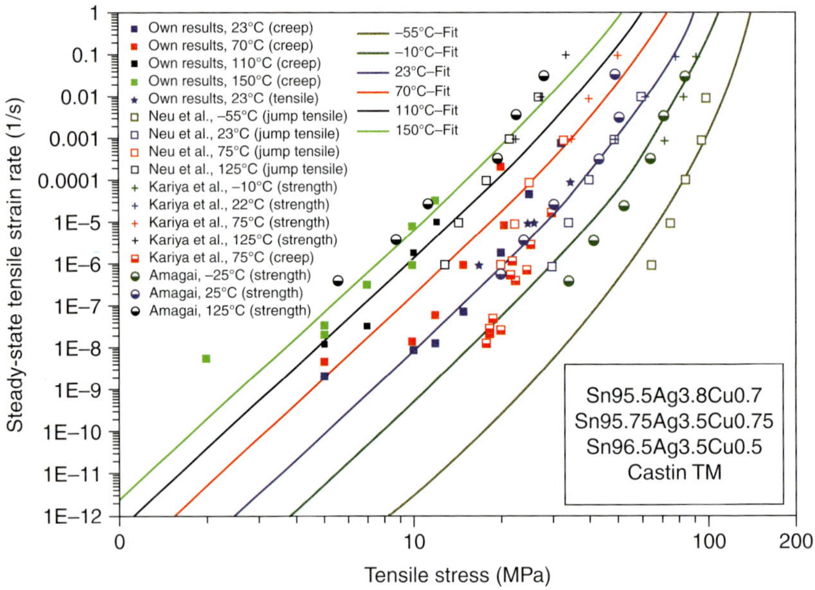

FIGURE 2.26 Creep test data collected by IZM.

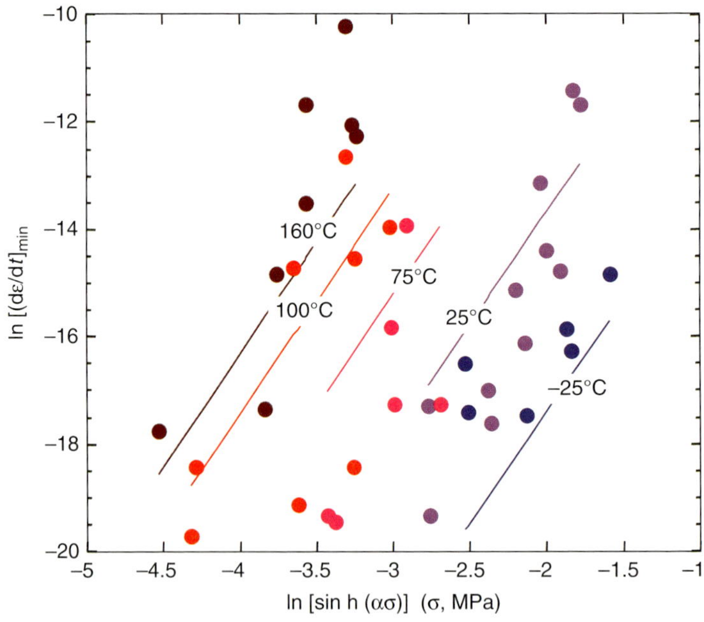

FIGURE 2.27 Creep test data by Sandia.

Constitutive Equations of SnAgCu Lead-Free Solders

Garofalo-Arrhenius creep constitutive equation expressed by [35]:

$$\frac{\mathrm{d}\gamma}{\mathrm{d}t} = C\left(\frac{G}{\Theta}\right)\left[\sinh\left(\omega\frac{\tau}{G}\right)\right]^{n}\exp\left(\frac{Q}{k\Theta}\right) \tag{2.37}$$

where γ is the creep shear strain, $\mathrm{d}\gamma/\mathrm{d}t$ is the creep shear strain rate, t is the time, C is a material constant, G is the modulus, Θ is the absolute temperature (K), Ω defines the stress level at which the power law stress dependence breaks down, τ is the shear stress, n is the stress exponent, Q is the activation energy for a specific diffusion mechanism (e.g., dislocation diffusion, solute diffusion, lattice self-diffusion, and grain boundary diffusion), and k is the Boltzmann's constant (8.617×10^{-5}eV/K). Equation (2.37) can be rearranged by lumping certain coefficients and expressed as:

$$\frac{\mathrm{d}\varepsilon}{\mathrm{d}t} = C_1[\sinh(C_2\sigma)]^{C_3}\exp\left(-\frac{C_4}{T}\right) \tag{2.38}$$

where C_1, C_2, C_3, and C_4 are determined by creep tests (curves) at various sets of constant stress and temperature. It should be noted that the last equation is exactly the same as the input form of the implicit creep model (TBOPT = 8) in ANSYS [36]. In the previous two equations, σ is the effective normal stress, τ is the shear stress, and $\mathrm{d}\varepsilon/\mathrm{d}t$ is the effective normal creep strain rate. The unit for both σ and τ is MPa.

Based on 32 samples (Fig. 2.27), Vianco et al. [1, 2] obtained the creep constitutive equation for the Sn3.9Ag0.6%Cu solder alloy (Alloy No. 1 in Table 2.4). Schubert and his colleagues [7], based on 108 measurement data (Fig. 2.26), obtained a new creep constitutive equation for the Sn3.8Ag0.7Cu, Sn3.5Ag0.75Cu, Sn-3.5Ag0.75Cu, and Castin alloys. These two equations are plotted in Fig. 2.28. It can be seen that for the temperatures (–25°C, 50°C, and 125°C), Sandia's data tend to predict faster creep rates at lower stresses and slower creep rates at higher stresses. Averaging the values of these two sets of data at different temperatures (by Lau and Dauksher of Agilent) leads to a new creep constitutive equation [3], called Alloy No. 3, as shown in Fig. 2.28 and described in Table 2.4. It can be seen that this new constitutive equation [Eq. (2.39)] for Sn(3-3.8)Ag0.5Cu is based on 140 creep test data and should be useful for the input of finite element analyses.

$$\frac{\partial\varepsilon}{\partial t} = 500000\sinh^{5.0}\left(\frac{\sigma}{1\times10^{8}}\right)\exp\left(\frac{-5807}{T(\mathrm{K})}\right) \tag{2.39}$$

The creep constitutive equations for other solders have been determined by Lau and his colleagues for:

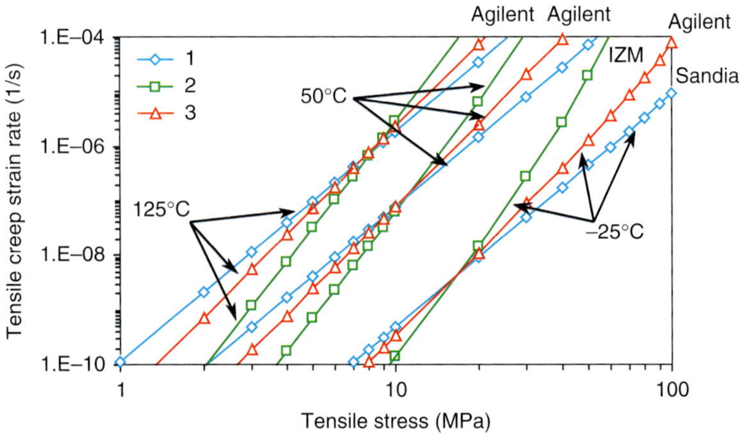

FIGURE 2.28 Average creep data (from Sandia and IZM) by Lau and Walter of Agilent [3].

80Au-20Sn:

$$\frac{\partial \varepsilon}{\partial t} = 1.3977 \times 10^{-21} \sigma^{2.55} \exp\left(\frac{-9516}{T}\right) \tag{2.40}$$

42Sn-58Bi:

$$\frac{\partial \varepsilon}{\partial t} = 9.82 \times 10^{-24} \sigma^{4.05} \exp\left(\frac{-8483}{T}\right) \tag{2.41}$$

48Sn-52In:

$$\frac{\partial \varepsilon}{\partial t} = \frac{2.54 \times 10^{11}(593 - T)}{T} \sinh^{3.11}\left(\frac{\sigma}{36.09 \times 10^{6} - 59628T}\right) \exp\left(\frac{-9705}{T}\right) \tag{2.42}$$

63Sn-37Pb:

$$\frac{\partial \varepsilon}{\partial t} = \frac{926(508 - T)}{T} \sinh^{3.3}\left(\frac{\sigma}{37.78 \times 10^{6} - 74414T}\right) \exp\left(\frac{-6360}{T}\right) \tag{2.43}$$

In Eqs. (2.39) through (2.43), the unit of σ is Pa and of T is K.

Figure 2.29 shows the Young's modulus versus temperature plots of the lead-free solders reported in the literature. Again, it can be seen that data are quite varied. The closest data are Alloy No. 1 and Alloy No. 2. Again, by taking the average of these two sets of data, the modulus of the proposed Alloy No. 3 is obtained.

The coefficients of thermal expansion (CTE) of various lead-free solders reported in the literature are shown in Fig. 2.30. Since most of the data agree with Sandia's data, this value was chosen for Alloy No. 3. In summary, the Young's modulus, CTE, Poisson's ratio, and

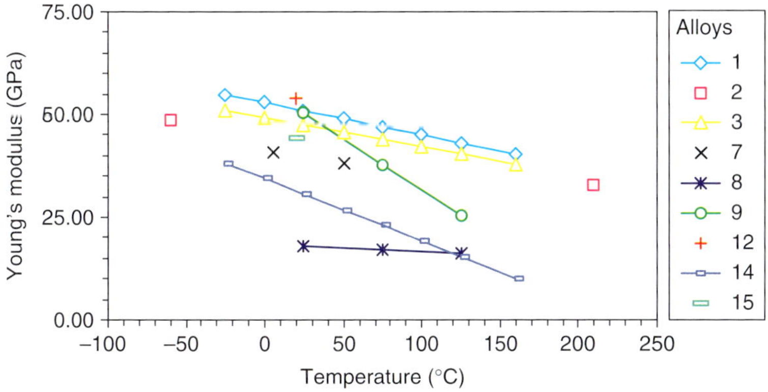

FIGURE 2.29 Young's Modulus.

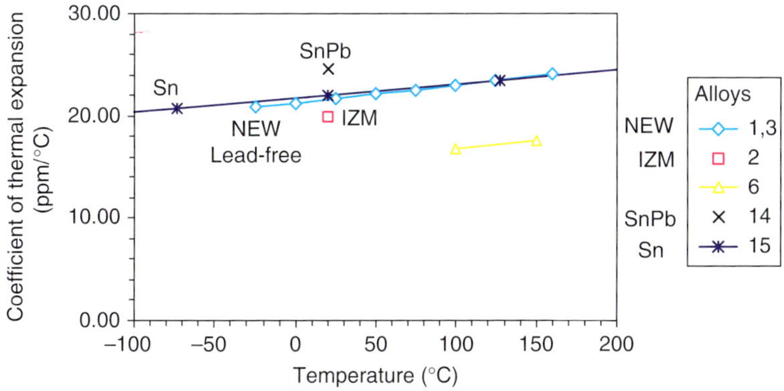

FIGURE 2.30 Coefficient of thermal expansion (CTE).

thermal conductivity of SAC and other solders 80Au20Sn, 42Sn58Bi, 48Sn52In, and 63Sn37Pb are shown in Table 2.5.

2.6.6 Isothermal Fatigue Tests of WLCSP Lead-Free Interconnects

The thermal fatigue life of solder interconnects (which are dominated by creep responses) may be estimated by the following equation:

$$N_f = \psi(\Delta W)^{\phi} \tag{2.44}$$

where N_f is the number of cycles to failure of the solder interconnect, ΔW is the creep strain energy density per cycle in the solder interconnect, and Ψ(always positive) and ϕ (always negative) are constants for the solder interconnects. In this equation, ΔW can be determined by creep analysis of the package or 3-D IC structure, which consists of the IC, substrate, solder interconnects, PCB, etc.,

Material	E (GPa)	α (ppm/K)	ν	k (W/m·K)
Si	167	2.54	0.28	154
Cu	76	17	0.35	398
95.5Sn-3.9 Ag-0.6Cu	68.11 – 0.07T	16.66 + 0.017T	0.3	70
80Au-20Sn	59	16	0.3	251
48Sn-52In	30.5	28	0.36	32.8
63Sn-37Pb	75.9–0.152T	24.5	0.35	50.6
Sn-58Bi	381.5–3.13T + 0.009T^2	14	0.35	18.3
AlGaAs	86	5.8	0.31	33.67
GaAs	86	5.8	0.31	33.7

Note: T in degrees Kelvin (K); E = Young's modules; α = thermal coefficient of expansion; ν = Poisson ratio; and k = thermal conductivity

TABLE 2.5 Material Properties

subjected to a specified thermal loading condition. In performing the creep analysis, the material properties of each part of the structure, such as the Young's modulus, CTE, Poisson ratio, and creep constitutive equations discussed in the previous section, are needed. The constants Ψ and ϕ are usually determined by isothermal fatigue tests of real solder interconnects, which are the focus of this section.

The test vehicle consists of the μSMD [37, 38] WLCSP with 95.5wt%Sn4.0wt%Ag0.5wt%Cu solder bumps, and the PCB with electroless nickel-immersion gold (ENIG or Ni-Au) surface finish. The lead-free solder paste used for the current surface mount assembly is 95.5wt%Sn3.9wt%Ag0.6wt%Cu.

Test Board Assembly

Figure 2.31 shows the 1.29 × 1.29 × 0.85 mm wafer level chip scale package (WLCSP) under consideration. There are eight solder bumps at 0.5 mm pitch and each with a bump diameter of 0.17 mm. The solder alloys are Sn4Ag0.5Cu.

The PCB is made of FR-4 fiberglass laminate material and is 1.6 mm thick. The surface finish of the copper pad (30–44 μm thick) is electroless Ni (4.5–5.6 μm thick) and immersion Au (~0.08 μm thick). The round copper pad (0.3 mm in diameter) for the WLCSP is non-solder mask defined (NSMD), and the solder mask opening is 0.45 mm in diameter. The pads on both WLCSP and PCB are interconnected in an alternating pattern so as to provide a daisy chain connection (for continuous measurement during testing) after they are assembled (Fig. 2.31).

The assembly process of the test boards is robust and is with regular equipments. For solder paste printing, the stainless steel stencil

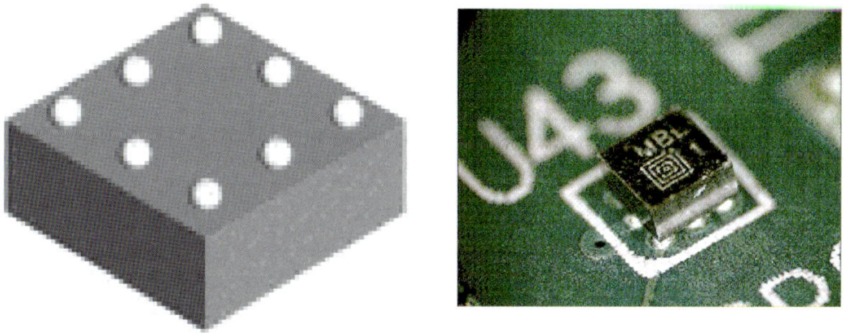

*1.29 × 1.29 × 0.85 mm WLCSP
*8 bumps at 0.5 mm pitch
*Bump diameter is 0.17 mm
*Solder alloys are Sn4Ag0.5Cu

*PCB is 1.6 mm thick
*FR-4 fiber-glass laminate
*Cu pads are 30~44 μm thick
*Cu pads are 0.3 mm in diameter
*Cu pads are NSMD (0.45 mm)
*EN(4.5~5.6 μm) IG (~0.08 μm)

FIGURE 2.31 μSMD WLCSP with eight lead-free solder bumps on PCB.

is manufactured with a laser-cutting process followed by electropolishing. The stencil thickness is 127 μm, and the aperture is 0.3 mm square, which gives an area ratio (of the aperture opening area to the aperture wall area) of 0.6.

The paste used in this study is a no-clean solder paste with Sn3.9Ag0.6Cu composition and 88.4wt% metal loading. The powder (particle) size is Type 3 (i.e., more than 80wt% of the nominal diameter is between 20 μm and 45 μm, less than 1wt% of the nominal diameter is larger than 45 μm, no more than 10wt% of the nominal diameter is less than 20 μm, and none of the powder size is larger than 50 μm). The solder paste is printed with the following parameters: print speed = 75 mm/s, squeegee pressure = 10 kg, and separation speed = 10 mm/s. The WLCSPs are picked and placed using a production fine-pitch placer.

A convection reflow oven with eight heating zones is used for reflowing the lead-free solder bumps and paste in air. The reflow temperature profile on the PCB is shown in Fig. 2.32. It can be seen that the peak temperature is 247°C and the x-ray image of an assembled WLCSP lead-free test board has some voids. The setup of the thermocouple probe design is also shown in Fig. 2.32.

Isothermal Fatigue Test and Results

The isothermal fatigue test is a force-controlled test, that is, the force amplitude is constant and the displacement is measured for each cycle. The very first step of the test is to perform an overload test of the sample to fail in order to determine the ultimate load (F) and displacement of the real solder interconnects. Use one-half of the ultimate load, that is, $F/2$, to perform the isothermal fatigue test and record

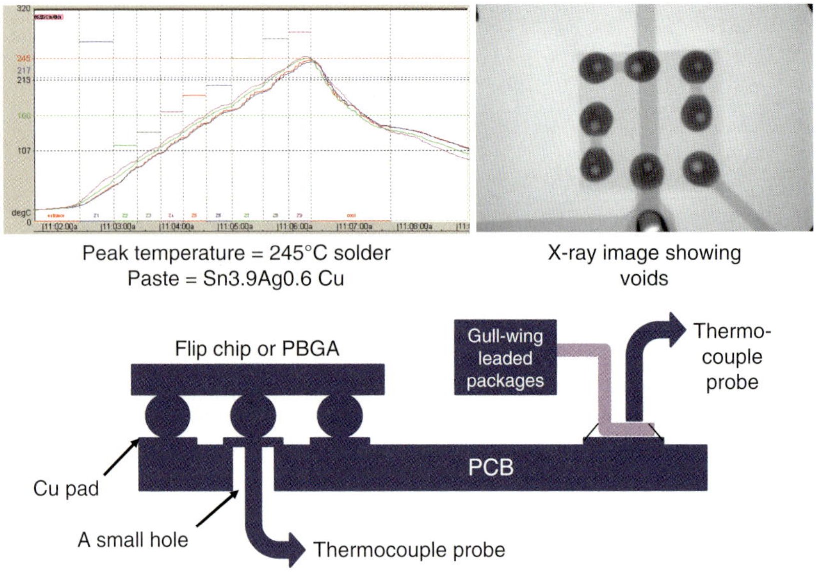

FIGURE 2.32 Lead-free reflow profile and x-ray image of the assembly. Also, the correct way to attach the thermal couple for area array and gull-wing interconnects is shown.

the displacement amplitude (Δ). In this study, the average value of the ultimate load is ~10 N, that is, $F/2 \sim 5$ N.

During the force-controlled fatigue test, record and plot the load-displacement hysteresis loops and the displacement-increase curves. The area within the load-displacement hysteresis loop is the strain energy per cycle. The number of the cycle after which the cyclic displacement amplitude is increased to 100% of that of the first cycle, that is, 2Δ, is defined as the fatigue life (failure), or number of cycle-to-failure.

The test setup and procedure of the lead-free WLCSP test board assembly are shown in Fig. 2.33 and highlighted in the following [37, 38]. All tests are performed at room temperature and at 12 cycles per minute, that is, 0.2 Hz.

(1) Samples are cut into the appropriate size (3 × 12 mm).

(2) The cut PCB with the WLCSP is attached on the bottom aluminum block by instant glue (Fig. 2.33*a*).

(3) The bottom aluminum block is then fixed on the shaft (Fig. 2.33*b*).

(4) Another aluminum block is placed on the top of the WLCSP (Fig. 2.33*c*).

(5) A thin metal plate with known thickness is inserted between the upper and the bottom aluminum blocks to ensure the gap distance is always kept constant.

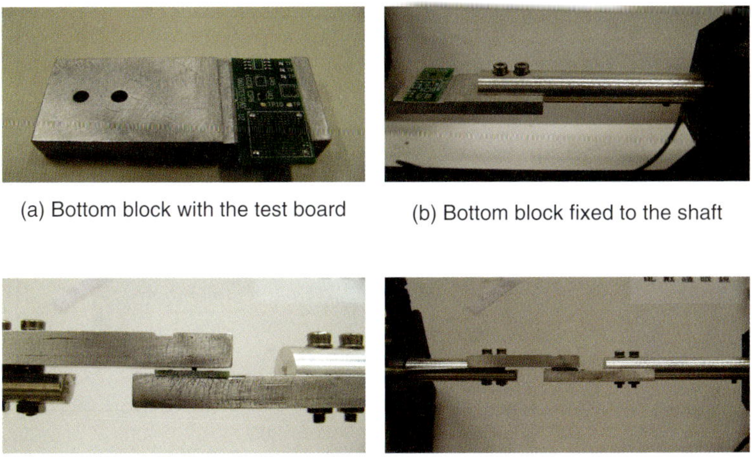

(a) Bottom block with the test board

(b) Bottom block fixed to the shaft

(c) Another block placed on top of the package

(d) Upper block connected to the load cell of Tytron

FIGURE 2.33 Isothermal fatigue test setup.

(6) The displacement and force value are set to zero.

(7) The upper block is connected to the load cell of the test equipment (MTS Tytron), and the instant glue is added between the block and the sample (Fig. 2.33*d* and Fig. 2.34).

(8) The program is executed after the instant glue is cured. The test is conducted under load control. The force amplitude is around 5 N and the cyclic frequency is 0.2 Hz.

FIGURE 2.34 Overall view of the Tytron microtesting system.

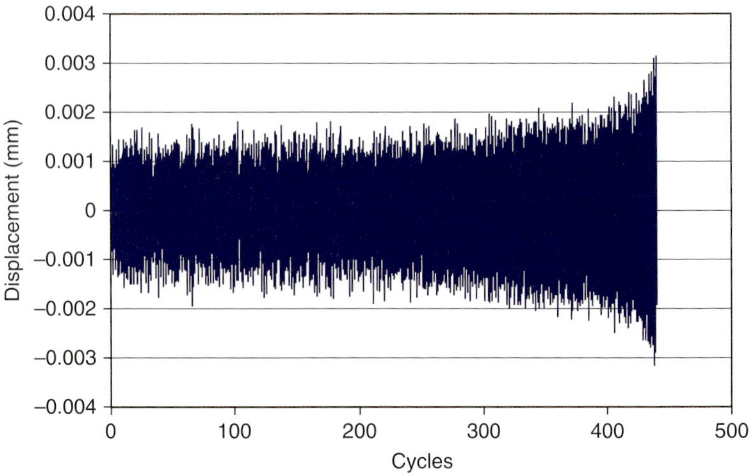

FIGURE 2.35 Displacement increase curve.

(9) The experiment is continued until the solder interconnect totally fails.

(10) The typical time history of displacement during the isothermal cyclic fatigue test is then reviewed.

Figure 2.35 shows a typical displacement increase curve versus time for one of the tested samples. In this case, it can be seen that the sample failed (total separation) at 443 cycles. The number of the cycles after which the cyclic displacement is increased to double of the initial displacement is 430 cycles, which is close to that (443 cycles) when the solder interconnects are totally separated. This shows that for the present force-controlled fatigue test, 100% displacement increase is a reasonable failure criterion. Similar results are observed for all other test samples, and their results are summarized in [37, 38].

A typical hysteresis loop (load vs. displacement curves) for one of the tested samples (Sample G1, F/2 = 5 N, Table 2.6) is shown in Fig. 2.36 for lower cycles as well as for higher cycles—the displacement amplitude is doubled. It can be seen that the initial load-displacement loops are kind of narrow and then they became broader as time goes by.

Figure 2.37 shows the plot for the applied force versus time (or N_f = number of cycle-to-failure) for all the test data in this study. It can be seen that (1) the N_f is not a linear function of the applied force, and (2) the larger the applied force, the smaller the life (N_f) of the lead-free solder interconnects.

As mentioned earlier, during the isothermal fatigue test, the sample is defined as failed when the initial displacement is doubled.

Cycles to Total Separation of the Package	Cycles to Double Displacement Amplitude	ΔW (Strain Energy Density per Cycle, MPa)	ΔW (Area of Hysteresis Loop of the Whole Package)	Picture Taken after Total Failure Package-Side	PCB-Side
1214	1098	0.109	0.0146		
1265	1240	0.133	0.0179	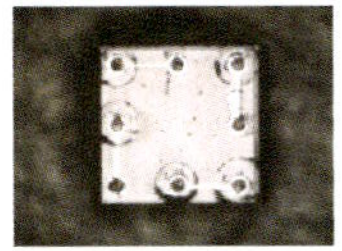	
1569	1550	0.122	0.0164		

TABLE 2.6 Isothermal Fatigue Test Results (WLCSP)

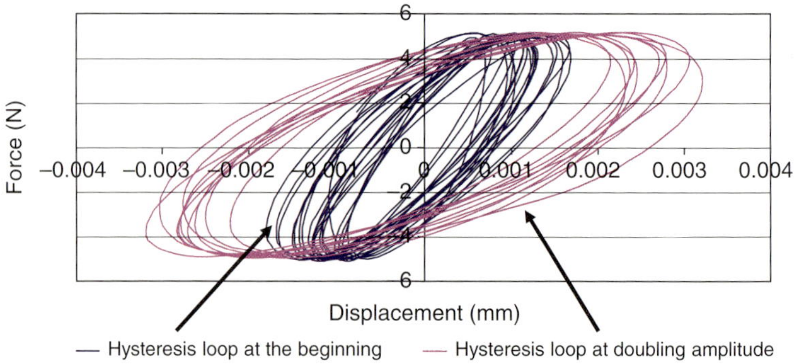

FIGURE 2.36 Force-displacement hysteresis loops.

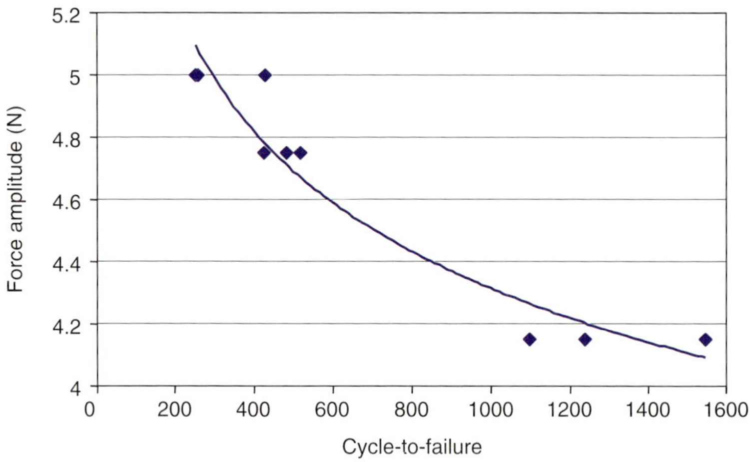

FIGURE 2.37 Number of cycle-to-failure vs. applied force.

The area under the hysteresis loop at this time (N_f) is defined as the strain energy per cycle at failure. When the strain energy per cycle is divided by the volume of the solder interconnects, the strain energy density per cycle (ΔW) is obtained.

The number of cycle-to-failure (N_f) is plotted against the strain energy density per cycle (ΔW) on a log-log scale, as shown in Fig. 2.38. A least-squares fit straight line is used to determine the coefficients ($\Psi = 18.15$ and $\phi = -1.96$) of the following thermal-fatigue equation for the SnAgCu solder alloys:

$$N_f = 18.15(\Delta W)^{-1.96} \tag{2.45}$$

Thus, once the strain energy density per cycle of the SnAgCu solder elements in a structure under thermal-cycling conditions is determined,

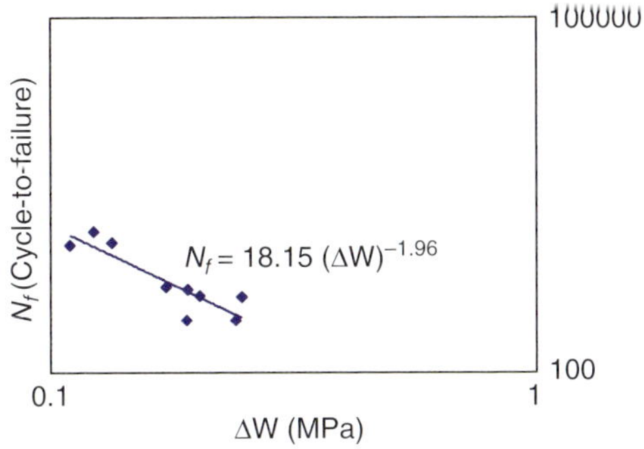

$N_f = 18.15\,(\Delta W)^{-1.96}$

N_f is the number of cycle-to-failure
ΔW is the strain energy density per cycle

FIGURE 2.38 Number of cycle-to-failure vs. strain energy density per cycle.

the average thermal-fatigue life of the solder interconnect can be estimated by Eq. (2.45).

The failure modes of the test samples are shown in Table 2.6. The close-up view of a solder interconnect is shown in Fig. 2.39. It can be seen that: (1) The failure modes of all the solder interconnects are similar and are subjected to shear-fatigue fracture in the solder interconnects. (2) The failure locations of all the solder interconnects are similar and are in the solder interconnects near the WLCSP (this is expected, since the cross-section of the solder interconnects near the chip is smaller than that near the PCB). (3) There are voids in the solder interconnects. The last two items will make the result presented herein more conservative.

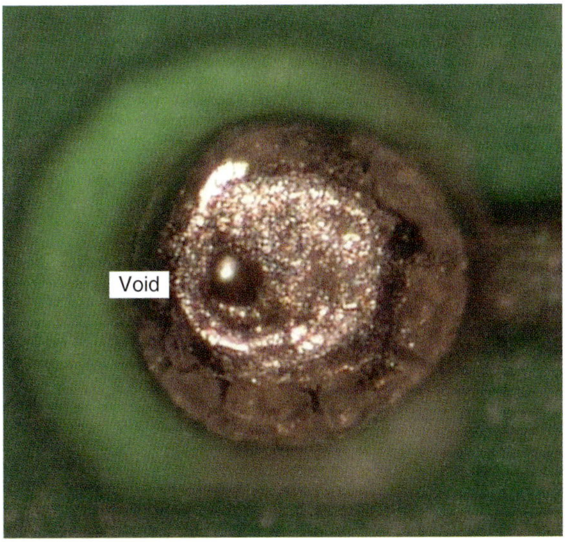

FIGURE 2.39 A typical shear-fatigue fracture surface with a large void.

Summary and Recommendations

Some important results and recommendations for the isothermal fatigue test of the WLCSP lead-free interconnect are summarized as follows:

(1) The number of cycle-to-failure of lead-free solder interconnects is a nonlinear function of the applied force.

(2) As expected, the larger the applied forces, the smaller the number of cycle-to-failure of the lead-free solder interconnects.

(3) The failure locations of the lead-free solder interconnects of all the samples are similar and are in the solder interconnects near the μSMD WLCSP side. (This is due to the poor pad design of the PCB.)

(4) The failure modes of the lead-free solder interconnects of all the samples are similar and are subjected to the shear-fatigue fracture in the solder interconnects.

(5) The thermal-fatigue life prediction equation presented herein should be conservative.

(6) The pad size of the PCB should be the same as that of the chip. In this case, the shear deformation due to the fatigue will be more even on both sides of the solder interconnect.

2.6.7 Isothermal Fatigue Tests of PBGA Lead-Free Interconnects at 60°C

In this section, the test vehicle consists of a PBGA with 95.5wt%Sn4.0wt%Ag0.5wt%Cu solder balls, and the PCB with ENIG surface finish. The lead-free solder paste used for the current surface mount assembly is 95.5wt%Sn3.9wt%Ag0.6wt%Cu [39, 40].

Test Board Assembly

Figures 2.40 through 2.42 show the 15 × 15-mm, 196-pin PBGA assembly under consideration. It can be seen that there are only eight connected solder joints at the four corners and the four middles of the outer edges. All the other solder balls are not soldered to the PCB. This is evidenced by viewing the x-ray images (Fig. 2.40) and noting that: (1) all solder balls (except those eight) are round, even though they are associated with daisy chains; and (2) the eight assembled solder-joint images are larger (even with solder wetting to the daisy chains) because of the additional solder paste. The solder alloys are Sn4Ag0.5Cu for the solder balls and Sn3.9Ag0.6Cu for the solder paste. The purpose of this special design is to have all eight solder joints fail at approximately the same time under the isothermal fatigue tests at 60°C.

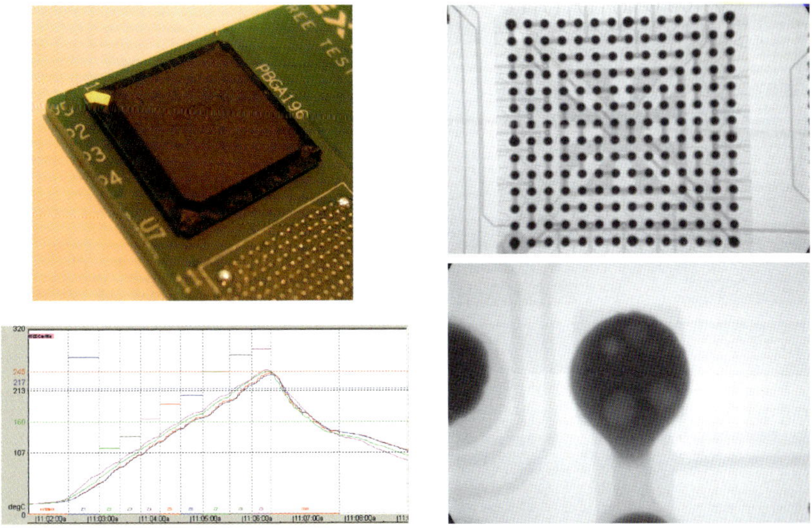

FIGURE 2.40 A PBGA with a lead-free assembly profile and their x-ray images showing the corner solder joint.

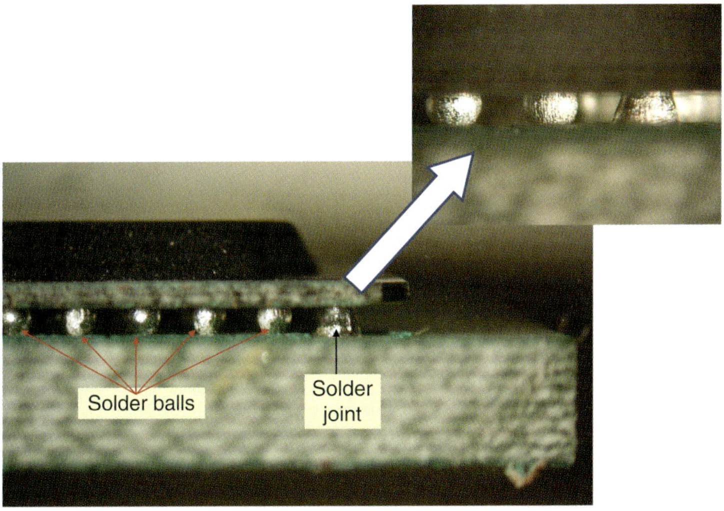

Only eight solder joints (four corners and four middle of the four outer edges). All other solder balls were not soldered onto the PCB.

FIGURE 2.41 Cross-section showing the corner solder joint.

The PCB is made of FR-4 fiberglass laminate materials and is 1.6 mm thick. The surface finish of the copper pad (30 to 44 μm thick) is ENIG. The 0.76-mm-diameter (averaging from 15) round pads are NSMD (Figs. 2.40 to 2.42). The PBGA substrate is made of BT materials, and

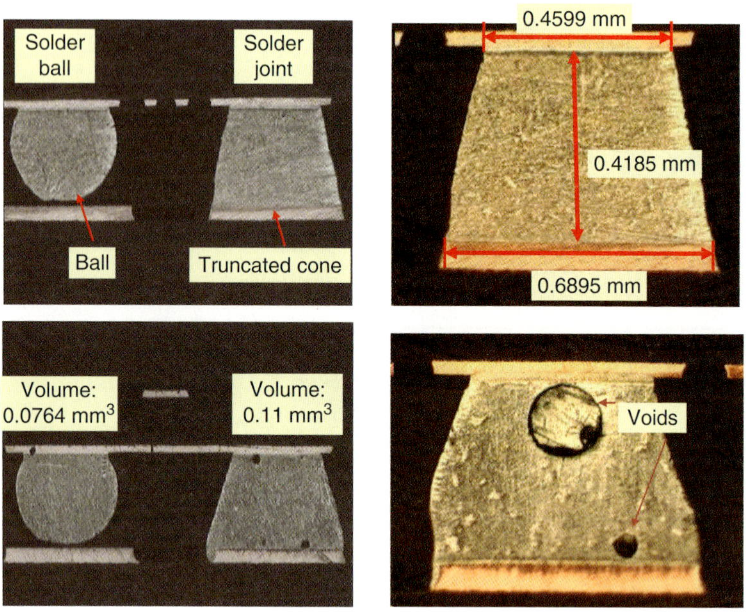

FIGURE 2.42 Cross-section and solder volume of the PBGA corner and middle (of 4-side) solder joints.

the 0.53-mm-diameter round copper pads are solder mask defined (SMD) (Figs. 2.40 to 2.42).

For solder paste printing, the stencil thickness is 127 μm and has only eight openings (at four corners and four middles of the outer edges) for this PBGA package. Why only eight solder joints? The reason is that we want all eight solder joints to fail at approximately the same time during the isothermal fatigue test; thus, the average strain energy density per cycle can be determined.

Isothermal Fatigue Test and Results

The test method and failure criterion are exactly the same as those in the previous section, except in the present case, the sample is inside a thermal chamber with a temperature equal to 60°C, as shown in Figs. 2.43 and 2.44. In this case, the average value of the ultimate load is ~70 N; hence, F/2 35 N. Figure 2.45 shows a typical displacement curve versus time for one of the tested samples (Sample S1, F/2 = 35 N, Table 2.7). In this case, it can be seen that the sample failed (total separation) at 251 cycles. The number of the cycles after which the cyclic displacement is increased to double that of the initial displacement is 217 cycles, which is close to that (251 cycles) when the solder interconnects are totally separated. This shows that for the present force-controlled fatigue test, the 100% displacement

Figure 2.43 Overall view of the Tytron microtesting system with a temperature chamber.

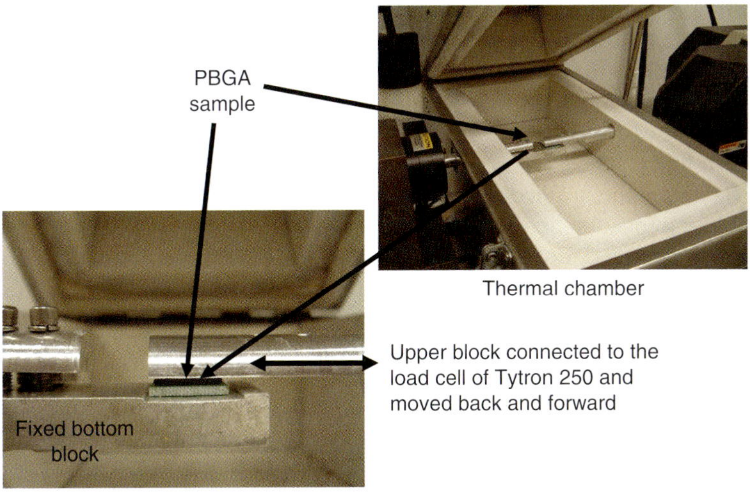

Figure 2.44 Close-up look at the temperature chamber with the PBGA sample (HKUST).

increase is a reasonable failure criterion. Similar results are observed for all other test samples, and the results are summarized in [40].

Typical hysteresis loops (load versus displacement curves) for one of the tested samples (Sample S1, $F/2 = 35$ N, Table 2.7) are shown in Fig. 2.46 (for lower cycles—just beginning—and for higher cycles—the displacement amplitude is doubled). It can be seen that the initial

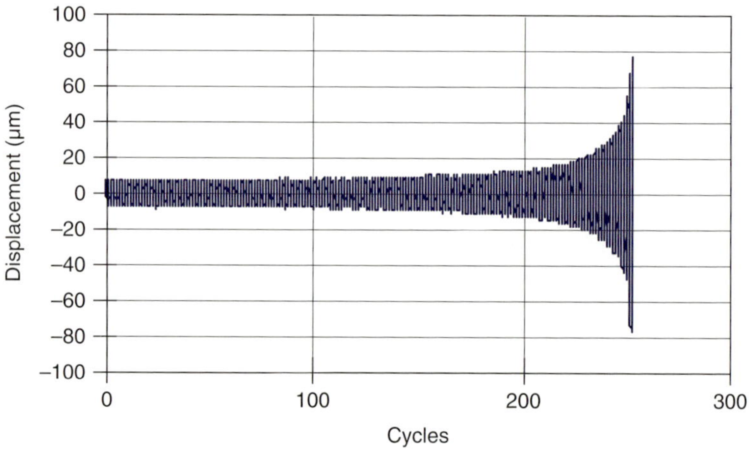

FIGURE 2.45 Displacement increase curve.

load-displacement loops are narrow and then they became broader as time goes by. The area within the hysteresis loop (when the displacement amplitude is double that of the initial displacement) of the whole package with eight solder joints per sample is shown in the fifth column of Table 2.7 with the units of N·mm. Similar results are obtained for all other test samples and are presented in [40].

Figure 2.47 shows the plot for the applied force versus time (or N_f = number of cycles-to-failure) for all the test data in this study. It can be seen that (1) the N_f is not a linear function of the applied force; and (2) the larger the applied force, the smaller the life (N_f) of the lead-free solder joints. As mentioned earlier, during the isothermal fatigue test, the sample is defined as failed when the displacement is increased to double the initial value. The area under the hysteresis loop at this time (W) is defined as the strain energy per cycle at failure. When the strain energy per cycle is divided by the total volume (8×0.119 mm^3 = 0.95 mm^3, Fig. 2.42) of those eight solder joints, the strain energy density per cycle (ΔW) is obtained. The unit of ΔW is MPa.

The number of cycles-to-failure (N_f) is plotted against the strain energy density per cycle (ΔW), as shown in Fig. 2.48. A least-squares fit straight line is used to determine the coefficients ($\Psi = 463$ and $\phi = -2.23$) of the following thermal-fatigue equation for the SnAgCu solder alloys at 60°C:

$$N_f = 463(\Delta W)^{-2.23} \tag{2.46}$$

Thus, once the strain energy density per cycle of the SnAgCu solder elements in a structure under thermal-cycling conditions is determined by creep structural analysis, the average thermal-fatigue life of the PBGA solder joints can be estimated by Eq. (2.46) [39, 40]. It should

Sample #/ Force Amplitude	Cycles to Total Separation of the Package	Cycles to Double Displacement Amplitude	ΔW(Strain Energy Density per Cycle, MPa)	Area of Hysteresis Loop of the Whole Package, N·mm	Picture Taken after Total Failure Package-Side	PCB-Side
S1/35 N	251	217	1.231	1.170	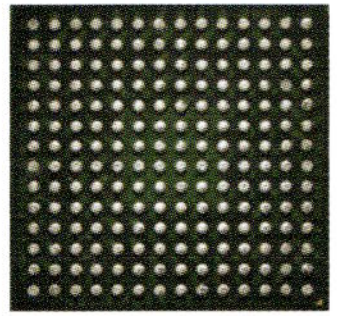	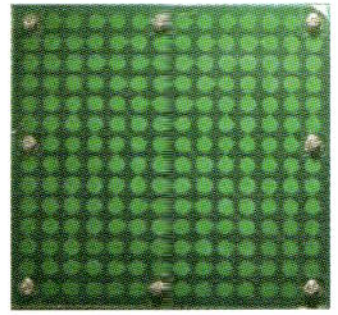
S2/35 N	278	261	1.270	1.208	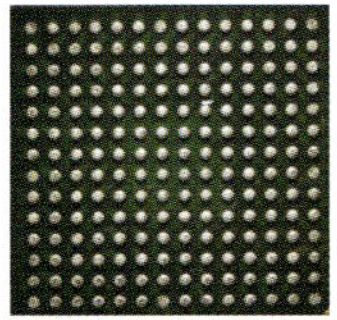	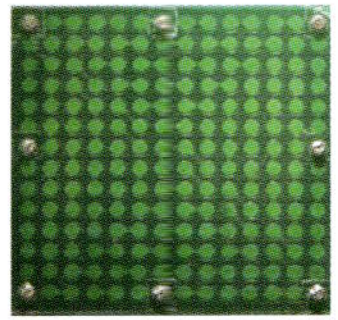
S3/35 N	382	343	1.165	1.108	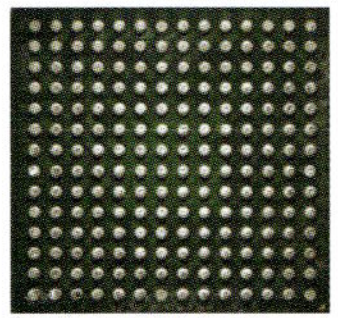	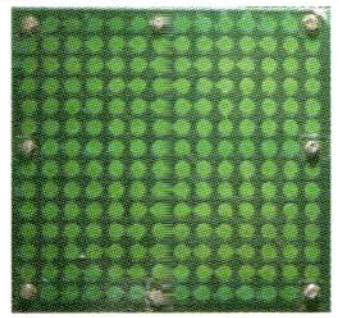

TABLE 2.7 Isothermal Fatigue Test Results (PBGA)

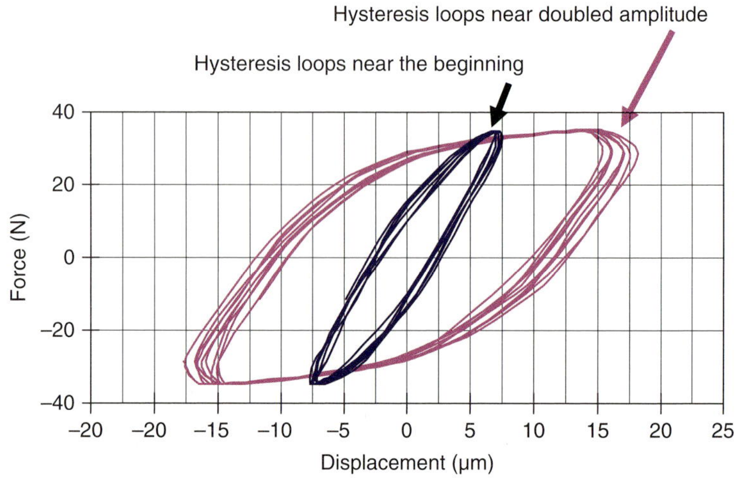

FIGURE 2.46 Force-displacement hysteresis loops.

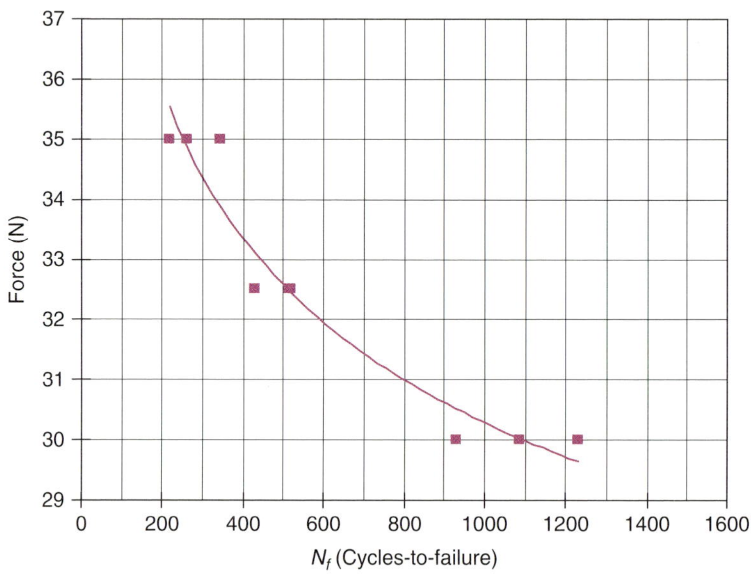

FIGURE 2.47 Number of cycles-to-failure (N_f) vs. applied force (F) at 60°C.

be pointed out that in most of the operating conditions, the temperature range (ΔT) during thermal cycling is 25°C to 85°C, that is, $\Delta T = 60$°C. Thus, Eq. (2.46) should be useful for predicting the average thermal-fatigue life of SnAgCu lead-free solder joints.

The failure modes of the test samples are shown in Table 2.7, and the close-up view of a solder joint is shown in Figs. 2.49 and 2.50.

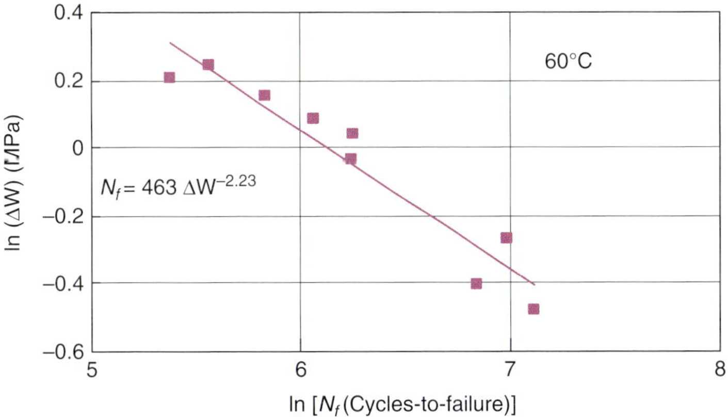

FIGURE 2.48 Number of cycles-to-failure (N_f) vs. strain energy density per cycle (ΔW) at 60°C on ENIG PCB.

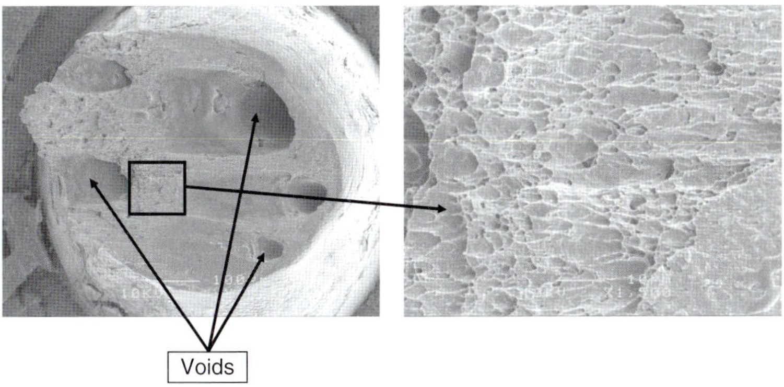

FIGURE 2.49 Typical shear-fatigue fracture surface near the substrate of the PBGA package (looking down to the PCB).

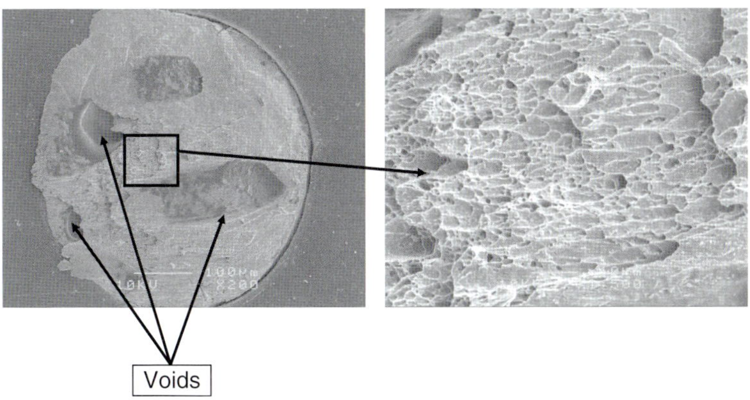

FIGURE 2.50 Typical shear-fatigue fracture surface near the substrate of the PBGA package (looking up to the substrate of the package).

It can be seen that: (1) The failure modes of all the solder joints are similar and are subjected to shear-fatigue fracture in the solder joints. (2) The failure locations of all the solder joints are similar and are in the solder joints near the PBGA package (this is expected, since the cross-section of the solder joints near the package is smaller than that near the PCB). (3) There are many voids in the solder joints.

2.6.8 Thermal-Fatigue Life Predictions of Lead-Free Interconnects

The DFR process is shown in Fig. 2.51, which shows a simple flow chart for the DFR to determine the thermal-fatigue life of solder interconnects. Basically, there are two steps—namely, (1) to determine the creep strain energy density per cycle (ΔW) of the solder interconnect by finite element methods, and (2) to substitute ΔW into the fatigue crack growth equation [$N_f = \psi(\Delta W)^{\phi}$] for determining the thermal-fatigue life (N_f), for example, Eq. (2.45) for WLCSP and Eq. (2.46) for PBGA.

In determining ΔW, we need to input the geometry and material property (constitutive equation) of the whole structure (which includes the IC chip, TSV, package, solder interconnect, PCB, etc.) to a finite element program (e.g., ANSYS [36]). Proper boundary

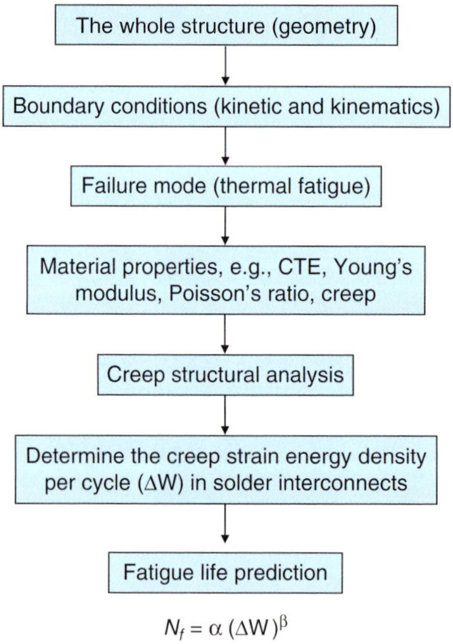

FIGURE 2.51 Structure analysis and life prediction of solder interconnects under low-cycle thermal fatigue.

conditions, such as the kinetic and kinematics, are necessary for performing the fatigue creep structural analysis of the assembly. Useful creep constitutive equations for various SnAgCu alloys have been given in Sec. 2.6.5.

The use of Eq. (2.45) has been shown in [41] and the use of Eq. (2.46) is shown herein. The 256-pin PBGA package under consideration (as shown in Figs. 2.52 and 2.53) has a 27 × 27 × 0.36 mm BT substrate with a 12.5 × 12.5 × 0.3 mm silicon IC chip and a 1.17-mm-thick overmold. The package has 0.635-mm-diameter pads (with a pitch of 1.27 mm) on the substrate, and the assembled lead-free (95.5wt%Sn-3.9wt%Ag-0.6wt%Cu) solder joints are assumed to have a height of 0.5 mm and a maximum diameter of 0.9 mm. The PCB thickness is 1.6 mm, and the pad diameter is 0.635 mm. Figure 2.52 shows the x-ray image of one-quarter of the PBGA assembly.

Thermal-Fatigue Life Prediction of 256-Pin PBGA Lead-Free Solder Interconnects

When the 256-pin PBGA PCB assembly is subjected to temperature cycle loadings, the solder joint closest to the chip edge will likely fail first. This is due to (1) the very large local thermal expansion mismatch between the silicon chip, the BT substrate, and the molding compound; and (2) the relatively high chip stiffness and high flexibility of the BT

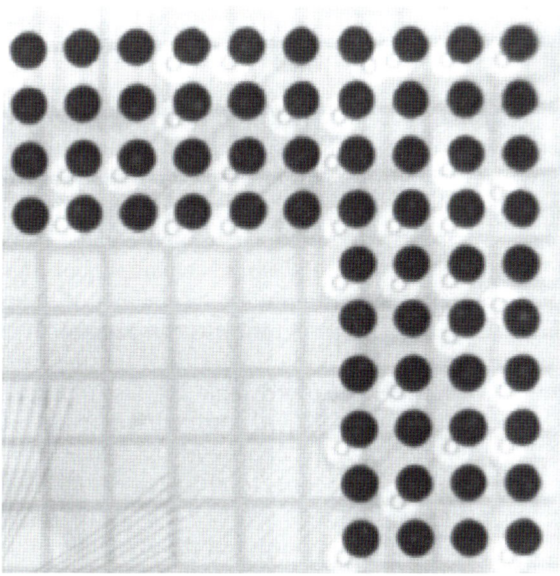

FIGURE 2.52 X-ray image of one-quarter of a lead-free 256-PBGA PCB assemble.

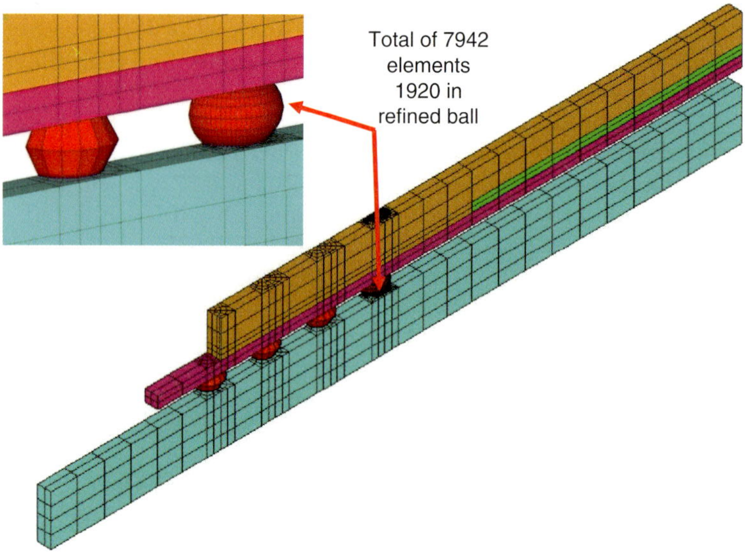

Figure 2.53 Finite-element analysis model of the lead-free 256-pin PBGA assembly.

laminate substrate. Thus, a finer mesh is deliberately applied to the lead-free solder joint located near the chip edge.

Figure 2.53 shows the 3-D finite-element model that captures the construction along a diagonal strip from the 256-pin PBGA lead-free assembly's geometric center to a corner. Figure 2.53 also provides a close-up view of the solder joint with the fine mesh, along with one with the coarser mesh due to its location further away from the chip. Because of the midplane symmetry, the mesh actually models a one-half strip (with one half of a solder joint) using the appropriate in-plane constraints placed on one symmetry plane. Coupled in-plane translations are applied to the other symmetry plane to produce a state of generalized plane strain. Using exclusively hexahedral solid elements, the model can capture the precise shape of the packages' solder joints while retaining significant computational efficiency over full octant models. The solder material properties of the PBGA assembly are shown in Sec. 2.6.5 and Table 2.5. The Young's modules, Poisson's ratio, and CTE of PCB (and laminate substrate) are, respectively 27 GPa, 0.39, and 18 and of overmold are, respectively 13 GPa, 0.3, and 15.

The temperature loading imposed to the lead-free PBGA assembly is –25°C to 125°C. The ramp-up, ramp-down, dwell-at-hot, and dwell-at-cold are each 15 minutes and shown in Fig. 2.54. Five thermal cycles are executed. The maximum creep strains and stresses occur at the solder joint interface (at a local volume from the outside)

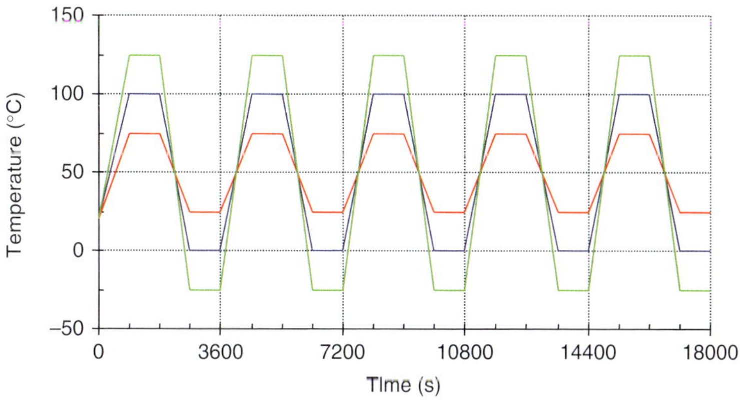

ΔT_1 = 25–75°C; ΔT_2 = 0°C–100°C; ΔT_3 = –25–125°C

FIGURE 2.54 Temperature cycling boundary conditions.

near the package substrate pad (slightly larger values) and PCB pad (slightly smaller values). Also, the maximum creep shear strain energy density per cycle near the package substrate pad is 0.316 MPa, as shown in Fig. 2.55 (i.e., the area within the fifth hysteresis loop). Thus, based on Eq. (2.46), we have the average thermal-fatigue life of the 256-pin solder interconnects as $N_f = 463(0.316)^{-2.23} = 6043$ cycles (–25°C–125°C).

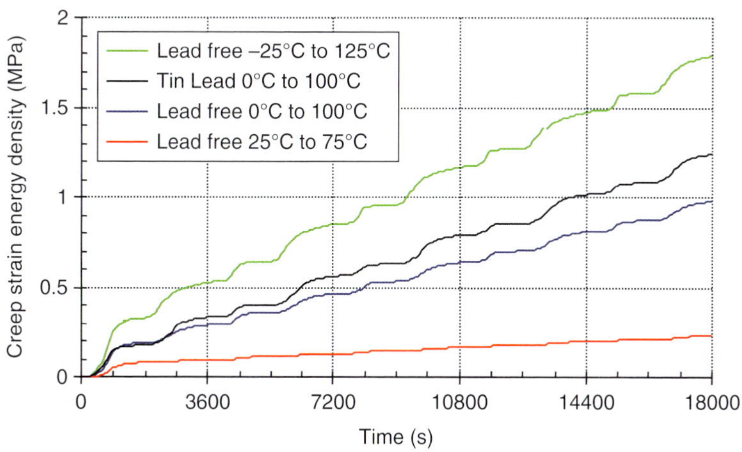

Creep strain energy density per cycle, ΔW (MPa):

LF 25–75 = 0.033
LF 0–100 = 0.171 (EU 0–100 = 0.226)
LF –25–125 = 0.316

FIGURE 2.55 Creep strain energy density at critical location of the lead-free 256-pin solder joint.

It should be pointed out that the direct application of fatigue crack-growth in Eq. (2.46) presumes that the creep strain energy density per cycle is uniformly distributed through the whole cross-sectional area near the interface of the solder joint with the PBGA substrate pad. Also, the material properties of some of the elements of the structure may not be accurate. However, the objective of this example is to present a systematic approach to estimate life rather than actually predicting the life of the 256-pin PBGA lead-free solder joints. Also, the other objective of this study is to show that it is impossible to compare the DFR results with the reliability test results, which depend on sample size, confidence levels, etc.

2.7 Quality of Lead-Free Solder Interconnects

During lead-free assembly process development and manufacturing, the following tests are usually performed:

- Wetting balance test
- Solder-spreading test
- Solderability test
- Cleanability test
- Printability test
- Temperature-profiling test
- Surface insulation resistance (SIR) test
- Automated optical inspection (AOI) test
- X-ray inspection test
- In-circuit test (ICT)
- Functional test

These tests are useful for ensuring and enhancing the quality of the lead-free solder interconnects. The quality will affect the lead-free solder interconnects, but cannot estimate the reliability of the interconnects (only a reliability test can.) Thus, it should be pointed out that the high quality of a solder interconnect doesn't mean this interconnect is reliable for any products under any intended function, for any specified period, under any given operating condition. For example, a high-quality lead-free PBGA is used for a mobile phone and a telecommunication router. Let's assume both products have the same intended function and operating condition, but different life times (e.g., mobile phone is 3 years and telecommunication router is 20 years). The probability (reliability) of the mobile phone with this PBGA for surviving three years is very high (in layman's terms: it is very reliable). On the other hand, the probability of the telecommunication router with the same PBGA surviving for 20 years is very low.

2.8 References

1. Lau, J., W. Dauksher, and P. Vianco, "Acceleration Models, Constitutive Equations and Reliability of Lead-Free Solders and Joints," *IEEE Electronic Components and Technology Conference Proceedings*, New Orleans, Louisiana, May 2003, pp. 229–236.
2. Viance, P., J. Rejent, and A. Kilgo, "Creep Behavior of the Ternary 95.5Sn-3.9Ag-0.6Cu Solder: Part I—As-Cast Condition," *Journal of Electronics Material*, Vol. 33, 2004, pp. 1389–1400.
3. Lau, J. H., and W. Dauksher, "Creep Constitutive Equations of Sn(3.5–3.9)wt%Ag(0.5–0.8)wt%Cu Lead-Free Solder Joints," *MicroMaterials and Nanomaterials*, Berlin, Germany, Editor: B. Michel, 2004, pp. 54–62.
4. Zhang, Q., A. Dasgupta, and P. Haswell, "Viscoplastic Constitutive Properties and Energy-Partitioning Model of Lead-Free Sn3.9Ag0.6Cu Solder Alloy," *IEEE Electronic Components and Technology Conference Proceedings*, New Orleans, Louisiana, June 2003, pp. 1862–1868.
5. Pang, J. H. L., B. S. Xiong, C. C. Neo, X. R. Zhang, and T. H. Low, "Bulk Solder and Solder Joint Properties for Lead Free 95.5Sn-3.8Ag-0.7Cu Solder Alloy," *IEEE Electronic Components and Technology Conference Proceedings*, New Orleans, Louisiana, May 2003, pp. 673–679.
6. Pao, Y. H., R. Govila, and E. Jih, "An Experimental and Modeling Study of Thermal Cyclic Behavior of Sn-Cu and Sn-Pb Joints," *Proceedings of Materials Science Society Symposium*, Vol. 323, 1994, pp. 153–158.
7. Schubert, A., R. Dudek, H. Walter, E. Jung, A. Gollhardt, B. Michel, et al., "Reliability Assessment of Flip-Chip Assemblies with Lead-Free Solder Alloys," *IEEE Electronic Components and Technology Conference Proceedings*, San Diego, CA, May 2002, pp. 1246–1255.
8. Schubert, A., R. Dudek, E. Auerswald, A. Gollhardt, B. Michel, and H. Reichl, "Fatigue Life Models for SnAgCu and SnPb Solder Joints Evaluated by Experiments and Simulation," *IEEE Electronic Components and Technology Conference Proceedings*, New Orleans, Louisiana, May 2003, pp. 603–610.
9. Wiese, S., A. Schubert, H. Walter, R. Dudek, F. Feustal, E. Meuesel, and B. Michel, "Constitutive Behavior of Lead-Free Solders vs. Lead-Containing Solders—Experiments on Bulk Specimens and Flip-Chip Joints," *IEEE Electronic Components and Technology Conference Proceedings*, Orlando, FL, pp. 890–902.
10. Wiese, S., S. Rzepka, and E. Meusel, "Time Dependent Plastic Behavior of Solders and Its Effect on FEM Simulations for Electronic Packages," *Proceedings of the 8th International Symposium on Advanced Packaging Materials*, 2002, pp. 101–111.
11. Song, H. J., J. W. Morris, and F. Hua, "The Creep Properties of Lead-Free Solder Joints," *Journal of Materials*, June 2002, pp. 30–32.
12. Hua, F., "Pb-Free Solder Challenges in Electronic Packaging and Assembly," *IEEE Electronic Components and Technology Conference Proceedings*, New Orleans, Louisiana, May 2003, pp. 58–63.
13. Kanchanomai, C., Y. Miyashita, and Y. Mutoh, "Low Cycle Fatigue Behaviors of Sn-Ag, Sn-Ag-Cu and Sn-Ag-Cu-Bi Lead-Free Solders," *Journal of Electronic Materials*, Vol. 31, No. 5, pp. 456–465.
14. Siewert, T., S. Liu, D. Smith, and J. Madeni, 2002, "Properties of Lead-Free Solders, Release 4.0," NIST and Colorado School of Mines.
15. Lau, J. H., *Low Cost Flip Chip Technologies for DCA, WLCSP, and PBGA Assemblies*, McGraw-Hill, New York, 2000.
16. Touloukian, Y. S., and C. Y. Ho, *Thermophysical Properties of Matter*, The TPRC Data Series, Plenum, New York, 1970.
17. Lau, J. H., and Y. Pao, *Solder Joint Reliability of BGA, CSP, Flip Chip, and Fine Pitch SMT Assemblies*, McGraw-Hill, New York, 1997.
18. Lau, J. H., *Design for Reliability, Reliability Testing and Data Analysis, and Failure Analysis of Solder Joints*, NEPCON West Workshops, Anaheim, CA, February 1990.

19. Lau, J., N. Hoo, R. Horsley, J. Smetana, D. Shangguan, W. Dauksher, et al., "Reliability Testing and Data Analysis of Lead-Free Solder Joints for High-Density Packages," *Journal of Soldering & Surface Mount Technology*, Vol. 16, No. 2, 2004, pp. 46–68.
20. Lau, J. H., "Reliability of Lead-Free Solder Joints," *ASME Transactions, Journal of Electronic Packaging*, Vol. 128, September 2006, pp. 297–301.
21. Lau, J., and W. Dauksher, "Reliability of an 1657CCGA (Ceramic Column Grid Array) Package with 96.5Sn3.9Ag0.6Cu Lead-Free Solder Pastie on PCBs (Printed Circuit Boards), *ASME Transactions, Journal of Electronic Packaging*, Vol. 127, June 2005, pp. 96–105.
22. Weibull, W., "A Statistical Distribution Function of Wide Applicability," *ASME Transactions, Journal of Applied Mechanics*, Vol. 18, No. 3, pp. 293–297, 1951.
23. Johnson, L. G., *The Statistical Treatment of Fatigue Experiments*, Elsevier Publishing Company, Amsterdam, 1964.
24. Johnson, L. G., *Theory and Technique of Variation Research*, Elsevier Publishing Company, Amsterdam, 1964.
25. Johnson, L. G., "The Median Tanks, of Sample Values in Their Population with an Application to Certain Fatigue Studies," *Industry Mathematics*, Vol. 2, 1951, pp. 1–9.
26. Mood, A. M., F. Graybill, and D. Boss, *Introduction to the Theory of Statistics*, 3rd edition, McGraw-Hill, New York, 1974.
27. Tobias, P. A., and D. Trindade, *Applied Reliability*, Van Nostrand Reinhold, New York, 1986.
28. Grant, E. L., and R. Leavenworth, *Statistical Quality Control*, McGraw-Hill, New York, 1980.
29. Lipson, C., and N. Sheth, *Statistical Design and Analysis of Engineering Experiments*, McGraw-Hill, New York, 1973.
30. Ang, A. H., and W. H., Tang, *Probability Concepts in Engineering Planning and Design*, John Wiley & Sons, New York, 1981.
31. Norris, K. C., and A. Landzberg, "Reliability of Controlled Collapse Interconnections," *IBM Journal of Research and Development*, Vol. 13, May 1969, pp. 534–544.
32. Puttlitz, K., "Flip Chip Solder Bump (FCSB) Technology: An Example," in *Multichip Module Technologies and Alternatives*, edited by D. Doane and P. Franzon, Van Nostrand Reinhold, New York, 1993, pp. 450–476.
33. Solomon, H., "The Solder Joint Fatigue Life Acceleration Factor," *ASME Transactions, Journal of Electronic Packaging*, Vol. 113, June 1991, pp. 186–190.
34. Lau, J. H., and J. Chu, "3D X-ray Tests and Analyses of Lead-Free PBGA (Plastic Ball Grid Array) Package Solder Joints," *Proceedings of SMT International Conference*, Chaicago, IL, October 2003, pp. J1–13.
35. Lau, J. H., *Thermal Stress and Strain in Microelectronics Packaging*, Van Nostrand Reinhold, New York, 1992.
36. ANSYS, *User's Manual*, Release 6.1, 2002.
37. Lau, J. H, D. Shangguan, D. Lau, T. Kung, and R. Lee, "Thermal-Fatigue Life Prediction Equation for Wafer-Level Chip Scale Package (WLCSP) Lead-Free Solder Joints on Lead-Free Printed Circuit Board (PCB)," *IEEE Proceedings of Electronics and Component Technology Conference*, Las Vegas, Nevada, 2004, pp. 1563–1569.
38. Lau, J., R. Lee, and D. Shangguan, "Thermal-Fatigue Life Prediction Equations for Real Lead-Free Solder Joints on Real Lead-Free Printed Circuit Board (PCB). *ASME Paper No. IMECE2004 – 62492*, November 2004.
39. Lau, J. H., F. Song, D. Lau, R. Lee, D. Shangguan, and W. Dauksher, "Thermal-Fatigue Life Prediction Equation for Plastic Ball Grid Array (PBGA) SnAgCu Lead-Free Solder Joints," *ASME Paper No. IPACK2005-73362*, July 2005.
40. Lau, J. H., R. Lee, F. Song, D. Lau, and D. Shangguan, "Isothermal Fatigue Tests of Plastic Ball Grid Array (PBGA) SnAgCu Lead-Free Solder Joints at 60°C," *IEEE Electronic Components and Technology Conference Proceedings*, New Orleans, Louisiana, June 2006, pp. 229–236.
41. Lau, J. H., R. Lee, and D. Shangguan, "Thermal Fatigue-Life Prediction of Lead-Free Solder Joints," *ASME Paper No. IMECE2004-62493*, November 2004.

CHAPTER 3

Notes on Failure Criteria

3.1 Introduction

As mentioned in Chap. 2, the one and only way to determine the reliability, failure rate, mean life, etc., of lead-free interconnects is by reliability tests and statistical analyses. The most important factors in reliability tests are the failure criteria and data extractions. In this chapter, these two key elements in reliability will be briefly presented and discussed.

A variety of environmental stress factors, for example, temperature; voltage; humidity; corrosion; current density (electromigration); and mechanical bending, shearing, twisting, shocking (drop), and vibrating loads, may lead to structural failures [1–128]. The most common failure modes in practice are *overload* and *fatigue.*

Overload failure occurs whenever the stress in a solder interconnect, brought about by the imposed stress factors, is greater than the capacity of the solder material. An example would be excessive bending and twisting or dropping of a printed circuit board (PCB) with solder-bumped flip chip, chip scale package (CSP), ball grid array (BGA), or 3-D IC integration packages soldered to its surface. In this case, the bending and twisting stresses are larger than the strength or fracture toughness of the lead-free solder materials.

Fatigue failure takes place via the initiation and propagation of a crack in a solder interconnect until it becomes unstable. The stress factors that typically cause failure by fatigue are far below the overload failure levels. Examples of fatigue failure include the temperature cycling and/or vibration of a solder-bumped flip chip, CSP, BGA, or 3-D IC integration package that is soldered to the surface of a PCB. In this book, the focuses are on the temperature, current density, shock (drop), and vibration fatigue stress factors.

3.2 Failure Criteria

During reliability tests, we continuously perform the resistance measurements. The failure criterion is defined as the resistance increases to certain (e.g., 1 to ∞) percentages of the original resistance. Usually, daisy chains, which connect the solder interconnects of the chip/package and substrate/PCB, allow the measurements. Since most solder interconnects under fatigue loadings will go through crack initiation, crack propagation, and crack rupture sooner or later, the daisy chains' resistance will increase accordingly, that is, from small to large, and become ∞ when the solder interconnect is totally cracked (opened). (Most people unintentionally pick the last one.) Since the exact "resistance vs. crack length" relations of various lead-free interconnects don't exist today, people just randomly pick a number, which is from 1 to ∞ percent of the initial resistance. That's one of the reasons why there are so many different Weibull plots in the lectures, even with the same package, PCB, solder paste, sample size, test condition, and test period.

Figures 3.1 through 3.4 show the Weibull plots of the 196-pin plastic ball grid array (PBGA), 8-pin wafer level chip scale package (WLCSP), 28-pin BBC, and 208-pin plastic quad flat pack (PQFP) with an SACC (SnAgCuCe) solder paste (which will be discussed

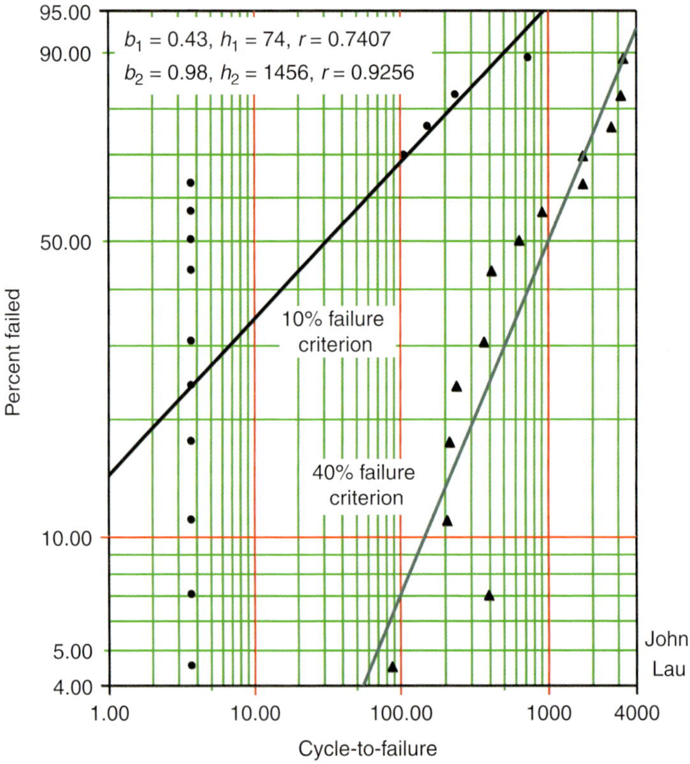

FIGURE 3.1 Thermal cycling test (−40°C–125°C) of PBGA196 with 10% and 40% failure criteria.

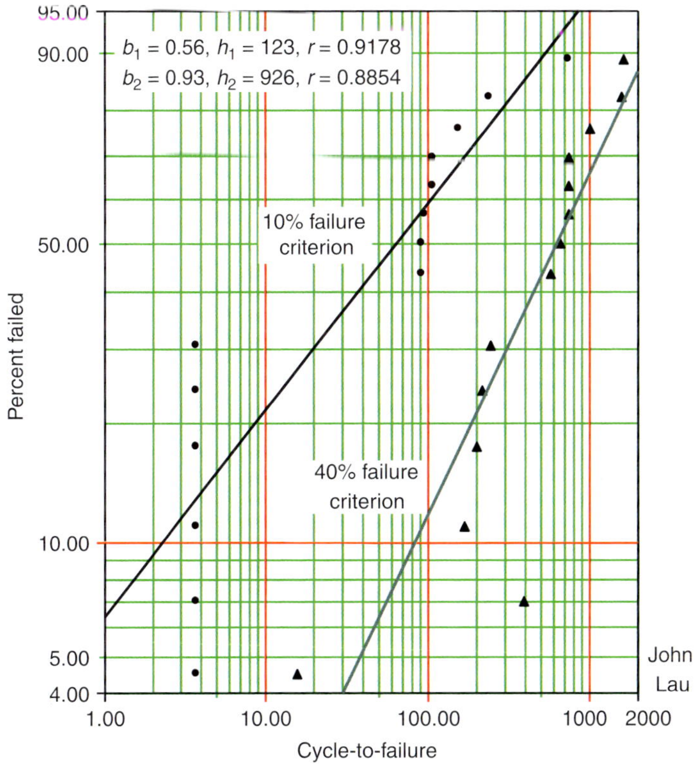

FIGURE 3.2 Thermal cycling test (−40°C–125°C) of WLCSP with 10% and 40% failure criteria.

in Chap. 9 of this book) on the same PCB and subjected to the same thermal cycling condition. (A computer stored all the resistance measurements.) The only difference is the way to define failure: one is based on 10% resistance increase, and the other is based on 40% resistance increase. It can be seen from these figures that: (1) the life distribution is many times different from the failure criterion based on the 10% resistance increase and the 40% resistance increase; (2) the effect of failure criterion on various packaging is different, for example, the characteristic life of the 208-pin PQFP based on the 40% failure criterion is more than 10 times of that based on the 10% failure criterion, but the characteristic life of the 28-pin BCC based on the 40% failure criterion is more than 100 times of that based on the 10% failure criterion; and (3) as expected, there are more failures with the failure criterion based on lower resistance increases.

What can we learn from these examples? Next time we read a Weibull plot, the first thing is to find out the failure criterion because

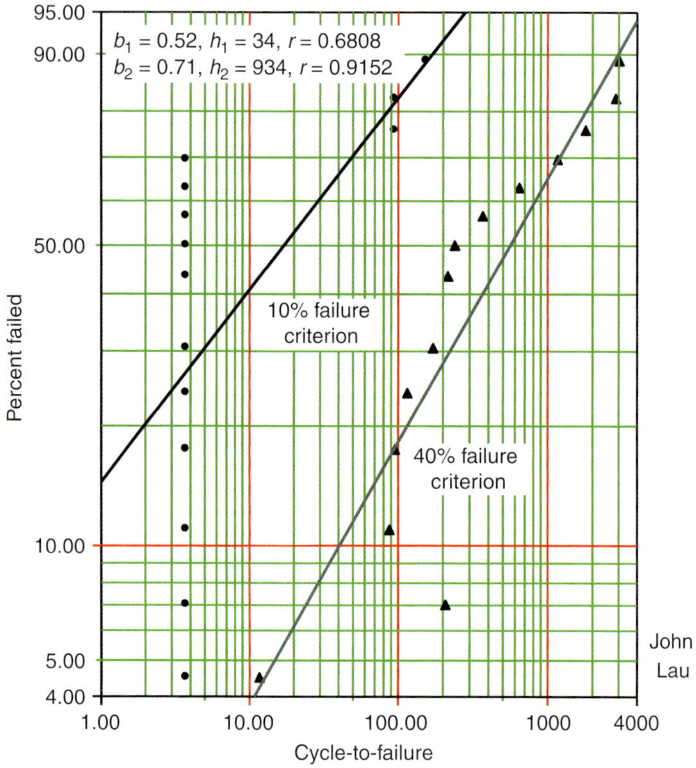

FIGURE 3.3 Thermal cycling test (−40°C–125°C) of BCC with 10% and 40% failure criteria.

it affects the life distribution, reliability, failure rate, etc. Then we should find out the sample size, test condition, etc., because they also affect the reliability. Many people like to compare and combine a few Weibull plots from the literature without even knowing their failure criterion, test condition, sample size, etc., and then draw conclusions, for example, regarding the acceleration factors. These are dangerous and unintentionally mislead others. Thus, there should be an industry standard that defines a few failure criteria for different products and applications. Before that, we should save all the measurement data and perform statistical analyses with different resistance changes, then compare their life distribution, reliability, failure rate, mean life, etc.

In general practices, after people define their failure criterion, they call the failure criterion "threshold resistance." In this case, it represents the pass or fail criteria. Either the resistance is below the threshold resistance (pass) or the resistance is above the threshold resistance (fail).

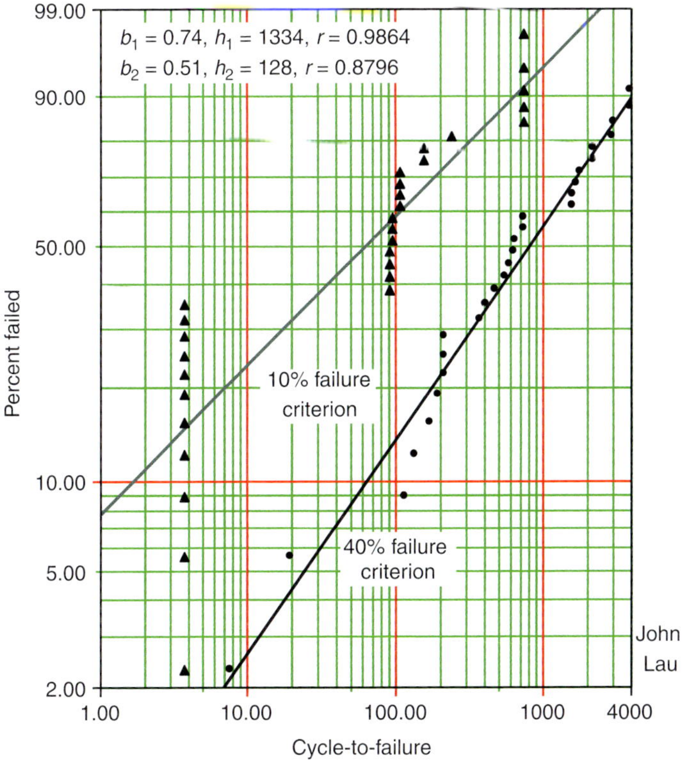

FIGURE 3.4 Thermal cycling test (–40°C–125°C) of PQFP208 with 10% and 40% failure criteria.

3.3 Resistance Measurements

Figure 3.5 shows a typical thermal cycling profile, for example, –40°C to 125°C and 15 minutes for ramp-up, ramp-down, and dwell at cold and hot (one-hour cycle). During tests, we may take four measurements (points 2, 4, 6, and 8) or all eight-point measurements at each cycle.

In order to avoid false failures, it is recommended to compare the measurements (no matter if it is four-point or eight-point) of every channel (daisy chain) at every two consequence cycles. If all the measurements of a particular channel show they are above the threshold in two consequence cycles, this channel is considered failed. If all the measurements of a channel show they are above the threshold in the n cycle and below the threshold in the $n + 1$ cycle, then we don't consider this channel failed. We will compare the measurements of this channel at $n + 1$ cycle with that at $n + 2$ cycle until we find another two consequence measurements above the threshold. It should be

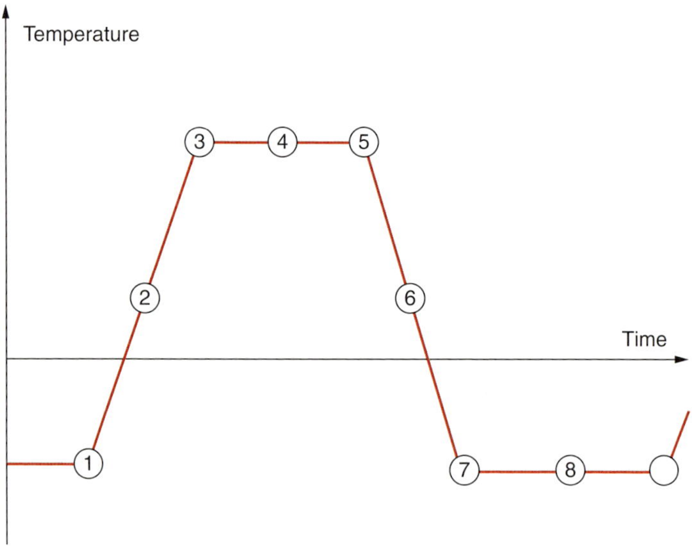

FIGURE 3.5 Thermal cycling profile and points of measurements.

pointed out that slightly better reliability would be estimated with the eight-point measurements, however it is not recommended (it does not worth the efforts).

3.4 Three-Cycle Moving Average (TCMA) Method

Another method to avoid false failures is the TCMA [16]. This method is especially suitable for judging the failures caused by the RF (radio frequency) interference, which induces a temporary increase in circuit resistance in the Anatech event detector.

The TCMA is useful for displaying the frequency at which events are being recorded and how the frequency varies with time. Moving averages are often used to assess how values change over time, such as share prices, batting average, etc. The method works by breaking a long period into a large number of smaller periods. For example, over the course of a year it may rain on 22% of the days. This would mean, on average, that it rained on 80.3 days. It would not identify when the rainy season occurs. If the year were to be analyzed in smaller portions, such as weeks or months, some portions would have more rainy than others. By analyzing when the highest proportion of rainy days per week or month occurred, the rainy season could be identified.

In this exercise a period of three thermal cycles is chosen over which to extract the data. Over the period of three thermal cycles, 12 (three at points 2, 4, 6, and 8, and from now on please forget points 1, 3, 5, and 7) evenly spaced resistance measurements are taken for

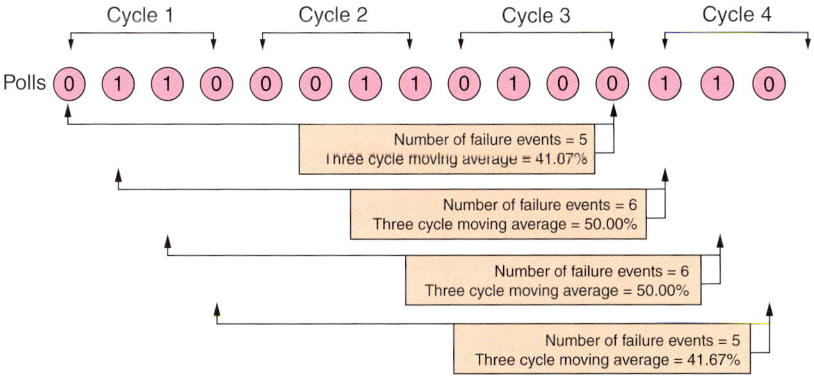

FIGURE 3.6 Schematic of the construction of the three cycle moving average (TCMA).

each test circuit (channel or daisy chain). These measurements are taken using the Anatech event detectors. If the measurement is above threshold resistance, the test is deemed to be a potential failure. The TCMA is calculated by tallying the number of potential failed test results from the first three thermal cycles and dividing by the total number of measurements taken in the three cycles. For each additional measurement, the new measurement is counted and the earliest (of the previous three cycles) discarded. For example, if the first 12 results are tallied, that is the first TCMA value. When the thirteenth result is taken, it is added to the test group but the first measurement is left out. When the fourteenth result is recorded, the second is discounted, and so on. This process is shown in Fig. 3.6.

A computer program can be written to identify the time it takes for each circuit to reach a genuine circuit resistance of threshold resistance. The program is designed to distinguish between genuine failures and RF interference events. This is done by analyzing the original Anatech data file and identifying the time of each recorded measurement above threshold resistance. The first stage of data analysis is to calculate the TCMA for each time when a measurement is taken and represent this frequency data as a percentage. The program then calculates the average (mean) frequency up to that point and the standard deviation (which is a measure of the variation of frequency values). Therefore, for each measurement that has been taken, the current TCMA value has been calculated, along with the average TCMA value for all preceding measurements and the standard deviation to that time.

A comparison has to be made to evaluate whether the circuit resistance is actually above the threshold resistance during the thermal cycling. As has been mentioned, the RF behavior is generally consistent over time and does not tend to change dramatically. When genuine failures occur, the frequency of failure reporting increases by a significant margin and remains at this elevated level. For each calculated TCMA value, the program compares it to the mean average and standard deviation of all the preceding values. The program is basically looking for dramatic changes of frequency behavior. When it finds one, it triggers a higher level of investigation. The change necessary to trigger this next stage is a value that is larger than the mean average TCMA + 2σ (standard deviation) for all the preceding TCMA values.

If an increase in the frequency is found that is larger than the mean average TCMA + 2σ, the program looks to see if this behavior is sustained. The program calculates the average values for the next 20 thermal cycles, that is, 80 TCMA values for four polls/cycle or 160 for eight polls/cycle. If the average TCMA value over the next 20 cycles is below the mean average TCMA + 2σ of all the TCMAs calculated prior to the trigger value, then the time is not recognized as the genuine failure point and the program continues its search. If the average TCMA for the next 20 cycles is larger than the mean average TCMA + 2σ, then the point is recognized as the failure point.

The event detectors record every instance of a channel circuit exceeding the threshold resistance value. This takes into account both genuine and RF-generated events. Neither the detectors nor the Anatech software can distinguish between a genuine event and an RF-generated effect.

In an ideal world, no RF-generated events would be detected. Therefore, prior to fatigue cracks increasing the resistance of a circuit, no measurement will exceed threshold resistance. As fatigue crack growth is the only active mechanism by which the resistance of the channel circuits can increase, the time at which the event detector records the first measurement above threshold resistance can be defined as the "failure" point of the circuit. This type of behavior is illustrated in Fig. 3.7.

On some circuits, RF-generated events are recorded. In order to make a judgment on the "failure" time of a circuit, it is necessary to filter these false events from the data. RF interference is random in severity and occurrence. In affected circuits, RF interference creates a "background" level of false events that, if severe enough, can mask the onset of genuine failures. Figure 3.8 shows the background level of noise generated by RF interference in a circuit whose resistance has not exceeded threshold resistance.

The best way to identify and filter the RF-generated events is through frequency analysis. This shows the proportion of "failure" events as a percentage of the number of measurements taken over a short period. If a change in the electrical resistance of the circuit

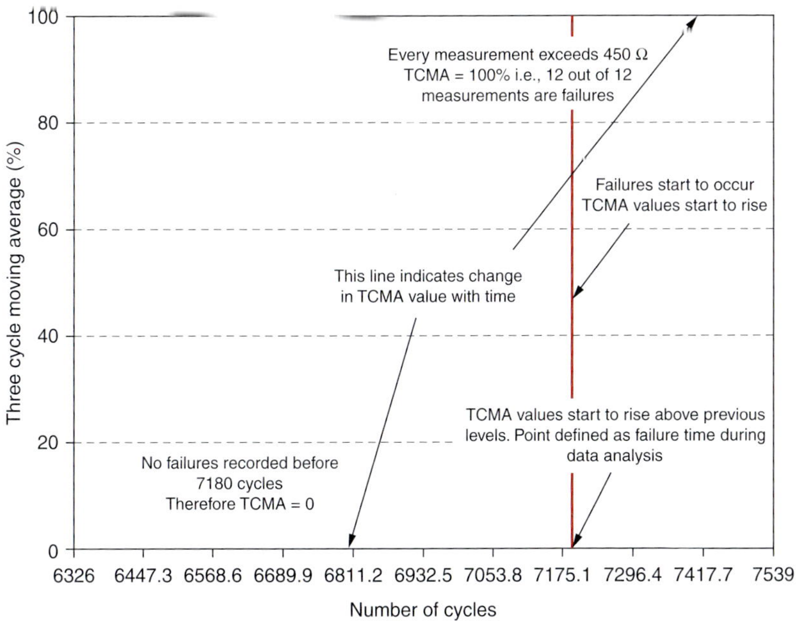

FIGURE 3.7 Ideal fatigue behavior. Prior to 7189 cycles, no measurements above the threshold resistance (450 Ω) were recorded. After this point, all measurements were above 450 Ω. The failure point can be easily identified.

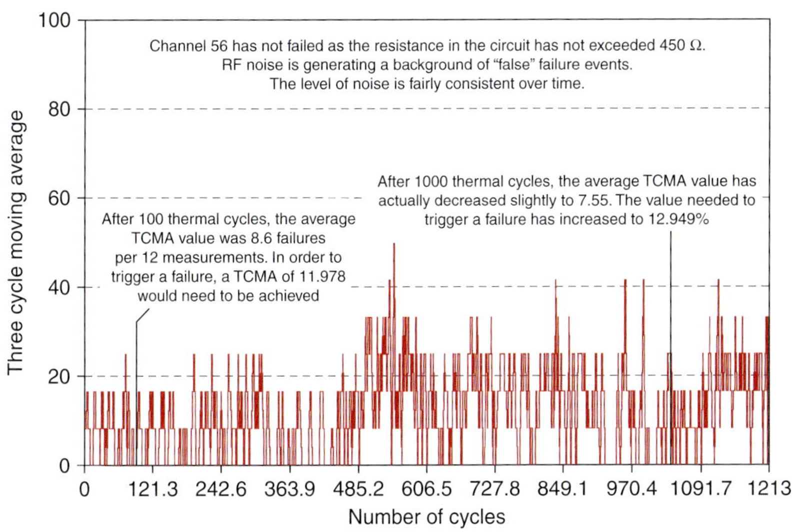

FIGURE 3.8 This channel is experiencing the effects of RF noise. This temporarily increases the effective resistance of the test circuit so that the event detector believes the threshold resistance has been exceeded. In fact, the events are false and the circuit has not failed. Note x-axis origin should be at 6326 cycles not 0.

occurs, it would be expected that there would be a corresponding change in the percentage of measurements that record values over threshold resistance. The response must be measured over a fairly short period, as an increase or decrease averaged over a long period would be difficult to resolve. Identification of changes in the frequency data is optimum when the percentage of failed events over three thermal cycles (during each cycle 4 measurements) are plotted versus time for a given channel. Using this method, the RF signal varies randomly, but over time, tends to average out at a specific level. The level of RF noise can be determined, as it is known that no fatigue cracks are present in the solder joints early on in testing and that the circuit resistances are all below threshold resistance. The level of RF noise can increase and decrease with time, but when it does so, the change usually occurs gradually over a large number of cycles. In some cases, RF spikes are detected where the frequency of false events increases dramatically but then returns to its initial background level after a short period.

When genuine failures occur, the frequency of failed events increases. The rise is usually large, sudden, and, to a large extent "irreversible." In many cases, the frequency of failed events rises to 100%, that is, every measurement taken by the event detectors yields a value in excess of the threshold resistance. In some cases, the frequency of failed measurements does not reach 100%. In all cases, though, an increase in the frequency is observed. This means that the average frequency of resistance measurements above threshold resistance is higher after the genuine failure point is reached than before. This difference in frequency behavior can be observed when a graph of the TCMA is plotted, as shown in Fig. 3.9.

Figures 3.10 and 3.11 show variations on the behavior of different circuits that eventually do fail. The circuit shown in Fig. 3.10 is subject to RF interference. A relatively low level of failure events is generated during thermal cycling before failure. Once the inherent circuit resistance does rise above the threshold resistance, all measurements exceed the threshold resistance value. Similar failure behavior is seen in Fig. 3.11. Before failure, channel 552 displayed a far higher frequency of background events. Even so, the failure point is still readily identifiable.

Figure 3.12 shows the TCMA plot for a 1657 CCGA (which will be discussed in Chap. 4) that was soldered using SnPb solder paste. It can be seen that based on the present failure criterion and data extraction method, this is not a failure! As a matter of fact, none of the 1657 CCGA packages soldered with SnPb paste suffered failure, and only three of nine registered any recorded resistance failures in the 1212 thermal cycles of the experiment.

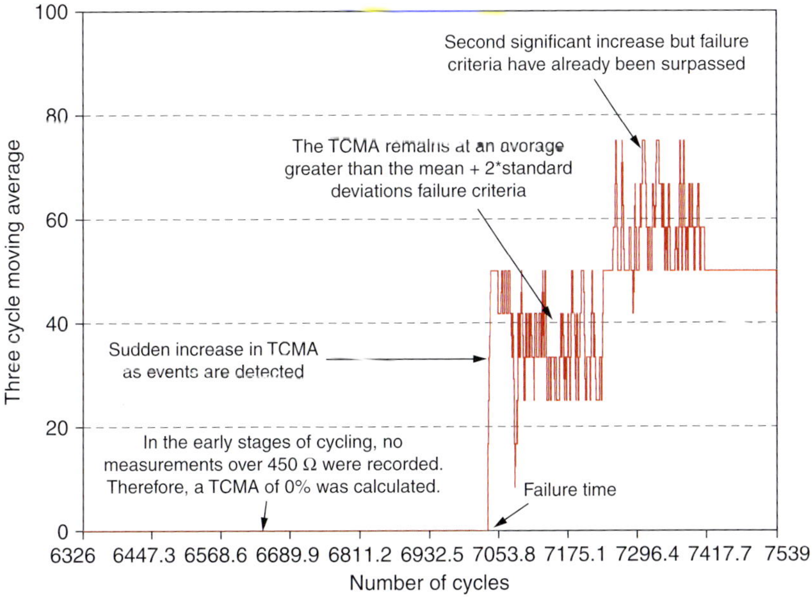

FIGURE 3.9 There is no RF noise prior to failure but when failure does occur, the TCMA does not increase to 100%. The significant and prolonged change in behavior (for more than 20 thermal cycles) achieves the failure criterion. A second prolonged rise in TCMA is also recorded but failure can only be recorded once.

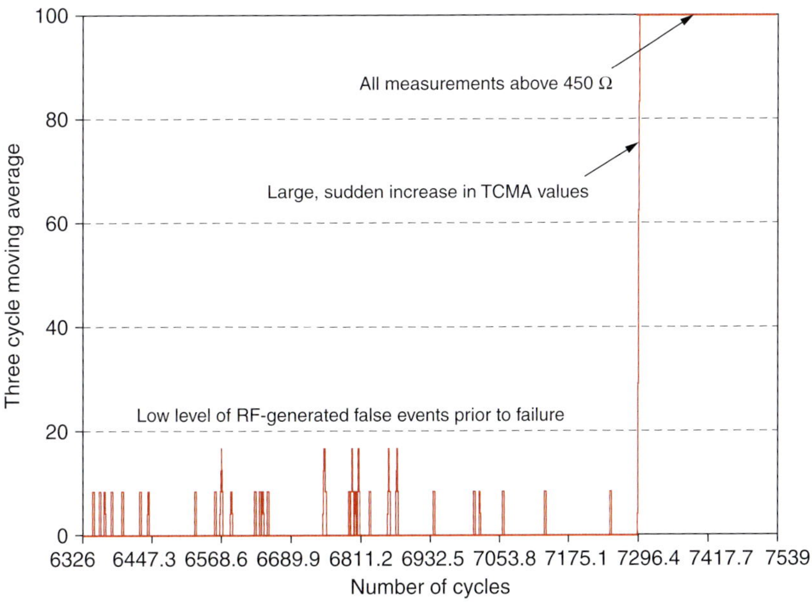

FIGURE 3.10 Low-level background of false events prior to a rapid deterioration in circuit integrity.

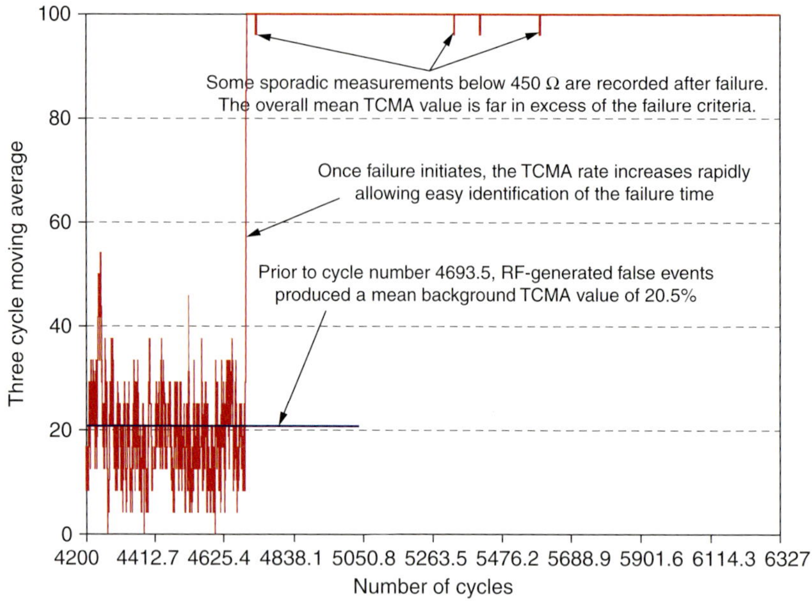

FIGURE 3.11 A substantial, but consistent level of RF interference is followed by a dramatic rise in TCMA values when genuine failures start to occur.

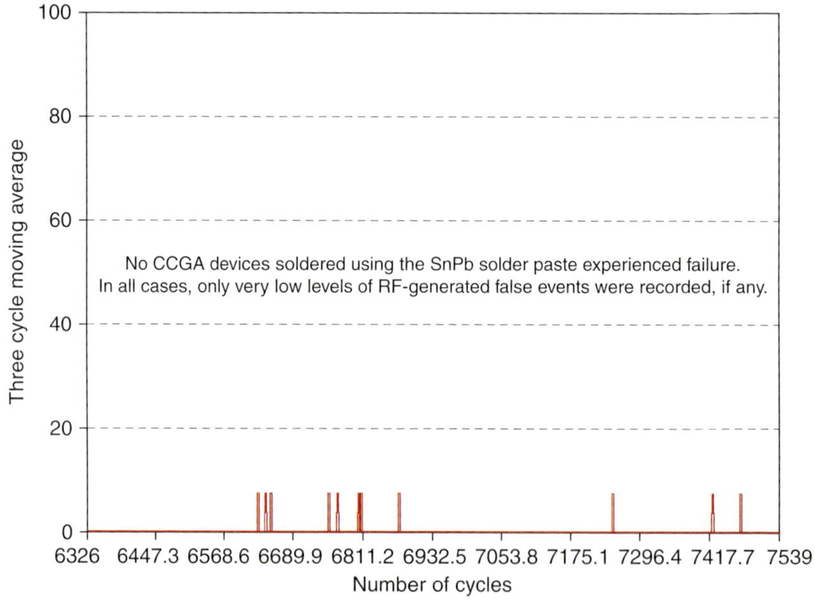

FIGURE 3.12 1657CCGA with SnPb solder paste on PCB.

3.5 Summary and Recommendations

Some important results and recommendations are summarized in the following:

1. Failure criterion is one of the most important factors in reliability testing and it strongly affects the reliability results.
2. The effect of failure criterion on the reliability results depends on the types of package.
3. The industry should establish standards that define a few failure criteria for different products and applications.
4. Next time when you read a Weibull plot, the first thing is to find out the failure criterion, otherwise you don't even know what you are look at.
5. Besides the failure criterion, the sample size, number of failures, test condition, data acquisition system, and data extraction method need to be known, because they all affect the reliability results.

3.6 References

1. Lau, J. H., "State-of-the-art and Trends in 3D Integration," *Chip Scale Review*, March/April, 2010, pp. 22–28.
2. Lau, J. H., Lee, R., Yuen, M., and Chan, P., "3D LED and IC Wafer Level Packaging," *Journal of Microelectronics International*, Vol. 27, No. 2, 2010, pp. 98–105.
3. Lau, J. H., "Critical Issues of 3D IC Integrations," *Proceedings of IMAPS International Symposium on Microelectronics*, San Jose, CA, November 2009, pp. 585–592. Also, to be published in the *IMAPS Transactions, Journal of Microelectronics and Electronic Packaging*.
4. Lau, J. H., "Design and Process of 3D MEMS Packaging," *Proceedings of IMAPS International Symposium on Microelectronics*, San Jose, CA, November 2009, pp. 1–9. Also, to be published in the *IMAPS Transactions, Journal of Microelectronics and Electronic Packaging*.
5. Lau, J. H., et al., "Optical, Electrical, Thermal, and Mechanical Design and Analysis of 3D Stacked Optoelectronics on Optical Printed Circuit Boards," accepted for publication in the *IEEE Transactions on Advanced Packaging*.
6. Lau, J. H., et al., "Characterization of Bump Chip Carrier (BCC++) Package Assembly and Solder Joint Reliability for Fiber-Optic Transceiver Applications," Accepted for publication in the *IEEE Transaction on Components, Packaging, and Technology*.
7. Lau, J. H., et al., "Effects of Underfills on the Thermal-Cycling Tests of SnAgCu PBGA (Plastic Ball Grid Array) Packages on ImAg PCB (Printed Circuit Board)," accepted for publication in the *IEEE Transactions on Advanced Packaging*.
8. Lau, J. H., "Thermal Stress Analysis of a Flip-Chip Parallel VCSEL (Vertical-Cavity Surface-Emitting Laser) Package with Low-Temperature Lead-Free (48Sn-52In) Solder Joints," accepted for publication in the *IEEE Transactions on Advanced Packaging*.
9. Lau, J. H., "Material Properties and Intermetallic Compounds of a New Lead-Free Solder: Sn3wt%Ag0.5wt%Cu0.019wt%Ce (SACC)," accepted for publication in the *IEEE Transactions on Advanced Packaging*.

10. Lau, J. H., et al., "Design, Materials, and Assembly Process of High-Density Packages with a Low-Temperature Lead-Free Solder (SnBiAg)," *Soldering & Surface Mount Technology*, Vol. 20, No. 2, 2008, pp. 11–20.
11. Lau, J. H., et al., "Reliability Test and Failure Analysis of High-Density Packages Assembled with a Low-Temperature Lead-Free Solder (SnBiAg)," *Soldering & Surface Mount Technology*, Vol. 20, No. 2, 2008, pp. 21–29.
12. Lau, J. H., Todd Castello, Dongkai Shangguan, Walter Dauksher, Joe Smetana, Rob Horsley, Dave Love, Irv Menis, and Bob Sullivan, "Failure Analysis of Lead-Free Solder Joints of an 1657CCGA (Ceramic Column Grid Array) Package," *Journal of Microelectronics and Electronics Packaging*, Vol. 4, No. 3, Third Quarter, 2007, pp. 189–213.
13. Lau, J. H., "Reliability of Lead-Free Solder Joints," *ASME Transactions, Journal of Electronic Packaging*, Vol. 128, September 2006, pp. 297–301.
14. Lau, J., and W. Dauksher, "Reliability of an 1657CCGA (Ceramic Column Grid Array) Package with 96.5Sn3.9Ag0.6Cu Lead-Free Solder Paste on PCBs (Printed Circuit Boards)," *ASME Transactions, Journal of Electronic Packaging*, Vol. 127, June 2005, pp. 96–105.
15. Lau, J., W. Dauksher, J. Smetana, R. Horsley, D. Shangguan, T. Castello, I. Menis, D. Love, and B. Sullivan, "Design for Lead-Free Solder Joint Reliability of High-Density Packages," *Journal of Soldering & Surface Mount Technology*, Vol. 16, No. 1, 2004, pp. 12–26.
16. Lau, J., N. Hoo, R. Horsley, J. Smetana, D. Shangguan, W. Dauksher, D. Love, I. Menis, and B. Sullivan, "Reliability Testing and Data Analysis of Lead-Free Solder Joints for High-Density Packages," *Journal of Soldering & Surface Mount Technology*, Vol. 16, No. 2, 2004, pp. 46–68.
17. Lau, J., J. Smetana, R. Horsley, K. Snowdon, D. Shangguan, J. Gleason, I. Memis, D. Love, W. Dauksher, and B. Sullivan, "Design, Materials, and Process for Lead-Free Assembly of High-Density Packages," *Journal of Soldering & Surface Mount Technology*, Vol. 16, No. 1, 2004, pp. 53–62.
18. Lau, J., D. Shangguan, T. Castello, R. Horsley, J. Smetana, W. Dauksher, D. Love, I. Menis, and B. Sullivan, "Failure Analysis of Lead-Free Solder Joints for High-Density Packages," *Journal of Soldering & Surface Mount Technology*, Vol. 16, No. 2, 2004, pp. 69–76.
19. Lau, J. H., et al., "3D Nonlinear Stress Analysis of Tin Whisker Initiation on Lead-Free Components," *ASME Transactions, Journal of Electronic Packaging*, Vol. 125, December 2003, pp. 621–624.
20. Lau, J. H., et al., "Thermal Stress and Deflection Analysis of a Glass Window (Circular Plate) Elastically Restrained Along Its Edge in a Photonic Device," *ASME Transactions, Journal of Electronic Packaging*, Vol. 125, December 2003, pp. 603–608.
21. Lau, J. H., et al., "Creep Analysis and Thermal-Fatigue Life Prediction of the Lead-free Solder Sealing Ring of a Photonic Switch," *ASME Transactions, Journal of Electronic Packaging*, Vol. 124, December 2002, pp. 403–410.
22. Lau, J. H., and C. Chang, "Characteristics and Reliability of Fast-Flow, Snap-Cure, and Reworkable Underfills for Solder Bumped Flip Chip on Low-Cost Substrates," *IEEE Transactions on Electronics Packaging Manufacturing*, Vol. 25, No. 3, July 2002, pp. 231–230.
23. Lau, J. H., et al., "Corrosion Analysis of a Fiber-Optic Transceivers Housing," *ASME Transactions, Journal of Electronic Packaging*, Vol. 124, December 2002, pp. 379–384.
24. Lau, J. H., "Critical Issues of Wafer Level Chip Scale Package (WLCSP) with Emphasis on Cost Analysis and Solder Joint Reliability," *IEEE Transactions on Electronics Packaging Manufacturing*, Vol. 25, No. 1, 2002, pp. 42–50.
25. Lau, J. H., and R. Lee, "Effects of Build-Up Printed Circuit Board Thickness on the Solder Joint Reliability of a Wafer Level Chip Scale Package (WLCSP)," *IEEE Transactions on Components & Packaging Technologies*, Vol. 25, No. 1, March 2002, pp. 3–14.
26. Lau, J. H., S. Pan, and C. Chang, "A New Thermal-Fatigue Life Prediction Model for Wafer Level Chip Scale Package (WLCSP) Solder Joints," *ASME*

Transactions, Journal of Electronic Packaging, Vol. 124, September 2002, pp. 212–220.
27. Lau, J. H., and R. Lee, "Modeling and Analysis of 96.5Sn-3.5Ag Lead-Free Solder Joints of Wafer Level Chip Scale Package (WLCSP) on Build-Up Microvia Printed Circuit Board," *IEEE Transactions on Electronics Packaging Manufacturing,* Vol. 25, No. 1, 2002, pp. 51–58.
28. Lau, J. H., R. Lee, S. Pan, and C. Chang, "Nonlinear Time Dependent Analysis of Micro Via-in-Pad Substrates for Solder Bumped Flip Chip Applications," *ASME Transactions, Journal of Electronic Packaging,* Vol. 124, September 2002, pp. 205–211.
29. Lau, J. H., C. Chang, and R. Lee, "Solder Joint Crack Propagation Analysis of Wafer-Level Chip Scale Package on Printed Circuit Board Assemblies," *IEEE Transactions on Components & Packaging Technologies,* Vol. 24, No. 2, 2001, pp. 285–292.
30. Lau, J. H., and R. Lee, "Computational Analysis on the Effects of Double-Layer Build-Up Printed Circuit Board on the Wafer Level Chip Scale Package (WLCSP) Assembly with Pb-Free Solder Joints," *International Journal of Microcircuits & Electronic Packaging, IMAPS Transactions,* Vol. 24, No. 2, 2001, pp. 89–104.
31. Lau, J. H., R. Lee, and C. Chang, "Solder Joint Reliability of Wafer Level Chip Scale Package (WLCSP): A Time-Temperature-Dependent Creep Analysis," *ASME Transactions, Journal of Electronic Packaging,* Vol. 122, No. 4, December 2000, pp. 311–316.
32. Lau, J. H., and R. Lee, "Fracture Mechanics Analysis of Low Cost Solder Bumped Flip Chip Assemblies with Imperfect Underfills," *ASME Transactions, Journal of Electronic Packaging,* Vol. 122, No. 4, December 2000, pp. 306–310.
33. Lau, J. H., "Cost Analysis: Solder Bumped Flip Chip Versus Wire Bonding," *IEEE Transactions on Electronics Packaging Manufacturing,* Vol. 23, March 2000, pp. 4–11.
34. Lau, J. H., C. Chang, and R. Lee, "Failure Analysis of Solder Bumped Flip Chip on Low-Cost Substrates," *IEEE Transactions on Electronics Packaging Manufacturing,* Vol. 23, No. 1, March 2000, pp. 19–27.
35. Lau, J. H., and C. Chang, "Taguchi Design of Experiment for Wafer Bumping by Stencil Printing," *IEEE Transactions on Electronics Packaging Manufacturing,* Vol. 21, No. 3, July 2000, pp. 219–225.
36. Lau, J. H., R. Lee, and C. Chang, "Effects of Underfill Material Properties on the Reliability of Solder Bumped Flip Chip on Board with Imperfect Underfill Encapsulants," *IEEE Transactions on CPMT,* June 2000, pp. 323–333.
37. Lau, J. H., Chang, C., Lee, S. W. R., Chen, T. Y., Cheng, D., Tseng, T. J. and Lin, D., 2000, "Design and Manufacturing of Micro Via-In-Pad Substrates for Solder Bumped Flip Chip Applications," *Journal of Electronics Manufacturing,* Vol. 10, No. 1, pp. 79–87.
38. Lau, J. H., and S. W. Lee, "Effects of Underfill Delamination and Chip Size on the Reliability of Solder Bumped Flip Chip on Board," *International Journal of Microcircuit & Electronic Packaging, IMAPS Transactions,* Vol. 23, No. 1, First Quarter, 2000, pp. 33–39.
39. Lau, J. H., and R. Lee, "Temperature-Dependent Popcorning Analysis of Plastic Ball Grid Array Package During Solder Reflow with Fracture Mechanics Method," *ASME Transactions, Journal of Electronic Packaging,* Vol. 122, March 2000, pp. 34–41.
40. Lau, J. H., C. Chang, R. Lee, T. Chen, D. Cheng, T. Tseng, and D. Lin, "Design and Manufacturing of Micro Via-In-Pad Substrates for Solder Bumped Flip Chip Applications," *Journal of Electronics Manufacturing,* Vol. 10, No. 1, March 2000, pp. 79–87.
41. Lau, J. H., R. Lee, and H. Chao, "Assembly of Large PBGAs on Printed Circuit Board with Large PQFPs Directly on the Opposite Side," *Journal of Electronics Manufacturing,* Vol. 9, No. 4, December 1999, pp. 293–298.
42. Lau, J. H., C. Chang, and C. Chen, "Characteristics and Reliability of No-Flow Underfills for Solder Bumped Flip Chip Assemblies," *International Journal of Microcircuit & Electronic Packaging, IMAPS Transactions,* Vol. 22, No. 4, 1999, pp. 370–381.

43. Lau, J. H., and C. Chang, "Delamination Analysis of Plastic Packages During Solder Reflow by Strain-Gage Measurement and Tomographic Acoustic Micro Imaging Methods," *Journal of Electronics Manufacturing,* Vol. 8, September & December, 1999, pp. 165–172.
44. Lau, J. H., C. Chang, M. Lee, D. Cheng, and T Tseng, "Printed Circuit Board Manufacturing and Testing of RIMM," *Journal of Electronics Manufacturing,* Vol. 9, No. 3, September, 1999, pp. 215–222.
45. Lau, J. H., and C. Chang, "How to Select Underfill Materials for Solder Bumped Flip Chips on Low Cost Substrates?" *International Journal of Microelectronics & Electronic Packaging, IMAPS Transactions,* Vol. 22, No. 1, 1999, pp. 20–28.
46. Lau, J. H., "Design, Manufacturing, and Testing of a Novel Plastic Ball Grid Array Package," *Journal of Electronics Manufacturing,* Vol. 9, No. 4, December 1999, pp. 283–291.
47. Lau, J. H., and C. Chang, "Characterization of Underfill Materials for Functional Solder Bumped Flip Chips on Board Applications," *IEEE Transactions on CPMT, Part A,* Vol. 22, No. 1, March 1999, pp. 111–119.
48. Lau, J. H., T. Chen, and R. Lee, "Effect of Heat Spreader Sizes on the Thermal Performance of Large Cavity-Down Plastic Ball Grid Array Packages," *ASME Transactions, Journal of Electronic Packaging,* Vol. 121, No. 4, 1999, pp. 242–248.
49. Lau, J. H., and T. Chen, "Low-Cost Thermal and Electrical Enhanced Plastic Ball Grid Array Package – NuBGA," *Microelectronics International,* 1999.
50. Lau, J. H, and C. Chang, "TMA, DMA, DSC, and TGA of Lead Free Solders," *Journal of Soldering & Surface Mount Technology,* Vol. 11, No. 2, 1999, pp. 17–24.
51. Lau, J. H., C. Chang, and C. Ouyang, "SMT Compatible No-Flow Underfill for Solder Bumped Flip Chip on Low-Cost Substrates," *Journal of Electronics Manufacturing,* Vol. 8, Nos. 3 & 4, September & December, 1998, pp. 151–164.
52. Lau, J. H., and T. Chou, "Electrical Design of a Cost-Effective Thermal Enhanced Plastic Ball Grid Array Package—NuBGA," *IEEE Transactions on CPMT, Part B,* Vol. 21, No. 1, February 1998, pp. 35–42.
53. Lau, J. H, "Solder Joint Reliability of Low Cost Chip Size Package—NuBGA," *Journal of Microelectronics Reliability,* Vol. 38, 1998, pp. 1519–1529.
54. Lau, J. H., "Flip Chip on PCBs with Anisotropic Conductive Film," *Advanced Packaging,* July/August 1998, pp. 44–48.
55. Lau, J. H., and T. Chen, "Cooling Assessment and Distribution of Heat Dissipation of a Cavity Down Plastic Ball Grid Array Package—NuBGA," *International Journal of Microelectronics & Electronic Packaging, IMAPS Transactions,* Vol. 21, No. 1, 1998, pp. 20–28.
56. Lau, et al., "A Low-Cost Solder-Bumped Chip Scale Package—NuCSP," *Journal of the Institute of Interconnection Technology,* Vol. 24, No. 3, April 1998, pp. 11–25.
57. Lau, J. H., and K. L. Chen, "Thermal and Mechanical Evaluations of a Cost-Effective Plastic Ball Grid Array Package," *ASME Transactions, Journal of Electronic Packaging,* Vol. 119, September 1997, pp. 208–212.
58. Lau, J. H., with R. Lee, "Solder Joint Reliability of Cavity-Down Plastic Ball Grid Array Assemblies," *Journal of Soldering & Surface Mount Technology,* Vol. 10, No. 1, February 1998, pp. 26–31.
59. Lau, J. H., "Nonlinear Analysis of Full-Matrix and Perimeter Plastic Ball Grid Array Solder Joints," *ASME Transactions, Journal of Electronic Packaging,* September 1997, pp. 163–170.
60. Lau, J. H., C. Chang, and R. Chen, "Effects of Underfill Encapsulant on the Mechanical and Electrical Performance of a Functional Flip Chip Device," *Journal of Electronics Manufacturing,* Vol. 7, No. 4, December 1997, pp. 269–277.
61. Lau, J. H., and K. L. Chen, "Thermal and Mechanical Evaluation of a Cost-Effective Plastic Ball Grid Array Package," *ASME Transactions, Journal of Electronic Packaging,* Vol. 119, September 1997, pp. 208–212.
62. Lau, J. H., with T. Chou, "A Low-Cost Chip Size Package—NuCSP," *Journal of the Institute of Interconnection Technology,* Vol. 24, No. 1, 1997, pp. 34–38.

63. Lau, J. H., with W. Jung and Y. Pao, "Nonlinear Analysis of Full-Matrix and Perimeter Plastic Ball Grid Array Solder Joints," *ASME Transactions, Journal of Electronic Packaging*, Vol. 119, September 1997, pp. 163–170.
64. Lau, J. H., "The Roles of DNP (Distance to Neutral Point) on Solder Joint Reliability of Area Array Assemblies," *Journal of Soldering & Surface Mount Technology*, No. 20, July 1997, pp. 58–60.
65. Lau, J. H., with R. Lee, "Design for Plastic Ball Grid Array Solder Joint Reliability," *Journal of the Institute of Interconnection Technology*, Vol. 23, No. 2, January 1997, pp. 11–13.
66. Lau, J. H., "Solder Joint Reliability of Flip Chip and Plastic Ball Grid Array Assemblies Under Thermal, Mechanical, and Vibration Conditions,," *IEEE Transaction on CPMT, Part B*, Vol. 19, No. 4, November 1996, pp. 728–735.
67. Lau, J. H., with R. Lee, "Effect of Chip Dimension and Substrate Thickness on the Solder Joint Reliability of Plastic Ball Grid Array Package," *Journal of the Institute of Interconnection Technology*, Vol. 23, October 1996, pp. 16–19.
68. Lau, J. H., E. Schneider, and T. Baker, "Shock and Vibration of Solder Bumped Flip Chip on Organic Coated Copper Boards," *Journal of Electronic Packaging, Transactions of ASME*, Vol. 118, June 1996, pp. 101–104.
69. Lau, J. H., "Temperature and Stress Time History Responses in Electronic Packaging," *IEEE Transactions on CPMT, Part B*, February 1996, pp. 248–254.
70. Lau, J. H., "On Murphy's Integrated Circuit Yield Integral," *Journal of Electronic Packaging, Transactions of ASME*, Vol. 117, June 1995, pp. 159–164.
71. Lau, J. H., K. Gratalo, E. Schneider, T. Marcotte, and T. Baker, "Solder Joint Reliability of Large Plastic Ball Grid Array Assemblies," *Journal of the Institute of Interconnection Technology*, Vol. 22, October 1995, pp. 27–32.
72. Lau, J. H., and M. Kelley, "Low Cost Solder Bumped Flip Chip MCM-L Demonstration," *Journal of the Institute of Interconnection Technology*, Vol. 21, June 1995, pp. 159–164.
73. Lau, J. H., M. Heydinger, J. Glazer, and D. Uno, "Design and Procurement of Eutectic Sn/Pb Solder-Bumped Flip Chip Test Die and Organic Substrates," *Journal of the Institute of Interconnection Technology*, Vol. 21, March 1995, pp. 20–24.
74. Lau, J. H., with B. Wun, "Characterization and Evaluation of the Underfill Encapsulants for Flip Chip Assembly," *Journal of the Institute of Interconnection Technology*, Vol. 21, March 1995, pp. 25–27.
75. Lau, J. H., S. Golwalkar, and S. Erasmus, "Advantages and Disadvantages of Thin Small Outline Packages (TSOP) with Copper Gull-Wing Leads," *Journal of Electronic Packaging, Transactions of ASME*, September 1994, pp. 154–157.
76. Lau, J. H., "Bending and Twisting of 63Sn37Pb Solder Interconnects with Creep," *Journal of Electronic Packaging, Transactions of ASME*, June 1994, pp. 154–157.
77. Lau, J. H., J. Miremadi, J. Gleason, R. Haven, S. Ottoboni, and S. Mimura, "No Clean Mass Reflow of Large Plastic Ball Grid Array Carriers," *Journal of the Institute of Interconnection Technology*, Vol. 20, March 1994, pp. 15–22.
78. Lau, J. H., Y. Pao, C. Larner, S. Twerefour, R. Govila, D. Gilbert, S. Eeasmus, and S. Dolot, "Reliability of 0.4 mm Pitch, 256-Pin Plastic Quad Flat Pack No-Clean and Water-Clean Solder Joints," *Journal of Soldering & Surface Mount Technology*, No. 16, February 1994, pp. 42–50.
79. Lau, J. H., "Creep of Solder Interconnections Under Combined Loads," *IEEE Transactions on CPMT*, December 1993, pp. 794–798.
80. Lau, J. H., and Erasmus, S., "Reliability of Fine Pitch Plastic Quad Flat Pack Leads and Solder Joints Under Bending, Twisting, and Thermal Conditions," *Journal of Electronic Packaging, Transactions of ASME*, Vol. 115, September 1993, pp. 322–328.
81. Lau, J. H., "Closed-Form Solutions for The Large Deflection of Curved Optical Glass Fibers Under Combined Loads," *Journal of Electronic Packaging, Transactions of ASME*, Vol. 115, September 1993, pp. 337–339.
82. Lau, J. H., Marcotte, T., Severine, J., Lee, A., Erasmus, S., Baker, T., Moldaschel, J., Sporer, M., and Burward-Hoy, G., "Solder Joint Reliability of Surface Mount

Connectors," *Journal of Electronic Packaging, Transactions of ASME*, Vol. 115, June 1993, pp. 180–188.

83. Lau, J. H., S. Golwalkar, P. Boysan, R. Surratt, D. Rice, R. Forhringer, and S. Erasmus, "Solder Joint Reliability of a Thin Small Outline Package (TSOP)," *Journal of the Institute of Interconnection Technology*, Vol. 20, No. 1, November 1993, pp. 12–19.
84. Lau, J. H., "Creep of 96.5Sn3.5Ag Solder Interconnects," *Journal of Soldering & Surface Mount Technology*, No. 15, September 1993, pp. 45–49.
85. Lau, J. H., "Thermal Fatigue Life Prediction of Flip Chip Solder Joints by Fracture Mechanics Method," *International Journal of Engineering Fracture Mechanics*, Vol. 45, No. 5, July 1993, pp. 643–654.
86. Lau, J. H., T. Krulevitch, W. Schar, M. Heydinger, S. Erasmus, and J. Gleason, "Experimental and Analytical Studies of Encapsulated Flip Chip Solder Bumps on Surface Laminar Circuit Boards," *Journal of the Institute of Interconnection Technology*, Vol. 19, No. 3, March 1993, pp. 18–24.
87. Lau, J. H., "Elasto-Plastic Large Deflection Analysis of a Copper Film," *Journal of Electronic Packaging, Transactions of ASME*, Vol. 114, June 1992, pp. 221–225.
88. Lau, J. H., Golwalkar, S., Rice, D., Erasmus, S., and Foehringer, R., "Experimental and Analytical Studies of 28-Pin Thin Small Outline Package Solder-Joint Reliability," *Journal of Electronic Packaging, Transactions of ASME*, Vol. 114, June 1992, pp. 169–176.
89. Lau, J. H., "Thermoelastic Solutions for a Semi-Infinite Substrate with a Powered Electronic Device," *Journal of Electronic Packaging, Transactions of ASME*, Vol. 114, September 1992, pp. 353–358.
90. Lau, J. H., R. Govila, C. Larner, Y. Pao, S. Erasmus, S. Dolot, M. Jalilian, and M. Lancaster, "No-Clean and Solvent-Clean Mass Reflow Processes of 0.4mm Pitch, 256-Pin Fine Pitch Qual Flat Packs (QFP)," *Journal of the Institute of Interconnection Technology*, Vol. 19, No. 1, October 1992, pp. 19–26.
91. Lau, J. H., G. Dody, W. Chen, M. McShane, D. Rice, S. Erasmus, and W. Adamjee, "Experimental and Analytical Studies of 208-Pin Fine Pitch Quad Flat Pack Solder-Joint Reliability," *Journal of the Institute of Interconnection Technology*, Vol. 18, No. 2, January 1992, pp. 13–19.
92. Lau, J. H., Subrahmanyant, R., Rice, D., Erasmus, S., and Li, C., "Fatigue Analysis of a Ceramic Pin Grid Array Soldered to An Orthotropic Epoxy Substrate," *Journal of Electronic Packaging, Transactions of ASME*, Vol. 113, June 1991, pp. 138–148.
93. Lau, J. H., "Thermoelastic Solutions for a Finite Substrate with an Electronic Device," *Journal of Electronic Packaging, Transactions of ASME*, Vol. 113, March 1991, pp. 84–88.
94. Lau, J. H., S. Leung, R. Subrahmanyant, D. Rice, S. Erasmus, and C. Li, "Effects of Rework on the Solder Joint Reliability of Pin Grid Array Interconnects," *Journal of the Institute of Interconnection Technology*, July 1991, pp. 5–10.
95. Lau, J. H., "Thermoelastic Problems for Electronic Packaging," *Journal of the International Society for Hybrid Microelectronics*, No. 25, May 1991, pp. 11–15.
96. Lau, J. H., Rice, W. D., and Harkins, G., "Thermal Stress Analysis of TAB Packages and Interconnections," *IEEE Transactions on CHMT*, Vol. 13, No. 1, March 1990, pp. 183–188.
97. Lau, J. H., and Harkins, G., "Stiffness of 'Gull-Wing' Leads and Solder Joints for a Plastic Quad Flat Pack," *IEEE Transactions on CHMT*, Vol. 13, No. 1, March 1990, pp. 124–130.
98. Lau, J. H., Powers, L., Baker, J., Rice, D., and Shaw, W., "Solder Joint Reliability of Fine Pitch Surface Mount Technology Assemblies," *IEEE Transactions on CHMT*, Vol. 13, September 1990, pp. 534–544.
99. Lau, J. H., S. Erasmus, D. Rice, "Overview of Tape Automated Bonding Technology," *Journal of the Institute of Interconnection Technology*, Vol. 16, No. 2, January 1990, pp. 5–24.

100. Lau, J. H., and Keely, C. A., "Dynamic Characterization of Surface Mount Component Leads for Solder Joint Inspection," *IEEE Transactions on CHMT*, Vol. 12, December 1989, pp. 594–602.
101. Lau, J. H., "Thermal Stress Analysis of SMT PQFP Packages and Interconnections," *Journal of Electronic Packaging, Transactions of ASME*, Vol. 111, March 1989, pp. 2–8.
102. Lau, J. H., "A Note on the Calculation of Thermal Stresses in Electronic Packaging by Finite Element Methods," *Journal of Electronic Packaging, Transactions on ASME*, Vol. 111, December 1989, pp. 313–320.
103. Lau, J. H., and Jeans, A. H., "Nonlinear Analysis of Elastomeric Keyboard Domes," *Journal of Applied Mechanics, Transactions of ASME*, Vol. 56, December 1989, pp. 751–755.
104. Lau, J. H., and Lian-Mueller, L. B., "Finite Element Modeling for Optimizing Hermetic Package Reliability," *Journal of Electronic Packaging, Transactions of ASME*, Vol. 111, December 1989, pp. 255–260.
105. Lau, J. H., and Moresco, L. L., "Mechanical Behavior of Microstrip Structures Made of Y-Ba-Cu-O Superconducting Ceramics," *IEEE Transactions on CHMT*, Vol. CHMT-11, No. 4, December 1988, pp. 419–426.
106. Lau, J. H., and Harkins, G., "Thermal Stress Analysis of SOIC Packages and Interconnections," *IEEE Transactions on CHMT*, Vol. CHMT-11, December 1988, pp. 380–389.
107. Lau, J. H., Harkins, G., Rice, D., Kral, J., and Wells, B., "Experimental and Statistical Analyses of Surface-Mount Technology PLCC Solder-Joint Reliability," *IEEE Transactions on Reliability*, Vol. 37, No. 5, December 1988, pp. 524–530.
108. Lau, J. H., Rice, D. W., and Avery, P. A., "Elasto-Plastic Analysis of Surface-Mount Solder Joints," *IEEE Transactions on CHMT*, Vol. CHMT-10, No. 3, September 1987, pp. 346–357.
109. Lau, J. H., and Barrett, G. H., "Stress and Deflection Analysis of Partially Routed Panels for Depanelization," *IEEE Transactions on CHMT*, Vol. CHMT-10, September 1987, pp. 411–419.
110. Lau, J. H., and C. Hu, "Deformation of Curved Bars with Creep," *Journal of Engineering for Gas Turbines and Power, Transactions of the ASME*, Vol. 107, January 1985, pp. 225–230.
111. Lau, J. H., and D. Rice, "Solder Joint Fatigue in Surface Mount Technology: State of the Art," *Solid State Technology*, Vol. 28, October 1985, pp. 91–104.
112. Lau, J. H., "Vibration Frequencies for a Non-Uniform Beam with End Mass," *Journal of Sound and Vibration*, Academic Press, Vol. 97, December 1984, pp. 513–521.
113. Lau, J. H., and T. Lau, "Bending and Twisting of Pipes with Strain-Hardening," *Journal of Pressure Vessel Technology, Transactions of the ASME*, Vol. 106, May 1984, pp. 188–195.
114. Lau, J. H., "Vibration Frequencies of Tapered Bars with End Mass," *Journal of Applied Mechanics, Transactions of the ASME*, Vol. 51, March 1984, pp. 178–181.
115. Lau, J. H., "Vibration Frequencies and Mode Shape for a Constrained Cantilever," *Journal of Applied Mechanics, Transactions of the ASME*, Vol. 51, March 1984, pp. 182–187.
116. Lau, J. H., S. Jung, T. Lau, "Creep of Thin-Walled Circular Cylinder Under Axial Force, Bending, and Twisting Moments," *Journal of Energy for Power, Transactions of the ASME*, Vol. 106, January 1984, pp. 79–83.
117. Lau, J. H., and T. Lau, "Creep of Pipes Under Axial Force and Bending Moment," *Journal of Engineering, Proceedings of ASCE*, Vol. 108, February 1982, pp. 190–195.
118. Lau, J. H., "Large Deflections of Beam with Combined Loads," *Journal of Engineering Mechanics, Proceedings of ASCE*, Vol. 108, February 1982, pp. 180–185.
119. Lau, J. H., and T. Lau, "Creep of Pipe Under Axial Force and Twisting Moment," *Journal of Engineering Mechanics, Proceedings of ASCE*, Vol. 108, February 1982, pp. 174–179.

120. Lau, J. H., "Torsion of Hollow Circular Cylinder with Creep," *Journal of Engineering Mechanics, Proceedings of ASCE,* Vol. 108, February 1982, pp. 185–190.
121. Lau, J. H., "Tables of Dimensionless Central Moments," *Journal of Engineering Mechanics, Proceedings of ASCE,* Vol. 107, December 1981, pp. 1250–1252.
122. Lau, J. H., "Bending of Circular Thin-Walled Cylinder with Creep," *Journal of Engineering Mechanics, Proceedings of ASCE,* Vol. 107, August 1981, pp. 704–710.
123. Lau, J. H., and G. Listvinsky, "Bending and Twisting of Internally Pressurized Thin-Walled Cylinder with Creep," *Journal of Applied Mechanics, Transactions of the ASME,* Vol. 48, June 1981, pp. 439–441.
124. Lau, J. H., Bending and Twisting of Pipe with Creep," *International Journal of Nuclear Engineering and Design, Academic Press,* Vol. 66, June 1981, pp. 367–374.
125. Lau, J. H., Fundamental Frequency of a Constrained Beam," *Journal of Sound and Vibration, Academic Press,* Vol. 78, 1981, pp. 154–157.
126. Lau, J. H., "Large Deflection of Cantilever Beam," *Journal of Engineering Mechanics, Proceedings of ASCE,* Vol. 107, February 1981, pp. 259–264.
127. Lau, J. H., "Bending of Circular Cylinder with Creep," *Journal of Engineering Mechanics, Proceedings of ASCE,* Vol. 107, February 1981, pp. 265–270.
128. Lau, J. H., "Stiffness of Corrugated Plates," *Journal of Engineering Mechanics, Proceedings of ASCE,* Vol. 107, February 1981, pp. 271–275.

CHAPTER 4

Reliability of 1657-Pin CCGA Lead-Free Solder Joints

4.1 Introduction

In many applications, such as computers and telecommunications, the IC chip sizes are very big, the on-chip frequency and power dissipation are very high, and the number of chip I/Os is very large. The ceramic column grid array (CCGA) package developed by IBM is one of the best candidates for housing these kinds of chips [1–7]. There are three parts in this chapter. The first part is concerned with design for reliability (DFR) to determine the time-history creep strain energy density of the 1657-pin CCGA solder column with two different solder paste materials, namely, 95.5wt%Sn3.9wt%Ag0.6wt%Cu and 63wt%Sn37wt%Pb, and under three different thermal cycling profiles, namely, 25°C to 75°C, 0°C to 100°C, and −25°C to 125°C. The effects of these solder pastes and temperature conditions on the thermal-fatigue life of the high-lead (10wt%Sn90wt%Pb) solder columns of the CCGA package are provided and discussed [8–10].

The second part is concerned with the thermal cycling test of the 1657CCGA package with 95.5wt%Sn3.9wt%Ag0.6wt%Cu and 63wt%Sn37wt%Pb solder pastes on lead-free printed circuit boards (PCBs) with hot-air solder leveling (HASL) CuSn, NiAu (electroless Ni and immersion Au, ENIG), and Entek organic solderability preservative (OSP) surface finishes [10, 11]. The test (0°C–100°C, 40-min. cycle) stopped at 7500 cycles. The test data is best fitted to the Weibull life distribution. Also, the sample mean, population mean, sample characteristic life, true characteristic life, sample Weibull slope, and true Weibull slope for the 1657CCGA package on the lead-free PCBs are provided and discussed. Furthermore, the relationship between the reliability and the confidence intervals for the life distribution is established. Finally, the confidence level for comparing the quality (mean life) of the CCGA solder joints is determined.

The last part of this chapter is concerned with performing failure analysis [12, 13]. Emphasis is placed on determining the failure locations and failure modes of the solder joints of the 1657CCGA assemblies after they have been through 7000 cycles of temperature cycling. The present results, which are compared with those obtained from design for reliability analysis, indicate that the test was performed properly.

4.2 Design for Reliability of the 1657CCGA with Lead-Free Solder Paste

4.2.1 The Structure

The 1657-pin CCGA package (shown in Figs. 4.1 and 4.2) under consideration consists of 10Sn90Pb solder columns (0.5 mm in diameter and 2.21 mm tall), an Al_2O_3 substrate (42.5 × 42.5 × 3.75 mm), and a cavity lid with an edge length of 40.5 mm and an overall thickness of 3 mm. The actual thickness of the lid over the cavity ranges from 1.3 to 1.7 mm; a 1.3-mm thickness is used in the analyses. Since the

FIGURE 4.1 The IBM 1657CCGA package.

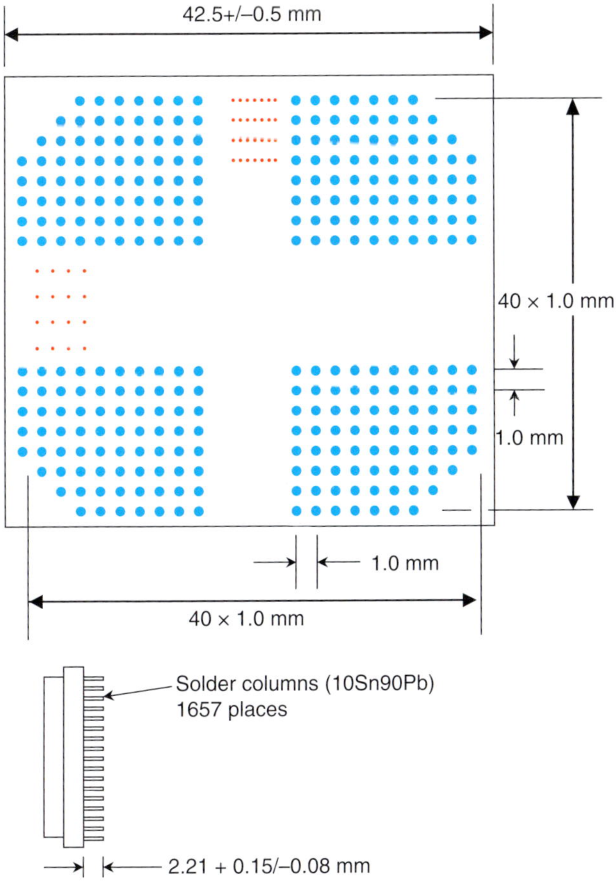

FIGURE 4.2 Dimensions and pin-out of the 1657CCGA.

stiffness and size of the IC chip are very small compared with those of the package, the chip is not modeled.

The second-level solder columns form a 41 × 41 array at 1 mm pitch with six pins depopulated per corner. This results in 1657 second-level pads. The CCGA PCB assembly is modeled as a three-dimensional strip that captures the construction along a diagonal path from the assembly's geometric center to a corner. Due to the symmetry about the vertical midplane of a full strip, the model is actually a half-strip with the appropriate in-plane restraints placed on one symmetry plane. Coupled in-plane translations are applied to the other symmetry plane to produce a state of generalized plane strain. Using exclusively hexahedral solid elements, the model can capture the precise shape of the packages' solder columns and potential distance to neutral point (DNP) effects while retaining significant computational efficiency over full octant models.

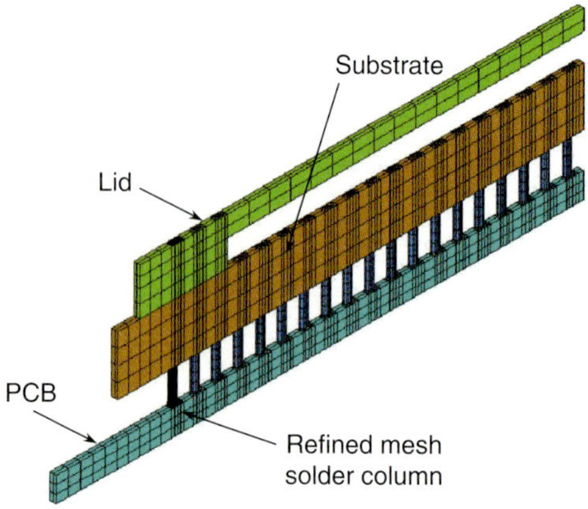

FIGURE 4.3 The finite element model.

Despite the overall economy of elements in the strip model, selective mesh refinement is used to concentrate highly refined elements in the solder columns where failure is anticipated. In the CCGA assembly, failure would be predicted in the column with the greatest DNP and close to the intersection of the solder column with the vertex of the solder fillet [12, 13].

The finite element model includes 19 columns, as shown in Fig. 4.3. The inner 18 columns are coarsely meshed, while the outermost column is constructed with a highly refined mesh, as shown in Fig. 4.4. While the pads on the 1.6-mm thick PCB are 0.673 mm in diameter, the diameter of those on the substrate is 0.8 mm. Since the strip model for this package should be 0.707 mm thick, incorporating the substrate pad necessitates a reasonable modification to the modeled geometry. The thickness of the strip is increased to 0.8 mm while the substrate's pad diameter is set to 0.75 mm. ANSYS [14] is the code selected for the modeling and analysis.

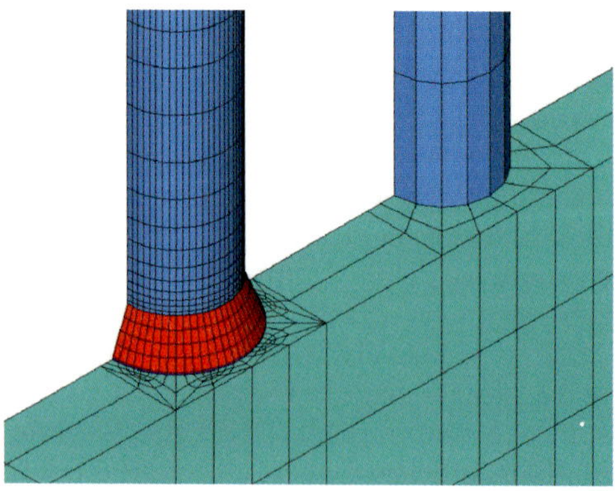

FIGURE 4.4 Detail of the refined mesh.

Solder Alloy	C_1 (1/s)	C_2 (1/Pa)	C_3	C_4 (K)
95.5Sn3.9Ag 0.6Cu	441000	5×10^{-9}	4.2	5412
SnPb	$926(508 - T)/T$	$1/(37.78 \times 106 - 74414T)$	3.3	6360
10Sn90Pb	$58.08/([1.732 \times 10^6]^{C_3})$	N.A	$8.0696 - 0.0122T$	7462

Note: All temperatures are in K

Table 4.1 Creep Material Properties

4.2.2 Material Properties

The material properties of the CCGA assembly are shown in Tables 4.1 and 4.2. All of the material properties are assumed to be constant except for those of the solders. Both the 95.5Sn3.9Ag0.6Cu and the 63Sn37Pb solders are assumed to follow the generalized Garofalo equation (sinh law) as discussed in Chap. 2, while the 10Sn90Pb solder is modeled with the Norton equation (power law), as follows:

$$\frac{\partial \varepsilon}{\partial t} = C_1[\sinh(C_2\sigma)]^{C_3} \exp\left(\frac{-C_4}{T}\right)$$

$$\frac{\partial \varepsilon}{\partial t} = C_1\sigma^{C_3} \exp\left(\frac{-C_4}{T}\right)$$

Material	Young's Modulus (GPa)	Poisson's Ratio	CTE (ppm/K)
PC board	27	0.39	18
Copper pad	76	0.35	17
95.5Sn3.9Ag0.6Cu	$74.84 - 0.08T$	0.3	$16.66 + 0.017T$
SnPb	$75.94 - 0.152T$	0.35	24.5
10Sn90Pb	$23.71 - 0.047T$	0.35	$24.495 + 0.015T$
Al_2O_3 substrate	270	0.3	5.5
Si	167	0.28	2.54
Lid	70	0.33	23
Lid seal	0.0024	0.35	270

Note: All temperatures are in K

Table 4.2 Material Properties of the Structure

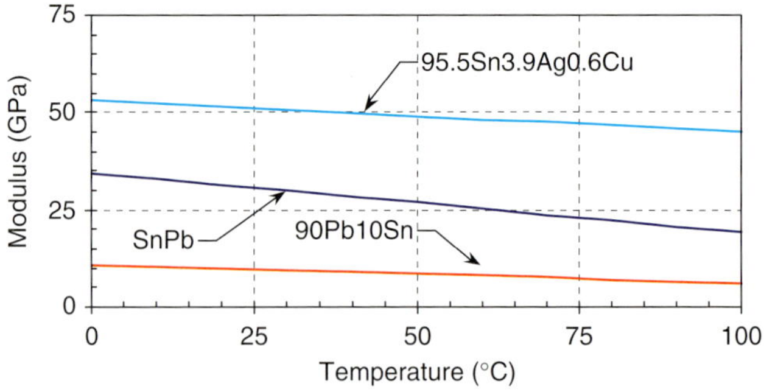

Figure 4.5 Young's modulus of the solders.

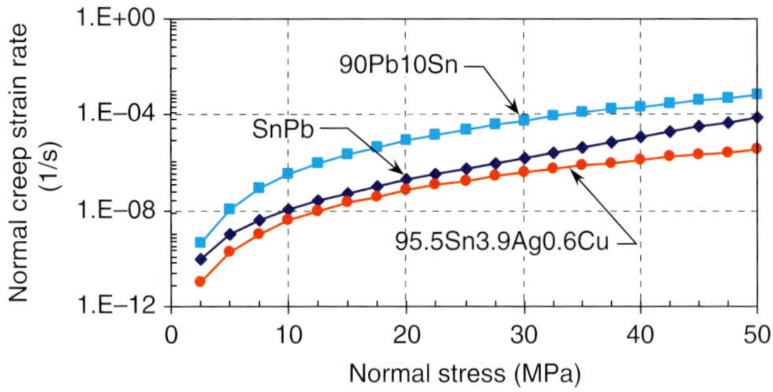

Figure 4.6 Solder creep rate at 0°C.

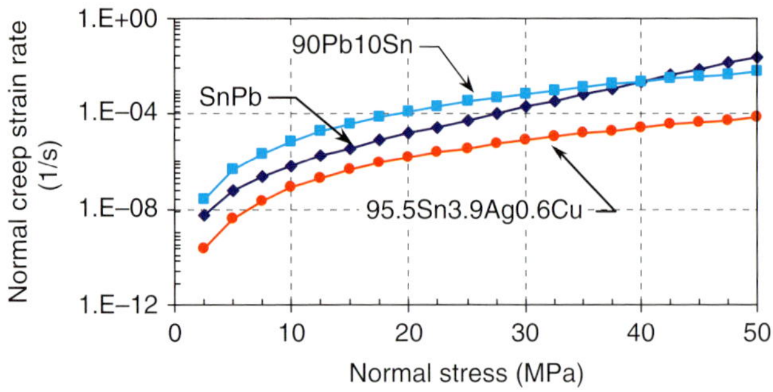

Figure 4.7 Solder creep rate at 50°C.

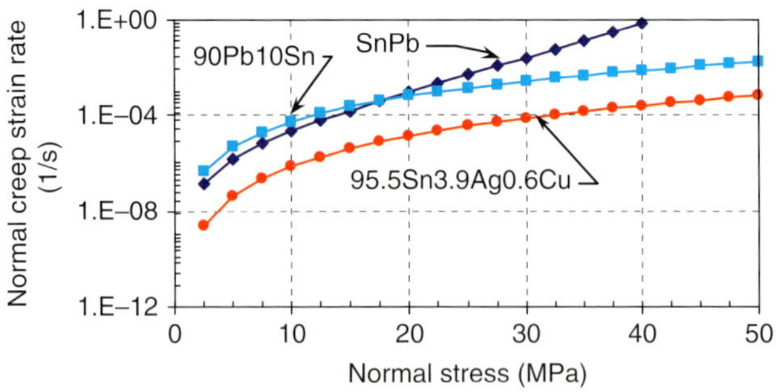

Figure 4.8 Solder creep rate at 100°C.

where the values of C_1 through C_4 are presented in Table 4.1. These creep formulations are currently available in ANSYS. The lead-free creep constitutive relation was obtained from compression samples tested over the temperature range −25°C to 160°C [15, 16], and the constitutive relations for the leaded solders were obtained on a number of configurations and represent generally accepted industry values [17]. While the creep constitutive relations are presented as one-dimensional, the finite element code uses von Mises stresses and strains during solution. The remaining material properties are specified in Table 4.2. The Young's modulus of these three solders is shown in Fig. 4.5. It can be seen that they are all temperature-dependent. Also, the Young's modulus of SnAgCu is the largest and that of 10Sn90Pb is the smallest. Figures 4.6 through 4.8 show the creep strain rate of these three solders at 0°C, 50°C, and 100°C. It can be seen that for all the temperatures, the SnAgCu solder has the smallest creep strain rate.

4.2.3 Temperature Boundary Conditions

Figure 4.9 shows the temperature profiles to be imposed on the CCGA PCB assembly. It can be seen that there are three temperature loading conditions: (1) 25°C to 75°C; (2) 0°C to 100°C; and (3) −25°C to 125°C. For all the cases, the cycle time is one hour and the ramp-up, ramp-down, dwell-at-hot, and dwell-at-cold are each 15 min. In each model simulation, the solder column/joint assembly is subjected to five cycles of one of the three aforementioned temperature regimes.

4.2.4 Simulation Results

The creep shear strain time-histories near the PCB side of the outermost column are shown in Figs. 4.10 and 4.11, respectively, for the SnPb and SnAgCu solder paste (fillet) CCGA assemblies exposed to the cited temperature ranges. As expected, for both solder

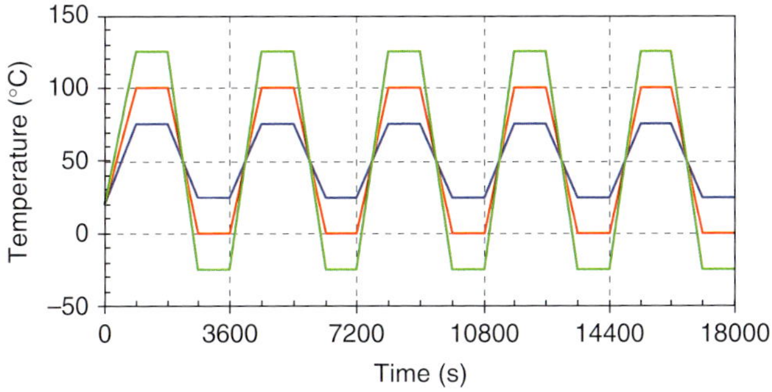

FIGURE 4.9 Input thermal profiles.

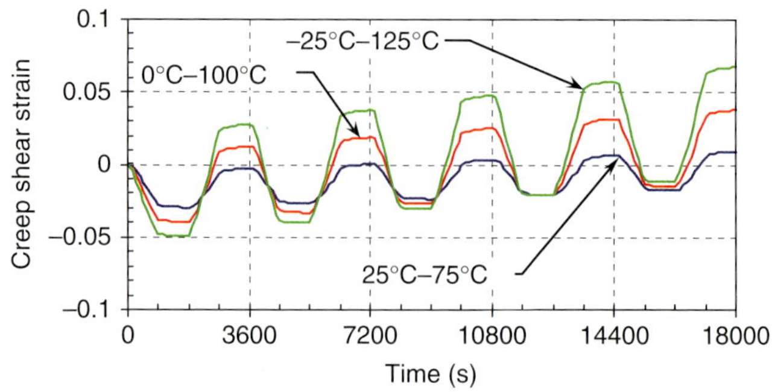

FIGURE 4.10 Maximum creep shear strain at the PCB side of the outermost column with eutectic PCB fillets.

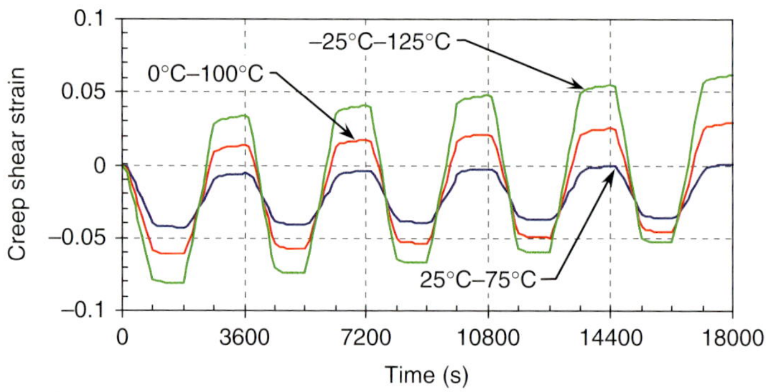

FIGURE 4.11 Maximum creep shear strain at the PCB side of the outermost column with lead-free PCB fillets.

fillets, the creep shear strain time-history is in phase with the thermal profile and the solder columns exhibit larger creep shear strain ranges with larger temperature spans. The creep shear strain distributions in the outermost column at the end of the fourth thermal cycle (i.e., 14,400 s) are shown in Figs. 4.12 and 4.13, respectively, for the SnPb and SnAgCu filleted CCGA assemblies exposed to ΔT = 100°C. It can be seen that for both cases, the maximum creep shear strain occurs near the fillet tips of the solder column and their distributions are similar, except the values are slightly different. The creep shear strain in the solder column with the SnPb solder fillet is slightly larger due to the larger creep strain rate of the SnPb solder alloys.

Figures 4.14 and 4.15 show, respectively, the shear stress time-history near the PCB side of the outermost column for the SnPb and SnAgCu solder pasted (filleted) CCGA assemblies exposed to

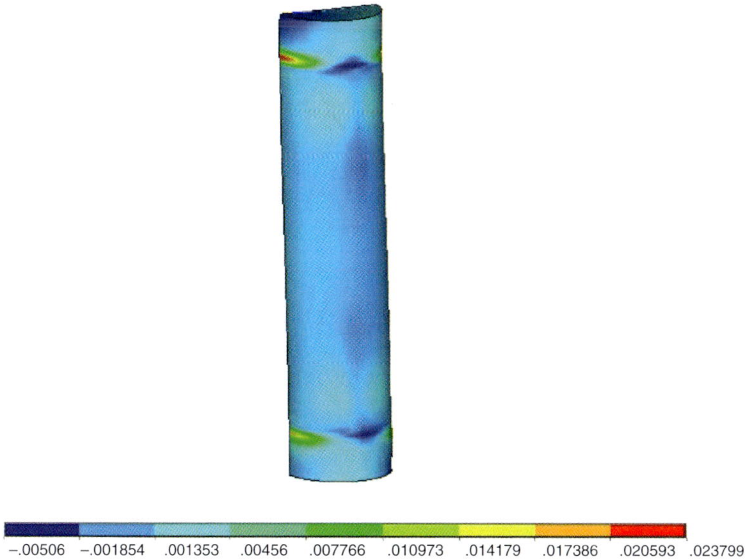

FIGURE 4.12 Creep shear strain in the outermost column with eutectic PCB fillets and subjected to a temperature range of 0°C to 100°C at 14,400 s.

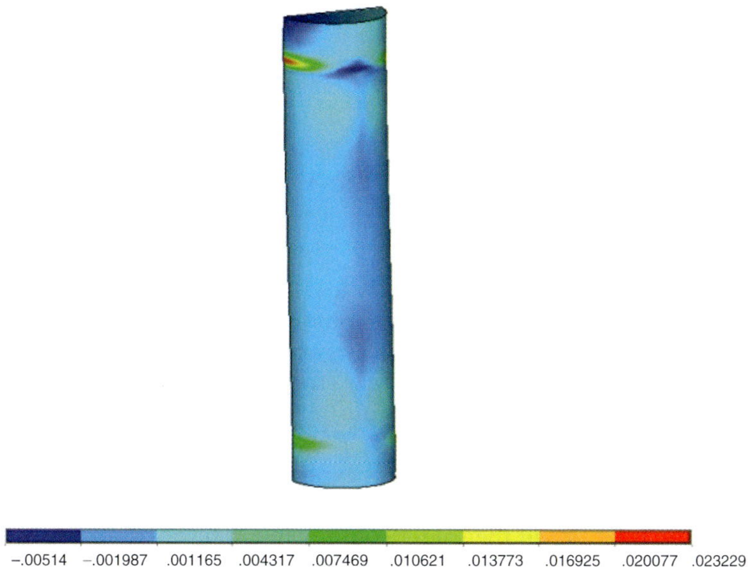

FIGURE 4.13 Creep shear strain in the outermost column with lead-free PCB fillets and subjected to a temperature range of 0°C to 100°C at 14,400 s.

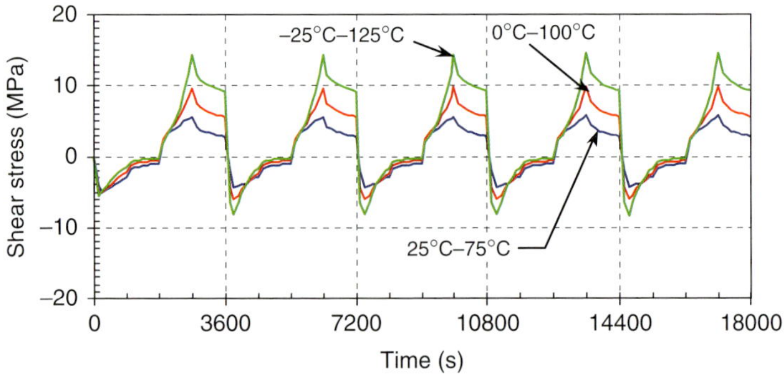

FIGURE 4.14 Maximum shear stress at the PCB side of the outermost column with eutectic PCB fillets.

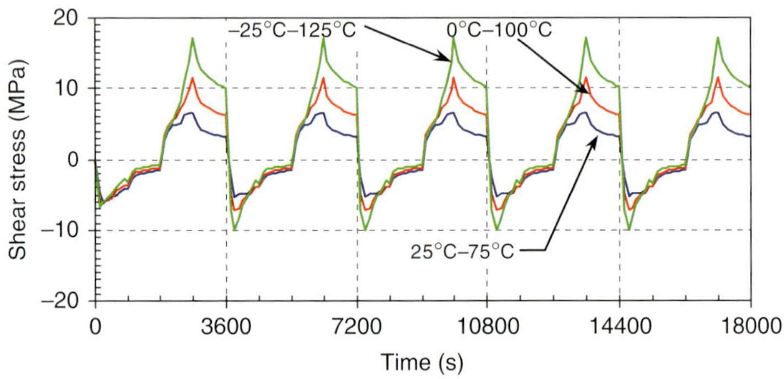

FIGURE 4.15 Maximum shear stress at the PCB side of the outermost column with lead-free PCB fillets.

the cited temperature ranges. As expected, for both solder fillets, the shear stress time-history is in phase with the thermal profile and the solder columns exhibit larger shear stress ranges with larger temperature spans. The shear stress distributions in the outermost column at the end of the fourth thermal cycle (i.e., 14,400 s) are shown in Figs. 4.16 and 4.17, respectively, for the SnPb and SnAgCu filleted CCGA assemblies exposed to $\Delta T = 100°C$. It can be seen that for both cases, the maximum creep shear stress occurs near the fillet tips of the solder column and their distributions are similar, except the values are slightly different. The shear stress in the solder column with the SnAgCu solder fillet is slightly larger due to the larger Young's modulus of the SnAgCu solder alloys. Thus, if any failures occur, they should be located near the solder fillet tips of the solder columns.

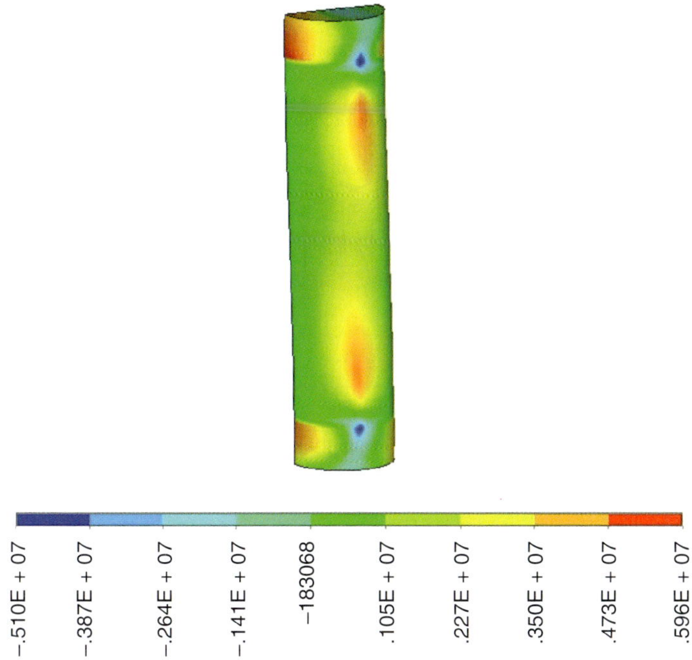

FIGURE 4.16 Shear stress in the outermost column with eutectic PCB fillets and subjected to a temperature range of 0°C to 100°C at 14,400 s.

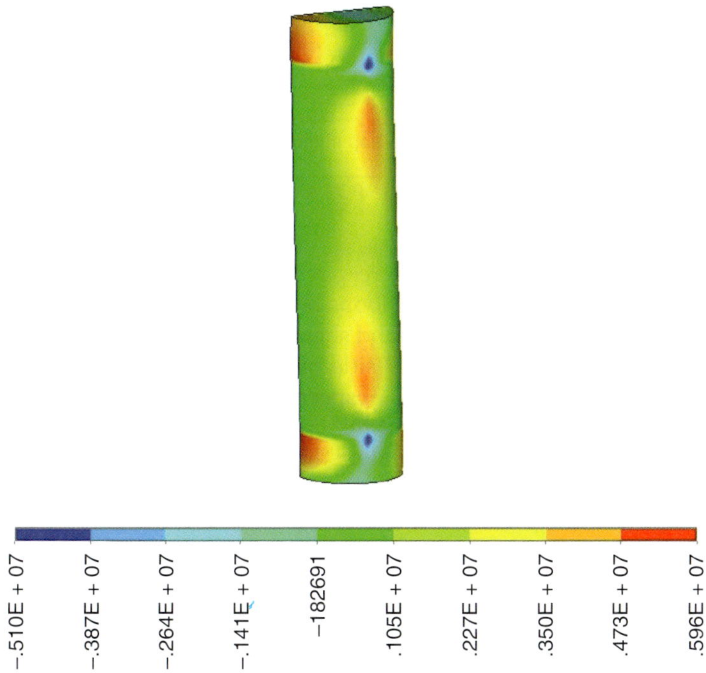

FIGURE 4.17 Shear stress in the outermost column with lead-free PCB fillets and subjected to a temperature range of 0°C to 100°C at 14,400 s.

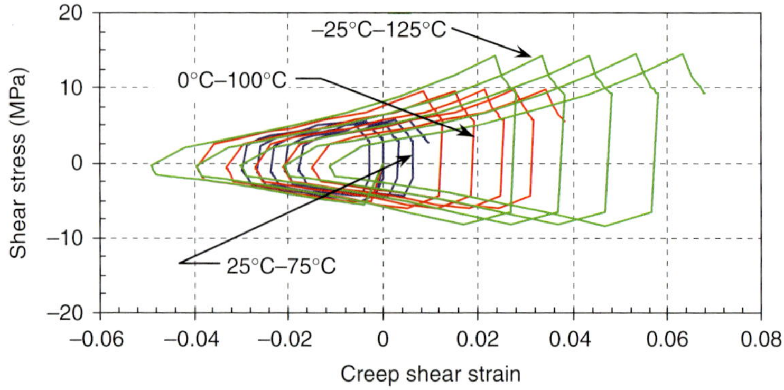

Figure 4.18 Hysteresis loops at the PCB side of the outermost column with eutectic PCB fillets.

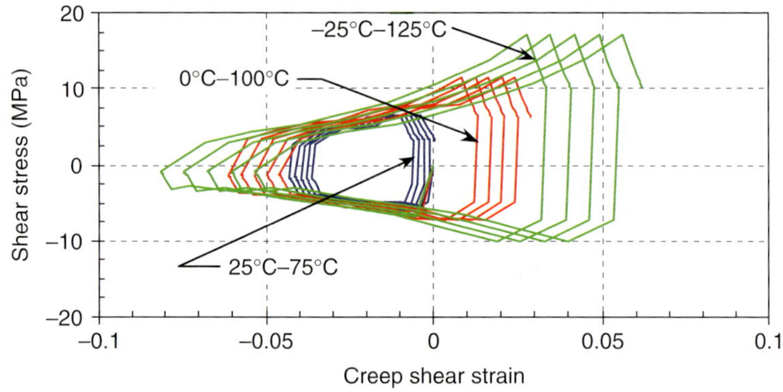

Figure 4.19 Hysteresis loops at the PCB side of the outermost column with lead-free PCB fillets.

It is important to study the creep responses for multiple cycles by observing when the hysteresis loops become stabilized. Figures 4.18 and 4.19 show the shear stress and shear creep strain hysteresis loops for multiple cycles at the PCB side of the outermost column with SnPb and SnAgCu solder fillets, respectively, under various temperature ranges. It can be seen that the creep shear strain versus shear stress loop is quite stabilized after the fourth cycle. Also, the area under the hysteresis loop is larger for higher ΔT.

Figures 4.20 and 4.21 show, respectively, the creep strain energy density time-history near the PCB side of the outermost column for the SnPb and SnAgCu solder pasted (filleted) CCGA assemblies exposed to the specified temperature ranges. The larger the temperature range, the larger the creep strain energy density range (per cycle) in both the assemblies. Also, it can be seen that for all ΔTs, the creep strain energy density in the solder column with the lead-free solder fillet is larger than that with the SnPb solder fillet.

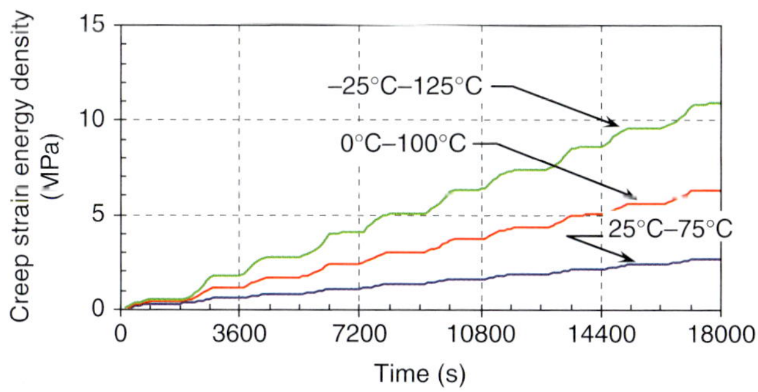

FIGURE 4.20 Critical creep strain energy density at the PCB side of the outermost column with eutectic PCB fillets.

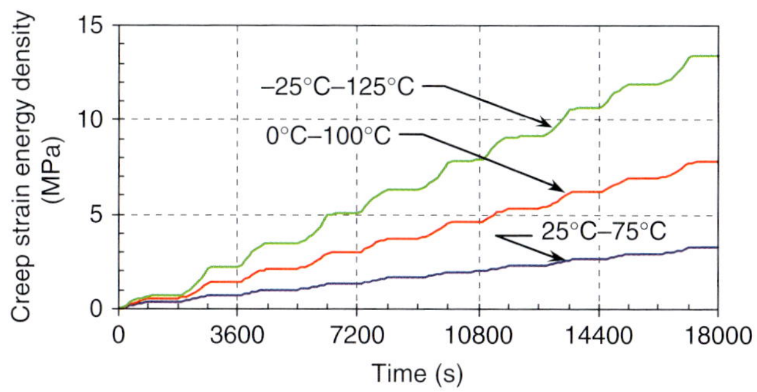

FIGURE 4.21 Critical creep strain energy density at the PCB side of the outermost column with lead-free PCB fillets.

4.2.5 Thermal-Fatigue Life Prediction

In 1965, Morrow [18] demonstrated that fatigue life can be correlated with the mechanical energy of the hysteresis loop. This approach recognizes that energy is required to initiate inelastic deformation and that the hysteresis loop represents the total energy dissipated during a loading cycle. While most of the imparted energy may be transformed into heat, a portion is assumed to damage the material, finally resulting in failure. The thermal fatigue life of solder columns (dominated by creep responses) may, then, be predicted by the following equation [19, 20] as shown in Chap. 2.

$$N_f = \psi(\Delta W)^{\phi}$$

where N_f is the number of cycles to failure, ΔW is the creep strain energy density per cycle, and ψ (always positive) and ϕ (always negative) are material constants for the solder columns. In this equation, ψ and ϕ are usually determined by isothermal fatigue tests of the 10Sn90Pb material. (Please see Chap. 2 for more details.) ΔW can be determined by creep analysis of the structure (IC package, solder columns, and PCB) subjected to the specified loading conditions. In this case, ΔW may be thought of as the area within the hysteresis loop that defines a cycle. ΔW is obtained at the point under consideration by integrating the product of the von Mises stress multiplied by the von Mises creep strain over the span of one thermal cycle. Since the material constants for the 10SnPb solder columns are not available at this moment, thermal fatigue life prediction of the solder columns is impossible. However, the calculated creep strain energy density per thermal cycle (the area under the creep hysteresis loop) can be used to provide some useful insights into the reliability of the solder columns.

Figure 4.22 shows the relationship between the creep strain energy density per cycle (ΔW) at the PCB side of the outermost column for the SnPb and SnAgCu solder pasted (filleted) CCGA assemblies exposed to the temperature ranges (ΔT) considered in this study. It can be seen that the ΔW of the solder column with the SnAgCu solder fillet is larger than that with the SnPb solder fillet (thus, the thermal fatigue life of the solder column with the lead-free solder fillet is smaller than that with the SnPb fillet). Also, the ΔW can be curve-fitted into a nonlinear function of ΔT for the solder columns with SnPb and SnAgCu solder fillets, respectively:

$$\Delta W = 40.122\Delta T^2 + 9485.3\Delta T - 54470$$

$$\Delta W = 48.138\Delta T^2 + 118181\Delta T - 72520$$

Thus, once the material constants (ψ, ϕ) of the solder columns are determined by experiments, the CCGA solder column thermal-fatigue life with either SnPb or SnAgCu solder paste can be predicted for any temperature range and the studied cyclic frequency without further finite element analyses and creep strain energy density calculations.

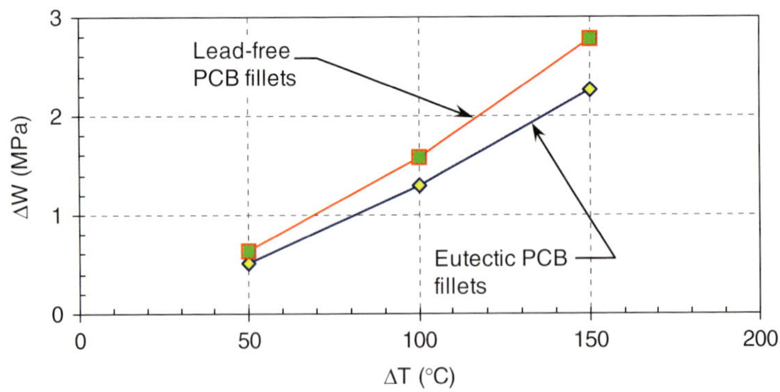

FIGURE 4.22 ΔW as a function of ΔT in the high lead column.

4.2.6 Summary and Recommendations

Some important results and recommendations for DFR are summarized in the following:

1. For the CCGA assembly, the maximum stress and creep strain in the solder column occur near the solder fillet tips with both the lead-free and SnPb solders.
2. For the CCGA assembly, the stresses in the solder columns with the lead-free solder fillets are larger than those with the SnPb solder fillets. This is because the stiffness of the lead-free solder is larger than that of the SnPb solder.
3. For the CCGA assembly, the creep strains in the solder columns with the lead-free solder fillet are smaller than those with the SnPb solder fillets. This is because the creep strain rate of the lead-free solder is smaller than that of the SnPb solder.
4. For the CCGA assembly with SnAgCu and SnPb solder fillets, the thermal-fatigue life of the solder columns with the lead-free solder fillets should be shorter than that with the SnPb solder fillets, since the maximum creep strain energy per cycle in the solder column with the lead-free solder paste is larger than that with the SnPb solder paste.
5. For the CCGA assembly with SnAgCu and SnPb solder fillets, the maximum creep strain energy density per cycle is a nonlinear function of the applied temperature ranges considered herein.
6. The fatigue crack-growth material constants (which can be determined by isothermal fatigue tests) for the 10Sn90Pb solder columns are desperately needed in order to make quantitative solder-column thermal-fatigue life predictions.

4.3 Reliability Testing and Data Analyses of the 1657CCGA with Lead-Free Solder Paste

4.3.1 Component and Test Boards

The 1657-pin CCGA packages (shown in Figs. 4.1 and 4.2) with Sn90Pb solder columns are joined to the PCB with both Sn37Pb and Sn3.9Ag0.6Cu solder pastes. The pad diameter on the PCB is 0.673 mm. The PCB (193 × 352 × 1.6 mm) is made of fiber-reinforced glass and has six copper layers with microvia fabrication. Three different lead-free surface finishes are considered: SnCu HASL, ENIG or NiAu, and Entek OSP, as shown in Table 4.3. Twenty boards (each with two components) are used for the temperature cycling tests. The sample size is 40.

Number of Boards	Board Finishes	Solder Paste	Number of CCGA/board
5	OSP	SnPb	2
5	OSP	SnAgCu	2
5	ENIG	SnAgCu	2
5	SnCu HASL	SnAgCu	2

Table 4.3 PCB and Package Sample Size

4.3.2 Test Chamber and Temperature Cycling Condition

Thermal cycling is performed in an Espec Corporation environmental chamber. This unit is capable of achieving chamber temperature as high as 150°C and as low as −35°C. Heating and cooling are achieved by forced convection, and the maximum ramp rate is 12°C per minute. All of the boards are placed vertically in the chamber and arranged in two parallel columns of 10 boards each. The relative position of each board is shown in Fig. 4.23. Each board is mounted so that the 50-pin connector attachment area (connectors are not populated) and ribbon cable are uppermost.

The temperature input to the air of the chamber is shown in Fig. 4.24. It can be seen that in order to achieve the temperature profile (0°C to 100°C) as planned on the test boards and the components, the input temperature range is much larger, that is, between −10°C and 110°C. This profile is composed of 10-min ramps, a 10-min hold at hot and a 10-min hold at cold. The actual result of measurement with a thermocouple attached on the PCB surface is shown in Fig. 4.25. The measurement results on the 1657CCGA at locations 2, 3, and 4, as shown in Fig. 4.23, are shown in Figs. 4.26, 4.27,

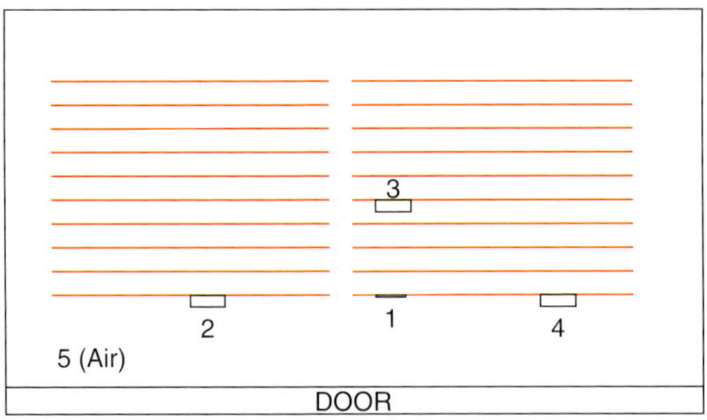

Figure 4.23 Test Boards (20) in Thermal cycling chamber (top-view) and Thermal couples' locations on the board (1), on the 1657CCGA (2, 3, 4), and in the air (5).

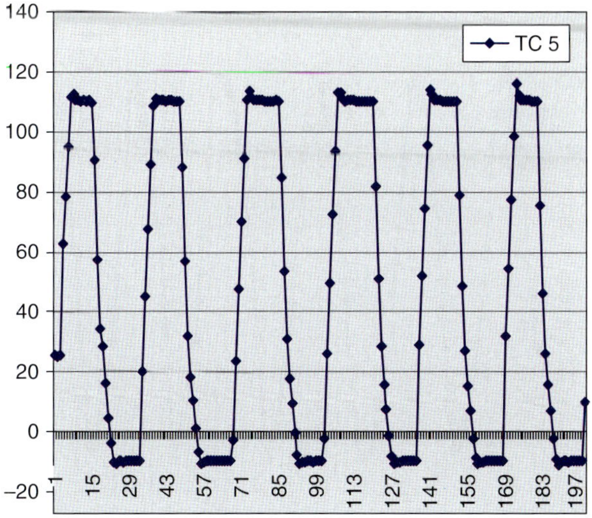

Figure 4.24 Thermal couple reading in the air of the thermal cycling chamber (–10°C–110°C, 40 min).

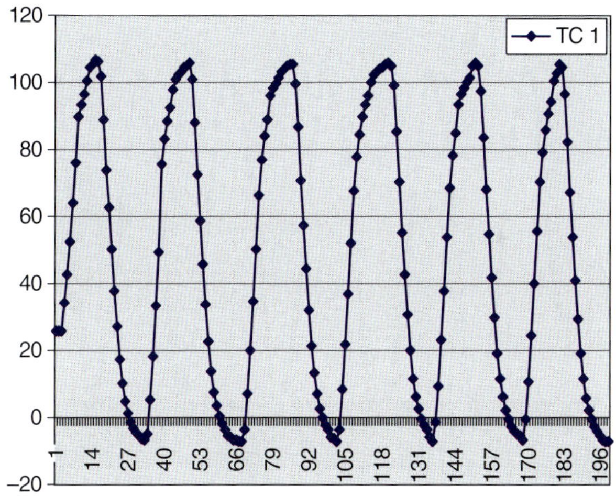

Figure 4.25 Thermal couple reading on the test board (–5°C–105°C, 40 min).

and 4.28, respectively. It can be seen that: (1) the measured temperature profiles on the boards and components have smaller temperature ranges than the input temperature profiles; and (2) the temperature profile range on the boards is slightly higher than that on the components.

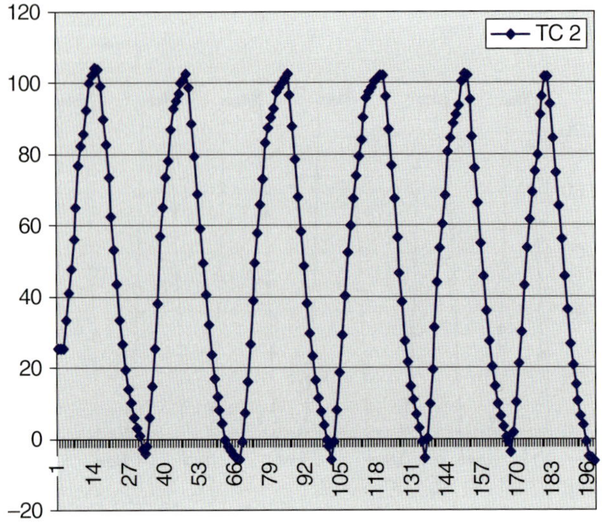

Figure 4.26 Thermal couple reading on the 1657CCGA at location 2 (0°C–100°C, 40 min).

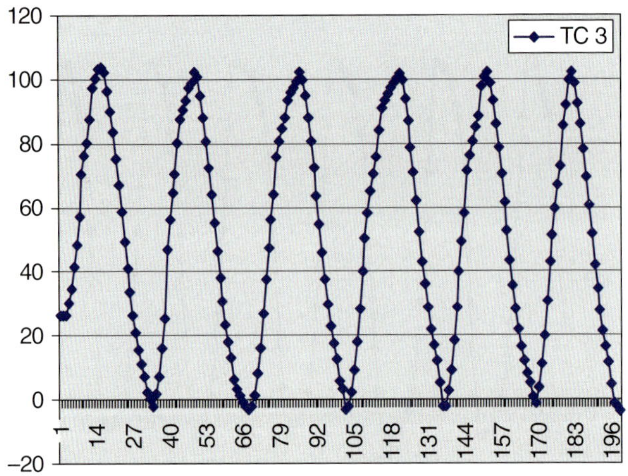

Figure 4.27 Thermal couple reading on the 1657CCGA at location 3 (0°C–100°C, 40 min).

4.3.3 Data Acquisition System and Failure Criterion

During the temperature cycling test, the electrical resistance monitoring in each electrical circuit is performance using the Agilent VXI Data Collection System (E8460A). With the software that is written for this project, the VXI system has the capability of collecting 1024+ (256 channels per card) temperature readings at a rate of

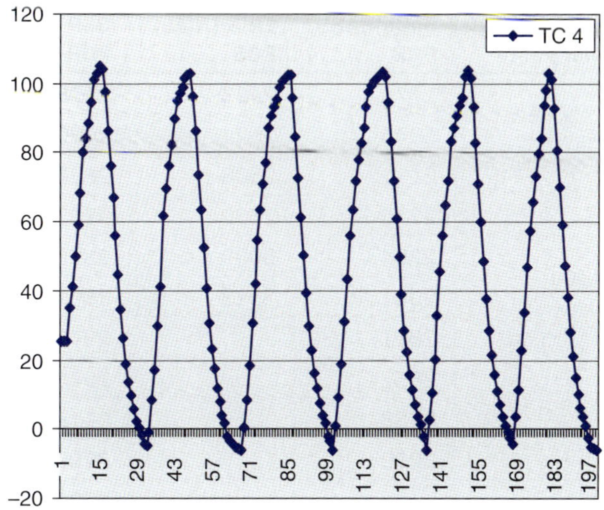

FIGURE 4.28 Thermal couple reading on the 1657CCGA at location 4 (0°C–100°C, 40 min).

1024 channels/20 s. All the data is collected on a 300 MHz (or better) PC (personal computer) in standard ASCII format. Resistance was measured continuously throughout the test. Part failure occurred when the resistance of any chain within the part exceeded 500 Ω for five consecutive cycles. The fifth cycle was reported as the failure cycle.

For a given confidence level (C) and the number of failures (N), the Weibull slope error (E) can be searched from Eq. (2.13). For example, given $N = 10$, $C = 90\%$, then $E = 37\%$.

One of the difficult tasks in statistical analysis of life-test data is to draw conclusions about a population from a small sample size. It is even more difficult to compare the populations of two products from the knowledge of their test data (with small sample size). If the quality (mean life) of one product lot is found to be superior to another, how confident (P) can it be that the same is true of the population? A simple yet useful equation given by Eq. (2.15) can be adopted for this study.

In order to determine the life distributions of lead-free solder joints at certain confidences (e.g., 90% confidence), certain rankings other than the median rank of the life-test data are necessary and can be determined by Eq. (2.12)

4.3.4 Statistical Analysis of 1657CCGA Solder Joints

Table 4.4 shows the test results for the 1657CCGA package with the SnPb and SnAgCu solder pastes on the SnCu HASL, NiAu, and OSP PCBs. It can be seen that: (1) there are failures on all three PCBs with the lead-free paste; (2) all of the samples with lead-free paste on the

Solder Paste	SnCu HASL PCB	ENIG (NiAu) PCB	OSP (Entek) PCB
SnAgCu	1771	2423	2128
	1945	2521	2480
	2111	2568	2620
	2276	2682	2722
	2394	3009	2826
	2581	3513	2899
	2611		2972
	2652		3045
	2688		3118
	3009		3200
SnPb	NA	NA	2303
			3451
			3537
			3555
			3684
			3694
			3851
			4404
			4507
			4595

TABLE 4.4 1657CCGA Thermal Cycling Test Results (0°C–100°C, 40 min Cycle)

SnCu HASL and OSP PCBs have failed; and (3) all of the samples with the SnPb solder paste on the OSP PCB have failed.

The test data of the CCGA with lead-free paste on the SnCu HASL, NiAu, and OSP PCBs are plotted with 90% confidence intervals in Figs. 4.29, 4.30, and 4.31, respectively. It can be seen that besides the median rank, the test data are plotted on the same paper against the 5% and 95% ranks. The resultant plot yields a 90% confidence interval within which the true mean and true characteristic values will lie. These values are shown in Table 4.5. It should be pointed out and expected that for a given percent failed, the life (number of cycles to failure) range (interval) is larger for the required higher confidences.

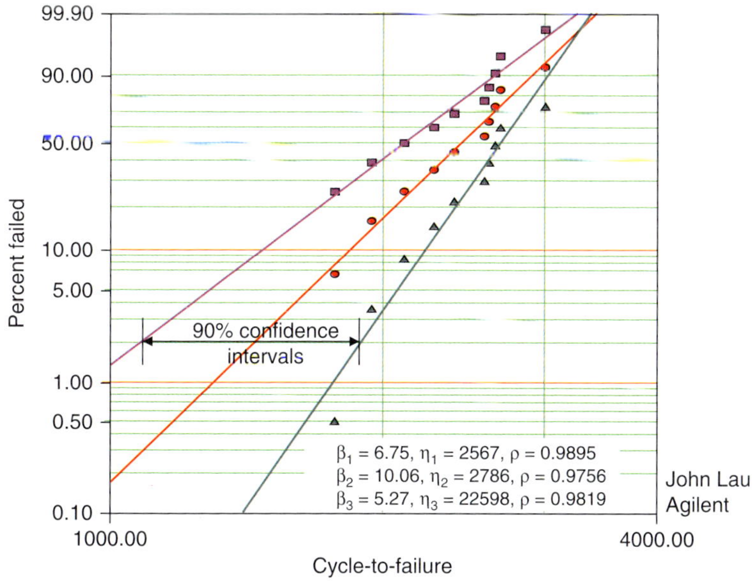

FIGURE 4.29 1657CCGA with SnAgCu paste on HASL (SnCu) PCB—90% confidence.

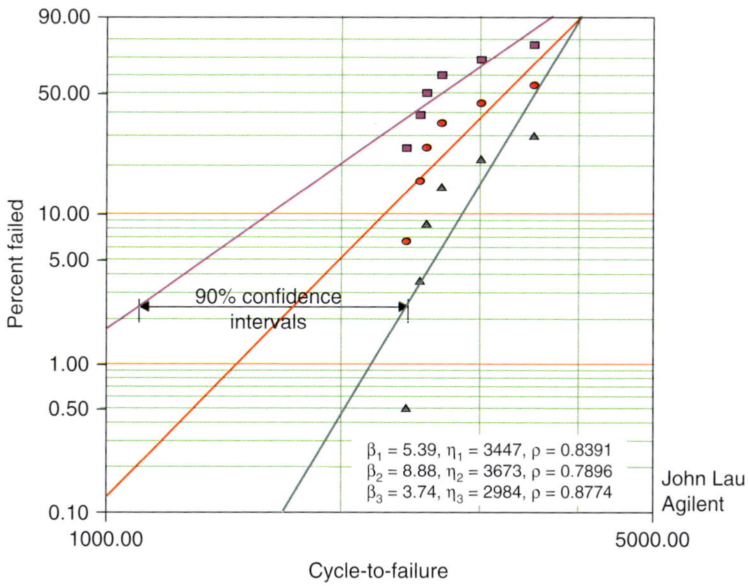

FIGURE 4.30 1657CCGA with SnAgCu paste on NiAu PCB—90% confidence.

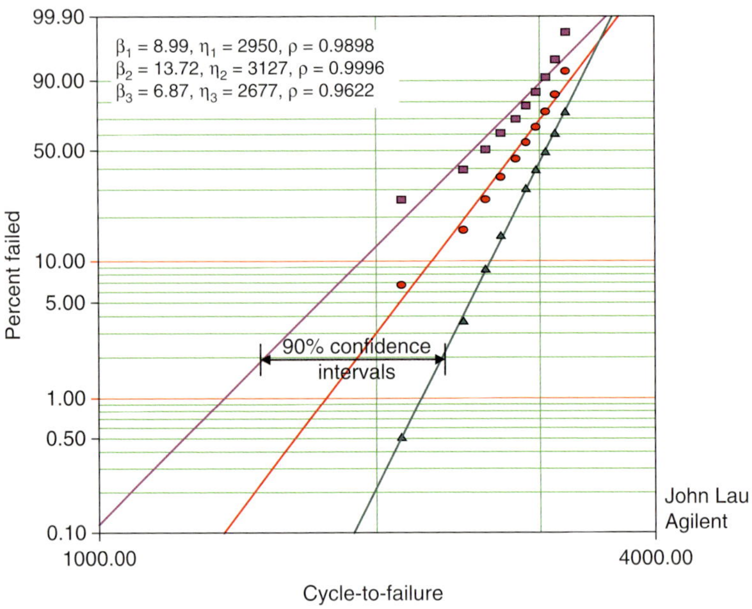

FIGURE 4.31 1657CCGA with SnAgCu paste on OSP PCB—90% confidence.

In the case of the CCGA solder columns with lead-free paste on a SnCu HASL PCB, from the test sample (Fig. 4.29) one would conclude that: (1) the characteristic life (at 63.2% failed) of the CCGA solder joint is 2567 cycles; and (2) the mean life, μ (quality), of the solder joints is 2397 cycles. However, the true characteristic life in 90 out of 100 cases (90% confidence) can be as low as 2259 cycles but not more than 2786 cycles. The percent failed at mean can be determined as F (2397) = 46.72%. Thus, the true mean life of the CCGA solder joint at 90% confidence can be at least 2069 cycles and as high as 2552 cycles. Similar conclusions can be made for other cases.

The Weibull slope error for all the cases has been determined from the chart provided in Chap. 2 and shown in Table 4.5. It can be seen that (for example, in the case of the CCGA with lead-free paste on SnCu HASL PCB) the sample Weibull slope is 6.75. However, in 90 out of 100 cases (C = 90% confidence), the Weibull slope error is E = 37%, and the true Weibull slope at 90% confidence can be as low as 4.25 and as high as 9.25.

Figure 4.32 shows the median rank plots of the CCGA with lead-free solder paste on the SnCu HASL, NiAu, and OSP PCBs. It should be emphasized that this kind of plot is absolutely meaningless from a statistical point of view, unless the confidence level for each pair of mean life ratios (comparisons) is determined

		SnAgCu Paste on SnCu HASL PCB	SnAgCu Paste on NiAu PCB	SnAgCu Paste on OSP PCB	SnPb Paste on OSP PCB
Mean Life (μ), cycle	Sample	2397	3179	2793	3760
	Percent failed at mean	46.72	47.61	45.75	47.66
	Population (μ) at 90% confidence	$2069 \le \mu \le 2552$	$2655 \le \mu \le 3497$	$2492 \le \mu \le 3017$	$3107 \le \mu \le 4267$
Characteristic Life (θ), cycle	Sample	2567	3447	2950	4082
	True θ (cycle) at 90% confidence	$2259 \le \theta \le 2786$	$2984 \le \theta \le 3673$	$2677 \le \theta \le 3127$	$3461 \le \theta \le 4530$
Weibull Slope (β)	Sample	6.75	5.39	8.99	5.29
	β error (%) at 90% confidence	37	47	37	37
	True β at 90% confidence	$4.25 \le \beta \le 9.25$	$2.86 \le \beta \le 7.92$	$5.66 \le \beta \le 12.32$	$3.33 \le \beta \le 7.25$

Table 4.5 Mean Life, Characteristic Life, and Weibull Slope of the 1657CCGA Solder Columns with SnAgCu and SnPb Solder Paste on Various PCBs (90% Confidence Intervals)

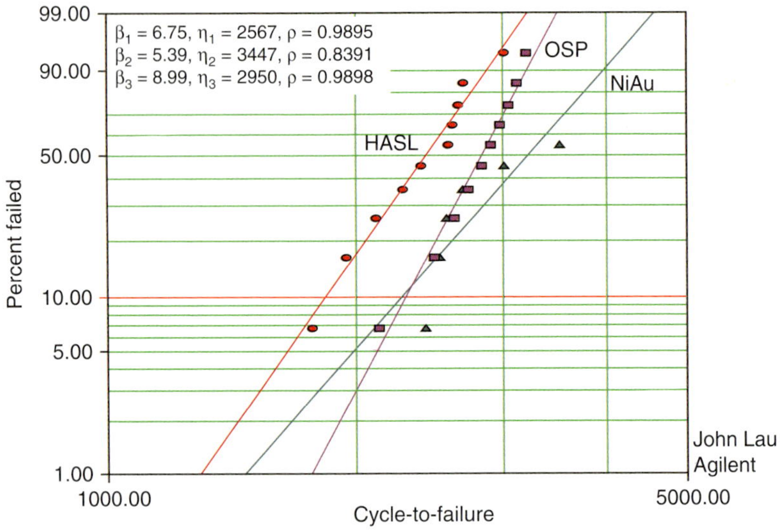

FIGURE 4.32 1657CCGA with SnAgCu paste on HASL, NiAu, and OSP PCBs (median rank).

for example, by Eq. (2.15). Table 4.6 shows the meanlife ratios of the CCGA on these PCBs. A number of conclusions can be drawn. First, the quality (mean life) of CCGA solder joints on a NiAu PCB is superior to those on a SnCu HASL PCB with 99% confidence, that is, in 99 out of 100 cases, the solder joints on a NiAu PCB are better than those on a SnCu HASL PCB (this could be due to the flatness/printability and consistency of the solder pastes on NiAu vs. an uneven SnCu HASL finish). Second, at 89% confidence, the

Different Assembly	Assigned Symbol	Mean Life, Cycles	Mean Life Ratio (Comparison)	Determined Confidence
SnAgCu paste on HASL PCB	A	2397	B/A = 1.33	99%
SnAgCu paste on NiAu PCB	B	3179	B/C = 1.14	89%
SnAgCu paste on OSP PCB	C	2793	C/A = 1.17	94%
SnPb paste on OSP PCB	D	3760	D/C = 1.35	99%

TABLE 4.6 Mean Life Ratios of the 1657CCGA Columns with SnAgCu and SnPb Solder Pastes on Various PCBs

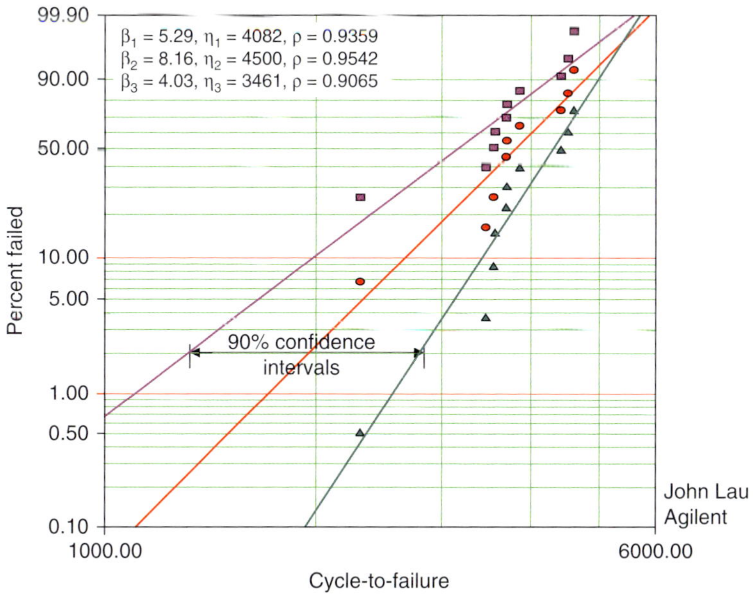

FIGURE 4.33 1657CCGA with SnPb paste on OSP PCB—90% confidence.

solder joints on NiAu PCBs are better than those on OSP PCBs (this could be due to better wetting to the NiAu finish than the OSP finish and thus, the solder columns are more centered). Finally, in 94 out of 100 cases, the solder-joint mean life on OSP PCBs is better than that on the SnCu HASL PCBs (this could be due to the more even OSP finish and the fact that it has more centered columns).

Figure 4.33 shows the life distributions of the 1657CCGA solder joints on the OSP PCB with SnPb solder paste at 90% confidence limits. It can be seen that: (1) the sample characteristic life is 4082 cycles; (2) the sample mean life is 2397 cycles; (3) the true characteristic life in 90 out of 100 cases can be as low as 3461 cycles but not more than 4500 cycles; (4) the percent failed at mean is 47.66%; (5) the true mean life at 90% confidence can be at least 3107 cycles and as high as 4267 cycles; (6) the sample Weibull slope is 5.29; (7) the Weibull error at 90% confidence is 37%; and (8) the true Weibull slope is no less than 3.33 and no more than 7.25.

Figure 4.34 shows the median rank plots of the CCGA solder columns with both tin-lead and lead-free solder pastes on the OSP PCB. It can be seen in Table 4.6 that the CCGA solder joint with SnPb solder paste on the OSP PCB is better than that with SnAgCu solder paste with 99% confidence, that is, in 99 out of 100 cases, the mean life of CCGA solder joints with tin-lead solder paste is superior to those with lead-free solder paste.

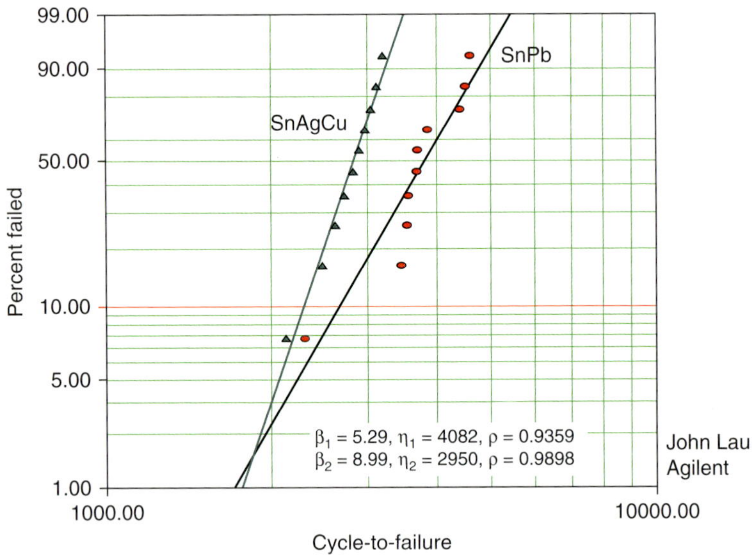

FIGURE 4.34 1657CCGA with SnAgCu and SnPb pastes on OSP PCB (median rank).

4.3.5 Summary and Recommendations

Some important results and recommendations for reliability testing and data analyses are summarized in the following:

1. At 90% confidence (that is, in 9 out of 10 cases), the true characteristic life of the CCGA solder columns with SnAgCu solder paste on the HASL (SnCu) PCB is larger than 2259 cycles and less than 2786 cycles. The true mean life can be as low as 2069 cycles and as high as 2552 cycles. The true Weibull slope is larger than 4.25 and less than 9.25.
2. Similarly, at 90% confidence, the true characteristic life of the CCGA solder columns with SnAgCu solder paste on the NiAu PCB can be as low as 2984 cycles and as high as 3673 cycles. The true mean life can be no less than 2655 cycles and no more than 3497 cycles. The true Weibull slope is larger than 2.86 and less than 7.92.
3. At 90% confidence, the true characteristic life of the CCGA solder columns with SnAgCu solder paste on the OSP (Entek) PCB is larger than 2677 cycles and less than 3127 cycles. The true mean life can be as low as 2492 cycles and as high as 3017 cycles. The true Weibull slope is larger than 5.66 and less than 12.32.
4. At 90% confidence, the true characteristic life of the CCGA solder columns with SnPb solder paste on OSP (Entek) PCB

is larger than 3461 cycles and less than 4500 cycles. The true mean life can be as low as 3107 cycles and as high as 4267 cycles. The true Weibull slope is larger than 3.33 and less than 7.25.

5. The quality (mean life) of the CCGA solder columns with SnAgCu solder paste on the NiAu PCB is better than that on the OSP, with 89% confidence, and that on the SnCu HASL, with 99% confidence.
6. In 94 out of 100 cases, the quality of the CCGA solder columns with SnAgCu solder paste on the OSP PCB is better than that on the SnCu HASL PCB.
7. The quality of the CCGA solder columns with SnPb solder paste on the OSP PCB is better than those with the SnAgCu solder paste with 99% confidence. This agreed very well with finite element simulation results reported in Sec. 4.2.4.
8. More thermal cycling tests and finite element simulations should be done on lead-free assemblies with different dwell times.

4.4 Failure Analyses of 1657CCGA Assemblies

4.4.1 Failure Analyses of CCGA with SAC and SnPb Solder Pastes

Figures 4.35 through 4.46 show the images of failed samples. Figures 4.35 through 4.40 and 4.41 through 4.46, respectively, are for the 1657CCGA package with the SnAgCu and SnPb solder pastes. Those with

FIGURE 4.35 Photograph showing portion of the outer-row columns with SnAgCu solder paste.

FIGURE 4.36 Close-up photograph showing the outer-row columns with SnAgCu solder paste.

FIGURE 4.37 Close-up of fracture (totally) in the Sn90Pb solder column near the ceramic substrate with SnAgCu solder paste.

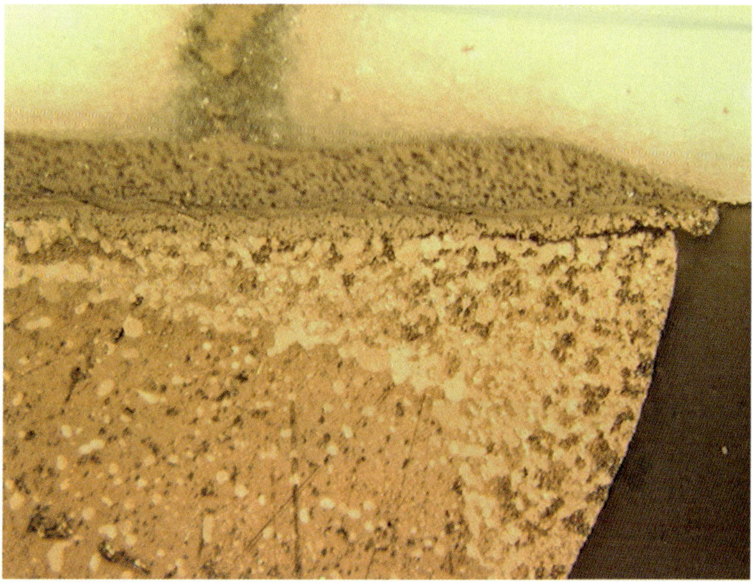

Figure 4.38 Close-up of cracks in the solder fillet near the interface between the solder fillet and the IMC of the ceramic substrate with SnAgCu solder paste.

Figure 4.39 Close-up of fracture (totally) in the Sn90Pb solder column near the PCB with SnAgCu solder paste.

FIGURE 4.40 Sn90Pb Solder column on microvia pad PCB showing the dissolution of Pb into the SnAgCu solder joint with SnAgCu solder paste.

FIGURE 4.41 Photograph showing portion of the outer-row columns with Sn37Pb solder paste.

lead-free solder paste failed at 2128 and 2722 cycles (the left line of Fig. 4.34) and those with tin-lead solder paste failed at 3451 and 4507 cycles (the right line of Fig. 4.34). It should be pointed out that failure analysis was performed on parts subsequent to 7000 cycles, that is, samples were not taken out of the thermal cycling chamber until 7000 cycles.

In all cases, fractured solder columns are common on the perimeter rows of the package, as shown in Figs. 4.35, 4.36, 4.41, and 4.42.

FIGURE 4.42 Close-up photograph showing the outer-row columns with Sn37Pb solder paste.

FIGURE 4.43 Close-up of fracture (totally) in the Sn90Pb solder column near the ceramic substrate with Sn37Pb solder paste.

FIGURE 4.44 Close-up of fracture (totally) in the Sn90Pb solder column near the PCB with Sn37Pb solder paste.

FIGURE 4.45 Close-up of fracture (totally) in the solder joint near the PCB with Sn37Pb solder paste.

FIGURE 4.46 Close-up of crack in the solder joint near the interface between the solder joint and the IMC of the ceramic substrate with Sn37Pb solder paste.

It can be seen that nearly all perimeter-row columns have fractures near the interface between the solder columns and the solder fillets on the package side and the interface between the solder columns and the solder fillets on the PCB side. This compares very well with the DFR results. Also, in general, more solder columns failed on the package side. DFR also predicts higher creep strain and stress at this location.

Figures 4.37 and 4.39 (with the lead-free solder paste) and Figs. 4.43 and 4.44 (with the tin-lead solder paste) show the typical cross-sections of total fractures (separations) of the solder columns near the package side and near the PCB side. It can be seen that the failure mode is bending fatigue, which is due to the global thermal expansion mismatch between the ceramic package and the organic PCB during thermal cycling. Most of the failures are at these locations. This compares very well with the simulation results. Also, in general, more solder columns failed in the 1657CCGA package with lead-free solder paste. This agrees with the DFR results.

Figs. 4.37, 4.38, and 4.46 show cracks near the interface between the solder fillet and the IMC (intermetallic compound) of the package substrate. Only a few columns failed at this location.

Figure 4.45 shows a total fracture near the interface between the solder fillet and the IMC of the PCB. This failure location is rare and most likely it is caused by either pick and place error, solder column misalignment, or not enough solder paste at the first place.

Figure 4.40 shows a typical solder joint on a copper pad with microvia with SnAgCu lead-free solder paste. It can be seen that the lead from the Sn90Pb solder column is dissolved into the SnAgCu solder during reflow, even though the maximum reflow temperature is below the melting temperature (302°C) of the Sn90Pb solder alloy, this is by dissolution, not by melting. The lead dendrites can be seen throughout the solder fillet connecting the Sn90Pb to the PCB.

4.4.2 Summary and Recommendations

Some important results and recommendations for failure analyses are summarized in the following:

1. With both SnAgCu and SnPb solder pastes, fractured solder columns are common on the perimeter rows of the package.
2. With both SnAgCu and SnPb solder pastes, most of the failures occur near the fillet tips of the solder column.
3. With both SnAgCu and SnPb solder paste, a few failures occur in the solder fillet near the interface between the solder and the IMC of the package substrate.
4. With both SnAgCu and SnPb solder pastes, the failure mode of the solder columns is bending fatigue and the failure mode of the solder fillet (joint) near the package side is shear fatigue.
5. With both SnAgCu and SnPb solder pastes, there are more failures at the fillet tip near the package side.
6. There are more failures in the solder columns with the SnAgCu solder paste than with the SnPb solder paste.
7. The failure locations and failure modes of the CCGA solder joints compare very well with the DFR results and indicate the thermal cycling test was properly performed.
8. The lead from the Sn90Pb solder columns dissolves into the SnAgCu solder joint during reflow, even though the reflow temperature is below the melting temperature of the Sn90Pb solder alloy.
9. In order to have a good quality of the 1657CCGA solder joints, the solder columns must meet the specifications, the solder paste must be sufficient, and the pick and place must be properly done.

4.5 References

1. Cole, M., DeSousa, I., Interrante, M., Jozwiak, J., June, M., Martin, G., Milkovich, C., Ross, J., and Shannon, M., "The New Millennium for CCGA—Beyond 2000 I/O," *Proceedings of SMI*, Chicago, IL, 2001, pp. 297–306.

2. Caulfield T., Cole, M., Cappo, F., Zitz, J., and Benenati, J., "An Overview of Ceramic Ball and Column Grid Array Packaging," in *Ball Grid Array Technology*, Lau, J., ed., McGraw-Hill, New York, 1995, pp. 131–169.
3. Banks, D., Hoebener, K., and Viswanadham, P., "Ceramic Ball Grid Array Assembly," in *Ball Grid Array Technology*, Lau, J., ed., McGraw-Hill, New York, 1995, pp. 171–192.
4. Master, R., Cole, M., and Martin, G., "Ceramic Column Grid Array for Flip Chip Applications," *IEEE Proceedings of Electronic Components & Technology Conference*, 1995, pp. 925–929.
5. Ray, S., et. al., "Ceramic Column Grid Array for a High Performance Workstation Application," *IEEE Proceedings of Electronic Components and Technology Conference*, 1997, pp. 319–324.
6. Ray, S., et. al., "CLASP Ceramic Column Grid Array Technology for Flip Chip Carriers," *Proceedings of SEMICON West*, 1999, pp. 1–7.
7. Interrante, M. J., et. al., "High Density Column Grid Array Connections and Method Thereof," *US Patent No. 6,429,388*, 2002.
8. Lau, J., Smetana, J., R. Horsley, K. Snowdon, D. Shangguan, J. Gleason, I. Memis, D. Love, W. Dauksher, and B. Sullivan, "Design, Materials, and Process for Lead-Free Assembly of High-Density Packages," *Journal of Soldering & Surface Mount Technology*, Vol. 16, No. 1, 2004, pp. 53–62.
9. Lau, J., W. Dauksher, J. Smetana, R. Horsley, D. Shangguan, T. Castello, I. Menis, D. Love, and B. Sullivan, "Design for Lead-Free Solder Joint Reliability of High-Density Packages," *Journal of Soldering & Surface Mount Technology*, Vol. 16, No. 1, 2004, pp. 12–26.
10. Lau, J., and W. Dauksher, "Reliability of a 1657CCGA (Ceramic Column Grid Array) Package with 96.5Sn3.9Ag0.6Cu Lead-Free Solder Paste on PCBs (Printed Circuit Boards)," *ASME Transactions, Journal of Electronic Packaging*, Vol. 127, June 2005, pp. 96–105.
11. Lau, J., N. Hoo, R. Horsley, J. Smetana, D. Shangguan, W. Dauksher, D. Love, I. Menis, and B. Sullivan, "Reliability Testing and Data Analysis of Lead-Free Solder Joints for High-Density Packages," *Journal of Soldering & Surface Mount Technology*, Vol. 16, No. 2, 2004, pp. 46–68.
12. Lau, J., D. Shangguan, T. Castello, R. Horsley, J. Smetana, W. Dauksher, D. Love, I. Menis, and B. Sullivan, "Failure Analysis of Lead-Free Solder Joints for High-Density Packages," *Journal of Soldering & Surface Mount Technology*, Vol. 16, No. 2, 2004, pp. 69–76.
13. Lau, J. H., Todd Castello, Dongkai Shangguan, Walter Dauksher, Joe Smetana, Rob Horsley, Dave Love, Irv Menis, and Bob Sullivan, "Failure Analysis of Lead-Free Solder Joints of an 1657CCGA (Ceramic Column Grid Array) Package," *IMAPS Transactions, Journal of Microelectronics and Electronics Packaging*, Vol. 4, Number 3, Third Quarter, 2007, pp. 189–213.
14. ANSYS, *User's Manual*, Release 6.1, 2002.
15. Vianco, P. and Rejent, J., "Compression Deformation Response of 95.5Sn-3.9Ag-0.6Cu Solder," *UCLA Lead-Free Workshop*, 2002.
16. Lau, J. H., Dauksher, W., and Vianco, P., "Acceleration Models, Constitutive Equations, and Reliability of Lead-Free Solders," *IEEE Proceedings of Electronic Components & Technology Conference*, pp. 229–236, 2003.
17. Lau, J. H., and Pao, Y., Solder Joint Reliability of BGA, CSP, Flip Chip, and Fine Pitch SMT Assemblies, McGraw-Hill, New York, 1997.
18. Morrow, J., "Cyclic Plastic Strain Energy and Fatigue of Metals," ASTM STP-378, 1965, pp. 45–87.
19. Lau, J. H., Mei, Z., Pang, S., Amsden, C., Rayner, J., and Pan, S., "Creep Analysis and Thermal-Fatigue Life Prediction of the Lead-Free Solder Sealing Ring of a Photonic Switch," *ASME Transactions, Journal of Electronic Packaging*, Vol. 124, 2002, pp. 403–410.
20. Lau, J. H., "Reliability of Lead-Free Solder Joints," *ASME Transactions, Journal of Electronic Packaging*, Vol. 128, September 2006, pp. 297–301.

CHAPTER 5

Reliability of PBGA Lead-Free Solder Joints (With and Without Underfills)

5.1 Introduction

The plastic ball grid array (PBGA) package was invented by Motorola in the early 1990s, and has had significant effects on semiconductor packaging assemblies and tests [1]. Compared to the lead-frame technology, PBGA uses an organic, for example, bismaleimide triazine (BT) substrate to fan out the circuitry from the chip on its top side, through the vias (or wrap around the edges [2, 3]) of the substrate, and connect to the area-array solder balls on its bottom side, which are then soldered to the printed circuit board (PCB). Because the coefficient of thermal expansion (CTE) mismatch between the BT substrate ($\sim 15 \times 10^{-6}/°C$) and PCB ($\sim 18 \times 10^{-6}/°C$) is not very large, most solder joints are reliable for most operating conditions, especially for the SnPb solder joints.

However for lead-free solder joints [4–42] and very high reliability requirements, for example, more than 10 years of life, due to lack of reliability test data, the confidence level of the lead-free solder joints to survive is not very high and thus methods to enhance their reliability are needed. One method is to use underfills to cement the PBGA carrier to the PCB, to reduce the relative displacement between the carrier and the PCB, and thus reduce the stresses and strains at the lead-free solder joints. Figure 5.1 shows an example that uses underfills between the PBGAs and the PCB to enhance the solder joint reliability. Underfills also help in drop conditions.

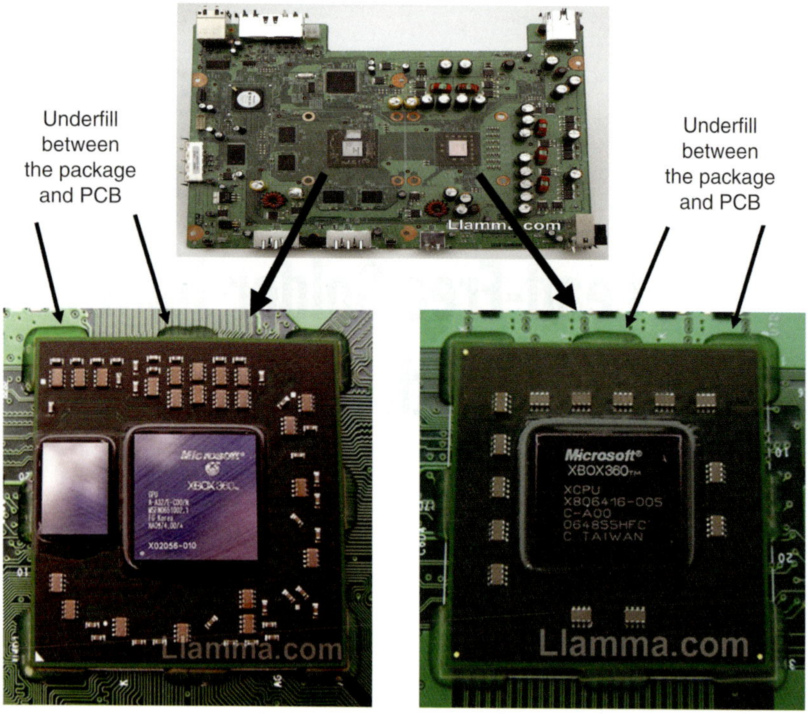

FIGURE 5.1 Microsoft's Xbox360 packages with underfills.

Underfills were first used in the late 1980s by Hitachi, Citizen Watch, and IBM at Yasu for solder-bumped flip chips on low-cost organic substrates or PCBs. This is because the CTE mismatch between the silicon chip ($2.5 \times 10^{-6}/°C$) and the organic substrate or PCB is very large and the flip-chip solder joints don't have a chance to survive under most operating conditions. Thus, underfill is a must for low-cost flip-chip applications [9], except for very small chip size.

In this chapter, the effects of underfills on the thermal cycling test and drop test of lead-free (SnAgCu) PBGA packages on ImAg (immersion silver) PCB are investigated. Three different underfills are studied. These underfills are dispensed on the PCB to fill the gaps between the BT substrate of the PBGA, the Sn3wt%Ag0.5wt%Cu (SAC) solder joints, and the PCB.

In addition, two more sets of samples (without underfill) are tested (one is with SAC solder balls and paste, and the other is with Sn37wt%Pb solder balls and paste), which serve as controls. The sample size for each set of samples is 48. The life distribution, reliability function, failure rate, and mean-time-to-failure (MTTF) of the PBGA solder joints under thermal cycling and drop tests for these five different sets of samples have been determined. The true characteristic life, true MTTF, and true Weibull slope of the PBGA solder joints under certain confidence requirements have also been determined. Failure analysis has been performed and examined on the failed samples. The results presented in this chapter should be useful for understanding the effects of underfills on the lead-free PBGA solder joint reliability.

5.2 Reliability of PBGA SAC Solder Joints With and Without Underfills under Thermal Cycling Condition

5.2.1 Components, Test Board, and Underfills

The package used for this study is shown in Fig. 5.2 [43]. It is Amkor's A-CABGA160.8 mm–12 mm. It is a chip-size package and has 160 lead-free SAC solder balls on a 0.8-mm pitch. The dummy chip dimensions are 10×10 mm, body dimensions are 12×12 mm, and for other dimensions such as molding, substrate, and solder balls, please refer to Amkor's product catalog.

FIGURE 5.2 Amkor's daisy-chain A-CABGA 160-.8 × -12 mm package and thermal cycling test board with underfill.

The test board is also shown in Fig. 5.2. It is made of fiber-reinforced glass with high glass transition temperature (T_g) and with a dimension of 265.5 × 165.5 × 1.6 mm. It is an eight-layer laminate board with 1 oz copper for the inner layers and 0.5 oz for the outer layers. The PCB copper pad is ImAg, and its diameter (300 μm) is non-solder mask defined (NSMD) with a 400-μm solder mask opening. It can be seen that there are 48 components on the test board.

In a test board of this type, routing is largely a matter of connecting the traces described by the schematic. As such, it is important to understand the schematic and its requirements. The component used in this test is interconnected between pins inside the package in an alternating pattern so as to provide a daisy chain connection when the component is soldered to the PCB. The internal silicon die on this device is typically a dummy die, placed in the package to provide the physical characteristics of functional die. The pads on the PCB are interconnected in an alternating pattern so as to provide a daisy chain connection with the component. These are the test chains that are monitored during thermal cycling to evaluate the solder joint. A typical schematic on the test board for the component is shown in Fig. 5.16.

There are five different sets of test boards, as shown in Table 5.1, namely: (1) PBGA with Sn37Pb solder balls and paste (no underfill) or just called SnPb; (2) PBGA with SAC solder balls and paste (no underfill) or just called SAC; (3) PBGA with SAC solder balls and paste (with no-filler underfill-X) or called SACX; (4) PBGA with SAC solder balls and paste (with no-filler underfill-Y) or called SACY; and (5) PBGA with SAC solder balls and paste (with filler underfill-Z) or called SACZ. The material properties and characteristics of these underfills are shown in Table 5.2. It should be noted that underfill-Z has the minimum CTE because of its 40% filler contents. The CTE of BT substrate, PCB, Sn37Pb, and SAC are [9], respectively, $15 \times 10^{-6}/°C$, $18.5 \times 10^{-6}/°C$, $24.5 \times 10^{-6}/°C$, and $20 \times 10^{-6}/°C$.

Symbols	Descriptions
SnPb	SnPb PBGA with SnPb paste (no underfill)
SAC	SAC PBGA with SAC paste (no underfill)
SACX	SAC PBGA with SAC paste (with no-filler underfill-X)
SACY	SAC PBGA with SAC paste (with no-filler underfill-Y)
SACZ	SAC PBGA with SAC paste (with filler underfill-Z)

SAC = Sn3wt%Ag.5wt%Cu; SnPb = Sn37wt%Pb

TABLE 5.1 Symbols and Descriptions

Properties	Underfill-X	Underfill-Y	Underfill-Z
Reworkable	No	No	No
Filler content	No	No	Yes (40%)
CTE (ppm/°C)	50	60	38
T_g (°C)	110	135	115
Viscosity (cps)	5500	7000	5000
Cure time @165°C	3 min	5 min	5 min
Pot life/Work life	7 days @ 22°C	14 days @ 25°C	7 days @ 25°C
Shelf life	6 m @ 2 to 8°C	6 m @ –5°C	6 m @ –5°C

TABLE 5.2 Underfills Material Characteristics

The most desirable features of underfills are [43–58]: (a) low viscosity (fast flow), which can increase throughput; (b) low curing temperature/fast curing time, which can reduce cost and be less harmful to other components; (c) low TCE, which can reduce thermal expansion mismatch between the chip, solder bumps, and PCB; (d) high modulus, which leads to good mechanical properties; (e) high glass transition temperatures, which enables endurance of higher temperature environments; (f) low moisture absorption, which can extend shelf life; (g) good adhesion, which can improve product life time; (h) reworkable; and (i) lead-free–process (260°C) compatible.

Before testing, all the test boards are inspected by x-ray and C-SAM (C-mode scanning acoustic microscope) for voids, opens, shorts, misalignments, insufficient solders, excess solders, and delaminations. In this study, all test boards are properly assembled and the typical x-ray (left) and C-SAM (right) images are shown in Fig. 5.3.

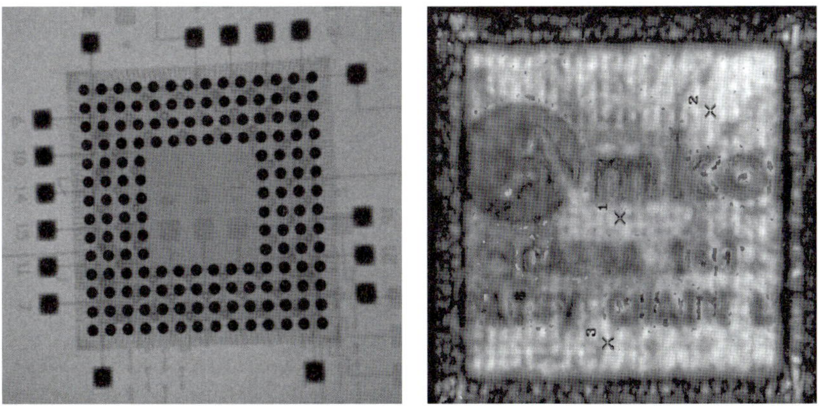

FIGURE 5.3 X-ray (left) and C-SAM (right) images of PBGA after assembly.

5.2.2 Design for Reliability

The methods and procedures of finite-element modeling and analyses of the current structures are exactly the same as those shown in Sec. 2.6. in Chapter 2. Figure 5.4 shows the typical stress and strain contours in the SnPb and SAC solder joints when they are subjected to temperature-cycling loading. The maximum von Mises (equivalent) stress and strain are located in the solder joint near the interface between the BT substrate and the solder, and near the interface between the solder and the PCB. Thus, if any failure occurs, it should be in one of these locations.

5.2.3 Test Chamber and Temperature-Cycling Profile

Thermal cycling is performed in a Votsch VTS 7027-15 environmental chamber. Heating and cooling are achieved by forced air convection. In order for the chamber to operate at peak efficiency, it is important that the airflow in the chamber be disrupted as little as possible. Airflow in the chamber is directed from top to bottom, and for this reason the boards are placed vertically in the chamber as shown in Fig. 5.5. The temperature-cycling profile measured (with the method of thermocouple probe shown in Fig. 2.32) on the solder joints is also shown in Fig. 5.5. It can be seen that the temperature ranges from −40°C to 125°C and the cycle time is 60 min with 15 min ramp up and down, 15 min hold at cold (−40°C), and 15 min hold at hot (125°C).

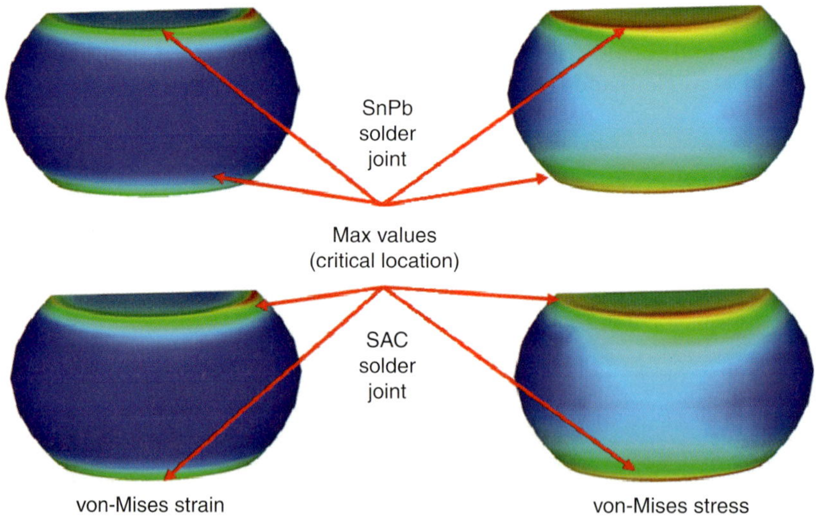

FIGURE 5.4 Finite-element analysis results showing the maximum stress/strain locations.

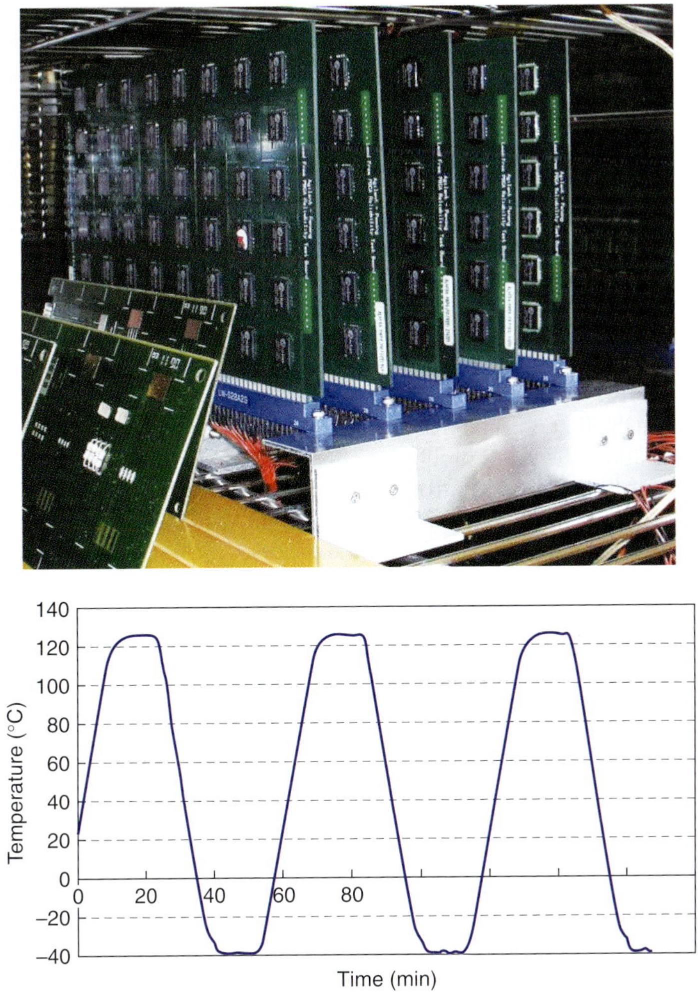

Figure 5.5 Temperature cycling setup and profile.

5.2.4 Data Acquisition, Failure Criterion, and Data Extraction

The test boards are inserted into sockets as shown in Fig. 5.5, which are connected to two data loggers (Agilent 37940A) for real-time monitoring. By monitoring the daisy chain resistance, the time-to-failure of each package can be recorded. In this case, the solder joint of a package is considered to have failed when the daisy chain resistance reaches infinity. However, the resistance may vary with the temperature changes. In order to catch the correct cycle-to-failure, the resistance of each daisy chain is recorded at 15-min intervals. By taking the data in this sampling rate, the condition of the daisy chain in each package at high temperature dwell, low temperature dwell, and room temperature are monitored.

5.2.5 Weibull Slope, Characteristic Life, and MTTF from Test Samples

Table 5.3 shows the number of failures, Weibull slope (β), characteristic life (θ), and MTTF = $\theta\Gamma(1+1/\beta)$ (where Γ is the gamma function) for the thermal-cycling test samples. It can be seen that the number of failures, Weibull slope, characteristic life, and MTTF are, respectively, 43, 3.19, 2114, and 1893 for SAC; 47, 3.01, 1963, and 1753 for SACX; 48, 3.78, 2107, and 1905 for SACY; 30, 3.63, 3555, and 3207 for SACZ; and 48, 16.94, 2192, and 2124 for SnPb. There are only 47 failures for SACX because the first one failed at 82 cycles, which is considered a assembly defect and thus is eliminated from the total samples. The Weibull slope of SnPb is unusually high (~17). This shows that the SnPb solder joints we made are very consistent, because the higher the Weibull slope, the more uniform the product.

5.2.6 True Characteristic Life

The thermal cycling test results at 90% confidence for SAC, SACX, SACY, SACZ, and SnPn, as shown in Figs. 5.6 through 5.10, respectively. In 90 out of 100 cases (the other 10 cases no one but God knows about), the characteristic life is from 1886 cycles to 2281 cycles for the SAC solder joints; from 1729 to 2126 cycles for the SACX solder joints; from 1898 to 2256 cycles for the SACY solder joints; from 3350 to 3559 cycles for the SACZ solder joints; and from 2141 to 2232 cycles for the SnPb solder joints (Table 5.4). For example, the true characteristic life of SAC solder joints without underfill can be at least 1886 cycles but no more than 2281 cycles at 90% confidence. In this study, the sample sizes are 47/48 and thus, the intervals are not very large. However, they are still sizable. Of course, at higher (e.g., 95%) confidence or smaller sample size, the intervals should be wider!

	No. of Failures/ No. of Samples	Weibull Slope	Characteristic Life (cycles)	MTTF (cycles)
SAC	43/48	3.19	2114	1893
SACX	47/47	3.01	1963	1753
SACY	48/48	3.78	2107	1905
SACZ	30/48	3.63	3555	3207
SnPb	48/48	16.94	2192	2124

TABLE 5.3 Weibull Slope, Characteristic Life, and MTTF for Test Samples

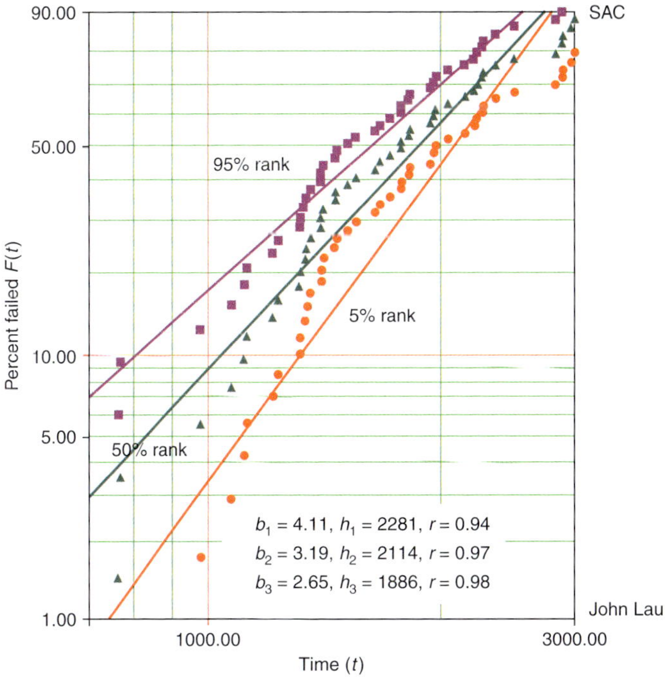

FIGURE 5.6 Thermal cycling test results of SAC solder joints with 90% confidence.

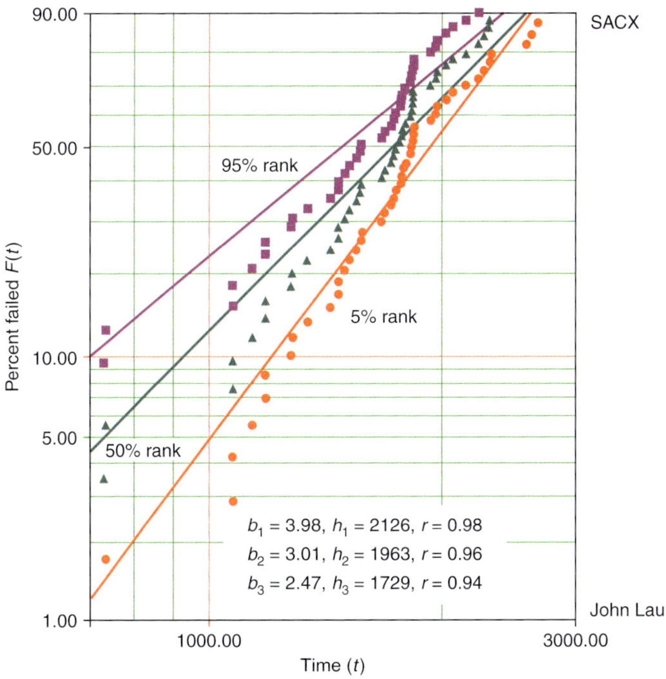

FIGURE 5.7 Thermal cycling test results of SACX solder joints with 90% confidence.

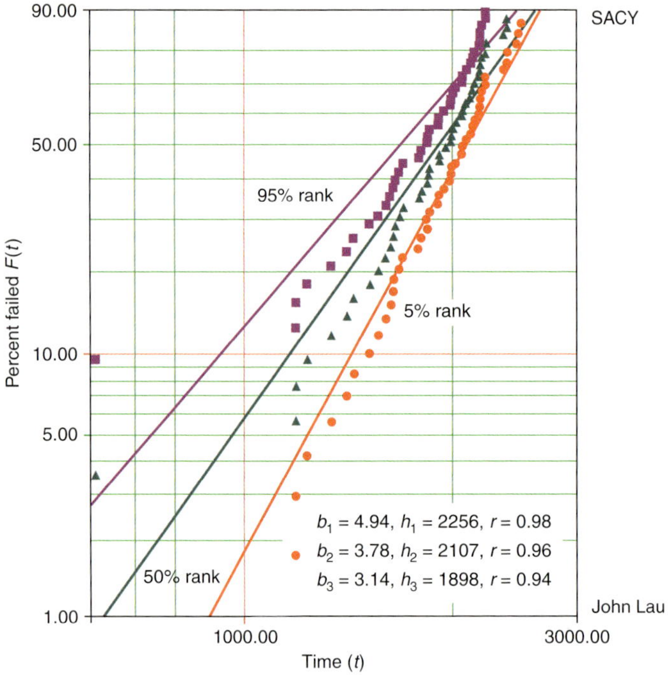

FIGURE 5.8 Thermal cycling test results of SACY solder joints with 90% confidence.

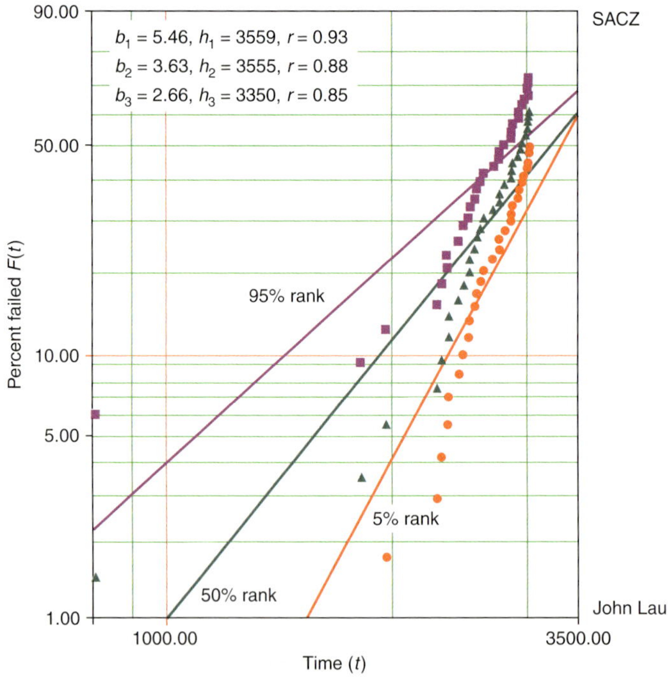

FIGURE 5.9 Thermal cycling test results of SACZ solder joints with 90% confidence.

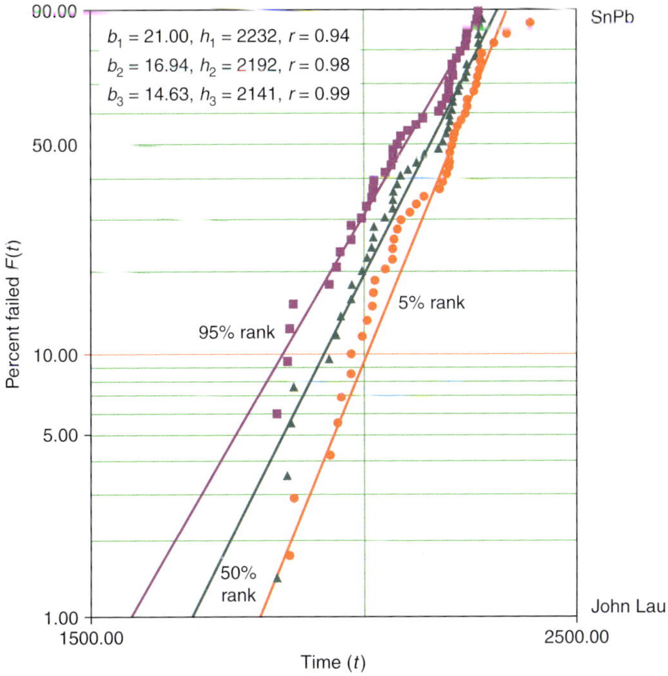

FIGURE 5.10 Thermal cycling test results of SnPb solder joints with 90% confidence.

	True Characteristic Life (cycles) at 90% Confidence
SAC	$1886 \leq \theta \leq 2281$
SACX	$1729 \leq \theta \leq 2126$
SACY	$1898 \leq \theta \leq 2256$
SACZ	$3350 \leq \theta \leq 3559$
SnPb	$2141 \leq \theta \leq 2232$

TABLE 5.4 True Characteristic Life (θ) at 90% Confidence

5.2.7 True MTTF

The percent failed at the sample mean can be determined by $F(x = mean)$ and is equal to 51%, 51%, 50%, 49%, and 44% for SAC, SACX, SACY, SACZ, and SnPb solder joints, respectively. The true expected (or MTTF) values at 90% confidence for SAC, SACX, SACY, SACZ, and SnPn can be determined, respectively, from Figs. 5.6 through 5.10, with the determined percent failed and the 95% rank and 5% rank probabilities, and are shown in Table 5.5. It can be seen that their expected life in 90 out of 100 cases is from 1660 cycles to 2101 cycles for the SAC solder joints; from 1510 to 1954 cycles for the SACX

	True MMTF (μ) (cycles) at 90% Confidence
SAC	$1660 \le \mu \le 2101$
SACX	$1510 \le \mu \le 1954$
SACY	$1688 \le \mu \le 2097$
SACZ	$2883 \le \mu \le 3314$
SnPb	$2063 \le \mu \le 2175$

TABLE 5.5 True MTTF (Mean-Time-to-Failure) at 90% Confidence

solder joints; from 1688 cycles to 2097 cycles for the SACY solder joints; from 2883 cycles to 3314 cycles for the SACZ solder joints; and from 2063 cycles to 2175 cycles for the SnPb solder joints. For example, the true mean life of SACZ solder joints in 90 out of 100 cases can be as low as 2883 cycles and as high as 3314 cycles. Again, the mean life intervals (ranges) are larger for higher confidence levels or smaller sample sizes.

5.2.8 True Weibull Slope

For a given 90% confidence (C = 0.9), the Weibull slope error is searched from Eq. (2.13) for a given number of failures and plotted in Fig. 5.11 (a special case of Fig. 2.2) for engineering practice convenience. Based on this curve, the true Weibull slope at 90% confidence for SAC, SACX, SACY, SACZ, and SnPb can be determined, and is shown in Table 5.6. It can be seen that in 90 out of 100 cases, the true Weibull slopes of SAC, SACX, SACY, SACZ, and SnPb solder joints range, respectively, from 2.62 to 3.76, 2.5 to 3.52, 3.14 to 4.42, 2.83 to 4.43, and 14.06 to 19.82. Again, the ranges of the intervals for the true Weibull slope depend on the required confidence and sample size (more precisely, the number of failures).

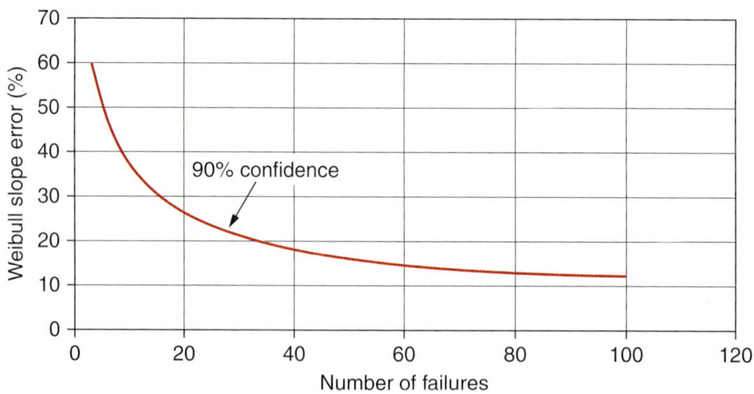

FIGURE 5.11 Weibull slope error (%) at 90% confidence.

	True Weibull Slopes (β) at 90% Confidence
SAC	$2.62 \leq \beta \leq 3.76$
SACX	$2.50 \leq \beta \leq 3.52$
SACY	$3.14 \leq \beta \leq 4.42$
SACZ	$2.83 \leq \beta \leq 4.43$
SnPb	$14.06 \leq \beta \leq 19.82$

TABLE 5.6 True Weibull Slope at 90% Confidence

5.2.9 Comparison of MTTF from Two Sets of Test Samples

One of the difficult tasks in statistical analysis of life-test data is to draw conclusions about a population from a small sample size. It is even more difficult to compare the populations of two products from the knowledge of their test data (with small sample size). If the quality (e.g., mean life) of one product lot is found to be superior to another, how *confident* (P) can it be that the same is true of their populations?

Figure 5.12 plots the life distributions of SAC, SACX, SACY, SACZ, and SnPb samples and their number of failures; sample Weibull slopes and mean lives (MTTF) have been given in Table 5.3. Thus, based Eq. (2.15), the confidence levels for comparing every two of the samples can be determined, as shown in Table 5.7. From Fig. 5.12 and Table 5.7 it can be seen that the solder-joint quality (mean life) of SACZ is better than that of SAC with 99.99% confidence. This means that in 9999 out of 10,000 cases, the expected (mean) life of the lead-free solder joint with underfill-Z is better than that of the lead-free solder joint without underfill. The underfill-Z cements the BT substrate of the package to the PCB. Thus, when the whole structure is subjected to thermal cycling, the package and PCB deform (bending, shearing, and twisting) as a unit and the relative displacement between the BT substrate and the PCB is very small. Consequently, the stresses and strains in the solder joints are small.

Similarly, from Fig. 5.12 and Table 5.7, the quality of SACZ solder joints is better than that of SnPb solder joints without underfill with very high confidence (99.99%) at this particular test condition. Also, from Fig. 5.12 and Table 5.7 it can be seen that the mean life of SnPb solder joints without underfill is better than that of the SAC lead-free solder joints without underfill at 98.9% confidence (at least at the present test condition).

Finally, the mean life of the SACY solder joint with underfill-Y without fillers is *better* than the SAC solder joints without underfill at 51.45% confidence (i.e., in 51 out of 100 cases). Since the confidence is so low, thus the last statement is not statistically significant. Essentially the mean life of the SAC and SACY solder joints are statistically the same! This means that the underfill-Y without fillers doesn't help

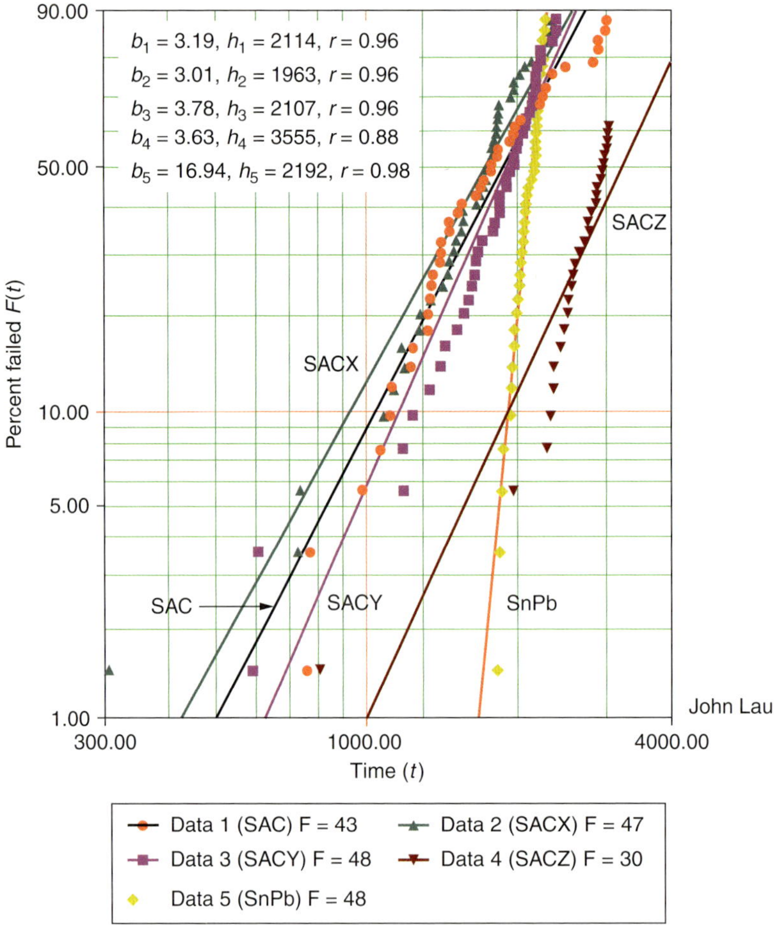

FIGURE 5.12 Thermal cycling results of SAC, SACX, SACY, SACZ, and SnPb solder joints.

Comparing	MTTF Ratio	Determined Confidence Levels
SACZ/SAC	3207/1893 = 1.69	99.99%
SACZ/SnPb	3207/2124 = 1.51	99.99%
SnPb/SAC	2124/1893 = 1.12	98.90%
SACY/SAC	1905/1893 = 1.01	51.45%

TABLE 5.7 Calculated Confidences for Comparing the MTTF of Two Populations

to reduce the stresses and strains much in the lead-free solder joints and is not as good as underfill-Z. One of the reasons for this could be that there are no fillers in underfill-Y and thus its CTE is very high compared with that of the BT substrate, solder joints, and PCB. Also, since there are no fillers, the Young's modulus (stiffness) of the underfill-Y is lower than that of underfill-Z. Consequently, there is a large local thermal expansion mismatch among the underfill-Y, the BT substrate, solder joints, and PCB—even the global thermal expansion mismatch between the package and PCB is reduced.

The effect of underfill-X without fillers on SAC solder joint reliability is very similar to underfill-Y, as shown in Fig. 5.12 and Table 5.7, and the material properties (large CTE and no filler) of these two underfills. It should be pointed out that all the results in these sections are for thermal cycling at a particular condition (−40°C to 125°C). Since the acceleration factor for SAC is not known, no conclusion could be made at other temperature conditions.

5.2.10 Failure Analyses

Based on design for reliability (DFR), the maximum stress and strain are located in the solder joint near the interface between the BT substrate and the solder, and near the interface between the solder and the PCB. Thus, if any failure occurs, it should be in one of these locations (Fig. 5.4).

Figures 5.13 through 5.15 show some typical cross-sectional areas of the failed samples. First of all, it can be seen that all the failure locations occurred at the solder joint near the interface areas. For all the SnPb solder-joint failures, the crack location is near the interface between the SnPb solder and the BT substrate. However, for all the SAC solder-joint failures (with and without underfills), the crack location is near the interface between the SAC solder and the PCB. This could be due to the more Sn (~96wt%) in the SAC solder than that (63wt%) in the SnPb solder. The IMCs such as Cu_3Sn and Cu_6Sn_5

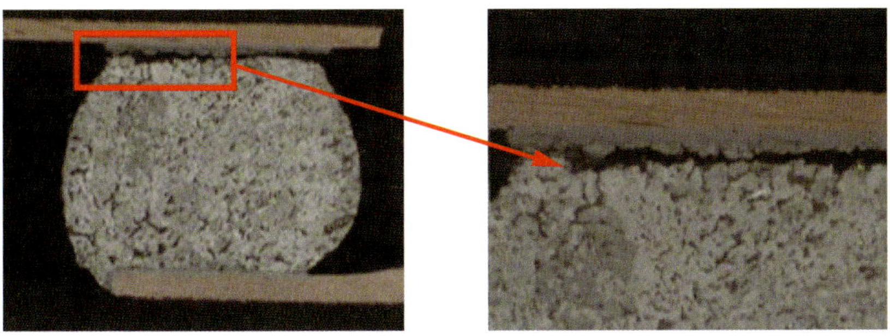

100X 200X

FIGURE 5.13 Cross-section of an SnPb solder joint without underfill (failed at 2189 cycles).

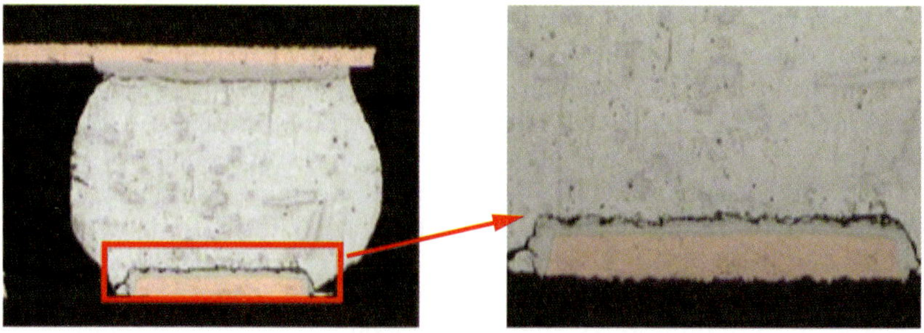

FIGURE 5.14 Cross-section of an SAC305 solder joint without underfill (failed at 2046 cycles).

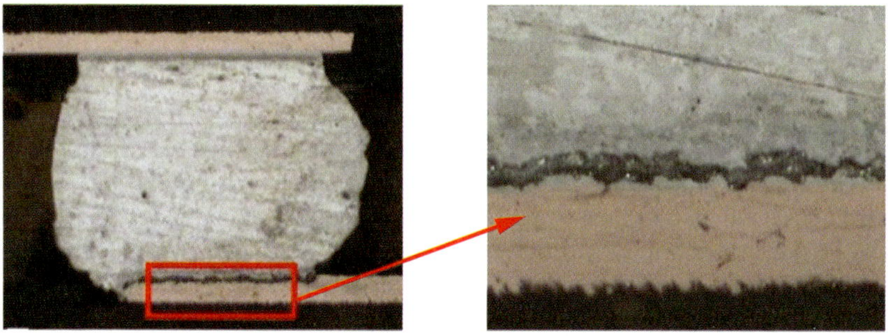

FIGURE 5.15 Cross-section of an SAC305 solder joint with underfill (failed at 2088 cycles).

between the SAC solder and the Cu pad are thicker than those between the SnPb and the Cu pad, that is, the SAC consumes more Cu than the SnPb does. The immersion Ag on the Cu pad dissolves into the SAC or SnPb solder joints and cannot even be found with an energy-dispersive x-ray spectrometry (EDX).

5.2.11 Summary

Some important results are summarized in the following:

1. At 90% confidence (in 90 out of 100 cases), the true Weibull slope, true characteristic life, and true MTTF of the SnPb PBGA solder joints without underfill are, respectively, in the ranges of 14.06 and 19.82, 2141 and 2232, and 2063 and 2175.
2. At 90% confidence, the true Weibull slope, true characteristic life, and true MTTF of the SAC PBGA solder joints without underfill are, respectively, in the ranges of 2.62 and 3.76, 1886 and 2281, and 1660 and 2101.

3. At 90% confidence, the true Weibull slope, true characteristic life, and true MTTF of the SAC PBGA solder joints with underfill-X are, respectively, in the ranges of 2.5 and 3.52, 1729 and 2126, and 1510 and 1954.
4. At 90% confidence, the true Weibull slope, true characteristic life, and true MTTF of the SAC PBGA solder joints with underfill-Y are, respectively, in the ranges of 3.14 and 4.42, 1898 and 2256, and 1688 and 2097.
5. At 90% confidence, the true Weibull slope, true characteristic life, and true MTTF of the SAC PBGA solder joints with underfill-Z are, respectively, in the ranges of 2.83 and 4.43, 3350 and 3559, and 2883 and 3314.
6. At 98.9% confidence, the MTTF of SnPb PBGA solder joints without underfill is better than that of SAC PBGA solder joints without underfill at the present temperature condition (−40°C to 125°C, one-hour cycle).
7. At 99.99% confidence, the MTTF of SAC solder joints with underfill-Z is better than that of the SnPb solder joints without underfill at the present temperature condition (−40°C to 125°C, one-hour cycle).
8. Essentially, the MTTF of SAC solder joints without underfill and with underfill-X and underfill-Y is statistically the same!
9. At 99.99% confidence, the MTTF of ASC solder joints with underfill-Z is better than that of the SAC solder joints without underfill. Thus, the underfill-Z performs better than underfill-X and underfill-Y.
10. The failure locations of the failed samples are near the interface between the BT substrate and solder, and near the interface between the solder and PCB.
11. For SnPb PBGA failed samples, all the solder-joint cracks occur near the interface between the BT substrate and the solder joint.
12. For SAC PBGA failed samples with and without underfills, all the solder-joint cracks occur near the interface between the solder joint and the PCB.

5.3 Reliability of PBGA SAC Solder Joints With and Without Underfills under Drop Condition

Recently, many research results show that under certain loading conditions and for certain components, the SnAgCu interconnects perform worse than those of the SnPb. For example, in [28], Amagai and his colleagues show that for a 10- × 10-mm PBGA with 6- × 6-mm die size and mounted on 160- × 64-mm FR-4 two-metal-layer PCB, the SnAgCu solder joints don't perform as well as the SnPb solder joints under drop conditions.

The effects of aging and underfills on the drop test of lead-free (SAC) PBGA packages on ImAg PCB are investigated [44]. Only two different underfills (underfill-Y and underfill-Z) are studied. These underfills are dispensed on the PCB to fill the gaps between the BT substrate of the PBGA, the Sn3wt%Ag0.5wt%Cu solder balls and paste, and the PCB. In addition, two more sets of samples (without underfill) are tested (one is with SAC solder balls and paste, and the other is with Sn37wt%Pb solder balls and paste), which serve as controls.

The sample size for each of the four different sets of samples is 24, and half of them (12) are subjected to temperature aging (125°C for 500 hours). The test condition is based on JESD-B111 (the acceleration = 2900 g and the impact duration = 0.3 ms). A strain gauge is used to measure the deformation of the test board during impacts. The true characteristic life, true mean-time-to-failure (MTTF), and true Weibull slope of the PBGA interconnects under certain confidence requirements have also been determined. Failure analysis has been performed and examined on some of the failed samples. The results presented in this section should be useful for understanding the effects of aging and underfills on the lead-free PBGA interconnect reliability under drop conditions.

Figure 5.16 shows the test board for the drop experiments. There are four different sets of test boards, as shown in Table 5.8, namely, (1) PBGA with Sn37Pb solder balls and paste (no underfill); (2) PBGA with SAC solder balls and paste (no underfill); (3) PBGA with SAC solder balls and paste (with no-filler underfill-Y); and (4) PBGA with SAC solder balls and paste (with filler underfill-Z). For each set of test boards, half of the samples are subjected to thermal aging at 125°C for 500 h. The material properties and characteristics of these underfills have been shown in Table 5.2. The chip size, package, daisy chains, pad geometry, and characteristics are exactly the same as those in Sec. 5.2 for the thermal cycling test.

5.3.1 Drop Test Setup, Data Acquisition, and Drop Profile

A drop tower with guiding rods is used for the drop tests, as shown in Fig. 5.17. The PBGA packages are facing downward during the test. Four standoffs on the fixture provide the support and the spacing for deflections of the PCB during impacts.

Two high-speed oscilloscopes are used to record the drop profile and monitor the daisy chain resistance and strain in real time during the drop test. Since the high-speed oscilloscope can only measure voltage, an additional setup is required so the electrical continuity of the daisy chain can be monitored. A 100-Ω resistor is connected to each daisy chain in series, and a 5-V potential difference is applied. The voltage across the resistor is measured by the

FIGURE 5.16 Drop test board configuration and daisy-chain design.

high-speed oscilloscope. If the solder joint fails during the drop test, the resistance of the daisy chain will increase dramatically. As a result, the voltage across the 100-Ω resistor will decrease. The corresponding schematic diagram is shown in Fig. 5.18. The drop profile is shown in Fig. 5.19. It can be seen that according to

Sample Type	Thermal Aging Condition	Test Board	PBGA Qty	PBGA Location (#1 to #6) and Number of Drops to Failure (x means no failure at 30 drops)					
				#1	#2	#3	#4	#5	#6
SnPb solder joint w/o underfill	Without aging	#A	6	8	2	8	2	4	18
		#B	6	6	4	13	5	4	24
	Thermal aging @65°C, 500 h	#A	6	x	6	x	x	28	x
		#B	6	x	22	x	30	x	23
SAC solder joint w/o underfill	Without aging	#A	6	2	1	4	1	1	7
		#B	6	3	1	2	2	1	3
	Thermal aging @65°C, 500 h	#A	6	17	5	9	8	5	6
		#B	6	9	2	4	8	3	4
SAC solder joint with underfill-Y	Without aging	#A	6	x	x	x	x	x	x
		#B	6	x	x	x	x	x	x
	Thermal aging @65°C, 500 h	#A	6	x	x	x	x	x	x
		#B	6	x	x	x	x	x	x
SAC solder joint with underfill-Z	Without aging	#A	6	x	x	x	x	x	x
		#B	6	x	x	x	x	x	x
	Thermal aging @65°C, 500 h	#A	6	x	x	x	x	x	x
		#B	6	x	x	x	x	x	x

TABLE 5.8 Experimental Results of Drop Tests

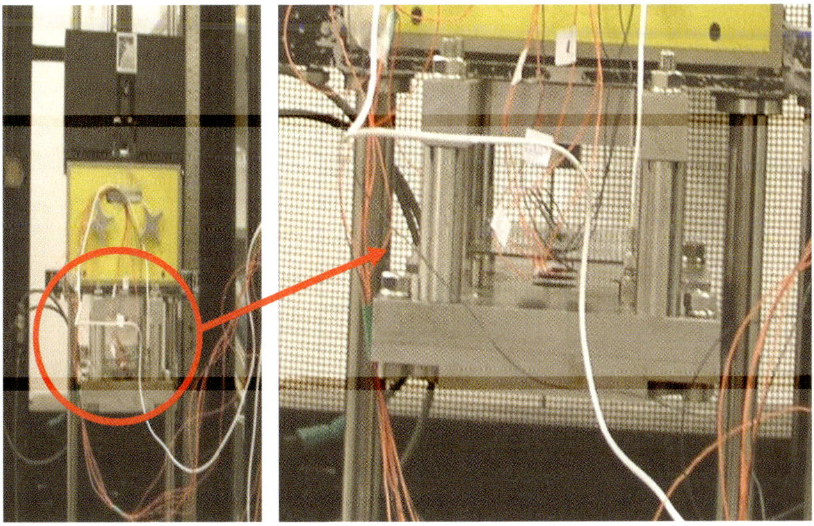

FIGURE 5.17 Drop tower with guiding rods for drop test.

JESD22-B111, the g-level required is 2900 g and the duration time is 0.3 ms.

5.3.2 Failure Criterion

The resistance of each component is monitored in real time during the drop test. The component is defined as a failure when there is a more than 20% increase in resistance. Every test board is dropped a maximum 30 times if there is no failure observed. Table 5.8 summarizes the experimental results.

5.3.3 Strain Time-History During Drop Impact

A strain gauge is attached to the middle of the back side of all the test boards to measure their deformation during impacts. A typical strain time-history is shown in Fig. 5.20. It can be seen that the maximum strain occurs right after the impact and then decreases to zero. The average maximum strain of these four sets of test boards is shown in Table 5.9. It can be seen that for all the cases considered, the maximum strain (deformation) of the thermal aging (125°C for 500 h) samples is smaller than those without aging. This could be due to the fact that aging bakes out the moisture and thus yields a stronger PCB and lighter assembly.

Also, it can be seen from Table 5.9 that for all the cases considered, the maximum deformation of the samples with underfills is smaller than those without underfills. This could be due to the fact that underfills "glue" the PBGA and the PCB together and lead to a stronger and higher bending-stiffness assembly.

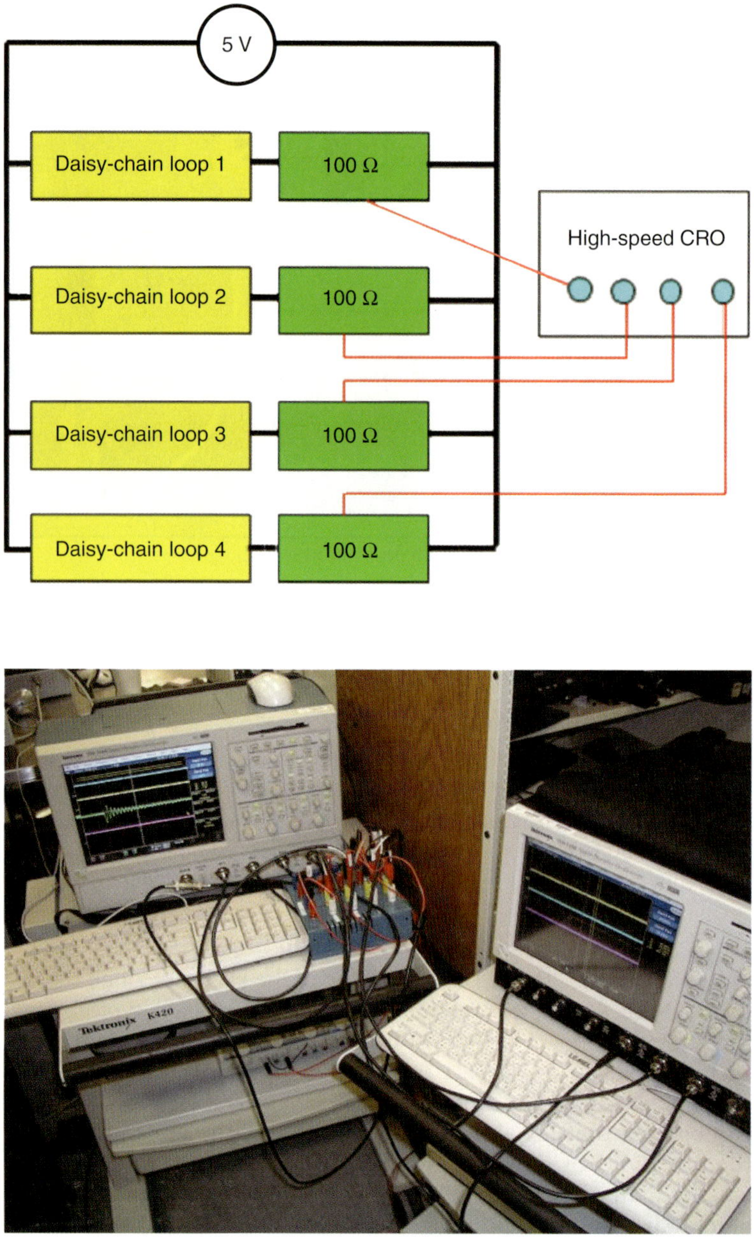

FIGURE 5.18 Data acquisition setup and system for drop test.

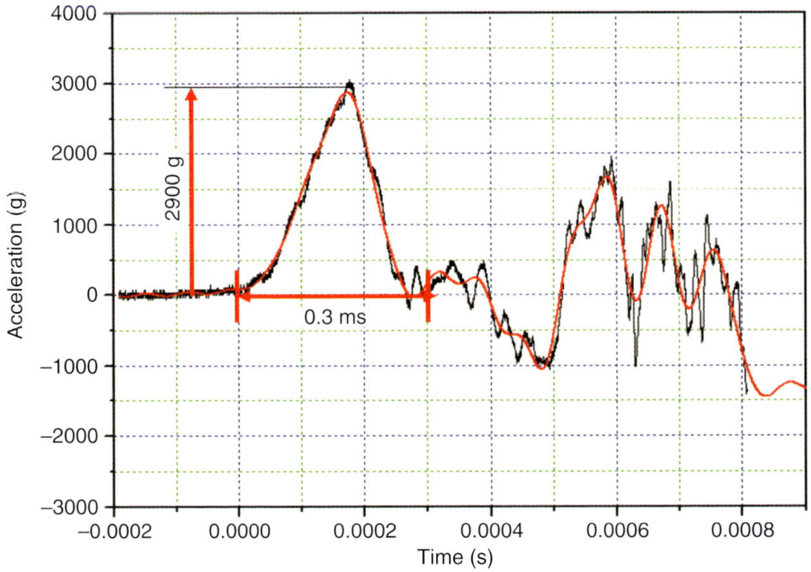

Figure 5.19 Drop profile according to JESD22-B111.

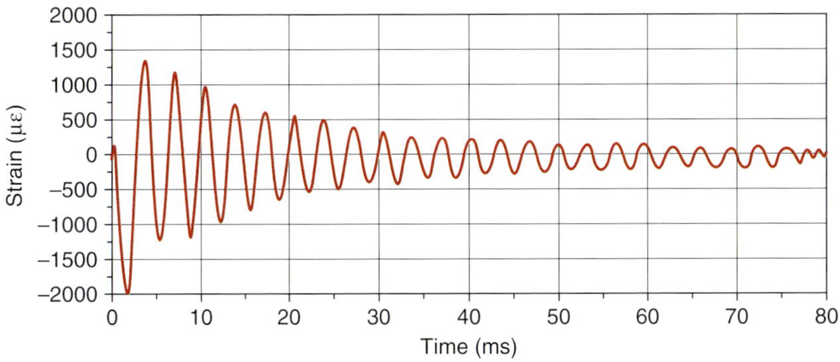

Figure 5.20 A typical strain time-history during drop test.

5.3.4 Weibull Slope, Characteristic Life, and MTTF from Test Samples

Table 5.10 shows the number of failures, Weibull slope (β), characteristic life (θ), and MTTF = $\theta\Gamma(1+1/\beta)$ (where Γ is the gamma function) for the drop test samples. It can be seen that the number of failures, Weibull slope, characteristic life, and MTTF are, respectively, 12, 1.34, 10, and 9 for the SnPb PBGA assembly; 5, 2.23, 45, and 40 for the aged SnPb PBGA assembly; 12, 1.52, 3, and 3 for the SAC PBGA assembly; and 12, 1.82, 8, and 7 for the aged SAC PBGA

Sample Type	Thermal Aging Condition	Average Max. Strain (Micro-strain)
SnPb solder joint w/o underfill	Without aging	−1895
	Thermal aging @125°C, 500 h	−1780
SAC solder joint w/o underfill	Without aging	−2011
	Thermal aging @125°C, 500 h	−1956
SAC solder joint with underfill-Y	Without aging	−1179
	Thermal aging @125°C, 500 h	−161
SAC solder joint with underfill-Z	Without aging	−1041
	Thermal aging @125°C, 500 h	−977

TABLE 5.9 Average Maximum Strain Under Drop Tests

Sample Types	No. of Failures/ No. of Samples	Weibull Slope	Characteristic Life (Drops)	MTTF (Drops)
SnPb	12/12	1.34	10	9
Aged SnPb	5/12	2.23	45	40
SAC	12/12	1.52	3	3
Aged SAC	12/12	1.82	8	7

TABLE 5.10 Weibull Slope, Characteristic Life and MTFF from Test Samples

assembly. In this study, the life (number of drops) of drop tests has been rounded to integers.

5.3.5 True Characteristic Life

The drop test results at 90% confidence for SnPb, aged SnPb, SAC, and aged SAC are shown in Figs. 5.21 through 5.24, respectively. It can be seen that in 90 out of 100 cases (the other 10 cases are unknowns) the characteristic life ranges from 5 to 15 drops for the SnPb PBGA assembly, from 35 to 45 drops for the aged SnPb PBGA assembly, from 2 to 4 drops for the SAC PBGA assembly and from 5 to 11 drops for the aged SAC assembly (Table 5.11). For example, the true characteristic life of the SnPb PBGA assembly can be at least 5 drops but no more than 15 drops at 90% confidence. In this study, the sample sizes are 12 and thus, the intervals are quite large. Of course, at higher (e.g., 95%) confidence, the intervals are even wider!

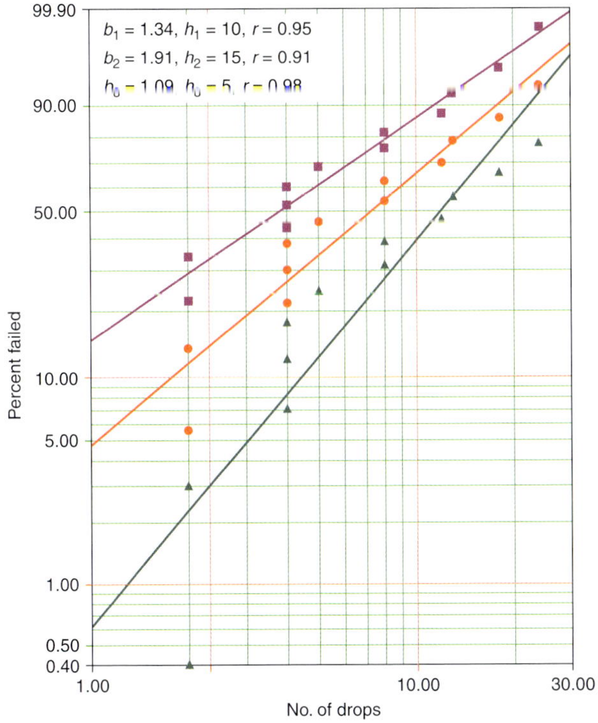

FIGURE 5.21 SnPb solder joint samples at 90% confidence.

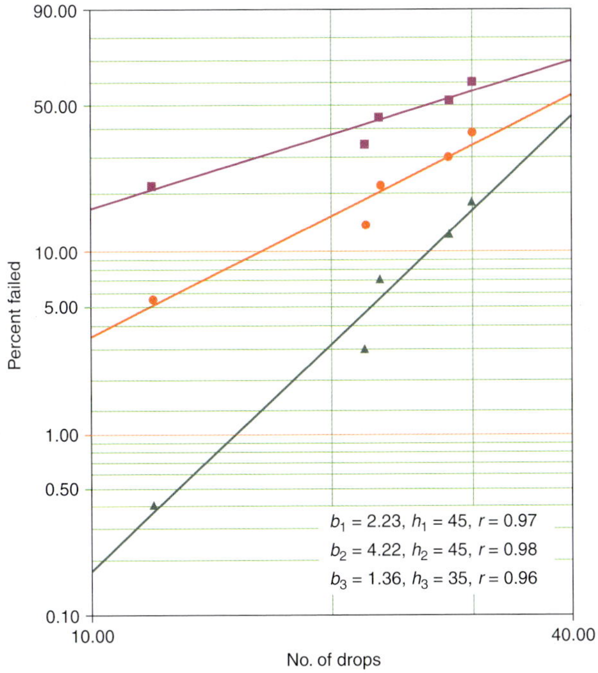

FIGURE 5.22 Aged SnPb solder joint samples at 90% confidence.

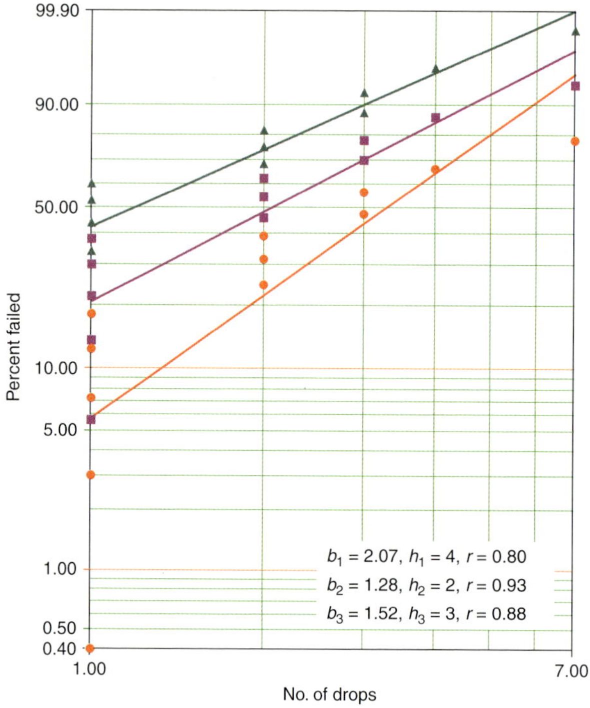

FIGURE 5.23 SAC solder joint samples at 90% confidence.

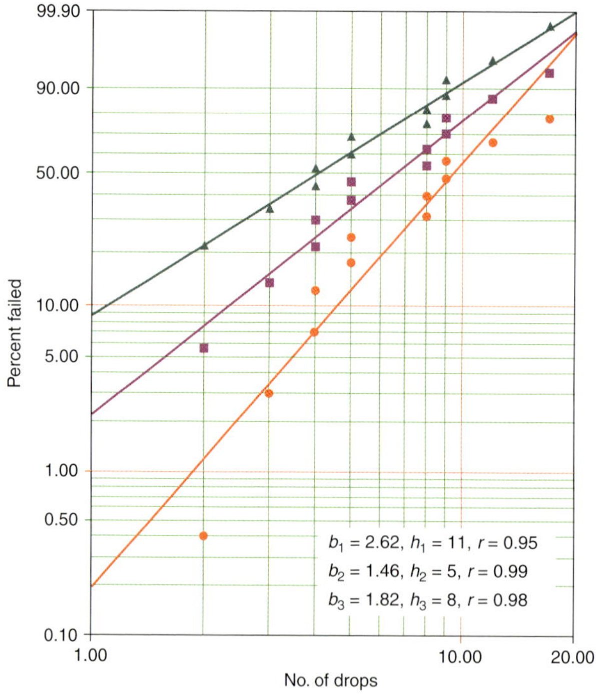

FIGURE 5.24 Aged SAC solder joint samples at 90% confidence.

Sample Type	True Characteristic Life (No. of Drops) at 90% Confidence
SnPb	$5 \leq \theta \leq 15$
SnPb with aging	$35 \leq \theta \leq 45$
SAC	$2 \leq \theta \leq 4$
SAC with aging	$5 \leq \theta \leq 11$

TABLE 5.11 True Characteristic Life (θ) at 90% Confidence

5.3.6 True MTTF

The percent failed at the sample mean can be determined by $F(x = mean)$ and is equal to 58%, 63%, 54%, and 63% for, respectively, SnPb, aged SnPb, SAC, and aged SAC PBGA assemblies. The true expected (or MTTF) values at 90% confidence for SnPb, aged SnPb, SAC, and aged SAC PBGA assemblies can be determined, respectively, from Figs. 5.21 through 5.24, with the determined percent failed (i.e., 58, 63, 54, and 63%) and the 95% rank and 5% rank probabilities, and are shown in Table 5.12. It can be seen that their expected life in 90 out of 100 cases is from 4 to 14 drops for the SnPb PBGA assembly, from 29 to 42 drops for the aged SnPb PBGA assembly, from 2 to 4 drops for the SAC PBGA assembly, and from 4 to 10 drops for the aged SAC assembly (Table 5.12). For example, the true mean life of the SAC PBGA assembly in 90 out of 100 cases can be as low as two drops and as high as four drops. Again, the mean life intervals (ranges) are larger for higher confidence levels.

5.3.7 True Weibull Slope

For a given 90% confidence ($C = 0.9$), the true Weibull slope for SnPb, aged SnPb, SAC, and aged SAC PBGA assemblies can be determined by the curve shown in Fig. 5.11, and is shown in Table 5.13. It can be seen that in 90 out of 100 cases, the true Weibull slopes of SnPb, aged SnPb, SAC, and aged SAC PBGA assemblies are, respectively, ranging from 0.88 to 1.8, 1.07 to 3.39, 1 to 2.04, and 1.2 to 2.44. Again, the ranges of the intervals for the true Weibull slope depend on the required confidence and the number of failures.

Sample Type	True MTTF (μ) (No. of Drops) at 90% Confidence
SnPb	$4 \leq \mu \leq 14$
SnPb with aging	$29 \leq \mu \leq 42$
SAC	$2 \leq \mu \leq 4$
SAC with aging	$4 \leq \mu \leq 10$

TABLE 5.12 True MTTF (μ) at 90% Confidence

Sample Type	True Weibull Slope at 90% Confidence
SnPb	$0.88 \le \beta \le 1.8$
SnPb with aging	$1.07 \le \beta \le 3.39$
SAC	$1.00 \le \beta \le 2.04$
SAC with aging	$1.20 \le \beta \le 2.44$

TABLE 5.13 True Weibull Slope at 90% Confidence

5.3.8 Effects of Aging and Underfill on the Reliability of PBGA Interconnects under Drop Impacts

Figure 5.25 plots the life distributions of SnPb, aged SnPb, SAC, and aged SAC PBGA assemblies and their number of failures, and sample Weibull slope and mean lives (MTTF) are given in Table 5.10. Thus,

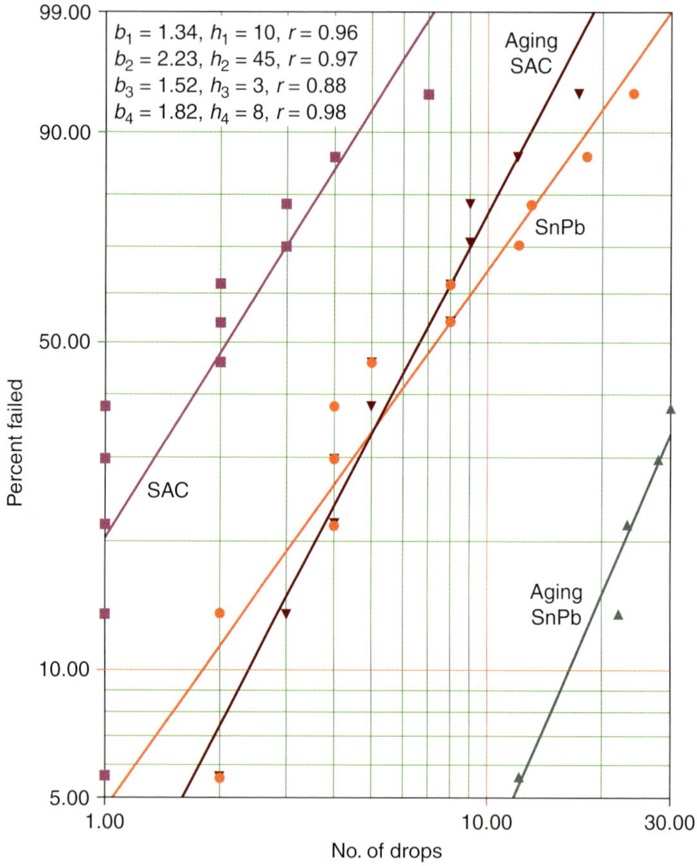

FIGURE 5.25 Solder joint life distribution of SnPb, aged SnPb, SAC, and aged SAC PBGA assemblies.

Comparing	MTTF Ratio	Determined Confidence Levels
SnPb/SAC	9/3 – 3	99.99%
Aged-SnPb/SnPb	40/7 = 5.7	99.99%
Aged-SnPb/Aged SAC	40/7 = 5.7	99.99%
Aged-SAC/SAC	7/3 = 2.3	99.99%

TABLE 5.14 Calculated Confidences for Comparing the MTTF of Two Populations

based on Eq. (2.15) shown in Sec. 2.2.10, the confidence levels for comparing every two of the samples can be determined as shown in Table 5.14.

From Fig. 5.25 and Table 5.14, it can be seen that the aged SnPb PBGA interconnect quality (mean life) is better than that without aging, with 99.99% confidence. This means that in 9999 out of 10,000 cases, the expected (mean) life of the aged SnPb PBGA interconnect is better than that of the SnPb PBGA interconnect without aging. This is because the aging process bakes out the moisture of the PCB and the PBGA component, and leads to a stronger PCB and lighter assembly. Similar aging effects have been observed for the SAC PBGA assemblies as shown in Fig. 5.25 and Table 5.14.

Since the acceleration factor of SnPb and SAC PBGA under drop conditions is not known, it is not possible to compare their populations for any operating conditions. However, just consider the present particular test condition (acceleration = 2900 g, impact duration = 0.3 ms, and room temperature), it can be seen from Fig. 5.25 and Table 5.14 that the mean life (both with and without aging) of the SnPb PBGA interconnects are better than that of the SAC PBGA interconnects, with 99.99% confidence. Also, in view of Table 5.9, the measured average maximum strain of the SnPb assembly (1895 μ-strain without aging and 1780 μ-strain with aging) is smaller than those of the SAC assembly (2011 μ-strain without aging and 1956 μ-strain with aging). This could be due to the higher Young's modulus (stiffness) of the SAC solder joints than that of the SnPb solder joints, see, for example [33]. During impact, higher stress should be expected in the SAC solder joint and its surrounding elements, for example, copper traces.

The effect of underfills is to reduce the impact deformation and enhance the drop life. These can be seen from Tables 5.8 and 5.9. For all the cases considered, there is no failure after 30 drops of the aged and non-aged SAC PBGA assemblies with both underfill-Y and underfill-Z (Table 5.8). These underfills glue the PBGA and the PCB together, and increase the bending stiffness of the assemblies. Consequently, the deformations of the underfilled SAC are much smaller than those without underfills, as shown in Table 5.9.

5.3.9 Failure Analysis

The failure mode of all the failed samples (both SnPb and SAC PBGA assemblies without underfill) due to drop impact (acceleration = 2900 g, impact duration = 0.3 ms, and room temperature) is the cracking of the copper trace (daisy chain) between solder joints, as shown in Figs. 5.26 through 5.29. Excess bending of the PCB due to the present drop impact is obvious. Even in some cases, the cracks initiate at both SnPb and SAC solder joints; however, they only propagate at a very short

FIGURE 5.26 SnPb PBGA assembly without underfill (failed at the 18th drop).

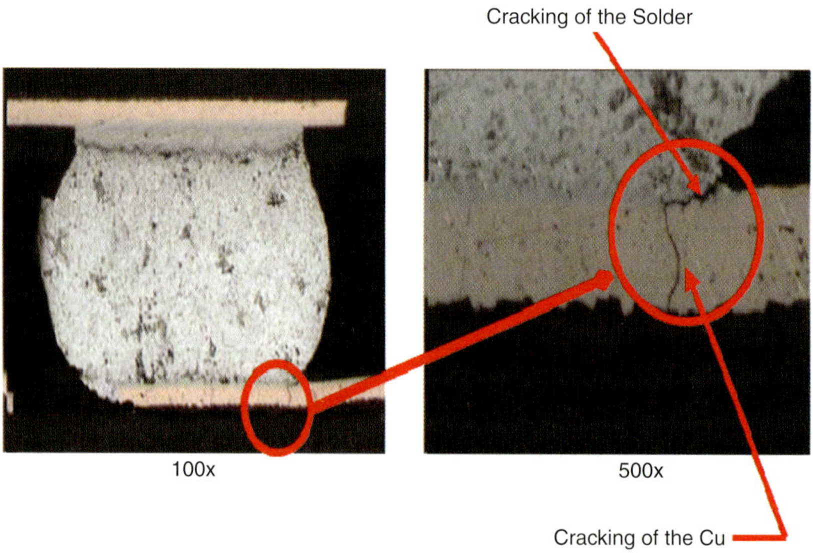

FIGURE 5.27 SnPb PBGA assembly without underfill (failed at the fourth drop).

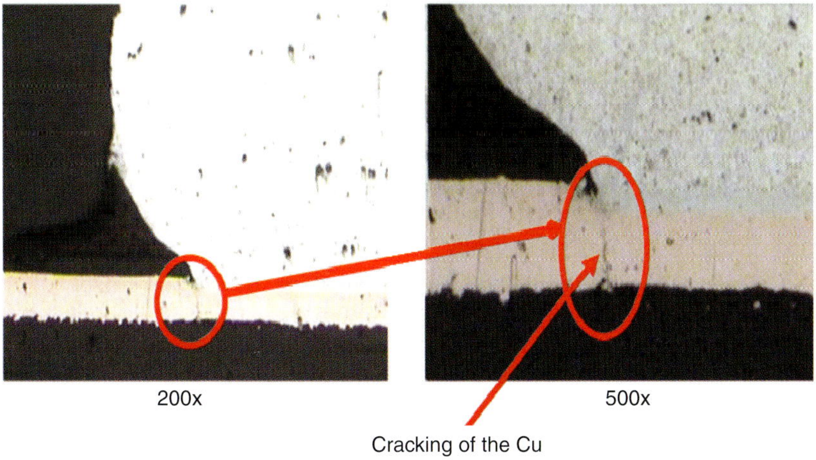

FIGURE 5.28 SAC PBGA assembly without underfill (failed at the first drop).

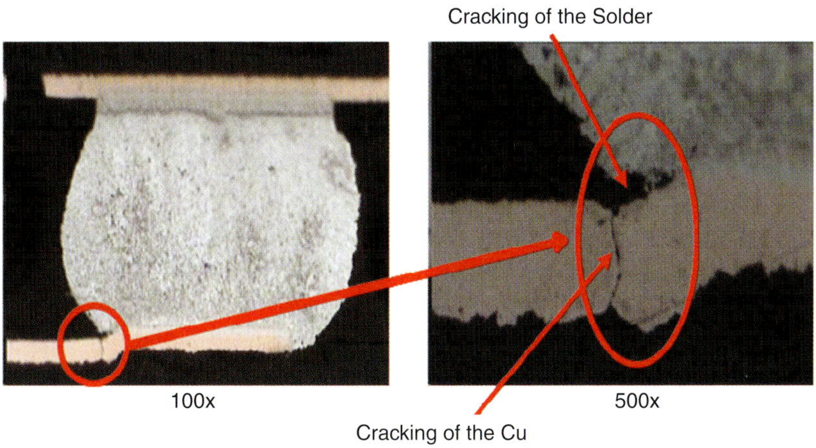

FIGURE 5.29 SAC PBGA assembly without underfill (failed at the fourth drop).

distance and then start cracking through the copper trace. There is no crack-through solder joint observed. For both SnPb and SAC assemblies with underfills, there are no crack-through copper traces.

5.3.10 Summary

Some important results are summarized in the following:

1. At 90% confidence (in 90 out of 100 cases), the true Weibull slope, true characteristic life, and true MTTF of the SnPb PBGA interconnects without underfill are, respectively, in the ranges of 0.88 and 1.8, 5 and 15 drops, and 4 and 14 drops.

2. At 90% confidence, the true Weibull slope, true characteristic life, and true MTTF of the SAC PBGA interconnects without underfill are, respectively, in the ranges of 1 and 2.04, 2 and 4 drops, and 2 and 4 drops.
3. At 90% confidence, the true Weibull slope, true characteristic life, and true MTTF of the SnPb PBGA interconnects without underfill but with aging are, respectively, in the ranges of 1.07 and 3.39, 35 and 45 drops, and 29 and 42 drops.
4. At 90% confidence, the true Weibull slope, true characteristic life, and true MTTF of the SAC PBGA interconnects without underfill but with aging are, respectively, in the ranges of 1.2 and 2.44, 5, and 11 drops, and 4 and 10 drops.
5. At 99.99% confidence, the MTTF of SnPb PBGA interconnects without underfill but with aging (125°C for 500 h) is better than that without aging.
6. At 99.99% confidence, the MTTF of SAC PBGA interconnects without underfill but with aging (125°C for 500 h) is better than that without aging.
7. For both SnPb and SAC PBGA, the aging process (125°C for 500 h) will (a) bake out the moisture from the PCB and PBGA, and yield a stronger PCB and lighter assembly; (b) reduce the maximum deformation of the assembly; and (c) increase the number of drops.
8. At 99.99% confidence, the MTTF of SnPb PBGA interconnects without underfill is better than that of SAC PBGA interconnects without underfill at the present drop condition (acceleration = 2900 g, impact duration = 0.3 ms, and room temperature).
9. At 99.99% confidence, the MTTF of SnPb PBGA interconnects without underfill but with aging (125°C for 500 h) is better than that of the SAC PBGA interconnects without underfill but with the same aging at the present drop condition (acceleration = 2900 g, impact duration = 0.3 ms, and room temperature).
10. There are no failures after 30 drops of both the SnPb and SAC PBGA interconnects with underfills. Thus, the effects of underfill are to (a) make the assembly stronger and with higher bending stiffness, (b) reduce the deformation of the solder joints, and (c) increase the number of drops.
11. The failure mode of both aged and non-aged SnPb and SAC PBGA assemblies without underfills is the cracking of the copper trace between the solder joints. There are no crack-through solder joints.

5.4 References

1. Lin, P., McShane, M., and H. Wilson, "Semiconductor device having a pad array carrier package," *US Patent No. 5216278*. Filed: March 2, 1992; Granted: June 1, 1993.
2. Lau, J. H., Y. Chen, T. Chou, F. Wu, K. Chen, W. Koh, "Single-Core Two-Side Substrate with U-Strip and Co-Planar Signal Traces, and Power and Ground Planes Through Split-Wrap-Around (SWA) or Split-Via-Connections (SVC) for Packaging IC Devices," *US Patent No. 5825084*, Filed: February 11, 1997; Granted: October 20, 1998.
3. Lau, J. H., T. Tseng, and D. Cheng, "Heat Spreader with a Placement Recess and Bottom Saw-Teeth for Connection to Ground planes on a This Two-Side Single-Core BGA Substrate," *US Patent No. 6,057,601*, Filed: November 27, 1998; Granted: May 2, 2000.
4. Lee, N. C., *Reflow Soldering Processes and Troubleshooting: SMT, BGA, CSP and Flip Chip Technologies*, Elsevier Science, New York, 2004.
5. Shangguan, D., *Lead-Free Solder Interconnect Reliability*, ASM International, Materials Park, Ohio, 2005.
6. Schubert, A., R. Dudek, E. Auerswald, A. Gollhardt, B. Michel, and H. Reichl, "Fatigue Life Models for SnAgCu and SnPb Solder Joints Evaluated by Experiments and Simulation," *IEEE Electronic Components and Technology Conference Proceedings*, New Orleans, Louisiana, June 2003, pp. 603–610.
7. Song, H. J., J. W. Morris, and F. Hua, "The Creep Properties of Lead-Free Solder Joints," *Journal of Materials*, June 2002, pp. 30–32.
8. Lau, J. H., Wong, C. P., Lee, N. C., and Lee, R., *Electronics Manufacturing with Lead-Free, Halogen-Free, and Conductive-Adhesive Materials*, McGraw-Hill, New York, 2003.
9. Lau, J. H., *Low-Cost Flip Chip Technologies*, McGraw-Hill, New York, 2000.
10. Anson, S. J., Slezak, J. G., and Srihari, K. "Investigation of Enhanced Solder Wetting in 63Sn/37Pb and Sn-Ag-Cu Lead Free Alloy." In *IEEE Proceedings of Electronic & Components Technology Conference*, Lake Buena Vista, FL, May 27–30, 2008, pp. 1737–1744.
11. Zhao, X. J., Caers, J. F. J. M., Wong, E. H., Seah, S. K. W., van Driel, W. D., Owen, N., Lai, Y. S. "Frequency Dependent S-N Curves for Predicting Drop Impact Robustness of Pb-Free Solder Interconnects." In *IEEE Proceedings of Electronic & Components Technology Conference*, San Diego, CA, May 25–29, 2009, pp. 93–99.
12. Xie, D., Geiger, D., and Gektin, V. "Reliability Study of High-end Pb-Free CBGA Solder Joint under Various Thermal Cycling Test Conditions." In *IEEE Proceedings of Electronic & Components Technology Conference*, San Diego, CA, May 25–29, 2009, pp. 109–116.
13. Chen, H. Y., and Chen, C. "In-Situ Observation of the Failure Induced by Thermomigration of Interstitial Cu in Pb-Free, Flip-Chip Solder Joints." In *IEEE Proceedings of Electronic & Components Technology Conference*, San Diego, CA, May 25–29, 2009, pp. 319–324.
14. Coyle, R., Popowich, R., Read, P., Fleming, D., Reid, M., Ryan, C., Collins, M., Punch, J., and Chatterji, I. "The Influence of the Pb-Free Solder Alloy Composition and Processing Parameters on Thermal Fatigue Performance of a Ceramic Chip Resistor." In *IEEE Proceedings of Electronic & Components Technology Conference*, San Diego, CA, May 25–29, 2009, pp. 423–430.
15. Logterman, M., and Gopalakrishnan, L. "A Product Feasibility Study of Assembling Pb-Free BGAs in a Eutectic Sn/Pb Process." In *IEEE Proceedings of Electronic & Components Technology Conference*, San Diego, CA, May 25–29, 2009, pp. 742–751.
16. Su, P., Li, L., Lai, Y. S., Chiu, Y. T., and Kao, C. L. "A Comparison Study of Electromigration Performance of Pb-Free, Flip-Chip Solder Bumps." In *IEEE Proceedings of Electronic & Components Technology Conference*, San Diego, CA, May 25v29, 2009, pp. 903–908.

17. Nicholls, L., Darveaux, R., Syed, A., Loo, S., Tee, T. Y., Wassick, T. A., and Batchelor, B. "Comparative Electromigration Performance of Pb-Free Flip Chip Joints with Varying Board Surface Condition." In *IEEE Proceedings of Electronic & Components Technology Conference*, San Diego, CA, May 25–29, 2009, pp. 914–921.
18. Lu, M., Shih, D. Y., Lauro, P., Kang, S., Goldsmith, C., and Seo, S. K. "The Effects of Ag, Cu Compositions and Zn Doping on the Electromigration Performance of Pb-Free Solders." In *IEEE Proceedings of Electronic & Components Technology Conference*, San Diego, CA, May 25–29, 2009, pp. 922–929.
19. Gadepalli, H., Dhanasekaran, R., Ramkumar, S. M., Jensen, T., and Briggs, E. "Influence of Reflow Profile and Pb-Free Solder Paste in Minimizing Voids for Quad Flat Pack No-Lead (QFN) Assembly." In *IEEE Proceedings of Electronic & Components Technology Conference*, San Diego, CA, May 25–29, 2009, pp. 2016–2024.
20. Ma, H., Lee, T. K., Kim, D. H., Liu, K. C., Kim, S. H., and Park, H. G. "Isothermal Aging Effects on the Dynamic Performance of Lead-Free Solder Joints." In *IEEE Proceedings of Electronic & Components Technology Conference*, San Diego, CA, May 25–29, 2009, pp. 390–397.
21. Kwon, D., Azarian, M. H., and Pecht, M. "Detection of Solder Joint Failure Precursors on Tin-Lead and Lead-Free Assemblies Using RF Impedance Analysis." In *IEEE Proceedings of Electronic & Components Technology Conference*, San Diego, CA, May 25–29, 2009, pp. 663–667.
22. Ahmad, M., Xie, W., Liu, K. C., Xue, J., and Towne, D. "Parametric Acceleration Transforms for Lead-Free Solder Joint Reliability under Thermal Cycling Conditions." In *IEEE Proceedings of Electronic & Components Technology Conference*, San Diego, CA, May 25–29, 2009, pp. 682–691.
23. Liu, W., and Lee, N. C., Porras, A., Ding, M., Gallagher, A., Huang, A., Chen, S., and Lee, J. C. B. "Achieving High Reliability Low Cost Lead-Free SAC Solder Joints Via Mn or Ce Doping." In *IEEE Proceedings of Electronic & Components Technology Conference*, San Diego, CA, May 25–29, 2009, pp. 994–1007.
24. Suganuma, K., Kim, K. S., Kim, S. S., Kim, D. S., Kang, M., and Kim, S. J. "Joining Characteristics of Various High Temperature Lead-Free Interconnection Materials." In *IEEE Proceedings of Electronic & Components Technology Conference*, San Diego, CA, May 25–29, 2009, pp. 1764–1768.
25. Sawada, Y., Sato, M., Abe, T., Tokunaga, M., Baba, S., and Hatanaka, Y. "Low Thermal Stress Flip-Chip Package for Ultra Low-k Die and Lead-Free Bumps." In *IEEE Proceedings of Electronic & Components Technology Conference*, San Diego, CA, May 25–29, 2009, pp. 1775–1780.
26. Meier, K., Wolter, K. J., Wiese, S., and Roellig, M. "Mechanical Characterization of Lead-Free Solder Alloys under High Strain Rate Loads." In *IEEE Proceedings of Electronic & Components Technology Conference*, San Diego, CA, May 25–29, 2009, pp. 1781–1787.
27. Geiger, D., Shangguan, D., and Yi, S., "Thermal Study of Lead-Free Reflow Soldering Processes," *Proceedings of the 3rd IPC/JEDEC Annual Conference on Lead-Free Electronic Assemblies and Components*, San Jose, CA, 2003, pp. 95–98.
28. Amagai, M., Toyoda, Y., Tajima, T., "High Solder Joint Reliability with Lead Free Solders," *IEEE Proceedings of Electronic Components and Technology Conference*, New Orleans, LA, May 2003, pp. 317–322.
29. Lau, J. H., and Pao, Y. H., *Solder Joint Reliability*, McGraw-Hill, New York, 1997.
30. Tan, K., Xu, L,. Lye, H., Pang, J., Lau, J., and Zhang. X., "Creep Properties of Sn3wt%Ag0.5wt%Cu0.019wt%Ce (SACC) Lead-Free Solder." In *IEEE Proceedings of Electronics Packaging Technology Conference*, Singapore, December 9–12, 2008, pp. 521–526.
31. Yamashita, M., Hidaka, N., and Shohji, I., "The Effect of Ag, Ni and Ge Elements in Lead-free Sn Base Solder Alloy." In *IEEE Proceedings of Electronics Packaging Technology Conference*, Singapore, December 9–12, 2008, pp. 582–587.
32. Eu, P. H., Ding, M., Wong, T. L., Amin, N., Ahmad, I., and Mok, Y. L., "A Study of SnAgNiCo vs Sn3.8Ag0.7Cu C5 Lead Free Solder Alloy on Mechanical

Strength of BGA Solder Joint." In *IEEE Proceedings of Electronics Packaging Technology Conference*, Singapore, December 9–12, 2008, pp. 588–594.

33. Lau, J. H., "Reliability of Lead-Free Solder Joints," ASME Transactions, Journal of Electronic Packaging, Vol. 128, September 2006, pp. 297–301.
34. Lee, N., "Future Lead-Free Solder Alloys and Fluxes—Meeting Challenges of Miniaturization." In *IEEE Proceedings of Electronics Packaging Technology Conference*, Singapore, December 9–12, 2008, pp. 864–872.
35. Eu, P., Wong, T. L., Amin, N., and Ahmad, I., "FCPBGA with SOP Pad Finishing: A Study of Lead-Free Solder Ball Attach Improvement." In *IEEE Proceedings of Electronics Packaging Technology Conference*, Singapore, December 9–12, 2008, pp. 1040–1045.
36. Zhang, L., Wang, D., and Tong W., "A Numerical Simulation Study of Wetting and Bridging of SnAgCu Solder Liquid." In *IEEE Proceedings of Electronics Packaging Technology Conference*, Singapore, December 9–12, 2008, pp. 1347–1350.
37. Xu, H., Li, M., Kim, J. M. and Kim, D. W., "Lead-Free Soldering Technique by using Medium-Frequency Electromagnetic Field." In *IEEE Proceedings of Electronics Packaging Technology Conference*, Singapore, December 9–12, 2008, pp. 1351–1357.
38. Song, J. M., Lin, L. W., Lee, N. C. and Lai, Y. S, and Chiu, Y. T., "Metallurgical Perspective on Alloying Modification of Sn-Ag-Cu Solders." In *IEEE Proceedings of Electronics Packaging Technology Conference*, Singapore, December 9–12, 2008, pp. 1357–1363.
39. Ibitayo, D., Everhart, L., Morgenstern, M., Geil, B., and Mitchell, J., "Reliability Assessment of Lead-Free Universal Solder for Direct Bonding in Power Electronics Packaging." In *IEEE Proceedings of Electronics Packaging Technology Conference*, Singapore, December 9–12, 2008, pp. 1364–1368.
40. Huang, Y., Lin, K. L., and Liu, D., "Effect of Impact Rate on the Strength of Lead-free Solder Joints." In *IEEE Proceedings of Electronics Packaging Technology Conference*, Singapore, December 9–12, 2008, pp. 1369–1374.
41. Xu, J., Hu, Q., He, H., Zhang, F., and Zhao, Z., "Study of Sn-Bi-Cu Lead-free Solder." In *IEEE Proceedings of Electronics Packaging Technology Conference*, Singapore, December 9–12, 2008, pp. 1375–1380.
42. Lau, J., W. Dauksher, and P. Vianco, "Acceleration Models, Constitutive Equations and Reliability of Lead-Free Solders and Joints," *IEEE Electronic Components and Technology Conference Proceedings*, New Orleans, Louisiana, June 2003, pp. 229–236.
43. Lau, J. H., J. Lo, J. Lam, E. Soon, W. Chow, and R. Lee, "Effects of Underfills on the Thermal-Cycling Tests of SnAgCu PBGA Packages on ImAg PCB," *IEEE Proceedings of Electronic Packaging and Technology Conference*, Singapore, December 2007, pp. 785–790.
44. Lau, J. H., J. Lo, J. Lam, E. Soon, W. Chow, and Ricky Lee, "Effects of Aging and Underfills on Mechanical-Drop Tests of SnAgCu PBGA Packages on ImAg PCB," *IEEE Proceedings of Electronics Materials and Packaging Conference*, October 2007, pp. 1–8.
45. Lau, J. H., and C. Chang, "Characteristics and Reliability of Fast-Flow, Snap-Cure, and Reworkable Underfills for Solder Bumped Flip Chip on Low-Cost Substrates," *IEEE Transactions on Electronics Packaging Manufacturing*, Vol. 25, No. 3, July 2002, pp. 231–230.
46. Lau, J. H., and R. Lee, "Fracture Mechanics Analysis of Low Cost Solder Bumped Flip Chip Assemblies with Imperfect Underfills," *ASME Transactions, Journal of Electronic Packaging*, December 2000, Vol. 122, No. 4, pp. 306–310.
47. Lau, J. H., R. Lee, and C. Chang, "Effects of Underfill Material Properties on the Reliability of Solder Bumped Flip Chip on Board with Imperfect Underfill Encapsulants," *IEEE Transactions on Components & Packaging Manufacturing Technology*, June 2000, pp. 323–333.
48. Lau, J. H., and S. W. Lee, "Effects of Underfill Delamination and Chip Size on the Reliability of Solder Bumped Flip Chip on Board," *IMAPS Transactions, International Journal of Microcircuit & Electronic Packaging*, Vol. 23, No. 1, First Quarter, 2000, pp. 33–39.

49. Lau, J. H., C. Chang, and C. Chen, "Characteristics and Reliability of No-Flow Underfills for Solder Bumped Flip Chip Assemblies," *IMAPS Transactions, International Journal of Microcircuit & Electronic Packaging*, Vol. 22, No. 4, 1999, pp. 370–381.
50. Lau, J. H., and C. Chang, "How to Select Underfill Materials for Solder Bumped Flip Chips on Low Cost Substrates?" *IMAPS Transactions, International Journal of Microelectronics & Electronic Packaging,* Vol. 22, No. 1, 1999, pp. 20–28.
51. Lau, J. H., and C. Chang, "Characterization of Underfill Materials for Functional Solder Bumped Flip Chips on Board Applications," *IEEE Transactions on Components & Packaging Manufacturing Technology, Part A,* Vol. 22, No. 1, March 1999, pp. 111–119.
52. Lau, J. H., C. Chang, and C. Ouyang, "SMT Compatible No-Flow Underfill for Solder Bumped Flip Chip on Low-Cost Substrates," *Journal of Electronics Manufacturing,* Vol. 8, Nos. 3 & 4, September & December, 1998, pp. 151–164.
53. Lau, J. H., C. Chang, and R. Chen, "Effects of Underfill Encapsulant on the Mechanical and Electrical Performance of a Functional Flip Chip Device," *Journal of Electronics Manufacturing*, Vol. 7, No. 4, December 1997, pp. 269–277.
54. Lau, J. H., "Solder Joint Reliability of Flip Chip and Plastic Ball Grid Array Assemblies Under Thermal, Mechanical, and Vibration Conditions", *IEEE Transaction on Components & Packaging Manufacturing Technology, Part B*, Vol. 19, No. 4, November 1996, pp. 728–735.
55. Wun, B., and J. H. Lau, "Characterization and Evaluation of the Underfill Encapsulants for Flip Chip Assembly," *Journal of the Institute of Interconnection Technology*, Vol. 21, March 1995, pp. 25–27.
56. Kelly, M., and J. H. Lau, "Low Cost Solder Bumped Flip Chip MCM-L Demonstration", *Journal of the Institute of Interconnection Technology*, Vol. 21, March 1995, pp. 14–17.
57. Lau, J. H., T. Krulevitch, W. Schar, M. Heydinger, S. Erasmus, and J. Gleason, "Experimental and Analytical Studies of Encapsulated Flip Chip Solder Bumps on Surface Laminar Circuit Boards," *Journal of the Institute of Interconnection Technology*, Vol. 19, March 1993, pp. 18–24.
58. Lau, J. H., "Thermal Fatigue Life Prediction of Encapsulated Flip Chip Solder Joints for Surface Laminar Circuit Packaging," *ASME Paper No. 92W/EEP-34*, 1992, ASME Winter Annual Meeting.

CHAPTER 6

Reliability of LED Lead-Free Interconnects

6.1 Introduction

The global consumption (usage) of component-level light-emitting diodes (LEDs) in 2008 reached over 1.5 billion units, and it is forecast to increase to 3.8 billion LEDs by 2015. Exterior applications, driven by maintenance/labor, safety/security, and "Green-Tech" issues will continue to dominate the use of LEDs in solid-state general lighting. LED has come a long way!

One hundred years ago, the discovery of electroluminescence by Henry Joseph Round in 1907 [1] and the invention of the first LED by Oleg Vladimirovich Losev in the 1920s [2, 3] *foreshadowed* the development of generations of LED to come. The invention of the first practical visible-spectrum red LED by Nick Holonyak, Jr. in 1962 [4], the first yellow LED by M. George Craford in 1972 [5], and the first high-brightness (HB) blue LED by Shuji Nakamura in 1993 [6–21] *excited* the development of LEDs in a broader range of technologies. The use of a transparent contact made of indium-tin-oxide (ITO, e.g., Sn_2O_3:In_2O_3 = 1:9) on (e.g., AlGaInP/GaAs) LED to enhance the efficiency of high-brightness LEDs was demonstrated by Alberto Barbieri in 1995. The existence of the HB blue LEDs and high-efficiency HB LEDs quickly led to the development of the first white LED by Nakamura, who employed a $Y_3Al_5O_{12}$:Ce, or YAG:Ce yellow phosphor coating on blue LEDs to produce light that appears white [22–24].

Nakamura's invention has been the most important and powerful driver for the development of the LED industry in the past 10+ years (as shown in Fig. 6.1 with the sharp turn) and opened the doors for LEDs in many applications, such as (1) indicators and signs, for example, traffic lights and vehicle lighting; (2) lighting, for example, street lamps and backlighting for LCD (liquid crystal display) televisions; (3) smart lighting, for example, to transmit broadband data and wireless routers for data transport; (4) environmentally sustainable lighting, for example, non-toxic; (5) economically sustainable lighting,

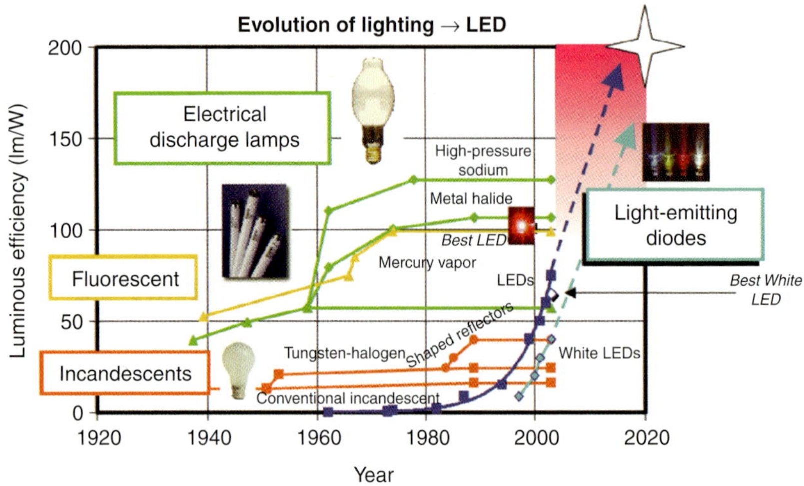

FIGURE 6.1 Evolution of lighting—all pointing to LED (Lumileds).

for example, reduce cost; (6) nonvisual applications, for example, sensors and light matter interaction and communication; and (7) light sources for machine vision systems, for example, barcode scanners. A few years later, the demonstration of high-power (HP) AlGaInN flip-chip (FC) LEDs by Jonathan Wierer and his colleagues at Lumileds [25] pushed the doors even wider for LEDs in more new applications. 3-D LED and IC integration with through-silicon-via (TSV) technology will make LED products with better performance, lower cost, less footprint, lighter weight, and smaller form factor [26]. LED is a clean (green) product because LED lighting is less harmful to human beings. Also, LED lighting can last for a very long time. Thus, unlike the current lighting system, much less LEDs will be end up in the land-field to harm the environment. It is an exciting time for LEDs.

In this chapter, a workable lead-free wave-soldering assembly process of a LED display on a printed circuit board (PCB) is presented. Also, these LED displays on PCBs are subjected to a temperature cycling test and the 85°C/85%RH with 5 mA test. The objective is to show that the current LED displays will survive the lead-free wave-soldering assembly process as well as the reliability tests, since the reliability of lead-free interconnects are of great concern [27–55]. Furthermore, a time- and temperature-dependent (creep) finite-element analysis of the LED display lead-free solder joint is performed and its thermal fatigue life is discussed. Finally, the future of 3-D LED and integrated circuit (IC) integration with TSV and their thermal management system will be discussed.

6.2 LED Seven-Segment Display Family

The 20-mm (0.8-inch) LED seven-segment displays are designed for viewing distances up to 10 m (33 feet) [56]. These devices have excellent appearance, high light output, and high peak current, and use an industry standard size package and pin-out (15.24-mm leads on 2.54 mm centers). All devices are available as either common anode or common cathode, and in various colors, for example, AlGaAs red, high efficiency red, yellow, and green (as shown in the upper-left corner of Fig. 6.2). These displays are excellent for long digit string multiplexing intensity and color selection option. The key dimensions of the displays are 27.94 × 20 × 7.36 mm (Fig. 6.3).

As mentioned earlier, today the leading lead-free solder candidate for both reflow and wave-soldering flow is SnAgCu; even SnCu has been recommended for wave soldering by International Electronics Manufacturing Initiative (iNEMI). The melting point for SnAgCu is 217°C and for SnCu is 227°C. These are much higher than that (183°C) for the eutectic SnPb solder. Thus, the lead-free PCB assembly (either by reflow or wave soldering) and reliability of conventional components and PCBs are of great concern. This is especially true for optoelectronic components.

In this chapter, an SnAgCu lead-free wave-soldering assembly process for the leaded LED display on a plated through hole (PTH) PCB is developed. The assembled LED-PCBs are subjected to thermal cycling test (–55°C to 100°C) and a biased temperature and humidity test

FIGURE 6.2 LED display and its PCB assembly.

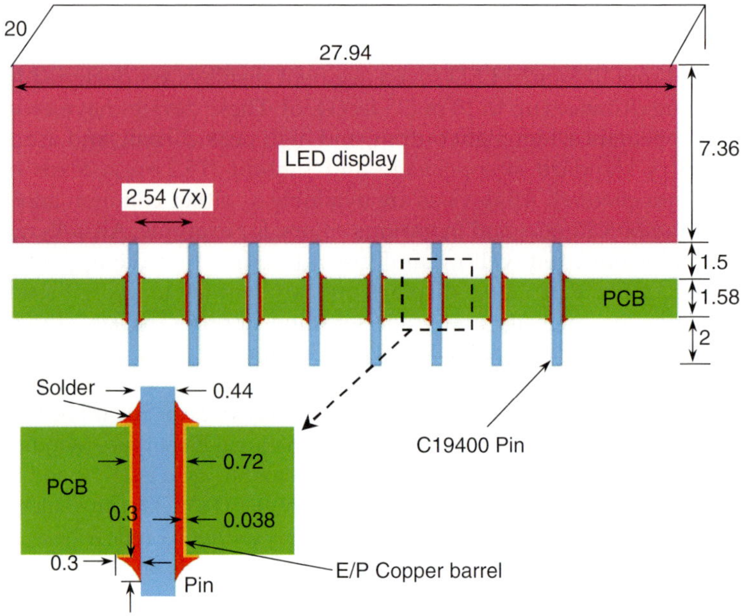

FIGURE 6.3 Schematic of the LED display PCB assembly.

(85°C/85%RH test with 5 mA). After 100 thermal cycles and 1000 hours of 85/85 tests, if the light output performances of the LED displays are still well within the specifications, then a workable lead-free wave-soldering assembly process for this optoelectronic component will have been successfully developed.

Another concern regarding the LED display assembly is its solder joint reliability under operating conditions. In this study, nonlinear stress and strain analyses of the LED display with lead-free solder joints are performed to understand its thermal-mechanical behavior. LED displays with SnPb solder joints are also considered for comparison purposes.

6.3 Lead-Free Wave Soldering of LED Displays

6.3.1 PCB

The PCB is made of fiber-reinforced (FR) glass epoxy. The dimensions of the PCB are 80 × 45 × 1.58 mm, as shown in Figs. 6.2 and 6.3.

6.3.2 Flux and Solder Bar

Flux is the substance (chemical/material) that prepares and protects the surface to be soldered by removing surface oxides. This provides a clean metallic surface and prevents further oxidation during the soldering process. This cleaning function is called "fluxing." Flux also assists in the transfer of heat from the molten solder to the joint area so that the base metals reach a high enough temperature to be wetted by the solder. Furthermore, flux reduces the interfacial surface tension between the solder and the base metal, enabling the solder to flow over and metallurgically wet the solderable surface.

Since the wetting characteristics of SnAgCu alloys are not as good as those typically exhibited by SnPb alloys at lower process temperatures, the selection of flux is very important to assure good solderability [28]. For surface mount technology (SMT), the flux is mixed with the solder powder (metal alloys) in a solder paste [27–30]. However, for wave soldering of PTH components, the flux and the solder are applied separately. The standard wave-soldering process is (after inserting the PTH components on the PCB): (1) apply the flux on the PCB; (2) preheat the PCB (to limit the thermal shock as the PCB contacts the much higher temperature of the molten solder); and (3) wave-solder flow the PCB. Thus, for wave soldering, the flux should be able to withstand: (1) higher PCB topside preheat temperature (130°C) and (2) higher solder temperature (280°C) for longer contact time (3 s).

In this study, a volatile organic compound (VOC) free and water soluble flux is used. This flux is designed to withstand high temperatures and works very well with the SnAgCu lead-free solder alloys. A fluxer (with a nozzle diameter of 1 mm) sprays the flux on the PCB with a 60° angle to achieve the optimal effects.

The solder bar used in this study is the 95wt%Sn3.5wt%Ag 0.7wt%Cu. It will take at least 4 h to melt the solder bars in the solder pot, which is maintained at 270°C to improve solderability and shorten contact time. Usually, it will take at least 8 h for the solder and pot to cool down to room temperature. During the lead-free transition period, this could decrease the throughput if SnPb and lead-free solders are used on the same wave-soldering machine.

6.3.3 Lead-Free Wave-Soldering Thermal Profiles

The main purpose of the wave-soldering process (fluxing, preheat, and wave) is to wet the surfaces to be jointed and to solidify the solder into a metallurgical bond. The molten solder is assisted by hydrodynamic forces up the holes and into intimate contact with the surfaces to be joined.

The exposure of optoelectronic components to high temperatures for relatively long periods can cause internal damage, that is, delamination between the plastic housing and leads, deformation and shrinkage of adhesives, thermal instability of the epoxy, and damage to plastic lenses. Using thermocouples and a recording device, precise temperature levels are recorded throughout the flow cycle. Through several experiments, the settings are established for the desired flow

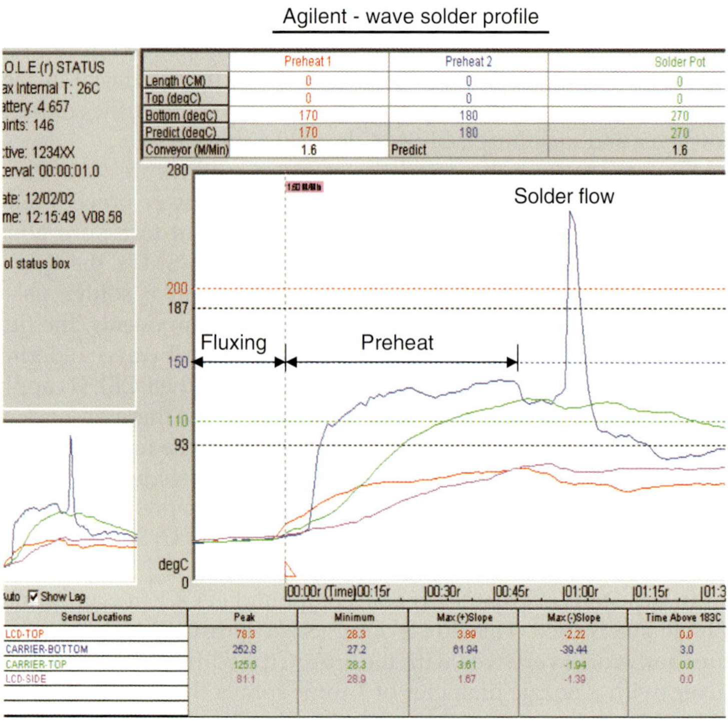

FIGURE 6.4 Lead-free wave-soldering flow thermal profile.

thermal profile. The wave-soldering setting is: (1) preheat temperature = 170°C/180°C; (2) solder pot temperature = 270°C; and (3) conveyor speed = 1.6 m/min.

Figure 6.4 shows the optimized wave-soldering flow temperature profiles of the LED displays. The four thermocouples (as shown in Fig. 6.5) are placed on the top of the LED, one side of the LED, the top of the PCB, and the bottom of the PCB. Important results are: (1) the peak temperature at the bottom of the PCB is 252.8°C; (2) the peak temperature at the top of the LED is 78.3°C; (3) the peak temperature at the side of the LED is 81.1°C; and (4) the peak temperature at the top of the PCB is 125.6°C. Figure 6.2 shows the LED-display PCB assemblies.

6.4 Thermal Cycling Test of LED Display Assemblies

Thirty lead-free wave-soldered LED display assemblies are subjected to the thermal cycling test. The test condition is from −55°C to 100°C, with 15-min dwells at both the hot and cold limits. At select test points (5, 20, 50, and 100 cycles), the LED display assemblies are subjected to the hot functional test. The hot functional test is carried out with the display placed on a hot plate at an elevated temperature (100°C) and soaked for a specific period (10 min) before the test starts using a functional

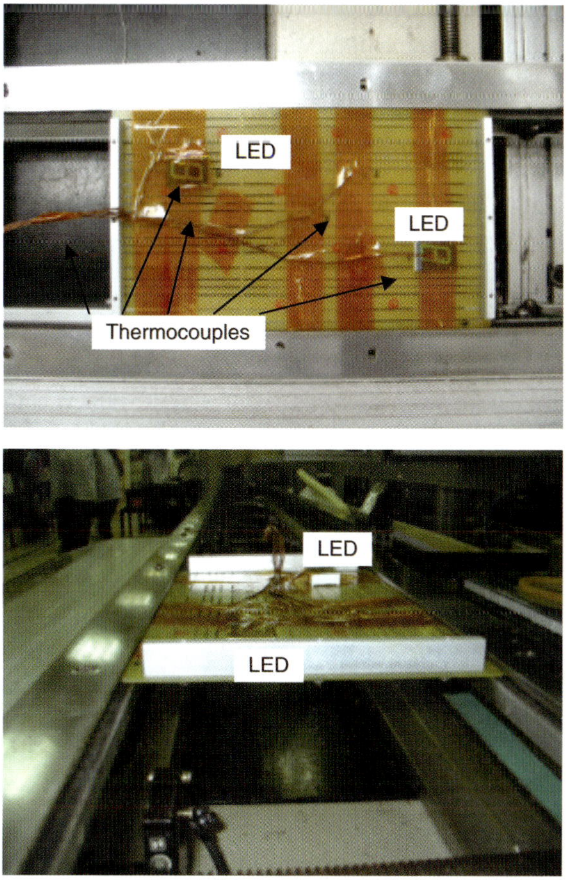

FIGURE 6.5 Thermocouples on the LED and PCB during wave soldering.

lit-up visual box. The operator will visually inspect the lit-up unit, and any segment of the display observed with no light will be considered a failure. In this study, all 30 samples passed these test points.

6.5 85°C/85%RH Test of LED Display Assemblies

Another 28 samples are subjected to 1000 hours of the 85°C/85%RH (with a stress current of 5 mA) test. The test time is 1000 hours. At test points 24, 168, 500, and 1000 h, the intensity (brightness) of the display is measured with a tester. The unit of measurement is millicandela. The measured value is compared with the initial value. Any drop of brightness is calculated in delta percentage. This delta percentage will be compared to the specification values, average < 35% and maximum < 50%. This maximum actually refers to the individual segment intensity delta value, which should be less than 50%. The hot functional test and functional lit-up visual box mentioned earlier can also be used to determine if the units pass or not. The tests show that all 28 samples passed these test points.

6.6 Creep Analysis of the LED Display Lead-Free Solder Joints

The thermal-fatigue life of the LED display lead-free solder joints is discussed in this section. The results are compared to those with SnPb solder joints.

6.6.1 The Structure

The 20-mm, LED seven-segment display (with green color) under consideration has a 27.94 × 20 × 7.36-mm body, as shown in Figs. 6.2 and 6.3. It consists of the diode, C19400 lead frame, epoxy, and housing. The pin diameter is 0.44 mm and is longer than 5.8 mm. The pin pitch is 2.54 mm. The PCB is 1.58 mm thick with a hole diameter of 0.72 mm. The electroplated copper thickness is 0.038 mm.

6.6.2 Material Properties—Constitutive Equations

The material properties of the LED PCB assembly are tabulated in Tables 6.1 and 6.2. It can be seen that the effective Poisson's ratio and coefficient of thermal expansion (CTE) of the LED are, respectively, 0.3 and 60×10^{-6}/K (the latter is a conservative value). The effective

	Young's Modulus (GPa)	Poisson's Ratio	CTE (ppm/K)
PC board	28 (in-plane) 12 (out-of-plane)	0.11 (in-plane) 0.39 (transverse)	18 (in-plane) 70 (out-of-plane)
E/P copper barrel	76	0.35	17
C19400 pin	121	0.35	16.3
LED	11 (218 K), 10 (298 K), 8 (328 K), 6 (398 K)	0.3	60
95.5Sn3.9Ag0.6Cu	$74.84 - 0.08T$	0.3	$16.66 + 0.017T$
Sn-Pb	$75.94 - 0.152T$	0.35	24.5

All temperatures in K

TABLE 6.1 Material Properties of the LED PCB Assembly

Solder Alloy	C1 (1/s)	C2 (1/Pa)	C3	C4 (K)
95.5Sn3.9Ag0.6Cu	441000	5×10^{-9}	4.2	5412
SnPb	926(508−*T*)/*T*	1/(37.78 × 106 − 74414*T*)	3.3	6360

All temperatures in K

$$\frac{\partial \varepsilon}{\partial t} = C_1[\sinh(C_2\sigma)]^{C_3} \exp\left(\frac{C_4}{T}\right)$$

TABLE 6.2 Creep Material Properties of the SnAgCu and SnPb Solder Alloys

Young's modulus of the LED is assumed to be temperature-dependent, as shown in Table 6.1. While the Poisson's ratio and CTE of the copper pin (C19400) on the LED and the electroplated copper on the PCB are similar, their Young's moduli (E) are very different. The Young's modulus of the electroplated copper (76 GPa) is significantly less than that of the C19400 copper (121 GPa).

The constitutive equations of the 95.5wt%Sn-3.9wt%Ag-0.6wt%Cu solder alloys have been given in [51] and are used for the present study:

$$E = -0.08T(^{\circ}\mathrm{C}) + 53$$

$$CTE = 0.017T(^{\circ}\mathrm{C}) + 21.3 \times 10^{-6}/^{\circ}\mathrm{C}$$

$$\frac{d\varepsilon}{dt} = 4.41 \times 10^5 \exp\left[-45(\mathrm{kJ/mol})/RT(\mathrm{K})\right] \sinh^{4.2}(0.005\sigma)$$

In these equations, E is the Young's modulus (GPa), R is the gas constant, ε is the normal creep strain, and σ is the normal stress (MPa). The last equation can be rewritten so that it is suitable for the input of ANSYS [57], and is shown here:

$$\frac{d\varepsilon}{dt} = C_1[\sinh(C_2\sigma)]^{C_3} \exp\left(\frac{-C_4}{T(\mathrm{K})}\right)$$

The constants C_1, C_2, C_3, and C_4 [please be noted that these set of coefficients are an earlier version than those give in Eq. (2.39)] are given in Table 6.2, where the material constants for the eutectic SnPb solder [49, 54] are also provided.

6.6.3 Boundary Conditions

Figure 6.6 shows the finite-element model that captures the construction along the diagonal from the LED PCB assembly's geometric center to a corner pin. It can be seen from the close-up view that the solder joint has finer meshes than the others, since the focus is in determining the stresses and creep strains at the solder joint. The model has

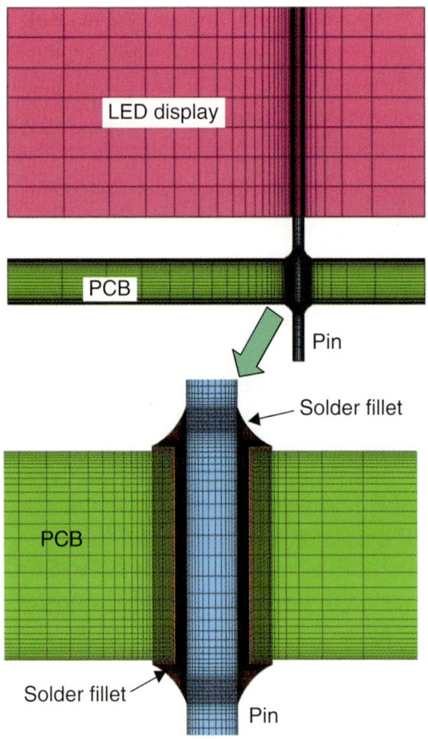

Figure 6.6 Finite-element model for the analysis of the LED PCB assembly.

approximately 10,000 elements and assumes a state of plane strain in the direction that is not being modeled. It should be noted that the present model neglects the interior pins' contribution to resist the thermal expansion mismatch between the LED and the PCB. Thus, the results calculated with the present model are conservative.

Figure 6.7 shows the temperature profiles to be imposed on the LED PCB assembly. It can be seen that the temperature loading condition is: (1) −25°C to 125°C; and (2) the cycle time is 1 h and the time for ramp-up, ramp-down, dwell-at-hot, and dwell-at-cold is 15 min each.

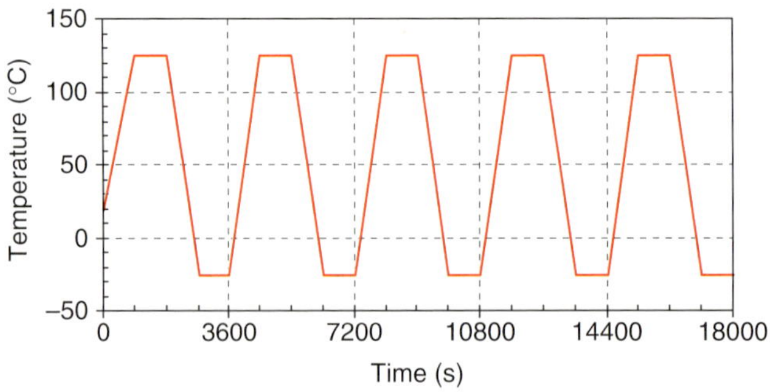

Figure 6.7 Temperature input to the LED PCB assembly.

6.6.4 Deformation of the LED Assembly

Figure 6.8 shows the deformations of the LED PCB assemblies. The top one is with SnAgCu solder, and the bottom one is with SnPb solder. It can be seen that: (1) globally, they look the same and have similar magnitudes; and (2) locally, the SnAgCu solder joint seems to have slightly

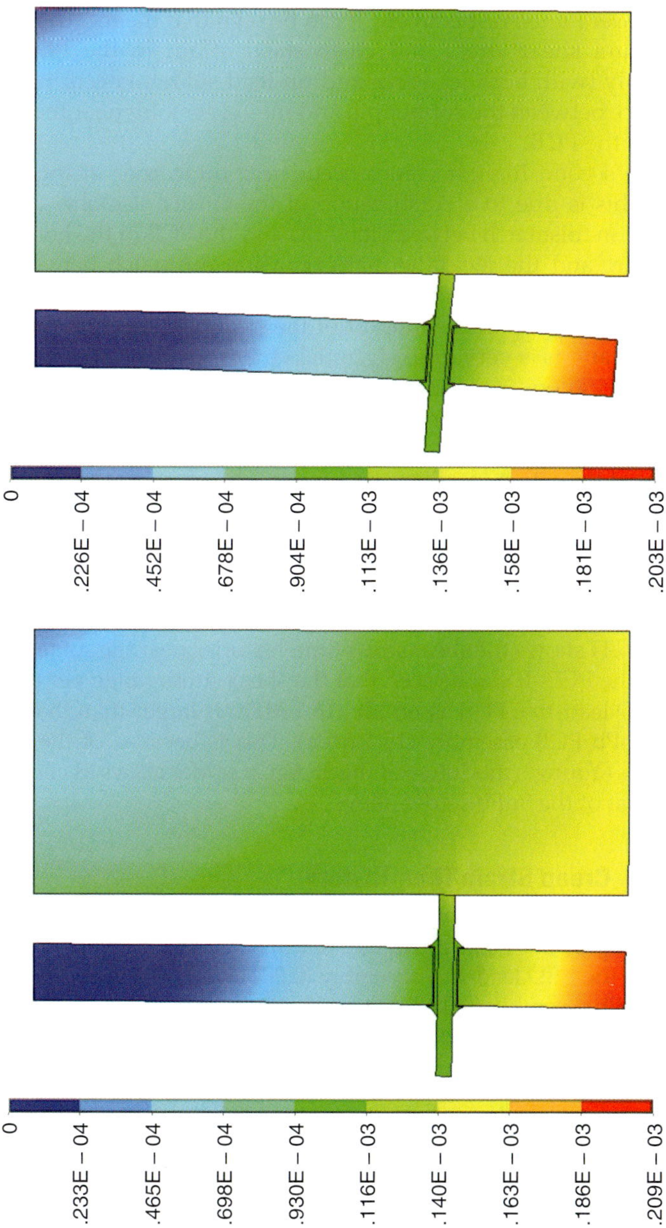

Figure 6.8 Deformation of the LED PCB assembly (top: SAC, bottom: SnPb).

larger rotation, which could be due to its higher stiffness and lower creep strain rate.

6.6.5 Stress and Creep Strain Distributions

Figures 6.9 and 6.10 show the maximum shear stress (top) and creep shear strain (bottom) distributions in the tin-lead and lead-free solder joints, respectively. It can be seen that the location of the maximum shear stress and creep shear strain in the LED PCB assembly (with both lead-free and tin-lead solders) occur near the interface between the copper pin and the solder joint near the upper PTH of the PCB.

The second highest values occur near the corner of the solder joint. This is due to a combination of the effects of global thermal expansion mismatch between the LED and the PCB in the horizontal direction, and the local thermal expansion mismatch between the PCB and the solder joint in the vertical direction. Since these high values are located in the interior of the structure, crack initiation at these locations is very unlikely, unless there are voids to begin with at these locations of the solder joint. Thus, the focus in this study is on the fillet tips of the solder joint. Also, it can be seen that for both lead-free and tin-lead solders, the shear stress (as well as the creep shear strain) at these two locations are similar. Therefore, in the following, only the stress and strain near the solder fillet tip on top of the PCB will be shown.

6.6.6 Stress Time-History

Figure 6.11 shows the shear stress time-histories near the solder tip on top of the PCB. It can be seen that the shear stress range per cycle in the LED lead-free PCB assembly (16.1 MPa) is larger than that in the LED SnPb PCB assembly (15.1 MPa). This is because of the higher stiffness (Young's modulus) of the SnAgCu solder alloys as compared with that of the SnPb solder alloys.

6.6.7 Creep Strain Time-History

Figure 6.12 shows the shear creep strain time-histories near the solder tip on top of the PCB. It can be seen that the shear creep strain per cycle in the LED lead-free assembly (0.002) is smaller than that in the LED SnPb assembly (0.0025). This is because of the lower creep strain rate of the SnAgCu solder alloys as compared with that of the SnPb solder alloys.

6.6.8 Hysteresis Loops and Creep Strain Energy Density

Figure 6.13 shows the shear stress and shear creep strain hysteresis loops near the solder tip on top of the PCB. It can be seen that the shapes (as well as the areas) of the loops for the SnAgCu and SnPb are

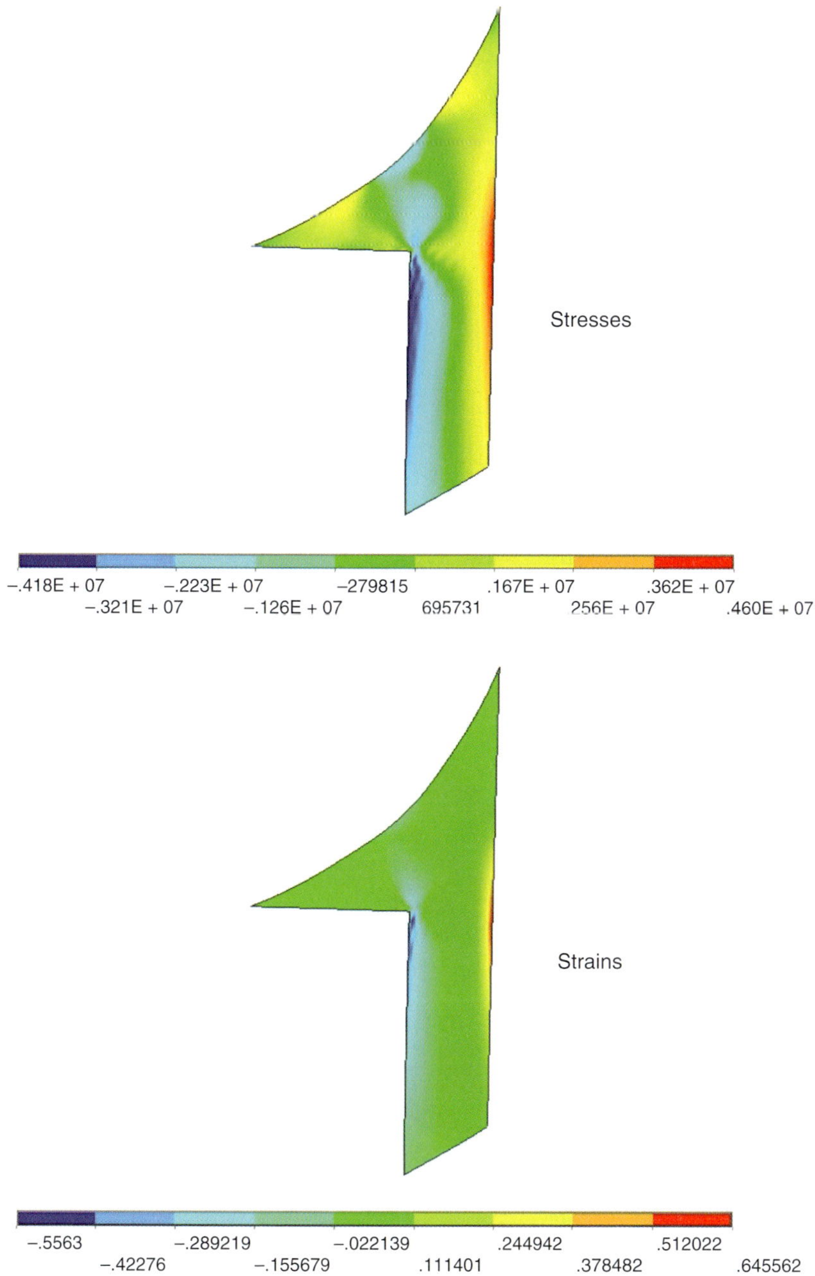

FIGURE 6.9 Shear stress (top) and creep strain (bottom) distributions in the SnPb LED PCB assembly.

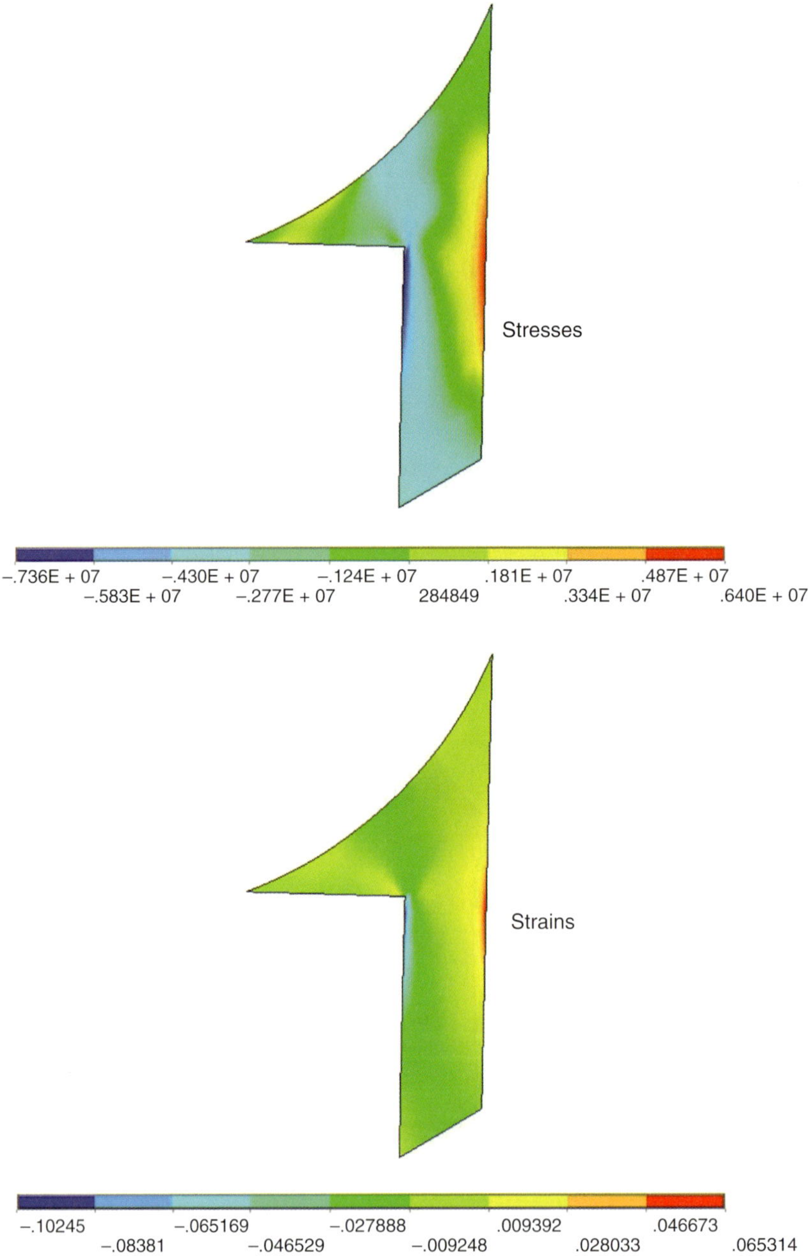

Figure 6.10 Shear stress (top) and creep strain (bottom) distributions in the SAC LED PCB assembly.

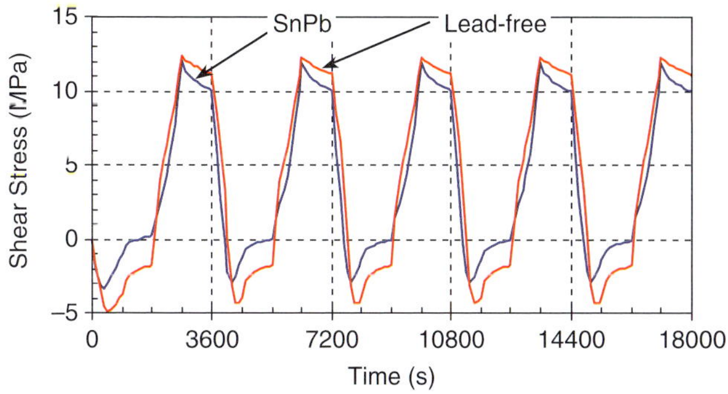

FIGURE 6.11 Shear stress time-history in the LED solder joints.

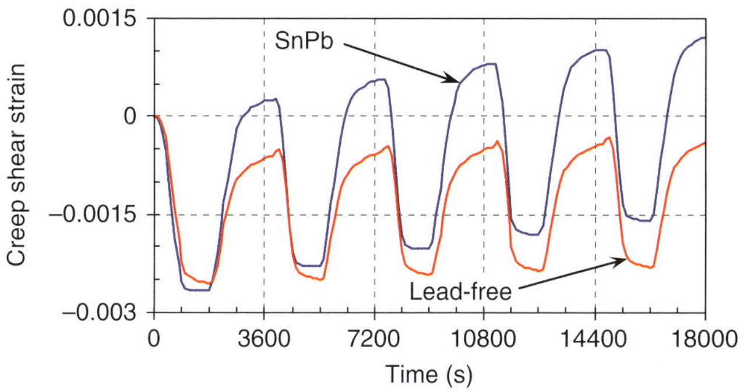

FIGURE 6.12 Shear creep strain time-history in the LED solder joints.

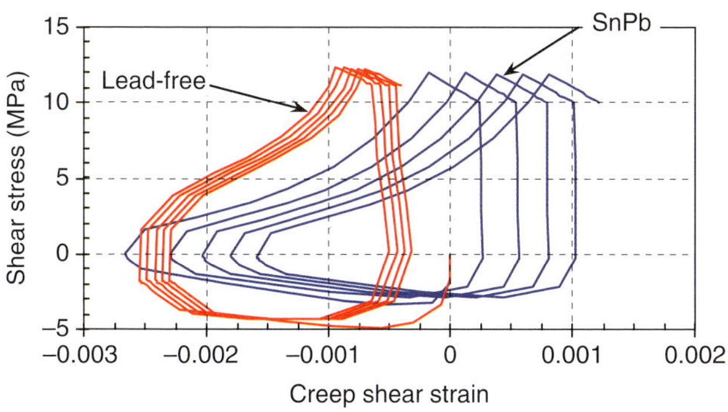

FIGURE 6.13 Shear stress and creep shear strain hysteresis loops.

not exactly the same. The creep strain energy density per cycle (area under the hysteresis loop) in the LED lead-free assembly (0.0175 MPa) is slightly smaller than that in the LED SnPb assembly (0.0181 MPa).

6.6.9 Thermal-Fatigue Life of LED Solder Joints

Thermal-fatigue life of solder joints (dominated by creep responses) may be predicted by the following equation, for examples, [58] and Chap. 2 of this book.

$$N_f = \psi(\Delta W)^{\phi}$$

where N_f is the number of cycles-to-failure, ΔW is the creep strain energy density per cycle, and ψ (always positive) and ϕ (always negative) are material constants for the solder. In this equation, ψ and ϕ are usually determined by isothermal fatigue tests of the solder joint. (Thus, ψ and ϕ will be different for lead-free and tin-lead solders.)

ΔW can be determined by creep analysis of the structure (LED, solder joints, and PCB) subjected to the specified loading conditions. Since the material constants for the SnAgCu lead-free solder are not available at this moment, thermal-fatigue life prediction of the LED display lead-free solder joints is impossible. However, since the creep shear strain energy density per cycle (for both lead-free and tin-lead solders) is small, they should be reliable for use under normal operating conditions. Also, it has been experimentally shown in [59] that PTH solder joints are very reliable under fatigue loadings because of the solder-joint crack-closure effects [59].

6.7 Summary and Recommendations

Some important results are summarized in the following:

1. With the workable wave-soldering temperature profile developed: (1) the peak temperature at the bottom of the PCB is 252.8°C; (2) the peak temperature at the top of the LED is 78.3°C; (3) the peak temperature at the side of the LED is 81.1°C; and (4) the peak temperature at the top of the PCB is 125.6°C.
2. Thirty samples passed the thermal cycling tests at 5, 20, 50, and 100 cycles. The test condition is from –55°C to 100°C, and the dwell time at hot and at cold for each cycle = 15 min.
3. Twenty-eight samples passed the 85°C/85%RH (with a stress current of 5 mA) tests at 24, 168, 500, and 1000 hours.
4. The locations of the maximum stress and strain in the solder joint are at the interface between the copper pin and the

solder joint near the upper PTH of the PCB and at the corner of the copper barrel. However, cracking initiations are unlikely to occur at these locations since they are in the interior of the structure, unless there are voids to begin with.

5. For both lead-free and tin-lead solders, the shear stress (as well as the shear creep strain) at both tips of the solder-joint fillet are similar. Solder-joint cracking should initiate at these locations, that is, (1) the interface between the copper pin and the tip of the solder fillet, and (2) the interface between the copper pad on the PCB and the other tip of the solder fillet.
6. The shear stress near the tips of the solder fillet with the SnAgCu solder is slightly larger than that with the SnPb solder.
7. The shear creep strain near the tips of the solder fillet with the SnAgCu solder is slightly smaller than that with the SnPb solder.
8. The shear creep strain energy density per cycle (ΔW) near the tips of the solder fillet with the SnAgCu solder is slightly smaller than that with the eutectic SnPb solder. However, because both solders' ΔW is very small, their solder joints should be reliable for most operating conditions.
9. It should be pointed out that the proposed wave-soldering temperature profile is only good for this particular LED display. In order to broaden the process window for other components, it is necessary to improve the material properties (e.g., thermal stability of epoxies) of optoelectronic components so they can withstand higher temperatures at longer times.

6.8 3-D LED and IC Integration

Figure 6.14 schematically shows the HP FCLED sitting on an Si submount. It can be seen that the blue light comes out from the multiple quantum wells (MQW), and the GaN is extracted through the transparent sapphire substrate, which is coated with the yellow phosphor. This yellow phosphor absorbs a portion of the blue light and is stimulated to emit yellow light, which mixes with the rest of the blue light to generate light that appears white to human eyes. Also, the power FCLED has a large emitting area and an optimized contacting scheme, allowing high current operation with low forward voltages, and therefore, high power conversion efficiencies [25]. This technology has been in mass production, and some of the products are shown in Fig. 6.15.

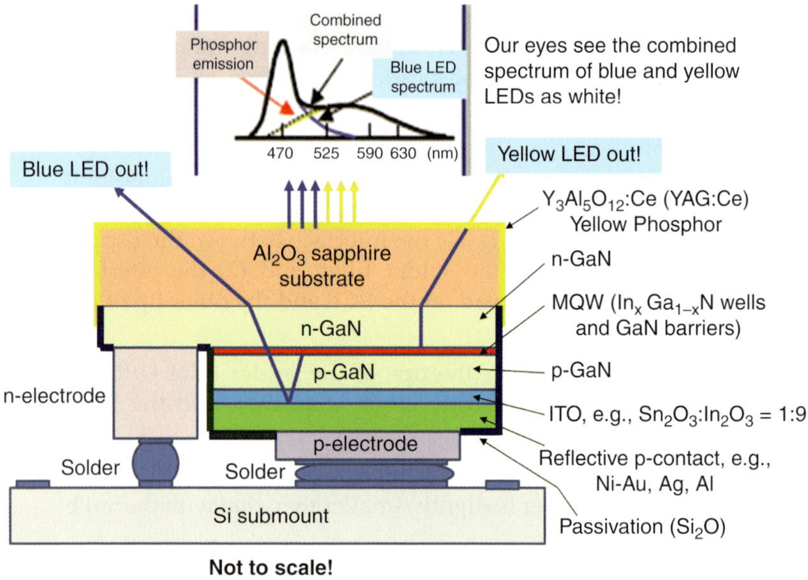

FIGURE 6.14 High Power (HP) Flip Chip (FC) LED (FCLED) on a Si submount.

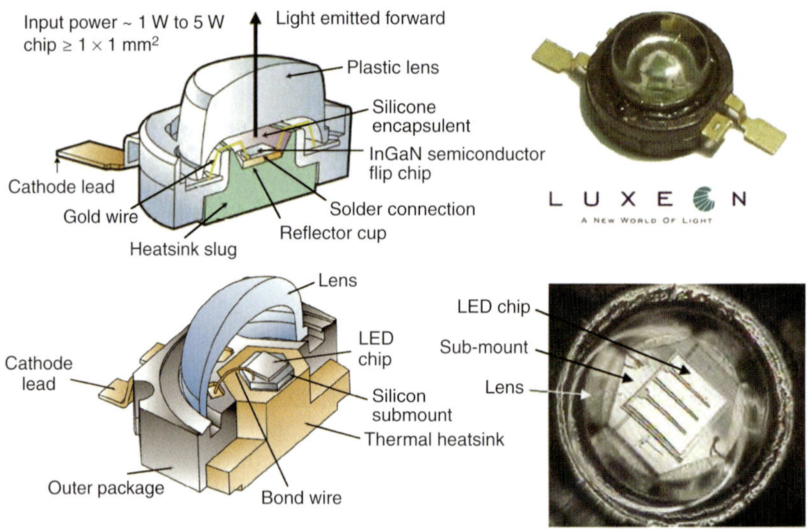

FIGURE 6.15 Lumileds' HP FCLED is in volume production.

Usually, the HP FCLED with the Si submount is packaged in a reflector cup with metal coating (but some are packaged without the cup, as shown in the lower-left corner of Fig. 6.15), attached to a interconnect, for example, a housing and/or PCB, filled with a transparent silicone encapsulent and then covered by a plastic lens (either curved

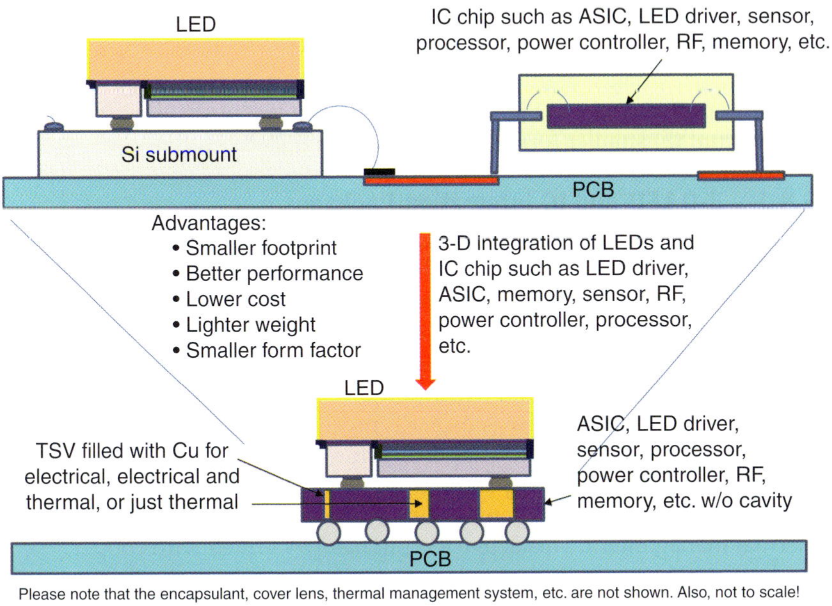

FIGURE 6.16 Three-dimensional LED and IC (e.g., LED driver, controller, RF, integration driver, ASIC, memory, processor, sensor, power).

or flat). The input power ranges from 1 W to 5 W, and the LED chip size is $\geq 1 \times 1$ mm^2.

Most of the system houses design and assemble the LED modules (components) on one location of the PCB and the application specific integrated circuit (ASIC), LED driver, sensor, radio frequency (RF), power controller, processor, memory, etc., on another location of the PCB, as shown in Fig. 6.16. (For simplicity, the cover lens, encapsulant, thermal management system, etc., of the LED module are not shown.)

The major trend in the electronic industry today is to make products more personal by making them smarter, lighter, smaller, thinner, shorter, and faster, while simultaneously making them more friendly, functional, powerful, reliable, robust, innovative, and economical. As the trend toward smaller and faster products continues, the introduction of innovative products that are more user-friendly and contain a wider variety of functions will be a driving factor in providing growth in the market. One of the key technologies that will help to make these product design goals possible is 3-D IC integration [60, 61].

The bottom schematic of Fig. 6.16, for example, shows the passive Si submount replaced by an IC chip, such as the ASIC, LED driver, processor, power controller, sensor, RF, etc., that is, integrating the LEDs and the IC chip together in a 3-D manner [26]. Their electrical feed-through and some of the thermal paths can be effectively achieved by through-silicon-vias (TSV) filled with copper on the IC chip. The advantages of this 3-D LED and IC integration package are: (1) better performance, (2) lower cost, (3) smaller footprint, (4) lighter weight, and (5) smaller form factor.

6.8.1 3-D LED and IC Integration Packages

Figures 6.17 and 6.18 show two 3-D LED and IC integration packages. The one in Fig. 6.18 is a package with a reflector cup and a flat plastic lens. The other one (Fig. 6.17) shows a package without a cup but a plastic curved lens. Both packages are able to support multi-LEDs. The passive Si submount (shown in Figs. 6.14 and 6.15) has been replaced by an active IC chip, which could be, for example, the ASIC, LED driver, power controller, sensor, RF, processor, and memory. This active IC chip can be used to support the multi-LEDs for many functions, for example, dimming and lighting control, step-up and step-down topologies, power conversion, electrical feed-through, and thermal management. This active IC chip has TSVs filled with copper (and most likely with RDL) for pure electrical feed-through, electrical feed-through as well as thermal path, and pure thermal path

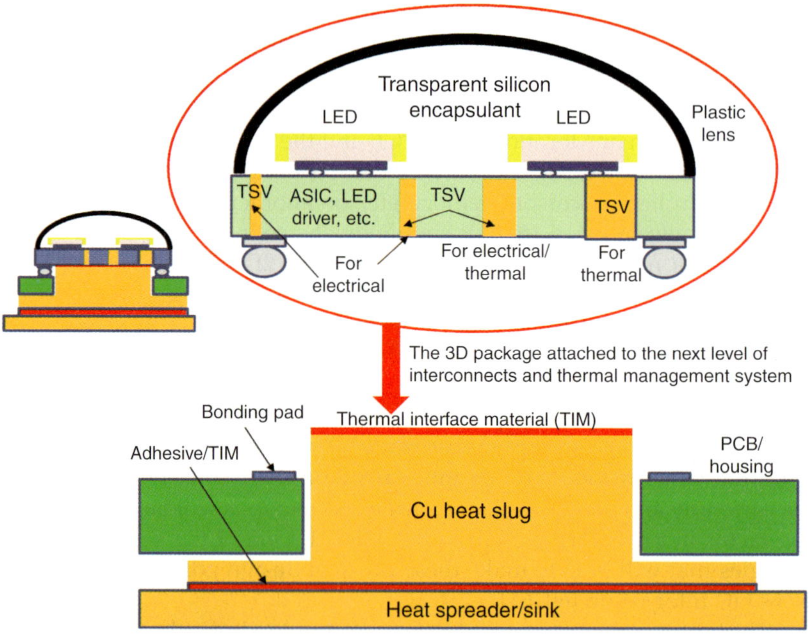

FIGURE 6.17 Three-dimensional LED and IC integration package (without a cavity) on an ordinary thermal management system.

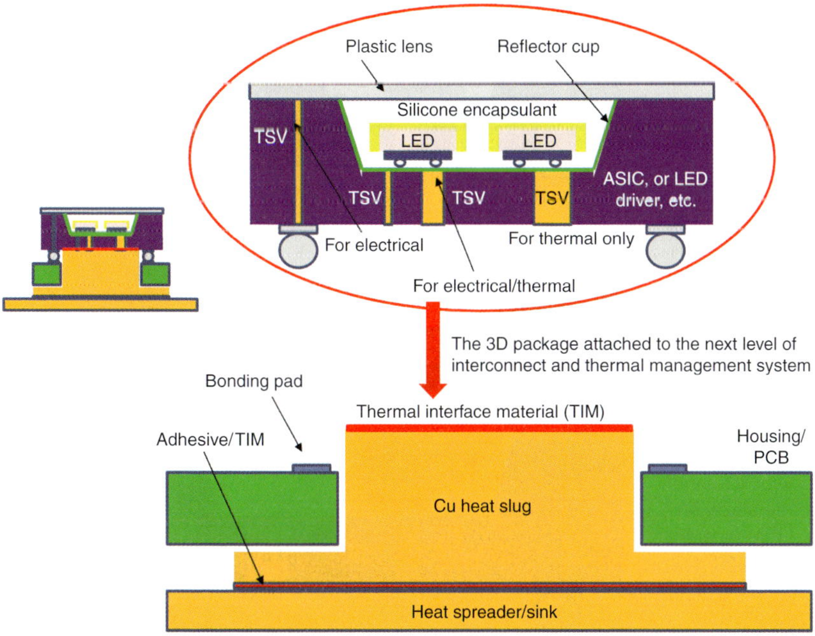

Figure 6.18 Three-dimensional LED and IC integration package on an ordinary thermal management system.

(i.e., large dummy vias filled with copper). The thermal conductivity of Si is 150 W/m/K and of Cu is 390 W/m/K.

6.8.2 Manufacturing Process of 3-D LED and IC WLP

The manufacturing processes of these two 3-D LED and IC integration packages shown in Figs. 6.17 and 6.18 are basically the same, except the one in Fig. 6.18 has the cavity and its manufacturing processes are shown in Fig. 6.19. The fabrication of the reflector cup (cavity) of the active silicon IC wafer is by wet anisotropic etched with products (solutions) such as potassium hydroxide solution (KOH), ethylenediamine pyrocatechol solution (EDP), and tetramethylammonium hydroxide (TMAH). Depending on the dimensions of the mask opening, a V-groove or trapezoidal basin can be formed in the active IC wafer. KOH is the most commonly used etchant. It is much less dangerous (toxic) than the others, easy to handle, readily available, and etches fast. The greatest disadvantages are that KOH is IC incompatible and that the selectivity to plasma-enhanced chemical vapor deposition (PECVD) oxide is rather poor. TMAH is nontoxic and IC compatible, but fewer studies exist on this system. EDP is not easy to handle. It is toxic, and the solution degrades if it comes in contact with oxygen. Thus, EDP is mainly used in research laboratories and is not used in mainstream semiconductor fabrications.

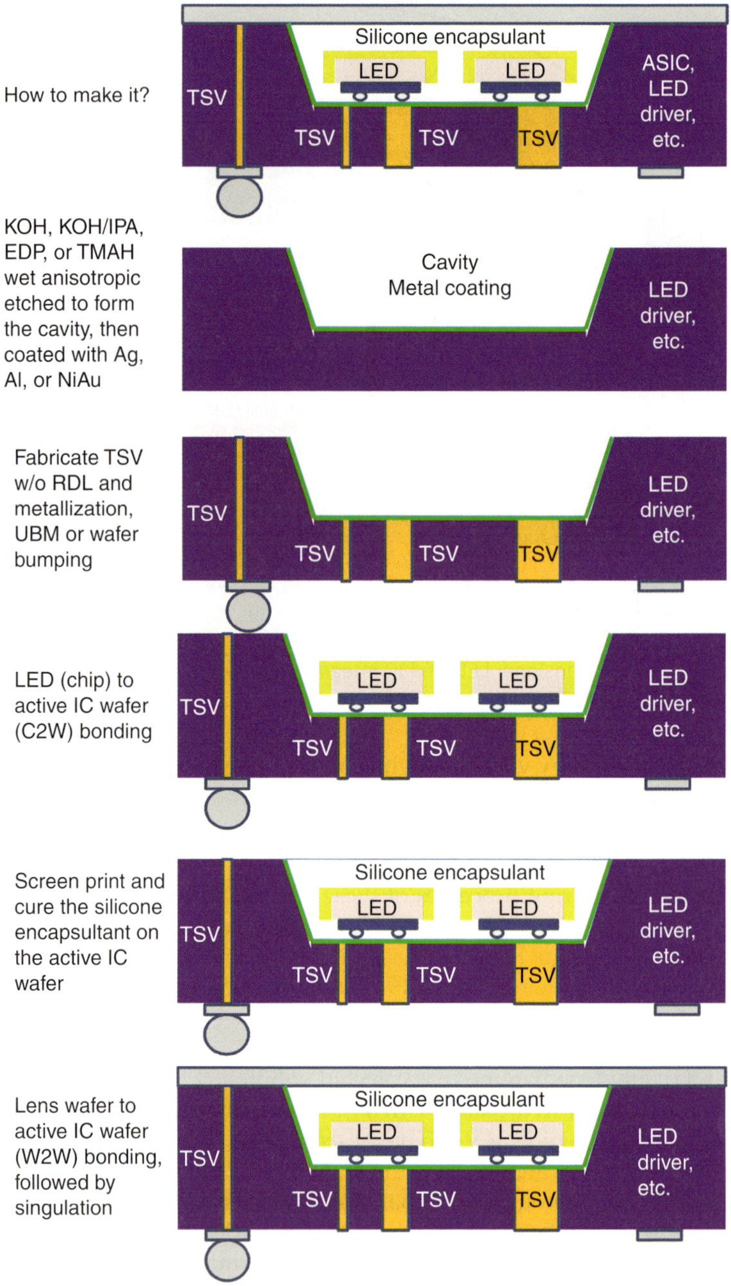

FIGURE 6.19 Manufacturing process of 3-D LED and IC integration package with TSV.

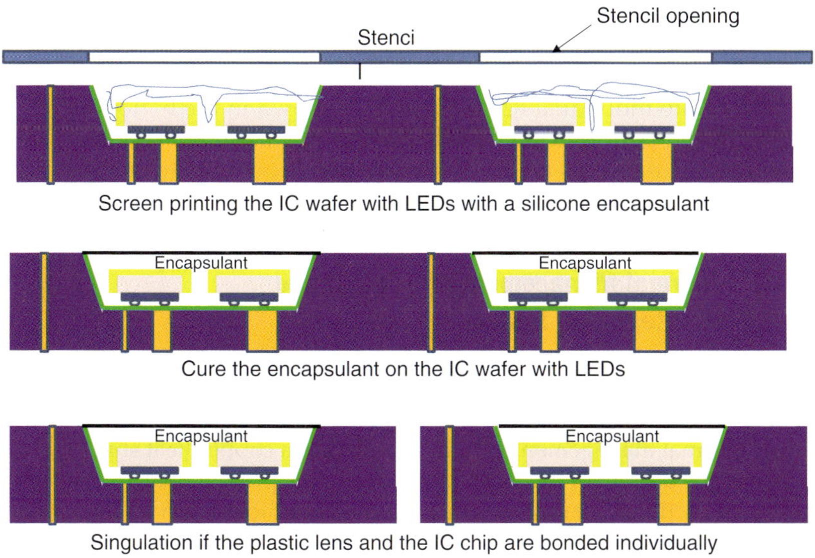

Figure 6.20 Screen print of encapsulant on IC wafer with LEDs.

TSV with RDL is the most important key enabling technology for 3-D IC integration [60–72]. There are five major steps (via formation, dielectric isolation layer deposition, barrier/adhesion and seed metal layer, via filling with copper, and copper polishing) to make just the TSV, and more steps to make the RDL, under bump metallurgy (UBM), solder bumps, and metallization, as shown in Fig. 1.20.

There are at least two ways to bond the LEDs (chips) to the IC wafer (C2W) with TSV. One is to use the pick and place (P&P) machine to place all the LEDs on the IC wafer and then reflow all the LEDs at once. The other is to use a flip-chip bonder to P&P and bond the LEDs, one at a time, on the IC wafer. Both methods have been demonstrated in [62–67] and are shown in the last step of the most likely 3-D IC integration process (Fig. 1.20).

The C2W bonding is followed by silicone encapsulation. Figure 6.20 shows the simple process steps. A stainless steel stencil with openings that are slightly larger than the top opening of the cavity on the IC wafer is used to screen print the silicone encapsulant into the cavity. Then, the encapsulant is cured.

After the cavity of the IC wafer is filled with encapsulant, it is ready for the lens wafer to IC wafer (W2W) bonding. This step is similar to the W2W bonding of the cavity-cap wafer and the ASIC wafer (with the MEMS devices already attached on its top surface), which is detailed in [61]. Then, singulate the bonded wafers by either mechanical dicing saw or lasers.

If the plastic lens is not in a wafer form, that is, an individual lens, there are two ways to seal the package. One is by chip (lens) to IC wafer bonding and then singulation. The other is to singulate the IC wafer (into individual IC chips) first, as shown in the bottom of Fig. 6.20, and perform chip (plastic lens) to IC chip (C2C) bonding. Now the 3-D packages are ready for the next level of interconnects and thermal management.

6.8.3 Thermal Management of 3-D LED and IC Integration System

One of the key challenges of HP LED packaging is to take the heat out from a very tiny LED chip "area" before it burns itself out or affects its light performance. Figures 6.17 and 6.18 show the most conventional thermal management system, which uses a copper heat slug to attach to the bottom of the Si submount (or in the present case, to one side of the active IC chip) through a thermal interface material (TIM). If that is not enough to take the heat out, add a copper heat spreader, and if that is still not enough, add a heat sink.

In this chapter, a thermal management system that consists of a microchannel carrier, a mini-pump, a heat exchanger, and an adapter, as shown in Figs. 6.21 and 6.22, for HP LEDs and IC integration

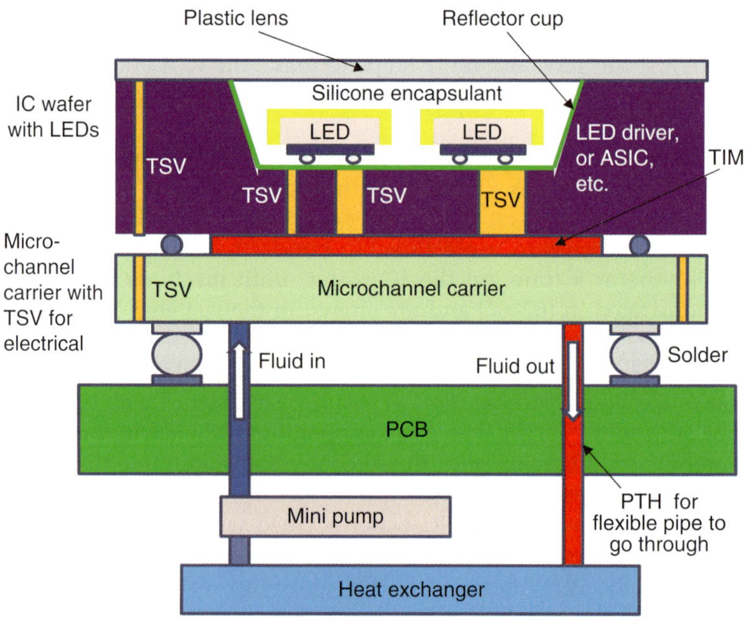

Figure 6.21 Thermal management of 3-D HP LEDs and IC (with a cavity) integration system.

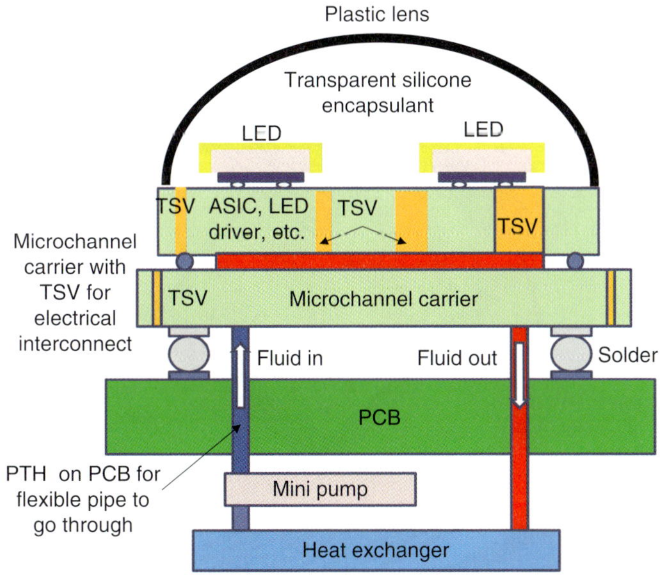

Figure 6.22 Thermal management of 3-D HP LEDs and IC integration system.

is proposed. The silicon microchannel carrier consists of two silicon chips, and an optimized liquid cooling channel structure is embedded in between the chips, as shown in Fig. 6.23. Silicon is chosen as carrier material because it is a suitable material for the integration of both electrical and fluidic structures in the same substrate with microfabrication process. The difference between these two chips is that the bottom Si chip does not have any outlets.

TSVs can be designed along the periphery of the carrier. After Si wafer to Si wafer (W2W) bonding, electrical interconnection through the carrier is made by the TSV with on-wall metallization (in this case, the copper filling may not be necessary). The fluidic channels are connected out through the inlet and outlet. There are sealing rings around both the fluidic path and the individual TSV to isolate the fluid from the electrical interconnection. (Please see Chap. 11 for more details.) The mini-pump is used to pump the cold water into the microchannel from the heat exchanger. This technology has been demonstrated [66, 68, 69, 70, 72], specifically, that a 15 × 15 mm^2 microchannel carrier is able to manage 100 W of heat from a 10 × 10 mm^2 chip. The microchannel and TSV are made by the deep reactive ion etch (DRIE) method; for examples, see [60, 61]. The LEDs/IC/thermal management system can be made even more miniature if the mini-pump is replaced by an electro-osmotic "pump" as shown in Fig. 6.24.

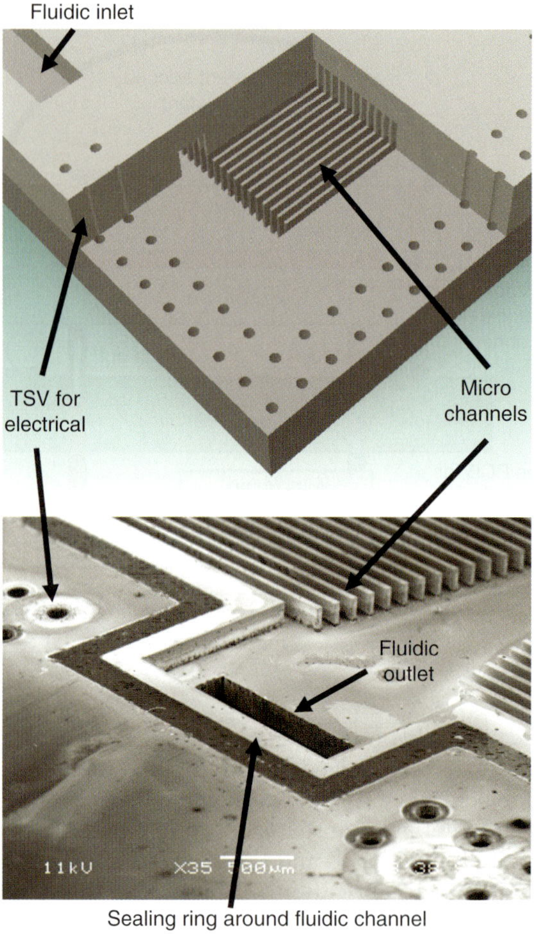

FIGURE 6.23 Microchannel carrier (with TSV for electrical feed-through) to cool high-power chips.

6.8.4 Summary and Recommendations

Some important results and recommendations are summarized in the following:

1. By replacing the Si submount of the HP FCLEDs with an active IC chip such as the ASIC, LED driver, processor, memory, RF, sensor, and power controller, that is, integrating the LEDs and the active chip together into a three-dimensional package. The advantages of these 3-D integration packages are: (1) better performance, (2) lower cost, (3) smaller footprint, (4) lighter weight, and (5) smaller form factor.

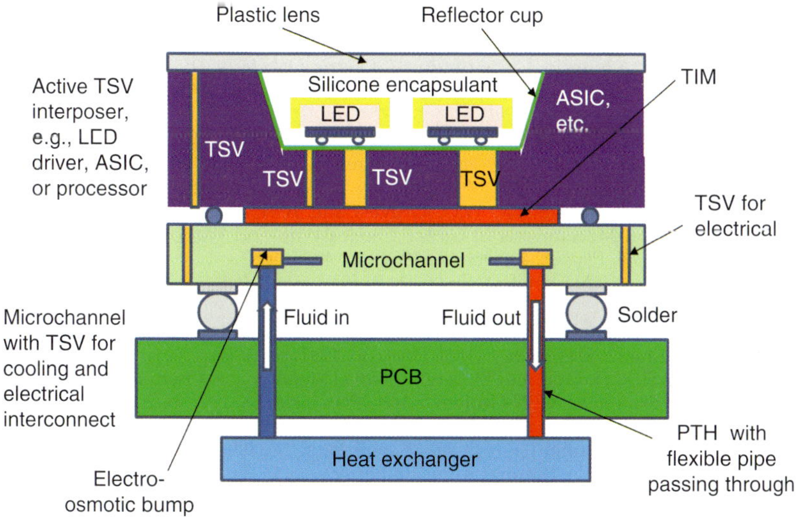

Figure 6-24 Thermal management of 3-D HP LEDs and IC osmotic "pump."

2. By utilizing the knowledge of the etch rates for different crystallographic planes of the starting wafers, a variety of micromachined structures (reflector cups) can be formed by properly selecting the masking materials, such as the Si_3N_4, SiO_2, Au, Pt, Cr, Cu, Ag, and Ta, and etchants such as the EDP, KOH/IPA, or TMAH.
3. The copper-filled TSV in the active IC chip can be used for not only electrical feed-through, but also as a thermal path. Since the thermal conductivity of copper is more than two times that of silicon, some large dummy TSVs filled with copper are recommended.
4. It is recommended that the LEDs be picked and placed on the IC wafer (C2W) and reflowed all at once to increase throughputs and reduce costs.
5. The transparent silicon encapsulant can be screen-printed with a stainless steel stencil with openings slightly larger than the opening of the cavity of the IC wafer.
6. It is recommended that the plastic lens be in wafer form, which allows IC wafer to plastic lens wafer (W2W) bonding to increase throughputs and save costs.
7. The proposed microchannel carrier is not only cooling the IC chip with LEDs, but also allowing electrical feed-through.
8. The thermal performance of the microchannel carrier is much better than that of the conventional thermal management method such as the heat slug/spreader/sink.

6.9 References

1. Round, H. J., "A Note on Carborundum." *Electrical World* 19: (1907), p. 309.
2. Losev, O. V., "CII. Luminous carborundum detector and detection effect and oscillations with crystals," *Phil. Mag*. 6, 1928, pp. 1024–1044.
3. Soviet patent #12191 granted in 1929.
4. Holonyak, Jr. H., and S. F. Bevaqua, "Coherent (visible) light emission from Ga(As1-xPx) junctions," *Appl. Phys. Lett.*, Vol. 1, No. 4, 1962, pp. 82–83.
5. Craford, M. G., R.W. Shaw, W.O. Groves, and A.H. Herzog, "Radiative Recombination Mechanisms in GaAsP Diodes With and Without Nitrogen Doping." *J. Appl. Physics,* 43, October 1972, pp. 4075–4083.
6. Nakamura, S., "GaN Growth Using GaN Buffer Layer," *Jpn. J. Appl. Phys.,* 30, 1991, p. L1705.
7. Nakamura, S., T. Mukai, and M. Senoh, "High-Power GaN p-n Junction Blue-Light-Emitting Diodes," *Jpn. J. Appl. Phys.*, 30, 1991, p. L1998.
8. Nakamura, S., Y. Harada, and M. Senoh, "Novel Metalorganic Chemical Vapor Deposition System for GaN Growth," *Appl. Phys. Lett.*, 58, 1991, p. 2021.
9. Nakamura, S., M. Senoh, and T. Mukai, "Highly P-Type Mg-Doped GaN Films Grown with GaN Buffer Layers," *Jpn. J. Appl. Phys.,* 30, 1991, p. L1708.
10. Nakamura, S., et al., "Thermal Annealing Effects on P-Type Mg-Doped GaN Films," *Jpn. J. Appl. Phys.*, 31, 1992, p. L139.
11. Nakamura, S. and T. Mukai, "High-Quality InGaN Films Grown on GaN Films," *Jpn. J. Appl. Phys.*, 31, 1992, p. L1457.
12. Nakamura, S., T.Mukai, and M. Senoh, "Si- and Ge-Doped GaN Films Grown with GaN Buffer Layers," *Jpn. J. Appl. Phys.,* 31, 1992, p. 2883.
13. Nakamura, S., et al., "Hole Compensation Mechanism of P-Type GaN Films," *Jpn. J. Appl. Phys.,* 31, 1992, p. 1258.
14. Nakamura, S., M. Senoh, and T. Mukai, "P-GaN/N-InGaN/N-GaN Double-Heterostructure Blue-Light-Emitting Diodes," *Jpn. J. Appl. Phys.,* 32, 1993, p. L8.
15. Nakamura, S., T. Mukai, and M. Senoh, "Candela-Class High-Brightness InGaN/AlGaN Double-Heterostructure Blue-Light-Emitting-Diodes," *Appl. Phys. Lett.,* 64, 1994, p. 1687.
16. Nakamura, S., "Growth of $In_XGa_{1-X}N$ Compound Semiconductors and High-Power InGaN/AlGaN Double Heterostructure Violet-Light-Emitting Diodes," *Microelectronics,* 25, 1994, p. 651.
17. Nakamura, S., et al., "High-Power InGaN Single-Quantum Well-Structure Blue and Violet Light-Emitting Diodes," *Appl. Phys. Lett.*, 67, 1995, p. 1868.
18. Nakamura, S., et al., "Superbright Green InGaN Single-Quantum-Well-Structure Light-Emitting Diodes," *Jpn. J. Appl. Phys.*, 34, 1995, p. L1332.
19. Nakamura, S., et al., "High-Brightness InGaN Blue, Green and Yellow Light-Emitting Diodes with Quantum Well Structures," *Jpn. J. Appl. Phys.*, 34, 1995, p. L797.
20. Nakamura, S., et al., "Room-Temperature Continuous-Wave Operation of InGaN Multi-Quantum-Well Structure Laser Diodes," *Appl. Phys. Lett.*, 69, 1996, p. 4056.
21. Nakamura, S., "Characteristics of Room Temperature-cw Operated InGaN Multi-Quantum-Well-Structure Laser Diodes," *MRS Internet J. Nitride Semicond. Res.*, 2 (5), 1997.
22. Japan Patent Application No. 10-194156 (194156, 1998) filed on Jul. 9, 1998.
23. Japan Patent Application No. 10-316169 (316169, 1998) filed on Nov. 6, 1998.
24. Japan Patent Application No. 10-321605 (321605, 1998) filed on Nov. 12, 1998.
25. Wierer, J., D. A. Steigerwald, M. R. Krames, J. J. O'Shea, M. J. Ludowise, G. Christenson, Y.-C. Shen, C. Lowery, P. S. Martin, S. Subramanya, W. Götz, N. F. Gardner, R. S. Kern, and S. A. Stockman, "High-power AlGaInN flip-chip light-emitting diodes," *Applied physics Letters*, V. 78, no. 22, May 28, p. 2001.
26. Lau, J. H., Lee, R., Yuen, M., and Chan, P., "3D LED and IC Wafer Level Packaging", IEEE Proceedings of EP09-20-05, 2009, also *Journal of Microelectronics International*, Vol. 27, Issue 2, 2010, pp. 98–105.

27. Hwang, J. S., *Environmentally Friendly Electronic Lead-Free Technology*, Electrochemical Publications, Ayr, Scotland, 2001.
28. Lau, J. H., C. P. Wong, N. C. Lee, and S.W. R. Lee, *Electronics Manufacturing with Lead-Free, Halogen-Free, and Conductive-Adhesive Materials*, McGraw-Hill, New York, 2003.
29. Lee, N. C., *Reflow Soldering Processes and Troubleshooting for SMT, BGA, CSP, and Flip Chip Technologies*, Newnes, Boston, MA, 2002.
30. Lau, J. H., *Low Cost Flip Chip Technologies for DCA, WLCSP, and PBGA Assemblies*, McGraw-Hill, New York, 2000.
31. Smetana, J., R. Horsley, J. Lau, K. Snowdon, D. Shangguan, J. Gleason, I. Memis, D. Love, W. Dauksher, and B. Sullivan, "Lead-Free Design, Materials, and Process of High Density Packages," *Proceedings of APEX*, Anaheim, CA, March 2003, pp. S42–1–1/13.
32. Lau, J., W. Dauksher, J. Smetana, R. Horsley, D. Shangguan, T. Castello, I. Memis, D. Love, and B. Sullivan, "Design for Lead-Free Solder Joints reliability of High-Density Packages," *Proceedings of APEX*, Anaheim, CA, March 2003, pp. S42–2–1/13.
33. Lau, J., N. Hoo, R. Horsley, J. Smetana, D. Shangguan, W. Dauksher, D. Love, I. Memis, and B. Sullivan, "Reliability Test and Data Analysis of High-Density Packages' Lead-Free Solder Joints," *Proceedings of APEX*, Anaheim, CA, March 2003, pp. S42–3–1/24.
34. Lau, J., D. Shangguan, T. Castello, R. Horsley, J. Smetana, W. Dauksher, D. Love, I. Memis, and B. Sullivan, "Failure Analysis of High-Density Packages' Lead-Free Solder Joints," *Proceedings of APEX*, Anaheim, CA, March 2003, pp. S42–4–1/11.
35. Deepak, G., T. Lane, P. Kinzie, C. Panichas, K. Chong, and O. Villalobos, "Failure Mechanism of Brittle Solder Joint Fracture in the Presence of Electroless Nickel Immersion Gold (ENIG) Interface," *IEEE Proceedings on Electronic Components and Technology Conference*, 2002, pp. 732–739.
36. Jeon, Y., K. Paik, K. Bok, W. Choi, and C. Cho, "Studies on Ni-Au Intermetallic Compound and P-Rich Ni layer at the Electroless Ni UBM – Solder Interface and Their Effects on Flip Chip Solder Joint Reliability," *IEEE Proceedings on Electronic Components and Technology Conference*, 2001, pp. 69–75.
37. Chan, Y., P. Yu, C. Tang, K. Hung, J. Lai, "Reliability Studies of μBGA Solder Joints—Effects of Ni-Sn Intermetallic Compound," *IEEE Transactions on Advanced Packaging*, Vol. 24, 2001, pp. 25–31.
38. Beelen-Hendrikx, C., J. Klerk, and J. van de Water, "Lead-free Semiconductor Packaging," *IEEE Proceedings of Electronics Packaging Technology Conference*, Singapore, December 2002, pp. 11–19.
39. Sy, H., P. Arulvanan, and P. Collier, "Rework and Reliability of QFP and BGA Lead-Free Assemblies," *IEEE Proceedings of Electronics Packaging Technology Conference*, Singapore, December 2002, pp. 194–199.
40. Chung, C., R. Aspandiar, K. Leong, and C. Tay, "The Interactions of Lead (Pb) in Lead Free Solder (Sn/Ag/Cu) Systems," *IEEE Proceedings of Electronic Components & Technology Conference*, June 2002, pp. 168–175.
41. Kujala, A., T. Reinikainen, and W. Ren, "Transition to Pb-Free Manufacturing using Land Grid Array Packaging Technology," *IEEE Proceedings of Electronic Components & Technology Conference*, June 2002, pp. 359–364.
42. Schubert, A., R. Dudek, H. Walter, E. Jung, A. Gollhardt, B. Michel, and H. Reichl, "Reliability Assessment of Flip-Chip Assemblies with Lead-free Solder Joints," *IEEE Proceedings of Electronic Components & Technology Conference*, June 2002, pp. 1246–1255.
43. Lee, R., B. Lui, Y. Kong, B. Bayton, T. Leung, P. Umali, and H. Agtarap, "Assessment of board level solder joint reliability for PBGA assemblies with lead-free solders," *Soldering & Surface Mount Technology*, Vol. 14, No. 2, 2002, pp. 46–50.
44. Guo, Y., J. Lin, and A. De Silva, "Reliability Evaluations of Chip Interconnect in Lead-Free Solder Systems,, *IEEE Proceedings of Electronic Components & Technology Conference*, June 2002, pp. 1275–1280.

45. Lau, J. H., "Creep of 96.5Sn-3.5Ag Solder Interconnects," *Soldering & Surface Mount Technology*, Vol. 5, 1993, pp. 45–49.
46. Lau, J. H., "Creep of Solder Interconnects Under Combined Loads," *IEEE Transactions on Components, Hybrids, and Manufacturing Technology*, December 1993, pp. 794–798.
47. Lau, J. H., and C. Chang, "TMA, DMA, DSC, and TGA of Lead Free Solders," *Soldering & Surface Mount Technology*, Vol. 11, No. 2, 1999, pp. 17–24.
48. Arra, M., D. Xie, and D. Shangguan, "Performance of Lead-Free Solder Joints Under Dynamic Mechanical Loading," *IEEE Proceedings of Electronic Components & Technology Conference*, June 2002, pp. 1256–1262.
49. Lau, J. H., S. Pan, C. Chang, "Creep Analysis of Wafer Level Chip Scale Package (WLCSP) with 96.5Sn-3.5Ag and 100In Lead-Free Solder Joint and Microvia Build-Up Printed Circuit," *ASME Transactions, Journal of Electronic Packaging*, Vol. 124, June 2002, pp. 69–76.
50. Park, T., and S. Lee, "Isothermal Low Cycle Fatigue Tests of Sn/3.5Ag/0.75Cu and 63Sn/37Pb Solder Joints under Mixed-Mode Loading Cases," *IEEE Proceedings of Electronic Components & Technology Conference*, June 2002, pp. 979–984.
51. Lau, J. H., W. Dauksher, and P. Vianco, "Acceleration Models, Constitutive Equations, and Reliability of Lead-Free Solders," *IEEE Proceedings of Electronic Components & Technology Conference*, May 2003, pp. 229–236.
52. Lau, J. H., S. Pan, and C. Xu, "3D Large Deformation and Nonlinear Stress Analysis of Tin Whisker Initiation and Growth on Lead-Free Components," *IEEE Proceedings of Electronic Components & Technology Conference*, May 2003, pp. 692–697.
53. Lau, J. H., and S. Pan, "3D Nonlinear Stress Analysis of Tin Whisker Initiation on Lead-Free Components," *NEPCON West Proceedings*, CD-ROM, 2002.
54. Lau, J. H., and S. W. R. Lee, *Microvias for Low Cost, High Density Interconnects*, McGraw-Hill, New York, 2001.
55. Choi, W., T. Lee, K. Tu, N. Tamura, R. Celestre, A. MacDowell, Y. Bong, L. Nguyen, and G. Sheng, "Structure and Kinetics of Sn Whisker Growth on Pb-free Solder Finish," *Proceedings of Electronic Components and Technology Conference*, 2002, pp. 628–633.
56. Agilent, "20 m (0.8 inch) Seven Segment Displays," *Technical Data*, July 26, 2001.
57. ANSYS, *User's Manual*, Release 5.7, 2000.
58. Lau, J. H., Z. Mei, S. Pang, C. Amsden, J. Rayner, and S. Pan, "Creep Analysis and Thermal-Fatigue Life Prediction of the Lead-Free Solder Sealing Ring of a Photonic Switch," *ASME Transactions, Journal of Electronic Packaging*, Vol. 124, December 2002, pp. 403–410.
59. Lau, J. H., R. Subrahmanyan, D. Rice, S. Erasmus, and C. Y. Li, "Fatigue Analysis of a Ceramic Pin Grid Array Soldered to An Orthotropic Epoxy Substrate," *ASME Transactions, Journal of Electronic Packaging*, Vol. 113, June 1991, pp. 13–20.
60. Garrou, P., C. Bower, and P. Ramm, *3D Integration: Technology and Applications*, Wiley, John & Sons, 2009.
61. Lau, J., C. Lee, C. Premachandran, and A. Yu, *3D MEMS Packaging*, McGraw-Hill, 2010.
62. Zhang, X., T. Chai, J. Lau, C. Selvanayagam, K. Biswas, S. Liu, D. Pinjala, G. Tang, Y. Ong, S. Vempati, E. Wai, H. Li, B. Liao, N. Ranganathan, V. Kripesh, J. Sun, J. Doricko, and C. Vath. "Development of through silicon via (TSV) interposer technology for large die (21 × 21 mm) fine-pitch Cu/low-*k* FCBGA package." *IEEE Proceedings of Electronic Components and Technology Conference*, San Diego, CA, May 2009, pp. 305–312. Also, accepted for publication in IEEE Transactions in Advanced Packaging.
63. Yu, A., J. Lau, S. Ho, A. Kumar, W. Hnin, D. Yu, M. Jong, V. Kripesh, D. Pinjala, and D. Kwong, "Study of 15μm Pitch Solder Microbumps for 3D IC Integration," *IEEE Proceedings of Electronic Components and Technology Conference*, San Diego, CA, May 2009, pp. 6–10.

64. Yu, A., J. Lau, S. Ho, A. Kumar, H. Yin, J. Ching, V. Kripesh, D. Pinjala, S. Chen, C. Chan, C. Chao, C. Chiu, M. Huang, and C. Chen, "Three dimensional interconnects with high aspect ratio TSVs and fine pitch solder microbumps." *IEEE Proceedings of Electronic Components and Technology Conference,* San Diego, CA, May 2009, pp. 350–354. Also, accepted for publication in IEEE Transactions in Advanced Packaging.
65. Ho, S., S. Yoon, Q. Zhou, K. Pasad, V. Kripesh, and J. Lau, "High RF performance TSV for silicon carrier for high frequency application." *IEEE Proceedings of Electronic Components and Technology Conference,* Orlando, FL, May 27–30, 2008, pp. 1956–1952.
66. Khan, N., V. Rao, S. Lim, S. Ho, V. Lee, X. Zhang, R. Yang, E. Liao, N. Ranganathan, T. Chai, V. Kripesh, and J. Lau, "Development of 3D silicon module with TSV for system in packaging." *IEEE Proceedings of Electronic Components and Technology Conference,* Orlando, FL, May 27–30, 2008, pp. 550–555.
67. Selvanayagam, C., J. Lau, X. Zhang, S. Seah, K. Vaidyanathan, and T. Chai, "Nonlinear thermal stress/strain analysis of copper filled TSV (through silicon via) and their flip-chip microbumps," *IEEE Transactions on Advanced Packaging,*Vol. 32, No. 4, 2009, pp. 720–728.
68. Khan, N., H. Li, S. Tan, S. Ho, N. Su, W. Hnin, V. Kripesh, D., Pinjala, J. Lau, and T. Chuan, "3D Packaging with Through Silicon Via (TSV) for Electrical and Fluidic Interconnections," *IEEE Proceedings of Electronic Components and Technology Conference,* San Diego, CA, May 2009, p. 1153–1158.
69. Tang, G., S. Tan, N. Khan, D. Pinjala, J. Lau, A. Yu, V. Kripesh, and C. Toh, "Fluidic Interconnects in Integrated Liquid Cooling Systems for 3-D Stacked TSV Module," *IEEE Proceedings of Electronic Packaging and Technology Conference,* December 2008, pp. 552–558.
70. Tang, G., O. Navas, D. Pinjala, J. Lau, A. Yu, and V. Kripesh, "Integrated Liquid Cooling Systems for 3-D Stacked TSV Modules," *IEEE Transactions on Components and Packaging Technologies*, Vol. 33, No. 1, 2010, pp. 184–195.
71. Lau, J. H., and G. Tang, "Thermal management of 3D IC integration with TSV (through silicon via)." *IEEE Proceedings of Electronic Components and Technology Conference,* San Diego, CA, May 2009, pp. 635–640.
72. Yu, A., N. Khan, G. Archit, D. Pinjalal, K. Toh, V. Kripesh, S. Yoon, and J. Lau, "Development of silicon carriers with embedded thermal solutions for high power 3-D package." *IEEE Transactions on Components and Packaging Technology*, Vol. 32, No. 3, September 2009, pp. 566–571.

CHAPTER 7

Reliability of VCSEL Lead-Free Interconnects

7.1 Introduction

Most of the optical fiber communication systems nowadays use intensity modulation (IM) and direct detection (DD). The basic IM-DD system consists of a light source, a fiber, and an optical receiver. In general, the electrical signal is converted into the optical signal through the light source and the optical signal is transmitted through the optical fiber or polymer waveguide. Then, the optical signal is converted back into the electrical signal through the optical receiver. This is the fundamental principle of optical-fiber transceivers [1]. For optical fiber communication applications, most of the light sources are based on the electron-hole recombination in semiconductor materials. This recombination results in the release of energy (in the form of another photon), which can take place either spontaneously, such as the light-emitting diodes (LEDs), or as a result of an external stimulus, such as the semiconductor lasers. Usually, the LEDs are used for very short distance, such as chip-to-chip communications with plastic fibers. On the other hand, semiconductor lasers are used for longer distances, such as telecommunications [2].

Both LEDs and semiconductor lasers can collect the light in the direction parallel to the p–n junction, and are called edge-emitting diodes and edge-emitting lasers, respectively. They can also collect the light in the direction perpendicular to the p–n junction, and are called the surface-emitting diodes and vertical-cavity surface-emitting lasers (VCSELs), respectively. One of the advantages of surface-emitting light sources is that they can be used in parallel optics that offer low-cost interconnection with perhaps the best use of backplane space in a network system [3–6].

Unlike older edge-emitting diodes that emit infrared (IR) radiation in the plane of the semiconductor, the VCSEL produces a nearly circular beam of laser light that is perpendicular to the plane of the

FIGURE 7.1 Honeywell's 4 × 4 VCSELs. (*Courtesy of Honeywell.*)

semiconductor. These devices may be constructed on the surface of a fabricated wafer and have the advantage of combining large 2-D emitter arrays (such as the one shown in Fig. 7.1) and active devices such as complementary metal-oxide-semiconductors (CMOS) drivers with conventional flip-chip technology. In addition, VCSELs offer ease of fiber (or polymer waveguide) coupling and of physical conformity with vertical laser cavity assemblies such as the one shown in Fig. 7.2 (Fig. 10.47 of [7]).

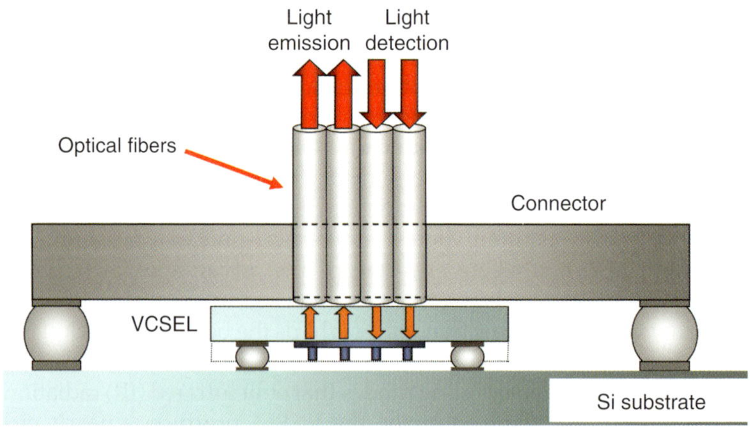

FIGURE 7.2 Optical communication system with VCSELs.

How does one package the surface-emitting light sources with a good balance between cost, performance, and reliability? This is a difficult question to answer. Although the light source (chip) may be the most sophisticated and important component in an optoelectronic package, the weight, size, performance, reliability, and cost of the package are dominated by the auxiliary components required to run the light source chip. These components usually include, among others: (1) the thermistor used as a temperature sensor, (2) the photodiode sensing the light source power, (3) the lens collimating the light source output field, (4) the optical isolator isolating the light source from reflection, (5) the thermoelectrical cooler needed to keep the light source temperature at a constant value, (6) the fiber coupling lens, (7) the fiber mount, (8) the substrate, (9) the heat sink, and (10) the hermetic window. In general, the cost of the light source chip is only a fraction of the total auxiliary component cost.

It is well known that the temperature around the surface-emitting light source can change the wavelength of the light emission. In order to make the light source stable, the junction temperature near the cavity needs to be kept below a constant temperature. The typical acceptable temperature is below 85°C, even though some light sources may be able to tolerate up to 100°C [8–10]. In principle, the wavelength in the infrared (IR) region has the advantage of integrating large 2-D emitter arrays with active (such as CMOS driver) components. This is because the semiconductor becomes transparent in this wavelength region and, therefore, the device can emit lights through the substrate. This feature yields to the possibility of applying the conventional flip-chip technology [11, 12] with wafer-level redistribution to the LED/VCSEL assembly such as the one shown in Figs. 7.3 and 7.4 [13].

The choice of a package configuration requires recognition not only of critical cost issues, but also those of thermal performance and mechanical reliability. In particular, a stable thermal environment is required. Excursions from the desired operating temperature may alter the wavelength of the emitted light, consequently degrading the VCSEL's performance. Since the choice of bump material is critical to the conduction of heat from the light source, the judicious choice of the solder bump material is an important component of the VCSEL design. Therefore, the integrity of the solder bump is a vital design component. Since the bump provides the electrical/thermal path, loss of the bump due to creep damage accumulation during in-service operation is unacceptable. Also, the stresses acting at the AlGaAs pad should be lower than its allowable value.

As such, this chapter compares the effects of thermal conductivity of the 48wt%Sn-52wt%In, 42wt%Sn-58wt%Bi, 63wt%Sn-37wt%Pb, Sn(3–4)wt%Ag(0.5–0.7)wt%Cu, and 80wt%Au-20wt%Sn solder alloys to simulate potential in-service operating temperature distributions.

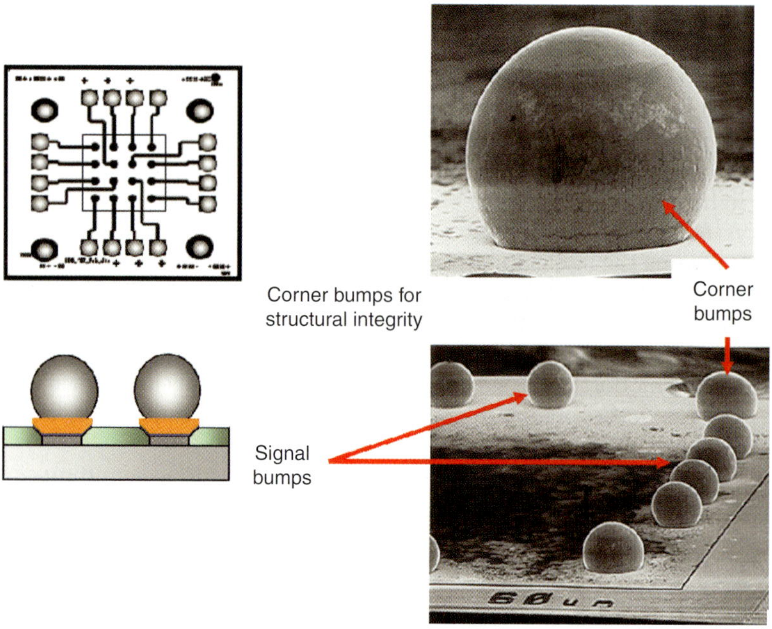

Figure 7.3 Wafer-level redistribution with 4 × 4 VCSELs.

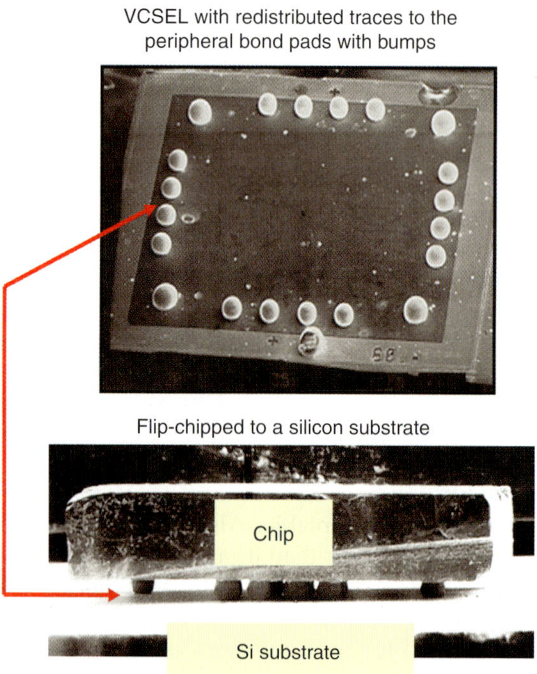

Figure 7.4 Wafer-level redistribution with 4 × 4 VCSELs on Si substrate.

These temperature distributions are then applied to the VCSEL assembly over a 24-h period to assess the stresses in the AlGaAs pads and the accumulated creep damage at the solder joints for each candidate material. In addition, a 3-D optoelectronics package on a printed circuit board with polymer waveguide is presented.

7.2 Lead-Free Interconnect Reliability of a VCSEL under Transient and Steady-State Loadings

7.2.1 The Structure

Figures 7.5 and 7.6 detail the construction of the VCSEL [14–19]. Solder bumping, with its inherent self-aligning capability, becomes a natural packaging construction. The GaAs chip is mounted on a silicon substrate with four flip-chip solder joints. Unlike those shown in Figs. 7.3 and 7.4, in the present case, the bumps are directly attached to the VSCEL. Since the GaAs is transparent to the IR wavelength, transmission occurs through the chip. The 2 × 2 array of lights sources have a 250 μm pitch. Both the AlGaAs light source and the copper pad have 80 μm diameters and 5 μm thicknesses. The assembled bump height is 75 μm.

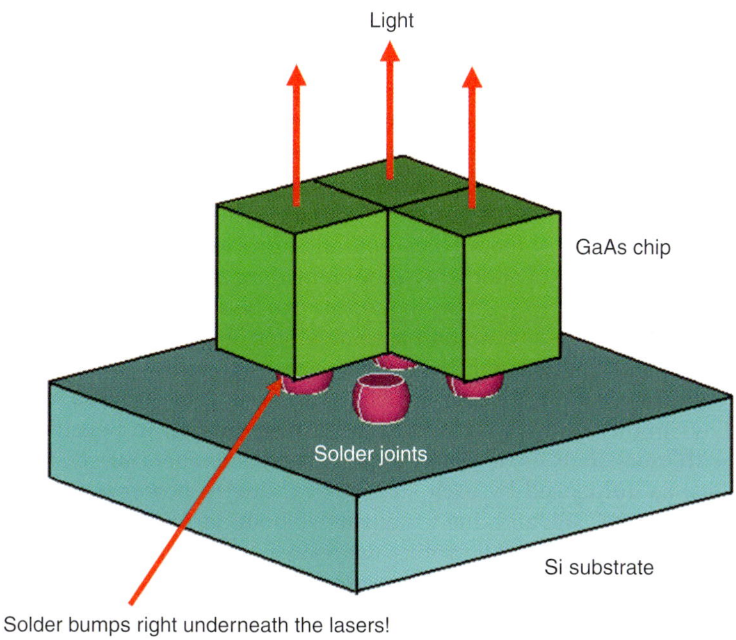

FIGURE 7.5 The construction of the VCSEL.

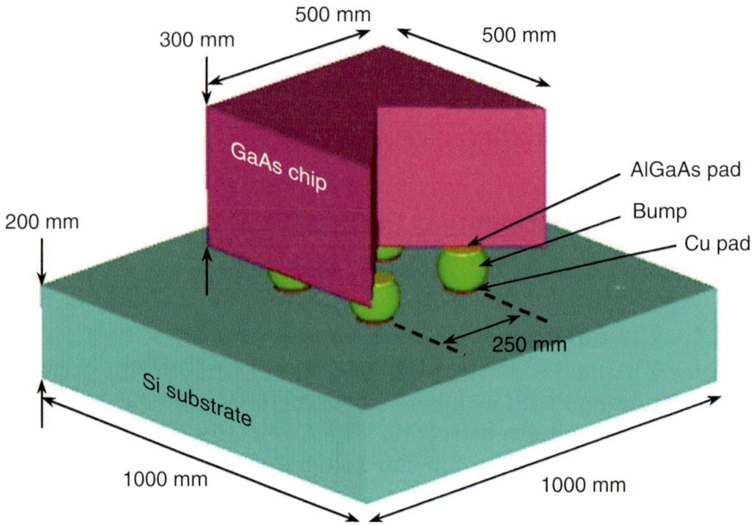

FIGURE 7.6 The dimensions of the VCSEL.

7.2.2 Materials

The melting points of 48wt%Sn-52wt%In, 42wt%Sn-58wt%Bi, 63wt%Sn-37wt%Pb, Sn(3–4)wt%Ag(0.5–0.7)wt%Cu, and 80wt%Au-20wt%Sn solder alloys are, respectively 118°C, 138°C, 183°C, 217°C, and 280°C. For optoelectronics applications, lower melting solders are preferred, for example, optical components with plastic lenses depend on the opacity and stability of epoxies to meet their optical transmission and alignment specifications. Nonglass optical materials tend to darken or crack when exposed to high processing temperatures. Also, low-temperature solders such as Sn-In and Sn-Bi allow a two-step or hierarchical soldering process of one component or module on top of another without remelting already soldered connections, for example, an Sn-In module can be soldered on top of an SnAgCu module as a second assembly step without remelting the solder joints on the latter.

In this chapter, thermal-structural analysis evaluates the solder alloys in the context of in-service operating conditions. Specifically, the thermal analysis determines the in-service temperature distributions, as influenced by each solder alloy, due to power generation within the VCSEL and due to convective boundary conditions. Subsequently, these thermal profiles are used as the thermal loads in an evaluation of the stress and creep response in the laser pads and solder joints. Emphasis is placed on the relaxation of stresses at the laser pads and on the creep strains developed at the solder joints during a 24-h power-on condition.

Material	E (GPa)	CTE (ppm/K)	ν	k (W/mK)
Si	167	2.54	0.28	154
Cu	76	17	0.35	398
95.5Sn-3.9Ag-0.6Cu	$68.11 - 0.07T$	$16.66 + 0.017T$	0.3	70
80Au-20Sn	59	16	0.3	251
48Sn-52In	30.5	28	0.36	32.8
63Sn-37Pb	$75.9 - 0.152T$	24.5	0.35	50.6
Sn-58Bi	$381.5\text{-}3.13T + 0.009T^2$	14	0.35	18.3
AlGaAs	86	5.8	0.31	33.67
GaAs	86	5.8	0.31	33.7

Note: T in degrees Kelvin

Table 7.1 Material Properties

Table 7.1 presents the material properties used in this study. With a significantly higher thermal conductivity, the 80Au-20Sn solder is anticipated to provide superior thermal performance. The conductivity of the SnAgCu (SAC) alloy exceeds that of the SnPb and SnIn materials, and the thermal conductivity of SnBi is the smallest!

The solders exhibit creep deformations when under thermal loads as shown in Sec. 2.6.5 of this book. The AuSn and SnBi materials are assumed to obey the power law creep model as, respectively [7, 17]:

$$\frac{\partial \varepsilon}{\partial t} = 1.3977 \times 10^{-21} \sigma^{2.55} \exp\left(\frac{-9516}{T}\right)$$

$$\frac{\partial \varepsilon}{\partial t} = 9.82 \times 10^{-24} \sigma^{4.05} \exp\left(\frac{-8483}{T}\right)$$

The SnPb, SAC, and SnIn solders follow, respectively, the Garofalo-Arrhenius equation [7, 19, 20, 21]:

$$\frac{\partial \varepsilon}{\partial t} = \frac{926(508-T)}{T} \sinh^{3.3}\left(\frac{\sigma}{37.78 \times 10^6 - 74414T}\right) \exp\left(\frac{-6360}{T}\right)$$

$$\frac{\partial \varepsilon}{\partial t} = 500000 \sinh^{5.0}\left(\frac{\sigma}{1 \times 10^8}\right) \exp\left(\frac{-5807}{T}\right)$$

$$\frac{\partial \varepsilon}{\partial t} = \frac{2.54 \times 10^{11}(593-T)}{T} \sinh^{3.11}\left(\frac{\sigma}{36.09 \times 10^6 - 59628T}\right) \exp\left(\frac{-9705}{T}\right)$$

In these equations, ε is the normal strain, σ is the normal stress in Pa, t is time in seconds, and T is temperature in K. The creep responses of the solders at 0°C, 50°C, and 100°C are presented in Figs. 7.7 through 7.9, respectively. The SnIn and SnPb solders have a creep rate that is more than 10 orders of magnitude greater than that of the AuSn solder and approximately 1 to 2 orders of magnitude greater than the SAC alloy's creep rate. As a result, for a given stress, the SnIn and SnPb solder joints should have both a faster creep response and higher creep strains.

When comparing the SnBi solder alloy with the SnPb and SAC solder alloys, it can be seen that: (1) at low temperatures, for example, at 0°C, the creep rate of SnBi is slightly slower than those of SnPb and SnAgCu at higher stress conditions, and is slightly faster than that of

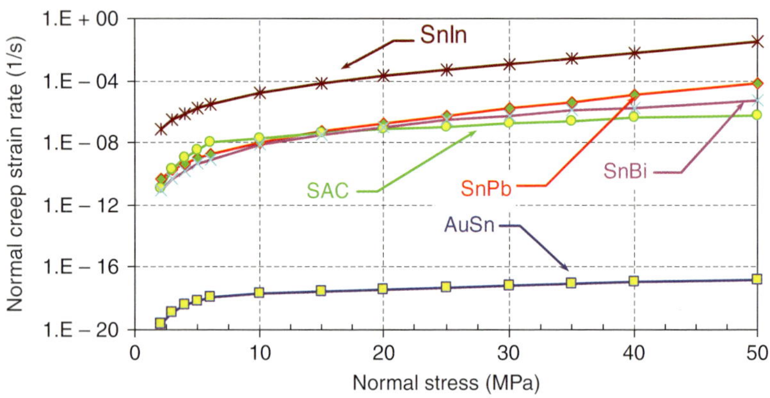

FIGURE 7.7 Creep rate of solder alloys at 0°C.

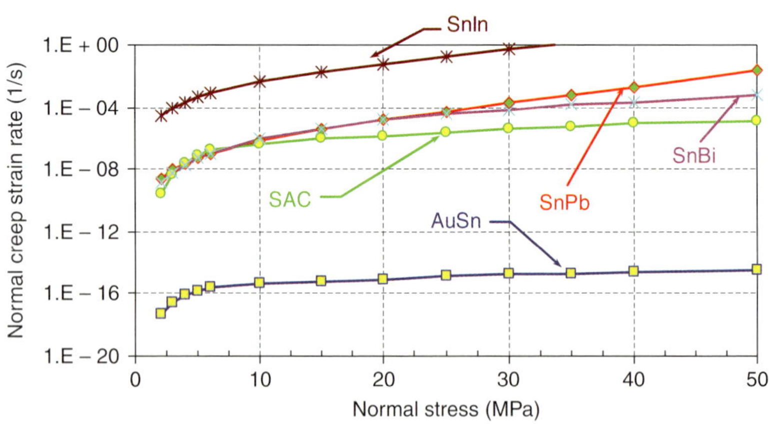

FIGURE 7.8 Creep rate of solder alloys at 50°C.

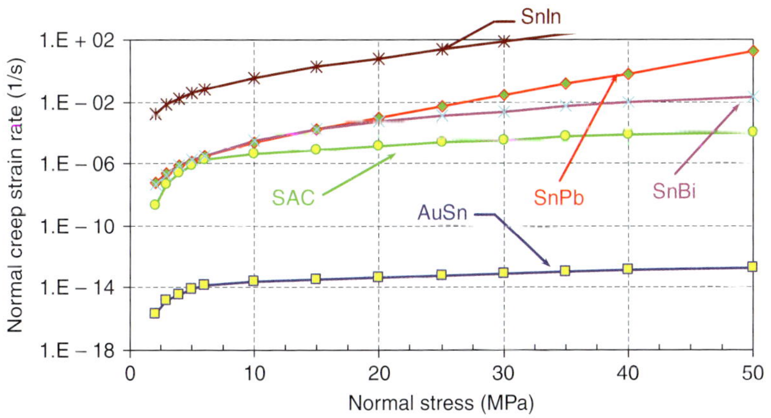

FIGURE 7.9 Creep rate of solder alloys at 100°C.

SnAgCu, and is about the same as SnPb's at lower stress conditions; and (2) at higher temperatures, for example, 50°C and 100°C, the creep rate of SnAgCu is the slowest and the creep rate of SnPb is faster than that of SnBi. The creep rate of SnIn is the fastest and that of AuSn is the slowest among all the solders considered.

Conversely, the moduli of all the lead-free solder alloys (except SnIn) are larger than that of the SnPb solder alloy (see Fig. 7.10). Therefore, a prescribed displacement should produce higher stresses in all the lead-free (except SnIn) solder joints considered herein when compared to the SnPb eutectic solder joints. The modulus of SnBi is less than that of AuSn and SAC. Below room temperature, the modulus of SnIn is slightly lower than that of SnPb. However, above room temperature, the modulus of SnIn is larger than that of SnPb.

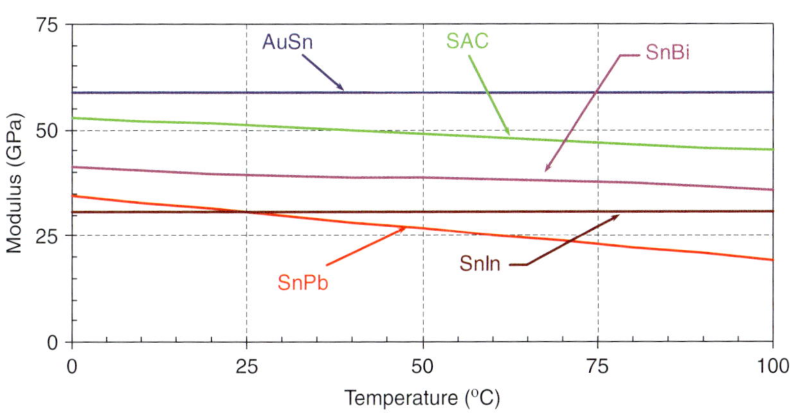

FIGURE 7.10 Solder alloys' modulii.

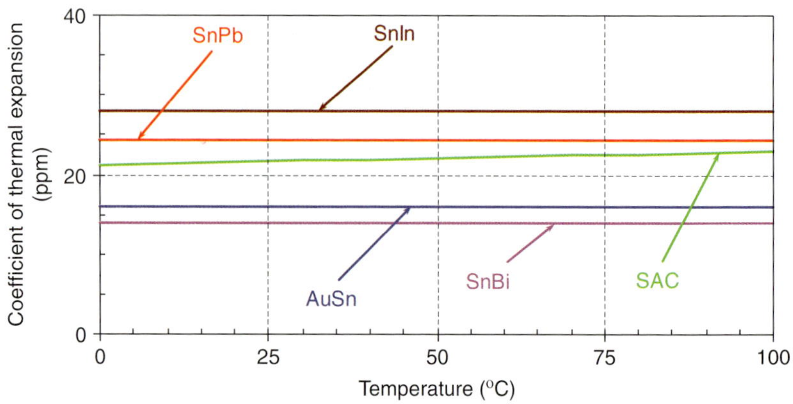

FIGURE 7.11 Solder alloys' coefficient of thermal expansion.

Finally, following Fig. 7.11, the coefficient of thermal expansion (CTE) of all the lead-free solders (except SnIn) considered herein is lower than that of the SnPb. It can also be seen from Fig. 7.11 that the CTE of SnIn is the largest among all these solder alloys.

7.2.3 Heat-Transfer Analysis and Results

In this portion of the study, a thermal analysis determines the spatially nonuniform temperature distribution within a VCSEL package based on a localized heat source, the thermal conductivities of the package materials, and the convection coefficients on the VCSEL's surface. The use of convection coefficients as thermal boundary conditions, rather than imposed node temperatures on the external surfaces, differentiates this study from previous efforts.

In service, the VCSEL will be switched on and off at a frequency of millions (or more) of cycles per second. Since the thermal time constant of the VCSEL structure is significantly greater than the period of the power-on/power-off cycling, the structure will equilibrate to a near-uniform operating temperature. We conservatively presume that the VCSEL has continuous power generation of 1 mW that is uniformly distributed within the AlGaAs pad. Convection boundary conditions are applied to the external surfaces of the structure. These convection coefficients are derived from a computational fluid dynamics analysis. Recognizing that the local convection coefficient along any surface is likely nonuniform, the values used represent surface averaged values and, consequently, may be applied to the present thermal model. The ambient temperature is 55°C.

The finite-element model, constructed of 3200 eight-node brick elements, is shown in Fig. 7.12. Due to the geometric symmetry that exists in the package, only one octant was modeled. This simplification implies the simultaneous power generation by all AlGaAs pads.

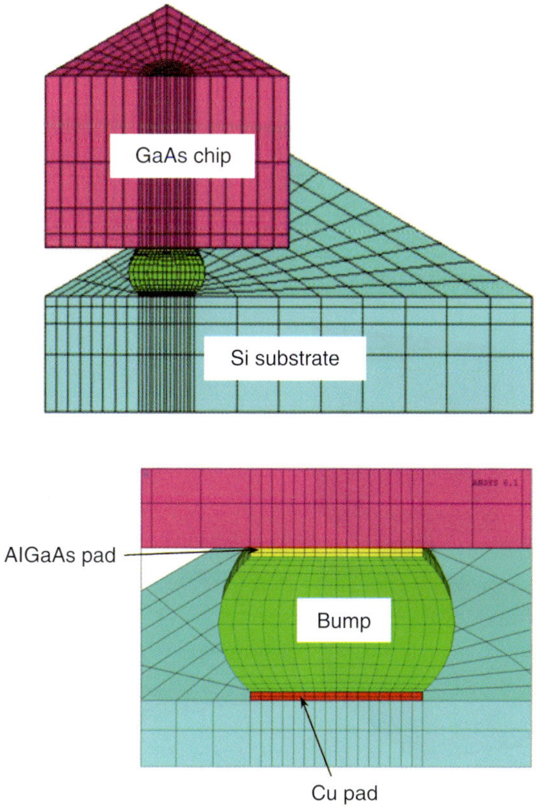

FIGURE 7.12 The finite-element model.

On the external surfaces, convection coefficients, as specified and depicted in Fig. 7.13, are applied. Heat flow is prohibited across the cut symmetry planes.

Initially, a transient analysis was performed to evaluate the time-temperature response of the VCSEL. Figure 7.14 presents the time-temperature response at the center of the AlGaAs pad with an SnPb bump during the first minute of the analysis. This representative data indicates that the VCSEL achieves a steady-state temperature distribution within approximately 20 s. Similar results may be found for the lead-free alloys.

Since the structural analysis that follows examines the VCSEL over a 24-h period, and due to the brevity of the time-dependent thermal response, the transient analysis was discarded in favor of a steady-state thermal analysis. The resulting temperature profiles are shown in Fig. 7.15. These temperatures are in line with normal application specific integrated circuits (ASIC) operating temperatures of approximately 100°C, but are believed to be conservative for two

Location	Convection coefficient (W/m^2 °C)
GaAs chip, inner surface	9.82
GaAs chip, edge	25.96
GaAs chip, outer surface	25.96
Si substrate, inner surface	9.82
Si substrate, edge	25.96
Si substrate, outer surface	21.42

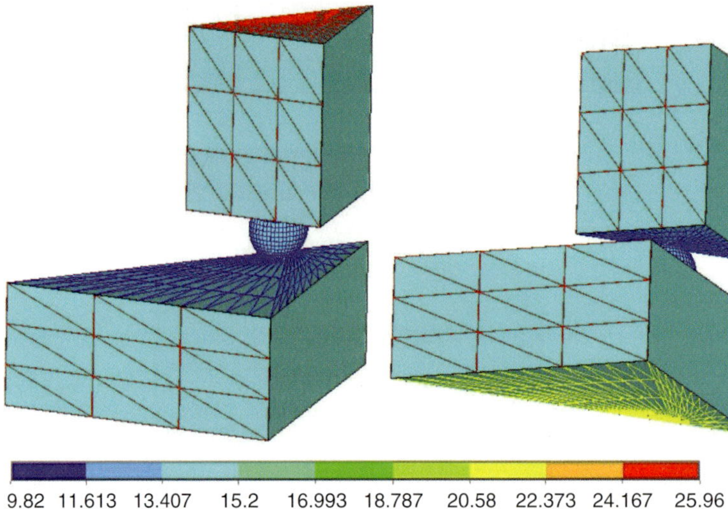

FIGURE 7.13 Convection coefficients (W/m^2 °C) at 55°C applied in the thermal analysis.

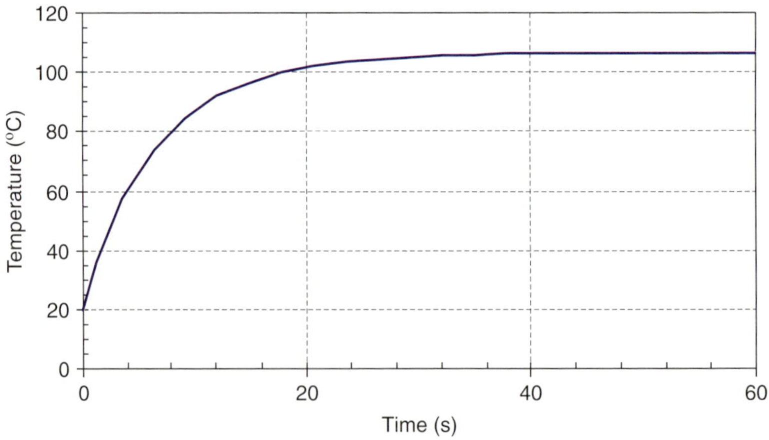

FIGURE 7.14 Time-history at the center node of the AlGaAs pad with an SnPb bump.

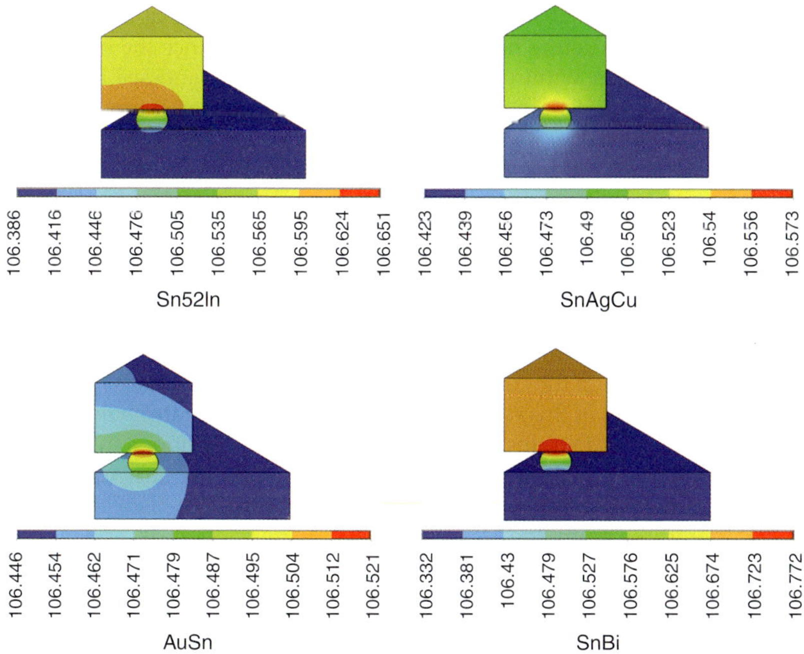

FIGURE 7.15 Temperature of the package with various solder joints (°C).

reasons. First, the analysis ignored the beneficial thermal path offered by the printed circuit board (PCB). Second, the boundary condition of continuous power generation within all diodes is clearly an upper limit on the true, switched case.

While the results presented in Fig. 7.15 do not demonstrate temperature differences of engineering significance based on solder alloy, a detailed analysis that examines the VCSEL temperature on the timescale of operational power-on/power-off frequencies is expected to do so. Still, it can be seen at the laser (AlGaAs pad) that the temperature of the device with AuSn (106.52°C) is smaller than that (106.57°C) with SAC, (106.60°C) with SnPb, (106.65°C) with SnIn, and (106.77°C) with SnBi. The present analysis is intended to be a conservative structural reliability evaluation.

7.2.4 Thermal-Stress Analysis and Results

A prevalent theory is that the fatigue life of the solder interconnects is governed by the accumulation of inelastic work within the solder joint. In particular, the relationship between creep strain or creep strain energy accumulation and solder joint life is well documented; see, for example, [7, 12]. Often, the finite-element assessment of creep strains results from thermal loading conditions that presume the package structure is subject to temperature changes but that at any point in time the structure has a uniform temperature. Since the temperature is uniform within the structure at any time, stresses and strains result solely from the effects of differences in material properties, the applied temperature, and the structure's geometry. It is in this context that the present analysis differs. It is well known that spatial nonuniformity in a temperature profile alone may introduce thermal stresses within a continuum. Therefore, the spatially nonuniform temperature profiles of the preceding thermal analysis are used as the loads in the structural analysis.

The mesh of the structural model is coincident with that of the thermal model. Note that the solder-joint mesh is graded from a coarse element size at the center of a meridian to a refined size at the pads. The intention of this refinement is to obtain an accurate evaluation of the stress and strains at the anticipated failure locations (near the pads) while maintaining an economy of elements. The structural analysis restricts nodal translations through the symmetry planes.

Temperatures from the thermal analysis are applied as nodal thermal loads in the structural analysis. Two load steps are evaluated. For both load steps, the package is assumed to be stress-free at 20°C. The first has a duration of 1 s, and the second has duration of 24 h. In the first load step, which is a reasonable approximation of powering up the VCSEL, the nodal temperatures ramp up from room temperature to the values determined in the thermal analysis. In the subsequent 24-h load step, the node temperatures are held constant at the thermal analysis values.

Thermal stresses and creep strains within the solder joint are evaluated during the course of the 24-h power-on case. Figure 7.16 details the creep shear strains at one second and six hours in the AuSn, SAC, SnBi, and SnIn solder joints, respectively. A number of trends are evident. The first is that the creep occurs almost instantly in the SnIn and SnPb solder joints, while the AuSn solder joints' creep strain develops rather slowly. The creep shear strains in the SAC and SnBi solder joints are similar to those in the SnPb solder joints, only slightly delayed in time and lower in magnitude due to the slower creep rates of SAC and SnBi. Finally, the magnitude of the creep shear strain is significantly lower in the AuSn solder joints than in the SnIn, SnPb, SAC, or SnBi solder joints.

Figure 7.17 presents the highest value of the creep shear strain in each solder joint at the AlGaAs pad interface over the 24-h period. The figure provides additional details of the time-strain progression. The dramatically lower creep strain value in the AuSn solder alloy, when compared with those in the SnIn, SnPb, SnBi, and SAC solder alloys, is particularly evident. While further increases in the AuSn creep strain appear likely beyond the 24-h period, it seems that an asymptote will be reached around 10^6 s. As will be discussed

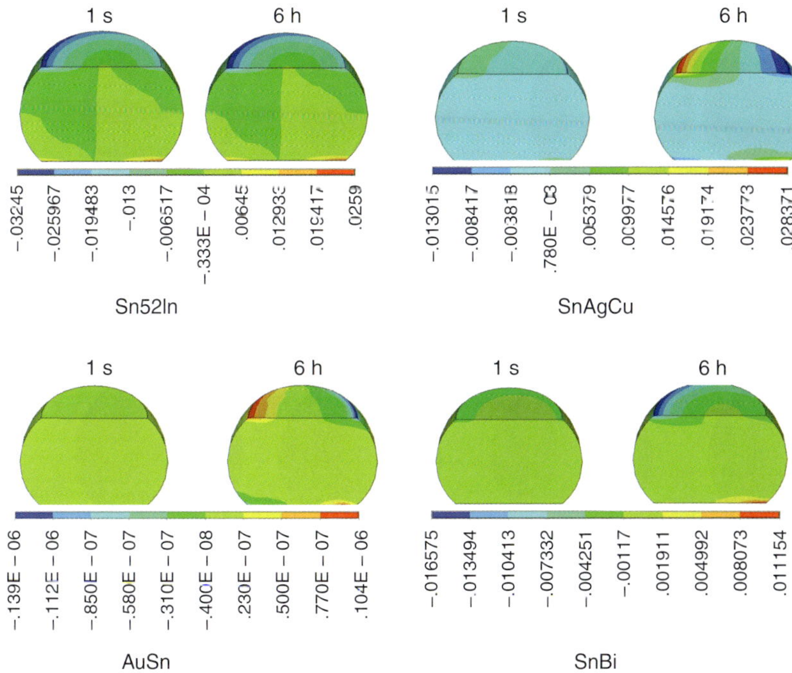

FIGURE 7.16 Creep shear strain in various solder joints at 1 s and 6 h.

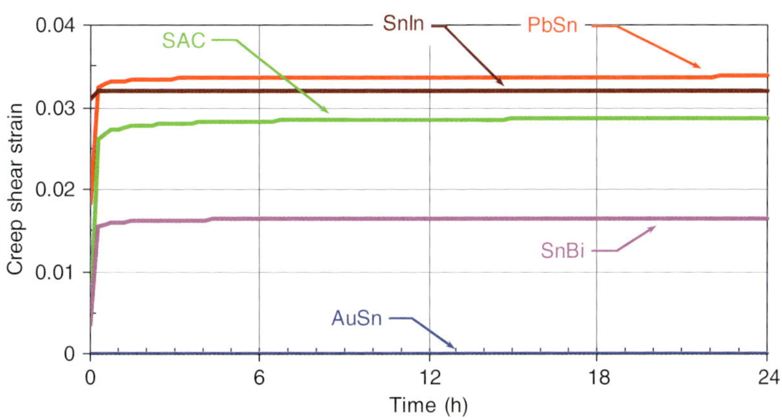

FIGURE 7.17 Creep shear strain at the critical bump location as a function of time.

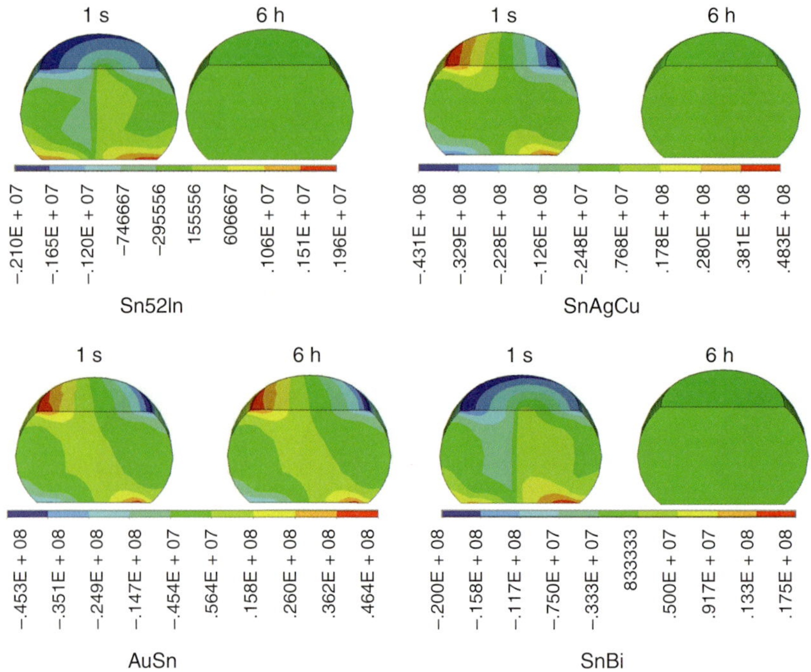

FIGURE 7.18 Shear stress (Pa) in various solder joints at 1 s and 6 h.

in the next paragraph, the creep strains cause a significant stress relief in the SnIn, SnPb, SnBi, and SAC solder joints. Conversely, the creep-resistant AuSn solder experiences negligible stress relief. The creep deformation in the SnIn and SnPb solder joints is about the same.

The corresponding shear stress distributions at 1 s and 6 h are shown in Fig. 7.18. As may be expected from an examination of the creep strains, the shear stress within the SnIn and SnPb solder joints drops rapidly from high initial values to values of less than 0.5 MPa at six hours. This stress relief is evident in the SAC and SnBi alloys also, where high initial stresses also drop to negligible values at 6 h. The creep-resistant AuSn solder joints have a higher shear stress value than that of the SnIn, SnPb, SAC, and SnBi alloys, and this high value appears unabated after 6 h.

Figure 7.19 presents the stress time-history of the critical value at the solder near the AlGaAs pad. From these figures, the beneficial decrease in shear stress due to creep strain relief is evident for the SnIn, SnPb, SnBi, and SAC solders. Also, it can be seen that the SnIn solder-joint stresses relax more than those of the SnPb, SnBi, and SAC solder joints.

Based on these analyses and results, it is concluded that: (1) the order (from small to large) of creep deformations in solder joints is

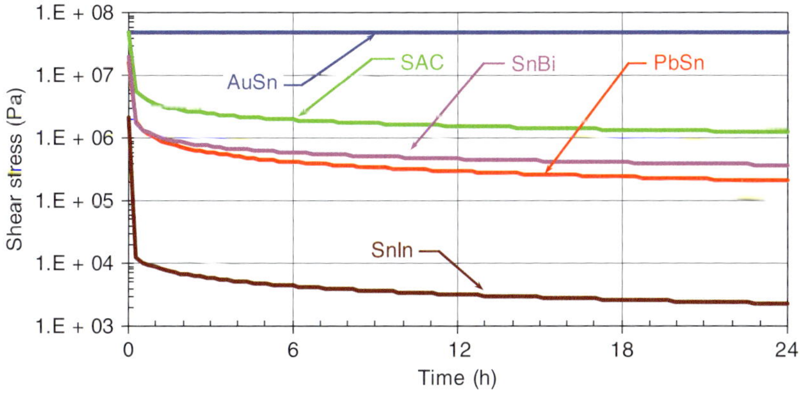

FIGURE 7.19 Shear stress at the critical bump location as a function of time.

from AuSn, SnBi, and SAC to SnPb/SnIn; and (2) the order (from small to large) of stresses in solder joints is from SnIn, SnPb, SnBi, SAC, and AuSn. In addition to the examination of stresses and creep strains, the accumulated creep strain energy density in the solder joints has been used in the examination of lead-free solder joints; see, for example, [7, 18]. In general, for the same solder material, the larger the creep strain energy density, the lower the thermal-fatigue life of the solder joints.

Figure 7.20 presents the creep strain energy density at the critical location as a further basis of comparison of the four solder alloys. It can be seen that the creep strain energy density is more than four

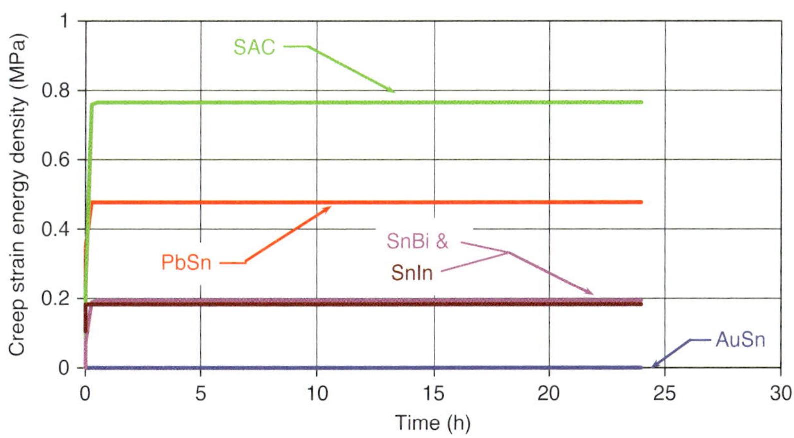

FIGURE 7.20 Creep strain energy density at the critical bump location as a function of time.

orders of magnitude lower for the AuSn solder joint when compared to the other solder joints. This difference results from the significantly slower creep rate of the AuSn solder, and is a strong indicator that the AuSn solder should provide solder-joint thermal life improvements over the other solder alloys under in-service conditions, even though they may not have the same fatigue crack growth material properties.

It should also be noted from Fig. 7.20 that the creep strain energy density in the SnIn and SnBi solder joints is almost two times and four times, respectively, lower than that in the SnPb and SAC solder joints. Again, even though these materials have differing properties, the results suggest that the SnIn and SnBi solder joints may last longer than the SnPb and SAC solder joints.

The structural response of the solders may have an effect on the reliability of other packaging components. Of particular interest is the AlGaAs pad. As a metric, Fig. 7.21 depicts the time history of the shear stress at the edge of the AlGaAs pad at the mid-thickness location. This stress is primarily due to the global (GaAs chip and Si substrate) and local (AlGaAs pad and the solder joint) thermal expansion mismatches. Note that the time history response is directly analogous to the trends seen in the solder-joint stresses. The pad shear stresses associated with the stiff and creep-resistant AuSn solder are relatively high and unabated through the 24-h period. The SAC, SnPb, SnBi, and SnIn alloys have a significantly greater creep rate than the AuSn alloy, and this results in pad stresses that rapidly decrease during the course of the analysis. Finally, the pad shear stresses are lowest and decrease most rapidly with the SnIn alloy. The consideration of the solder-joint influence on the AlGaAs pad stresses suggests that the SnIn alloy may be a better lead-free choice than all the other materials.

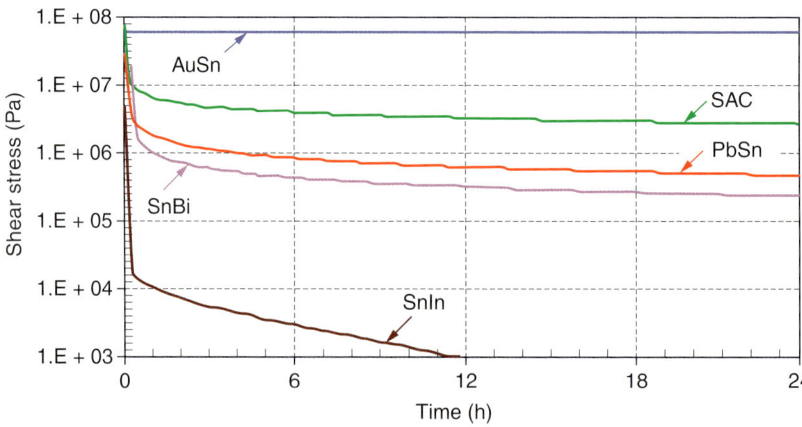

FIGURE 7.21 Shear stress at the edge of the AlGaAs pad as a function of time.

7.2.5 Summary and Recommendations

Some important results are summarized in the following:

1. The creep rate of AuSn is slower than that of SAC, SnBi, SnPb, and SnIn. Also, the creep rate of SAC is slower than SnBi, SnPb, and SnIn. In general, the creep rate of SnIn is the largest.
2. The modulus of AuSn is the largest, and the moduli of SnPb and SnIn are the smallest among the solder alloys considered herein. Above room temperature, the modulus of SnIn is larger than that of SnPb. The modulus of SnBi is smaller than that of SAC.
3. The CTE of SnIn is the largest and the CTE of SnBi is the smallest of the alloys studied.
4. The order (from small to large) of the steady-state temperature at the AlGaAs laser pad with different alloys is AuSn, SAC, SnPb, SnIn, and SnBi.
5. The order (from small to large) of the maximum creep strain at the solder joints with different alloys is AuSn, SnBi, SAC, and SnPb/SnIn.
6. The order (from small to large) of the maximum stress at the solder joints with different alloys is SnIn, SnPb, SnBi, SAC, and AuSn. The stress relaxation of AuSn solder joints is very slow!
7. The order (from small to large) of the maximum creep strain energy at the solder joints with different alloys is AuSn, SnIn/SnBi, SnPb, and SAC.
8. The order (from small to large) of the maximum stress at the AlGaAs laser pad with different alloys is SnIn, SnBi, SnPb, SAC, and AuSn. The stress relaxation at the AlGaAs laser pad with AuSn is very slow.
9. Based on the least creep damage of solder joints and the lower stress at the AlGaAs laser pads, SnIn is the best choice among the solder alloys considered.

7.3 Lead-Free Interconnect Reliability of a VSCEL on an Optical PCB

The major trend in the optoelectronic industry today is to make products more personal by making them smarter, lighter, smaller, thinner, shorter, and faster, while at the same time making them more friendly, functional, powerful, reliable, robust, innovative, creative, and less expensive. As the trend toward miniature and compact products continues, the introduction of cool products that are more user-friendly and contain a wider variety of functions will provide growth in the market. One of the key technologies that is helping to make these cool product design goals possible is 3-D bare chips stacked on optical PCB.

As IC technology advances, the speed and complexity of interconnects increase tremendously. As the number of devices per chip, the number of chips per board, the modulation speed, and the degrees of integration continue to increase, electrical interconnects are facing their fundamental bottlenecks, for example, speed, data rate, and power dissipation.

Optical interconnect is an attractive solution to the bandwidth limitation of the electrical transmission line for board-level interconnections. However, the implementation of optical waveguide in a commercial PCB imposes severe restrictions on the materials and processes, like low propagation loss, high thermal stability, cost-effective fabrication, etc. Currently, polymer-based waveguide is the most popular choice for optical interconnect due to its low cost, suitability for mass production, high thermal stability, and low optical loss [22–43].

When the VCSEL is integrated with the optical PCB, however, it requires an out-of-plane 45-degree reflector to couple the optical signal to the polymer waveguide due to its surface-emitting characteristic. This problem is also encountered by the surface-illuminated photo detector. The optical polymer waveguide is usually designed in multimode and has a larger aperture, for reasons of low cost and compatibility with the PCB process. An optical polymer waveguide is designed and analyzed for the present optoelectronic system.

The speed, data rate, power, and size of optoelectronic devices generally scale with the wavelength of light used. The complexity of the present-day 2-D photonic modules is limited by their substrate sizes and the difficulty in connecting large numbers of electrical connections. By progressing to multilayer interconnects, the IC density can be increased greatly. Since fewer connections are required between bare chips, a higher performance will result. Some devices can actually be made smaller when integrated vertically rather than laterally. Vertical stacking also allows more chips per wafer. Complete optoelectronic integration of devices of different materials onto a single substrate (like the optical PCB) will achieve a dramatic cost and labor reduction as well. Novel combinations of materials/geometries may allow for devices far superior than their in-plane counterparts. Additional flexibility can be afforded in package design, interconnect routing, and package placement through 3-D architectures. These factors have the potential to revolutionize the way 3-D hybrid modules are produced and used.

A low-cost, 3-D hybrid optoelectronic system with electrical and photonics ICs for board-level applications, as shown in Figs. 7.22 through 7.24, is designed and analyzed. The entire system consists of

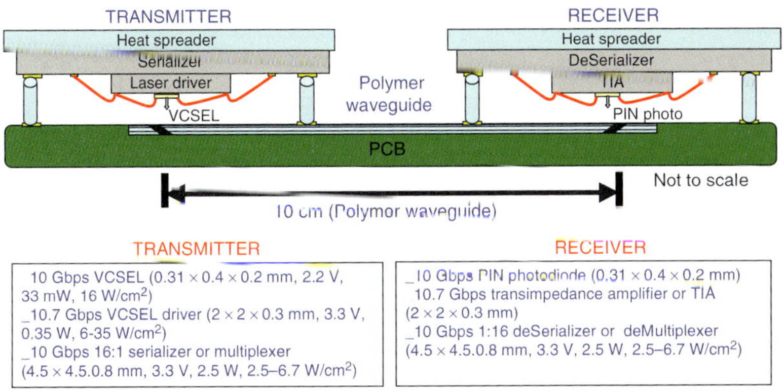

Figure 7.22 Three-dimensional stacking of optoelectronics on optical printed circuit board.

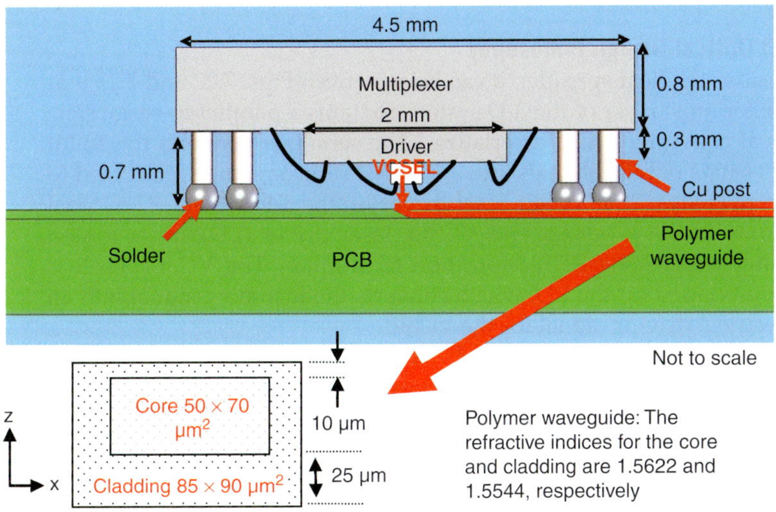

Figure 7.23 VCSEL, laser driver, serializer (mux), and polymer waveguide.

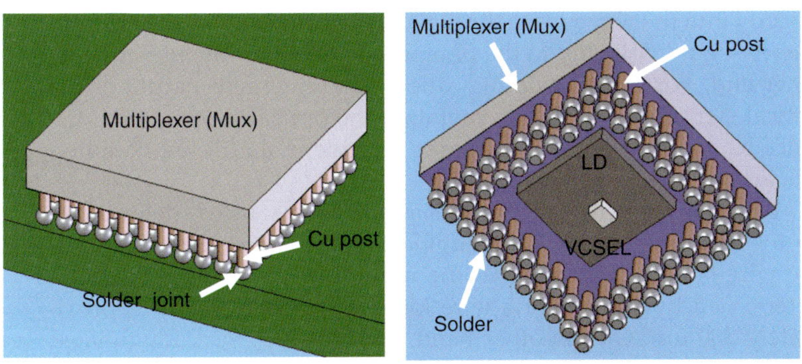

Figure 7.24 VCSEL, laser driver (LD), serializer (mux), and Cu post with solder.

a rigid PCB made of fiber reinforced (FR) material with an optical polymer waveguide, a VCSEL, a laser driver (LD), a 16:1 serializer, a photo-diode detector, a tans-impedance amplifier (TIA), a 1:16 deserializer, and heat spreaders. The bare VCSEL, LD, and serializer chips are stacked with wire bonds and then solder-joined on the optical PCB via copper posts. Similarly, the bare photo-diode detector, TIA, and deserializer chips are stacked with wire bonds and then solder-joined on the optical PCB via copper posts. The optical, thermal, and mechanical design philosophy and their metrics of the proposed 3-D optoelectronic system will be given and the analysis results will be discussed.

7.3.1 Optical Design and Analysis

(A) Optical Design Philosophy

Besides the heat spreader, it can be seen from Figs. 7.22 and 7.23 that the topmost layer of the 3-D system contains a parallel-to-serial electrical converter called a serializer. The serializer converts the 16-bit 622 Mbps parallel signals into a 10 Gbps serial signal, which is fed to the next layer (the VCSEL laser driver) in the 3-D system. The signal inputs to the serializer and VCSEL laser driver (LD) are re-routed using wafer level RDL deposition techniques. The VCSEL itself is wire-bonded on the LD, which converts the incoming signal into an electrical driving signal for the VCSEL.

Again, from Figs. 7.22 and 7.23, it can be seen that the receiver schematic is similar to that of the transmitter, except that the internal ICs are the receiving elements. The 3-D stacked transmitter and receiver can be used for board-level optical interconnects, as shown in Figs. 7.22 and 7.23, with the transmitter output coupling to one end of the polymer waveguide, and the output on the other end of the polymer waveguide coupling to the receiver element.

Direct coupling (without a lens) is used for optical coupling design in this study. The optical signal is coupled from the VCSEL directly into the waveguide through its mirror on one end, and from the waveguide directly to the photodiode through the mirror on its other end. For direct coupling, due to the divergent nature of the optical signal, it is critical to maintain the optical propagating distance as short as possible in order to achieve the desire coupling efficiency.

Based on a given structure, this minimum distance is determined by the assembly design of the optoelectronic components [33]. For the wire-bonding component, the minimum distance or height is determined by the bonding wire loop, which is typically approximately 200 μm. On the other hand, for flip-chip, back-illuminated optoelectronics components, the minimum height can be as close as possible (~50 μm for conventional solder bumps and ~25 μm for microbumps) without damaging the active area of the optoelectronic components.

(B) Optical Simulation

The design of the optical waveguide is shown in Figs. 7.22 and 7.23. The waveguide is embedded on the top surface of the FR4 PCB to give a complete planar design. The cross-sectional design of the waveguide is shown in Fig. 7.23, and the refractive indices for the core and cladding are, respectively 1.5622 and 1.5544. At both ends of the waveguide there is the out-of-plane reflector or mirror, which reflects the optical signal at 90 degrees.

(C) Results

The optical simulation is done using the commercial software, ASAP, and Fig. 7.25 shows the layout used in the simulation. The VCSEL used in the simulation has a beam divergence of 30 degrees, and the photodiode has an active area of 40 μm. The mirror on both sides of the waveguide is simulated with a loss of 0.6 dB, and the waveguide loss is not simulated. Using these parameters, the coupling losses are 8.5 dB and 5 dB, respectively, for 250 μm and 50 μm heights (distances

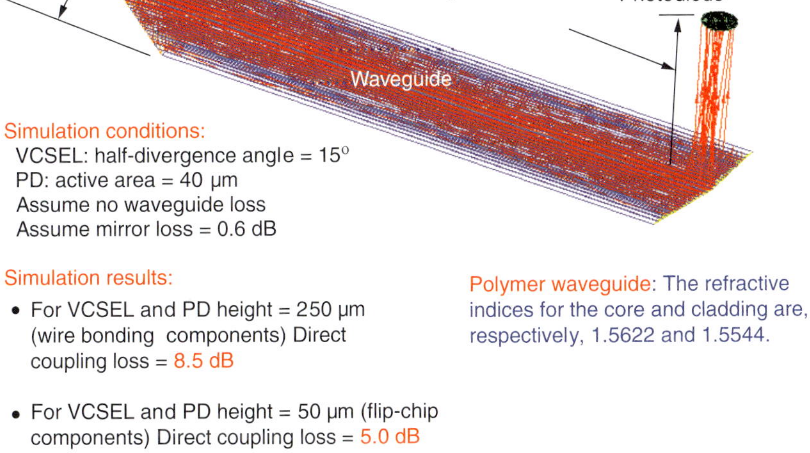

FIGURE 7.25 Simulated model showing the direct coupling of the optical signal.

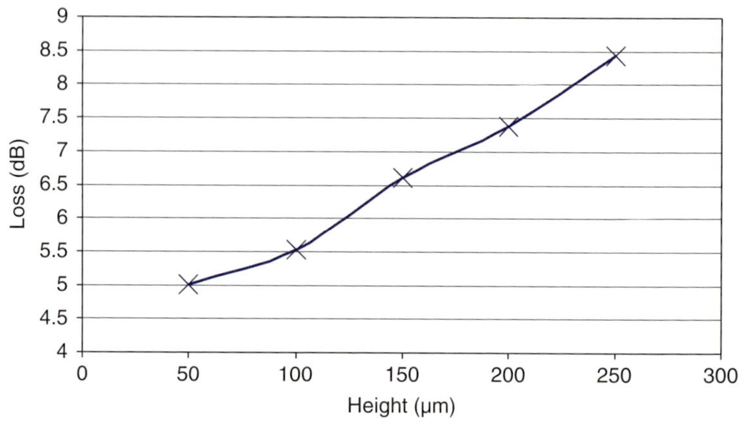

FIGURE 7.26 Coupling loss vs. the height of optoelectronic components.

of the VCSEL and photodiode from the waveguide). Figure 7.26 shows the coupling losses for other heights. As expected, the greater the height, the larger the coupling loss.

(D) Analysis and Summary

The typical optical power from the VCSEL is 3 dBm, and the responsitivity for the GaAs photodiode at 850 μm is 0.6 A/W. Assuming the waveguide loss is 0.1 dB/cm, the optical loss of a 10-cm-long waveguide will be 1 dB. Also assuming that the minimum input optical power required for a 10 Gbps receiver is –13 dBm, then the maximum optical acceptable coupling loss is approximately 15 dB. Therefore, the direct coupling schemes are feasible for the 10 Gbps opto-electronic circuit board (OECB) design.

7.3.2 Thermal Design and Analysis

(A) Design Philosophy and Metrics

The choice of a photonic package configuration requires recognition not only of critical cost issues and optical and electrical performances, but also those of thermal performance and mechanical reliability. Since low-cost bare chips (VCSEL, PIN, laser driver, TIA, serializer, and deserializer) are used to stack up the optoelectronic structure as shown in Fig. 7.22, thermal management is a critical issue. This is compounded with the high-power dissipation of the 10 Gbps laser (2.2 V, 33 mW, and 16 W/cm^2), the 10.7 Gbps laser driver and TIA chips (3.3 V, 0.35 W, and 6–35 W/cm^2), and the 10 Gbps 1:16 serializer and the 10 Gbps 16:1 deserializer (3.3 V, 2.5 W, and 2.5–6.7 W/cm^2); thus, how to take the heat out from these chips before they burn themselves is a challenging problem. In particular, a stable thermal environment is required for the VCSEL. Temperatures higher than the desired operating temperature may alter the wavelength of the emitted light, consequently degrading the VCSEL's performance. In this study, the target temperature at the laser of the 10 Gpbs VCSEL is 85°C.

(B) Heat-Spreader Design and Analysis

The full 3-D optoelectronic structure shown in Fig. 7.22 is modeled and analyzed by FlowTHERM, which is a computational fluid dynamic (CFD) thermal-modeling analysis tool to simulate the air flow and heat transfer in optoelectronic systems. Four different heat-spreader sizes are considered: 0 × 0 × 0 mm (no heat spreader), 15 × 15 × 0.8 mm, 30 × 30 × 0.8 mm, and 50 × 50 × 0.8 mm.

(C) Boundary Conditions and Material Properties

The 3-D package size is about: X = 175 mm, Y = 125 mm, and Z = 3.15 mm. The gravity direction is set in the negative Z direction. The overall dimensions of the volume or enclosure (domain for computation) are X = 200 mm, Y = 150 mm, and Z = 60 mm. Since the gravity is set in the negative Z direction, the air flow is mainly in the Z direction, and hence more space for free convection is given in the upper and lower parts of the package. The boundary conditions of all exposed surfaces are free convection with the ambient, that is, there is no fan to help cool the system. The boundary conditions of the inner boundaries between the components are by conduction.

The ambient dry air temperature is set to 25°C, the pressure is set to 1 atm, and the properties of the air at the specified condition (i.e., 25°C and 1 atm) are set as follows: density = 1.185 kg/m^3, viscosity = 18.35 × 10^{-6} kg/(m·s), specific heat = 1.005 kJ/(kg·°C), and thermal conductivity = 0.0263 W/(m·°C). All other material properties, power, and dimensions are shown in Table 7.2 [17, 18].

Component	PCB	HS	MUX	LD	VCSEL	De-MUX	TIA and LA	PD
Material	FR4	Copper	Si	Si	GaAs	Si	Si	GaAs
Thermal conductivity (W/m/°C)	// 0.8 ⊥ 0.3	390	150	150	68	150	150	68
Dimension detail (mm)	175 × 125 × 1.6	Variable	4.5 × 4.5 × 0.8	2.0 × 2 × 0.3	0.31 × 0.4 × 0.2	4.5 × 4.5 × 0.8	2.0 × 2.0 × 0.3	0.31 × 0.4 × 0.2
Power (W)	N.A.	N.A.	2.5	0.35	0.033	2.5	0.35	0.033

TABLE 7.2 Material Properties, Power, and Dimension for Thermal Modeling

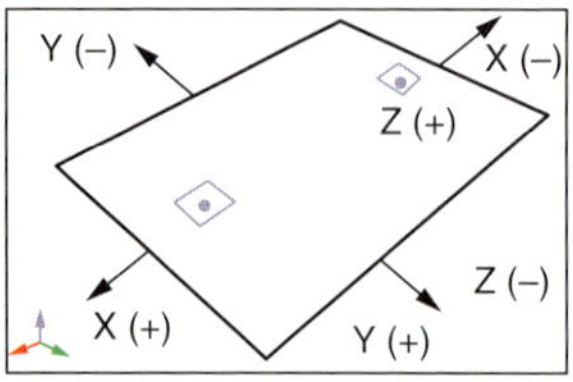

Face	PCB	HS	Mux	LD	VCSEL
X (+)	11.4	11.5	1.5	3.4	8.8
X (–)	11.3	12.6	2.5	3.6	7.4
Y (+)	9.1	10.4	1.5	3.4	7.7
Y (–)	9.1	10.4	2.5	3.6	12.4
Z (+)	1.2	5.8	n/a	n/a	n/a
Z (–)	3.4	7.9	2.7	17.7	39.1

Heat spreader size:
50 × 50 × 0.8 mm

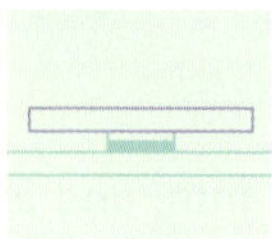

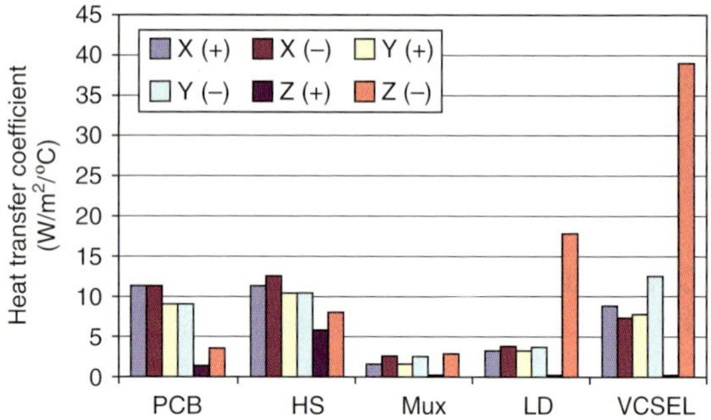

FIGURE 7.27 Heat transfer coefficient (heat spreader size: 50 × 50 × 0.8 mm.

(D) Results and Analyses

Figure 7.27 shows the calculated heat transfer coefficient at the faces of all the key components (e.g., PCB, laser of the VCSEL, serializer, driver, etc.) of the 3-D system, with a heat spreader of 50 × 50 × 0.8 mm. It can be seen and expected that the maximum heat transfer coefficient occurs at the laser of the VCSEL. Similarly, for 3-D stacked optoelectronic components with 15 × 15 × 0.8 mm, and 30 × 30 × 0.8 mm heat spreaders, the maximum heat transfer coefficient occurs at the laser of the VCSEL, and it decreases as the heat spreader size increases, as shown in Fig. 7.28. This means that the heat spreader helps to dissipate the heat from the system.

Figure 7.29 shows the temperature distributions in the 3-D system with a 50 × 50 × 0.8 mm heat spreader. It can be seen, and is expected, that the maximum temperature occurs at the laser of the VCSEL and is equal to 84.7°C, which meets the target (85°C). Figure 7.30 shows the effects of heat spreader size on the maximum temperature of the 3-D system. It can be seen, and is expected, that the larger the heat spreader, the less the maximum temperature at the laser of the VCSEL.

(E) Summary

Based on the design and analysis of a few heat spreaders for the 3-D optoelectronic system, it is found that the 50 × 50 × 0.8 mm heat spreader (without a fan) can help to lower the maximum temperature at the laser of the VCSEL to 85°C.

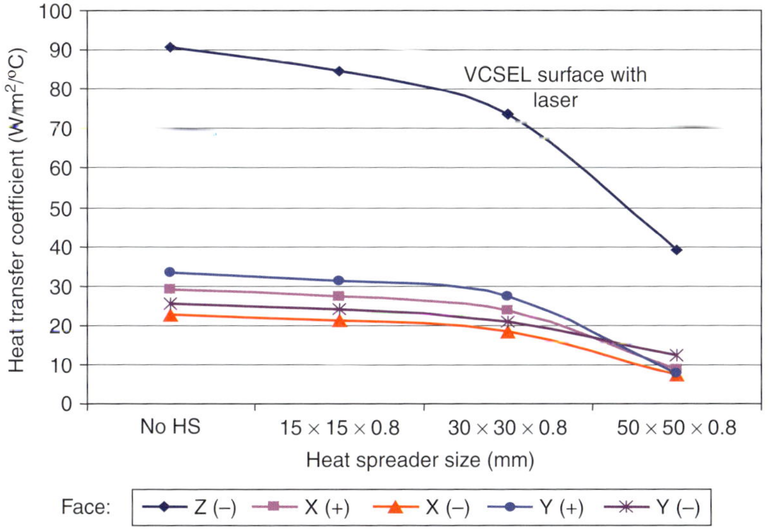

FIGURE 7.28 Heat transfer coefficient at the VCSEL for various heat spreader sizes.

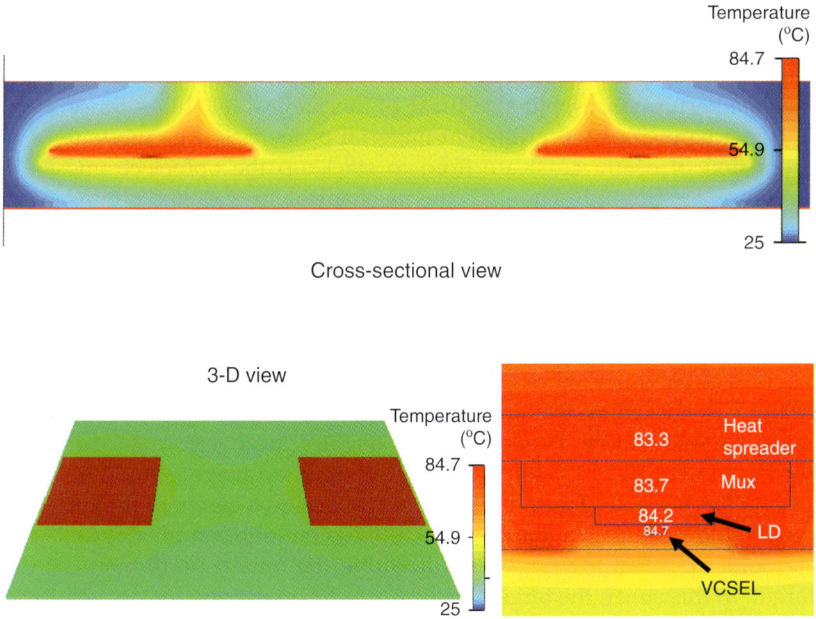

FIGURE 7.29 Temperature distribution of the 3-D stacked optoelectronics.

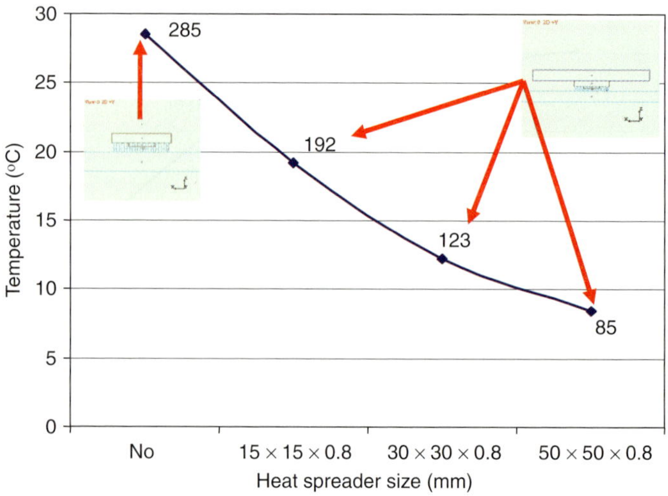

FIGURE 7.30 Effects of heat-spreader size on the maximum temperature at VCSEL.

7.3.3 Lead-Free Interconnect Reliability

(A) Design Philosophy and Metric

As mentioned earlier, the choice of a photonic package configuration requires recognition not only of critical cost issues and optical and thermal performances, but also those of mechanical reliability. Since low-cost bare chips with different material properties (e.g., thermal expansion of coefficients) are used to stack up the optoelectronic structure, as shown in Fig. 7.22, the mechanical stresses and strains (i.e., creep strain energy density per temperature cycle) in the solder joints and the mechanical stresses at the laser of the VCSEL due to the global and local thermal expansion mismatches under the most common environmental stress condition (i.e., temperature cycling with −40°C to 125°C, one hour per cycle) are critical issues.

One of the functions of Cu post is to increase the stand-off height of the 3-D stacked system and thus to relax the stress and strain acting at the solder joints. However, based on the optical performance of the laser of the VCSEL without a lens, the stand-off height should be as small as possible. Thus, it is necessary to design a height of the Cu post + solder joint such that the structure has enough compliance for the solder-joint reliability yet is still within the optical performance target. In this study, the target for the creep strain energy density per temperature cycle is 0.3 MPa and the stress at the laser is 30 MPa.

(B) Design and Analysis

The Cu post and the solder joint shown in Figs. 7.22 through 7.24 are designed with the following dimensions: 0.5 mm × 0.2 mmΦ, and

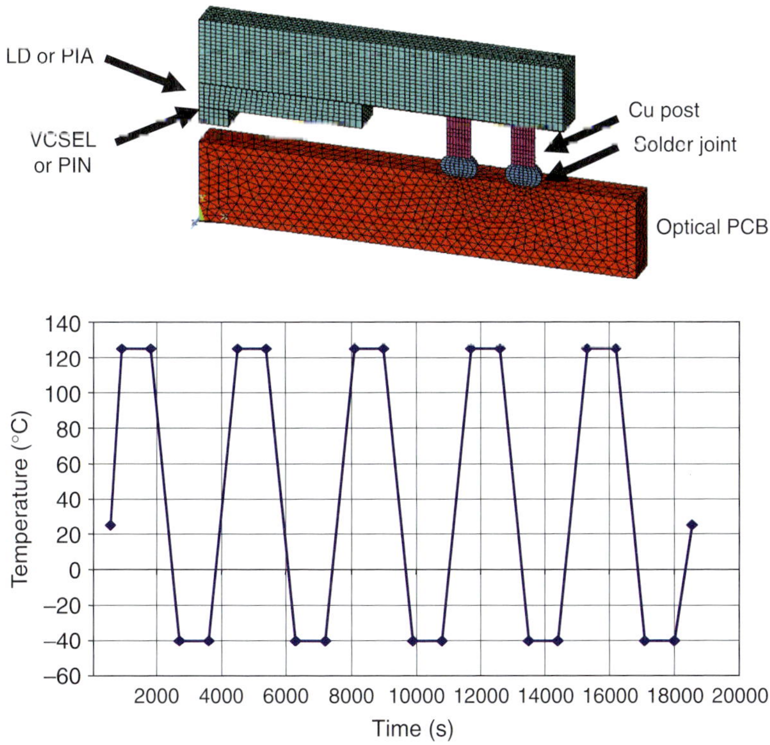

FIGURE 7.31 Mechanical analysis of the 3-D stacked optoelectronics on PCB.

0.2 mmΦ, respectively. Figure 7.31 shows the 3-D finite-element model that captures the construction along a diagonal strip from the 72-pin serializer lead-free assembly's geometric center to a corner. Because of the mid-plane symmetry, the mesh actually models a one-half strip (with one-half of a solder joint) using the appropriate in-plane constraints placed on one symmetry plane. Coupled in-plane translations are applied to the other symmetry plane to produce a state of generalized plane strain. Using exclusively hexahedral solid elements, the model can capture the precise shape of the packages' solder joints and potential distant to neutral point (DNP) effects while retaining significant computational efficiency over full octant models. ANSYS is the code selected for the modeling and analyses.

The material properties of the 3-D assembly are shown in Table 7.3. All of the material properties are assumed to be constant except for those of the solders, which creep.

(C) Temperature Boundary Conditions

The temperature loading imposed to the lead-free 3-D assembly is shown in Fig. 7.31. The temperature ranges are −40°C to 125°C and

Materials	CTE (ppm/°C)	E (GPa)	Poisson Ratio
VCSEL	5.6	85	0.30
Serializer	2.7	131	0.28
Laser driver	2.7	131	0.28
FR-4 PCB	15	21	0.18
Cu	17	110	0.34
RDL	17	110	0.34
Die attach film	40	0.69	0.3

TABLE 7.3 Material Properties for Environmental Stress Modeling

the ramp-up, ramp-down, dwell-at-hot, and dwell-at-cold are each 15 min.

(D) Maximum Stress/Strain Location

Figure 7.32 shows a typical von Mises stress and equivalent creep strain contour of the most-likely-to-fail (corner) lead-free solder joint in the 3-D assembly at the end of the fifth cycle. The location for the maximum von Mises stress and equivalent creep strain is near the corner interface between the solder and the pad of the PCB of the lead-free solder joint.

(E) Creep Hysteresis Loops

It is important to study the creep responses for multiple cycles by observing when the hysteresis loops become stabilized. Figure 7.33 shows a typical shear stress and creep shear strain hysteresis loops for multiple cycles at the corner of the most-likely-to-fail solder joint in the lead-free 3-D assembly subject to the –40°C to 125°C environmental temperature cycles. The creep shear strain versus shear stress loop is quite stabilized after the fourth cycle.

(F) Maximum Creep Strain Energy Density Time-History

Figure 7.34 shows the von Mises creep strain energy density (W) time-history for five complete cycles at the corner solder joint of the 3-D assembly. The creep strain energy density per temperature cycle (ΔW) can be determined from either the fourth or fifth cycle, and is equal to 0.123 MPa, which is less than the target (0.3 MPa).

(G) Maximum Stress in VCSEL

The maximum stress (Fig. 7.35) in the VCSEL due to the environmental stress condition (–40°C to 125°C) occurs at the laser in the bottom surface of the VCSEL, and is equal to 28.6 MPa. This value is less than the maximum allowable stress (30 MPa) of the target.

Maximum von mises stress = 0.994×10^7 Pa

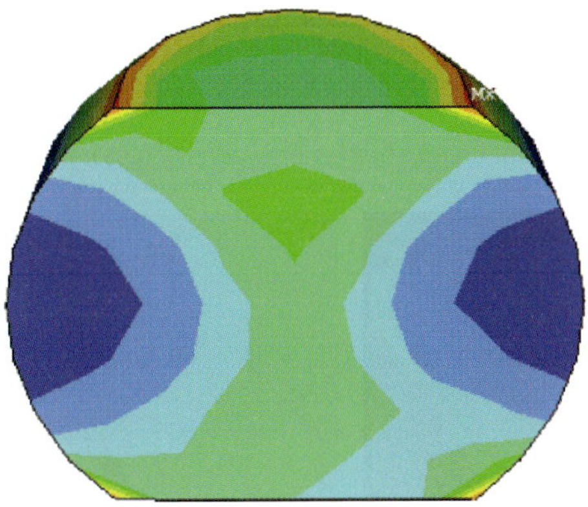

Maximum equivalent creep strain = 0.0121

FIGURE 7.32 Maximum von Mises stress and equivalent creep strain at the corner solder joint after the fifth temperature cycle.

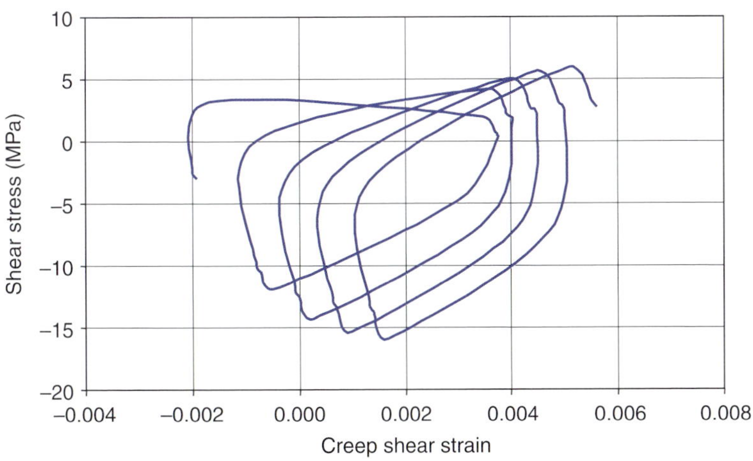

FIGURE 7.33 Maximum hysteresis loops at the corner solder joint.

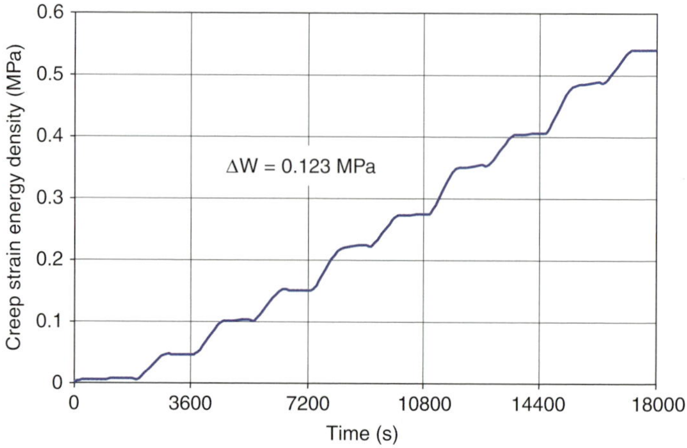

FIGURE 7.34 Maximum creep strain energy density history at the corner solder.

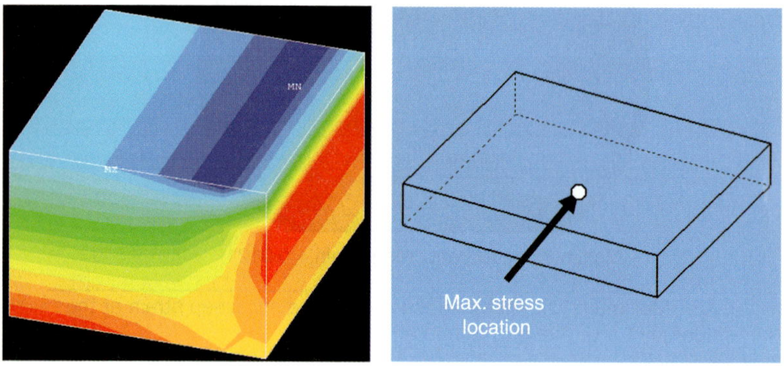

Max. Mises Stress = 28.6 MPa at the end of the 5th temperature cycle

The Max. Stress (28.6 MPa) occurs at the bottom-center (laser) of the VCSEL

FIGURE 7.35 Maximum von Mises stress in the VCSEL.

(H) Summary

Based on time and temperature-dependent analyses, it has been shown that the present design of the solder joint and Cu post in the 3-D optoelectronic system can withstand the most common environmental stress condition. Also, the maximum stress at the VCSEL is within the target.

7.3.4 Summary and Recommendation

Some important results are summarized in the following:

(1) Optical analyses showed that the present design of the cross-section of the polymer waveguide meets the optical acceptable coupling loss (15 dB). Therefore, the direct coupling

schemes (without lens) are feasible for the 10 Gbps optical board-level interconnect.

(2) Thermal analyses showed that the 50 × 50 × 0.8 mm heat spreader (without fans) can lower the maximum temperature at the laser of the VCSEL to below 85°C.

(3) Mechanical stress analyses showed that: (a) the maximum creep strain energy density per temperature cycle (0.123 MPa) in the corner lead-free solder joint is less than that of the allowable one (0.3 MPa); and (b) the maximum von Mises stress (28.6 MPa) at the laser of the VCSEL is less than the target (30 MPa).

(4) The assembly process of optical waveguides on PCB is out of the scope of this book, but is shown in [38–43].

7.4 References

1. Ebeling, K. J., *Integrated Optoelectronics*, Springer-Verlag, Berlin, 1993.
2. Strzelecka, E. M., D. A. Louderback, K. Bertilsson, B. J. Thibeault, M. Mondry and L. A. Coldren, "Free-spaceOptical Link Realized with Microlensed Components," *Proc. 47th Electronic Components & Technology Conference*, May 1997, pp. 376–381.
3. Chenevas-Paule, A., T. Collette, P. Scheer, J-P. Bouzinac, P. Churoux and S. Paineau, "From Inter-System to ICs Level, VCSEL as a Key Component for Optoelectronic Active Interconnects—a Survey," *Proc. 47th Electronic Components & Technology Conference*, May 1997, pp. 363–367.
4. Jiang, W., P. Claisse, C. Gaw, P. Kiely, B. Lawrence, M. Lebby and M. Roll, "Vertical Cavity Surface Emitting Laser Packaging with Auto Power Control," *Proc. 47th Electronic Components & Technology Conference*, May 1997, pp. 368–370.
5. Schnitzer, S., M. Grabherr, R. Jager, C. Jung, R. Michalzik, G. Reiner, W. Schmid, B. Weigl, D. Wiedenmann and K. J. Ebeling, "Bias-free 1 Gb/s Data Transmission Using High Efficient VCSELs," *Proc. 47th Electronic Components & Technology Conference*, May 1997, pp. 371–375.
6. Kosaka, H., M. Kajita, M. Yamada, Y. Sugimoto, K. Kurata, T. Tanabe and Y. Kasukawa, "Plastic-based Receptacle Type VCSEL-array Modules with One and Two Dimensions Fabricated Using the Self-alignment Mounting Technique," *Proc. 47th Electronic Components & Technology Conference*, May 1997, pp. 382–390.
7. Lau, J. H., *Low Cost Flip Chip Technologies*, McGraw-Hill, NY, 2000.
8. Schwartz, D. B., C. K. Y. Chun, B. M. Foley, D. H. Hartman, M. Lebby, H. C. Lee, S. C. Long, S. M. Kuo, S. G. Shook and B. Webb, "A Low-Cost High-Performance Optical Interconnect," *IEEE Trans. on Components, Packaging & Manufacturing, Part B: Advanced Packaging*, Vol. 19, 1996, pp. 532–539.
9. Hayashi, Y., T. Mukaihara, N. Hatori, N. Ohnoki, A. Matsutani, F. Koyama and K. Iga, "Lasing Characteristics of Low-threshold Oxide Confinement InGaAs-GaAlAs Vertical Cavity Surface Emitting Lasers," *IEEE Photon. Tech. Letter*, Vol. 7, 1995, pp. 1324–1326.
10. Yang, G. M., M. H. MacDougal and P. D. Dapkus, "Ultralow Threshold Current Vertical Cavity Surface Emitting Lasers Obtained with Selective Oxidation," *Electron. Lett.*, Vol. 31, 1995, pp. 886–888.
11. Lau, J.H., *Flip Chip Technologies*, McGraw-Hill, NY, 1996.
12. Lau , J.H. and Y. H. Pao, *Solder Joint Reliability of BGA,CSP, Flip Chip, and Fine Pitch SMT Assemblies*, McGraw-Hill, NY, 1997.

13. Jim, K., G. Faulkner, D. O'Brien, D. Edwards, and J. Lau, "Fabrication of Wafer Level Chip Scale Packaging for Optoelectronic Devices," *IEEE Proceedings of Electronic Components and Technology Conference*, June 1999, pp. 1145–1147.
14. Lau, J. H., and R. Lee, "Thermal Analyses of Vertical-Cavity Surface Emitting LED/VCSEL Assembly with Lead-Free Flip Chip Interconnects," *IEEE Proceedings on Photonic Devices and Systems Packaging Symposium*, Stanford University, July 2002, pp. 26–30.
15. Zou, Y., J. Lau and S. Camerlo, "3D Nonlinear Thermal Stress Analysis of VCSEL (Vertical-Cavity Surface-Emitting Laser) Assemblies with Lead-Free Flip-Chip Interconnects," *IEEE Proceedings of Electronic Components and Technology Conference*, June 2003, pp. 1684–1690.
16. Dauksher, W., and J. Lau, "Lead-Free Solder-Joint Reliability of a Photonic Device Under Transient and Steady State Loadings," *ASME Paper No. IMECE2003-42253*, November 2003.
17. Lau, J. H., and W. Dauksher, "Thermal Stress Analysis of a Flip-Chip Parallel VCSEL (Vertical-Cavity Surface-Emitting Laser) Package with Low-Temperature Lead-Free (48Sn-52In) Solder Joints," *IEEE Proceedings of Electronic Components & Technology Conference*, San Diego, CA, May 30-June 2, 2006, pp. 1009–1017.
18. Lau, J. H., and W. Dauksher, "Thermal-Mechanical Analysis of a Flip-Chip VCSEL (Vertical-Cavity Surface-Emitting Laser) Package with Low-Temperature Lead-Free (Sn-Bi) Solder Joints," *ASME Paper No. IMECE2005-79981*, November 2005.
19. Lau, J. H., and S. Pan, "Creep Analysis and Thermal-Fatigue Life Prediction of the Lead-free Solder Sealing Ring of a Photonic Switch," *ASME Transactions, Journal of Electronic Packaging*, Vol. 124, December 2002, pp. 403–410.
20. Garofalo, *Fundamentals of Creep and Creep-Rupture in Metals, Macmillan*, New York, 1965.
21. Lau, J., and W. Dauksher, "Creep Constitutive Equations of Sn(3.5-3.9) wt%Ag(0.5-0.8)wt%Cu Lead-Free Solder Alloys," in *Micromaterials and Nanomaterials*, edited by B. Michel, IZM, Berlin, German
22. Chen, Y. M., Cheng, L. Yang, Yao, L. C., Chen, H. H., Chen, Y. C., Chu, Y., Hsieh, T. E., "10Gbps Multi-Mode Waveguide for Optical Interconnect," *IEEE Proceedings of Electronic Components and Technology Conference*, Lade Buena Vista, FL, 2005, pp. 1739–1743.
23. Shioda T., "Recent Progress and Potential Markets for Optical Circuit Boards," *IEEE Proceedings of Polytronic Conference*, 2007, pp. 1–3.
24. Hiramatsu, S., Miura, K., and Hirao, K., "Optical Backplane Connectors Using Three-Dimensional Waveguide Arrays," *IEEE Journal of Lightwave Technology*, Vol. 25, No. 9, September 2007, pp. 2776–2782.
25. Chang, G. K., "High-speed, High-density Optical Interconnects for Next Generation Computing and Communications Systems," *IME Seminar*, April 13, 2006.
26. Glebov, A. L., Uchibori, C. J., and Lee, M. G., "Direct Attach of Photonic Components on Substrates with Optical Interconnects," *IEEE Photonics Technology Letters*, Vol. 19, No. 8, April 2007, pp. 547–549.
27. Pugliano, N., Chiarotto, N., Fisher, J., Heiks, N., Ho, T., Khanarian, G., Moynihan, M., Pawlowski, N., Shelnut, J., Sherrer, D., Sicard, B., and Zheng, H. B., "Progress Toward the Development of Manufacturable Integrated Optical Data Buses," *SPIE Proceedings of Photonics Packaging and Integration IV*, Vol. 5358, 2004, pp. 71–79.
28. Wang, L., Choi, J., Wang, X. L., Chen, R. T., Hass, D., and Magera, J., "Thin Film Optical Wavegide and Optoelectronic Device Integration for Fully Embedded Board Level Optical Interconnects," *SPIE Proceedings of Photonics Packaging and Integration IV*, Vol. 5556, 2004, pp. 1–13.
29. Lee, B., Pamidigantham, R., and Premachandran, C. S., "Prototype Development for Chip-Chip Interconnection by Multimode Waveguide," *IEEE Proceedings of Electronic Packaging and Technology Conference*, Lake Buena Vista, FL, 2005, pp. 488–491.

30. Schow, C. L., Doany, F. E., Liboiron-Ladouceur, O., Baks, C., Kuchta, D. M., Schares, L., John, R., and Kash, J. A., "160-Gb/s, 16-Channel Full-Duplex, Single-Chip CMOS Optical Transceiver," *Proceedings of OThG4*, 2007.
31. Dellmann, L., Berger, C., Beyeler, R., Dangel, R., Gmur, M., Hamelin, R., Horst, F., Lamprecht, T., Meier, N., Morf, T., Oggioni, S., Spreafico, M., Stevens, R., and Offrein, B. J., "120 Gb/s Optical Card-to-Card Interconnect Link Demonstrator with Embedded Waveguides," *IEEE Proceedings of Electronic Components and Technology Conference*, Reno, NV, 2007, pp. 1288–1293.
32. Mikawa, T., Kinoshita, M., Hiruma, K., Ishitsuka, T., Okabe, M., Hiramatsu, S., Furuyama, H., Matsui, T., Kumai, K., Ibaragi, O., and Bonkohara, M., "Implementation of Active Interposer for High-Speed and Low-Cost Chip Level Optical Interconnects," *IEEE Journal of Selected Topics in Quantum Electronics*, Vol. 9, No. 2, 2003, pp. 452–459.
33. Lau, J. H., Y. Lim, T. Lim, G. Tang, K. Houe, X. Zhang, P. Ramana, J. Zhang, C. Tan, J. Chandrappan, J. Chai, J. Li, G. Tangdiongga, and D. Kwong, "Design and Analysis of 3D Stacked Optoelectronics on Optical Printed Circuit Boards," *Proceedings of SPIE—Photonics Packaging, Integration, and Interconnects VIII*, San Jose, CA, January 19–24, 2008, Vol. 6899, pp. 07.1–07.20. Also, accepted for publication in the *IEEE Transactions on Advanced Packaging*.
34. Johnson, E., and A. Caletka, "Passive Alignment of VCSELs to Waveguides in Opto-Electronics Cards and Printed Circuit Boards," *US Patent No. 2005/0089264A1*.
35. Doany, F., et al., "300-Gbps 24-Channel Bidirectional Si Carrier Transceiver Optochip for Board-Level Interconnects," *2008 IEEE/ECTC Proceedings*.
36. Lu, D., "Chip-to-Chip Optical Interconnect," *U.S. Patent No. 7373033B2*. May 13, 2008.
37. Chen, R. T., "Packaging Enhanced Board Level Opto-Electronic Interconnects," *US Patent No. 6243509B1*, June 5, 2001.
38. Chai, J., G. Yap, T. Lim, C. Tan, Y. Khoo, C. Teo, L. Lim, H. Yee, P. Ramana, J. H. Lau, R. Chang, T. Tang, H. Chang, S. Chiang, D. Cheng, and T. J Tseng, "Electrical Interconnect Design Optimization for Fully Embedded Board-level Optical Interconnects," *IEEE/EPTC Proceedings*, December 2008, pp. 1126–1130.
39. Lim, L, C. Teo, H. Yee, C. Tan, O. Chai, Y. Jie, T. Lim, P. Ramana, J. H. Lau, R. Chang, H. Chang, T. Tang, S. Chiang, D. Cheng, and T. J. Tseng, "Optimization and Characterization of Flexible Polymeric Optical Waveguide Fabrication Process for Fully Embedded Board-level Optical Interconnects," *IEEE/EPTC Proceedings*, December 2008, pp. 1114–1120.
40. Teo, C., W. Liang, H. Yee, L. Lim, C. Tan, J. Chai, Y. Jie, T. Lim, P. Ramana, and J. H. Lau, R. Chang, T. Tang, S. Chiang, D. Cheng, and T. J. Tseng, "Fabrication and Optimization of the 45° Micro-mirrors for 3-D Optical Interconnections," IEEE/EPTC Proceedings, December 2009, pp. 1121–1125.
41. Lim, T. G., Lee, B., Shioda, T., Kuruveettil, H., Li, J., Suzuki, K., Fujita, K., Yamada, K., Pamidighantam, R., and J. H. Lau, "Demonstration of High Frequency Data Link on FR4 PCB Using Optical Waveguides," to be published in *IEEE Trans. Advanced Packaging*.
42. Ramana, P., H. Kuruveettil, B. Lee, K. Suzuki, T. Shioda, C. Tan, J. Chandrappan, T. Lim, C. Teo, J. Chai, Y. Jie, C. Sharmani, and J. H. Lau, "Bi-directional Optical Communication at 10 Gb/s on FR4 PCB Using Reflow Solderable SMT Transceiver," *IEEE/ECTC Proceedings*, May 2008.
43. Chang, C., J. Chang, J. H. Lau, A. Chang, T. Tang, S. Chiang, M. Lee, T. J. Tseng, C. Tan, L. Lim, Y. Jie, C. Teo, and J. Chai, "Fabrication of Fully Embedded Board-level Optical Interconnects and Optoelectronic Printed Circuit Boards," *IEEE/EPTC Proceedings*, December 2009, pp. 973–976.

CHAPTER 8

Reliability of Low-Temperature Lead-Free (SnBiAg) Solder Joints

8.1 Introduction

In this chapter, the practical aspects of using a lower-melting-point, lead-free solder paste, Sn-57wt%Bi-1wt%Ag, are presented. A number of niche assembly applications cannot withstand the processing temperature range of 230°C to 260°C required for SnAgCu assemblies [1–4]. For example, optical components with plastic lenses depend on the transparency and stability of epoxies to meet their optical transmission and alignment specifications. Nonglass optical materials tend to darken or crack when exposed to high processing temperatures. Lead-free solders for low-cost electronic packaging, including SnBi, have been investigated as early as 1994 [5] and others [6–14]. This chapter will describe the rationale and material set used for an experiment to test surface mount technology (SMT) assembly and reliability characteristics using the 42Sn-57Bi-1Ag alloy, which has a melting point of 139°C, to assess a low-temperature solder alternative. Emphases are placed on the design for reliability, reliability testing and data analysis, and failure analysis of various lead-free packages' solder joints on the printed circuit board (PCB) using the low-melting-temperature, lead-free (SnBiAg) solder paste.

8.2 Benefits of Using SnBiAg Low-Temperature Alloy

The benefits of low-temperature solder alloys are in three areas: cost reduction, increased product reliability, and broadened design flexibility.

8.2.1 Cost Reduction

The cost of materials and components generally increases with increasing temperature requirements. BiSnAg melts at ~139°C, which is 78°C lower than the melting temperature of SnAgCu. Lower-temperature manufacturing processes could allow the use of lower-cost materials and components. Second, the energy required for SMT reflow ovens would be reduced when using SnBiAg compared to SnAgCu. The peak reflow process oven temperatures for SnAgCu are 230°C to 250°C, which is ~50% higher than the 160°C to 180°C temperature range needed for SnBiAg, assuming the same SMT reflow oven designs and efficiencies. Last, wave soldering for temperature-sensitive and other through-hole components may be eliminated, assuming an SMT paste-in-hole design is practical.

8.2.2 Product Reliability

Lower assembly process temperatures impart lower stresses to semiconductor package structures, which generally improves long-term reliability. There is also less probability of package "popcorning" or laminate delamination at lower processing temperatures. The higher standoff of nonmelting Pb-free chip scale package (CSP) and ball grid array (BGA) balls should lead to longer solder-joint life when considering the coefficient of thermal expansion (CTE) mismatch between the substrate and the packages. Last, the rework process can be performed at a lower temperature, which may reduce possible component damage.

8.2.3 Design Flexibility

A low-temperature solder such as SnBiAg allows a two-step or hierarchical soldering process of one component or module on top of another without remelting already soldered connections. For example, an SnAgCu containing module/daughter board can be soldered to a board with SnBiAg without remelting the SnAgCu solder joints.

8.3 Concerns with Use of Bismuth-Based Solders

Concerns with the use of Bi-based solders tend to fall into categories of brittleness, availability, environmental impact, recyclability, and lower reliability performance when mixed with lead. We discuss each of these areas of concern in turn.

8.3.1 Brittleness

The SnBi eutectic alloy has good accelerated thermal-fatigue characteristics, as reported in the NCMS Lead-Free Solder Project Report [7]. McCormack et al. found that the addition of small amounts of Ag (less than ~0.5wt %) significantly improved the ductility of the binary SnBi solder. The ductility improvement, which is more than a threefold increase in tensile elongation, is observed even at relatively high strain rates of 0.01/s [7].

8.3.2 Availability

Our estimate of bismuth availability is based on a Prismark lead-free assembly report that predicted in 2003 the worldwide consumption of solder paste would be approximately 7000 metric tons [15]. Solder paste typically contains approximately 90% metal by weight, which implies a 63Sn37Pb alloy weight of 0.90×7000, or 6300 metric tons. The corresponding weight of lead consumed would be 0.37×6300, or 2330 metric tons/year, if use continues at this rate.

The amount of bismuth needed to fulfill SMT needs in the electronics industry is based on the weight-percent ratio of bismuth versus lead and the solder density ratio to achieve the same solder-joint volume. For 42Sn-57Bi-1Ag, approximately twice as bismuth is required as lead in 63Sn-37Pb solder because of the 57/37 increase in the Bi weight-percent in SnBiAg and the additional solder alloy needed to attach nonmelting SnAgCu BGA balls. As an upper bound, we consider converting 10% of 2003 worldwide production of SnPb solder paste to the SnBiAg alloy. This would require an additional $0.10 \times 2 \times 2330$, or 466 metric tons/year of bismuth.

According to the bismuth chapter of the 2004 USGS Minerals Yearbook [16], the worldwide mine production for bismuth for 2003 was 3750 metric tons. According to the January 2004 USGS Mineral Commodity Summaries, approximately one-third of bismuth mined in 2003 was a by-product of tungsten extraction/production from China. China had 73% of the world reserves of bismuth in 2003 [16]. The bismuth reserves for 2003, which represents bismuth that is known to be economically extractable, was 330,000 metric tons. The bismuth reserve base, which represents known deposits that could be extractable at some higher price, was 680,000 metric tons. Thus, there is a more-than-adequate supply of bismuth for 10% of SMT applications, which could use up to 466 metric tons in our example. The 2004 price for bismuth varied from \$2.95 to \$3.95 per pound, and the price of tin varied from \$3.00 to \$3.35 per pound, depending on market conditions. In contrast, the price of lead varied from \$0.41 to \$0.50 [16].

8.3.3 Environmental Impact

Bismuth is not considered a toxic material, and is even used in medicines such as Pepto-Bismol. The University of Tennessee Center for Clean Products, under a contract to the Environmental Protection Agency (EPA), is completing a report on the life cycle impact analysis of lead-free solders compared to tin-lead solders. According to Table 1 in a recent paper [17], SnBiAg has the lowest environmental and health impact score in each of sixteen life cycle impact categories compared to the tested lead-free solders, which included SnAgCu. The full EPA report, "Solders in Electronics: A Life Cycle Assessment," Geibig and Socolof, 2005, is at [18].

8.3.4 Recyclability

Bismuth "contamination" is a concern for copper refining operations because small amounts of bismuth (<100 ppm) in copper can cause copper to become brittle. This embrittlement is not tolerated for high-grade Cu that is extruded into wire. Several copper refiners prefer not to have bismuth in their concentrate-input stream because it requires special processing and the associated costs to remove it. However, options are available for bismuth recycling, which will be discussed next. Bismuth is mined as a by-product of lead, copper, or tungsten in China, which has the largest bismuth reserves [16].

Most primary metal mines treat bismuth as a by-product. For example, Noranda operates a copper mine in Peru that contains bismuth as a by-product. Noranda's copper smelter uses a two-stage smelting/refining process that is designed to separate bismuth along with other materials. In the first stage the smelter is used for separation to produce >98% pure copper anodes. The second stage is an electrolytic refining process that removes the rest of the bismuth (and other elements). The noncopper materials, such as bismuth, antimony, arsenic, and gold, end up in sludge. The refinery needs a small amount of bismuth to balance out the antimony and arsenic to optimize the electrolytic refining process. If the bismuth content is too low, more is added. In summary, bismuth has not been desirable in electronic circuit boards, but it is manageable up to certain limits at modern smelting/refining processes [19].

8.3.5 Lead Contamination

When BiSn-based solders are mixed with lead, the ternary eutectic of 18Sn-52Bi-30Pb forms, which melts at 96°C. This may be too low for nonconsumer applications. Therefore, it is not recommended to use SnBiAg solder with lead if this low melting phase is a concern. For some consumer applications where the environmental temperature does not exceed 75°C, some lead may be tolerated. Reference [8] describes a successful thermal cycling experiment running between 18°C and 75°C for seven months (7000 thermal cycles), using the SnBi alloy combined with lead. This report shows no noticeable change in the microstructure from the initial condition.

8.4 Design for Reliability

8.4.1 Problem Definition

In this section, emphasis is placed on the determination of the creep responses of Sn-57wt%Bi-1wt%Ag lead-free solder joints for a Sn-3.9wt%Ag-0.6wt%Cu solder-balled 256-plastic ball grid array (PBGA) package as shown in Fig. 8.1. Also, totally SnAgCu solder joints (i.e., both solder paste and balls are SnAgCu) are considered.

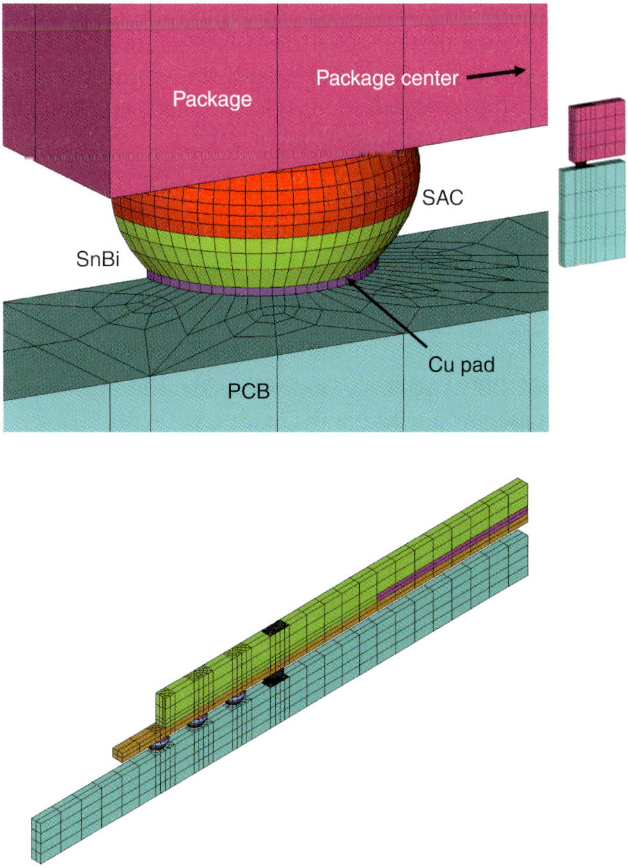

FIGURE 8.1 Finite-element model of a 256-pin PBGA assembly.

Furthermore, Sn-37wt%Pb components and solder paste are used as controls. Finally, the PCB design and assembly process for this and other high-density components assembled with either the SnBiAg or SnPb solder are also presented and discussed.

8.4.2 Material Properties

The material properties of the solder alloys and of the SnAgCu solder-balled 256-pin PBGA assembly are shown in Chaps. 2 and 7. All of the material properties are assumed to be time-independent except for those of the solders. It should be pointed out that since there is no creep material property data available for Sn-57Bi-1Ag, the properties of Sn-58Bi are used for this study.

As mentioned in Chap. 7, the Young's modulus of SnAgCu is the largest and that of SnPb is the smallest. Also, for the creep rate: (1) at low temperatures, for example, at 0°C, the creep rate of SnBi is slightly slower than that of SnPb and SnAgCu at higher stress conditions, and is slightly faster than that of SnAgCu but is about the same of SnPb's at lower stress conditions; and (2) at higher temperatures, for example, 50°C and 100°C, the creep rate of SnAgCu is the slowest and the creep rate of SnBi is slightly faster than that of SnPb.

8.4.3 Temperature Condition

Figure 8.2 shows the temperature profile imposed on the 256-pin PBGA assemblies. The temperature loading is from 0°C to 100°C, the cycle time is 44 min, the ramp-up and ramp-down times are each 12 min, and the dwell-at-hot and dwell-at-cold times are each 10 min.

The 256-pin PBGA package under consideration has a 27 × 27 × 0.36 mm bismaleimide triazene (BT) substrate with a 12.5 × 12.5 × 0.3 mm silicon integrated circuit (IC) chip and a 1.17-mm-thick overmold. The package has 0.635-mm-diameter pads (with a pitch of 1.27 mm) on the substrate, and the lead-free (95.5wt%Sn-3.9wt%Ag-0.6wt%Cu) solder balls are assumed to have a height of 0.5 mm, which are soldered on the PCB with the SnBiAg lead-free solder paste and the SnAgCu solder paste. The PCB thickness is 1.6 mm (0.062 in.), and the pad diameter is 0.635 mm. Again, the SnPb solder balls and the SnPb solder paste are used as controls.

Figure 8.1 shows the 3-D finite-element model that captures the construction along a diagonal strip from the 256-pin PBGA lead-free assembly's geometric center to a corner. Because of the mid-plane symmetry, the mesh actually models a one-half strip (with one-half of a solder joint) using the appropriate in-plane constraints placed on one symmetry plane. Coupled in-plane translations are applied to the other symmetry plane to produce a state of generalized plane strain. Using exclusively hexahedral solid elements, the model can capture

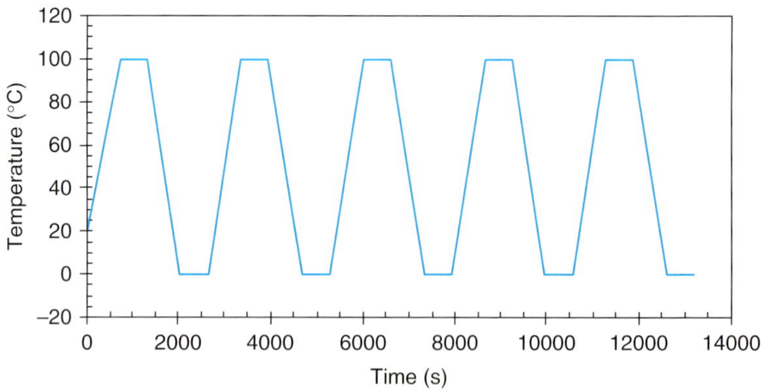

FIGURE 8.2 Temperature boundary condition.

the precise shape of the packages' solder joints and potential distant to neutral point (DNP) effects while retaining significant computational efficiency over full octant models. ANSYS [20] is the code selected for the modeling and analyses.

8.4.4 Results and Discussion

The creep shear strain contours at 13,200 s (five completed temperature cycles) in the SnBi, SnAgCu, and SnPb fillets are shown in Fig. 8.3 [21]. The creep shear strain time-history at the critical location in the SnBi, SnAgCu, and SnPb fillets are shown in Fig. 8.4. It can be seen that: (1) the maximum creep shear strain in the SnBiAg solder fillets is a few times larger than that in the SnAgCu and SnPb solder fillets and joints; and (2) the location of the maximum creep shear strain occurs at the center near the PCB and near the edge of solder fillets. Thus, during solder-joint fatigue, the SnBi solder fillets can be expected to break first (even the fatigue-crack growth behavior of

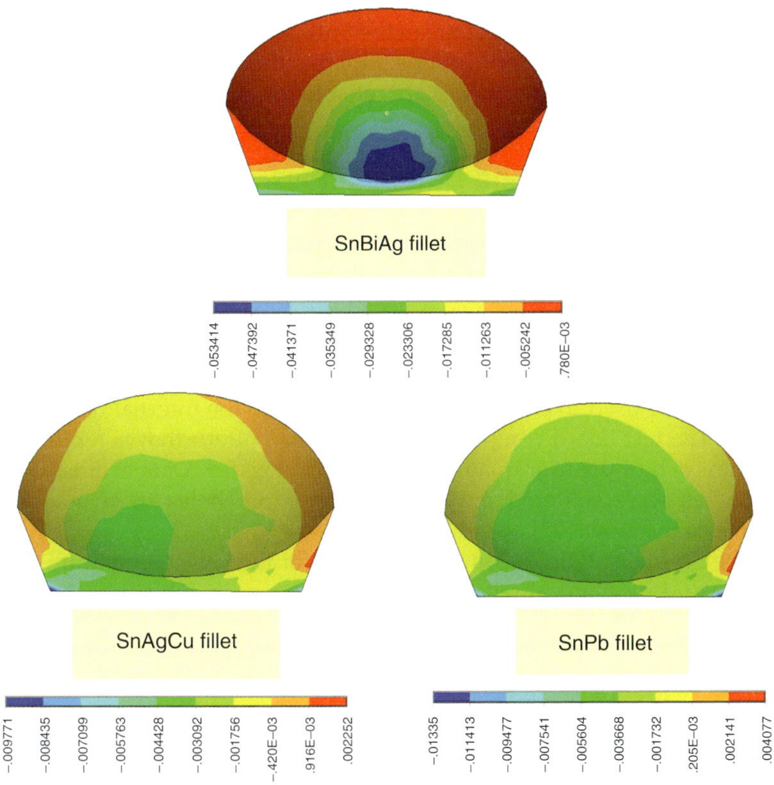

FIGURE 8.3 Creep shear strain in the SnBi, SnAgCu, and SnPb fillets at 13,200 s.

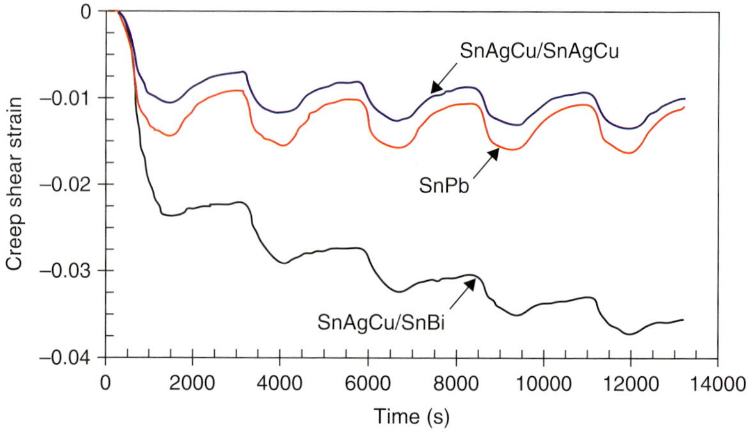

Figure 8.4 Creep shear strain at the critical location.

SnBi, SnPb, and SnAgCu solder alloys is not fully understood), and the failure location should be near the edge and the bottom side of the solder fillets. This doesn't mean the SnBi solder paste is not sufficiently reliable for all applications. Equally, this doesn't mean the SnPb and SnAgCu solder pastes are reliable for all applications.

8.5 Reliability Testing and Data Analyses

8.5.1 Components and Test Board

The size of the test board is 7.4 × 16.0 × 0.062 in (188 × 406 × 1.6 mm). It has six layers with standard FR-4 construction and uses resin-coated copper (RCC) foil for layers 1–2 and 5–6. The PCB finish is an organic solderability preservative (OSP) using Enthone Entek-106. Pad microvias are used to connect component pads to the internal daisy chains. The test board component layout has a pattern that is replicated in each of the four quadrants, as shown Fig. 8.5. Data acquisition wires were soldered into the connector pin holes to ensure reliable connections [22].

The Pb-free and SnPb components used for the low-temperature thermal and mechanical reliability tests are shown in Table 8.1. The simulated 256 PBGA is one of the components assembled on the test board.

8.5.2 Board Assembly Process

Twenty test boards were built: eight Pb-free, and eight SnPb "controls" plus two each for shock and vibration tests, which is described in [21]. The 42Sn57Bi1Ag solder paste was stencil-printed onto the eight Pb-free test boards with a conventional MPM-2000 paste stencil

FIGURE 8.5 Low-temperature solder reliability test board.

Part Type	Pin Count	Pitch (mm)	Ball Composition or Lead Finish for Pb-free/SnPb Components	Quantity in each ATC Test Group
PBGA	388	1.27	SnAgCu, SnPb	32
PBGA	256	1.27	SnAgCu, SnPb	32
PBGA	208	0.8	***SnPb, SnPb**	32
PBGA	196	1.0	SnAgCu, SnPb	32
PBGA	172	1.0	SnAgCu, SnPb	32
PBGA	144	0.8	SnAgCu, SnPb	32
PBGA	64	0.8	SnAgCu, SnPb	32
PQFP	80	0.5	NiPd	32
TSOP	48	0.5	Sn, SnPb	32
TSSOP	56	0.5	Sn, SnPb	32
Header	10	2.54	Sn, SnPb	32

*Not Pb-free!

TABLE 8.1 Pb-free and SnPb Test Components

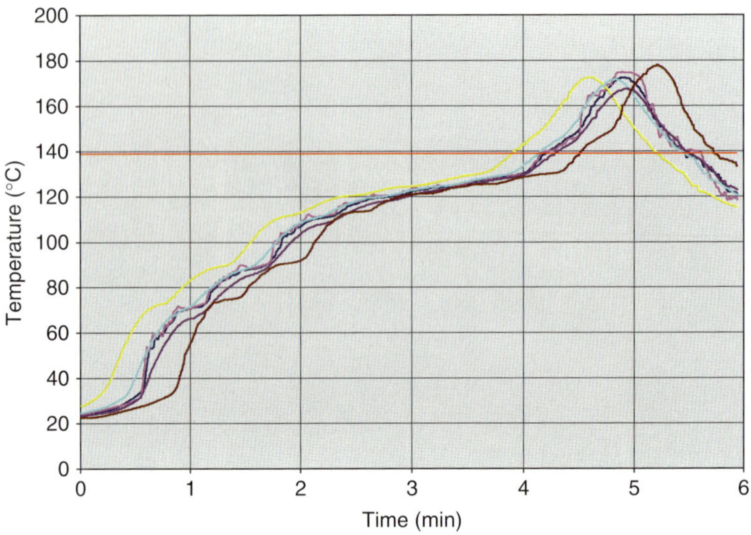

FIGURE 8.6 Reflow temperature profile used for 42Sn-57Bi-1Ag assemblies.

machine. The eight "control" boards were built with Alpha UP-78 SnPb solder paste. A standard Universal GSM pick and place machine was used to place the components on the test boards. The low-temperature SMT reflow process was run on a Heller 1800 EXL oven using a conveyor speed of 22 in (559 mm)/min in an air (not N_2) atmosphere. The temperature profile was set to have a peak reflow process temperature of 175°C, as shown in Fig. 8.6. For reference, the melting temperature of the SnBiAg alloy is shown by the line at 139°C.

8.5.3 Post-Assembly Inspection

Following the assembly process, the solder joints were inspected with a Feinfocus x-ray machine. An ERSA optical scope was used to inspect the solder joints. The solder joints on the leaded components appeared to be well formed (see Figs. 8.7 through 8.11).

After assembly, it became obvious that the BGA SnBiAg joints had insufficient solder paste volume, which was because the SnAgCu solder balls did not melt! The PBGA-388, PBGA-256, and other BGAs had hourglass-shaped solder joints that were concave, rather than convex.

One of the interesting challenges of building Pb-free assemblies is the "simple" logistics of keeping Pb-free and SnPb components separated. In one case, a "Pb-free" PBGA-208 was mislabeled and not found to be an SnPb part until after the assembly process! This demonstrates that factory controls need to be put in place to prevent

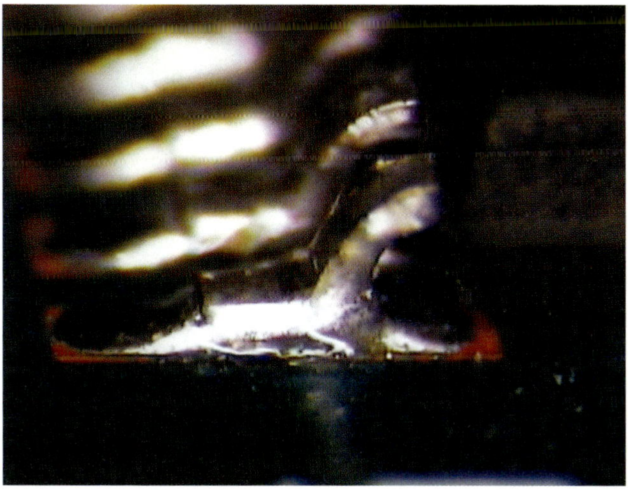

FIGURE 8.7 TSOP-48 solder joints.

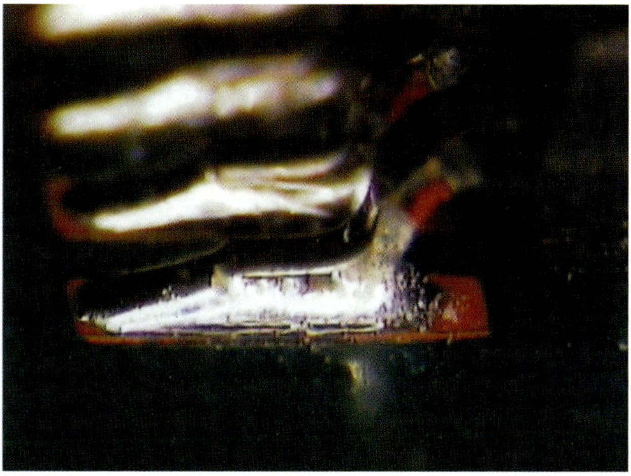

FIGURE 8.8 TSSOP-56 solder joints.

accidental mixing of Pb-free and SnPb components and materials. The new labeling standards JEDEC JESD-97 and IPC-1066 will help to keep Pb-free and non-Pb–free components separated.

8.5.4 Air-to-Air Temperature Cycling (ATC) Tests

Sixteen assembled test boards were placed in a Thermotron ATC test chamber, which is shown in Fig. 8.12. The data acquisition system was an HP/Agilent Model 8500 that continuously monitored the resistance of each component's daisy chain. When a component

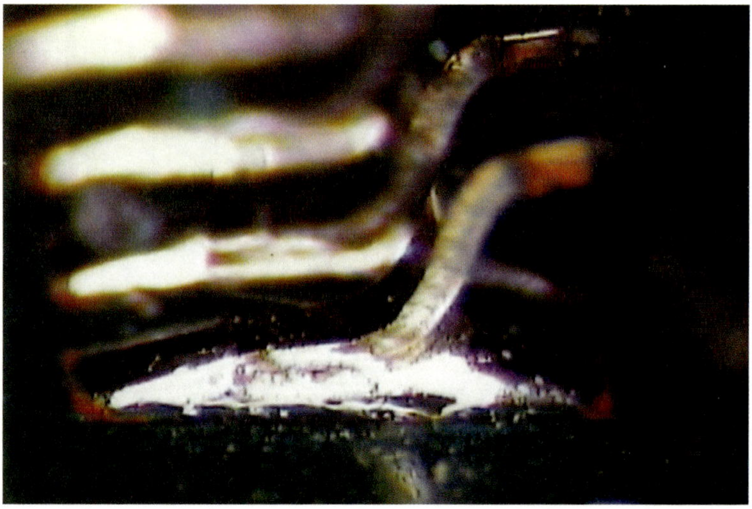

FIGURE 8.9 QFP-80 solder joints.

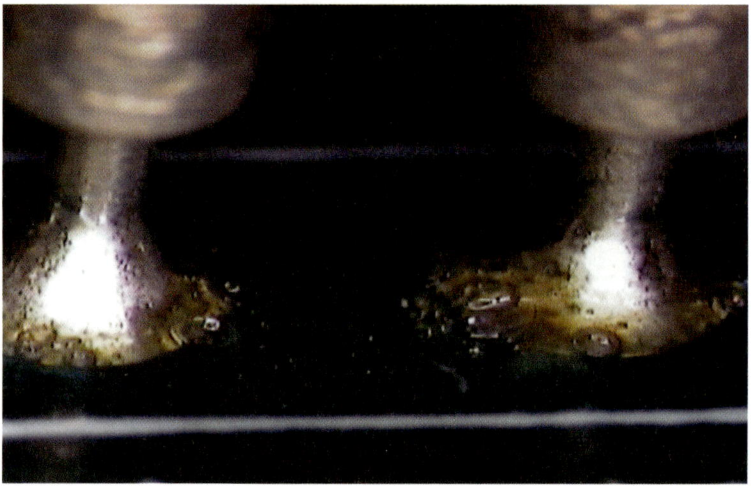

FIGURE 8.10 Through-hole in paste solder joints.

daisy-chain resistance exceeded 1.5 times its initial value (i.e., 50% increase), it was considered a "failure." The sample size for each component was 64, namely, 32 with SnPb solder balls + SnPb solder paste (controls) and 32 with SnAgCu solder balls + SnBiAg solder paste (lead-free). It should be noted that for the PBGA208 components, all the solder balls are made of SnPb.

The test was run for 8144 cycles from 0°C to 100°C, with 10-min dwells at the low and high temperatures. The rise and

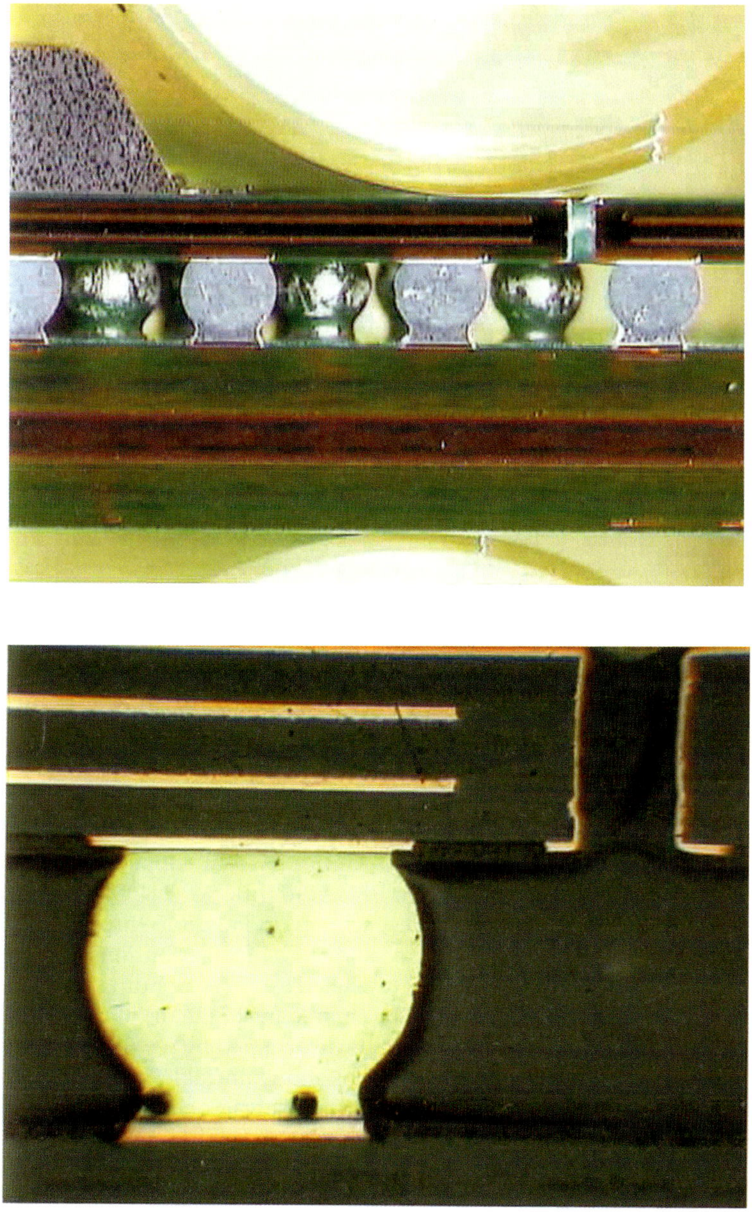

FIGURE 8.11 Cross-sections of concave PBGA-388 solder joints.

fall ramp times were both 12 min, which resulted in 33 cycles/day. Sixteen thermocouples (please see Fig. 2.32 for the set-up) were attached to the eight lead-free and eight control boards, and the average of these 16 measured values was used to set the 0°C to 100°C temperatures. All samples remained in the chamber until

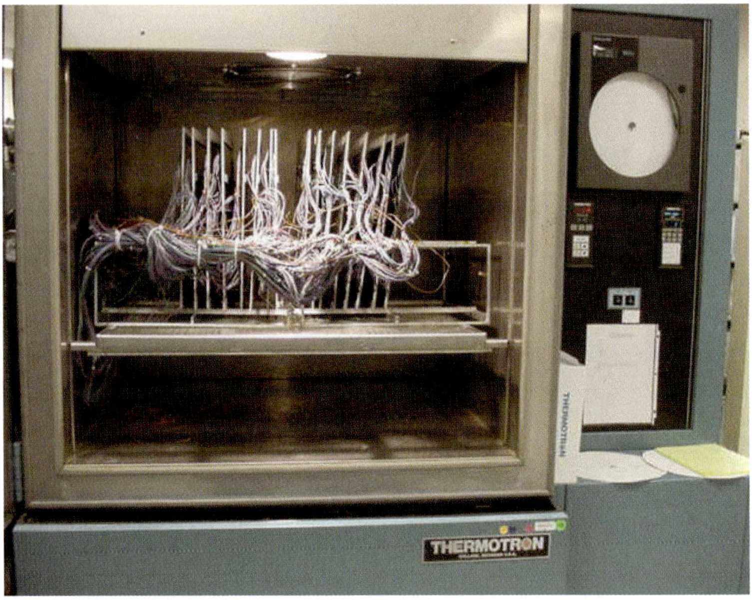

FIGURE 8.12 Thermal cycling chamber loaded with 16 test boards.

termination of the ATC test after 8144 cycles. Again, the failure criterion is defined as the 50% resistance increase.

8.5.5 Results of ATC Tests

The Weibull distribution was used to fit the test data to determine the sample characteristic life (θ) and sample Weibull slope (β) for the PBGA388, PBGA256, PBGA208, PBGA196, PBGA172, PQFP80, and TSSOP56 packages as shown, respectively, in Figs. 8.13 through 8.19. In each of the plots, both lead-free and SnPb life distributions are presented. Their solder-joint characteristic lives (corresponding to 63.2% failed) are shown in Table 8.2. The sample mean life (μ) of the solder joints of these seven packages can be determined by $\mu = \theta\Gamma\,(1 + 1/\beta)$, where Γ is the gamma function, and is shown in Table 8.2.

One of the difficult tasks in statistical analysis of life-test data is to draw conclusions about a population from a small sample size. It is even more difficult to compare the populations of two products from the knowledge of their test data (with small sample size). If the mean life (quality) of one product lot is found to be superior to another, how *confident* (P) can it be that the same is true of their populations?

By using Eq. (2.15), the calculated confidence level for comparing the quality of the SnPb solder joints and the lead-free solder joints is shown in Table 8.2. For the PBGA388, the mean life of SnPb solder joints is superior to that of lead-free solder joints, with 93.37% confidence, that is, in 93 out of 100 cases, the SnPb solder joints are better

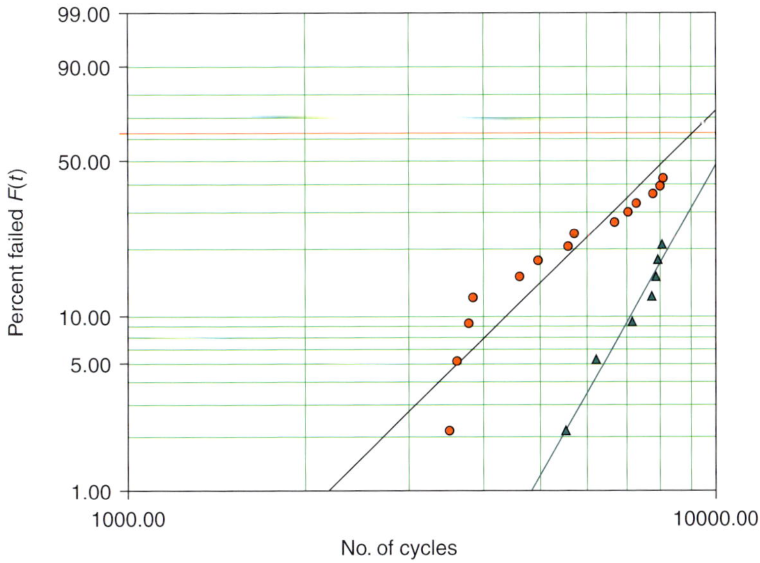

FIGURE 8.13 Life distributions of PBGA388 lead-free and SnPb solder joints (red data are for SAC solder paste).

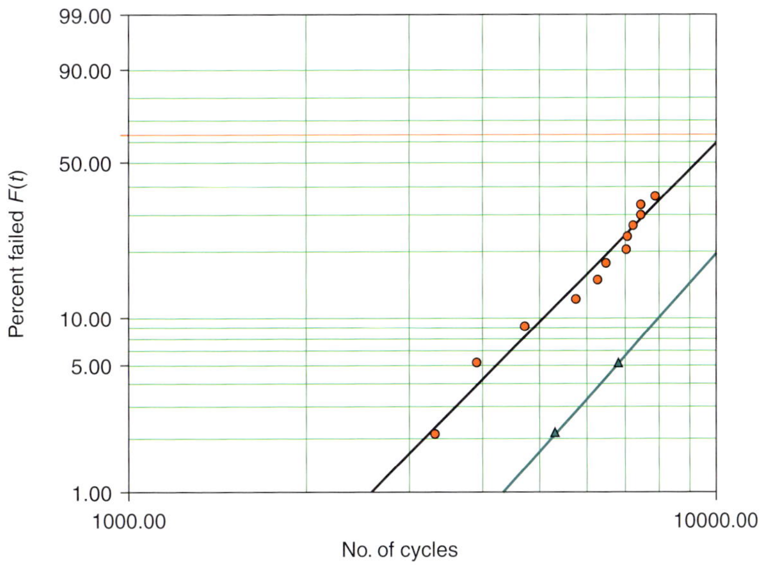

FIGURE 8.14 Life distributions of PBGA256 lead-free and SnPb solder joints (red data are for SAC solder paste).

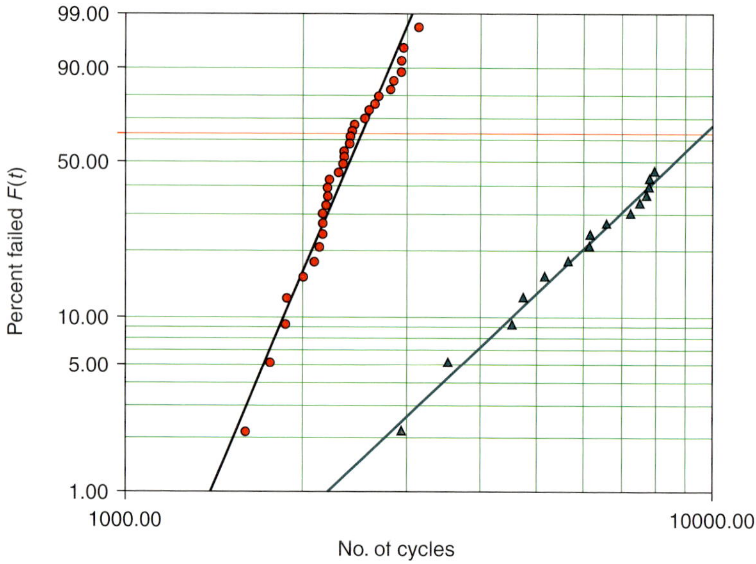

FIGURE 8.15 Life distributions of PBGA208 lead-free and SnPb solder joints (red data are for SAC solder paste).

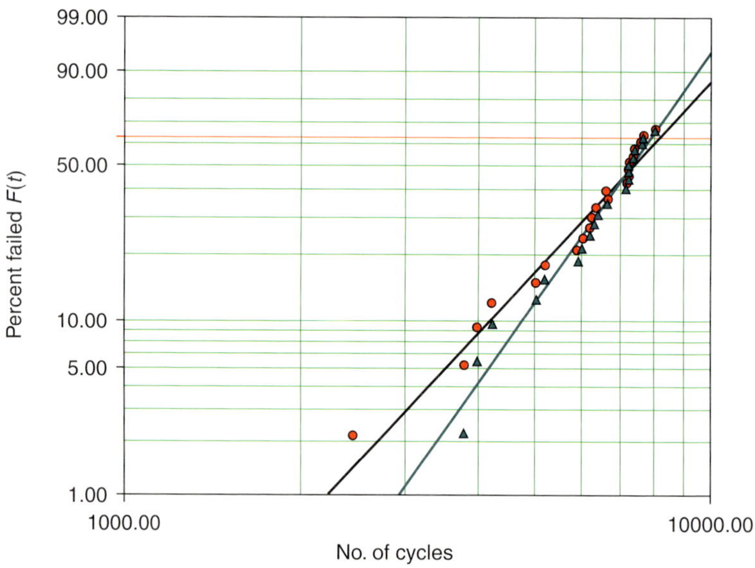

FIGURE 8.16 Life distributions of PBGA196 lead-free and SnPb solder joints (red data are for SAC solder paste).

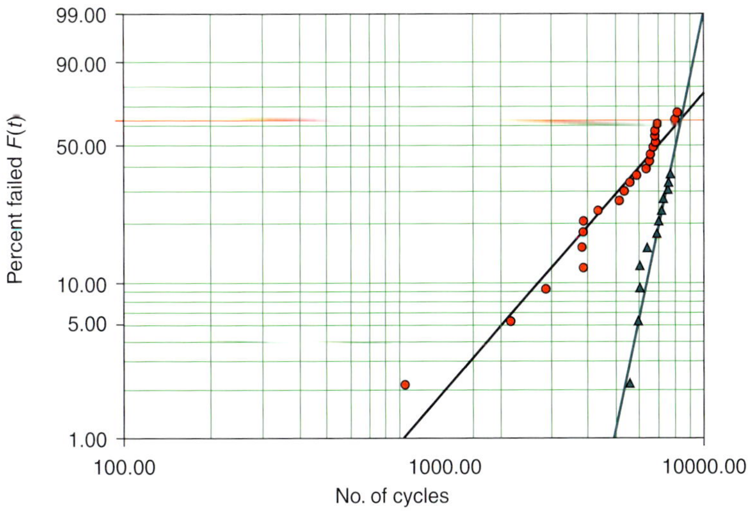

Figure 8.17 Life distributions of PBGA172 lead-free and SnPb solder joints (red data are for SAC solder paste).

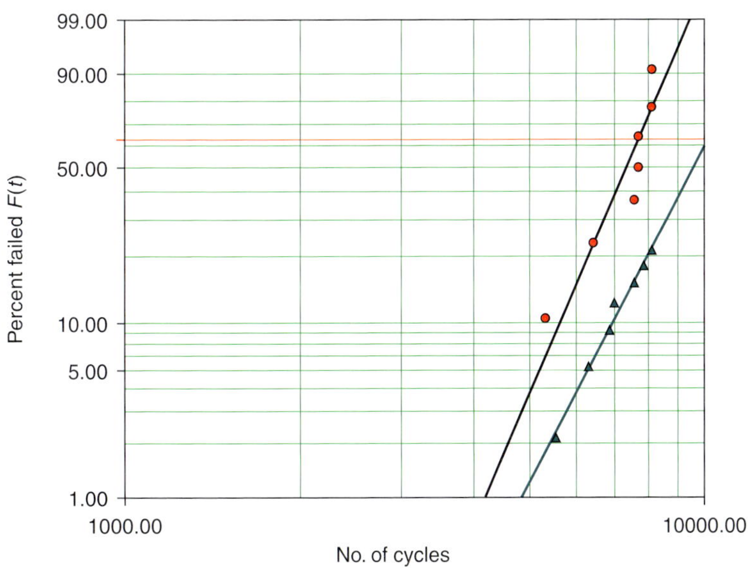

Figure 8.18 Life distributions of PQFP80 lead-free and SnPb solder joints (red data are for SAC solder paste).

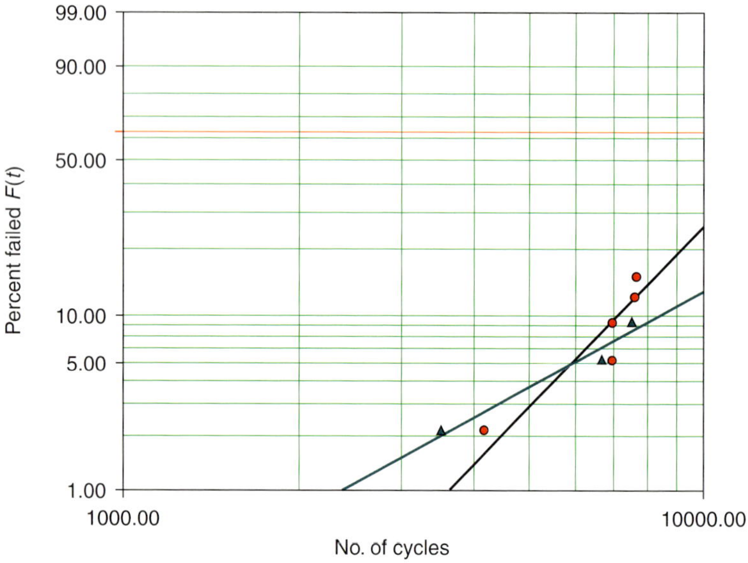

FIGURE 8.19 Life distributions of TSSOP56 lead-free and SnPb solder joints (red data are for SAC solder paste).

Package	Characteristic Life (Cycles)		Mean Life (Cycles)		Calculated Confidence (%)
	Lead-free	SnPb	Lead-free	SnPn	
PBGA388	9178	10700	8220	9908	93.37
PBGA256	10323	15125	9263	13644	95.54
PBGA208	2521*	9678	2371*	8656	99.999
PBGA196	8134	7901	7324	7226	56.36
PBGA172	8236	8269	7295	7816	73.00
PQFP80	7682	10102	7214	9391	99.40
TSSOP56	14408	32112	12928	28590	97.30

Lead-free means SnAgCu balls (Sn or NiPdAu leads) and SnBiAg paste. SnPb means SnPb balls (or SnPb leads) and SnPb paste. Lead-free* means the PBGA208 is with SnPb solder balls and SnBiAg paste. Test condition: 0°C–100°C, 40 min cycle.

TABLE 8.2 Confidence Level for Comparing the Quality (Mean Life) of Lead-Free and SnPb Solder Joints than the lead-free solder joints. Similarly, for the PBGA256 package, at 95.54% confidence level, the SnPb solder joints are better than the lead-free solder joints. It should be emphasized that these are only true for this particular temperature condition and component because the acceleration factor of SnPb and lead-free solder alloys may not be the same at other temperature conditions.

On the other hand, for the PBGA196 package, the quality of the lead-free solder joint is slightly superior to that of SnPb solder joints, with only 56.36% confidence level. This means in 56 out of 100 cases, the lead-free solder joints are better than the SnPb solder joints. But this confidence level is too low to be statistically significant. At best, we can only say that for the PBGA196 SnPb and lead-free solder joints, their thermal-fatigue lives are about the same! Again, this is only true for 0°C to 100°C.

The temperature cycling results show that for most of the PBGAs, their lead-free solder joints are not as good as the SnPb solder joints. This could be due to the SnBiAg solder volumes (fillets) for the nonmelting SnAgCu ball PBGAs being significantly smaller than they should be, as shown in Fig. 8.11. Also, the creep rate of SnBi solder is slightly faster than that of the SnPb solder during most of the temperature cycling. This "starved" PBGA solder joint volume could have limited the accelerated thermal-fatigue life of these components, and this negative effect was stronger on the larger PBGAs with 1.27-mm pitch than on the smaller balls on finer pitches (0.8 mm and 1.0 mm).

In order to make the lead-free solder joints more reliable, the copper pads on the PCB should be at least as large as the copper pads on the PBGA package substrate. In that case, more SnBiAg solder paste will form a bigger and stronger solder fillet for the nonmelting SnAgCu solder balls to resist the thermal expansion mismatch between the components and the PCB.

8.5.6 Shock (Drop) Tests and Results

Four 11 × 16 inch (279 × 406 mm) test boards were cut in half resulting in eight 11 × 8 inch (279 × 203 mm) sample assemblies. Four of these were assembled with lead-free components and SnBiAg solder, and four were assembled with conventional SnPb eutectic solder (for control). There are 20 daisy-chained components on each half-board, which were checked for continuity before the tests.

Four sample boards with no preconditioning were sequentially mounted on an Avco Shock Test machine, as shown in Fig. 8.20. The imposed shock levels were monitored by an accelerometer attached to the support table. Two vises were used to grip each board as shown. The test protocol was to shock each of the four board samples three times at each shock level, measure all of the daisy-chain continuities, and then progress to the next higher level of shock test. The test levels were done in sequence at 200 g, 400 g, 800 g, 1000 g, 1200 g, and

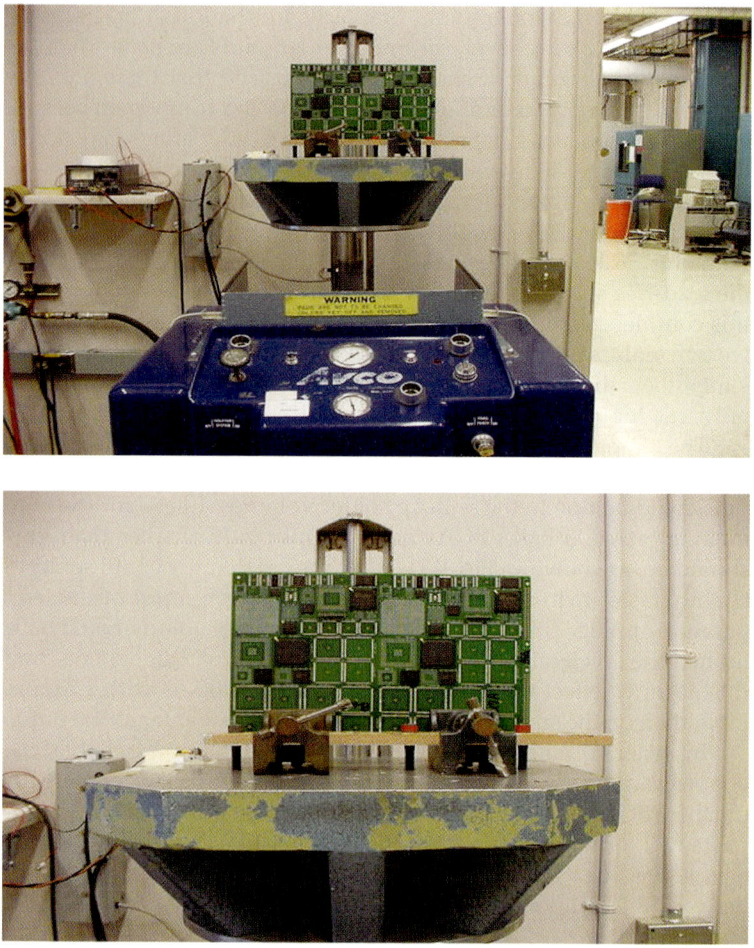

FIGURE 8.20 Test board mounted in (vertical) shock test fixture.

finally 1500 g. No open daisy-chain connections were found at each stage of the test. The bottom-line result is that no daisy chains were broken by this series of shock tests.

8.5.7 Vibration Tests and Results

The remaining four sample boards were shocked at 200 g prior to being vibration-tested. The Thermotron vibration table was set up for the HP Standard Class B1-II portable vibration test (5–500 Hz, 0.75 g survival/random scan), as shown in Fig. 8.21. Three tests were performed, and the 20 continuity measurements were made on each board after each run. First, the four boards were clamped flat on the vibration table and the class B1-II random vibration test was run for 10 min. Next the four boards were supported $^{3}/_{8}$ inch above the vibration

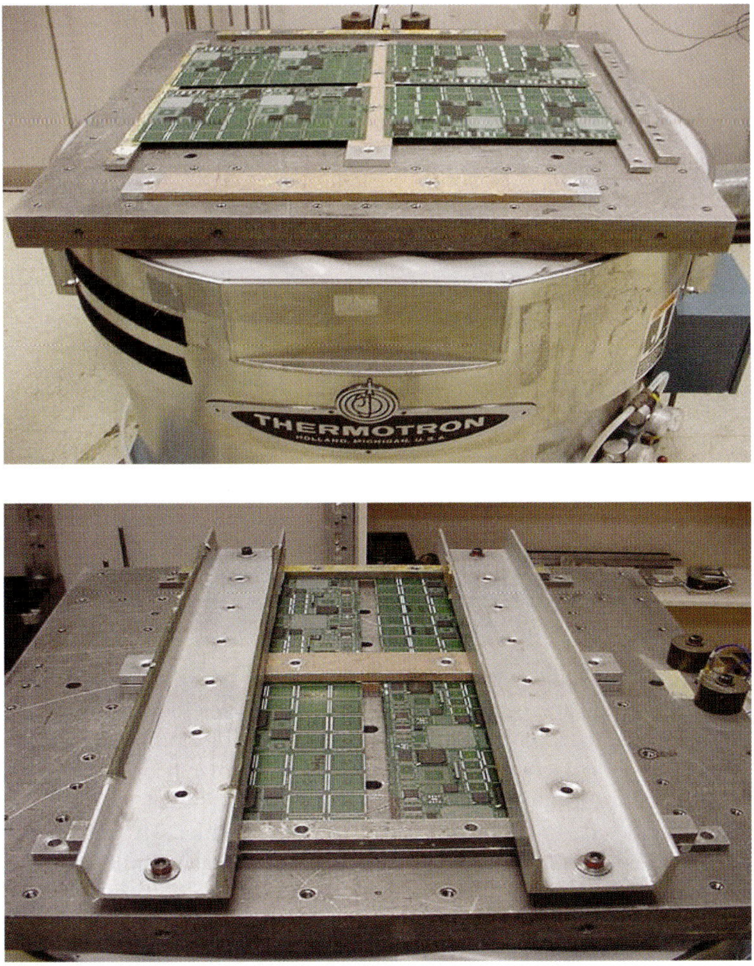

FIGURE 8.21 Test boards mounted on vibration test table. Top: Test boards are $^3/_8$ inch from the table. Bottom: Test boards are clamped on the table.

table and the 10-min test was repeated. Finally, an accelerometer was placed on a board and the vibration frequency was scanned for the dominant board resonance, which was 38 Hz. This test was run for five minutes with an 11 g acceleration measured on the board. The table acceleration was 0.75 g RMS. The final set of continuity measurements was made, and no opens were found.

8.6 Failure Analyses

Since there are no failures with the shock and vibration tests, failure analyses was only performed on the thermal-cycling failed samples. For all the PBGA solder joints (except the lead-free PBGA208 and

FIGURE 8.22 Typical lead-free PBGA failure mode (SnBiAg solder fillet cracked).

lead-free PBGA172, which will be discussed later) their very first solder-joint failure occurred at over 2500 cycles. Thus, based on these results, they (both lead-free and SnPb solder joints) should be reliable for most consumer products under normal operations. A typical failure mode of the lead-free PBGAs is shown in Fig. 8.22. The crack occurs in the SnBiAg lead-free solder fillet. This compares very well with the predicted failure location from the finite-element results.

One set of Pb-free components failed rather early (two-thirds failed by 1500 cycles). The Pb-free PBGA208 was analyzed and found to have SnPb solder balls instead of SnAgCu. The resulting coarse microstructure is shown in Fig. 8.23. Note also the large void, which was probably caused by out-gassing of trapped flux/air in the microvia in the copper pad. Since the ternary melting point of 52Bi-30Pb-18Sn is 96°C, these solder joints at least partially melted on every transition to 100°C, significantly limiting their thermal-fatigue life.

The Weibull plot shown on the left in Fig. 8.15 illustrates the effect of the early failures caused by the contamination of Pb from SnPb solder balls in the SnBiAg solder joints. All of the lead-free solder joints failed by approximately 3000 cycles. The Weibull plot on the right is for the PBGA208 SnPb control components.

The PBGA172 also exhibited early failures (the first solder joint failed at about 1000 cycles), as shown in Fig. 8.24. Dye and pry tests and cross-sections (Fig. 8.22) show large voids that were likely caused by the out-gassing of trapped flux/air in the microvias on the copper

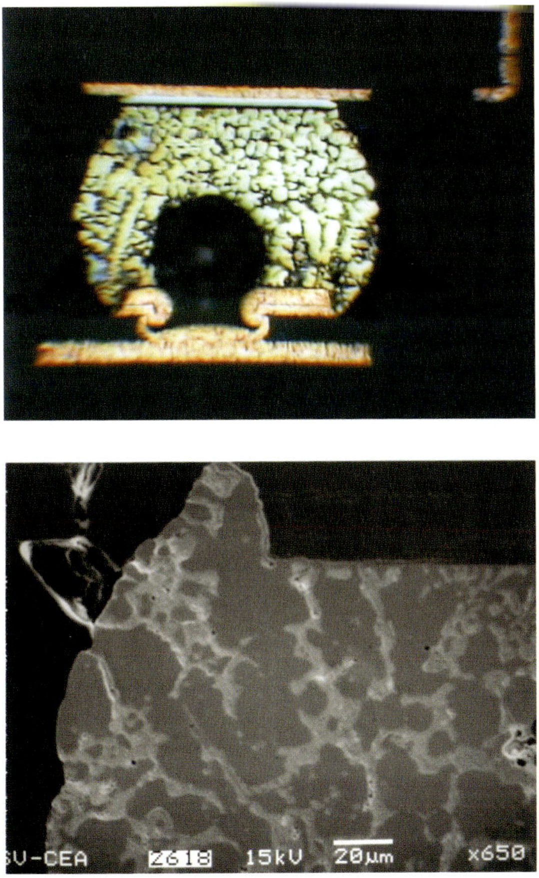

FIGURE 8.23 PBGA208 with SnPb ball and SnBiAg paste (52Bi30Pb18Sn, MP ~96°C).

pad. These voids degrade the thermal-fatigue life of the lead-free PBGA172 solder joints. Thus, in order to make the lead-free solder joints on microvias more reliable, the void size needs to be reduced. One way is to use N_2 in a reflow atmosphere to reduce the out-gassing of trapped flux/air in the microvias.

Temperature cycling test results for the SnPb and lead-free solder joints of the PQFP80 and TSSOP56 packages are shown in Figs. 8.18 and 8.19, respectively, and Table 8.2. It can be seen that: (1) for both packages, the SnPb solder joints are superior to the lead-free solder joints at very high confidence levels; and (2) for both packages and both SnPb and lead-free solder joints, their very first solder joint failure is larger than 3500 cycles. Also, these results show they are reliable for most of the consumer products under normal operations.

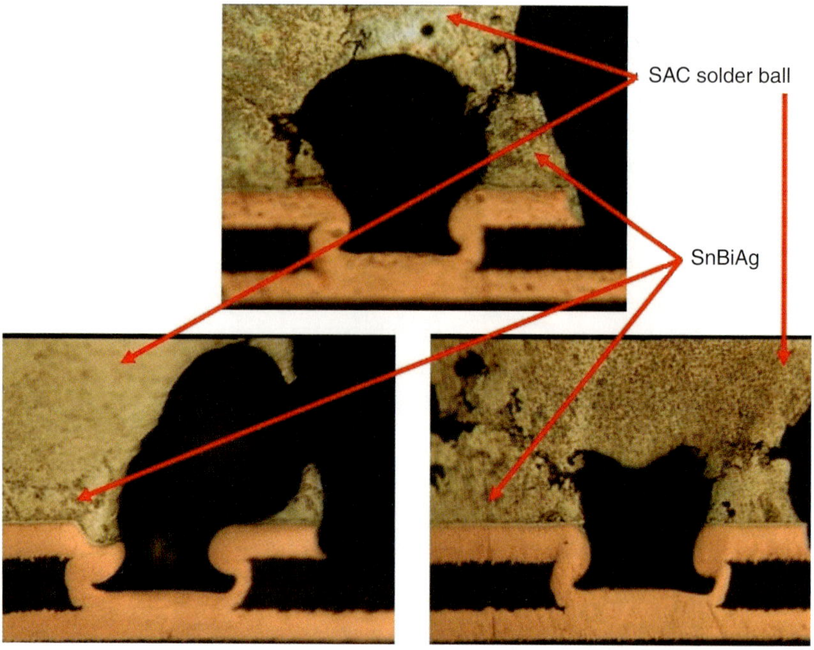

FIGURE 8.24 PBGA172 lead-free solder-joint failures with large voids on microvias.

8.7 Summary and Recommendations

Some important results and recommendations are summarized in the following:

1. For both the SnPb and SnBiAg assemblies, and for all the components, there were no solder-joint failures after the shock and vibration tests.
2. For the PBGA388 package, the SnPb solder joints are superior to the SnBiAg solder joints with 93.37% confidence level. The first lead-free solder joint failed at 3513 cycles.
3. For the PBGA256 package, in 95 out of 100 cases, the SnPb solder joints are better than the SnBiAg solder joints. The first lead-free solder joint failed at 3328 cycles.
4. For the PBGA196 package, the lead-free solder joints are slightly better than the SnPb solder joints with very low confidence level (56.36%). Basically, from a statistical point of view, there is almost no difference! The first lead-free solder joint failed at 2451 cycles.
5. For the PBGA172 package, in 73 out of 100 cases, the SnPb solder joints are better than the lead-free solder joints. The

first lead-free solder joint failed at 932 cycles. This could be due to large voids caused by the out-gassing of trapped flux/air in the microvia on the copper pads. Reflowing the SnBiAg with N_2 may help to reduce the number of voids and void size.

6. For the PBGA208 package, the SnPb solder joints are superior to the SnBiAg solder joints with 99.999% confidence level. In this case, the solder balls are SnPb and the lead-free solder paste is SnBiAg. The thermal-fatigue life of the SnPb-SnBiAg composite solder joints is very poor because of the ternary meting point (96°C) of 18Sn-42Bi-30Pb. All the lead-free solder joints (100%) failed before 3137 cycles.
7. One of the major reasons that the PBGA SnPb solder joints are superior to the SnBiAg solder joints is the SnBiAg solder volumes (fillets) for the nonmelting SnAgCu solder ball is significantly smaller than they should have been. In order to make the lead-free solder joints more reliable, the copper pads on the PCB should be at least as large as the copper pads on the PBGA package substrate. In that case, more SnBiAg solder paste will form a bigger and stronger solder fillet for the nonmelting SnAgCu solder balls.
8. Most of the lead-free solder joints of the PBGA packages failed at the SnBiAg solder fillets. This agreed very well with the finite-element simulation results.
9. For the leaded components, PQFP80 and TSSOP56, the SnPb solder joints are better than the lead-free solder joints with very high confidence levels. The first lead-free solder joint failed for the PQFP80 at 5,325 cycles, and for the TSSOP56 at 4,177 cycles. These are more than adequate for most of the consumer products under normal operating conditions.

8.8 References

1. Hwang, J. S., *Environment-Friendly Electronics: Lead-Free Technology*, Electrochemical Publications, Asahi House, British Isles, 2001.
2. Hwang, J. S., *Implementing Lead-Free Electronics*, McGraw-Hill, New York, 2004.
3. Lau, J. H., *Low Cost Flip Chip Technologies for DCA, WLCSP, and PBGA Assemblies*, McGraw-Hill, New York, 2000, Chapter 3, Lead-Free Solders.
4. Lau, J. H., C. P. Wong, N. C. Lee, and S. W. R. Lee, *Electronics Manufacturing with Lead-Free, Halogen-Free and Conductive-Adhesive Materials*, McGraw-Hill, New York, 2003.
5. Glazer, J., "Microstructure and Mechanical Properties of Pf-free Solder Alloys for Low-Cost Electronic Packaging," *Journal of Electronic Materials*, Vol. 23, No. 8, 1994, pp. 701–707.
6. Lau, J. H., and W. Dauksher, "Thermal-Mechanical Analysis of a Flip-Chip VCSEL (Vertical-Cavity Surface-Emitting Laser) Package with Low-Temperature

Lead-Free (Sn-Bi) Solder Joints," *ASME Paper No. IMECE2005-79981*, November 2005.

7. McCormack, M., H. S. Chen, G. W. Kammlott, and S. Jin, "Significantly improved mechanical properties of Bi-Sn solder alloys by Ag-doping," *Journal of Electronic Materials*, 26 (8), 1997, pp. 954–958.
8. Mei, Z., Fay Hua, and Judy Glazer, "Sn-Bi-X Solders," *SMTA Proceedings*, September, 1999. Chicago, IL.
9. Hua, F., Zequn Mei, Judy Glazer, and Ayn Lavagnino, "Eutectic Sn-Bi as an Alternative Pb-free Solder," *IPC Works*, 1999.
10. Lau, J. H, and C. Chang, "TMA, DMA, DSC, and TGA of Lead Free Solders," *Journal of Soldering & Surface Mount Technology*, Vol. 11, No. 2, 1999, pp. 17–24.
11. Schroeder, V., and G. Chew, "The Applicability of Bi-42Sn-1Ag Solder for Consumer Products: A Study of Pb Contamination Effects," *IPC Works*, 2001.
12. Schroeder, V., J. Gleason, F. Hua, "Strength and Fatigue Behavior of Joints Made with Bi42Sn-1Ag Solder Paste: An Alternative to Sn-3.5Ag-0.7Cu for Low Cost Consumer Products," *SMTA Proceedings*, 2001, Chicago, IL.
13. Schroeder, V., J. Gleason, F. Hua, "Applicability of Bi-42Sn-1Ag Solder for Consumer Products," *APEX Proceedings*, January 2002, Anaheim, CA.
14. Ferrer, E., and H. Holder, "57Bi-42Sn-1Ag: A lead Free, Low Temperature Solder for the Electronics Industry, *JEDEX Conference*, March 22–25, 2003, San Jose, CA.
15. Lead Free Assembly, *Prismark Report*, August 2001, New York.
16. USGS Mineral Industry Surveys: http://minerals.usgs.gov/minerals.
17. Geibig, J., M. Socolof, M., "Life-Cycle Impacts of Lead and Lead-Free Solder use in Reflow Soldering and Electronics," *Electronics goes Green Conference*, Berlin, June 2004.
18. Center for Clean Products, University of Tennessee, http://eerc.ra.utk.edu/ccpct/index.html.
19. Private conversation with Hai Teh, Noranda Corp., Anaheim, CA.
20. ANSYS, *User's Manual*, Release 6, 2003, Pittsburgh, PA.
21. Lau, J. H., Gleason, J., V. Schroeder, G. Henshall, W. Dauksher, and B. Sullivan, "Design, Materials, and Assembly Process of High-Density Packages with a Low-Temperature Lead-Free Solder (SnBiAg)," *Soldering & Surface Mount Technology*, Vol. 20, No. 2, 2008, pp. 11–20.
22. Lau, J. H., Gleason, J., V. Schroeder, G. Henshall, W. Dauksher, and B. Sullivan, "Reliability Test and Failure Analysis of High-Density Packages Assembled with a Low-Temperature Lead-Free Solder (SnBiAg)," *Soldering & Surface Mount Technology*, Vol. 20, No. 2, 2008, pp. 21–29.

CHAPTER 9

Reliability of Lead-Free (SACX) Solder Joints

9.1 Introduction

The effect of a very low level of doping of Ce (cerium) on the SnAgCu solder is presented. Emphasis is placed on the determination of the material properties (such as the temperature and strain-rate–dependent Young's modulus, yield stress, ultimate strength, and the temperature- and time-dependent creep) and the intermetallic compounds (IMC) on organic solderability preservative (OSP) and NiAu (electroless Ni and immersion Au) copper surface finishes of the new Sn3wt%Ag0.5wt%Cu0.019wt%Ce (SACC) lead-free solder. The PCB (printed circuit boards) assembly (PCA) process developments and the solder-joint reliability of plastic quad flat pack (PQFP), plastic ball grid array (PBGA), wafer-level chip scale package (WLCSP), and bumped chip carrier (BCC) components on PCB with the newly proposed lead-free SACC are also provided and discussed.

9.2 SACC

In the past decade, the candidate for lead-free solder materials has been SnAgCu [1–12]. Recently, many research results show that under certain loading conditions and for certain components, the SnAgCu solder joints perform worse than the SnPb solder joints. For example, in [13] Amagai and his colleagues show that for a 10- × 10-mm PBGA with a 6- × 6-mm die size mounted on 160- × 64-mm FR-4 two-metal layer PCB, the SnAgCu solder joints don't perform as well as the SnPb solder joints under drop conditions.

More recently, many researchers have been devoted to slightly modifying the existing SnAgCu (i.e., adding a very low level of doping—X, also called a nanoparticle) and developing new lead-free solder alloys SnAgCuX with almost the same melting point

(217°C) as SnAgCu, but with better thermal and drop performances. For example, Sousa and her colleagues at IBM show that [14] with a very low level of Zn (0.6wt%) added to the SAC305, that is, Sn3wt%Ag0.5wt%Cu0.6wt%Zn, it will: (1) drastically reduce the rate of formation in the interfacial Cu_3Sn phase; (2) significantly reduce the Cu pad consumption rate during reflow and thermal aging; and (3) dramatically lower the Kirkendall void formation at the interface during thermal aging. Also, Amagai and his colleagues at TI and Senju show that [15] SnAgCuP alloys reveal high reliability performance for dynamic loads like drop tests. Finally, for the effects of nanoparticles such as Co, Ni, Pt, Al, P, Cu, Zn, Ge, Ag, In, Sb, or Au in SnAg-based lead-free solders for intermetallic compounds and drop-test performance, please see [16].

In this study, a very low level of Ce (0.019wt%) is added to the SAC solder. It can be made by using a master batch alloy. First, make a batch of master alloy with 1.9wt%Ce and 98.1wt%Sn. Then add 1.0% of this master batch alloy to the SnAgCu solder. This way, 0.019wt%Ce in SnAgCu can be obtained.

9.3 Material Properties of SACC

The objectives of this section are to determine the fundamental material properties, such as the Young's modulus, yield strength, ultimate strength, and creep of the SACC.

9.3.1 Bulk Solder Test Specimen

Dog bone-shaped bulk solder specimens, as shown in Fig. 9.1 for the uniaxial tensile test, are prepared by machining from the extruded SACC solder bar. The specimens have a total length of 65 mm and a gage length of 15 mm. The diameter of the gage length is 3 mm, as shown in Fig. 9.1. The specimens are annealed for 24 h at 60°C in a vacuum to eliminate the surface residual stresses. The specimen preparation and test methods used in this study are basically the same as those reported in [17, 20].

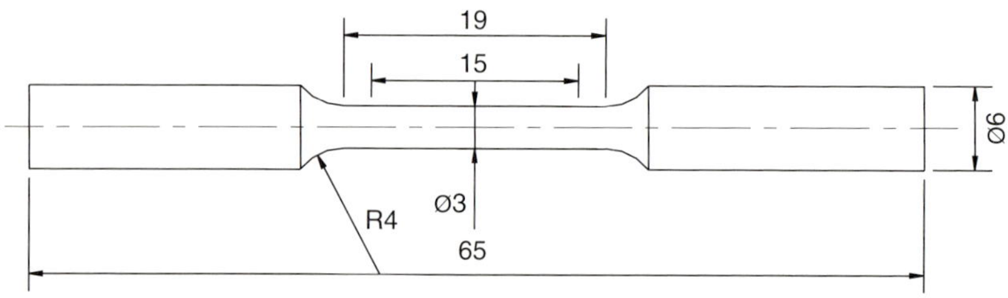

FIGURE 9.1 Bulk specimen for tensile tests.

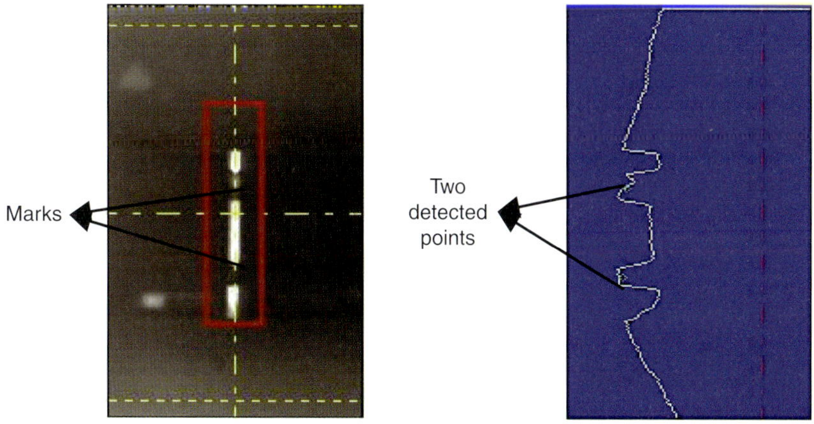

FIGURE 9.2 Video extensometer measurement of tensile strain.

9.3.2 Tensile Test Procedures

Tensile tests are carried out on a universal testing machine (Instron model 5569). Three different temperatures (25°C, 75°C, and 125°C) are investigated. At each temperature, three different strain rates are used (5.6×10^{-4}/s, 5.6×10^{-3}/s, and 5.6×10^{-2}/s). The tensile tester is fitted with a thermally insulated chamber. A video extensometer is employed in the specimen gauge section to measure the true extension of the specimen, as shown in Fig. 9.2. A data acquisition system that consists of a computer and some software is used to store and analyze the experimental data.

9.3.3 Tensile Test Results

The effects of strain rate and temperature on the SACC tensile material properties are evaluated. Figure 9.3 shows the stress-strain curves for the three different strain rates at a constant temperature of 25°C. The mechanical behavior of SACC is strongly dependent on the test strain rate. Figure 9.4 shows the stress-strain curves for the three different temperatures at a constant strain rate of 5.6×10^{-2}/s. Again, it can be seen, and is expected, that the mechanical behavior of SACC is strongly temperature-dependent.

The Young's modulus (or elastic modulus), E, of a material is determined by the initial slope of the stress-strain curve. The effects of test temperature and strain rate on the elastic modulus are shown in Figs. 9.5 and 9.6, respectively. It can be seen that: (1) for all the strain rates, the elastic modulus decreases as the temperature increases; and (2) for all the temperatures, the larger the strain rate, the larger the elastic modulus.

The yield stress, σ_y, of a material is computed from the stress-strain curve at 0.2% plastic strain (i.e., upon unloading, the permanent

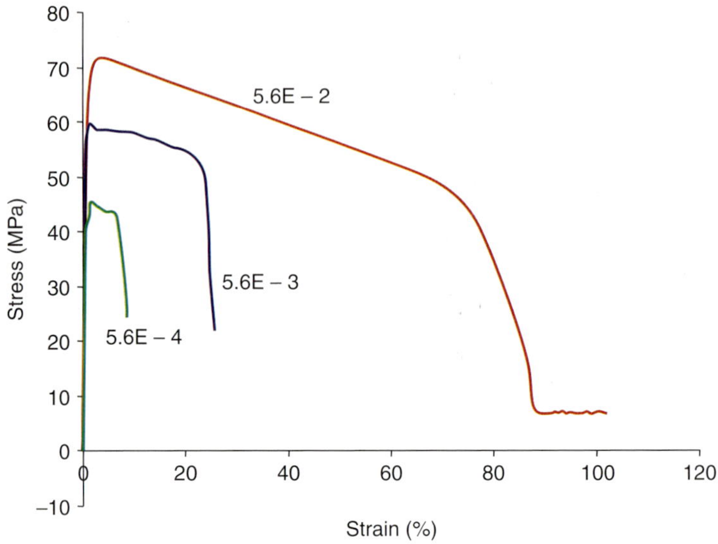

FIGURE 9.3 Strain rate effect at a constant temperature (25°C).

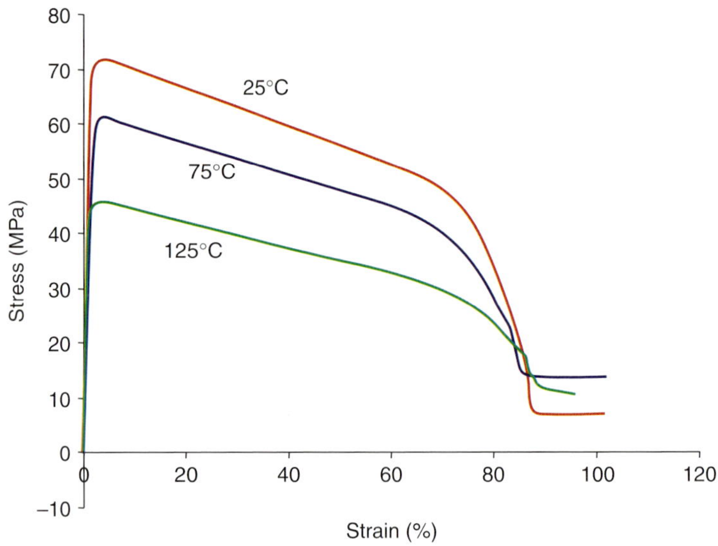

FIGURE 9.4 Temperature effect at a constant strain rate (0.056/s).

strain is equal to 0.002). The effects of test temperature and strain rate on the yield stress are given in Figs. 9.7 and 9.8, respectively. Again, it can be seen that: (1) for all the strain rates, the larger the temperature, the smaller the yield stress; and (2) for all the temperatures, the larger the strain rate, the larger the yield stress.

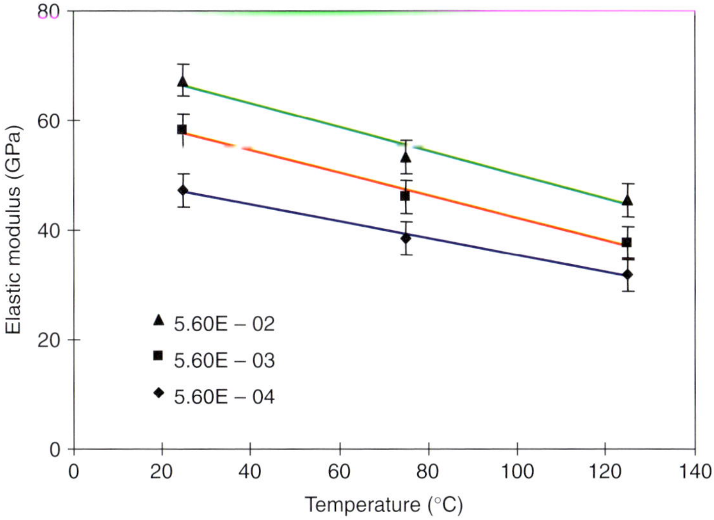

Figure 9.5 Effect of temperature on Young's modulus.

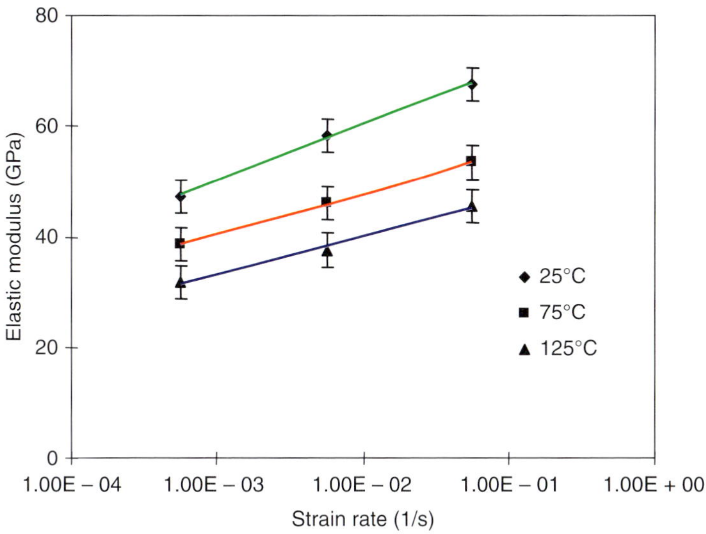

Figure 9.6 Effect of strain rate on Young's modulus.

The ultimate tensile stress (UTS) of a material is computed from the maximum stress realized in the stress-strain curve. The effects of test temperature and strain rate on the UTS are given in Figs. 9.9 and 9.10, respectively. The UTS results follow the similar trend as the elastic modulus and yield stress results. For all the temperatures UTS

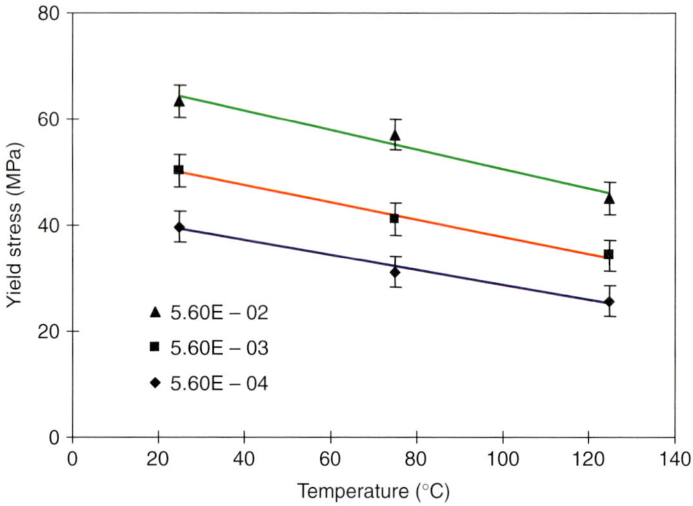

Figure 9.7 Effect of temperature on yield stress.

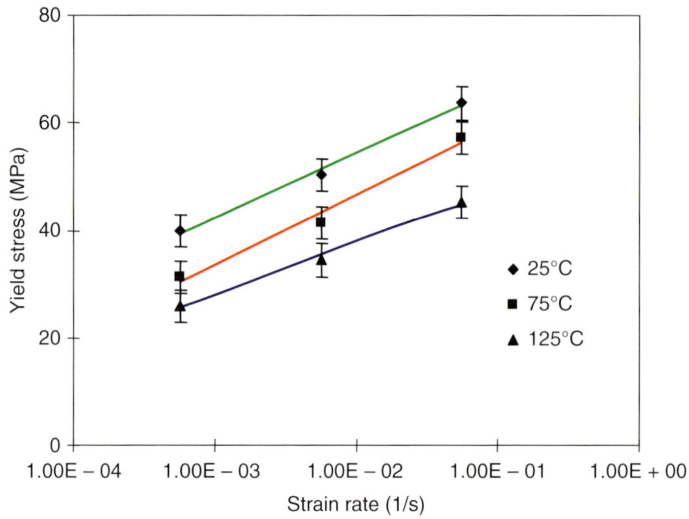

Figure 9.8 Effect of strain rate on yield stress.

increases with an increase in strain rate, and for all the strain rates, UTS decreases with an increase in temperature.

9.3.4 Equations for Young's Modulus and Yield Stress

The temperature- and strain-rate–dependent Young's (elastic) modulus, $E(T, \zeta)$, and yield stress, $\sigma_y(T, \zeta)$, of the SACC are summarized in Table 9.1. In design for reliability, usually the Young's modulus and yield stress are the required material inputs to a finite-element

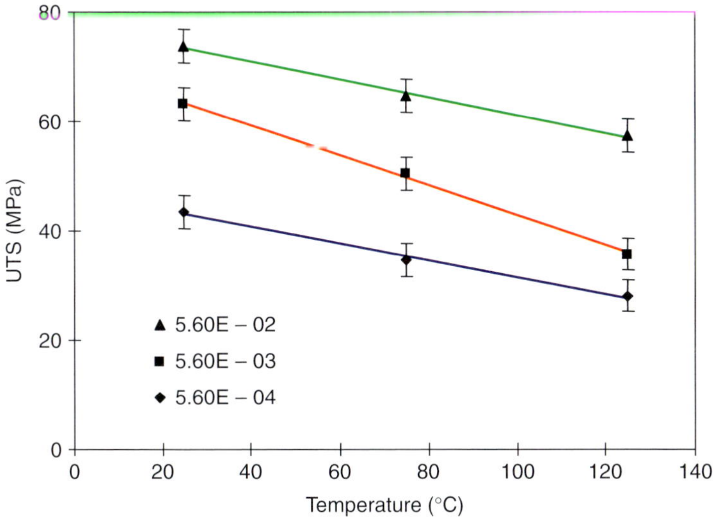

FIGURE 9.9 Effect of temperature on ultimate tensile strength.

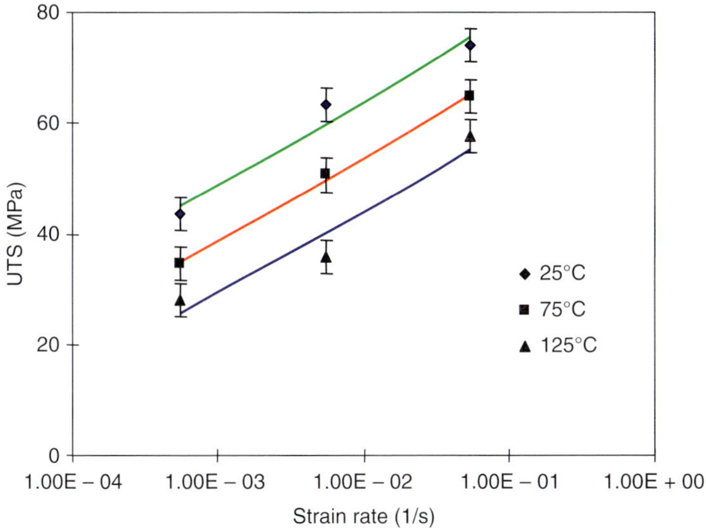

FIGURE 9.10 Effect of strain rate on ultimate tensile strength.

computer program. A statistical method with linear regression is employed to quantify the temperature (T) and strain rate (ζ)–dependent $E(T, \zeta)$ and $\sigma_y(T, \zeta)$ of the SACC solder as:

$$E(T, \zeta) = 55.1 - 0.1815T + 289.89\zeta - 0.96T\zeta$$

$$\sigma_y(T, \zeta) = 44.8 - 0.1328T + 522.34\zeta - 1.55T\zeta$$

Temperature/Strain Rate	E (GPa)	Yield (MPa)	UTS (MPa)
25°C			
5.6×10^{-4}/s	47.3	39.8	43.5
5.6×10^{-3}/s	58.3	50.3	63.3
5.6×10^{-2}/s	67.4	63.5	73.9
75°C			
5.6×10^{-4}/s	38.6	31.26	34.7
5.6×10^{-3}/s	46.0	41.29	50.6
5.6×10^{-2}/s	53.4	57.2	64.7
125°C			
5.6×10^{-4}/s	31.8	25.8	28.1
5.6×10^{-3}/s	37.6	34.3	35.7
5.6×10^{-2}/s	45.5	45.1	57.5

TABLE 9.1 Summary of Test Results (E, Yield Stress, and UTS) at Various Temperatures and Strain Rates

In these equations, the units of $E(T, \zeta)$, $\sigma_y(T, \zeta)$, T, and ζ are, respectively, GPa, MPa,°C, and 1/s. The calculated values and percentage error of the predicted Young's modulus and yield stress compared to the test results are given in Table 9.2. It can be seen that: (1) the best-fit Young's modulus equation is within 11.3% of the test results; and

Temperature/Strain Rate	E (GPa) (% error)	Yield (MPa) (% error)
25°C		
5.6×10^{-4}/s	50.7 (7.2)	41.7 (4.9)
5.6×10^{-3}/s	52.0 (–10.7)	44.2 (–12.2)
5.6×10^{-2}/s	65.5 (–2.9)	68.6 (8.0)
75°C		
5.6×10^{-4}/s	41.6 (7.8)	35.1 (12.2)
5.6×10^{-3}/s	42.7 (–7.2)	37.1 (–10.1)
5.6×10^{-2}/s	53.7 (0.6)	57.6 (0.7)
125°C		
5.6×10^{-4}/s	32.5 (2.2)	28.4 (10.0)
5.6×10^{-3}/s	33.4 (–11.3)	30.0 (–12.4)
5.6×10^{-2}/s	42.0 (–7.8)	46.6 (3.3)

TABLE 9.2 Curve-Fit Equations-Results and Percentage Errors

(2) the calculated yield stress is within 12.4% of the experimental results. Thus, the present curve-fit equations lead to satisfactory estimation of the temperature and strain-rate–dependent mechanical properties of SACC, and are in the input format of most commercially available, finite-element computer codes.

Sometimes, instead of the tensile yield stress, it is necessary to input the shear yield stress to a finite-element analysis. In this case, the equivalent shear yield stress, $\tau_y(T, \gamma)$, can be converted from the tensile yield stress, $\sigma_y(T, \zeta)$, by the von Mises relationships: $\sigma = \sqrt{3}\tau$ and $\zeta = \gamma/\sqrt{3}$. Then, we have

$$\tau_y(T, \gamma) = 25.87 - 0.077T + 174.12\gamma - 0.52T\gamma$$

In this equation, γ is the shear strain rate and the units of $\tau_y(T, \gamma)$, T, and γ are, respectively, MPa,°C, and 1/s.

9.3.5 SnAgCu (SAC) versus SnAgCuCe (SACC)

The yield stress of the SACC solder and the SAC solder reported in [20] are plotted in Figs. 9.11 and 9.12, respectively, for different strain rates and temperatures. It can be seen that for all the tested temperatures and strain rates, the yield stress of the SACC is higher than that of the SAC solder. By comparing the test data in Table 9.1 with those of [20], it can also be shown that for all the strain rates and temperatures, the Young's modulus and UTS of the SACC solder are larger than those of the SAC solder.

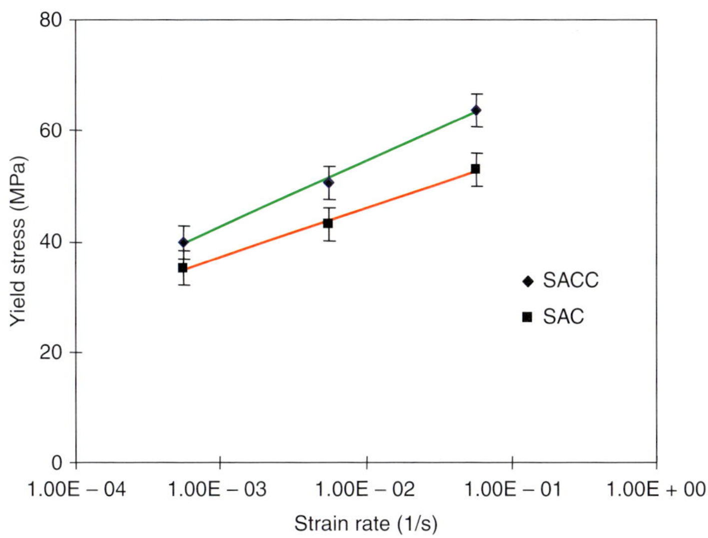

FIGURE 9.11 SAC versus SACC at different strain rates.

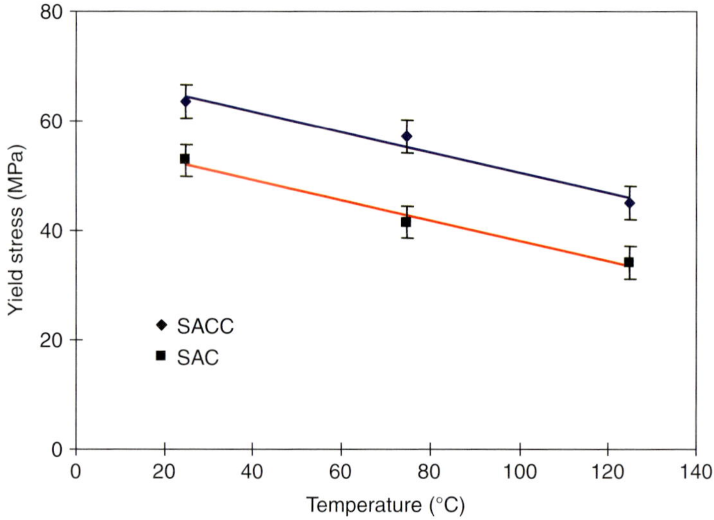

FIGURE 9.12 SAC versus SACC at different temperatures.

9.3.6 Creep Test and Data

Creep Test Specimen and Procedure

The size and preparation of the dog bone-shaped bulk solder specimens for creep testing are exactly the same as those in Sec. 9.3.2. The creep tests were carried out on a universal testing machine. A thermal chamber was used for the creep test over a wide range of test temperatures. In the creep test, the specimen was deformed in tension at constant load. Tests were carried out at three different temperatures (25°C, 75°C, and 125°C), with constant stress set at different values. After the specimen was griped and the chamber had reached thermal equilibrium, the specimen was placed into the chamber for a minimum of half an hour prior to testing. The specimen temperature was monitored by a thermocouple, which is used to regulate the testing temperature in the chamber.

Creep Test Results

Creep is the time-dependent plastic deformation of a material at a high homologous temperature of the solder material in this study. The effect of increasing the stress and temperature on creep behavior is shown in Fig. 9.13. The creep strain curves at three different stresses (7, 10, and 20 MPa) at 125°C show increasing creep strain at higher applied nominal stresses.

The creep strain rates of SACC are plotted against the applied normal stresses, as shown in Fig. 9.14. The figure shows the creep data of SACC corresponding to different stress levels (5 MPa, 7 MPa, 10 MPa, 20 MPa) applied on the dog bone–shaped specimens at three

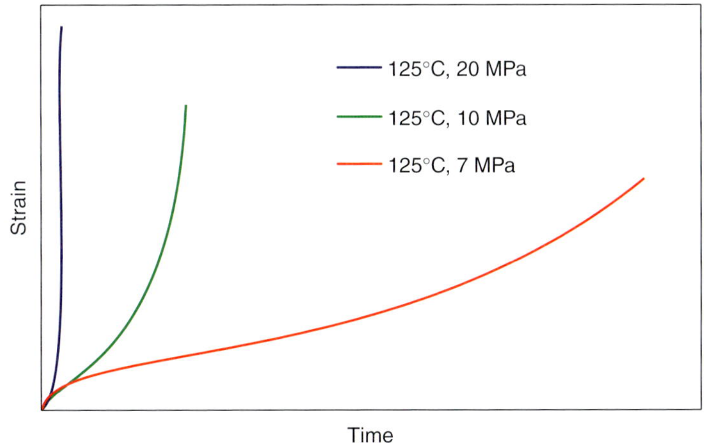

FIGURE 9.13 Creep behavior of SACC solder at 125°C.

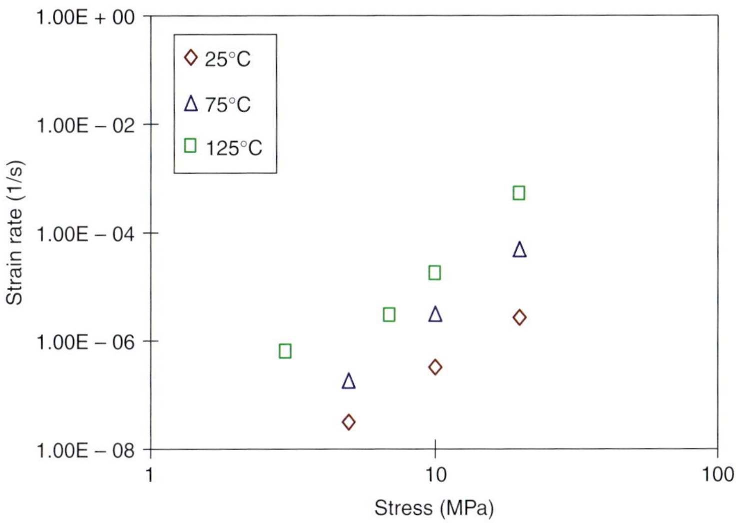

FIGURE 9.14 Steady-state creep strain rate of SACC solder.

test temperature (25°C, 75°C, and 125°C). The numerical results of the creep strain rates at different stresses and temperatures are given in Table 9.3.

From the creep test results, it can be seen that the creep strain rate of SACC solder increases with increases in applied normal stress and test temperature. Every increase of 50°C in temperature leads to an increase of one order of magnitude in creep strain rate. Compared to creep data of SAC387 reported by Pang et al. [17], SACC shows different

Stress (MPa)	Steady State Creep Rate (1/s)
25°C	
5	3.03E–08
10	3.28E–07
20	2.54E–06
75°C	
5	1.88E–07
10	3.04E–06
20	5.01E–05
125°C	
3	6.03E–07
7	2.88E–06
10	1.67E–05
20	5.21E–04

Table 9.3 Steady-State Creep Data for SACC Solder

creep strain rate behavior at lower temperatures at 25°C, but fairly good agreement with SAC387 creep data shown in Fig. 9.15. SACC solder has better creep resistance than SAC387 at high stress levels above 10 MPa. SAC387 has better creep resistance at low stress levels below 5 MPa.

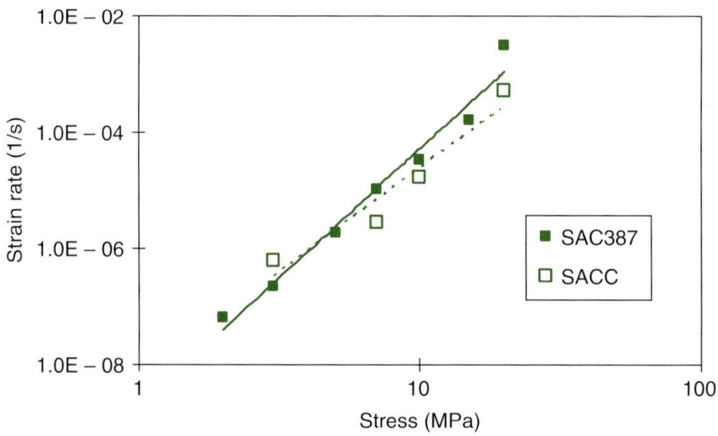

Figure 9.15 Comparison of SACC with SAC387 at 125°C.

Creep Constitutive Model

The creep law can be characterized by Eq. (9.1), where the creep strain rate, $\dot{\varepsilon}$, can be simply expressed as a power law relation with stress and an exponential term with activation energy and temperature:

$$\dot{\varepsilon} = A\sigma^n e^{\frac{Q}{kT}} \tag{9.1}$$

where A is a material constant, σ is applied stress, n is the stress exponent, Q is creep activation energy, k is Boltzmann's constant, and T is absolute temperature. The activation energy, Q, and the stress exponent, n, change with the dominant creep mechanism, and may have different values at different regimes of the applied normal stress magnitude.

Curve fitting of the SACC creep test data to Eq. (9.1) is shown in Fig. 9.16. The curves fit the experimental data very well. The creep model for SAC is given in Eq. (9.2) by the regression curve-fitting method. The creep model has a constant, A, of 2.2×10^{-3} with stress exponent, n, equal to 3.51 and activation energy $-Q/k$ equal to -5082.25.

$$\dot{\varepsilon} = 2.2 \times 10^{-3} \cdot \sigma^{3.51} \exp\left(-\frac{5082.25}{T}\right) \tag{9.2}$$

Alternatively, the creep strain rate at different regimes of stress and temperature can be expressed as a hyperbolic-sine model, given in Eq. (9.3):

$$\dot{\varepsilon} = C[\sinh(\alpha\sigma)]^n \exp\left(-\frac{Q}{kT}\right) \tag{9.3}$$

where C is a material constant, σ is applied stress, n is the stress exponent, Q is creep activation energy, k is Boltzmann's constant, and

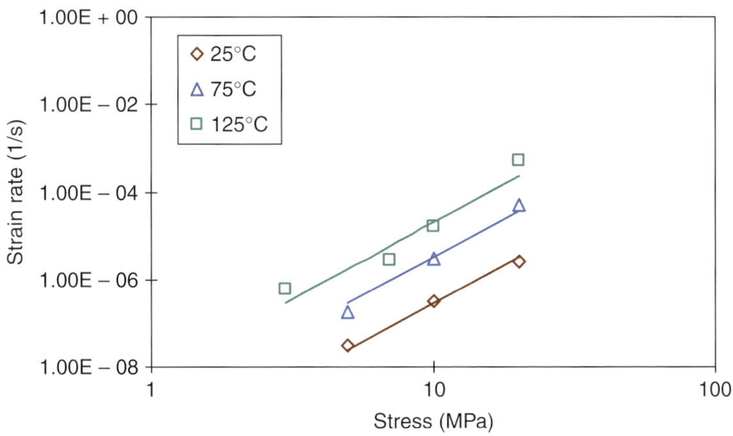

FIGURE 9.16 Power law creep model for SACC.

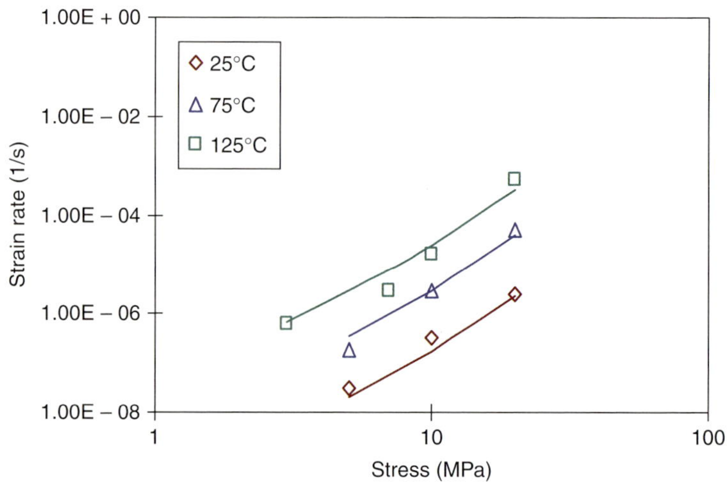

FIGURE 9.17 Hyperbolic-sine creep model for SACC.

T is absolute temperature. The stress- and temperature-dependent creep strain rate data for SACC were curve-fitted to the model, as in Eq. (9.3) and shown in Fig. 9.17, at temperatures of 25°C, 75°C, and 125°C. Compared to Fig. 9.16, the hyperbolic-sine creep model fits the creep data better than the power-law creep model over the creep test data region.

The hyperbolic-sine creep model of SACC is as follows:

$$\dot{\varepsilon} = 120[\sinh(0.072\sigma)]^{2.8} \exp\left(-\frac{5856.02}{T}\right) \tag{9.4}$$

where the constant C is 120, the stress constant α is 0.072, the stress exponent n is 2.8, and the activation energy $-Q/k$ is -5856.02.

Pang et al. [18, 19] reported creep test data for SAC387 for tests carried out at similar test temperatures (25°C, 75°C, and 125°C) but wider stress levels (2, 3, 5, 7, 10, 15, 20, 25, 30, 35, and 40 MPa). The creep model for SAC397 reported by Pang et al. is given in Eq. (9.5).

$$\dot{\varepsilon} = 501.3[\sinh(0.0316\sigma)]^{4.96} \exp\left(-\frac{5433.5}{T}\right) \tag{9.5}$$

Comparing Eqs. (9.4) and (9.5), both the creep models for SACC and SAC387 have approximately comparable activation energy. The stress exponent for SACC is lower compared to SAC387. The curve-fitted parameters for SACC derived from curve fitting based on power-law and hyperbolic-sine models are summarized in Table 9.4.

Power Law Model			
A	n		$-Q/k$
2.2×10^{-3}	3.51		−5082.25
Hyperbolic-sine Model			
C	α	n	Q/k
120	0.072	2.8	−5856.02

TABLE 9.4 Parameters for Different Creep Models for SACC

Iso-Strain Rate Tensile Test

The effects of strain rate and temperature on the SACC tensile mechanical properties were reported by Lau et al. [20] earlier. Figure 9.3 shows the stress-strain curves for the three different strain rates at a constant temperature of 25°C. It can be seen that the mechanical behavior of SACC is strongly dependent on the strain rate.

The yield stress of a material is computed from the stress-strain curve at 0.2% plastic strain. The yield stress of SACC at 25°C, 75°C, and 125°C may be plotted against the iso-strain rate, respectively, as shown in Fig. 9.18. At the instant when the material starts to yield during the iso-strain rate tensile test, it is subjected to a short state of creep strain. The yield stress derived from the tensile test can be treated as the equivalent creep stress, and the iso-strain rate is equivalent to the creep strain rate. The iso-strain rate tensile test data are plotted together with the current creep test data for SACC solder, as shown in Fig. 9.19. The hyperbolic-sine creep model for Eq. (9.4) fits the combined creep test data and iso-strain rate tensile test data.

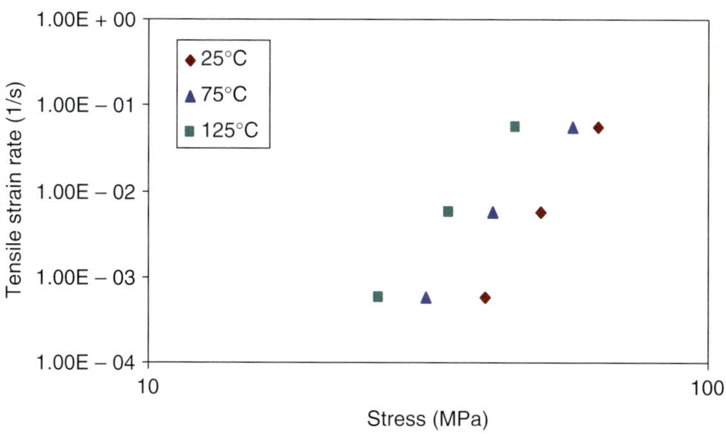

FIGURE 9.18 Iso-strain rate tensile test data of SACC.

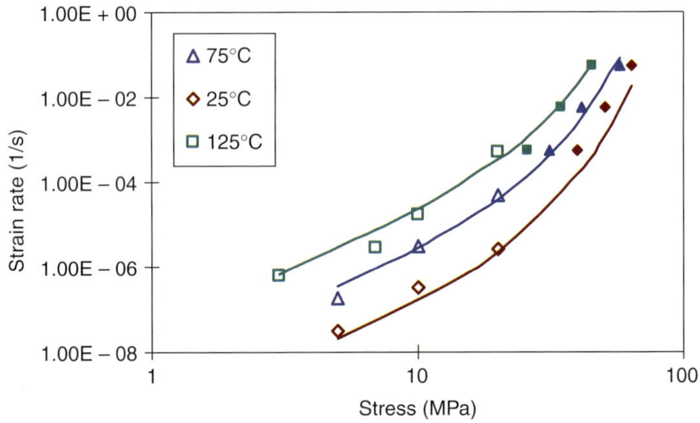

FIGURE 9.19 Hyperbolic-sine creep model for SACC (iso-strain rate tensile test data included).

Iso-Strain Rate Nano-Indentation Test

The indentation strain rate is defined as the indentation tip moving-in rate divided by the depth into the surface [21, 22, and 23]:

$$\varepsilon_h = \dot{h}/h \tag{9.6}$$

In a conventional constant load or displacement control indentation test, the indentation strain rate is not constant. Using a similar indentation creep test method with a Berkovich tip reported by Xu et al. [4], the indentation strain rate and hardness can be kept constant. The indentation hardness of a material is given by:

$$H = \frac{P}{A} = \frac{P}{ch^2} \tag{9.7}$$

where H is the hardness, P is the load, A is the projected contact area, h is the displacement (contact depth), and c is a constant that depends upon the geometry of the indenter tip (for a perfect Berkovich tip, c = 24.56). The relationship between yield stress and hardness is given by:

$$H = 3\sigma_y \tag{9.8}$$

Hence, the strain-rate–dependent yield stress can be determined from the hardness measurements from nano-indentation tests at iso-strain rate conditions. The summary of hardness and yield stress of SACC is shown in Table 9.5.

The nano-indentation tests were carried out at a constant strain rate; the test data can be collaborated with creep data for creep behavior analysis. During the indentation test, when the tip indents to a certain

Indentation Strain Rate (1/s)	0.01	0.1	1
Hardness (GPa)	0.193	0.279	0.453
Yield Stress (MPa)	64.33	93	151

TABLE 9.5 Hardness and Yield Stress of SACC

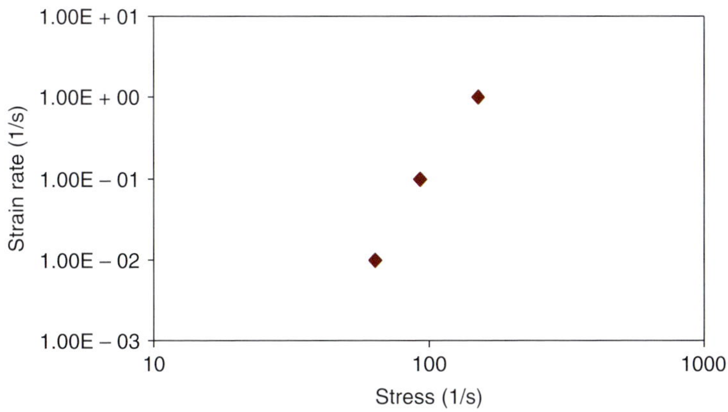

FIGURE 9.20 Iso-strain rate nano-indentation data at 25°C.

depth of the material, hardness can be obtained and the yield stress can be calculated. The calculated yield stress and the test strain rate of SACC at 25°C can be plotted as equivalent creep strain rate versus applied stress, as shown in Fig. 9.20.

A hyperbolic-sine model, as shown in Eq. (9.9) fits the combined creep test data and iso-strain rate nano-indentation test data at room temperature of 25°C. The constant C is 6.58×10^4, the stress coefficient α is 0.019, the stress exponent is 4.2, and the activation energy $-Q/k$ is -5856.02. Figure 9.21 shows that Eq. (9.9) fits well with both the creep test and nano-indentation test data at 25°C.

$$\dot{\varepsilon} = 6.58 \times 10^4 [\sinh(0.019\sigma)]^{4.2} \exp\left(-\frac{5856.02}{T}\right) \tag{9.9}$$

9.3.7 Summary and Recommendations

Some important results and recommendations are summarized in the following:

(1) The Young's modulus, yield stress, and ultimate strength of the SACC solder are strongly temperature and strain rate dependent.

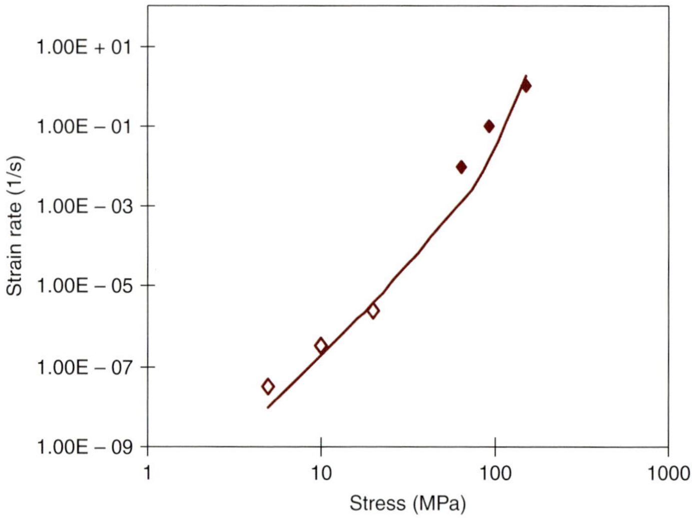

FIGURE 9.21 Hyperbolic-sine creep model for SACC at 25°C (iso-strain rate nano-indentation test data included).

(2) For all the temperatures considered, the higher the strain rate, the higher the Young's modulus, yield stress, and ultimate strength of the SACC solder.

(3) For all the strain rates considered, the higher the temperature, the lower the Young's modulus, yield stress, and ultimate strength of the SACC solder.

(4) A set of temperature- and strain-rate–dependent equations for the Young's modulus and yield stress with satisfactory accuracy has been provided for the SACC solder, which can be directly input to most of the commercially available finite-element computer codes.

(5) For all the temperatures and strain rates considered, the Young's modulus, yield stress, and ultimate strength of the SACC solder are larger than those of the SAC solder.

(6) The creep constitutive equation of SACC is strongly temperature and time dependent.

(7) The creep data for the SACC solder is comparable to the SAC387 solder at 125°C with a crossover behavior seen at a stress of ~10 MPa.

(8) Creep data can be extracted from iso-strain rate tensile tests and iso-strain rate nano-indentation tests and combined with creep test data to provide a wider range of applied stress and strain rates. The hyperbolic-sine creep model is shown to be suitable for fitting these combined creep data.

9.4 Intermetallic Compounds of SACC

The objectives of this section are to determine the intermetallic compounds of the SACC on various PCB surface finishes under aging conditions [20].

9.4.1 Sample Preparation

Two different surface finishes, namely, OSP and NiAu on FR-4 copper test coupons, are used for the intermetallics study. The copper thickness is 17 μm and the Au is 3 to 8 μin and Ni is 100 to 200 μin. Two different solder pastes are used. One is the Sn3wt% Ag0.5wt%Cu0.019wt%Ce alloy (metal/flux ratio = 89.5/10.5) and the other is a standard SAC305 solder paste.

For each surface finish, the solder pastes are printed onto three sets of coupons. The stencil thickness is 127 μm, and the opening is a circle with a diameter of 2.16 mm. The pastes are reflowed under air with a temperature profile shown in Fig. 9.22.

9.4.2 Tests and Measurements

The first set of samples is cross-sectioned as unaged fresh samples. The second set of samples is aged at 150°C for 10 days and then cross-sectioned. The third set of samples is aged at 150°C for 20 days and then cross-sectioned.

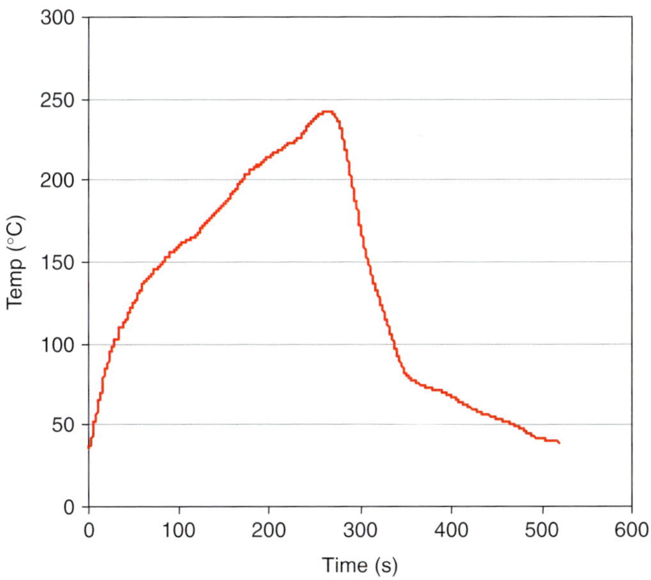

FIGURE 9.22 Reflow profile used for solder pastes IMC study.

All cross-sectioned samples are examined with scanning electron microscopy (SEM) and energy dispersive x-ray spectrometry (EDX). The composition and thickness of intermetallic layers at the interface of solder and base metal are determined. The IMC thickness is determined by taking the average of eight thickness readings evenly spaced on the SEM pictures.

9.4.3 IMC on OSP Surface Finish

The formation of IMC at the solder- interface on the copper substrate with the OSP surface finish as a function of aging time at 150°C is shown in Fig. 9.23 for both SAC305 solder paste and SACC solder paste. There

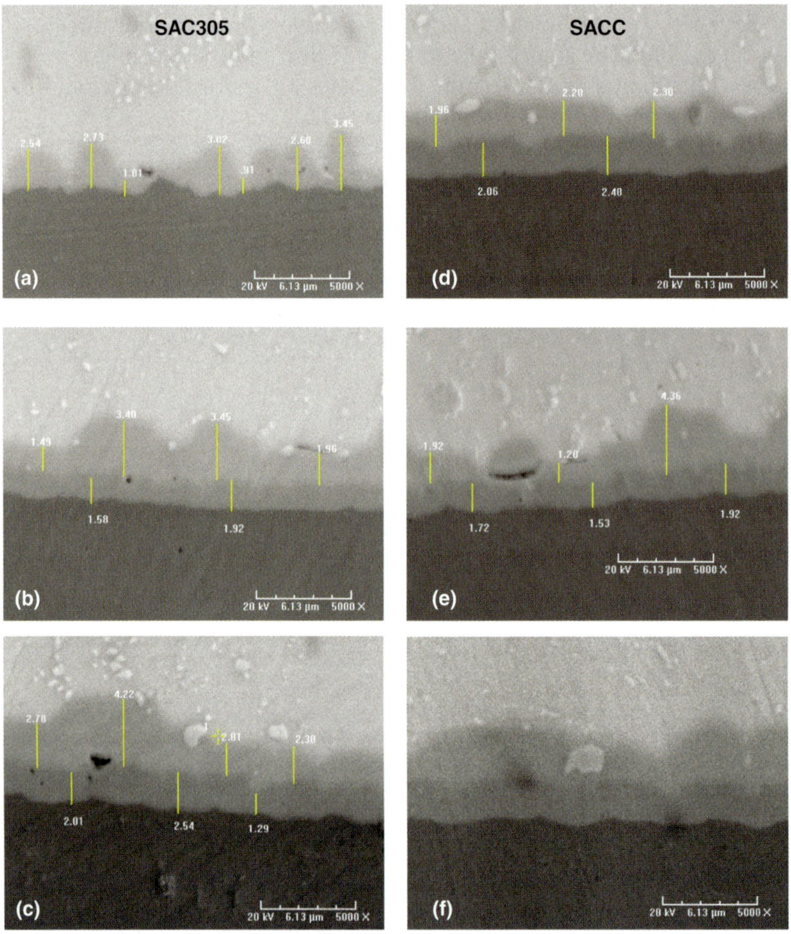

FIGURE 9.23 SEM images of solder joints on copper substrate with OSP: (a) SAC305 as reflowed, (b) SAC305 after aging at 150°C for 10 days, (c) SAC305 after aging at 150°C for 20 days, (d) SACC as reflowed, (e) SACC after aging at 150°C for 10 days, (f) SACC after aging at 150°C for 20 days.

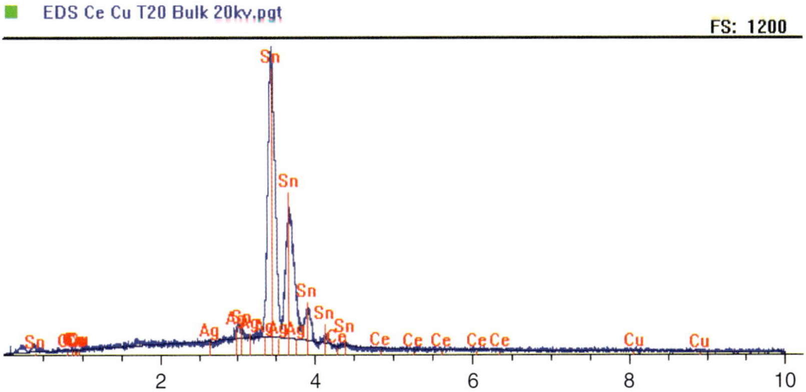

FIGURE 9.24 EDX spectra for SACC bulk solder on Cu substrate with OSP (after 150°C for 20 days).

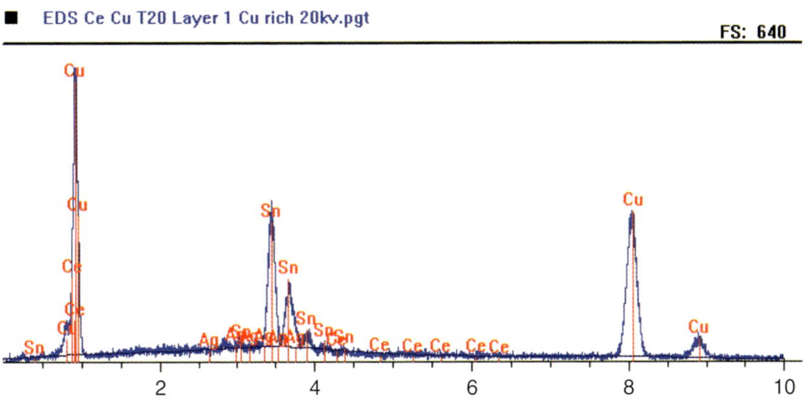

FIGURE 9.25 EDX spectra for Cu-rich IMC layer with SACC on Cu substrate with OSP (after 150°C for 20 days).

are two layers of IMC: one is an Sn-rich (Cu_6Sn_5) layer on the solder side, and the other is a Cu-rich (Cu_3Sn) layer on the copper side. Some of the typical EDX spectra for the SACC on Cu substrate with OSP surface finish after 150°C for 20 days are shown in Figs. 9.24 (in bulk solder), 9.25 (in the Cu-rich IMC layer), and 9.26 (in the Sn-rich IMC layer). Examples of the elemental compositions of each IMC layer for samples aged at 150°C for 20 days are shown in Table 9.6. No Ce can be discerned within either IMC layer.

The growth rate of IMC layers for solders aged at 150°C is shown in Fig. 9.27 for the Sn-rich layer and in Fig. 9.28 for the Cu-rich layer. The thickness of the Sn-rich layer is virtually unchanged through the thermal aging process, although a slight increase may be discerned for the SAC305 system. However, for the Cu-rich IMC layer, the thickness

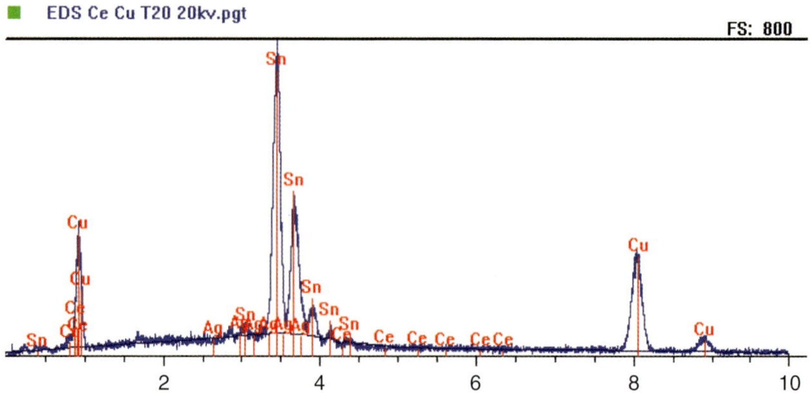

FIGURE 9.26 EDX spectra for Sn-rich IMC layer with SACC on Cu substrate with OSP (after 150°C for 20 days).

	SACC		SAC305	
Element	**Sn-rich**	**Cu-rich**	**Sn-rich**	**Cu-rich**
Sn	54.8	28.10	72.62	31.25
Ag	0.00	0.00	0.06	0.00
Cu	45.2	71.90	27.33	68.75
Ce	0.00	0.00	0.00	0.00

TABLE 9.6 Examples of Elemental Composition (Atomic %) of Intermetallic Layers formed on Copper Substrate with OSP for Samples Aged at 150°C for 20 Days

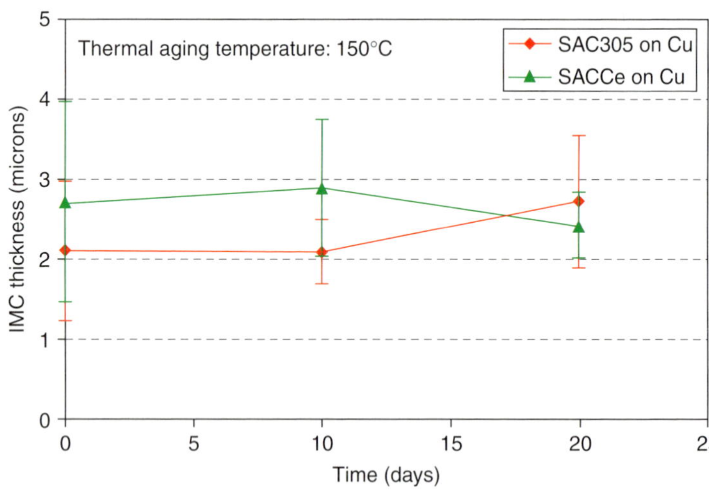

FIGURE 9.27 Growth rate of Sn-rich IMC layer for joints of SAC305 and SACC alloys on Cu substrate with OSP aged at 150°C.

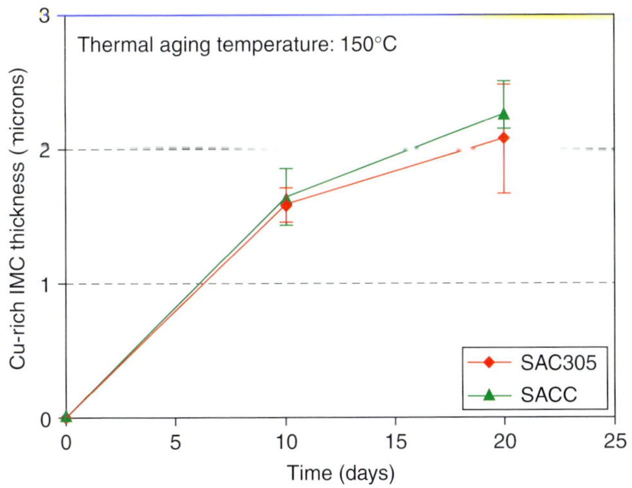

FIGURE 9.28 Growth rate of Cu-rich IMC layer for joints of SAC305 and SACC alloys on Cu substrate with OSP aged at 150°C.

increases as the aging time increases, reflecting the Cu consumption rate. The effect of Ce on the IMC growth rate on Cu is not significant.

9.4.4 IMC on NiAu Surface Finish

The formation of IMC at the solder-interface on the Cu substrate with the NiAu surface finish as a function of aging time at 150°C is shown in Fig. 9.29 for both the SAC305 solder paste and the SACC solder paste. Some of the typical EDX spectra for the SACC on the Cu substrate with the NiAu surface finish after 150°C for 10 days are shown in Figs. 9.30 (in IMC layer—Site 1) and 9.31 (in IMC layer—Site 2). The elemental compositions of the IMC layer for samples aged at 150°C for 10 days are shown in Table 9.7. Again, no Ce can be discerned within the IMC layer.

The growth rate of the IMC layer for joints aged at 150°C is shown in Fig. 9.32. The thickness of the IMC layer is virtually unchanged through the thermal aging process. The effect of Ce on the IMC growth rate on NiAu is also not significant.

9.4.5 SnAgCu (SAC) versus SnAgCuCe (SACC)

Based on the previous data, it can be seen that there is no significant difference in the IMC thickness and growth between the SAC and SACC solders on either the OSP or the NiAu surface finishes. Also, based on the EDX data, Ce is not present in the IMC layer between the SACC solder and the Cu-OSP or Cu-NiAu surface finishes. Furthermore, no significant effect on the IMC composition can be

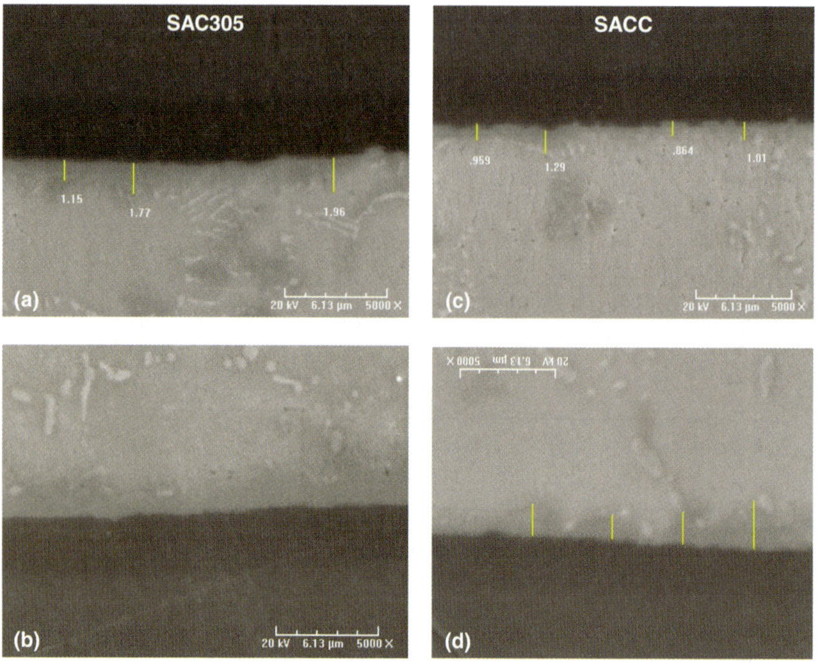

FIGURE 9.29 SEM images of cross-sectioned solder joints on Cu substrate with NiAu surface finish: (a) SAC305 as reflowed, (b) SAC305 after aging at 150°C for 10 days, (c) SACC as reflowed, (d) SACC after aging at 150°C for 10 days.

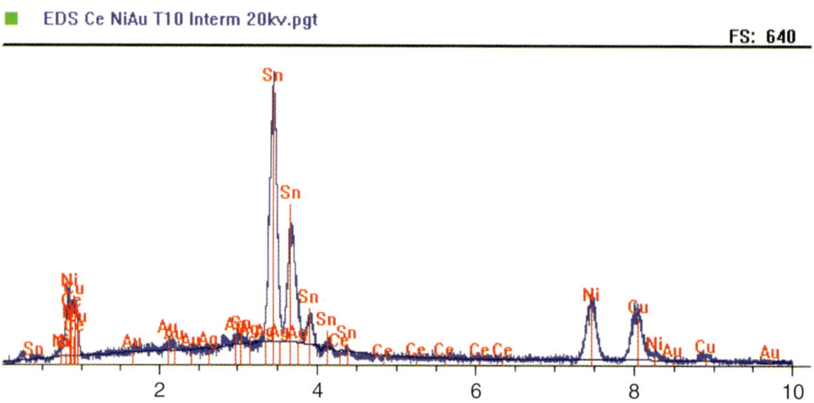

FIGURE 9.30 EDX spectra for IMC layer with SACC on Cu substrate with NiAu—Site 1 (after 150°C for 10 days).

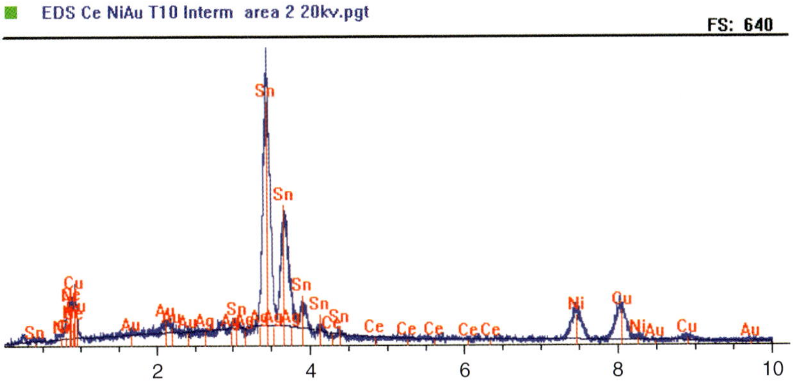

FIGURE 9.31 EDX spectra for IMC layer with SACC on Cu substrate with NiAu—Site 2 (after 150°C for 10 days).

Element	SACC		SAC305	
	Site 1	Site 2	Site 1	Site 2
Sn	55.50	65.66	66.87	64.12
Ag	0.50	0.00	1.13	0.00
Cu	22.02	18.90	4.96	20.63
Ce	0.00	0.00	0.00	0.00
Ni	21.19	13.91	26.36	14.39
Au	0.79	1.52	0.68	0.87

TABLE 9.7 Examples of Elemental Composition (Atomic %) of Intermetallic Layer formed on Cu Substrate with NiAu for Samples Aged at 150°C for 10 Days

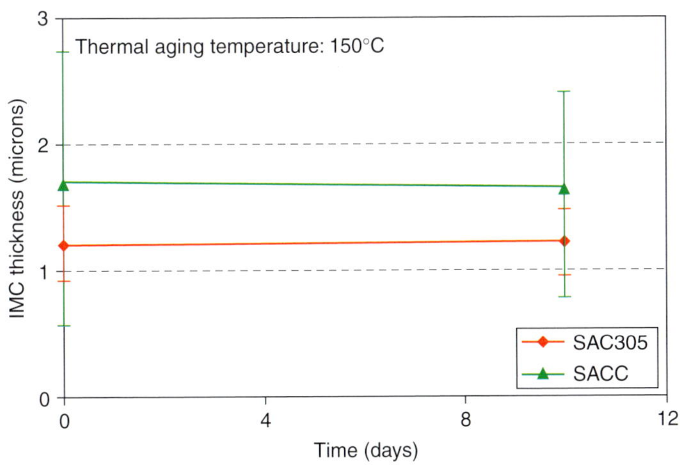

FIGURE 9.32 Growth rate of IMC layer for joints of SAC305 and SACC alloys on Cu substrate with NiAu aged at 150°C.

observed for both surface finishes. This implies that any possible effect of Ce on the solder-joint reliability would be related to bulk solder properties.

9.4.6 Summary and Recommendations

Some important results and recommendations are summarized in the following:

(1) The Sn-rich IMC thickness on the Cu substrate with OSP is basically constant (~2.5 μm) for both the SACC and SAC solders, even aged at 150°C for 20 days.

(2) The Cu-rich IMC thickness on the Cu substrate with OSP is increasing (from 0 to 2 μm) with increasing aging time (150°C for 20 days), which reflects the Cu consumption rate. Again, there is no significant difference between the SACC and SAC solders.

(3) The IMC thickness on the Cu substrate with NiAu is virtually unchanged (~1.5 μm) through the thermal aging process for both the SACC and SAC solders.

(4) For the OSP and NiAu surface finishes considered, the Ce is not present in the IMC layers.

(5) Since there is no significant effect on the IMC, this implies that any possible effect of Ce on the solder-joint reliability would be related to the bulk SACC material properties.

9.5 Process Development of SACC with Various Packages

The objectives of this section are: (1) to develop an assembly process of the SACC for various lead-free components (PQFP, PBGA, WLCSP, and BCC); and (2) to determine the four-point bending fatigue of SACC solder joints with lead-free (PQFP, PBGA, WLCSP, and BCC) packages on a lead-free PCB [24].

9.5.1 Components and PCBs

The components are shown in Table 9.8. They are the 208-pin PQFP208, 196-ball PBGA, 8-bump CSP8, and 24-pin BCC24. They are lead-free components, that is, the PQFPs are with matte-Sn plating; the PBGA and CSP8 are, respectively, with SAC balls and bumps; and the BCCs are with NiAu. Also, the components are internally connected in such a way that when they are soldered on the test board there will be a daisy chain for continuity measurements during reliability tests.

The two-layer test board is shown in Fig. 9.33 and Table 9.8. The location of the PQFPs, PBGA, CSP, and BCC is marked, respectively, with U3, U4, U1, U43, and U7. The package size, pad size, and pitch are, respectively, 30.6 × 30.6 mm, 0.254 × 1.65 mm, and 0.5 mm for PQFPs;

	Type	Package Size (mm)	Pitch (mm)	Location on PCB	Pad Size (mm) on PCB
PQFP 208	Matte Sn Leads	30.6 × 30.6	0.5	U3 & U4	0.254 × 1.65
BCC24	LCC (NiAu)	4 × 4 × 0.8	0.5	U7	0.4 × 0.4
PBGA 196	SAC (0.51 mm)	15 × 15	1.0	U1	0.7
WLCSP8	SAC (0.17 mm)	1.29 × 1.29 × 0.85	0.5	U43	0.15

TABLE 9.8 Lead-Free Components and PCB Data

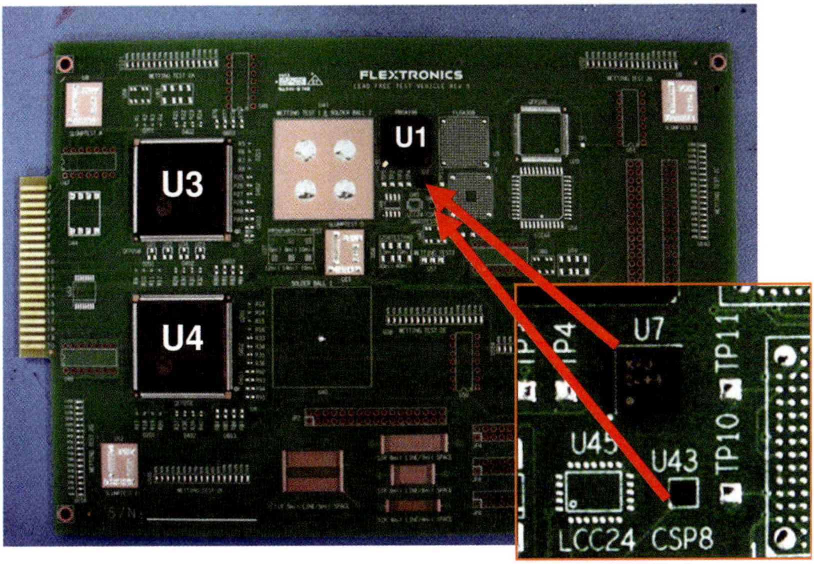

FIGURE 9.33 Test board with PQFP (U3 and U4), PBGA (U1), WLCSP (U43), and BCC (U7).

15 × 15 mm, 0.7 mm, and 1 mm for PBGA; 1.29 × 1.29 × .85 mm, 0.15 mm, and 0.5 mm for CSP; and 4 × 4 × 0.8 mm, 0.4 × 0.4 mm, and 0.5 mm for BCC. The solder ball diameter is 0.51 mm for the PBGA package, and the solder bump diameter is 0.17 mm for the wafer-level CSP8.

9.5.2 Printing Characteristics

All of the pastes are kept refrigerated upon arrival and are allowed to warm completely and naturally to ambient temperature for a minimum of 12 h prior to use. Solder paste mixing is performed prior to

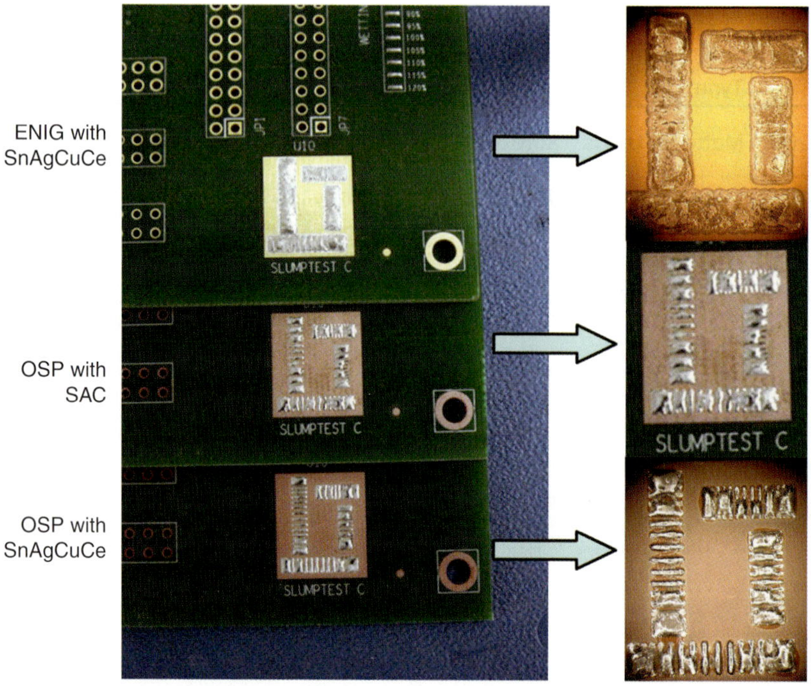

FIGURE 9.34 Spread tests.

application onto a stencil. Wetting (spread test) is performed on ENIG (electroless Ni and immersion Au) and OSP test coupons with the SACC and SAC solder pastes. At reflow, the total bridges formed are used to represent the wetting capability of the solder pastes. The more the bridges formed, the better the wetting. Figure 9.34 shows that: (1) there is no significant wetting difference on OSP between the SACC and SAC solder pastes; and (2) with SACC solder paste, the ENIG wets better than OSP.

The important parameters for stencil printing the SAC and SACC solder pastes on the ImAg test board are shown in Table 9.9, and some of the typical printing results for the CSP, PBGA, and PQFP packages are shown in Figs. 9.35 and 9.36, respectively, for the SAC and SACC solder pastes. It can be seen that the front and rear pressures, stencil thickness, and squeegee are the same for both solder pastes and are equal to, respectively, 12 kg, 5 mils, and metal with a 45-degree angle. However, due to the stickiness of the SACC solder paste, the separation speed (12 mm/s) is faster (to allow the solder paste to release better from the stencil) than that (10 mm/s) of the SAC. Also, the front and rear speeds of the SACC (25 mm/s) are slower than that (40 mm/s) of the SAC. Furthermore, it can be seen that with the printing parameters of these two lead-free solder pastes given in Table 9.9,

	SACC	SAC
Front & Rear Speed	25 mm/s	40 mm/s
Front & Rear Pressure	12 kg	12 kg
Separation Speed	12 mm/s	10 mm/s
Stencil Thickness	5 mils	5 mils
U43-WLCSP8	12 mils square aperture	12 mils square aperture
U1-PBGA196	27 mils round aperture	27 mils round aperture
Squeegee	Metal (45° angle)	Metal (45° angle)

Table 9.9 Stencil Printing Parameters for SAC & SACC

PQFP

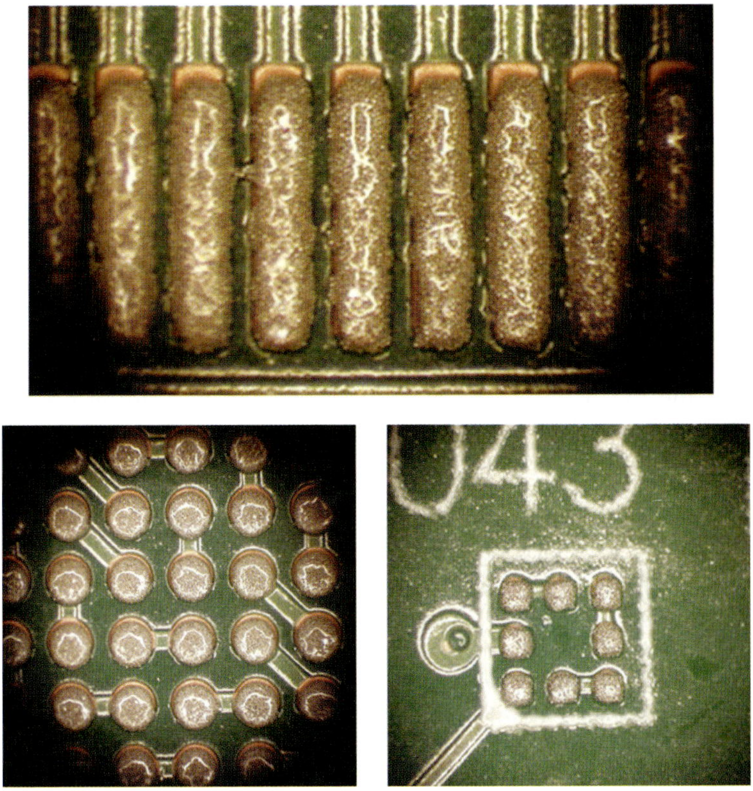

PBGA WLCSP

Figure 9.35 Solder paste printing results with SACC.

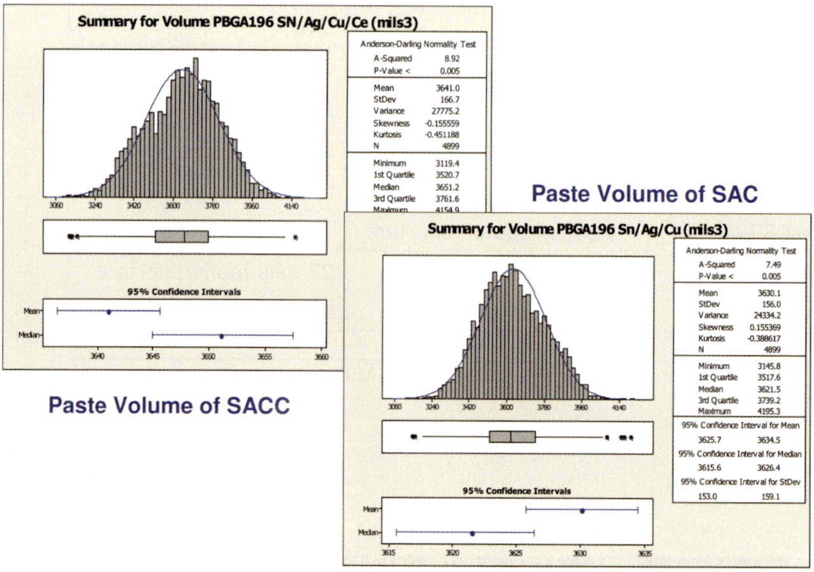

FIGURE 9.36 Paste volume measurements of SAC and SACC for the PBGA.

they yield almost the same amount of solder paste volume as shown in Fig. 9.36 for the PBGA package.

9.5.3 Reflow Temperatures

The same reflow temperature profiles are used to reflow the SAC and SACC solder pastes. One typical set of the reflow temperatures is shown in Fig. 9.37. The thermocouples are located at one corner of U3 (PQFP), another corner of U3 (PQFP), the center of U1 (PBGA), and the top of the PCB. It can be seen that, depending on the thermal mass of the components, the peak temperature of one corner of U3 is 245°C, of another corner of U3 is 244.4°C, and of the PBGA is 250.6°C. Also, the time above 217°C is 45 s for one corner of U3, 43 s for another corner of U3, and 49 s for the PBGA.

9.5.4 Voids

All the assemblies are inspected per IPC-610-D, and there is no significant difference between the SAC and SACC solder joints. However, x-ray inspections indicate that there are more voids in the SACC solder joints for all the packages considered—even the volume of voids in SACC solder joints is very small (5–8%) and acceptable (e.g., Fig. 9.38 shows the PBGA voids with SACC). It should be pointed out that the amount of voids can be significantly influenced by the solder paste formulation and the reflow profile. The effects of voids on solder joint reliability are shown in Chap. 16 of this book.

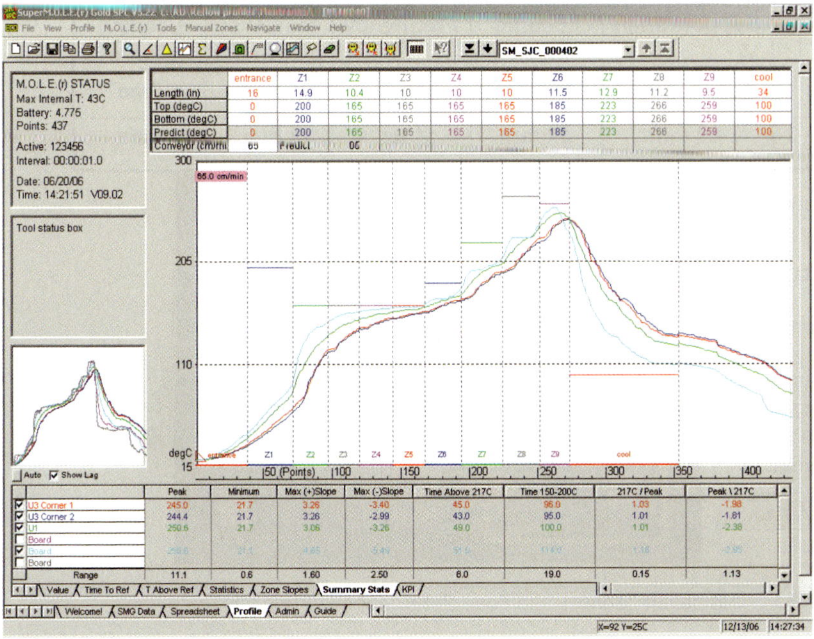

FIGURE 9.37 Reflow temperature profiles for both SAC and SACC.

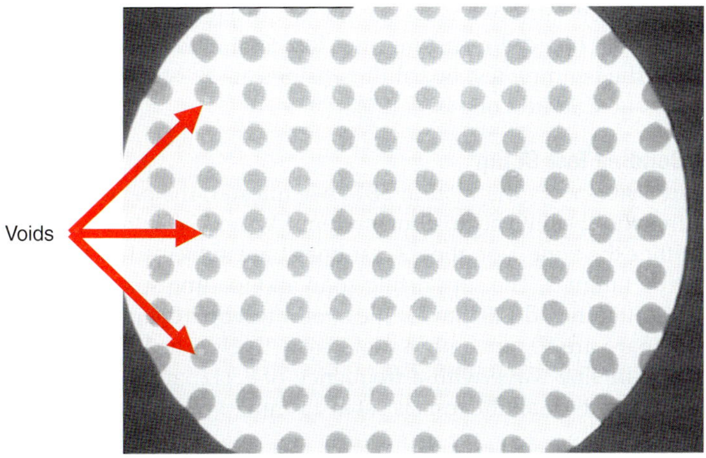

FIGURE 9.38 Voids in a PBGA solder joints with SACC.

9.5.5 Four-Point Bending Fatigue Tests

(A) Sample Preparation

The samples are divided into five sets, one at room temperature, the second set is aged at 150°C for one hour, the third set is aged at 150°C

	Support Span	Loading Span	Frequency	Max. Displacement	Loading Curve
Zone 1	130 mm	110 mm	1 Hz	1 mm	Sine
Zone 2	80 mm	54 mm	1 Hz	1 mm	Sine

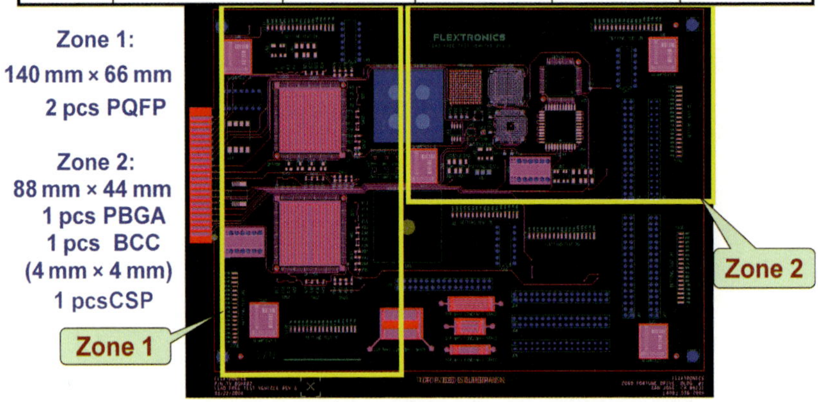

FIGURE 9.39 Test panels (Zone 1 and Zone 2) for four-point bending tests.

for 10 days, the fourth set is aged at 150°C for 20 days, and the fifth set is aged at 150°C for 30 days. The assembled board shown in Fig. 9.33 is cut into two zones, as shown in Fig. 9.39. In zone 1, the bending test board dimensions are 140 × 66 mm with two PQFPs. In zone 2, the bending test board dimensions are 88 × 44 mm with one PBGA, one CSP, and one BCC [24].

(B) Bending Test Setup

Figure 9.40 shows the Instron 5848 Microtester, the bending fatigue fixture, and the data acquisition system (Agilent 34970). The samples on either the zone 1 or zone 2 test board are subjected to four-point bending fatigues. The supporting span and loading span are, respectively, 130 mm and 110 mm for the zone 1 test board, and for the zone 2 test board are 80 mm and 54 mm, respectively. The bending fatigue curve is in sine wave with a frequency of 1 Hz, and the maximum displacement for all the tests is set at ±1 mm.

(C) Bending Fatigue Test Results

Figure 9.41 shows the four-point bending fatigue test results of the PQFP with SAC and SACC solders after 10-day, 20-day, and 30-day aging at 150°C. It can be seen that: (1) for both solders, the effect of aging is to reduce their number of cycles-to-failure—the longer the aging time, the smaller the number of cycles-to-failure; and (2) for all the aging conditions, the number of cycles-to-failure of the SAC solder joints is larger than that of the SACC solder joints. This could be due

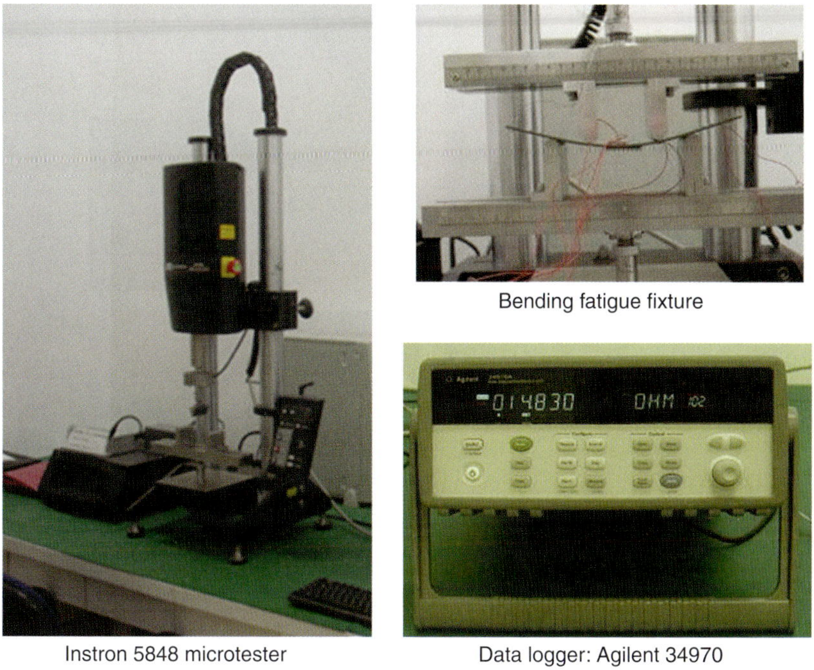

FIGURE 9.40 Four-point bending fatigue test setup.

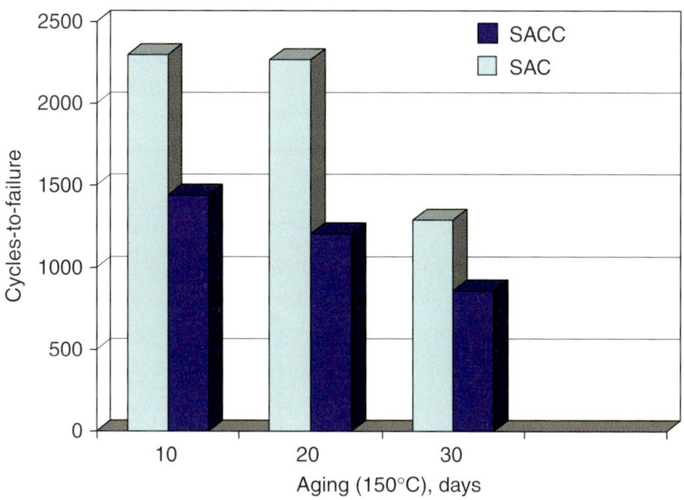

FIGURE 9.41 Four-point bending fatigue test results of PQFP.

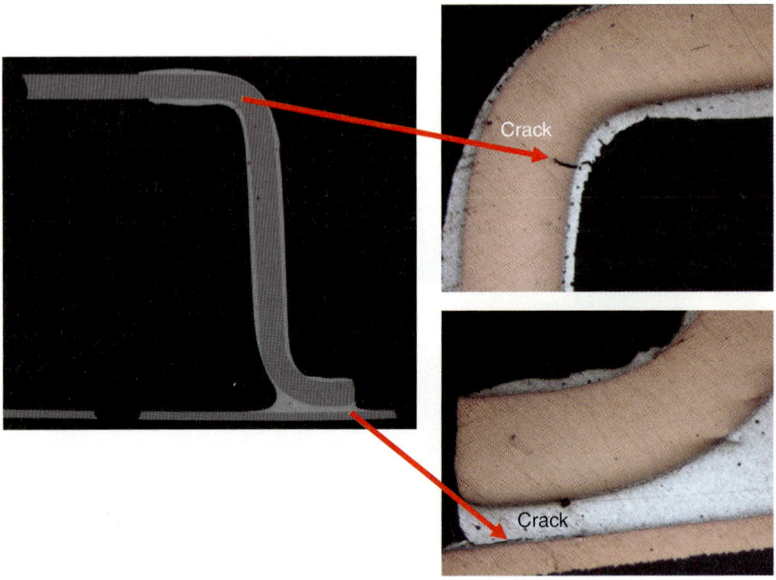

FIGURE 9.42 Failure locations of PQFP under four-point bending fatigue test.

to the higher Young's modulus (stiffness) of the SACC than that of the SAC [8]. The failure modes of the PQFP with both solders are similar and are shown in Fig. 9.42. The failure locations are at the elbow and the solder joint.

Figure 9.43 shows the four-point bending fatigue test results of the PBGA with SAC and SACC solders after 10-day, 20-day, and

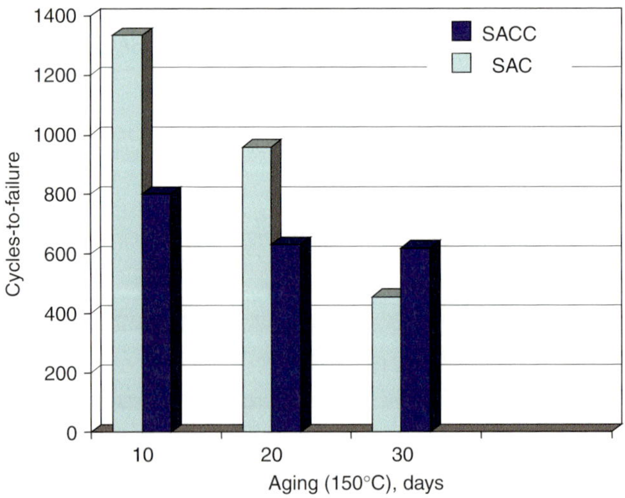

FIGURE 9.43 Four-point bending fatigue test results of PBGA.

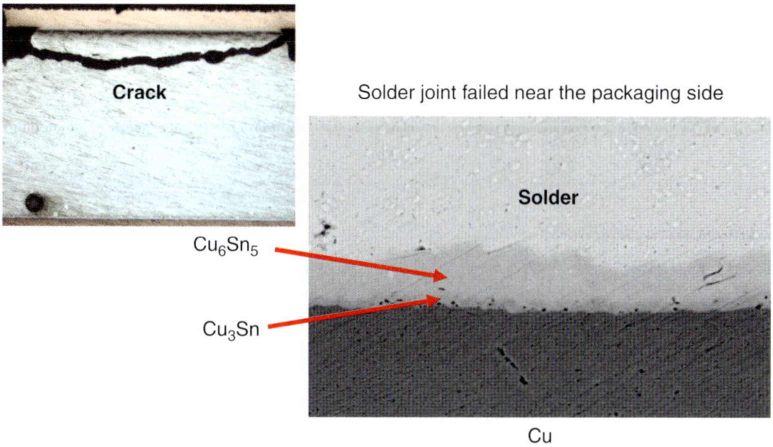

FIGURE 9.44 Failure location and IMC of PBGA under four-point bending fatigue test.

30-day aging at 150°C. It can be seen that: (1) for both solders, the longer the aging time, the smaller the number of cycles-to-failure; and (2) in general, the number of cycles-to-failure of the SAC solder joints is larger than that of the SACC solder joints. Again, the SACC is stiffer than the SAC. The failure mode of the PBGA with both solders is similar and shown in Fig. 9.44. It cracks at the solder joint near the package side.

Figure 9.45 shows the four-point bending fatigue test results of the WLCSP with SAC and SACC solders after 10-day, 20-day, and 30-day aging at 150°C. It can be seen that: (1) for both solders, the

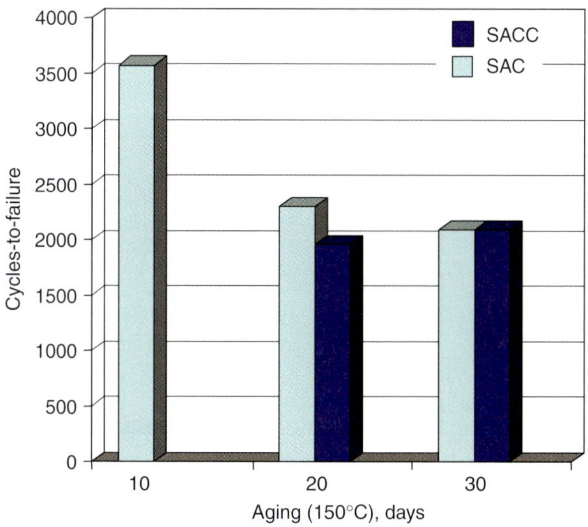

FIGURE 9.45 Four-point bending fatigue test results of WLCSP.

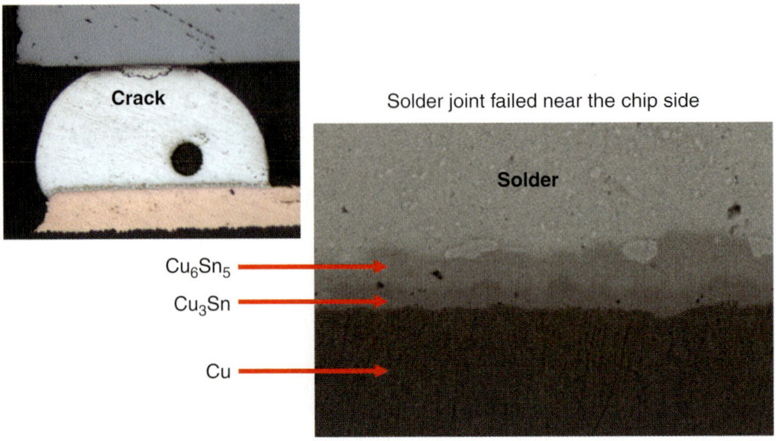

FIGURE 9.46 Failure location and IMC of WLCSP under four-point bending fatigue test.

longer the aging time, the smaller the number of cycles-to-failure; and (2) in general, the number of cycles-to-failure of the SAC solder joints is larger than that of the SACC solder joints. The failure mode of the CSP with both solders is similar and shown in Fig. 9.46. It cracks at the solder joint near the chip side.

Figure 9.47 shows the four-point bending fatigue test results of the BCC with SAC and SACC solders after 10-day, 20-day, and 30-day aging at 150°C. It can be seen that: (1) for both solders, the longer the

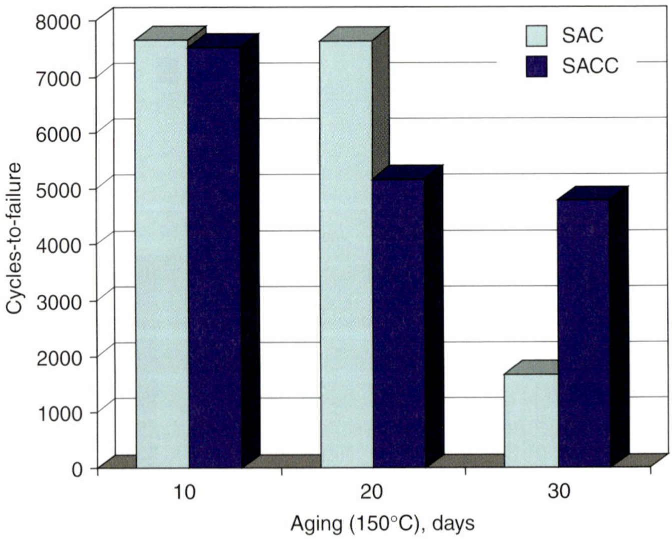

FIGURE 9.47 Four-point bending fatigue test results of BCC.

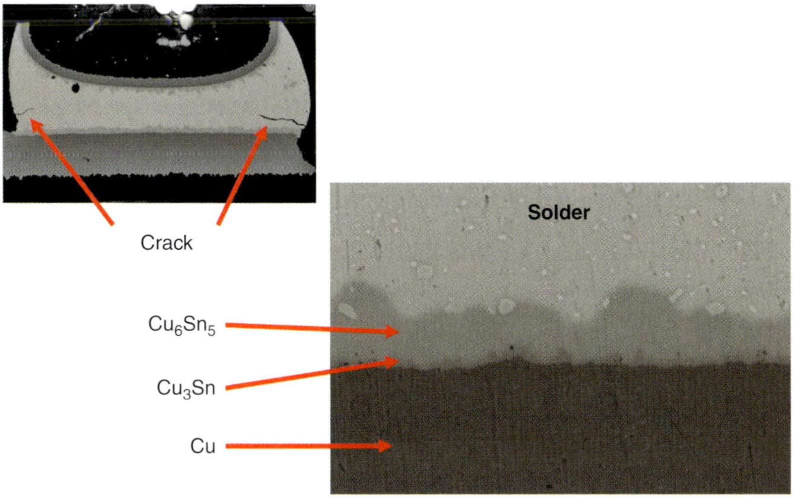

FIGURE 9.48 Failure locations IMC of BCC under four-point bending fatigue test.

aging time, the smaller the number of cycles-to-failure; and (2) in general, the number of cycles-to-failure of the SAC solder joints is larger than that of the SACC solder joints. The failure mode of the BCC with both solders is similar and shown in Fig. 9.48. It cracks at the solder joint near the PCB side.

(D) IMC Thickness

The IMC thickness on the PCB of the PBGA with the SAC and SACC solders is shown in Fig. 9.49. It can be seen that: (1) for both

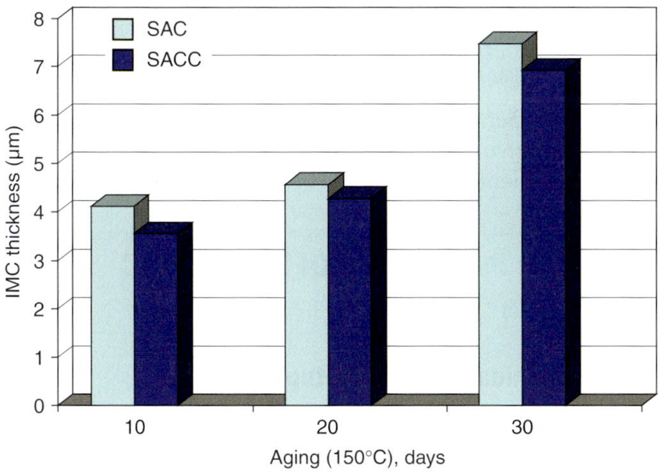

FIGURE 9.49 IMC thickness of PBGA as a function of aging.

SAC and SACC, the longer the aging, the thicker the IMC on the PCB, that is, the longer the aging time, the more coppers are consumed by the Sn in SAC or SACC; (2) the IMC thickness ($Cu_3Sn + Cu_6Sn_5$) for both solders is about the same; and (3) the IMC distribution for both solders is similar (a typical one is shown in Fig. 9.44). For both SAC and SACC cases, the immersion Ag on the Cu pads dissolves into their solder joints and cannot be found even with an EDX.

9.5.6 Summary and Recommendations

Some important results and recommendations are summarized in the following:

(1) There is no significant wetting difference between the SAC and SACC solder pastes on the OSP board.

(2) The SACC solder paste wets better on the ENIG board than on the OSP board.

(3) During stencil printing, because of the stickiness of the SACC, its separation speed is faster than that of SAC, and the front and rear speeds of the SACC are slower than that of SAC.

(4) Compared with SAC solder joints, there are more voids in the SACC solder joints; even the percentage of the volume of voids is small and acceptable.

(5) For both SAC and SACC solder pastes, in general, the effects of aging (at 150°C) are to reduce their number of cycles-to-failure and increase their IMC thickness (more Cu consumed by the Sn of SAC and SACC).

(6) In general, for the four-point bending fatigue tests, the number of cycles-to-failure of the SAC solder joints is larger (better) than that of the SACC solder joints.

(7) In general, the IMC thickness ($Cu_3Sn + Cu_6Sn_5$) of SAC and SACC solder joints is similar.

(8) More of bending, twisting, and shearing fatigue tests need to be done in order to increase the confidence on high-reliability requirement products.

9.6. Mechanical Pull Test of PQFP with SACC Solder Joints

9.6.1 Mechanical Pull Test Setup

The lead pull test is performed with an Instron 4411 by pulling the leads of the PQFP208 packages until failure, as shown in Fig. 9.50. Ten leads per sample are pulled with a 0.23-mm-diameter copper wire at a 45-degree angle using a pull speed of 10 mm/min. The

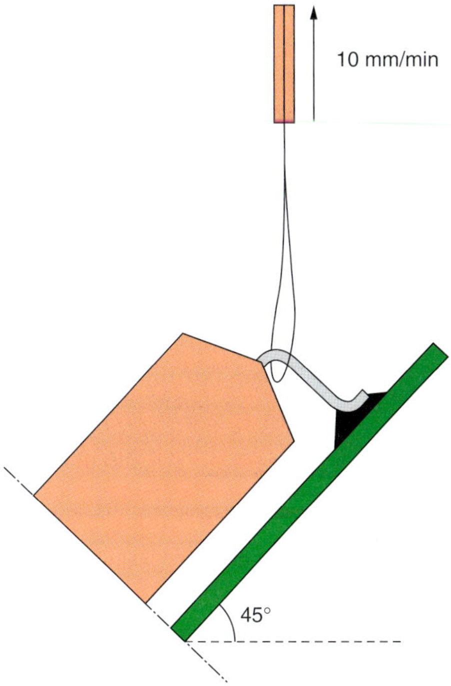

FIGURE 9.50 Test set-up diagram for 45-degree QFP lead pull test.

maximum pull force is recorded, and the failure modes are inspected [25].

9.6.2 Pull Test Results

All lead pull-force values exceed the minimum force of 5 N. The failure modes are recorded, and the SACC samples exhibited 60% to 70% PCB pad lifts without solder fractures, but the SAC samples exhibited 100% fractures within the solder joints [25]. This suggests that the SACC samples had stronger solder joints than the SAC samples. The SACC leads that do not exhibit pad lifts fractured at the interface between the lead and the bulk solder.

9.6.3 Failure Analyses

SEM/EDX analysis is performed to characterize the fracture surfaces on samples at time zero and following 30 days hot storage. The SACC samples (with the exception of the pad lifts) exhibit mostly ductile fracture morphology, whereas the SAC samples exhibit mostly brittle fracture morphology (Figs. 5.51 through 5.56). EDX analysis of the brittle-appearing fracture surfaces exhibit mostly copper and tin bearing intermetallics; however, EDX analysis of the ductile-appearing interfaces exhibit mostly bulk lead-free solder with some copper and tin intermetallics. The appearance of the fracture surfaces and the EDX spectra both indicate that the SAC samples exhibit brittle interfacial fractures within the intermetallic layer. On the other hand, the

Figure 9.51 SEM photo of SAC lead pull at time zero.

Figure 9.52 SEM photo of SAC lead pull at 30 days.

SACC samples fractured either within the bulk solder near the leads or at the interface between the bulk solder and the intermetallic layers. These differences in fracture surfaces between the SAC and SACC lead pull test samples are further discussed in terms of the observations of differences in the intermetallic formations and/or bulk solder microstructure found by cross-sectioning of the QFP leads [25].

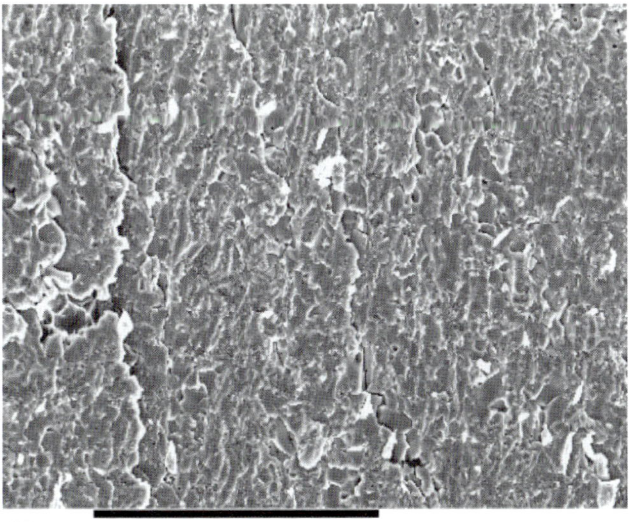

FIGURE 9.53 SEM photo of SAC lead pull at 30 days (Close-up view).

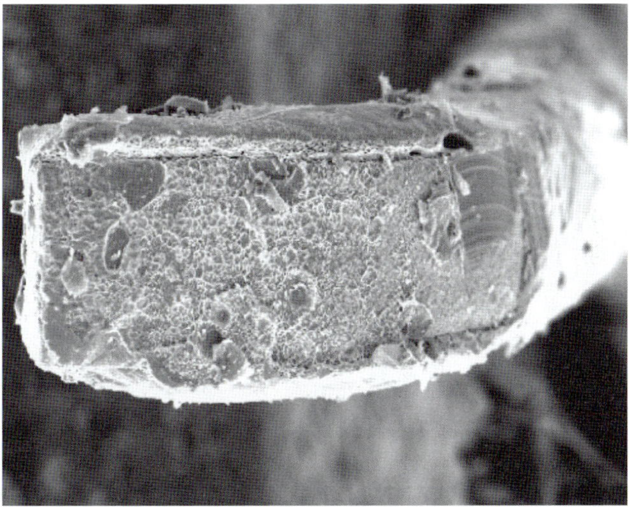

FIGURE 9.54 SEM photo of SACC lead pull at time zero.

Because of the brittle nature of thick intermetallic formations, the thicker copper-rich intermetallic formations that form in the SACC samples during hot storage would have been expected to have poorer mechanical integrity than thinner copper-rich intermetallic formations in the SAC samples. Assuming that intermetallic formations are the

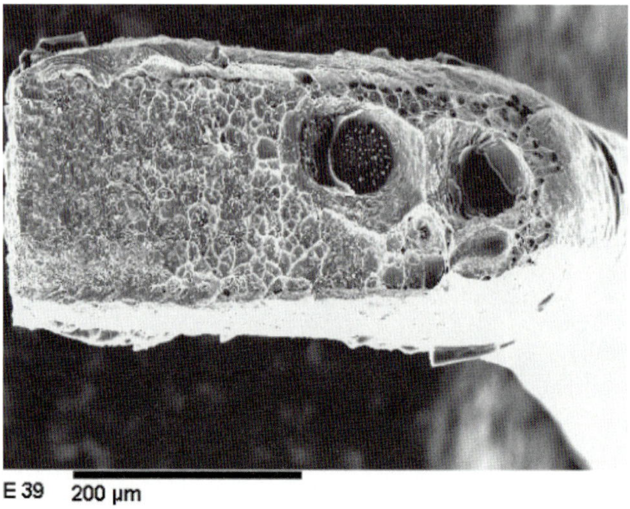

FIGURE 9.55 SEM photo of SACC lead pull at 30 days.

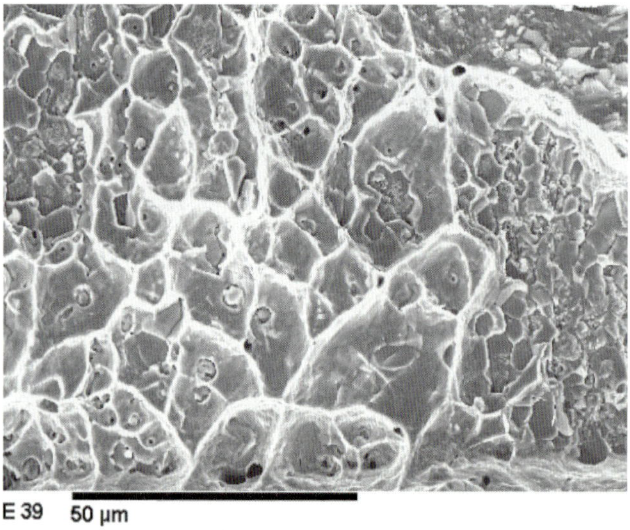

FIGURE 9.56 SEM photo of SACC lead pull at 30 days (Close-up view).

most dominant factor, the SACC samples should have exhibited more brittle fractures than in the SAC samples in lead pull-tests, because the SACC samples had thicker, more brittle, intermetallic formations. The fact that SACC samples had the thicker intermetallic formations but had stronger solder joints (as indicated by PCB pad lifts in lead pull testing) suggested that the material properties of the bulk solder were dominant over the intermetallic metallurgy in determining the failure mode in lead pull testing. Although we did observe a difference in intermetallic formation between SACC and SAC, the lead pull results indicate that the bulk solder properties of the SACC alloy are more important than intermetallic compound formations in determining the mechanical integrity of solder joints in lead pull tests.

9.7 Thermal Cycling Test of SACC Solder Joints

The objectives of this section are: (1) to perform a thermal cycling test of the ImAg PCB assemblies with the lead-free components with SAC and SACC solder pastes; (2) to estimate the life distribution, characteristic life, and mean-time-to-failure (MTTF) of various components' solder joints with SAC or SACC solder pastes; and (3) to compare the mean life of various components' solder joints with SAC and SACC solder pastes and determine the confidence levels of comparisons [26].

9.7.1 Components and Test Boards

The components and test boards are exactly the same as those in Sec. 9.5.

9.7.2 Thermal Cycling Test

Figure 9.57 shows the test boards in a thermal cycling chamber. Their commons and the end of each daisy chain in each component are connected to a data acquisition system. The resistance for each daisy

Figure 9.57 Thermal cycling chamber with test boards.

chain in each component is measured and stored in a computer. By monitoring the daisy-chain resistance, the time-to-failure of each component can be recorded. In this study, the failure criterion for the solder joints of a component is defined as a daisy-chain resistance increase of 40%.

In order for the chamber to operate at peak efficiency, it is important that the airflow in the chamber be disrupted as little as possible. Airflow in the chamber is directed from top to bottom, and for this reason the boards are placed vertically in the chamber as shown in Fig. 9.57. The cycling profile is from −40°C to 125°C, and the cycle time is 60 min with 15 min ramp up, 15 min ramp down, 15 min hold at cold (−40°C), and 15 min hold at hot (125°C).

9.7.3 Thermal Cycling Test Results

Table 9.10 shows the number of failures/sample size, Weibull slope (β or b in the Weibull plots), characteristic life (θ or h in the Weibull plots), and MTTF = $\theta\Gamma(1 + 1/\beta)$ (where Γ is the gamma function) for the thermal cycling test samples. The number of failures, Weibull slope, characteristic life (cycles), and MTTF (cycles) are, respectively: (1) 15, 0.92, 1328, 1381 for SAC and 23, 0.96, 1935, 1971 for SACC of the PBGA196; (2) 14, 0.78, 948, 1094 for SAC and 24, 1.07, 1763, 1717 for SACC of the BCC24; (3) 26, 0.97, 1447, 1467 for SAC and 43, 0.93, 2620, 2709 for SACC of PQFP208; and (4) 14, 1.38, 855, 781 for SAC and 29, 1.49, 472, 426 for SACC of WLCSP8.

There are 30 SACC test boards and each one has one PBGA, one BCC, one WLCSP, and two PQFPs. On the other hand, there are only 15 SAC test boards. Because of assembly defects, some samples failed very early and they are eliminated from the total samples and failures, as shown in Table 9.10, for example, for the PQFP, originally there were 60 samples and 45 failures with 2 that failed very early. As a result, the sample size is 58 and the number of failures is 43.

Figures 9.58 through 9.61, respectively, show the median-rank Weibull plots of the PBGA, BCC, PQFP, and WLCSP test samples with

	No. of Failures/ No. of Samples		Weibull Slope		Characteristic Life (cycles)		MTTF (cycles)	
	SAC	SACC	SAC	SACC	SAC	SACC	SAC	SACC
PQFP208	26/28	43/58	0.97	0.93	1447	2620	1467	2709
BCC24	14/14	24/28	0.78	1.07	948	1763	1094	1717
PBGA196	15/15	23/28	0.92	0.96	1328	1935	1381	1971
WLCSP	14/14	29/29	1.38	1.49	855	472	781	426

TABLE 9.10 Weibull Slope, Characteristic Life, and MTTF for the Test Samples

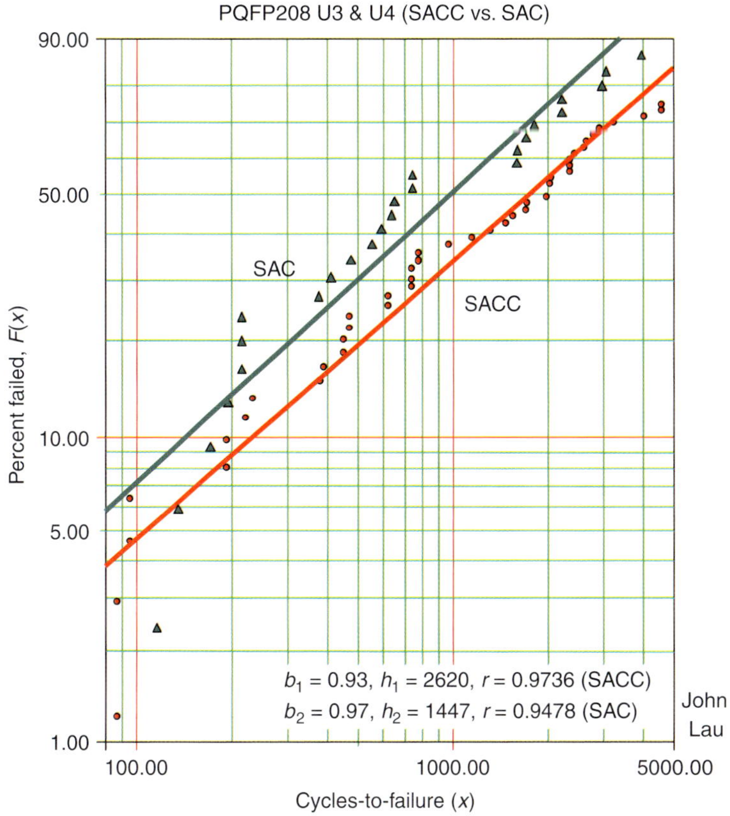

FIGURE 9.58 Thermal cycling (−40, +125°C, one-hour cycle) results for PQFP208 with SAC and SACC solder pastes.

the SAC and SACC solder pastes. Thus, based on the previous equations, the confidence levels for comparing pairs of the samples can be determined and are shown in Table 9.11. From Fig. 9.58 and Table 9.11, it can be seen that the solder-joint quality (mean life) of the matte-Sn–plated PQFP208 with SACC solder paste is better than that with SAC solder paste with 82% confidence. This means that in 82 out of 100 cases, the expected (mean) life of the (PQFP) SACC solder joints is better than that of the SAC solder joint. The other 18 cases are unknown, that is, they could be better or worse. Similarly, from Fig. 9.59 and Table 9.11, it can be seen that the solder-joint quality of the NiAu-plated BCC24 with SACC solder paste is better than that with SAC solder paste with 83% confidence.

In addition, as shown in Fig. 9.60 and Table 9.11, in 77 out of 100 cases, the expected life of the (SAC balled) PBGA solder joint with SACC solder paste is better than those with SAC solder paste. However, as shown in Fig. 9.61 and Table 9.11, the solder-joint quality of the

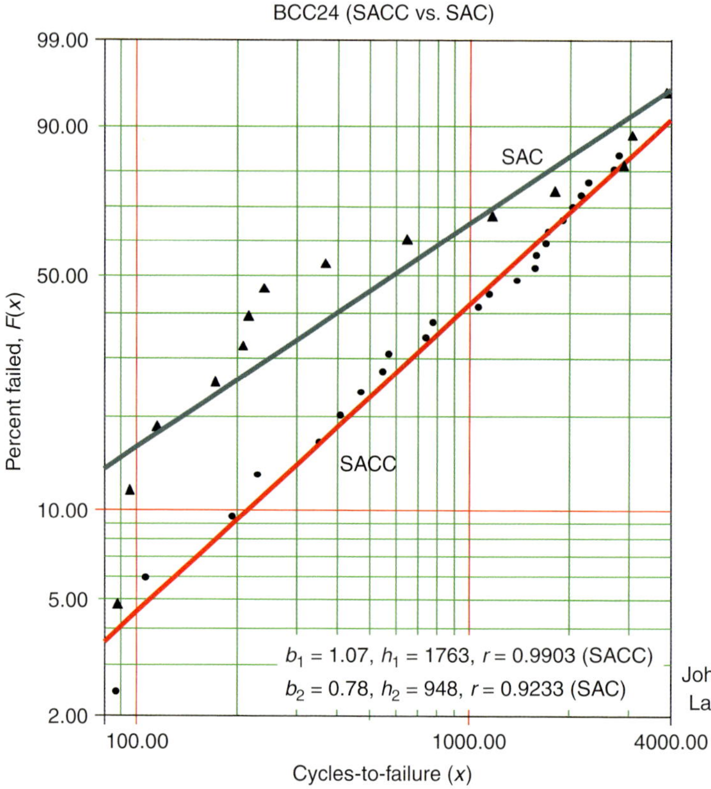

FIGURE 9.59 Thermal cycling (−40, +125°C, one-hour cycle) results for BCC24 with SAC and SACC solder pastes.

(SAC bumped) WLCSP8 with SAC solder paste is better than that with SACC solder paste at 93% confidence. (Finite-element modeling and analysis will be performed on this case in the next section.) It should be pointed out that: (1) in most of the cases, the Weibull slopes are rather small (0.78 for the BCC24 with SAC solder paste); and (2) except for the WLCSP8 case, the confidence levels for comparisons are rather low. Usually, they should be ≥90%.

Figures 9.62 and 9.63 show the Weibull plots of all the lead-free components with, respectively, SAC and SACC solder pastes. It can be seen that: (1) with the SACC solder paste, the quality of the BCC, PBGA, and PQFP solder joints is a few times better than that of the WLCSP solder joints with at least 98% confidences (based on the equation); and (2) with the SAC solder paste, the quality of the BCC, PBGA, and PQFP solder joints is less than twice that of the WLCSP solder joints with a confidence as low as 55%, which is not statistically significant.

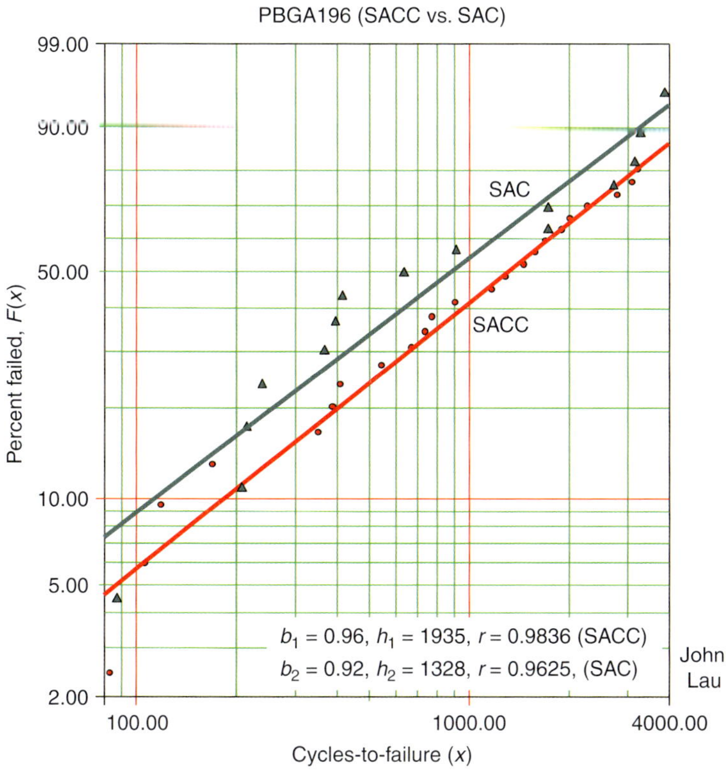

Figure 9.60 Thermal cycling (−40, +125°C, one-hour cycle) results for PBGA196 with SAC and SACC solder pastes.

9.7.4 Finite-Element Analysis and Results

Figure 9.64 shows the 3-D finite-element model that captures the construction along a diagonal strip from the peripheral WLCSP8 lead-free assembly's geometric center to a corner [27]. Because of the mid-plane symmetry, the mesh actually models a one-half strip (with one-half of a solder joint) using the appropriate in-plane constraints placed on one symmetry plane. Coupled in-plane translations are applied to the other symmetry plane to produce a state of generalized plane strain. Using exclusively hexahedral solid elements, the model can capture the precise shape of the packages' solder joints while retaining significant computational efficiency over full octant models. For the SAC solder paste case, the whole solder joint is made of SAC material. For the SACC case, one-third of the solder joint (SAC + SACC) is made of SACC material (Fig. 9.64). ANSYS [20] is the code selected for the modeling and analyses.

When the WLSCP PCB assembly is subjected to temperature cycle loadings, the solder joint at the chip corner will likely fail first. This is

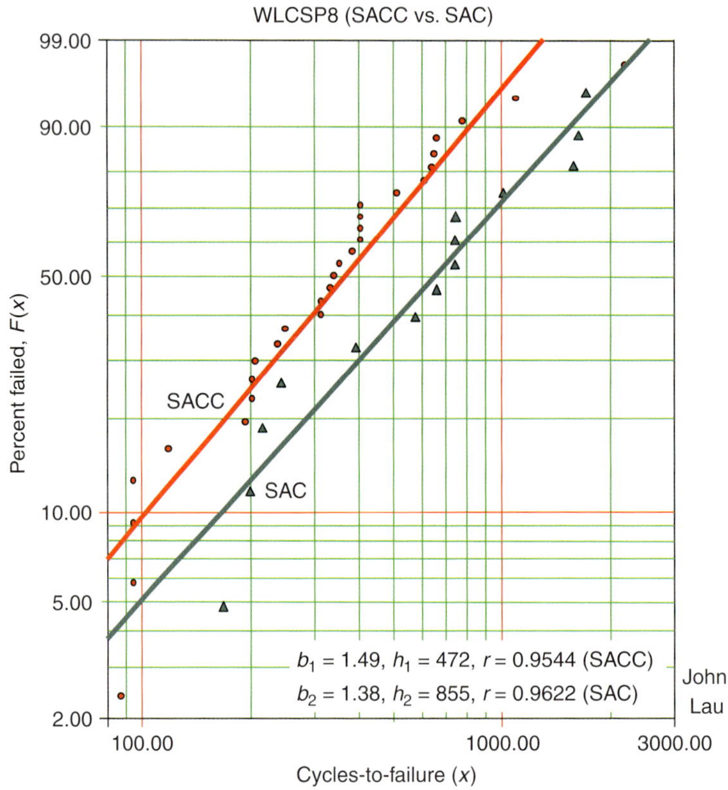

FIGURE 9.61 Thermal cycling (−40, +125°C, one-hour cycle) results for WLCSP8 with SAC and SACC solder pastes.

Components	MTTF Ratio	Determined Confidence (*P*)
PQFP208 SACC/SAC	2709/1467 = 1.85	82% (In 82 out of 100 cases, the mean life of PQFP208 solder joints with SACC solder paste is better than those with SAC solder paste)
BCC24 SACC/SAC	1717/1094 = 1.57	83% (In 83 out of 100 cases, the mean life of BCC24 solder joints with SACC solder paste is better than those with SAC solder paste)
PBGA196 SACC/SAC	1971/1381 = 1.43	77% (In 77 out of 100 cases, the mean life of PGBA196 solder joints with SACC solder paste is better than those with SAC solder paste)
WLCSP8 SAC/SACC	781/426 = 1.83	93% (In 93 out of 100 cases, the mean life of WLCSP8 solder joints with SAC solder paste is better than those with SACC solder paste)

TABLE 9.11 Calculated Confidences for Comparing the MTTF of SAC + SAC Paste and SAC + SACC Paste Populations

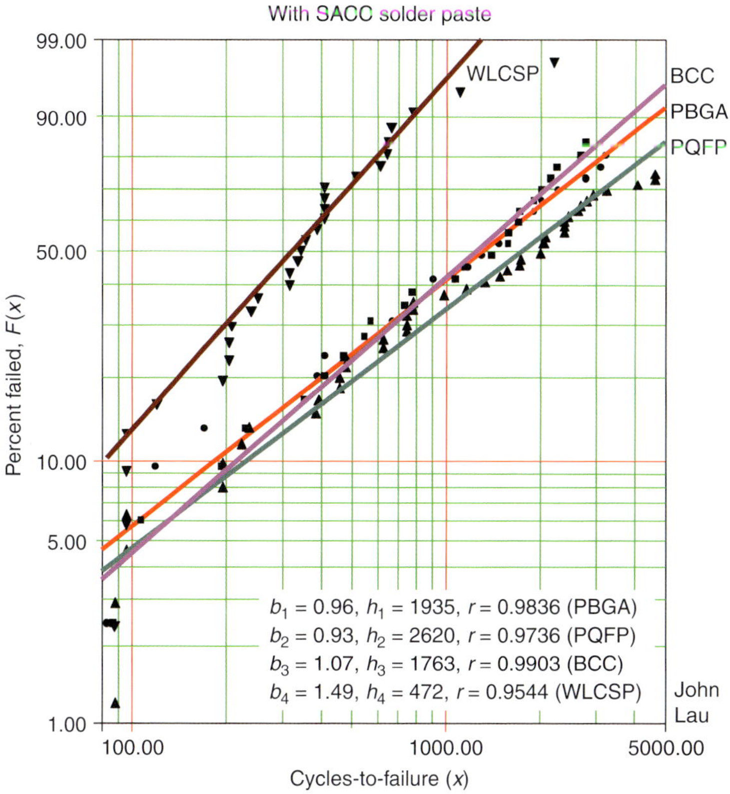

FIGURE 9.62 Thermal cycling (−40, +125°C, one-hour cycle) results for all components with SACC solder pastes.

due to: (1) the very large global thermal expansion mismatch between the chip and the PCB; and (2) the local thermal expansion mismatch between the silicon chip and the solder joint, and the solder joint and the PCB. Thus, a fine mesh is deliberately applied to the lead-free solder joint located at the chip corner as shown in Fig. 9.64.

The SAC material properties of the WLCSP assembly have been shown in Chap. 2, and the SACC creep material properties are shown in this chapter. Figure 9.65 shows the creep strain energy density (W) at critical locations of the all-SAC and SAC + SACC solder joints. For the all-SAC solder joint, the maximum W occurs near the interface between the chip and the solder joint. Thus, any failure of this solder joint should initiate there. On the other hand, for the SAC + SACC solder joint, the maximum W occurs near the interface between the solder joint and the PCB, and thus the failure should initiate here. According to the simulation results shown in Fig. 9.65, the maximum creep strain energy density per cycle (ΔW) that occurs in the SAC + SACC solder joint is larger than that in the SAC solder joint, as shown

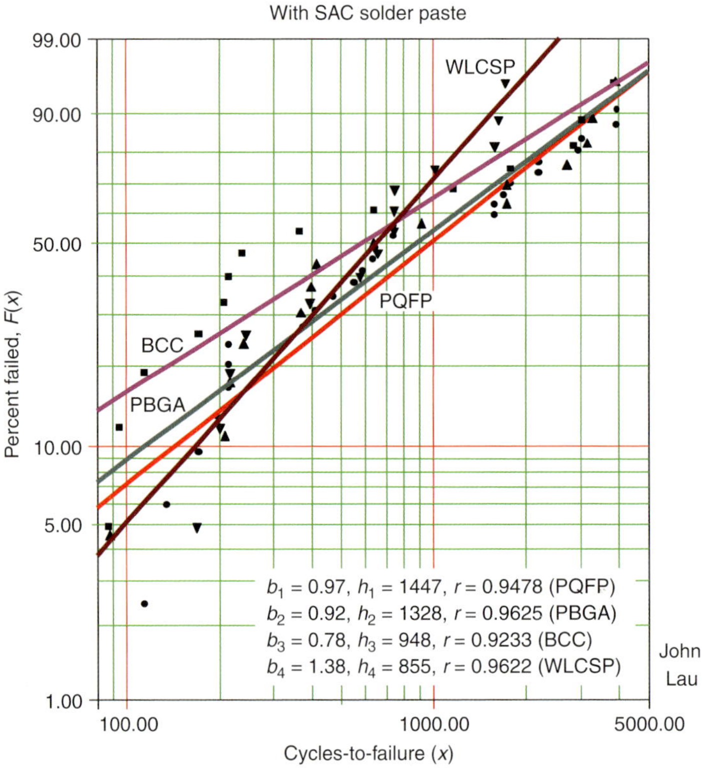

Figure 9.63 Thermal cycling (–40, +125°C, one-hour cycle) results for all components with SAC solder pastes.

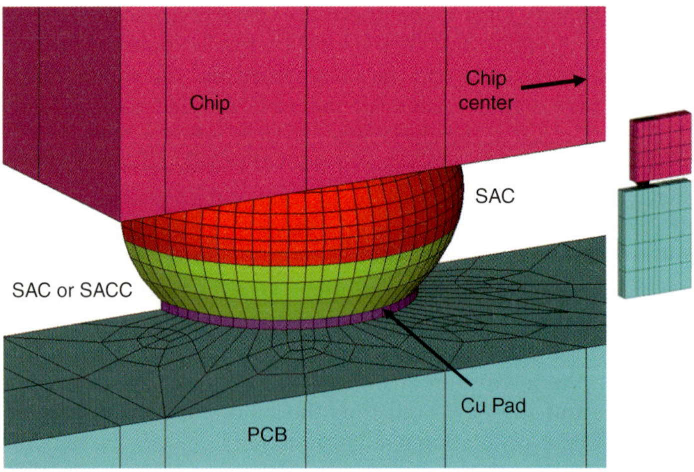

Figure 9.64 Finite-element analysis of the SAC solder-bumped WLCSP8 with SAC and SACC solder pastes.

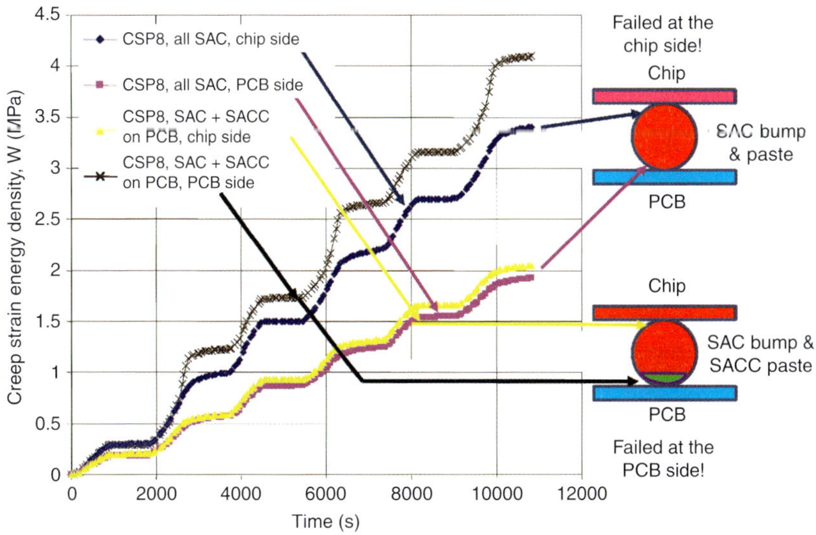

Figure 9.65 Creep strain energy density at critical locations of the SAC solder with SAC and SACC pastes.

in Fig. 9.64. Thus, one should expect the SAC + SACC solder joint to fail earlier than the all-SAC solder joint, assuming they have the same fatigue-crack growth rate.

9.7.5 Summary and Recommendations

Some important results and recommendations are summarized in the following:

(1) At 82% confidence, that is, in 82 out of 100 cases, the MTTF (quality) of the solder joints of the matted-Sn PQFP208 on the ImAg PCB with the SACC solder paste is better than that with the SAC solder paste at the present temperature condition (–40°C to 125°C, one-hour cycle).

(2) In 83 out of 100 cases, the solder-joint quality of the NiAu-plated BCC24 with the SACC solder paste is better than that with the SAC solder paste at the present temperature condition (–40°C to 125°C, one-hour cycle).

(3) At 77% confidence, the mean life of the solder joints of the SAC solder-balled PBGA196 on the ImAg PCB with the SACC solder paste is better than that with the SAC solder paste at the present temperature condition (–40°C to 125°C, one-hour cycle).

(4) In 93 out of 100 cases, the solder-joint MTTF of the SAC solder-bumped WLCSP8 on the ImAg PCB with the SAC solder paste is better than that with the SACC solder paste at the present temperature condition (–40°C to 125°C, one-hour cycle).

(5) For the SAC solder-bumped WLCSP8 with SAC solder paste (all-SAC) case, the maximum creep strain energy density per cycle (ΔW) occurs near the interface between the chip and the SAC solder joints.

(6) For the SAC solder-bumped WLCSP8 with SACC solder paste (SAC + SACC) case, the ΔW occurs near the interface between the solder joint and the PCB.

(7) Since the ΔW of the SAC + SACC case is larger than that of the all-SAC case, then the SAC + SACC solder joints should be expected to fail earlier than the all-SAC solder joints.

(8) The mechanical and creep materials properties of the reflowed SAC + SACC are unknown, and thus experimental testing of this new material is needed.

9.8 References

1. Lau, J. H., *Low-Cost Flip Chip Technologies*, McGraw-Hill, New York, 2000.
2. Hwang, J. S., *Environment-Friendly Electronics: Lead-Free Technology*, Electrochemical Publications, Asahi House, British Isles, 2001.
3. Lau, J. H., Wong, C. P., Lee, N. C., and Lee, R., *Electronics Manufacturing with Lead-Free, Halogen-Free, and Conductive-Adhesive Materials*, McGraw-Hill, New York, 2003.
4. Lee, N. C., *Reflow Soldering Processes and Troubleshooting: SMT, BGA, CSP and Flip Chip Technologies*, Elsevier Science, New York, 2004.
5. Hwang, J. S., *Implementing Lead-Free Electronics*, McGraw-Hill, New York, 2004.
6. Puttlitz, K. J., and K. A. Stalter, *Handbook of Lead-Free Solder Technology for Microelectronic Assemblies*, CRC Press—Marcel Dekker—Routledge, 2004.
7. Shangguan, D., *Lead-Free Solder Interconnect Reliability*, ASM International, Materials Park, Ohio, 2005.
8. Pecht, M., and S. Ganesan, *Lead-Free Electronics*, John Wiley and Sons, New York, 2006.
9. Evans, J., W. Engelmaier, and D. Kwon, *A Guide to Lead-free Solders: Physical Metallurgy and Reliability*, Springer, New York, 2007.
10. Bath, J. S., *Lead-Free Soldering*, Springer, New York, 2007.
11. Bradley, E., C. Handwerker, J. Bath, R. Parker, and R. Gedney, *Lead-Free Electronics: iNEMI Projects Lead to Successful Manufacturing*, John Wiley and Sons, New York, 2007.
12. Shina, S., *Green Electronics Design and Manufacturing*, McGraw-Hill, New York, 2008.
13. Amagai, M., Toyoda, Y., Tajima, T., "High Solder Joint Reliability with Lead Free Solders," *IEEE Proceedings of Electronic Components and Technology Conference*, New Orleans, LA, May 2003, pp. 317–322.
14. Sousa, I., Henderson, D., Patry, L., Kang, S., and Shih, D., "The Influence of Low Level Doping on Thermal Evolution of SAC Alloy Solder Joints with Cu Pad Structures," *IEEE Proceedings of Electronic Components and Technology Conference*, San Diego, CA, May-June 2006, pp. 1454–1461.
15. Amagai, M., Toyoda, Y., Tsukasa, O., and Akita, S., "High Drop Test Reliability: Lead-free Solders," *IEEE Proceedings of Electronic Components and Technology Conference*, Las Vegas, NV, June 2004, pp. 1304–1309.
16. Amagai, M., "A Study of Nano Particles in SnAg-Based Lead Free Solders for Intermetallic Compounds and Drop Test Performance," *IEEE Proceedings*

of Electronic Components and Technology Conference, San Diego, CA, May–June 2006, pp. 1170–1190.

17. Pang, J., Xiong, B., Neo, C., Zhang, X., and Low, T., "Bulk Solder and Solder Joint Properties for Lead Free 95.5Sn-3.8Ag-0.7Cu Solder Alloy," *IEEE Proceedings of Electronic Components and Technology Conference*, New Orleans, LA, May 2003, pp. 673–679.
18. Pang, J. H., and B. S. Xiong, "Mechanical Properties for 95.5Sn-3.8Ag-0.7Cu Lead-Free Solder Alloy," *IEEE Proceedings of TCPT*, Vol. 28, No. 4, December 2005.
19. Pang, J. H., B.S. Xiong, and T.H. Low, "Creep and Fatigue Characterization of Lead-Free 95.5Sn-3.8Ag-0.7Cu Solder," *IEEE Proceedings of Electronic Components and Technology Conference*, Las Vages, NV, May–June 2004, pp. 1333–1337.
20. Lau, J. H., J. H. Pang, N. C. Lee, and L. Xu, "Material Properties and Intermetallic Compounds of SACC," *IEEE Proceedings of Electronic Components and Technology Conference*, Reno, NE, May–June 2007, pp. 221–218.
21. Xu, L., K. Tan, and J. H. Pang, "Strain-Rate-Dependent Mechanical Properties for Sn-3.8Ag-0.7Cu and SAC-X Solder Alloy," IEEE Proceedings of *ITherm*, 2008, pp. 632–637
22. Xu, L., J. Pang, "Nano-Indentation Characterization of Lead-Free Solder and IMC Layer Properties," *Thin Solid Films*, Vol. 504, 2006, pp. 362.
23. Tan, K., L. Xu, J. Pang, J. H. Lau, and X. W. Zhang, "Creep Properties of SACC Lead-Free Solder," *IEEE Proceedings of Electronics Packaging Technology Conference*, Singapore, December 2008, pp. 521–526.
24. Lau, J. H., S. Liu, D. Shangguan, Z. Song, and D. Geiger, "Process Development and Solder Joint Reliability of a New Lead-Free Solder: SACC," *IEEE Proceedings of Electronic Packaging and Technology Conference*, Reno, NV, December 10–12, 2007, pp. 546–552.
25. Rooney, D., D. Geiger, D. Shangguan, and J. H. Lau, "Metallurgical Analysis and Hot Storage Testing of Lead-Free Solder Interconnects: SAC versus SACC," *IEEE Proceedings of Electronic Components & Technology Conference*, Lake Buena Vista, FL, May 27–30, 2008, pp. 89–98.
26. Lau, J. H., P. Tse, E. Richard, W. Dauksher, D. Shangguan, and J. Pang, "Reliability of Sn3wt%Ag0.5wt%Cu0.019wt%Ce (SACC) Solder Joints," *IEEE Proceedings of Electronic, Components & Technology Conference*, May, 2009, pp. 415–422.
27. Lau, J. H., and W. Dauksher, "Creep of Sn-(3.5-3.9)wt%Ag-(0.5-0.8)wt%Cu Lead-Free Solder," in *Micromaterials and Nanomaterials*, Edited by B. Michel, Fraunhofer Institute, IZM, Berlin, 2004, pp. 54–62.

CHAPTER 10

Chip-to-Wafer (C2W) Bonding and Lead-Free Interconnect Reliability

10.1 Introduction

As mentioned in Chap. 1, face-to-face silicon chips stacking is one of the package-on-package (PoP) 3-D integrated circuit (IC) packaging technologies [1–33]. In this chapter, the interconnect reliability of face-to-face Si chips stacked with AuSn lead-free material is presented. Wafer bumping and C2W bonding assembly process of the PoP are also discussed. In addition to AuSn, SnAg lead-free material is considered. Finally, low-temperature bonding [34–71] is discussed, and a special case of 3-D IC chip stacking with InSnAu C2W bonding is presented.

10.2 3-D Packaging (PoP) with AuSn Interconnects

AuSn solders are especially suited for optoelectronics and medical packaging. Most of the medical and optoelectronic devices have to be soldered in a fluxless process, which is possible using AuSn solders, as shown in the phase diagram in Fig. 10.1. Also, 80wt%Au20wt%Sn solder is used widely because of its advantages: (a) high strength, (b) high corrosion resistance, and (c) high fatigue resistance. Thus, the Au20Sn solder system is usually selected as the interconnect material for fluxless flip-chip applications.

10.3 Test Vehicle and Fabrication

10.3.1 Test Vehicle

The test vehicle is a silicon-stacked module, shown in Fig. 10.2 [32]. The assembled stacked module consists of a top silicon chip known

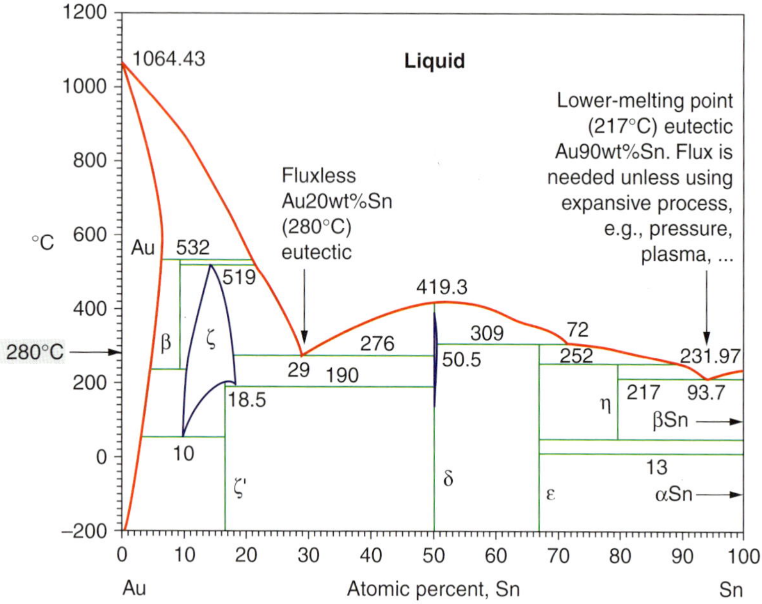

FIGURE 10.1 An-Sn solder phase diagram.

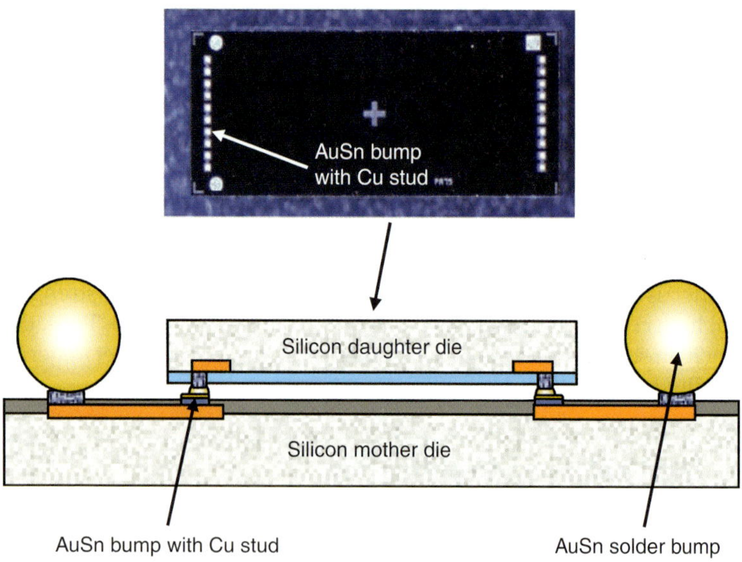

FIGURE 10.2 Stacked silicon module (daughter die and mother die) and a close look at the daughter die.

Package Type	Silicon Stacked Module	
Test Die	**Daughter**	**Mother**
Die Size (mm)	3.405 × 1.34 × 0.06	4.793 × 1.34 × 0.13
Pad opening	FC pad 30 μm	FC pad 30 μm
Pad pitch	100 μm	100 μm
Bump Type, (Height)	80Au20Sn (25 μm) on Cu stud	80Au20Sn (125 μm) on NiAu
Under Bump Metallization	Al	Electroless NiAu

TABLE 10.1 Materials and Geometry of Key Elements of the PoP Module

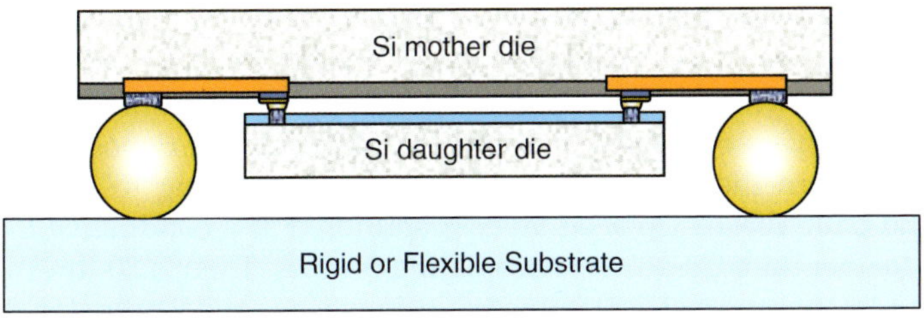

FIGURE 10.3 Stacked silicon module attached on a substrate.

as the daughter die and a bottom silicon chip known as the mother die, which are fabricated in wafers. On the daughter die, there are bumps made of Cu stud-coated with the AuSn solder. On the mother die, there are electroless NiAu under-bump metallurgy (UBM) for the bumps from the daughter die and AuSn solder bumps for the next level of interconnect, such as substrates. For both dies used in this test, the pads are interconnected in an alternating pattern so as to provide a daisy-chain connection when they are soldered together. There are 20 pads on the daughter die and they are arranged in two rows as shown in Fig. 10.2. The materials and dimensions of the key elements of the Si-stacked module is listed in Table 10.1. This PoP module is eventually attached to either a rigid or flexible substrate as shown in Fig. 10.3.

10.3.2 Test Vehicle Fabrication

AuSn solder can be manufactured by various methods, such as electron-beam evaporation, electron deposition, and paste and solder preform. Among these methods, evaporated solder allows more accurate control of dimensions and positions. Therefore, after UBM patterning, a sequence of alternating layers of Au and Sn are deposited to form an AuSn layered structure. Some of the advantages of using the electron-beam evaporation process for AuSn deposition are [32]:

(1) The amount of oxide formed after deposition will be reduced.

(2) The thickness and position of the solder can be precisely controlled.

(3) The gold deposition rate is high.

(4) Uniform AuSn thickness can be achieved across a 200-mm wafer.

The p-type silicon (100) wafers of 200-mm diameter with a thin electrical insulating film of 5000 Å of SiO_2 are used for fabricating the test vehicle. Figure 10.4 schematically shows the fabrication process for making the AuSn solder bumps on the daughter die. The first step is to deposit an AlCu layer on the Si wafer and then pattern the metal pads with daisy chains and etch the AlCu layer. Second, passivation layers of 5000 Å of SiO_2/5000 Å of SiN are deposited and then pattern the passivation layer using dry etching to open the metal pad. The Ti/Cu seed layer is sputtered as a conductive layer for further mini Cu-stud electroplating process. Then, a 20-μm-thick dry-film photoresist is laminated using a hot roll laminator and ultraviolet (UV) light using an EVG 640 contact mask aligner followed by developing the photoresist to define the UBM pads. These patterned vias are filled with copper up to 18 to 20 μm using the RENA 200-mm wafer electroplating tool, which is a fountain type (cup type) in which Sperolyte $CuSO_4$-based solution (Atotech Pte, Ltd.) is used as the electrolyte solution. Then, layers with a 2200 Å thickness of Au and a 2000 Å thickness of Sn are deposited alternatively using the electron-beam evaporator. A total of 16 layers are deposited to achieve a 3.5-μm-thick AuSn bump on the mini Cu stud. These AuSn layers are deposited on the whole wafer surface, but the AuSn on the dry-film layer can be removed using the lift-off process during dry-film stripping, and the AuSn will remain only on the Cu stud due to the adhesion of these layers with the Cu. After the dry-film stripping, the Ti/Cu seed layers are etched off one by one using a selective wet etching chemical to complete the AuSn solder-bump fabrication as shown in Fig. 10.4. These AuSn layers will melt and form homogenous Au80Sn20 solder joints during the reflow in the assembly process.

Figure 10.5 shows the AuSn solder bump where a 3.5 μm thickness of AuSn is deposited on a 20-μm-thick Cu stud by the evaporation process. The AuSn solder bumps are distributed at a 100-μm pitch along the two sides of the daughter die, and they are quite uniform, as the figure shows.

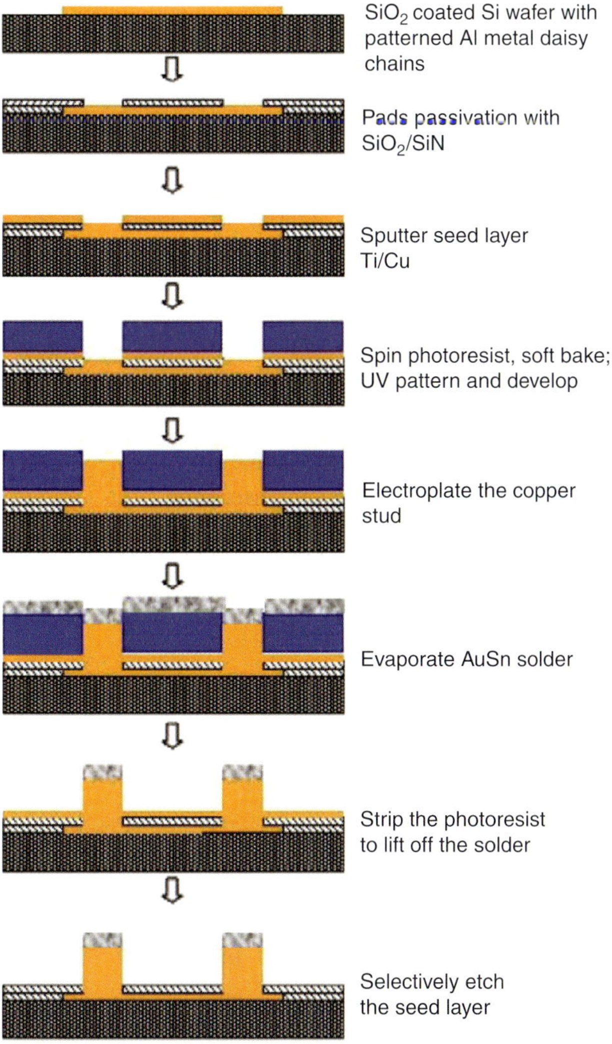

FIGURE 10.4 Fabrication process of the AuSn solder bump on the daughter-die wafer.

Figure 10.6 shows the bumps on the mother die—namely, the AuSn solder bump (which will not be discussed in this section) and the electroless NiAu UBM. The fabrication process for the mother die wafer is the same as the daughter die up to the AlCu metal pads being patterned for the daisy chain and metal pads' passivation opening. Then the electroless NiAu layers are deposited as the UBM as shown in Fig. 10.6.

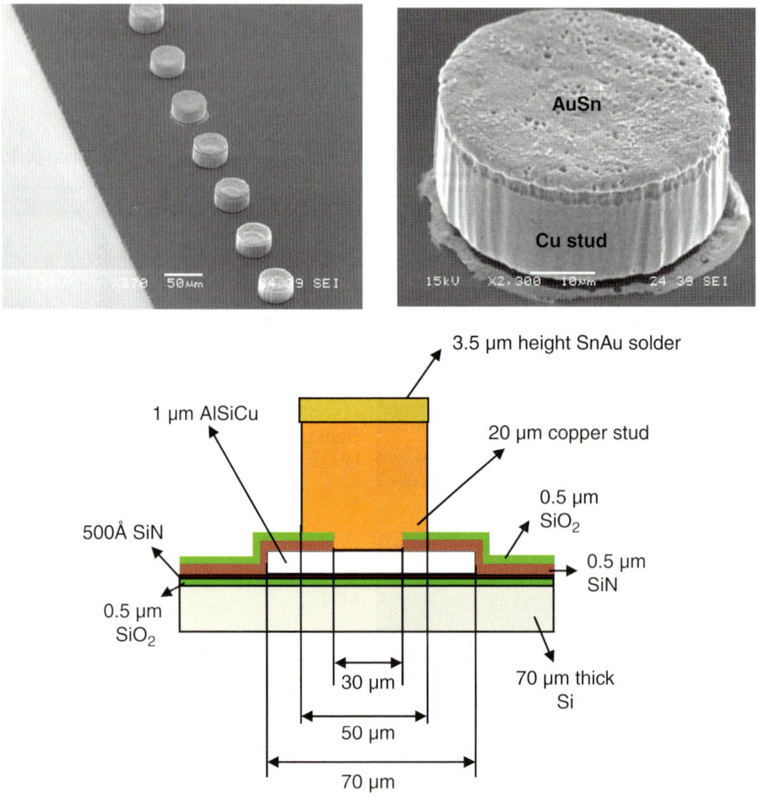

FIGURE 10.5 Dimensions and uniformity of the AuSn solder bump on the daughter-die wafer by electron beam evaporation process.

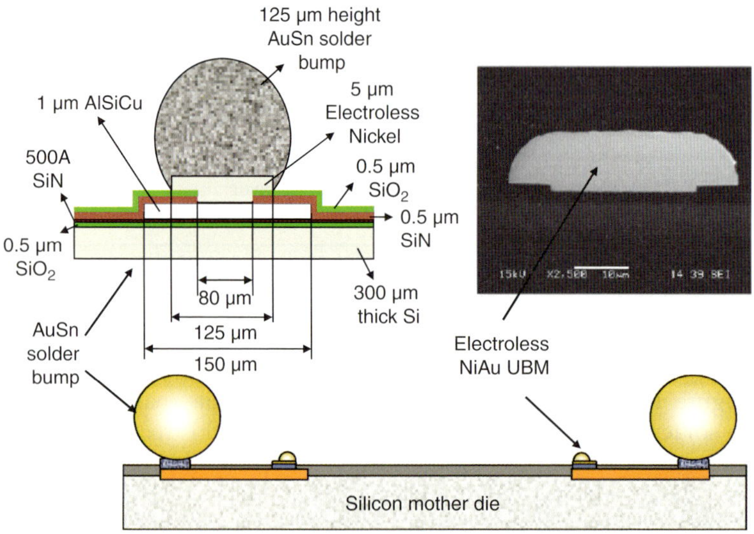

FIGURE 10.6 Dimensions and close-up look of the electronless NiAu UBM on the mother-die wafer.

10.4 C2W PoP Assembly

The overall assembly process of the AuSn solder-bumped daughter die on the mother die is by a chip-to-wafer (C2W) bonding method, that is, first dice the daughter-die wafer into individual dies and then pick up the good daughter dies and bond them on good mother dies of the mother-die wafer as shown in Fig. 10.7.

The bonding process is done with the flip-chip bonder FC150 from Karl Suss, as shown in Fig. 10.8. The assembly is carried out using nonreflowed bumps with N_2 purging during bonding. Argon sputter cleaning was performed on the daugther die and mother die (on a wafer) prior to the bonding process. The main challenge in AuSn flip-chip bonding is to achieve a good AuSn wetting on the electronless NiAu UBM bond pad with minimum AuSn solder squeezed between the bump and the UBM after bonding. In addition to bump height coplanarity, the assembly process optimization in terms of alignment accuracy, bond force, bond temperature and bond

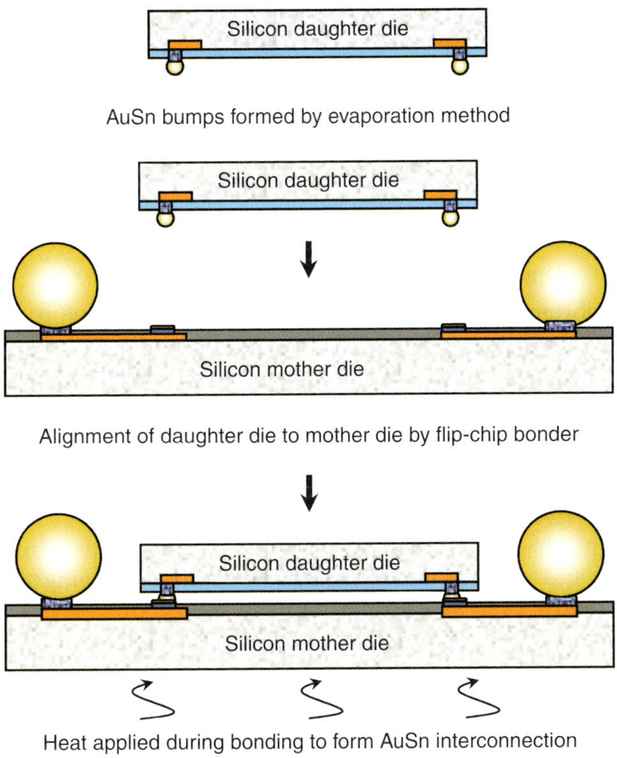

Figure 10.7 Overall PoP module assembly process.

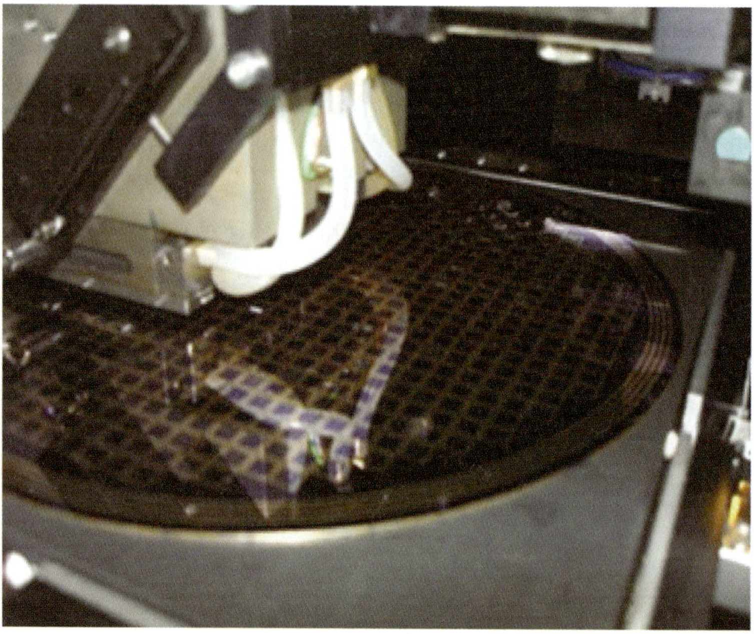

Figure 10.8 FC150 chip-to-wafer (C2W) bonder.

time on contact resistance, AuSn wetting, and AuSn solder-joint shape are investigated.

10.4.1 Bump Height Coplanarity

It is important to characterize the Au bump height before the flip-chip attachment process, as a nonuniform bump height will result in weak joints and even open continuity. Figure 10.9 shows the bump height measurements in an acceptable range of a maximum 2.5 μm at the die level and 1.5 μm at the wafer-to-wafer level.

10.4.2 Alignment Accuracy

In the flip-chip conventional soldering process, solder bumps are dipped into flux and attached onto the substrate pads before being subjected to the reflow process. During the oven reflow process, the solder bump on the die is able to self-align to the substrate pads and form good solder interconnections. However, with thermal compression bonding, the solder is unable to self-align during assembly. Hence, bonding alignment accuracy becomes critical, especially when using large dies with solder bumps. Precise bonding equipment with good alignment and bonding accuracy is required to form a good AuSn

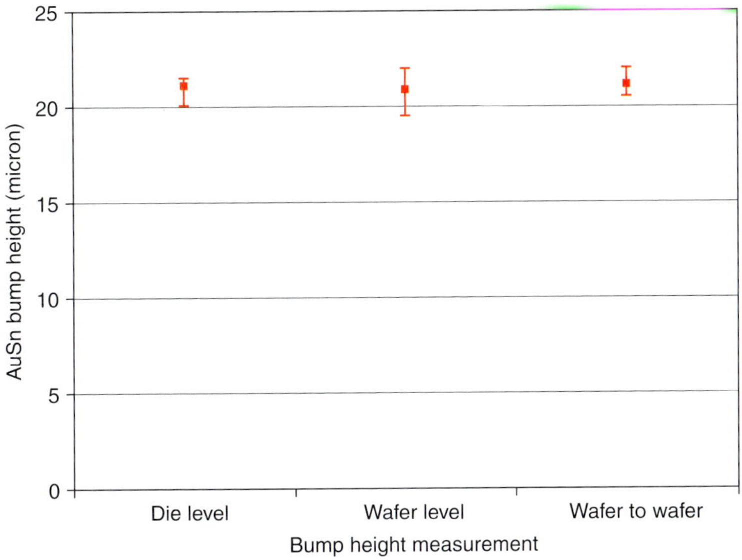

FIGURE 10.9 AuSn bump height measurement on various wafers.

interconnect with stable contact resistance and good AuSn wetting on the substrate pad.

The contact resistance of the interconnects is characterized at different bonding alignment accuracies, as shown in Fig. 10.10. A bond force of 0.8 kg is applied during the flip-chip bonding process. There is

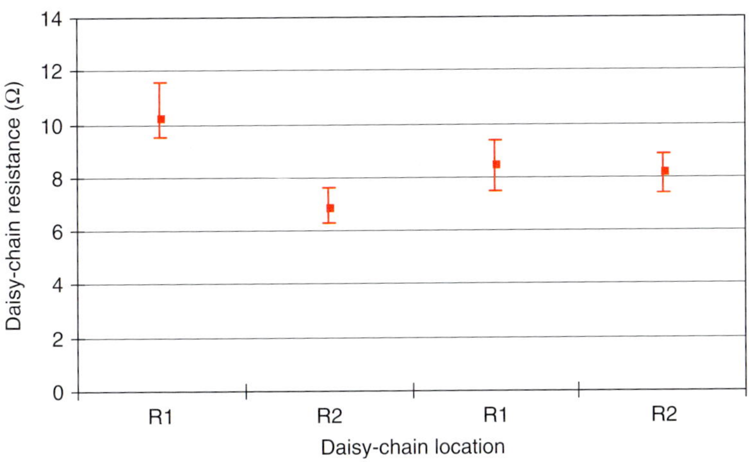

FIGURE 10.10 C2W bonding alignment accuracy vs. contact resistance measurements.

FIGURE 10.11 AuSn wetting failure modes. The top ones are for misaligned bonding alignment and the bottom ones are for 100% bonding alignment. Also, the left ones are for the daughter dies and the right ones are for the mother dies.

a significant difference in the contact resistance and the AuSn wetting on the electronless NiAu bond pads on misaligned AuSn bumps to electroless NiAu pads, compared to 100% aligned AuSn bumps to electroless NiAu pads, as shown in Fig. 10.11. This mode of joint configuration results in poor AuSn wetting on the bond pad and higher resistance readings. One of the reasons for this is that there is no self-alignment of the AuSn solder bumps to the electroless NiAu UBM after bonding. The AuSn solder is unable to wet the electroless NiAu UBM if there is a great difference in the misalignment offset. On the other hand, for the 100% aligned bonding, stable contact resistance and good AuSn wetting are observed on the electroless NiAu UBM. The contact resistances on the two daisy chains measured are lower and more consistent, compared to the resistance measured on the misaligned bonding. The cross-section of a good alignment interconnect is shown in Fig. 10.12.

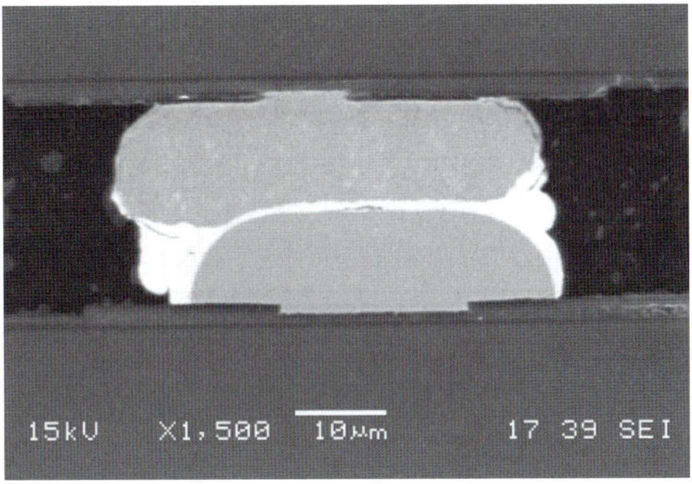

FIGURE 10.12 The cross-section of a good AuSn solder joint.

10.5 C2W Design of Experiments (DOE)

10.5.1 Three-Factor DOE

With the different assembly process issues addressed, actual assembly of the test vehicle is carried out without underfill according to the three-factor DOE (Table 10.2). Three different bond forces—two bond temperatures and time—are evaluated to determine the suitable bond force and bond temperature required to achieve consistent resistance measurements, good AuSn wetting, and AuSn joint formation. The three selected bond forces range from 0.4 kg to 1 kg (Table 10.3). The assembly yields are characterized and the contact resistances are monitored via a daisy-chain network, as shown in Fig. 10.13.

Batch	Bond Force (kg)	Bond Temp (°C)	Bond Time (s)
A	0.4	290	15
B		315	15
C	0.8	290	15
D		315	15
E	1	290	15
F		315	15

TABLE 10.2 Three-Factor DOE of C2W Bonding Evaluation

Factors	Variations			Response
Bond force (kg)	0.4	0.8	1	• Contact resistance • AuSn joint shape • AuSn wetting
Bond temperature (°C)	290		315	
Bond time (sec)	15			

Table 10.3 C2W Bonding with Various Combinations of Force, Temperature, and Speed

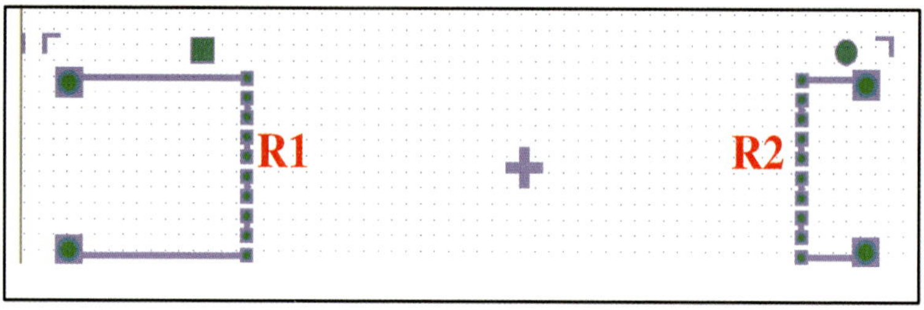

Figure 10.13 Resistance measurements through the chain on the mother die.

10.5.2 DOE Results

The bond force has a significant influence on the interconnection resistance and AuSn joint formation. Figure 10.14 shows stable contact resistance can be achieved at 0.8 kg and 1 kg bond forces. However, 1 kg is undesirable, as it results in chip passivation cracking at the Al pad/UBM interface, as shown in Fig. 10.15. Furthermore, a high bonding force tends to squeeze out the AuSn eutectic solder between

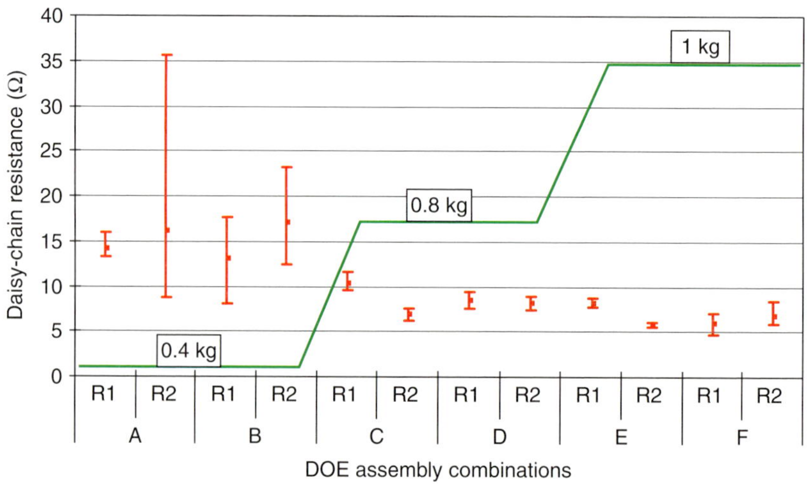

Figure 10.14 Contact resistance measurements from the DOE assemblies.

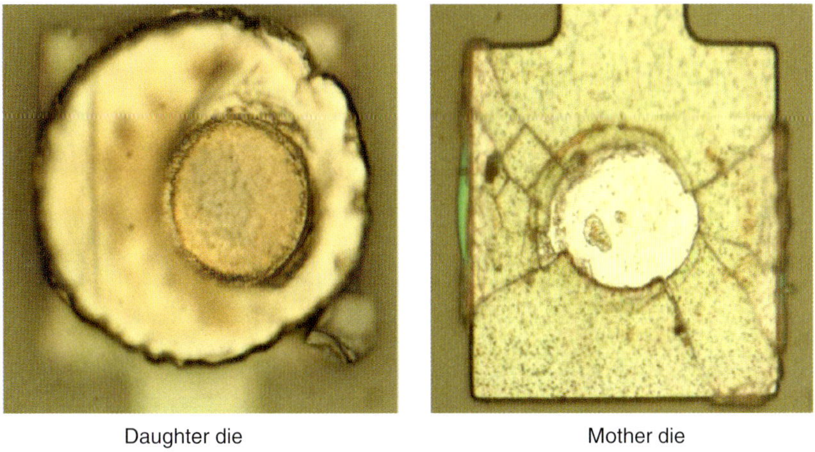

FIGURE 10.15 Passivation crack at high bonding force (1 kg).

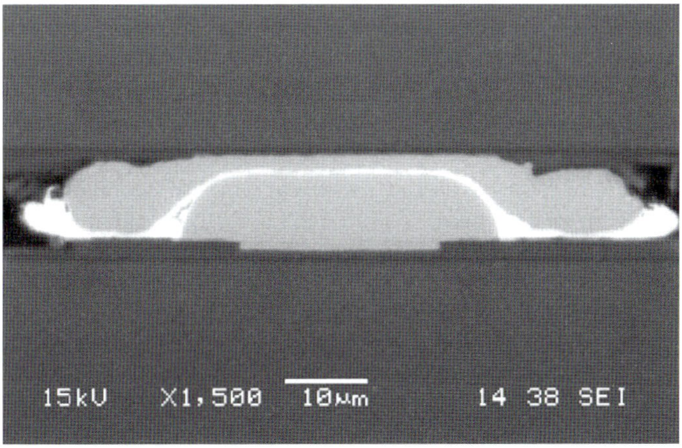

FIGURE 10.16 The cross-section of an AuSn assembly made by high bonding force (1 kg).

the bump and bond pad, and results in poor AuSn wetting on the bond pad. Figure 10.16 shows the cross-sectional view of the AuSn assembly at 1 kg bond force. The findings also suggest that the contact resistance is more sensitive to bond force than the bond temperature and bond time.

10.6 Reliability Tests and Results

A set (batch D) of the test vehicle assemblies (without underfill) is subjected to the temperature cycle test (–40°C to 125°C) to access the thermal-fatigue characteristics of the AuSn interconnects [32]. The electrical contact resistance is measured prior to the reliability test

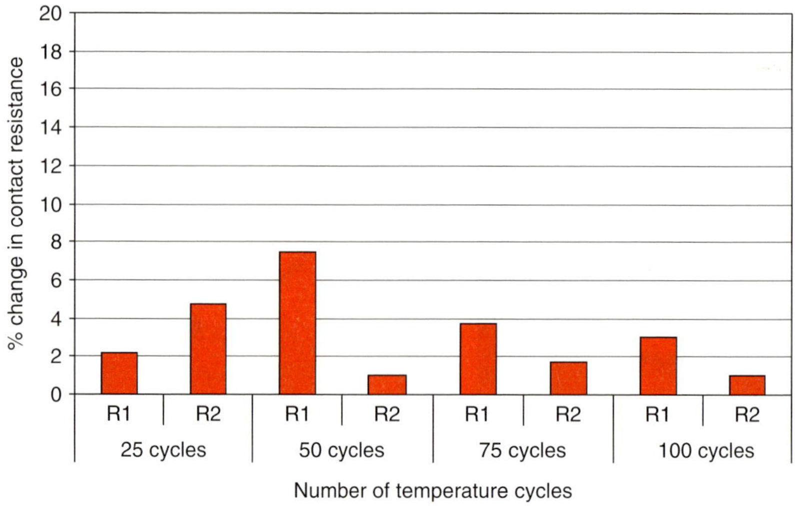

FIGURE 10.17 Percentage changes in contact resistance during temperature cycles (−40°C to 125°C).

and at every 25 cycles during the testing for 100 cycles. The criteria for failure would be an open continuity in the resistance value (∞) of the daisy chain.

This set of samples passed the thermal cycling test without open continuity up to 100 cycles, as shown in Fig. 10.17. A minimum change of less than 10% in the contact resistance is observed for bond force at 0.8 kg. This suggests that the interconnect configuration of a homogenous AuSn eutectic solder with direct contact to the electroless NiAu UBM bond pad is robust and able to withstand the temperature cycling conditions. The results indicate that stability in contact resistances and good reliability performance can be achieved in combination D (bonding temperature = 315°C and bonding time = 15 s) at 0.8-kg bond force.

10.7 Summary and Recommendations

Some of the important results are summarized are in the following:

1. Uniform bump coplanarity at the die level (±3.5 μm) is required to ensure that uniform pressure can be applied on the test die and form good AuSn interconnects.
2. Stable contact resistance (±2 Ω) and good AuSn wetting on the electroless NiAu pads is achieved on 100% aligned bonding at 0.8-kg bond force.
3. Assembly results show that samples assembled at high bonding force (1 kg) are undesirable as this force results in chip passivation cracking at the Al pad/UBM interface. Furthermore,

a high bonding force tends to squeeze out the AuSn eutectic solder between the bump and bond pad, and results in poor AuSn wetting on the bond pad.

4. Temperature cycle results show that stable contact resistances (±9 Ω) and good reliability performance (>100 temperature cycles) can be achieved without underfill at 0.8-kg bond force.
5. More importantly, the results of this work will serve as a reference for industries establishing flip-chip assemblies using AuSn fluxless solder-bump technology.

10.8 3-D Packaging (PoP) with SnAg Interconnects

Figure 10.18 shows the same daughter die and mother die, but with a different solder system. Instead of the AuSn solder, the Sn3wt%Ag + Cu-stud is considered on the daughter die as shown in Fig. 10.19. Also, the solder ball of the chip scale package (CSP) on the mother die is replaced by Sn37wt%Pb. The focus is, of course, on the reliability of the SnAg solder joints between the daughter die and mother die [33].

Prior to the assembly, the daughter die wafer with a 40-μm height of SnAg + Cu-stud microbumps is thinned down to 70 μm, and the mother die wafer with a 200-μm height of CSP SnPb + UBM microbumps is thinned down to 300 μm as shown in Fig. 10.20.

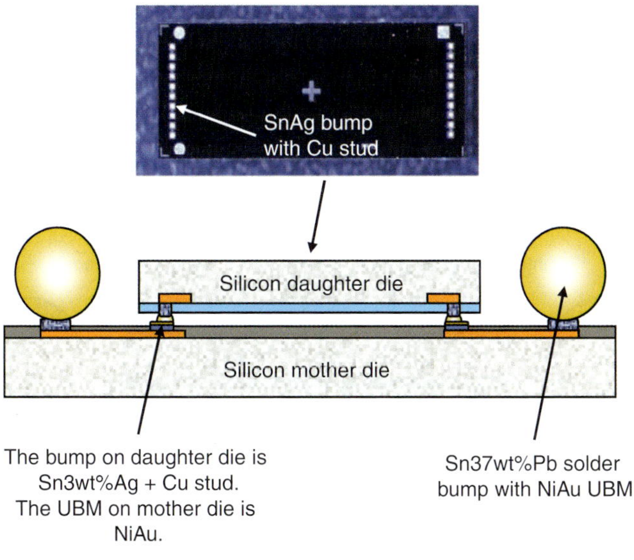

Figure 10.18 Stacked silicon module (daughter die and mother die) with SnAg lead-free solder.

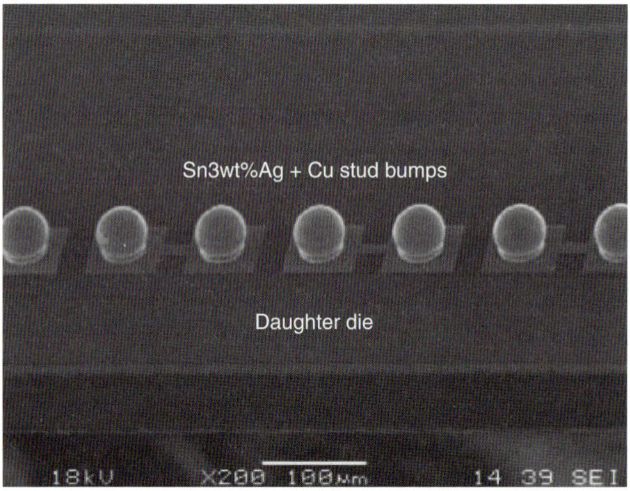

FIGURE 10.19 Lead-free Sn3wt%Ag + Cu-stud (~40 μm) bumps on daughter die.

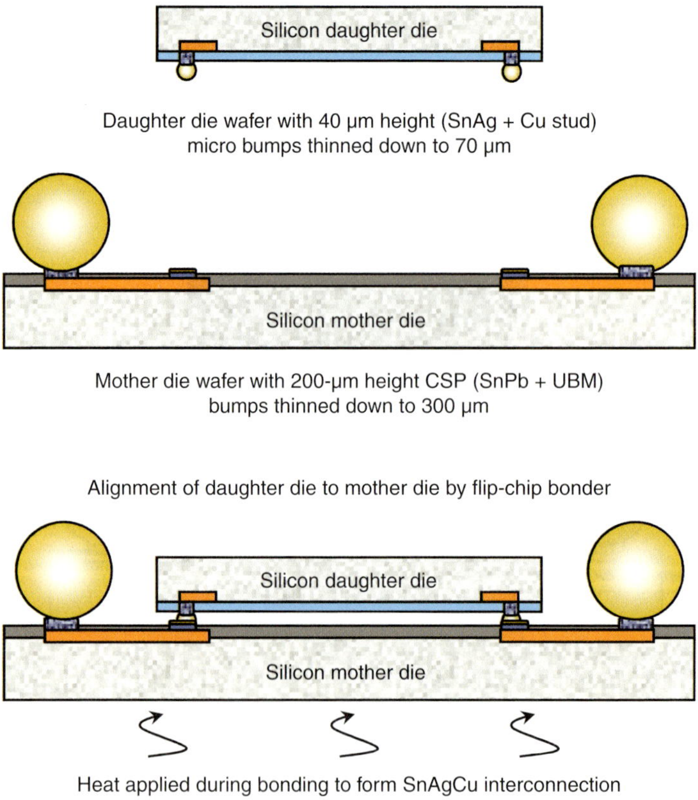

FIGURE 10.20 Overall PoP module assembly process with SnAg (not to scale).

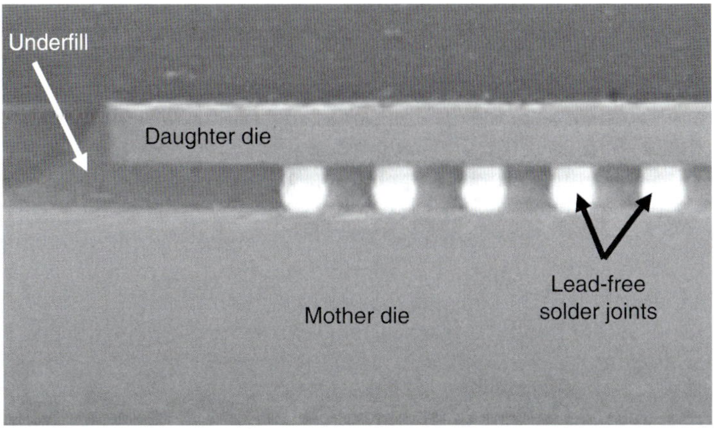

FIGURE 10.21 Daughter die assembled on mother die with SnAg solder joints.

Figure 10.21 shows a cross-section of the assembly where the underfill is applied.

After the assembly, the samples are subjected to unbiased highly accelerated stress test (UHAST): +130°C, 85%RH, 96 h; MST L3: 30°C, 60%RH, 192 h, 3-time reflow at 260°C; MST L1 85°C, 85%RH, 168 h, 3-time reflow at 260°C; HTS: +125°C, 1000 h; and TC: +125/–55°C, 15-min dwell, 15°C/min ramp, 1000 cycles. The typical image of failed samples after 1000 cyles is shown in Fig. 10.22. The crack initiates near the corner of the interface between the intermetallic compounds (IMC) (Cu_6Sn_5) and the bulk solder, and then propagates through the solder joint.

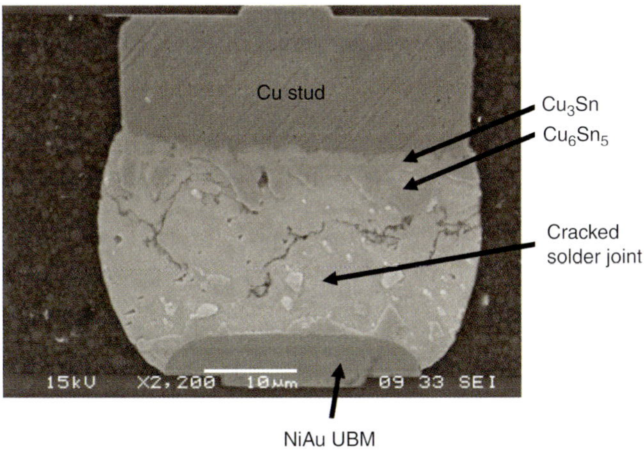

FIGURE 10.22 Failed sample after high temperature storage test, moisture sensitivity test level 1 and level 3, and 1000 thermal cycles test.

10.9 Low-Temperature C2W (InSnAu) Bonding for 3-D IC Chip Stacking

For 3-D IC chip stacking, as mentioned in Chap. 1, the chip is bonded with the base chip or other chips by either chip-to-chip (C2C) or chip-to-wafer (C2W) bonding methods. Due to IC chip yield and chip size differences, the wafer-to-wafer (W2W) bonding method is seldom used. Most current bonding methods use temperatures higher than 300°C [72–74]. Even the bonding temperature for silicon C2C, C2W, and W2W can be as low as room temperature [43]. However, for this kind of bonding, the bonding surfaces must be extremely flat and clean, which prevent it from being used for high-volume production. and thus it is beyond the scope of this book.

Low melting solders such as InAg, InCu, InSn, InNi, InAu, InSnAu, InCuNi, and InSnCu with a bonding temperature less than 200°C [34–71] are desired in order to reduce the damages (warpages) of the microstructures due to the thermal expansion mismatch of the bonding chips [one of them with copper-filled through-silicon-vias (TSVs)]. Also, the solder joints form robust IMC interconnects with high remelting temperatures, which allows stable 3-D IC chip stacking.

It should be emphasized that with the In-base solder (melting point ~156°C), the choice of the right composition of the solder and UBM to form nothing but IMC with a high remelting temperature is extremely important. Since the In atoms with additional elements tend to form a ternary eutectic phase, which has an even lower melting temperature than the In binary eutectic point [39], it is necessary to avoid forming the ternary eutectic phase by properly designing the suitable layer combination.

10.9.1 How Does Low-Temperature Bonding with Solder Work?

The basics of low-temperature bonding are shown in Fig. 10.23. The low-temperature solder (e.g., InAg, InCu, InSn, InNi, InAu, InSnAu, InCuNi, and InSnCu) is coated on the specially designed and fabricated UBM (e.g., Cu, Cu/Au, Ti/Au, Ni/Au, Ti/Cu, and Cu/Ti/Au). After aligning and bonding at a temperature of less than 200°C, all the solders will react with the UBM on the chips/wafers and become IMCs with melting points that are a couple of hundred degrees higher than that of the solders.

This feature is welcome by 3-D IC chip stacking and 3-D MEMS (microelectromechanical systems) packaging. For example, after bonding the first two chips at a low temperature, all the bonding interconnects are converted into IMC with a much higher remelting temperature. When the third chip is bonded on top of these two chips with the low temperature, the interconnects (IMC) between those two

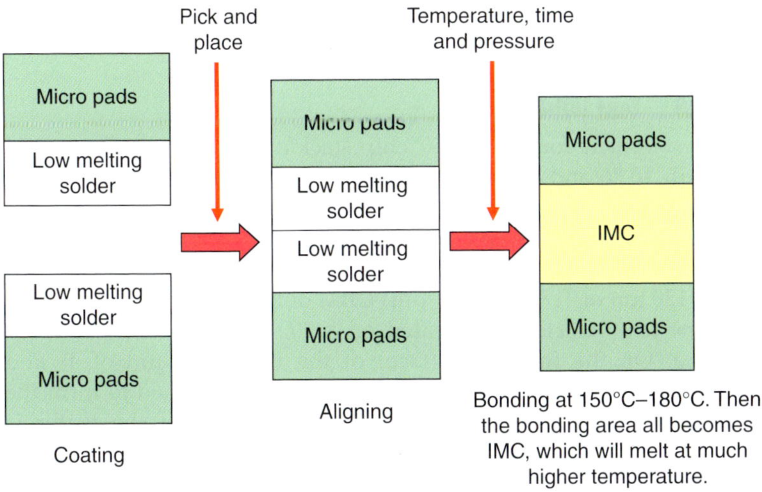

FIGURE 10.23 Fundamentals of low-temperature bonding.

chips will not be reflowed and moved. For another example, after the bonding of an MEMS device to an ASIC (application-specific IC) wafer with a low melting solder, all the bonding (solder-interconnect) areas become the IMC with a very high remelting temperature. When a cap wafer is bonded to the ASIC wafer (which is already bonded with the MEMS device), the interconnect between the ASIC and the MEMS devices will not be reflowed and moved. Furthermore, when the whole 3-D IC chip-stacking and/or 3-D MEMS packaging are attached to the printed circuit board (PCB) with surface mount technology (SMT) lead-free soldering (with a maximum reflow temperature as high as 260°C), the solder interconnects (IMC) between the chips for the 3-D chip stacking and between the ASIC and MEMS (for 3-D MENS) will not be reflowed and moved.

10.9.2 Solder Design

The use of indium with Ni, Cu, or Au has been studied for the low-temperature bonding in 3-D integrations [34–37]. Usually, a barrier layer such as Ti is inserted between the In layer and the other metal layers to prevent the excessive interdiffusion [38].

In this section, the In-based solder layer is chosen for the low-temperature solder and bonded to a thin Au layer to form AuIn-based IMC after bonding. Since the diffusion of Au into the In layer is fast enough to form AuIn intermetallics even at room temperature, a thin layer of Sn is inserted between the In and Au layers to minimize the interdiffusion before bonding, that is, the Sn reacts with the Au and temporarily forms a diffusion barrier layer before bonding so that the In layer is able to be preserved. On the other hand, this barrier layer should not interrupt the interfacial reaction between the In and Au layers during and after bonding to form the IMC completely.

10.9.3 Test Vehicle

The test vehicle for the 3-D IC chip stacking is shown in Figs. 10.24 through 10.26 and Table 10.4. The silicon test chip (8×8 mm^2) has ~1700 solder bumps and each consists of 100 μm in diameter of Ti/Au (0.1/1 μm) UBM and 80 μm in diameter of Sn/In (0.5/2 μm) solder with a flesh of Au (0.05 μm). The other side of the chip has 120×120 μm of Ti/Au (0.1/1 μm) UBM only.

The test vehicle is fabricated on a 200 mm Si wafer with SiO_2 coating. On the front side (face) of the wafer, 0.1 μm of Ti and 1 μm of Au are sputtered on the SiO_2 and are patterned to form the 120×120 μm UBM. Then the back side (back) of the wafer is

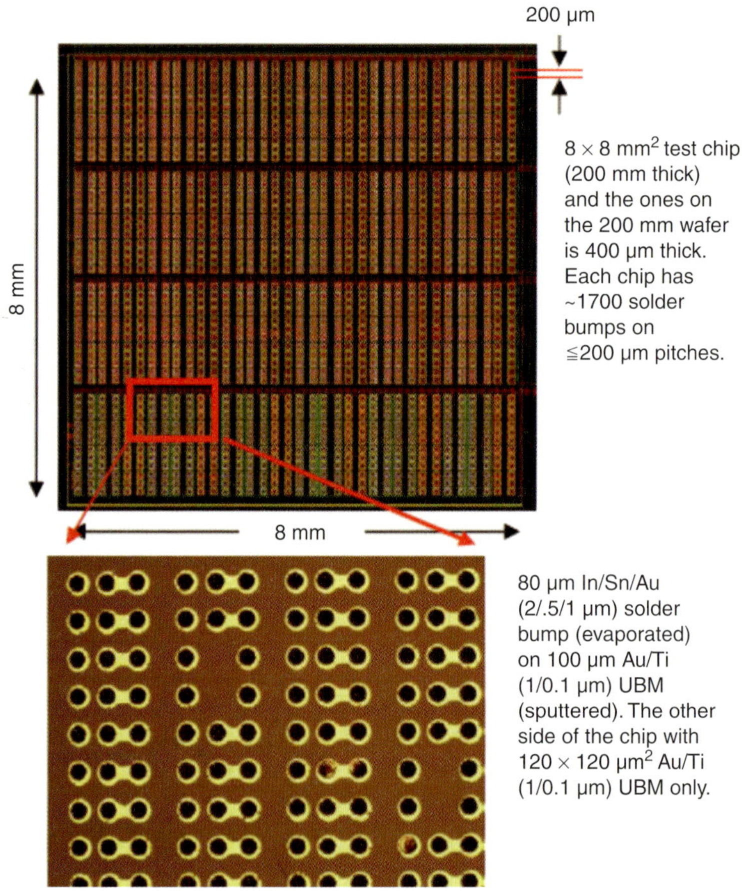

FIGURE 10.24 Solder bumps and pads on the top surface of the test vehicle.

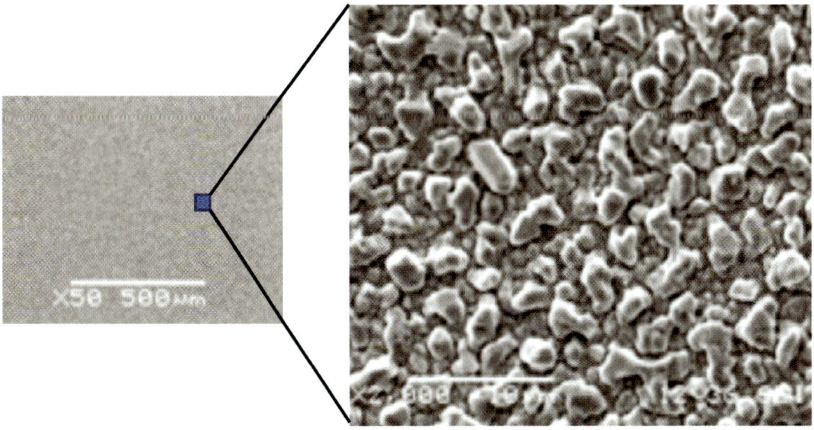

FIGURE 10.25 Scanning electron microscope (SEM) images (top view) of as-evaporated composite coating surface morphology.

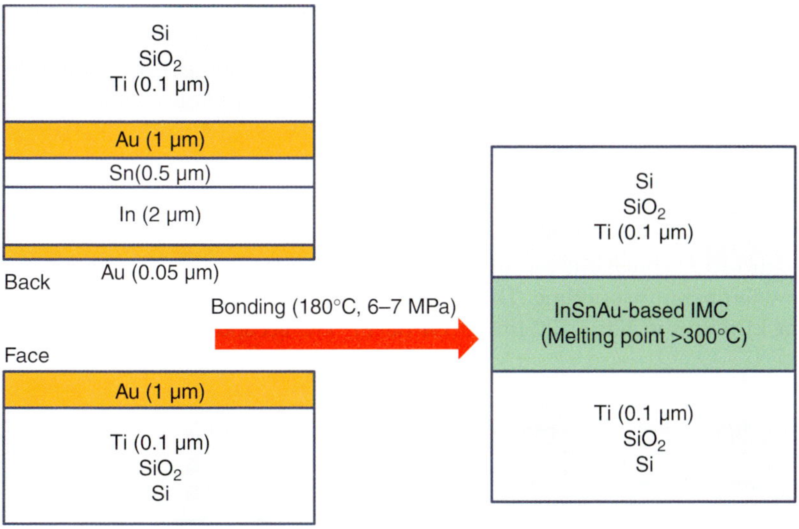

FIGURE 10.26 Schematic of the chip with UBM + solder bump and the chip with UBM only and the IMC after bonding.

sequentially (a) back-grinded to 200 μm thick, (b) coated with SiO_2 for the passivation layer, (c) deposited with 0.1 μm of Ti and 1 μm of Au (in a sputter chamber), (d) wet etched to pattern the 100 μm in diameter (Ti/Au) UBM pads, (e) laminated with a 20-μm dry film, (f) the openings for the solder bumps are developed, (g) the Sn (0.5 μm) is deposited on the Au (of the UBM) and then the In (2 μm) on the Sn in an evaporation chamber, and (h) the dry film is stripped (a lift-off process).

Items	Values
Wafer size	200 mm
Chip size	8 mm
Stacking chip thickness	200 mm
Base chip thickness	400 mm
Solder bump diameter	80 μm
Solder bump pitch	≤ 200 mm
Number of bumps	~ 1700
UBM (Ti/Au)	0.1/1 μm
Solder (In) bump	2.5 μm
Total bump height	3.6 μm

TABLE 10.4 Test Vehicle Dimensions

Figure 10.25 shows the top view of the as-evaporated composite coating surface morphology. The surface is rough and has a granular structure with a grain size of 2 to 3 μm. In order to bond surface with such rough features, a molten solder layer and high pressure are necessary to fill up the gaps between surfaces.

10.9.4 3-D IC Chip Stacking with InSnAu Low-Temperature Bonding

Figure 10.27 shows the assembly process of 3-D IC chip stacking with low-temperature bonding. The first layer is the 200-mm wafer (400 μm thick) with the 120 × 120 μm UBM (face). The second chip with the

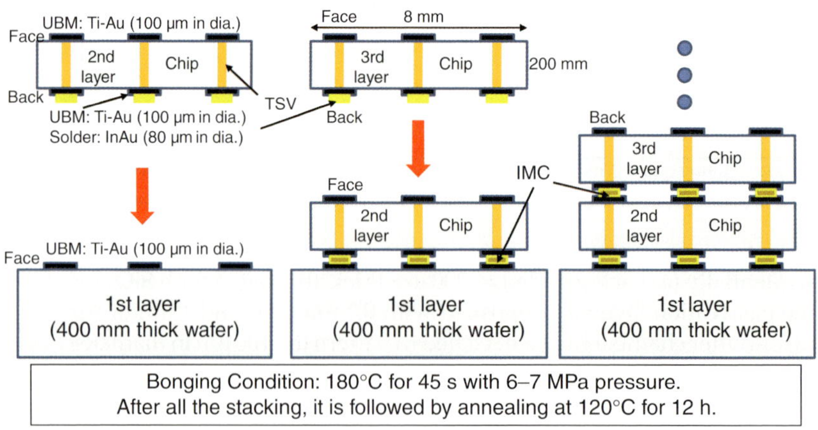

FIGURE 10.27 3-D IC stacking with low-temperature bonding.

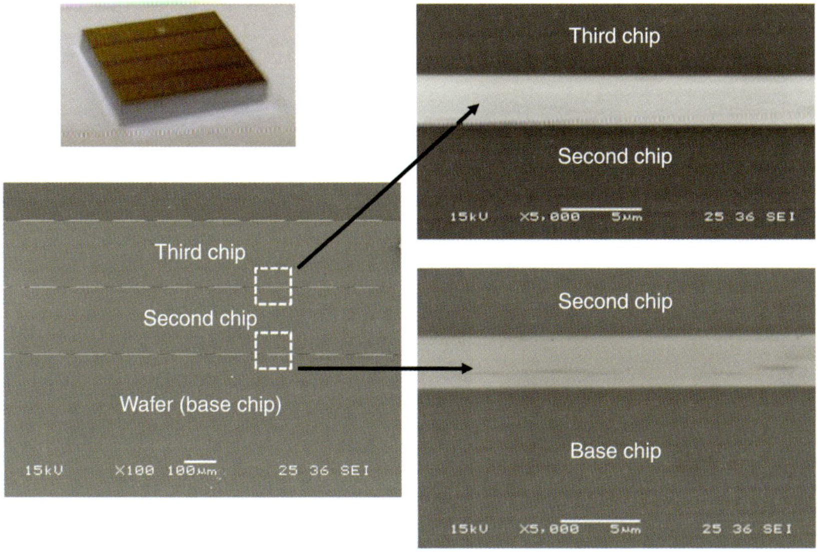

FIGURE 10.28 The assembled module and the SEM micrographs of the cross-sectional views of the three-layer stacked chips consisting of 1700 micro solder joints on each interconnect layer.

lead-free InSnAu solder bump (back) is bonded on the face of the wafer with 180°C and 6 MPa to 7 MPa for 45 s. After bonding, the lead-free solder interconnects become InSnAu IMC and the remelt temperature is a couple hundred degrees higher than the melting point of In. Then, the back of the third chip with solder bumps is bonded on the face of the second chip with the same conditions. The stacking is followed by annealing at 120°C for 12 h. Figure 10.28 shows the assembled three-chip stacking by the present low-temperature bonding process. The cross-sections show that the bonding is in a very good condition and there are no visible voids. The design of experiments leads to the present optimal bonding condition given in [42].

10.9.5 SEM, TEM, XDR, and DSC of the InSnAu IMC

Since the IMC layers are relatively thin, the transmission electron microscope (TEM) technique is used because it provides local chemical analysis. Figure 10.29 shows the TEM images of the region close to the base material silicon and interfacial reaction region between the AuTi and solder intermediate layers. The interfaces between each layer are void-free, uniform, and fully converted to IMC. The elemental composition results (at percent) of the selected points are labeled in Fig. 10.29. Two IMCs are identified along the main reaction zone, that is, InAu and InSnAu. This result is consistent with the x-ray diffraction (XRD) results, which confirmed the presence of these two IMCs in the bonding interface. Figure 10.30 shows the

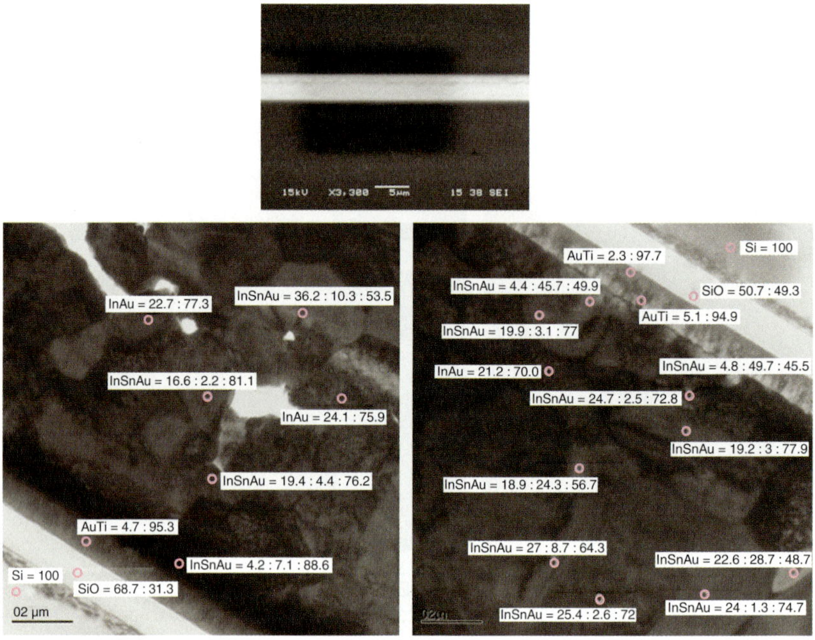

FIGURE 10.29 SEM (top) and TEM (transmission electron microscope) images at the interface region.

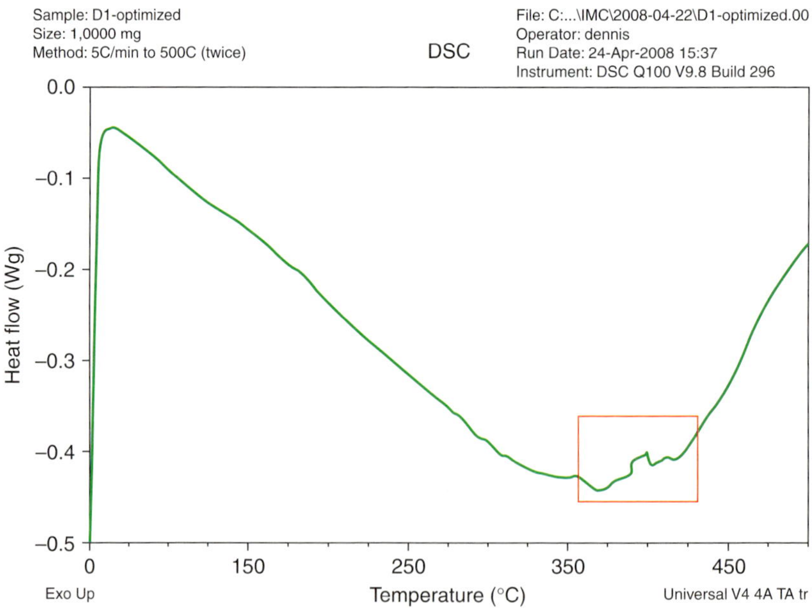

FIGURE 10.30 The DSC (differential scanning calorimetry) curve clearly shows the melting point ~360°C.

differential scanning calorimetry (DSC) curve indicating that the remelting temperature is higher than 365°C. The SEM image (Fig. 10.29) shows that uniform bonding has been made without any defects.

10.9.6 Young's Modulus and Hardness of the InSnAu IMC

The elastic modulus and hardness of the AuSn-In based IMC interconnects are determined by a nano-indentation method. Since the size of IMC layers from the cross-section of a sample is too thin for the tip of an optical microscope, the measurements are carried out along the fractured surfaces on the chips after shear tests. It is found that the Young's modulus of InSnAu is about 81 GPa, which is a little higher than that of Au (78 GPa) [40] and much lower than that of Cu_6Sn_5 (112.6 GPa) and Cu_3Sn (132.7 GPa) [41]. The measured hardness is ~1.5 GPa.

10.9.7 Three-Time Reflows of the InSnAu IMC

There are at least two different kinds of low-temperature assemblies—namely, one is for chip-level interconnects in a package and the other is for board-level interconnects. For board-level interconnects, since it is the final assembly, low-melting solders can be used to attach the packages on the board, as long as the solder joints are reliable (see, e.g., Chap. 8). However, for chip-level interconnects in a package, since the package has to be assembled on the board with lead-free solders, the package must be qualified before it is mounted on the PCB. Some of the qualifying tests, among others, are:

1. Temperature cycling test (−40°C to +125°C for 1000 cycles). The package should be preconditioned with humidity exposure and at least three lead-free SMT reflow cycles.
2. Biased HAST (130°C/85%RH, 1.8 V for 100 h)
3. Pressure-cook (steam) test (121°C, 100%RH, at 2 atm for 168 h)

In general, the quality and reliability requirements of the chip-level interconnects are always tougher than those of board-level interconnects, that is, the low-temperature solders that work for board-level interconnects may not work for chip-level interconnects. Figure 10.31 shows a typical reflow profile for qualifying packages. It also shows the cross-section at different magnifications of the IMC interconnects. After three times of reflow, there is no visible change of the InSnAu IMC interconnects of the 3-D IC chip stacking.

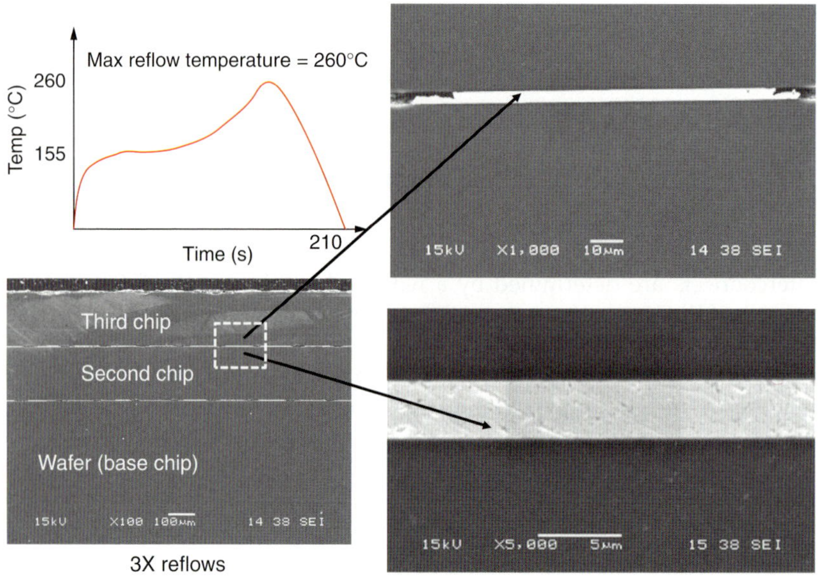

FIGURE 10.31 Cross-sections of the 3-D IC stacking after three-time reflows with a maximum temperature of 260°C.

10.9.8 Shear Strength of the InSnAu IMC

Since the InSnAu IMC is very strong and can resist shear, many of the 200 μm chips are broken during shear (horizontal push) tests. Thus, to determine the shear strength of the InSnAu IMC, 400 μm chips are used for the tests. Figure 10.32 shows the typical failure mode of the InSnAu IMC interconnects under shear tests. The fracture surfaces occur along the IMC joint and the interface between the TiAu UBM and the InSnAu IMC joint.

Figure 10.33 shows the shear strength test results of the InSnAu IMC interconnects as fabricated and after three times of reflows. It can be seen that: (1) the shear strength of the first IMC joint (between the base chip and the second chip) and the second IMC joint (between the second chip and the third chip) is reduced by ~15% because of the three-time reflows; (2) even with the reduction, however, their shear strength is still much higher than the required 20 MPa; and (3) for both cases, the shear strength of the first IMC joint is slightly higher (~8%) than that of the second IMC joint. This could be due to the shear test setup: (a) first shear (horizontal push) the third chip and hold the bottom two chips to obtain the second IMC strength; (b) then push the second chip and hold the base chip to get the first IMC strength. Since the holders are not 100% rigid, some of the push forces become relaxed. Also, the larger the holder, the more the push force relaxes.

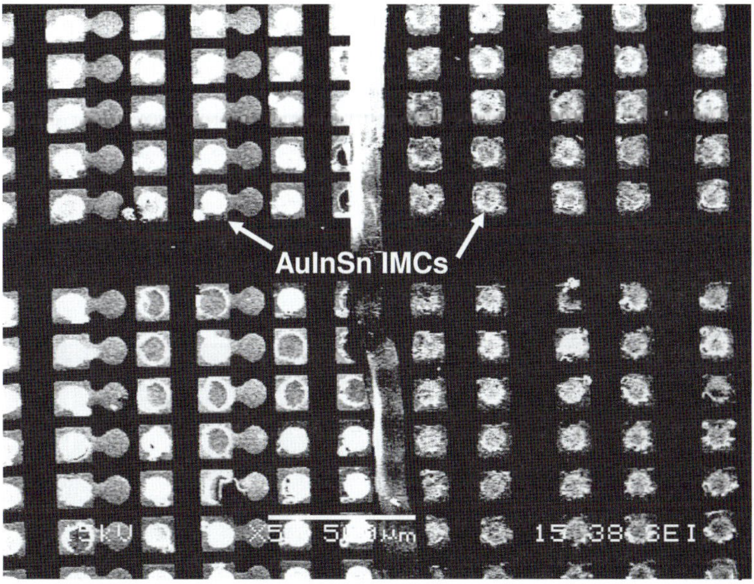

FIGURE 10.32 The fracture surfaces occur along the IMC joints and the interface between the UBM and IMC joints.

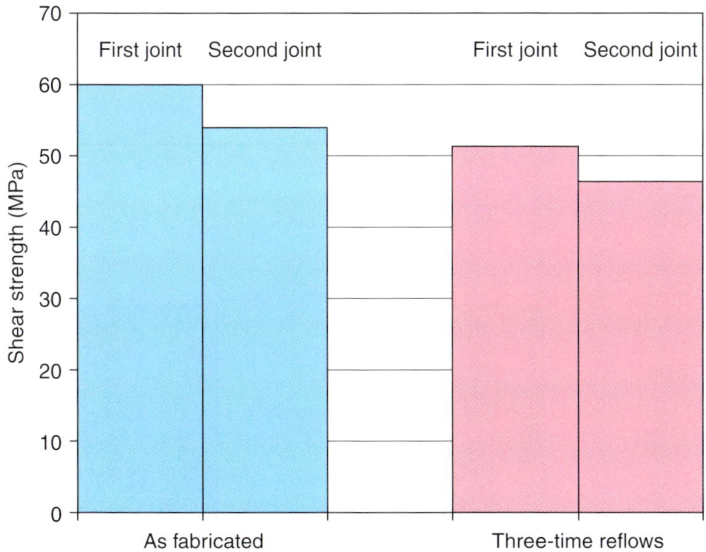

FIGURE 10.33 Bonding shearing strength of the three-layer stack: as fabricated and three-time reflows.

10.9.9 Electrical Resistance of the InSnAu IMC

The electrical resistance of the InSnAu IMC interconnect can be measured by Kevin's four-point measurement method, as shown in Fig. 10.34. Two bumps with 200 μm diameter are formed far away (800 μm) from each other. Figure 10.35 shows the measurement

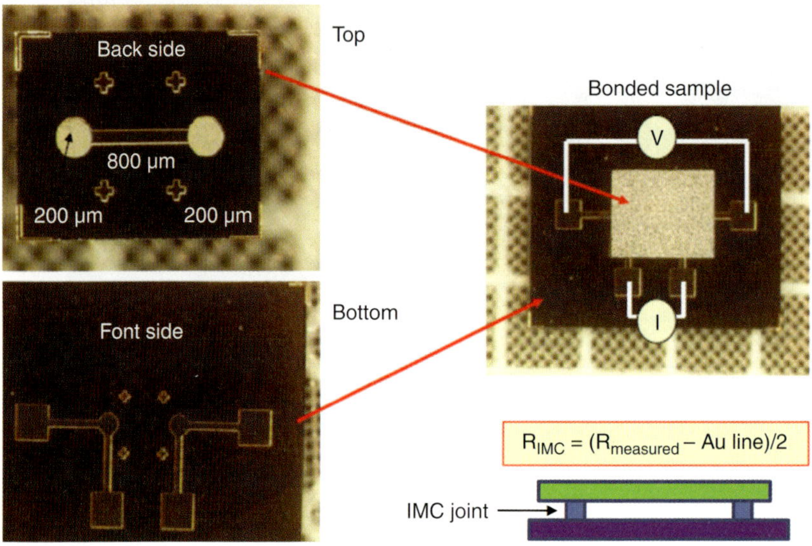

FIGURE 10.34 The test fixtures for the electrical resistance measurement.

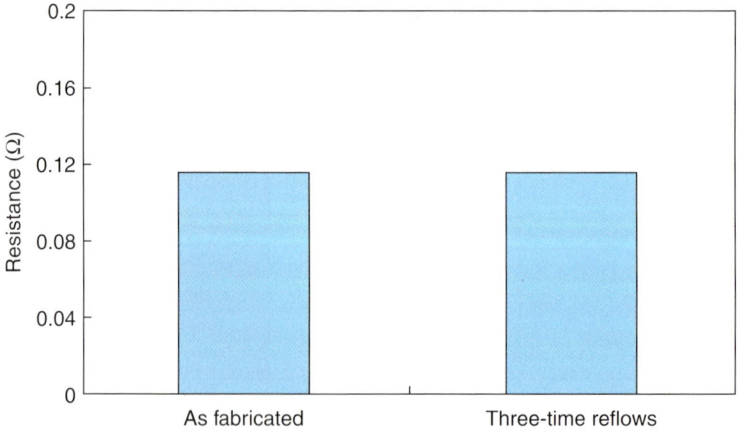

FIGURE 10.35 IMC joint electrical resistance: as fabricated and after three-time reflows.

results, and it can be seen that the electrical resistance of the InSnAu IMC joint is ~0.12 Ω and it is not degraded even after three times of reflow at 260°C.

10.9.10 When Does the InSnAu IMC Become Unstable?

It has been shown earlier by the DSC method that the remelting temperature of the InSnAu IMC is greater than 365°C. It would be interesting to find out the temperature when the IMC of the 3-D IC chip stack collapses. However, due to the platinum crucible size limit of the DSC equipment, the whole 3-D IC chip stack cannot be put in and measured by it. Instead, an indirect method is adopted, that employs the thermal mechanical analyzer (TMA) to measure the expansion of the whole 3-D IC chip stack versus temperature, as shown in Fig. 10.36. Before the InSnAu IMC starts to remelt, the curve should be moving upward (very close to a straight line) with the temperature. However, when the InSnAu IMC starts to remelt, the expansion curve starts to become unstable until it drops, which means the 3-D IC chip stack is totally collapsed. Thus, based on the three samples tested under TMA shown in Fig. 10.36, the IMC in the 3-D IC chip stack could start remelting at around 380°C to 400°C (these are higher than 365°C because the whole 3-D IC chip stacking is measured and absorbed more heat) and collapse at around 450°C to 480°C.

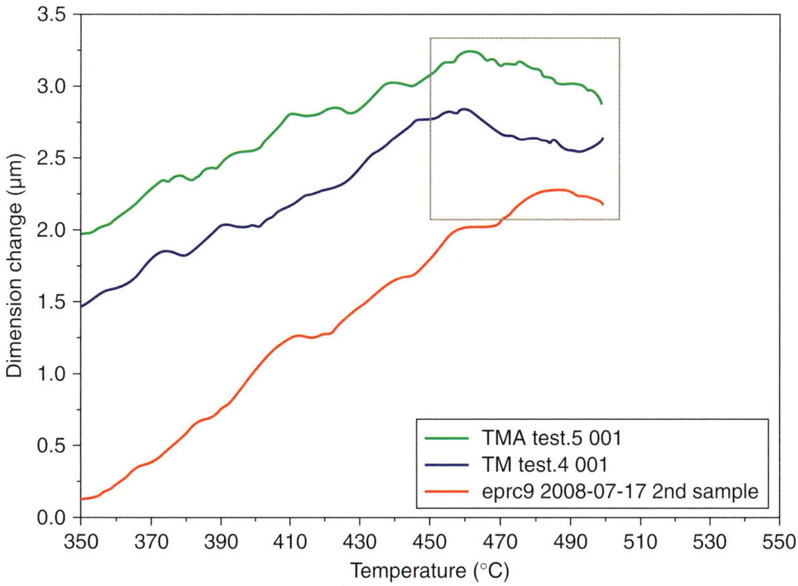

FIGURE 10.36 TMA results: remelting happens around 400°C.

10.9.11 Summary and Recommendations

Some important results and recommendations are summarized as follows:

1. A low-temperature solder system (InSnAu) has been designed for 3-D IC chip-stacking applications. The bonding condition is 180°C and 6 MPa for 45 s. After bonding, the whole assembly is annealed at 120°C for 12 h.
2. SEM, XDR, and TEM images show that uniform bonding has been made without any defects and all the bonding areas have been converted into InSnAu and InAu IMCs. Also, DSC results show that the remelting temperature of the IMC is greater than 365°C.
3. After three times of reflow (with a maximum temperature of 260°C), there is no (a) visible change of the InSnAu IMC of the 3-D IC chip stacking and (b) no change of the electrical contact resistance of the IMC.
4. Because of the three-time reflows (with a maximum temperature of 260°C), the shear strength of the InSnAu IMC is reduced by 15%. Even with this reduction, the IMC strength is still more than two times higher than the required 20 MPa.
5. The Young's modulus of the InSnAu IMC has been measured on the fracture face by a nano-indentation method, and is equal to 81 GPa. The measured hardness of the IMC is 1.5 GPa.
6. The failure modes (fracture surfaces) of the InSnAu IMC interconnects subjected to shear are along the IMC and at the interface between the TiAu UBM and the InSnAu IMC.
7. For 3-D IC chip stacking, since the chip thickness could be as thin as 20 μm, which makes the ordinary characterization methods very difficult or impossible, new methods and equipments are desperately needed.

10.10 References

1. Carson, F., K. Ishibashi, and Y. Kim, "Three-Tier PoP Configuration Utilizing Flip Chip Fan-In PoP Bottom Package," *IEEE Proceedings of Electronic Components & Technology Conference*, San Diego, CA, 2009, pp. 313–318.
2. Xie, D., D. Shangguan, D. Geiger, D. Gill, V. Vellppan, and K. Chinniah, "Head in Pillow (HIP) and Yield Study on SIP and PoP Assembly," *IEEE Proceedings of Electronic Components & Technology Conference*, San Diego, CA, 2009, pp. 752–758.
3. Sze, T., M. Giere, B. Guenin, N. Nettleton, D. Popovic, J. Shi, R. Ho, R. Drost, D. Douglas, and S. Bezuk, "Proximity Communication Flip-Chip Package with Micron Chip-to-Chip Alignment Tolerances," *IEEE Proceedings of Electronic Components & Technology Conference*, San Diego, CA, 2009, pp. 966–971.
4. Kumar, A., V. Sekhar, S. Lim, C. Keng, G. Sharma, S. Vempati, V. Kripesh, J. H. Lau, D. Kwong, and X. Dingwei, "Wafer Level Embedding Technology for 3D Wafer Level Embedded Package," *IEEE Proceedings of Electronic Components & Technology Conference*, San Diego, CA, 2009, pp. 1289–1296.

5. Kim, D., Y. Kim, K. Seong, J. Song, B. Kim, C. Hwang, and C. Lee, "Evaluation for UV Laser Dicing Process and Its Reliability for Various Designs of Stack Chip Scale Package," *IEEE Proceedings of Electronic Components & Technology Conference*, San Diego, CA, 2009, pp. 1531–1536.
6. Sharma, G., S. Vempati, A. Kumar, S. Nandar, Y. Lim, K. Houe, S. Lim, N. Vasarla, R. Ranjan, V. Kripesh, and J. H. Lau, "Embedded Wafer Level Packages with Laterally Placed and Vertically Stacked Thin Dies," *IEEE Proceedings of Electronic Components & Technology Conference*, San Diego, CA, 2009, pp. 1537–1543.
7. Pendse, R., "Flip Chip Package-in-Package (fcPiP): A New 3D Packaging Solution for Mobile Platforms," *IEEE Proceedings of Electronic Components & Technology Conference*, Reno, NV, May 2007, pp. 1425–1430.
8. Carson, F., Lee, S.M., Yoon, I.S., "The Development of the Fan-in Package-on-Package," *IEEE Proceedings of Electronic Components & Technology Conference*, Lake Buena Vista, FL, May 2008, pp. 956–963.
9. Kazuo, I., "PoP (Package-on-Package) Stacking Yield Loss Study," *IEEE Proceedings of Electronic Components & Technology Conference*, Reno, NV, May 2007, pp. 1403–1408.
10. Carson, F., S. Lee, and N. Vijayaragavan, "Controlling Top Package Warpage," *IEEE Proceedings of Electronic Components & Technology Conference*, CA, 2009, Reno, NV, May 2007, pp. 737–742.
11. Vijayaragavan, N., Carson, F., and Mistri, A., "Package on Package Warpage—Impact on Surface Mount Yields and Board Level Reliability," *IEEE Proceedings of Electronic Components & Technology Conference*, Lake Buena Vista, FL, May 2008, pp. 389–396.
12. IPC/JEDEC, "Moisture/Reflow Sensitivity Classification for Nonhermetic Solid State Surface Mounted Devices," *JSTD-020D*, June 2007.
13. Geiger, D., D. Shangguan, S. Tam, and D. Rooney, "Package Stacking in SMT for 3D PCB Assembly," *IEEE Proceedings of IEMT*, San Jose, CA, July 2003, pp. S207P6.
14. Carson, F., "Package-on-Package Variations on the Horizon," *Semiconductor International*, May 1, 2008.
15. Wei, L., A. Yoshida, and M. Dreiza, "Control of the Warpage for Package-on-Package (PoP) Design," *Proceedings of SMTAI*, Chicago, IL, 2007, pp. 320–327.
16. Dongji, X., D. Geiger, D. Shangguan, B. Hu, and J. Sjöberg, "Yield Study of Inline Package on Package (PoP) Assembly," *IEEE Proceedings of Electronic Packaging Technology Conference*, Singapore, Lake Buena Vista, FL, May 2008, pp. 1202–1208.
17. Kazuo, I., "PoP (Package-on-Package) Stacking Yield Loss Study," *IEEE Proceedings of Electronic Components & Technology Conference*, Reno, NV, 2007, pp. 1043–1048.
18. Jonas, S., D. Geiger and D. Shangguan, "Package on Package Process Development and Reliability Evaluation," *IEEE Proceedings of Electronic Components & Technology Conference*, Lake Beuna Vista, FL, May 2008, pp. 2005–2010.
19. Souriau, J., O. Lignier, M. Charrier, and G. Poupon, "Wafer Level Processing of 3D System in Package for RF and Data Applications," *IEEE Proceedings of Electronic Components & Technology Conference*, Lake Beuna Vista, FL, 2005, pp. 356–361.
20. Brunnbauer, M., E. Furgut, G. Beer, T. Meyer, H. Hedler, J. Belonio, E. Nomura, K. Kiuchi, K. Kobayashi, "An Embedded Device Technology Based on a Molded Reconfigured Wafer," *IEEE Proceedings of Electronic Components & Technology Conference*, San Diego, CA, 2006, pp. 547–551.
21. Keser, B., C. Amrine, T. Duong, Owen Fay, S. Hayes, G. Leal, W. Lytle, D. Mitchell, R. Wenzel "The Redistributed Chip Package: A Breakthrough for Advanced Packaging," *IEEE Proceedings of Electronic Components & Technology Conference*, Reno, NV, 2007, pp. 286–290.
22. Fillion, R., Charles Woychik, T. Zhang, D. Bitting, "Embedded Chip Build-Up Using Fine Line Interconnect," *IEEE Proceedings of Electronic Components & Technology Conference*, Reno, NV, 2007, pp. 49–53.

23. Kripesh, V., V. S. Rao, A. Kumar, G. Sharma, K. C. Houe, Z. Xiaowu, K. Y. Mong, N. Khan, and J. H. Lau, "Design and Development of a Multi-Die Embedded Micro Wafer Level Package," *IEEE Proceedings of Electronic Components & Technology Conference*, Lake Beuna Vista, FL, May 2008, pp. 1544–1549.
24. Hamano, T., T. Kawahara, J. I. Kasai, "Super CSPTM: WLCSP Solution for Memory and System LSI," *in proc. of International Symposium on Advanced Packaging Materials*, 1999, pp. 221–225.
25. Braun, T., K. Becker, M. Koch, V. Bader, U. Oestermann, D. Manessis, R. Aschenbrenner, H. Reichl, "Wafer Level Encapsulation—A transfer Molding Approach to System in Package Generation," *IEEE Proceedings of Electronic Components & Technology Conference*, San Diego, CA, 2002, pp. 235–244.
26. Ostmann, A., A. Neumann, S. Weser, E. Jung, L. Bottcher and H. Reichl, "Realization of a Stackable Package Using Chip in Polymer Technology," *IEEE Proceedings of Polytronic*, 2002, pp. 160–164.
27. Ko, C., S. Chen, C.W. Chiang, T.-Y. Kuo, Y.C. Shih and Y.-H. Chen, "Embedded Active Device Packaging Technology for Next-Generation Chip-in-Substrate Packages, CiSP," *IEEE Proceedings of Electronic Components & Technology Conference*, San Diego, CA, 2006, pp. 322–329.
28. Takyu, S., T. Kurosawa, N. Shimizu, S. Harada, "Novel Wafer Dicing and Chip Thinning Technologies Realizing High Chip Strength," *IEEE Proceedings of Electronic Components & Technology Conference*, San Diego, CA, 2006, pp. 1623–1627.
29. Werner K., F., Mariani, "Thinning and Singulation of Silicon: Root Causes of The Damage in Thin Chips," *IEEE Proceedings of Electronic Components & Technology Conference*, San Diego, CA, 2006, pp. 1317–1322.
30. Steen, M., *Laser Material Processing*, Third Edition, Springer, New York, pp. 107–125.
31. Li, J., H. Hwang, E.C. Ahn, Q. Chen, P.W. Kim, T.H. Lee, M.K Chung, T.G. Chung, " Laser Dicing and Subsequent Die Strength Enhancement Technologies for Ultra-thin Wafer," *IEEE Proceedings of Electronic Components & Technology Conference*, Reno, NV, 2007, pp. 761–766.
32. Lim, S., V. Rao, H. Yin, W. Ching, V. Kripesh, C. Lee, J. H. Lau, J. Milla and A. Fenner, "Process Development and Reliability of Microbumps," *IEEE Proceedings of Electronic Packaging Technology Conference*, December 2008, pp. 367–372. Also, accepted for the *IEEE Transactions on Components and Packaging Technology*.
33. Vempati, S., S. Nandar, C. Khong, Y. Lim, V. Kripesh, J. H. Lau, B. P. Liew, K. Y. Au, S. Tamary, A. Fenner, R. Erich, and J. Milla, "Development of 3-D Silicon Die Stacked Package Using Flip Chip Technology with Micro Bump Interconnects," *IEEE Proceedings of Electronic Components & Technology Conference*, San Diego, CA, 2009, pp. 980–987.
34. Fukushima, T., Y. Yamada, H. Kikuchi, T. Tanaka, and M. Koyanagi, "Self-Assembly Process for Chip-to-Wafer Three-Dimensional Integration," *IEEE Proceedings of Electronic Components & Technology Conference*, Reno, NV, 2007, pp. 836–841.
35. Sakuma, K., P. Andry, C. Tsang, K. Sueoka, Y. Oyama, C. Patcl, B. Dang, S. Wright, B. Webb, E. Sprogis, R. Polastre, R. Horton, and J. Knickerbocker," *IEEE Proceedings of Electronic Components & Technology Conference*, Lake Beuna Vista, FL, May 2008, pp. 18–23.
36. Wakiyama, S., H. Ozaki, Y. Nabe, T. Kume, T. Ezaki, and T. Ogawa, "Novel Low-Temperature CoC Interconnection Technology for Multichip LSI (MCL)," *IEEE Proceedings of Electronic Components & Technology Conference*, Reno, NV, 2007, pp. 610–615.
37. Liu, Y. M., and T. H. Chuang, "Interfacial Reactions between Liquid Indium and Au-Deposited Substrates," *J. of Electron. Mater.*, Vol. 29, No. 4, 2000, pp. 405–410.
38. Zhang, W., A. Matin, E. Beyne, and W. Ruythooren, "Optimizing Au and In Micro-Bumping for 3D Chip Stacking," *IEEE Proceedings of Electronic Components & Technology Conference*, Lake Beuna Vista, FL, May 2008, pp. 1984–1989.

39. Liu, H., C. Liu, K. Ishida, and Z. Jin, "Thermodynamic Modeling of the Au-In-Sn System," *J. of Electron. Mater.*, Vol. 32, No. 11, 2003, pp. 1290–1296.
40. http://www.webelements.com/WebElements: the periodic table on the website.
41. Kazumasa, T., "Micro Cu Bump Interconnection on 3D Chip Stacking Technology," *Jpn J. Appl. Phys.*, Vol. 43, No. 4B, 2004, pp. 2264–2270.
42. Choi, W., C. Premachandran, C. Ong, L. Xie, E. Liao, A. Khairyanto, B. Ratmin, K. Chen, P. Thaw, and J. H. Lau, "Development of Novel Intermetallic Joints Using Thin Film Indium Based Solder by Low Temperature Bonding Technology for 3D IC Stacking," In *IEEE Proceedings of Electronics and Components Technology Conference,* San Diego, CA, May 2009, pp. 333–338.
43. Shigetou, A. Itoh, T., Sawada, K., and Suga, T., "Bumpless Interconnect of 6-um pitch Cu Electrodes at Room Temperature", In *IEEE Proceedings of Electronic and Components Technology Conference,* Lake Buena Vista, FL, May 27-30, 2008, pp. 1405–1409.
44. Made, R., Gan, C. L., Yan, L., Yu, A.,Yoon, S. U., Lau, J. H., and Lee, C. "Study of low temperature thermocompression bonding in Ag-In solder for packaging applications." *J. Electron. Mater.* 38:365–371, 2009.
45. Yan, L.-L., Lee, C.-K., Yu, D.-Q., Yu, A.-B., Choi, W.-K., Lau, J. H., and Yoon, S.-U. "A Hermetic Seal Using Composite Thin Solder In/Sn as Intermediate Layer and Its Interdiffusion Reaction with Cu." *J. Electron. Mater.* 38:200–207, 2009.
46. Yan, L.-L., Lee, V., Yu, D.,Choi, W. K., Yu, A.,Yoon, S.-U., and Lau, J. H. "A Hermetic Chip to Chip Bonding at Low Temperature with Cu/In/Sn/Cu Joint." In *IEEE Proceedings of Electronics and Components Technology Conference,* Orlando, FL, May 2008, pp. 1844–1848.
47. Yu, A., Lee, C., Yan, L., Made, R., Gan, C., Zhang, Q., Yoon, S., and Lau, J. H. "Development of Wafer Level Packaged Scanning Micromirrors." *Proc. Photon. West* 6887:1–9, 2008.
48. Lee, C., Yu, A., Yan, L., Wang, H., Han, J., Zhang, Q., and Lau, J. H. "Characterization of Intermediate In/Ag Layers of Low Temperature Fluxless Solder Based Wafer Bonding for MEMS Packaging." *J. Sensors Actuators* (in press).
49. Yu, D.-Q., Lee, C., Yan, L. L., Choi, W. K., Yu, A., and Lau, J. H. "The Role of Ni Buffer Layer on High Yield Low Temperature Hermetic Wafer Bonding Using In/Sn/Cu Metallization." *Appl. Phys. Lett.* (in press).
50. Yu, D. Q., Yan, L. L., Lee, C., Choi, W. K., Yoon, S. U., and Lau, J. H. "Study on High Yield Wafer to Wafer Bonding Using In/Sn and Cu Metallization." In *Proceedings of the Eurosensors Conference*, Dresden, Germany, 2008, pp. 1242–1245.
51. Yu, D., Lee, C., and Lau, J. H. "The Role of Ni Buffer Layer Between InSn Solder and Cu Metallization for Hermetic Wafer Bonding." In *Proceedings of the International Conference on Electronics Materials and Packaging,* Taipei, Taiwan, October 22–24, 2008, pp. 335–338.
52. Yu, D., Yan, L., Lee, C., Choi, W., Thew, M., Foo, C., and Lau, J. H. "Wafer Level Hermetic Bonding Using Sn/In and Cu/Ti/Au Metallization." In *IEEE Proceeding of Electronics Packaging and Technology Conference,* Singapore, December 2008, pp. 1–6.
53. Choi, W., Yu, D., Lee, C., Yan, L., Yu, A., Yoon, S., Lau, J. H., Cho, M., Jo, Y., and Lee, H. "Development of Low Temperature Bonding Using In-based Solders." In *IEEE Proceedings* of *Electronics Component and Technology Conference*, Orlando, FL, May 2008, pp. 1294–1299.
54. Choi, W., Premachandran, C., Ong, C., Ling, X., Liao, E., Khairyanto, A., Chen, K., Thaw, P., and Lau, J. H. "Development of Novel Intermetallic Joints Using Thin Film Indium Based Solder by Low Temperature Bonding Technology for 3D IC Stacking." In *IEEE Proceedings* of *Electronics Component and Technology Conference,* San Diego, CA, May 2009, pp. 333–338.
55. Chen, K., Premachandran, C., Choi, K., Ong, C., Ling, X., Khairyanto, A., Ratmin, B., Myo, P., and Lau, J. H. "C2W Bonding Method for MEMS Applications." In *IEEE Proceedings of Electronics Packaging Technology Conference,* Singapore, December 2008, pp. 1283–1287.

56. Premachandran, C. S., Lau, J. H., Ling, X., Khairyanto, A., Chen, K., and Myo Ei Pa Pa. "A Novel, Wafer-Level Stacking Method for Low-Chip Yield and Non-Uniform, Chip-Size Wafers for MEMS and 3D SIP Applications." In *IEEE Proceedings of Electronic Components and Technology Conference,* Orlando, FL, May 27–30, 2008, pp. 314–318.
57. Simic, V., and Marinkovic, Z. "Room Temperature Interactions in Ag-metals Thin Film Couples." *Thin Solid Films* 61:149–160, 1979.
58. Lin, J.-C. "Solid-Liquid Interdiffusion Bonding Between In-coated Silver Thick Films." *Thin Solid Films* 61:212–221, 1979.
59. Roy, R., and Sen, S. K. "The Kinetics of Formation of Intermetallics in Ag/In Thin Film Couples." *Thin Solid Films* 61:303–318, 1979.
60. Chuang, R. W., and Lee, C. C. "Silver-Indium Joints Produced at Low Temperature for High Temperature Devices." *IEEE Trans. Components Packag. Technol. A* 25:453–458, 2002.
61. Chuang, R. W., and Lee, C. C. "High-Temperature Non-Eutectic Indium-Tin Joints Fabricated by a Fluxless Process." *Thin Solid Films* 414:175–179, 2002.
62. Lee, C. C., and Chuang, R. W. "Fluxless Non-Eutectic Joints Fabricated Using Gold-Tin Multilayer Composite." *IEEE Trans. Components Packag. Technol. A* 26:416–422, 2003.
63. Humpston, G., et al. *Principles of Soldering and Brazing,* pp. 128–132. Materials Park, MD: ASM International, 1993.
64. Chuang, T., Lin, H., and Tsao, C. "Intermetallic Compounds Formed During Diffusion Soldering of Au/Cu/Al2O3 and Cu/Ti/Si with Sn/In Interlayer." *J. Electron. Mater.* 35:1566–1570, 2006.
65. Lee, C., and Choe, S. "Fluxless In-Sn Bonding Process at 140°C." *Mater. Sci. Eng.* A333:45–50, 2002.
66. Lee, C. "Wafer Bonding by Low-Temperature Soldering." *Sensors & Actuators* 85:330–334, 2000.
67. Vianco, P. T. "Intermetallic Compound Layer Formation Between Copper and Hot-Dipped 100 In, 50In-50Sn, 100Sn and 63Sn-37Pb Coatings." *J. Electron. Mater.* 23:583–594, 1994.
68. Frear, D. R. "Intermetallic Growth and Mechanical-Behavior of Low and High Melting Temperature Solder Alloys." *Metall. Mater. Trans. A* 25:1509–1523, 1994.
69. Morris, J. W. "Microstructure and Mechancial Property of Sn-In and Sn-Bi Solders." *J. Miner. Metals Mater. Soc.* 45:25–27, 1993.
70. Mei, Z. "Superplastic Creep of Low Melting-Point Solder Joints." *J. Electron. Mater.* 21:401–407, 1992.
71. Chuang, T. H. "Phase Identification and Growth Kinetics of the Intermetallic Compounds Formed During In-49Sn/Cu Soldering Reactions." *J. Electron. Mater.* 31:640–645, 2002.
72. Klumpp, A. "Vertical System Integration by Using Inter-Chip Vias and Solid Liquid Interdiffusion Bonding." *Jpn. J. Appl. Phys.* 43:L829–L830, 2004.
73. Chen, K. "Microstructure Examination of Copper Wafer Bonding." *J. Electron. Mater.* 30:331–335, 2001.
74. Morrow, P. R. "Three-dimensional Wafer Stacking via Cu-Cu Bonding Integrated with 65-nm Strained-Si/low-*k* CMOS Technology." *IEEE Electron. Device Lett.* 27:335–337, 2006.

CHAPTER 11

Wafer-to-Wafer (W2W) Bonding and Lead-Free Interconnect Reliability

11.1 Introduction

W2W bonding is one of the most important technologies in high-throughput semiconductor manufacturing [1–53]. As discussed in Chaps. 1 and 6: (a) W2W bonding is the only assembly technology for 3-D Si integration; (b) for 3-D integrated circuit (IC) integration, W2W bonding is not very common except for examples, the microelectromechanical systems (MEMS) high-yield cap wafer to the MEMS/ASIC (application-specific IC) wafer [1, 3], and the high-yield transparent plastic lens wafer to the light-emitting diode (LED)/driver wafer [2]; and (c) for microchannel applications, W2W bonding is used for bonding two high-yield through silicon via (TSV) micro-channnel wafers together (see Sec. 6.8). Since the W2W bonding for 3-D Si integration is bumpless and far from volume production, only W2W bondings with solder bumps are presented in this chapter. Two different solder systems are considered: one is the SnIn solder and CuTiAu metallization (≤180°C, W2W bonding for MEMS applications), and the other is an AuSn solder and CuTiNiAu metallization (≤350°C, W2W bonding for TSV and micro-channel application).

11.2 Low-Temperature W2W Bonding with SnIn Solder on CuTiAu Metallization Pads

11.2.1 The Test Vehicle

Figure 11.1 [51, 52] shows the schematic view of the top and bottom wafers for bonding and the detailed schematic of metallization deposited on wafers. It can be seen that: (1) there are 119 square sealing rings

FIGURE 11.1 Schematic view of top wafer and bottom wafer for bonding (top) and metallization deposited on wafers for low-temperature bonding (bottom).

with a size of 11 × 11 mm and a width of 300 μm; (2) the sealing rings are constructed by SnIn solder, and its phase diagram is shown in Fig. 11.2; (3) the under-bump metallurgy (UBM) is TiCuTiAu; and (4) a cavity with an area of 6 × 6 mm and a depth of 250 μm.

11.2.2 Test Vehicle Fabrication

At first, a 300 Å thickness of SiO and a 1500 Å thickness of SiN are formed on a 200-mm silicon wafer by the thermal oxidation and low-pressure chemical vapor deposition process. They act as a hard mask for the cavity etching. The patterning process is done by photolithography using dry film as a photoresistant material. The cavity inside each bonding ring is formed by using a wet-etching KOH process on both the cap and bottom wafers. The sealing rings are constructed by SnIn solder on Cu-based high-temperature components. TiCuTiAu metallization with a thickness of 0.05, 2, 0.05, and 0.03 μm is sputtered on the Si/SiO/SiN substrate. Here, the first thin Ti layer acts as the adhesive layer and the second Ti layer as the buffer layer. The thin

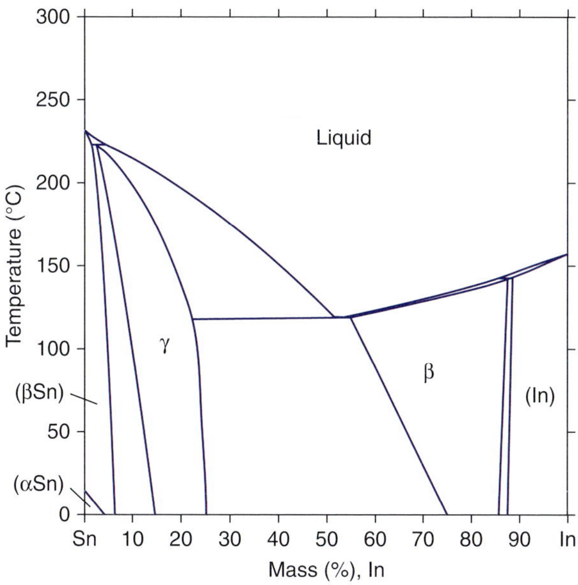

FIGURE 11.2 Phase diagram of InSn alloy.

Au layer is necessary for wetting and preventing metals from oxidizing before solder deposition.

In order to achieve an InSn alloy in a relative short time, four layers of Sn/In/Sn/In solders are deposited with a thickness of 1 μm, 1 μm, 0.7 μm, and 0.8 μm, respectively. According to this design, the atom percentage of Sn in InSn is about 50%, which is approximate to the eutectic composition of the InSn alloy. Lastly, a 0.03 μm thin layer of Au is deposited on its top to prevent oxidation during further process and storage. After materials deposition and dry film strip, O_2 plasma descum is conducted to remove the oxide layer and organic contaminants before the bonding process.

11.2.3 Low-Temperature W2W Bonding

The bonding is conducted in a controlled nitrogen atmosphere using the EVG520 wafer bonder with a pressure of 5.5 MPa upon the solder patterning area. The two bonded wafers are heated to a peak temperature of 180°C and 150°C, respectively. The reasons for selecting these two temperatures are: (1) the melting temperatures of Sn, In, and eutectic InSn alloys are, respectively, 232°C, 156°C, and 118°C; and (2) in order for solders to reflow properly the temperature must be 30 to 40°C higher than the melting temperature of the solders. Thus, the bonding temperature of 180°C is much higher than the melting temperature of both In and eutectic InSn alloys. By way of contrast, the bonding temperature of 150°C is mere a 30°C higher than the eutectic temperature of InSn. Therefore, these two bonding temperatures

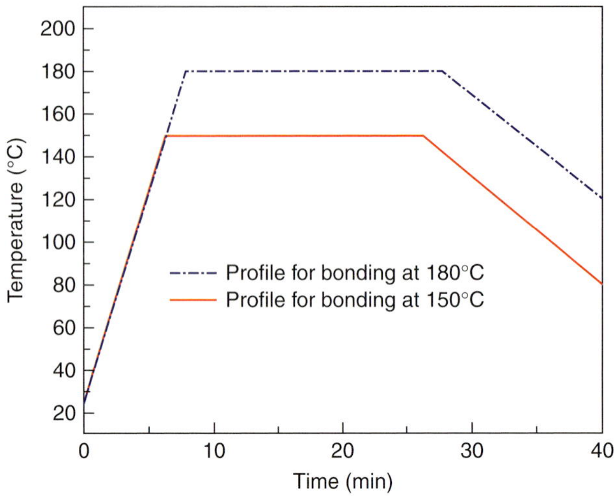

FIGURE 11.3 Temperature profiles for wafer bonding at 180 and 150°C.

would help determine the characteristics of W2W hermetic bonding using the InSn solder layers. Dwelling time at the peak temperature is 20 min to ensure sufficient diffusion between the solder and high-temperature components. Figure 11.3 shows the W2W bonding temperature profile, and Fig. 11.4 shows some typical C-mode scanning acoustic microscope (C-SAM) graphs for the bonded modules at 180°C (a) and 150°C (b). After bonding, the bonded wafers are diced into individual dies (13 × 13 mm) with a dicing speed of 2 mm/s.

11.2.4 C-SAM Inspection

After dicing, all the dies are nondestructively inspected by the C-mode-scanning acoustic microscopy (C-SAM) for the investigation of the W2W bonding interface. Some typical C-SAM images for the bonded modules are shown in Fig. 11.4. A die is defined as a "good die" if there are no detectable voids and/or cracks. The bonding yield is further defined as the percentage of good dies to the gross dies of a bonded wafer. Based on these definitions, 100% good seals (no voids and holes) throughout the 200-mm wafers bonded by both the 180°C and 150°C bonding temperatures have been obtained as shown in Fig. 11.4. However, since the resolution of C-SAM is about several micrometers, very tiny voids cannot be detected.

The effect of bonding temperature can also be observed by the C-SAM. More solder is squeezed out from the seal joint with a bonding temperature of 180°C than at 150°C. This is because at higher bonding temperatures, the flowability of liquid solder is better and, under the same pressure, more molten solders would flow over the seal ring, especially at the corner regions, as shown in the circles in Fig. 11.4*a*.

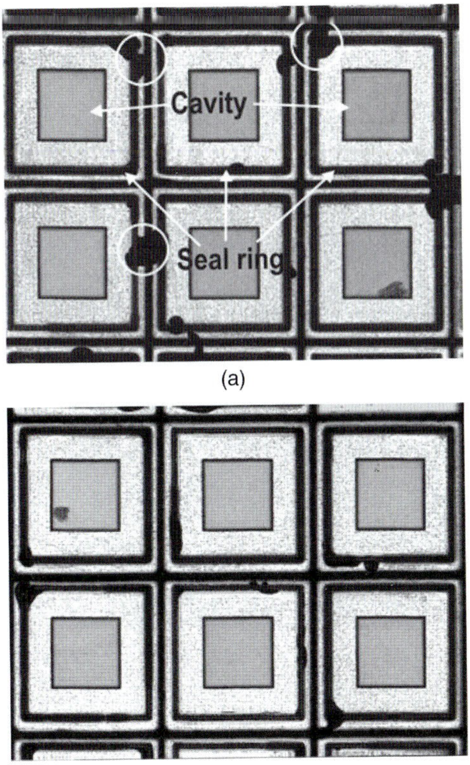

(a)

(b)

FIGURE 11.4 C-SAM graphs for bonded devices: bonding at (a) 180°C and (b) 150°C.

The maximum distance of the squeezed-out solder of the sealing ring is about 2 mm at the corners. For small packages, the flow-over solder may damage the device inside. A couple of simple yet useful methods are: (1) reduce the bonding pressure and/or solder thickness but at the same time ensure the yield; and (2) make grooves around the sealing ring to prevent the overflow of the solder.

11.2.5 Microstructure by SEM/EDX/FIB/TEM

For microstructure investigation, five samples for each bonding temperatures are mounted and cross-sectioned for SEM observation. These samples are grinded with SiC paper and polished with 1.0 μm diamond and 0.05 μm silica suspensions. The compositions of the bonded ring joints are analyzed by the energy dispersive x-ray (EDX). Focused ion beam (FIB) is employed to prepare the thin film for transmission electron microscopy (TEM) and EDX examinations of the solder joint.

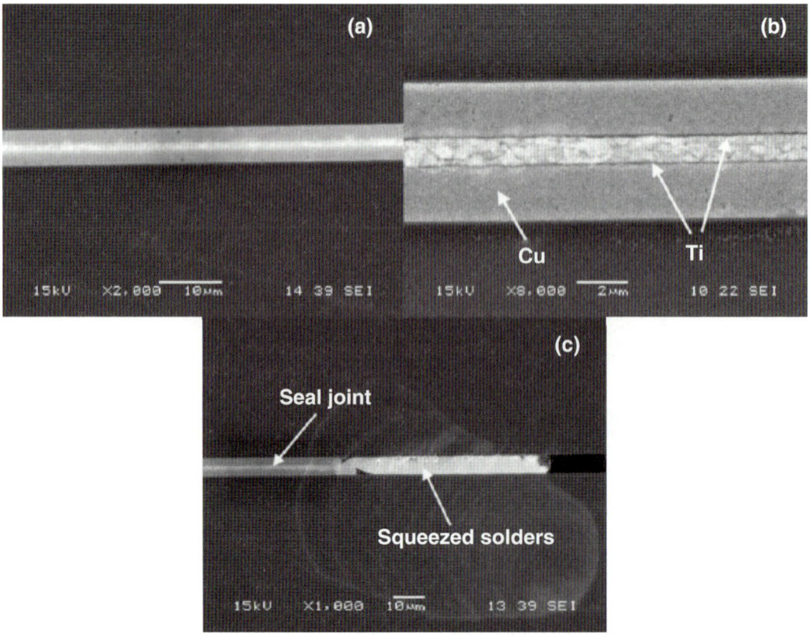

Figure 11.5 Interfacial microstructure of seal joint bonded at 180°C: (a) low magnification, (b) high magnification, and (c) solder squeezed out at the edge of the seal joint.

SEM images of sealing-ring joint bonding at 180°C are shown in Fig. 11.5. The thickness of the joint formed at 180°C is about 5.5 μm and that of the IMC is 1.5 μm, as shown in Figs. 11.5*a* and 11.5*b*. The whole joint after bonding is thinner, and there are no detectable voids and cracks. The overflow of solder is about 10 μm, as shown in Fig. 11.5*c*. The compositions of the solders are 55–60 at% Sn, 30–35 at% In, and 5–10 at% Au.

TEM/EDX is used to determine the accurate compositions along the sealing joint. A TEM microstructure is shown in Fig. 11.6, and the EDX results are listed in Table 11.1. It is quite obvious that two parallel Ti lines are embedded in the joint. According to the EDX results of **c** and **k**, a small amount of In, Cu, and Sn is found in the Ti layer, which means these elements can diffuse through the Ti layer. This proves the suitability of using Ti as a buffer layer for low-temperature W2W bonding. The compositions of **a**, **b**, and **l** adjacent to Cu sides show this phenomenon. For locations **a** and **b**, a small amount of Sn and In is found in Cu, which corresponds to the Cu-rich phase. For **l**, the ratio between Cu and SnIn is around 1:1. According to the thermodynamic calculation of phase equilibrium in a CuInSn system at 180°C as shown in Fig. 11.7 [54], an η phase exists in the system. In fact, the η phase exists in both the two binary systems—CuIn and CuSn,

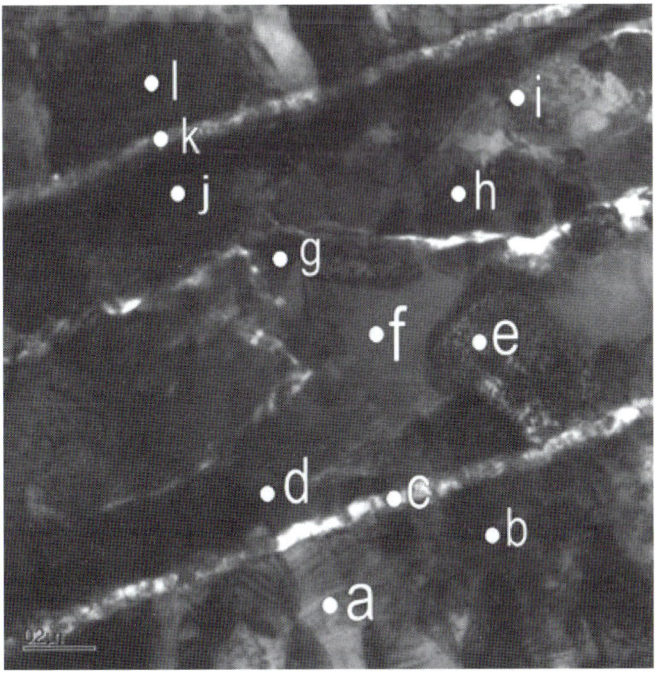

FIGURE 11.6 TEM/EDX analysis for the seal joint bonded at 180°C.

Position	Element (at%)				
	Cu	Ti	Sn	In	Au
a	88.1		7.1	4.7	
b	77		11.8	11.2	
c	5.3	79.6	9.9	5.1	
d	12.9	5.7	12.5	47.7	21.2
e	6.8		3	54.8	35.4
f			33.4	18.8	47.8
g	5.7		2.8	60.2	31.3
h	2.4		6.9	59.2	31.5
i	6.1		7.9	54.5	31.4
j	6	4	21.9	44.4	23.7
k	4.8	90.4	2.6	2.2	
l	50.3		30.7	19.1	

TABLE 11.1 Compositions of Selected Positions of TEM/EDX Analysis

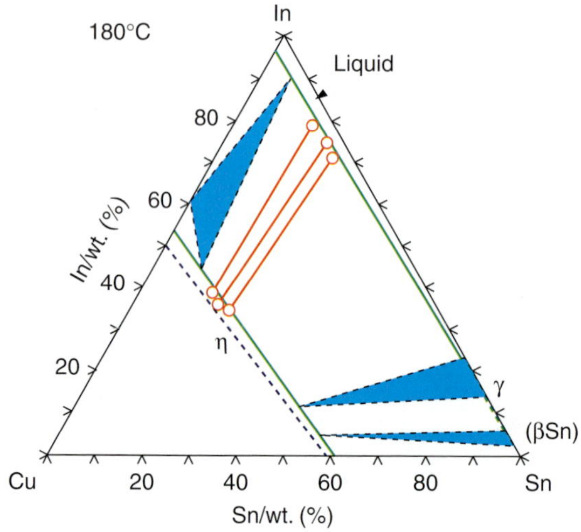

Figure 11.7 Isothermal section of CuInSn system at 180°C.

that is, η-Cu_2In and η-Cu_6Sn_5, with the same NiAs-type structure [55]. According to the study of Sommadossi et al. and the high solubility of Sn, the compounds should be η-$Cu_6(Sn\text{-}In)_5$ [55].

For **e**, **j**, **h**, **i** the compositions correspond to the $Au(InSn)_2$ phases. The small Cu content may be dissolved in the compounds during the liquid/solid diffusion bonding process. For location **d**, the composition of CuSnInAu comes from two phases: AuInSn and CuSnIn. For point **f**, the composition falls in the Au(SnIn) phase, which may be formed due to the absence of In atoms since a part of the solder is squeezed out.

SEM photos of the sealing-ring joint bonding at 150°C are shown in Fig. 11.8. The thickness of the joint is about 6.5 μm, in which the thickness of intermetallic compounds (IMCs) is around 2 μm. Comparing Fig. 11.5 with Fig. 11.8, we can also find that the thickness of the joint formed at 150°C was thicker than that formed at 180°C. The thickness differences of these two sealing joints are mainly caused by the different bonding temperatures. At higher temperature, solder material melts faster and has better flowability. Under the same pressure, more solder liquid is squeezed out at higher bonding temperatures. So a thin solder joint is formed with a thin IMC layer, since more solder alloy is kept at 150°C for diffusion bonding, that is, more solder materials are diffused through the Ti layer into Cu substrates, as shown in Fig. 11.8*b*.

When bonding at 150°C, due to the low bonding temperature, cracks at the center of the sealing joint are formed that may be harmful to its hermeticity and reliability properties. The reasons for the poor bonding qualities are: (1) at the surface of the solder materials,

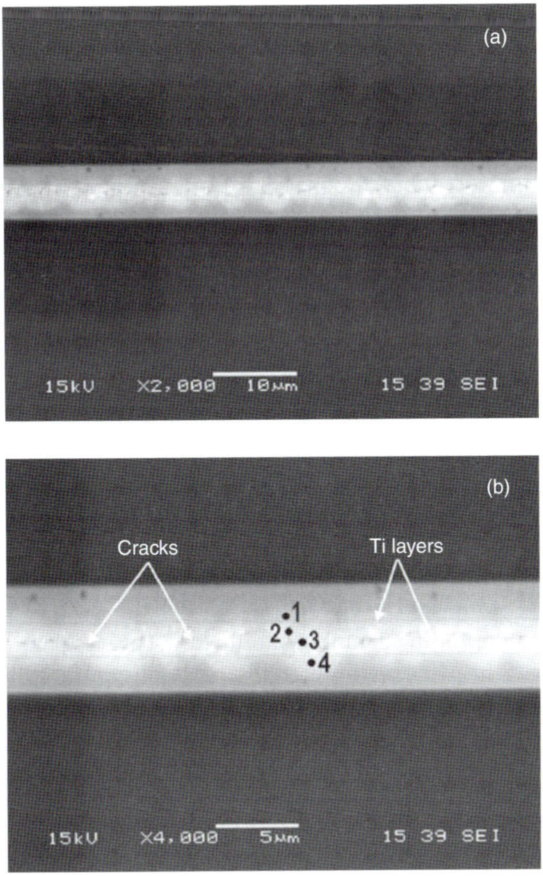

Figure 11.8 Interfacial microstructure of seal joint bonded at 150°C: (a) low magnification and (b) high magnification.

after the thin Au layer deposition, one thin AuIn layer may be formed; (2) during bonding, both bonding pairs must break through the AuIn layer and meet with each other; (3) the melting points of In and Sn are 156°C and 232°C, respectively; (4) ideally, it would be nice if the InSn layers could form the eutectic alloy at as low as 118°C; (5) unfortunately, it didn't and when the temperature reaches 150°C, it takes a relatively longer time for InSn interdiffusion to form a low-temperature alloy; (6) even if a eutectic alloy is finally obtained, the temperature difference is merely around 30°C; and (7) under such conditions, the flowability of liquid solder and the wetting property are poorer compared with that under higher bonding temperatures. Therefore, less liquid alloy is pushed through AuIn layers for interdiffusion and joining. As a result, voids along the AuIn layer are formed, and at some place, cracks are generated due to the continuity of the voids.

Position	Element (at%)				
	Cu	Ti	Sn	In	Au
1	78.01	5.05	3.48	8.94	4.52
2	72.51	5.49	13.60	8.40	0
3	60.17	3.31	22.88	13.64	0
4	51.21	2.02	20.86	18.21	7.70

TABLE 11.2 Compositions at the Interface in Fig. 11.8(*b*)

EDX analysis shows that Sn and In atoms diffuse through Ti into Cu and near the Cu regions, $Cu_3(SnIn)$ phases are formed. At the center region, both $Cu_6(SnIn)_5$ and AuIn phases are formed according to the EDX analysis listed in Table 11.2.

11.2.6 Helium Leak Rate Tests and Results

The hermeticity of the ring seals for patterned dies is evaluated by helium leak rate tests based on MIL-STD-883. According to the internal cavity volume of the bonded dies (~0.02 cm^3), the pressure of helium gas in bombing chamber should be set at 75 psi and the exposure time is 2 h. After bombing, the samples are put into the helium leak detector to measure the leak rates. For the leak rate test, the sample size is 21.

The results showed that the leak rate of the devices bonded at 180°C is smaller than 2×10^{-8} atm·cc/s, which is smaller than the rejected limit value (5×10^{-8} atm·cc/s) of MIL-STD-883E. However, for the packages bonded at 150°C, only 86% have a leak rate smaller than 5×10^{-8} atm·cc/s. This could be due to the voids/cracks in the middle of the sealing ring.

11.2.7 Reliability Tests and Results

Since the hermeticity of the packages bonded at 150°C is not very good, for the reliability study, only packages bonded at 180°C are considered. The reliability tests and results are listed in Table 11.3. The samples are subjected to the pressure cooker test (121°C, 2 atm for 300 h), high humidity test (85°C/85RH for 1000 h), high temperature storage test (125°C for 1000 h), and temperature cycling test (–40°C to 125°C for 1000 cycles). For each test, the sample size is 21. Before and after tests, these samples are examined by C-SAM, helium leak rate test, and shear test. The failure is defined as a leak rate smaller than 5×10^{-8} atm·cc/s or a large crack in the sealing ring. The ratios of packages with a leak rate smaller than 5×10^{-8} atm·cc/s after the pressure cooker test, high humidity storage, high temperature storage, and temperature cycling are 90.5%, 95.2%, 100%, and 90.5%, respectively. The shear strength requirement for this package is

Reliability Test Items	Results	
	Leak Rate Test (5×10^{-8} atm·cc/sec)	Shear Strength (MPa)
Pressure cooker test	$90.5\% < 5 \times 10^{-8}$	27.61
High humidity	$95.2\% < 5 \times 10^{-8}$	24.57
High temperature storage	$100\% < 5 \times 10^{-8}$	21.27
Temperature cycle	$90.5\% < 5 \times 10^{-8}$	25.62
*After bonding	$100\% < 2 \times 10^{-8}$	46.32

TABLE 11.3 Characterization Results after Reliability Tests

6 MPa. Thus, it can be seen from Table 11.3 that the samples' shearing stress results for all the tests are larger than the minimum requirement.

The interfacial microstructures are analyzed after reliability tests. Figure 11.9 shows that: (1) there are no significant changes on the microstructures; (2) with further SEM/EDX analysis, the compositions of the sealing ring also have no detectable changes, which is reasonable, since the reliability tests' temperatures are not very high compared with the melting temperatures of the IMCs; and (3) the Ti layer can also inhibit the diffusion between Cu and IMCs. However,

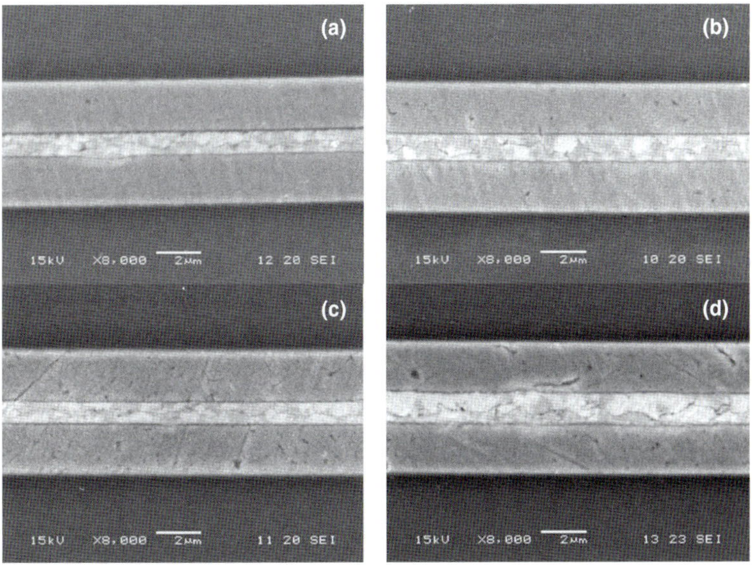

FIGURE 11.9 Interfacial microstructure of seal joint of good dies after reliability tests: (a) pressure cooker test (b) high humidity storage (c) high temperature (d) temperature cycling.

according to the shearing test results, the adhesions between different IMCs decrease after reliability tests.

Also, during long-term reliability tests, diffusion occurs between the overflow solder and high-temperature components (Cu, IMCs), as shown in Fig. 11.10 for the microstructure of the edges of the sealing ring after high-temperature storage. According to EDX analysis, all the overflowed solders become high-temperature compounds and tens of micrometer Cu substrates are consumed and turned into $Cu_6(SnIn)_5$. For the other three reliability tests, since the test temperatures are not high, the diffusion phenomena are not as active. For example, for the thermal cycling test, only 10–30 μm on each side of the sealing ring Cu substrates are consumed by about 20 μm of solders. Because the

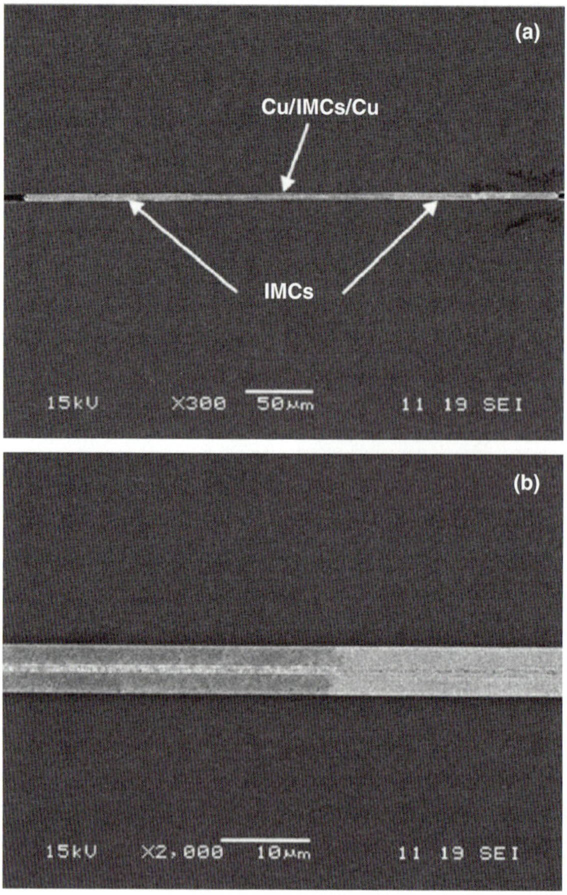

FIGURE 11.10 Diffusion between overflow solder and high temperature IMC joints after high temperature storage: (a) low magnification and (b) high magnification.

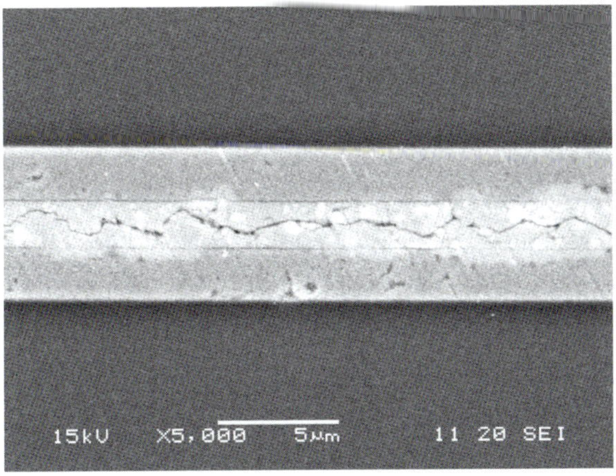

FIGURE 11.11 Interfacial microstructure of the failed (cracked) seal joint after pressure cooker test.

diffusion has not created new compounds and cracks, it will not cause the sealing ring to fail during reliability tests.

Samples with poor hermeticity are also analyzed to understand the failure mechanism of the sealing ring. Their failure modes are consistent. As shown in Fig. 11.11, a large crack throughout the sealing joint is found. The crack expands along the bonding line between the InAu compounds and CuSnIn compounds. When different IMCs are formed at the joints, under thermal cycling and pressure cooker tests, stresses may be generated. This is the reason why after these two tests, the leak rate test revealed a relative poorer hermeticity results, compared with the other tests. This means that the adhesion between the two phases is the key to determining the reliability of the sealing joint. Therefore, there are at least two ways to enhance the reliability of the present sealing joints: (1) one is to reduce AuIn compounds by a thin Au-layer deposition; and (2) the other is to reduce the Ti buffer layer thickness to allow fast diffusion between the solders and Cu substrate. In this case, less solder will flow out of the joint and thicker IMC joints will be formed with discontinuous AuIn phases.

11.2.8 Summary and Recommendations

Some important results are summarized in the following:

1) No detectable voids are found in the sealing rings of all packages after bonding at 150°C and 180°C based on C-SAM observation. When bonding at 180°C, all samples obtained a good leak rate (2×10^{-8} atm·cc/s). While bonding at 150°C, only 86% of packages can achieve acceptable hermeticity.

2) Under the two bonding temperatures (150°C and 180°C), In and Sn atoms can diffuse through the Ti layer to form joints with Cu. Sealing joints are composed of high-melting-point IMCs, which allow the package for the next level of interconnections with higher temperature.

3) More solders are squeezed out when bonding at 180°C.

4) Under bonding at lower temperatures, for example, 150°C, a thick sealing joint is obtained due to the low flowability of the liquid alloy. However, cracks are formed at the center of the sealing rings, which cause the hermeticity problem.

5) After bonding, high shear strength is obtained for the sealing rings by both bonding temperatures (150°C and 180°C).

6) After four kinds of reliability tests—namely, the pressure cooker test (121°C, 2 atm, 300 h), high humidity storage (85°C/85RH, 1000 h), high temperature storage (125°C, 1000 h), and temperature cycling test (–40°C to 125°C, 1000 h)—the package bonded at 180°C provides good reliability and hermeticity.

7) The propagation of a crack along different IMC layers is found in the failed packages after reliability tests. Therefore, the adhesion between different phases is the key for determining the reliability of the sealing joint. Two ways to enhance the reliability of the present sealing joints are recommended: (1) one is to reduce AuIn compounds by a thin Au-layer deposition; and (2) the other is to reduce the Ti buffer layer thickness to allow fast diffusion between the solders and Cu substrate.

8) The current material system (Cu/Ti/Au) with the bonding parameters (180°C for 20 min under 5.5 MPa pressure) is recommended to be used for MEMS packaging applications, which require low-temperature bonding and hermetic sealing.

11.3 W2W Bonding with AuSn Solder and TiCuNiAu Metallization Pads

11.3.1 The Test Vehicle

Figure 11.12 [53] schematically shows a silicon carrier with microchannels for thermal management and TSVs for electrical feedthrough. This carrier consists of two silicon chips, and the difference between these two chips is that the bottom Si chip does not have an outlet. There are 144 TSVs (at a pitch of 0.5 mm) and they are in rows along the periphery of the carrier. After bonding, electrical

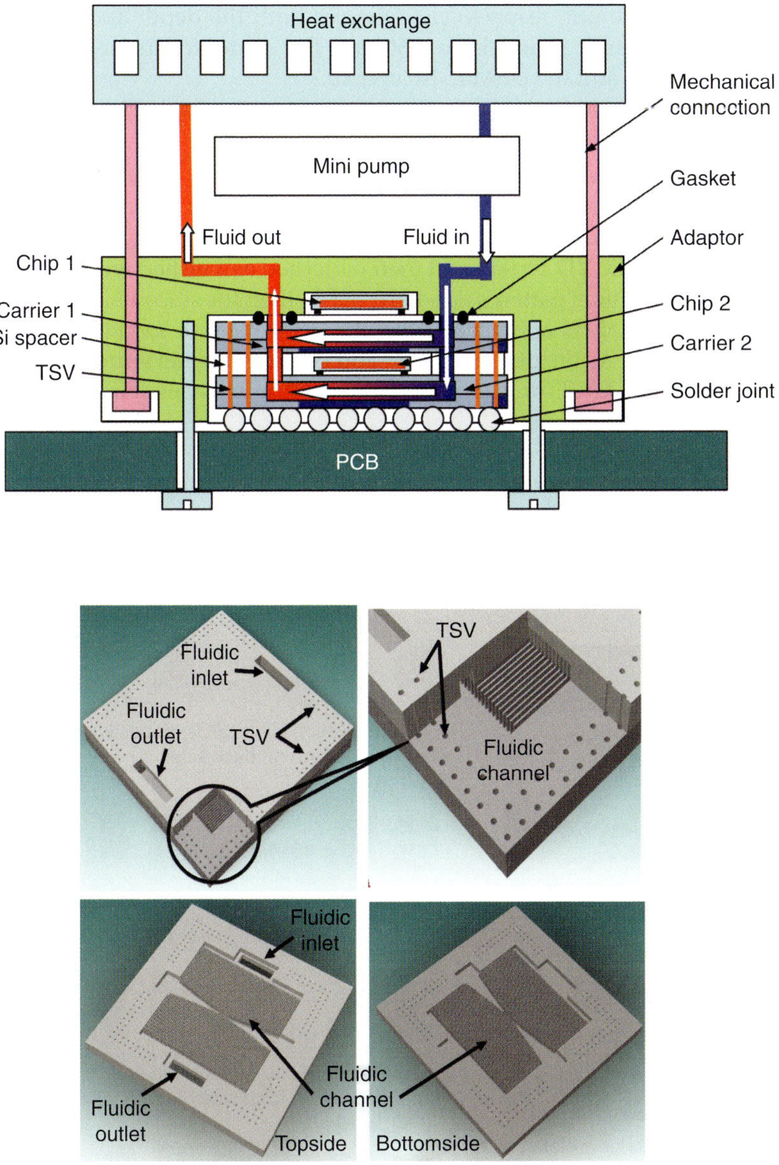

FIGURE 11.12 Schematic drawing of Si carrier with TSV for electrical feed-through and microchannel for thermal management.

interconnection through the carrier is made by TSV with on-wall metallization. The fluidic channels are connected through one/two inlets and one/two outlets. There are sealing rings around both the fluidic path and the individual TSV to isolate the fluid from the electrical interconnection. The depth and diameter of the silicon via are

400 μm and 150 μm, respectively, whereas both the depth and the width of the fluidic channels are 350 μm. The solder for the sealing ring is Au20Sn, and the UBM is TiCuNiAu.

11.3.2 Test Vehicle Fabrication

Figure 11.13 shows the fabrication process of the test vehicle on a 200-mm silicon wafer. First, a layer of SiO_2 of 3 μm thick is deposited by the plasma enhanced chemical vapor deposition (PECVD) process in a Novellus PECVD system and then patterned and etched with via and cooling channel patterns as a hard mask; then the silicon via and the cooling channels are etched with the deep reactive ion etching (DRIE) process. As the depth for TSV and cooling channel structures is different, the DRIE process is separated into two steps. First, via

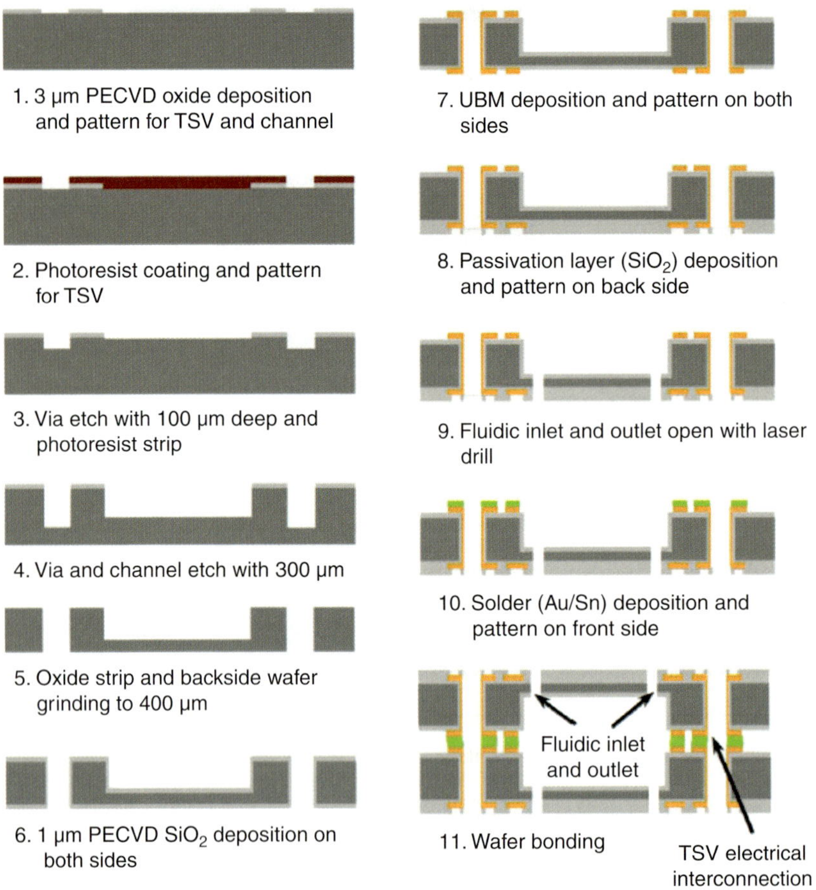

FIGURE 11.13 TSV and microchannel carrier fabrication process.

with 100 μm depth is etched using photoresist as etch mask; then the photoresist is stripped and the via and the channels are etched at the same time for 350 μm depth. SiO_2 patterned at step 1 is used as the etch mask. Since the DRIE process is restricted to blind silicon via, in order to achieve through silicon via (TSV), the silicon wafer is backgrinded and polished to 450 μm to expose the via after the DRIE process. Then a passivation layer is deposited on both sides of the wafer. The passivation layer is 1 μm of SiO_2 that is deposited by the Plasma-Therm SLR 720 system. After passivation, UBM is then sputtered and patterned around the TSV and the cooling channels on both sides of the wafer. The sputtering is done in the Balzer system. The UBM material used is Ti/Cu/Ni/Au with thickness of 0.1 μm, 2 μm, 0.5 μm, and 0.1 μm, respectively. In order to make sure that the side wall of the TSV is coated with a metal film for electrical connections, a thick copper layer (which is used in the UBM layer) is sputtered from both sides of the wafer. For this special case, the sputtering process is much simpler and less time-consuming than the copper plating process. Therefore, electrical interconnects are formed cost effectively on the TSV.

Usually, a thick UBM layer makes wet etching difficult. A workable process is shown in the following: (a) the top Au layer is etched with a solution of tri-iodide; (b) the Ni layer is etched with a solution of Ac-100; (c) the Cu layer is etched with a solution of A95; and (d) the Ti layer is etched with a 49% HF solution that is diluted with DI (deionized) water (49%HF:H_2O = 1:15). After UBM patterning, two or four rectangular holes (for inlets and outlets) are opened in the top silicon chip by laser drilling.

As mentioned earlier (Chap. 10) an Au20Sn solder can be manufactured by various methods including evaporation, electrodeposition, pasting, and solder preforms. Among these methods, evaporated solder allows more accurate control of dimension and position. Therefore, after UBM patterning, a sequence of Sn/Au layer (eight layers of Sn and eight layers of Au in total) is evaporated and simultaneously patterned by a lift-off process. The thickness for each Sn layer and Au layer is 0.2 μm and 0.24 μm, respectively. The evaporation can be done in a Temescal E-Beam Evaporator (Model VES-2550) from Semicore Equipment, Inc. The evaporation rate is 5 Å/s. During the lift-off process, a 20 μm thickness of dry film is used as the mold.

The fabricated wafer is shown in Fig. 11.14. The AuSn solder is 3.5 μm thick and the UBM, Ti(0.1 μm)/Cu(2 μm)/Ni(0.5 μm)/Au(0.1 μm), is 2.7 μm thick. The annular sealing ring around the TSV is 100 μm. The width of the sealing ring between the silicon wafers is 300 μm. The TSV is on a 500 μm pitch. The TSV is not totally filled with Cu; instead, the on-wall metallization is done by double-sided metal film sputtering.

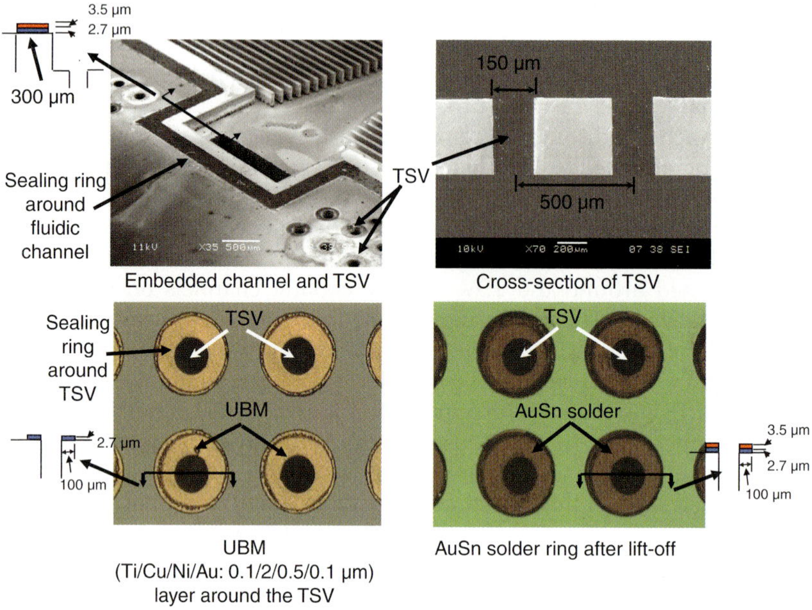

FIGURE 11.14 Fabricated TSV and microchannel carrier. The TSV, sealing ring for wafers, sealing ring for microchannel, Au20Sn solder bumps, and UBMs.

11.3.3 W2W Bonding

After solder layer formation, wafer-level bonding is conducted to bond the two wafers together. The bonding is carried out at 350°C, 15 min of peak temperature hold time, and 8 kN of compressive force (static pressure of 4.7 MPa). After bonding, the bonded wafer is diced into single silicon carriers with dimensions of 11.1 × 11.1 × 0.8 mm. The dicing machine used is the DAD 651 from Disco. The spindle speed is 30,000 rpm, and the feed rate is 5 mm/s.

11.3.4 Electrical Performance

The patterned UBM layer is shown in Fig. 11.14. It shows that the pattern edge is smooth. The electrical connectivity through the TSV with different diameters has been tested, and the results are listed in Table 11.4. When the diameter is larger than 150 μm, the resistance through TSV is less than 0.5 Ω. Therefore, a diameter of the TSV of 150 μm is used in order to have higher I/Os. Figure 11.15 shows the cross-section of a TSV after metallization with the diameter of 150 μm. It can be clearly seen that the side wall of the TSV is covered with a continuous metal layer.

11.3.5 Quality and Reliability

Shear strength measurements of the diced package samples are carried out by a commercially available shear tester (DAGE-SERIES-4000-T,

TSV Diameter	Resistance	Avg. Resistance
100 μm	1.5~12 kΩ	6.75 kΩ
125 μm	1.5~6.5 kΩ	4 kΩ
150 μm	0.46~0.52 Ω	0.49 Ω
200 μm	0.38~0.44 Ω	0.41 Ω
300 μm	0.38~0.44 Ω	0.41 Ω

TABLE 11.4 Electrical Resistance for Different Diameters of TSV

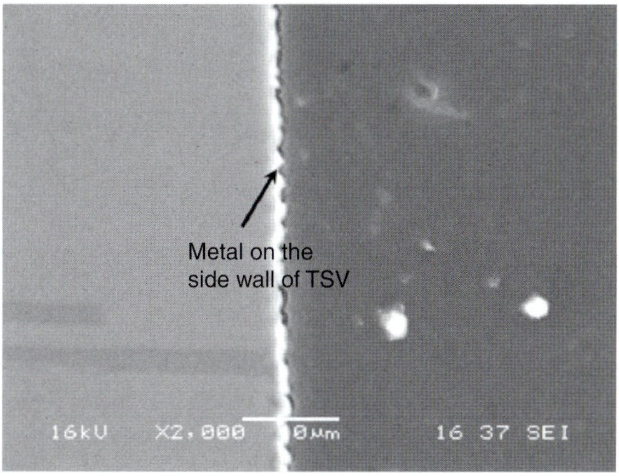

FIGURE 11.15 Thin metal film on the side wall of TSV.

Dage Precision Industries, Ltd., Aylesbury, UK). The shear strength value measured for 20 different samples ranged from 17.8 MPa to 35.1 MPa, and an average value of 27.2 MPa and a standard deviation of 2.2 are obtained. During the shear test, fracture always occurs at the UBM layer, which indicates that the bonding strength is stronger than the adhesion strength of the UBM layer. Figure 11.16 shows the fracture surface after the shear test, and it can be seen that the metal is peeled from the surface, showing good bonding strength. Figure 11.17 shows the cross-section of the bonding interface, and it can be seen that a uniform cross-section has been achieved.

During AuSn bonding, pressure is one of the critical parameters to obtain good bonding results. Figure 11.18 shows the cross-section of a poor AuSn solder ring after bonding at 350°C, 15 minutes of peak temperature hold time, and 3 kN of compression force (static pressure of 1.7 MPa). There is an obvious crack in the middle of the interface due to not enough bonding pressure. Higher pressure can improve the

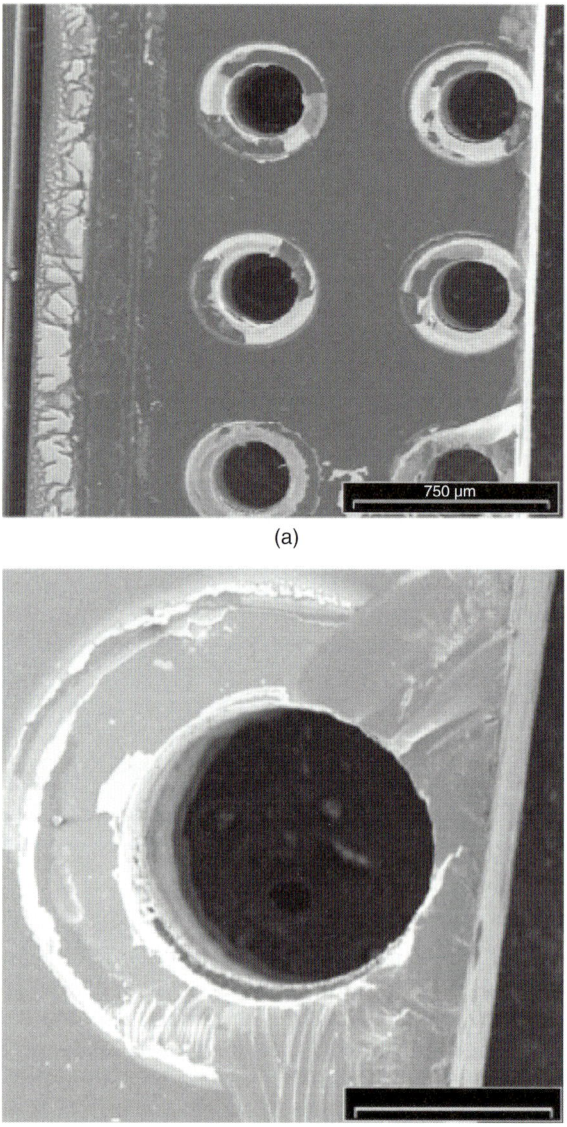

(a)

(b)

Figure 11.16 SEM images of solder surface after shear test (a) overview; (b) zoomed view.

contact area between two bonding surfaces from both macro and micro points of view. Normally, the silicon wafer is not ideally flat, especially after back-grinding and polishing and some microfabrication process. Therefore, high pressure can push (force) the two bonding surfaces against each other intimately. At the same time, the

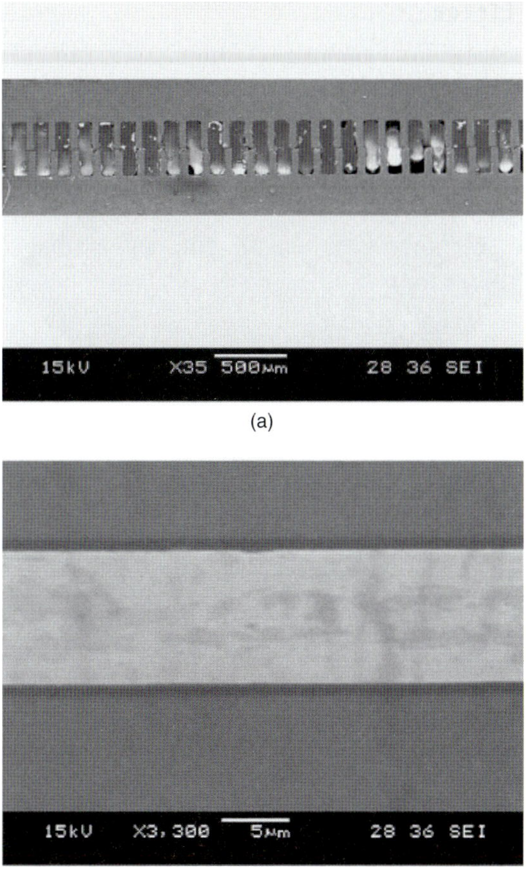

(a)

(b)

Figure 11.17 Cross-section of the bonding interface (a) cross-section of embedded cooling channels (b) cross-section of AuSn solder ring after bonding.

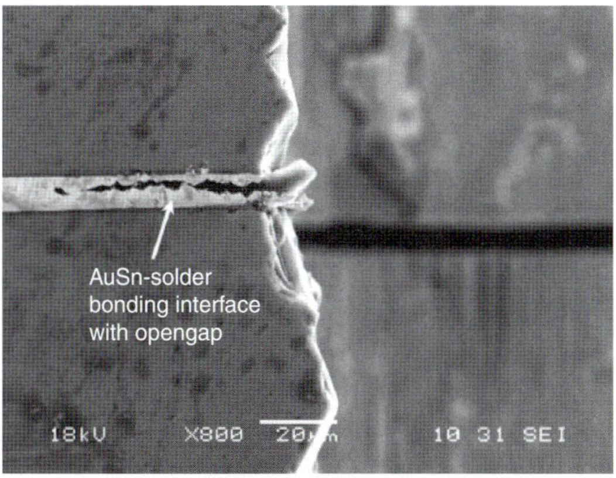

Figure 11.18 Crack in the bonding interface with small bonding pressure of 3 kN.

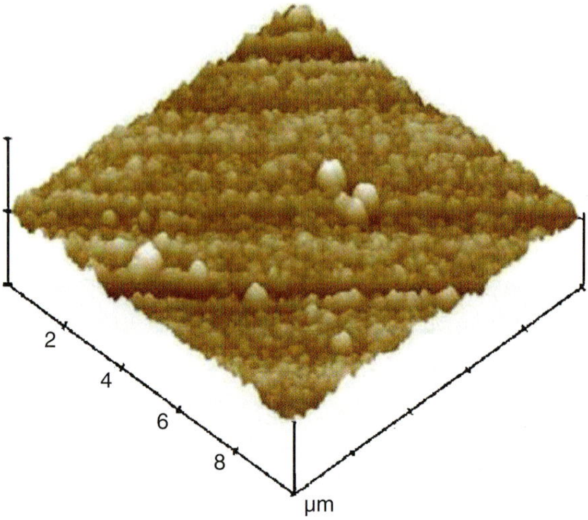

FIGURE 11.19 Surface roughness of AuSn solder after evaporation on UBM layer. The RMS roughness is 25 nm.

root-mean-square (RMS) roughness of the AuSn surface after evaporation is about 24 nm, as shown in Fig. 11.19. The rough surface will decrease the contact area between the two bonding surfaces. Therefore, high pressure is required to overcome the surface roughness in order to improve the contact area between the two bonding surfaces.

Figure 11.20 shows a bonded pair of silicon wafers (the top and bottom surfaces of a diced silicon carrier, respectively). The metal patterns on both sides provide electrical interconnection and rerouting. The yield after dicing is higher than 98%, which indicates the bonding strength is high enough to pass through the dicing step.

A thermal cycling test between –40°C (15 min) and 125°C (15 min) for 1000 cycles has been conducted. Temperature ramping rate is 15°C per minute. After the test, the average shear strength for 20 samples is 26 MPa, which is almost the same as that before the test.

Hermetic sealing of the bonded carrier is a key requirement of the project. A test setup has been developed to circulate water under high pressure through the carrier. The water flow rate is increased gradually up to 350 ml/min. No leakage is observed at a high flow rate. The maximum pressure drop in the carrier is 6×10^4 Pa.

11.3.6 Summary and Recommendations

Some important results are summarized as follows:

(1) An integration process for fabricating silicon carriers with both embedded fluidic channels and electrical interconnections has been developed. The silicon carriers can be stacked on top of each other, with silicon interposers in between to make up a stacked cooling module for high-power heat dissipation.

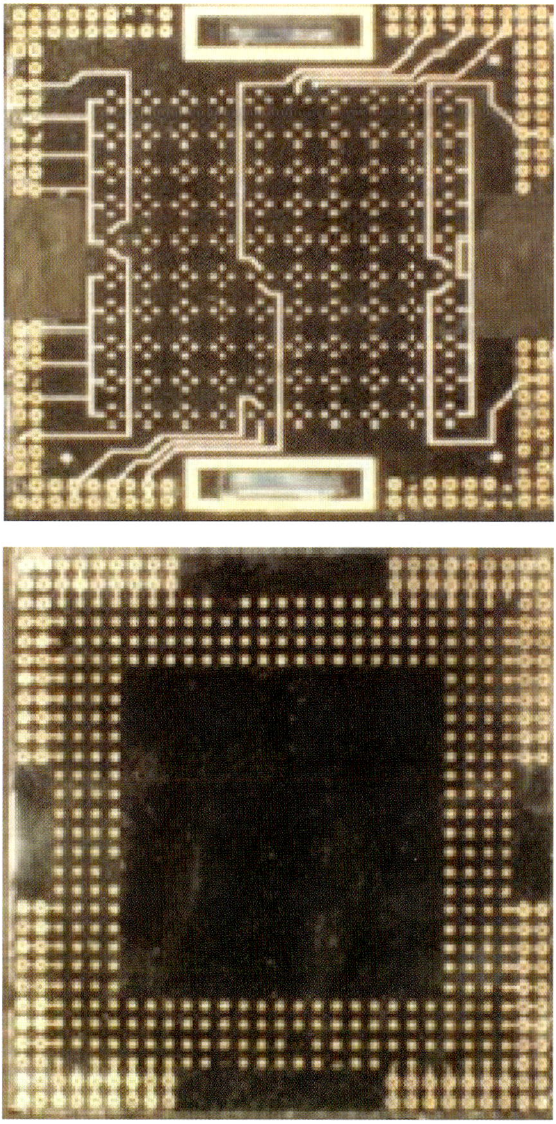

FIGURE 11.20 Optical images of a diced Si carrier: top surface (top); bottom surface (bottom).

(2) After on-wall metallization of TSV with a diameter larger than 150 μm and thickness less than 400 μm, the electrical resistance through TSV is 0.49 Ω. After bonding, the electrical resistance between the top and bottom surfaces of the silicon carrier is 1.16 Ω, which is 0.16 higher than the designed value. It can be reduced by using a thicker metal layer and/or a larger TSV diameter.

(3) For a single carrier, no leakage is observed at a high flow rate (350 ml/min). The maximum pressure drop in the carrier is 6×10^4 Pa.

(4) High bonding pressure can improve the bonding quality of evaporated AuSn solder. With a bonding pressure of 4.7 MPa, the die shear test strength is higher than 27 MPa and it is more than adequate for dicing/handling the silicon carrier and to provide a leakage-free fluidic path. After 1000 cycles of the thermal cycling test (–40°C to 125°C), the average die shear strength is 26 MPa and shows insignificant degradation after the thermal cycling test.

11.4 References

1. Lau, J. H., "Design and Process of 3D MEMS Packaging," *Proceedings of IMAPS International Symposium on Microelectronics*, San Jose, CA, November 2009, pp. 1–9. Also, IMAPS Transactions, *Journal of Microelectronics and Electronic Packaging*, First Quarter Issue, 2010, pp. 10–15.
2. Lau, J. H., Lee, R., Yuen, M., and Chan, P., "3D LED and IC Wafer Level Packaging," *Journal of Microelectronics International*, Vol. 27, Issue 2, 2010, pp. 98–105.
3. Lau, J. H., C. K. Lee, C. S. Premachandran, and A. Yu, *Advanced MEMS Packaging*, McGraw-Hill, New York, 2010.
4. Yu, R., Liu, F., Polastre, R., Chen, K., Liu, X., Shi, L., Perfecto, E., Klymko, N., Chace, M., Shaw, T., Dimilia, D., Kinser, E., Young, A., Purushothaman, S., Koester, S., and Haensch W., "Reliability of a 300-mm-compatible 3DI Technology Base on Hybrid Cu-adhesive Wafer Bonding," *Proceedings of Symposium on VLSI Technology Digest of Technical Papers*, 2009, pp. 170–171.
5. Wang, C., and Suga, T., "A Novel Moire Fringe Assisted Method for Nanoprecision Alignment in Wafer Bonding," In *IEEE Proceedings of Electronic Components & Technology Conference*, San Diego, CA, May 25–29, 2009, pp. 872–878.
6. McMahon, J. J. Lu, J. Q., and Gutmann, R. J. "Wafer Bonding of Damascene-Patterned Metal/Adhesive Redistribution Layers for Via-First Three-Dimensional (3D) Interconnect." In *IEEE Proceedings of Electronic Components & Technology Conference*, Lake Buena Vista, FL, May 31–June 3, 2005, pp. 331–336.
7. Wagner, M., and Zou, J. "Method for Creating Low Cost 3D MEMS Chemical Sensors by Using Through Wafer Vias & Wafer Bonding." In *IEEE Proceedings of Electronic Components & Technology Conference*, Reno, NV, May 29–June 1, 2007, pp. 1119–1126.
8. Fan, A., A. Rahman, and R. Reif, "Copper Wafer Bonding," *IEEE/ECS Electrochem. and Solid-State Lett.*, Vol. 2 (1999) pp. 534–536.
9. Chen, K. N., C. S. Tan, A. Fan, R. Reif, "Morphology and Bond Strength of Copper Wafer Bonding," *IEEE/ECS Electrochem. and Solid-State Lett.*, Vol. 7 (2004), pp. G14–G16.
10. Lin, C. W., S. C. L. Chiang, and T. K. A. Yang, "Bumpless flip chip packages," *Proc. IEEE 4th Int'l. Conf. on Electron. Mater. and Packag. (EMAP)*, 2002, pp. 173–177.
11. Towle, S., H. Braunisch, C. Hu, R. Emery, and G. Vandentop, "Bumpless Build-up Layer Packaging," *Proc. ASME Int'l. Mech. Eng. Congress and Exposition (IMECE)*, 2001, EPP–24703.
12. Shigetou, A., T. Itoh, K. Sawada, and T. Suga, "Bumpless Interconnect of 6 μm-pitch Cu electrodes at room temperature," *IEEE. Trans. on Adv. Packg.*, Vol. 31 (2008), pp. 473–478.
13. Takagi, H., K. Kikuchi, R. Maeda, T. R. Chung, and T. Suga, "Surface Activated Bonding of Silicon Wafers at Room Temperature," *Appl. Phys. Lett.*, Vol. 68 (1996), pp. 2222–2224.

14. Takahashi, K., and T. Onzawa, "The Effect of Surface Roughness on the Low Energy Bonding," *J. High Pressure Inst. of Japan,* Vol. 35 (1997), pp. 159–164.
15. Shigetou, A., T. Itoh, and T. Suga, "Direct Bonding of CMPCu Films by Surface Activated Bonding (SAB) Method," J. Mater. Sci., Vol. 40 (2005), pp. 3149–3154.
16. Shigetou, A., T. Itoh, and T. Suga, "Bumpless Interconnect of Cu Electrodes in Millions-Pins Level," *Proc. 56th Electronic Components and Technology Conf,* 2006, pp. 1223–1226.
17. Tong, Q.Y., and U. Gösele, *Semiconductor Wafer Bonding: Science and Technology,* Wiley, New York, 1999, pp. 1–13.
18. Esashi, M., "Wafer Level Packaging of MEMS", *J. Micromech. Microeng.,* Vol. 18, No. 073001, 2008, pp. 1–13.
19. Radu, I., I. Szafraniak, R. Scholz, M. Alexe, and U. Gösele, "GaAs on Si Heterostructures Obtained by He and/or H Implantation and Direct Wafer Bonding", *J. Appl. Phys.,* Vol. 94, 2002, pp. 7820–7825.
20. Pool, A. W., et. al., "Three-dimensional Integrated Circuits," *IBM J. Res. and Devel.,* Vol. 50, No. 4/5, 2006, pp. 491–506.
21. Tan, C. S., R. J. Gutmann, and L. R. Reif, Eds, *Wafer Level 3-D ICs Process Technology,* Springer, New York, 2008, pp. 59–63.
22. Christiansen, S. H., R. Singh, and U. Gösele, "Wafer Direct Bonding: From Advanced Substrate Engineering to Future Applications in Micro/Nanoelectronics," *Proc. IEEE,* Vol. 94, No.12, 2006, pp. 2060–2106.
23. Kawashima, S., M. Imada, K. Ishizaki and S. Noda, "High-Precision Alignment and Bonding System for the Fabrication of 3-D Nanostructures," *J. Micro electromech. Syst.,* Vol. 16, No. 5, 2007, 1140–1144.
24. Choa, S. H., "Reliability of Vacuum Packaged MEMS Gyroscopes," *Microelectron Reliability,* 45 (2005) 361–369.
25. Choa, S. H., "Reliability of MEMS Packaging: Vacuum Maintenance and Packaging Induced Stress," *Microsyst Technol* 11 (2005) 1187–1196.
26. Gosele, U., Stenzel, H., Martini, T., Steinkirchner, J., Conrad, D., and Scheerschmidt, K., "Self-Propagating Room Temperature Silicon Wafer Bonding in Ultrahigh Vacuum," *Appl. Phys. Lett.,* 67 (1995) 3614–3616.
27. Howlader, M. M. R., Suehara, S., Takagi, H., Kim, T. H., Maeda, R., and Suga, T., "Room-Temperature Microfluidics Packaging Using Sequential Plasma Activation Process," *IEEE Trans. Adv. Packag.,* 29(3), (2006) 448–456.
28. Howlader, M. M. R., Okada, H., Kim, T. H., Itoh, T., and Suga, T., "Wafer Level Surface Activated Bonding Tool for MEMS Packaging," *J. Electrochem. Soc.,* 151(7) (2004) G461–G467.
29. Takagi, H. and Maeda, R., "Aligned Room-Temperature Bonding of Silicon Wafers in Vacuum by Argon Beam Surface Activation," *J. Micromech. Microeng.* 15 (2005) 290–295.
30. Itoh, T. and Suga, T., "Necessary Load for Room Temperature Vacuum Sealing," *J. Micromech. Microeng.* 15 (2005) S281–S285.
31. Yu, W. B., Tan, C. M., Wei, J., Deng, S. S., and Nai, S. M. L., "Influence of Applied Load on Vacuum Wafer Bonding," *Sens. Actuators A* 115 (2004) 67–72.
32. Miki. N., X. Zhang, R. Khanna, A.A. Ayon, D. Ward, and S.M. Spearing, "Multi-stack Silicon-Direct Wafer Bonding for 3D MEMS Manufacturing," *Sens. Actuators A* 103 (2003) 194–201.
33. Daridon, A., Fascio, V., Lichtenberg, J., Wütrich, R., Langen, H., Verpoorte, E., and de Rooij, N. F., "Multilayer microfluidic glass chips for microanalytical applications," *Fresenius J Anal Chem* 371 (2001) 261–269.
34. Zhang, X. X. and Raskin, J. -P., "Low-Temperature Wafer Bonding: A Study of Void Formation and Influence on Bonding Strength," *J. Microelectromech. Sys.,* 14(2), (2005) 368–382.
35. Tang, Z., Shi, T., Liao, G. and Liu, S., "Modeling the Formation of Spontaneous Wafer Direct Bonding Under Low Temperature," *Microelectron. Engineer.,* 85 (2008) 1754–1757.
36. Turner, K. T. and Spearing, S. M., "Modeling of direct Wafer Bonding: Effect of Wafer Bow and Etch Patterns," *J. Appl. Phys.,* 92(12) (2002) 7658–7666.

37. Howlader, M. M. R., Watanabe, T., and Suga, T. "Investigation of the Bonding Strength and Interface Current of p-Si/n-GaAs Wafers Bonded by Surface Activated Bonding at Room Temperature," *J. Vac. Sci. Technol.*, B 19 (6) (2001) 2114–2118.
38. Smith, R. L. and S. D. Collins, "Micromachined Packaging for Chemical Microsensors," *IEEE Trans. Electron Devices*, vol. 35, no. 6 (1988), pp. 787–792.
39. AlSarawi, S. F., D. Abbott, and P. D. Franzon, "A Review of 3-D Packaging Technology," *IEEE Trans. Compon. Packag. Manufact. Technol., Part B—Adv. Packag.*, Vol. 21 (1998), pp. 2–14.
40. Heschel, M., J. F. Kuhmann, S. Bouwstra, and M. Amskov, "Stacking Technology for a Space Constrained Microsystem," *J. Intell. Mater. Syst. Struct.*, Vol. 9 (1998), pp. 749–754.
41. Fan, A., A. Rahman, and R. Reif, "Cu to Cu wafer bonding," *Electrochemical and Soild-State Letters* 2(10) (1999), 534–536.
42. Chow, E. M., Chandrasekaran, V., Partridge, A., Nishida, T., Sheplak, M., Quate, C. F., and Kenny, T. W., "Process Compatible Polysilicon-Based Electrical Through-Wafer Interconnects in Silicon Substrates," *Journal of Microelectromechanical systems*, Vol. 11, No. 6, 2002, 631–640.
43. Drost, A., G. Klink, S. Scherbaum, and M. Feil, "Simultaneous Fabrication of Dielectric and Electrical Joints by Wafer Bonding," *SPIE-Micromachined*, 2006, pp. 1–6.
44. Lu. J. Q., "Dielectric Adhesive Wafer Bonding for Back-End Wafer-Level 3D Hyper-integration," in *Dielectrics for Nanosystems: Materials Science, Processing, Reliability, and Manufacturing*, edited by R. Singh, H. Iwai, R.R. Tummala, and S.C. Sun, ECS PV 2004-04, 2004, pp. 312–323.
45. Guarini, K. W., "Electrical Integrity of State-of-the-Art 0.13 μm SOI CMOS Devices and Circuits Transferred for Three Dimensional (3D) Integrated Circuit (IC) Fabrication," in *Digest of International Electron Device Meeting*, (2002), pp. 943–945.
46. Kwon, Y., "Dielectric Glue Wafer Bonding for 3D ICs," in *Materials, Technology and Reliability of Advanced Interconnects*, MRS Proceedings Vol. 766, (2003), pp. 27–32.
47. Kwon, Y., *Wafer Bonding for Three Dimensional (3D) Integration*, Ph.D. Thesis, Rensselaer Polytechnic Institute, Troy, NY, 2003.
48. Price, D., *Dual Damascene Copper Interconnects with Low-k Polymer Dielectrics*, Ph.D. Thesis, Rensselaer Polytechnic Institute, Troy, NY, 1999.
49. Jindal, A., "Wafer Thinning for Monolithic 3D Interconnects." *Materials, Technology and Reliability of Advanced Interconnects* MRS Proceedings, Vol. 766 (2003), pp 21–26.
50. Cheng, K. N., "Morphology and Bond Strength of Copper Wafer Bonding," *Electrochemical and Solid-State Letters*, Vol. 7(1), 2004, pp. G14–G16.
51. Yu, D., L. Yan, C. Lee, W. Choi, S. Thew, C. Foo, and J. H. Lau, "Wafer-Level Hermetic Bonding Using Sn/In and Cu/Ti/Au Metallization," *IEEE Transactions on Components and Packaging Technologies*, Vol. 32 (4), 2009, pp. 926–934.
52. Yu, D., C. Lee, L. Yan, W. Choi, A. Yu, and J. H. Lau, "The Role of Ni Buffer Layer on high Yield Low Temperature Hermetic Wafer Bonding Using In/Sn/Cu Metallization," *Applied Physics Letters*, Vol. 94(3), 2009, pp. 34105.1–34105.3.
53. Yu, A., N. Khan, G. Archit, D. Pinjalal, K. Toh, V. Kripesh, S. Yoon, and J. H. Lau, "Development of Silicon Carriers with Embedded Thermal Solutions for High Power 3-D Package," *IEEE Transactions on Components and Packaging Technology*, Vol. 32 (3), 2009, pp. 566–571.
54. Liu, X. J., H. S. Liu, I. Ohnuma, R. Kainuma, K. Ishida, S. Itabashi, K. Kameda, and K.Yamaguchi, "Experimental determination and thermodynamic calculation of the phase equilibria in the Cu-In-Sn system," *J. Electron. Mater.*, Vol. 30, no. 9, 2001, p. 1093.
55. Sommadossi, S., and A. F. Guillermet, "Interface reaction systematic in the Cu/In-48Sn/Cu system bonded by diffusion soldering," *Intermetallics*, Vol. 15, 2007, p. 912.

CHAPTER 12

Through-Silicon-Via (TSV) Interposer Reliability

12.1 Introduction

As mentioned in Chap. 1, the through-silicon-via (TSV) [1–42], invented by 1956 Nobel Prize winner L. William Shockley in 1958 (Fig. 12.1), is the heart of three-dimensional (3-D) integrated circuit (IC) integration, and the only volume product with this technology today is Toshiba's CMOS (complementary metal oxide semiconductors) image sensor. By using their novel fabrication process (TCV technology), Toshiba is able to ship this product at very low cost. Their design for reliability (DFR) of the copper-plated TSV wall thickness is presented in this section, since it is one of the most important factors in their package reliability [1, 2].

In addition to Toshiba's simple case, for high pin-count and fine-pitch TSV interposers (either passive or active) the following topics are also presented in this chapter: (a) the DFR of the interface between the filled copper and the silicon, and between the copper and the silicon dioxide dielectric of TSVs, due to the local thermal expansion mismatch between the silicon and the filled copper; (b) the DFR of solder bumps between the IC chip and the TSVed interposer (either passive or active) due to the global thermal expansion mismatch between the silicon IC chip and the copper-filled TSV interposer [10, 20, 28]; (c) a new generic thermal-enhanced, electrical-sound, mechanical-integrity, and cost-effective 3-D IC SiP (System-in-Package) with a TSV interposer.

12.2 DFR of the Cu Thickness in Toshiba's TSV

The structure, materials, and assembly process of Toshiba's CMOS image sensor have been briefly mentioned in Chap. 1 and shown in Figs. 1.16 through 1.19. In this section, only finite-element modeling

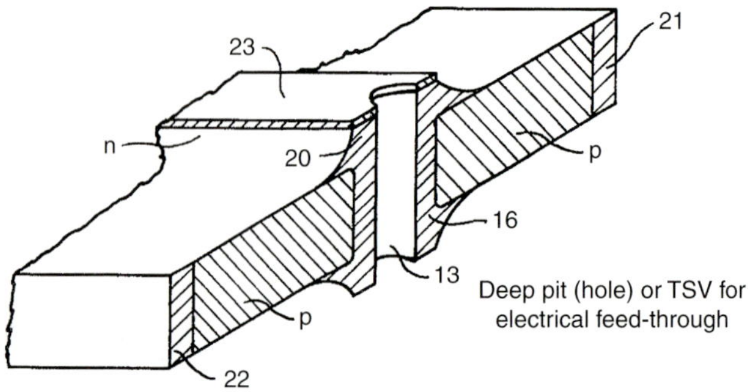

FIGURE 12.1 TSV invented by the 1956 Nobel Laureate William Shockley, U.S. Patent #3,044,909, filed on October 23, 1958, and granted on July 17, 1962.

and analyses for determining the stresses and strains at various Cu wall thicknesses of their TSV are presented.

12.2.1 Finite-Element Modeling

Figure 12.2*a* shows the finite-element model for analyzing a quarter (due to double symmetry) of the CMOS image sensor package (the Si chip is at the bottom and the glass chip is on top) [1]. The package size is 6.75 × 8.32 mm. The material properties of the model are shown in Table 12.1. Figure 12.2*b* shows the model around a via, which is located at the corner of the structure. This is because the maximum deflection (warpage) will occur at the corners of a package.

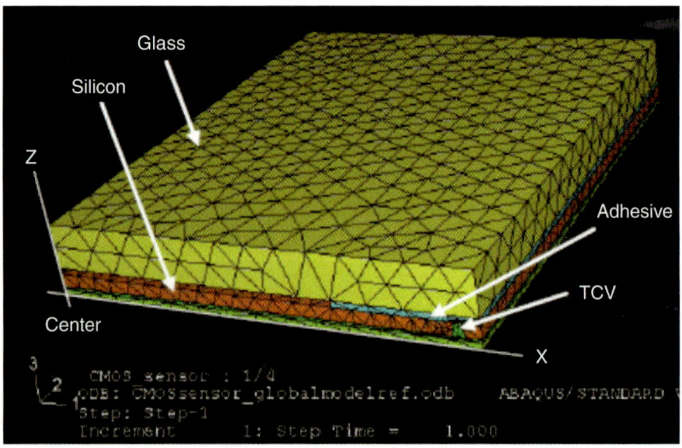

FIGURE 12.2a A quarter model of Toshiba's CMOS image sensor.

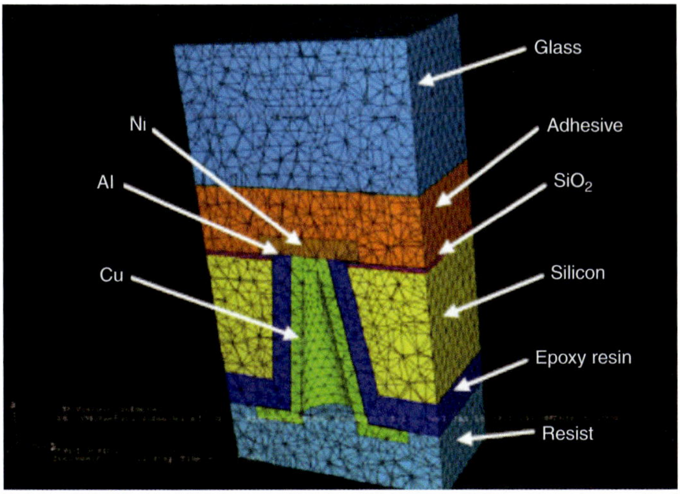

FIGURE 12.2*b* Submodel around a via.

Material	Young's Modulus (GPa)	Poisson's Ratio (Unitless)	CTE (ppm)
Silicon	170	0.28	3.5
SiO_2	73	0.27	0.4
SiN	98	0.3	2.2
Al	70	0.345	23.5
Cu	125	0.3	18
Ni	207	0.31	13.3
Adhesive	3	0.35	80
Epoxy resin	4	0.258	46
Glass	72.9	0.21	7.2
Resist	2.3	0.45	60

TABLE 12.1 A Quarter Model of the Toshiba's CMOS Image Sensor

The boundary condition is for the structure cooling from 120°C to 25°C (room temperature).

12.2.2 Modeling Results and Analyses

As expected, the maximum deflection does occur at the corners and the Si chip is convex due to the thermal expansion mismatch between the Si and glass chips (Fig. 12.3). The magnitude (~8 μm) is within their specification [1].

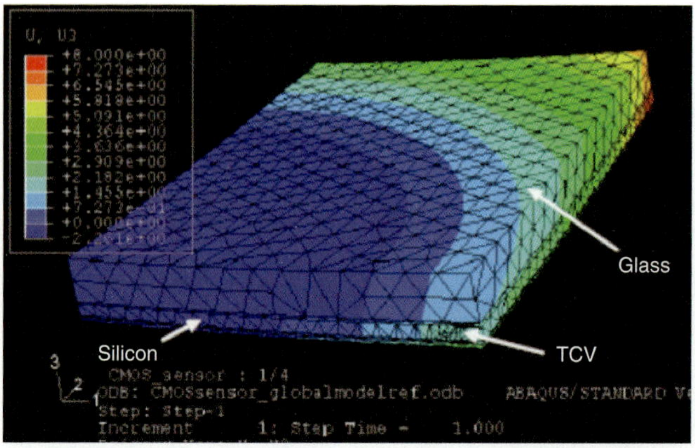

FIGURE 12.3 Result of simulation (package warpage × 50).

The von Mises stress contours for four different copper-filled thicknesses (5 μm, 10 μm, half-filled, and full-filled) are shown in Fig. 12.4. It can be seen that: (a) for all the cases, the high-stress regions are at the Cu landing edge on the epoxy resin side, the via center, and via bottom, which is the connection between the Ni and Cu; (b) for all

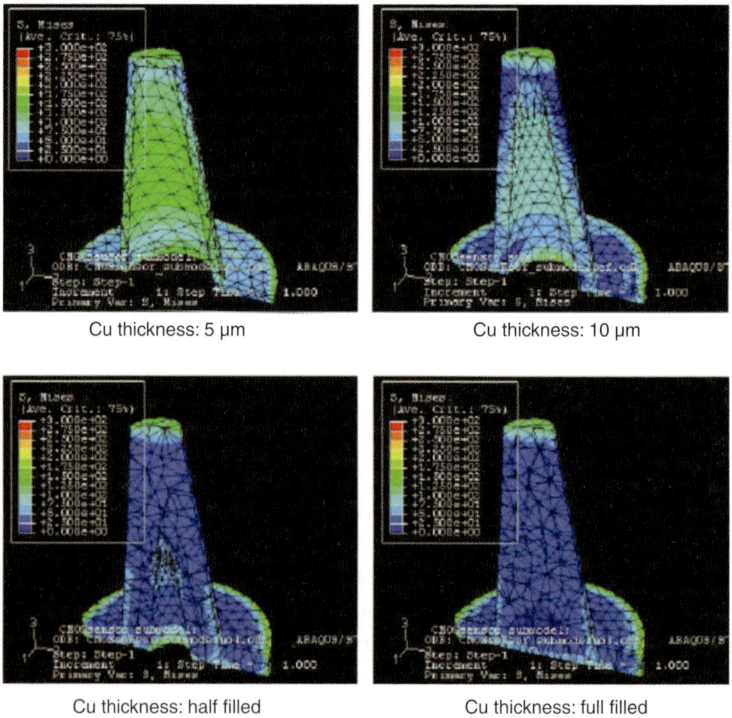

FIGURE 12.4 von Mises stress in the Cu.

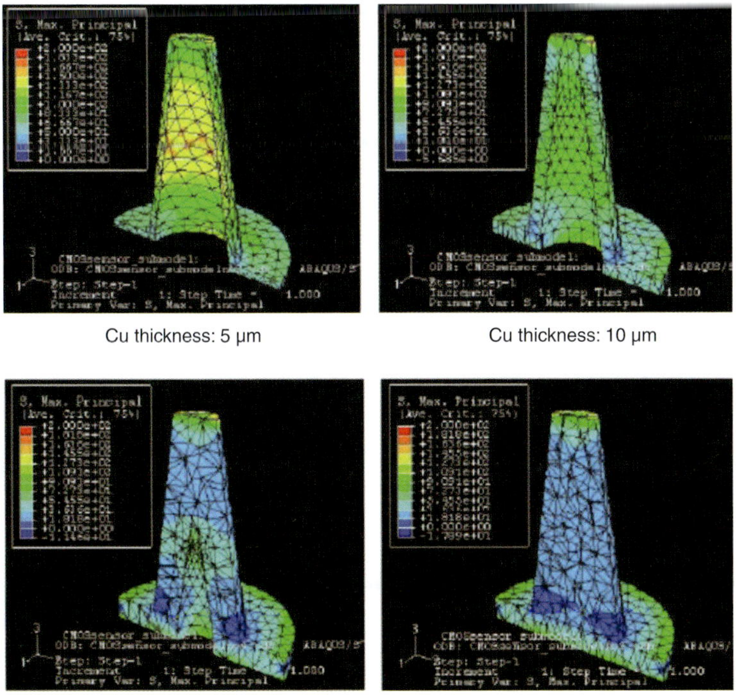

Cu thickness: 5 μm

Cu thickness: 10 μm

Cu thickness: half filled

Cu thickness: full filled

FIGURE 12.5 Maximum principal stress in the Cu.

the cases, the maximum stress occurs at the Cu landing edge on the epoxy resin side; and (c) the thicker the Cu wall thickness, the higher the von Mises stress at the Cu landing edge on the epoxy resin side.

Figure 12.5 shows the principal stress contours for those four different copper-filled thicknesses considered. It can be seen that: (a) for all the cases, the maximum principal stress occurs at the via bottom, which is the connection between the Ni and Cu; and (b) the largest maximum principal stress occurs at the bottom of the Cu half-filled via. Thus, by combining the results from von Mises stress and principal stress, one can conclude that a thin conformal Cu plating or full filling is better for the Cu-Ni interconnection.

12.2.3 Other Considerations

Since it takes a long time for the Cu plating to totally fill up the via, it increases the production cost and decreases the throughput. Also, the magnitudes of stresses for the 10 μm Cu wall thickness are only negligibly larger than those for the 5 μm Cu wall thickness. Furthermore, electroplating a 5 μm thickness of Cu along deep vias by a PCB fabrication process is marginal at this time. Thus, the 10-μm-thick Cu wall is chosen for the TSV metallization.

12.3 TSV Reliability Due to Local Thermal Expansion Mismatch

As mentioned earlier, the copper-filled TSV is one of the critical issues of 3-D IC integration. This is because: (1) the thermal coefficient of expansion (TCE) of silicon is $2.5 \times 10^{-6}/°C$; (2) the TCE of copper is $17.5 \times 10^{-6}/°C$; (3) there is a very large local thermal expansion mismatch between the silicon chip and the filled copper, which could create failures (delaminations) near the interface between the filled copper and the silicon and between the copper and the dielectric; and (4) there is a very large global thermal expansion mismatch between the fine-pitch IC silicon chip and the copper-filled TSV interposer (effective TCE ~ $10 \times 10^{-6}/°C$). The bumps (usually very small, e.g., microbumps [4, 10, 20, 28]) between them could fail (e.g., cracks) and may not be able to survive under thermal cycling conditions. In this section, the local thermal expansion mismatch is presented. The global thermal expansion mismatch will be presented in Sec. 12.4.

12.3.1 Boundary-Value Problems

In this section, DFR of the high-performance 2.5D IC integration system-in-package (SiP) (shown in Fig. 12.6) with copper-filled TSVs is performed by using the finite-element method. A very large chip (21 × 21 mm) with nine-metal layer and fine-pitch 65 nm Cu-low-k pads is supported by a silicon interposer (25 × 25 × 300 mm) with copper-filled TSVs. This TSV interposer is made with multilayer, double-sided metallization, CuNiAu under-bump metallurgy (UBM) on the top side, and SnAgCu (150 μm pitch) solder bumps on the bottom side [20]. The bismaleimide triazene (BT) substrate size is 45 × 45 mm square with 1-2-1 layer configuration (core thickness = 0.8 mm), and plastic ball grid array (PBGA) pad pitch of 1 mm. The package specification is shown in Table 12.2.

Nonlinear thermal stress due to the local thermal expansion mismatch at the interfaces between the copper, silicon, and dielectric are determined for a wide range of aspect ratios (of the TSV thickness and diameter). Figure 12.7*a* shows a simplified diagram of the TSV for finite-element analysis [10, 20, 28]. The assumptions are: (1) since the layer of tantalum (1 kÅ) is much thinner than the dielectric layer (1 μm), its effect is negligible; (2) the redistribution layer is very thin and therefore negligible; (3) silicon and silicon dioxide are not stressed beyond their elastic zone; (4) copper undergoes elastic deformation, followed by plastic deformation; and (5) the silicon interposer with TSV is stress-free at 125°C.

One-quarter of the TSV is modeled in an axisymmetric simulation. The boundary conditions used and the mesh at the critical interfaces are shown in Figs. 12.7*b* and 12.7*c*, respectively. Note that directions 1 and 2 represent the radial and axial directions, respectively. A static

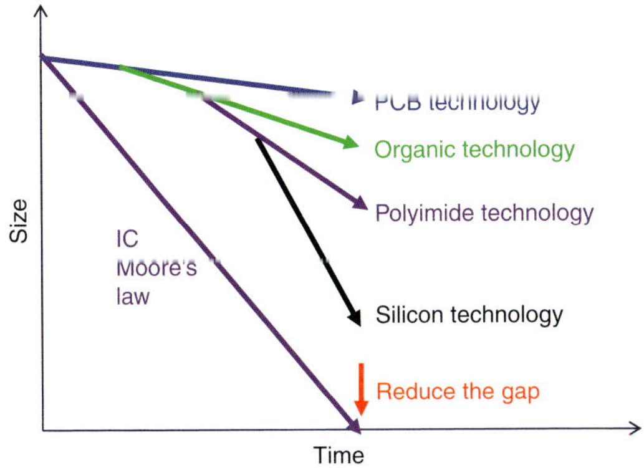

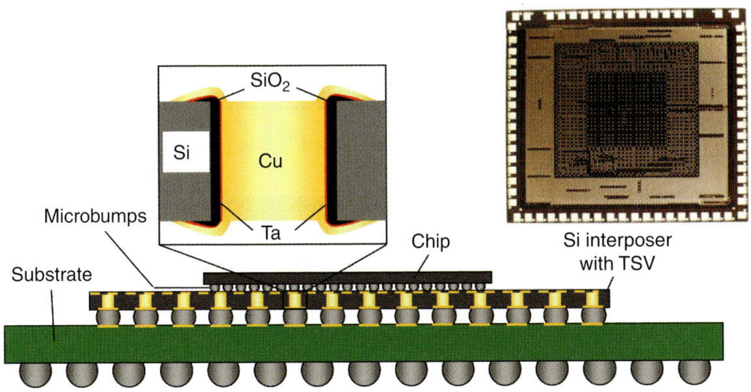

FIGURE 12.6 Schematic of the Cu-low-k chip on a TSV interposer (25 × 25 mm) (2.5D IC integration SiP).

Overall package	Body size	45 × 45 mm
	Thickness	1.8 mm
Top chip	Chip size	21 × 21 mm
	IC technology	Cu/low-k, 65nm technology
	I/O	11,000
	Pitch/Solder	150 μm/SnAg
TSV interposer	Outer/Inner pitch	300 μm/600 μm
	Tapered shape	100 μm/50 μm
Cu RDL	Layer	2
	Thickness	3 μm
	Via diameter	40 μm
	UBM/solder	CuNiAu/SnAgCu
Organic substrate	Layer	4
	Core thickness	800 μm
	Pitch	1 mm
	I/O	2000

TABLE 12.2 Package Dimensions and Configurations

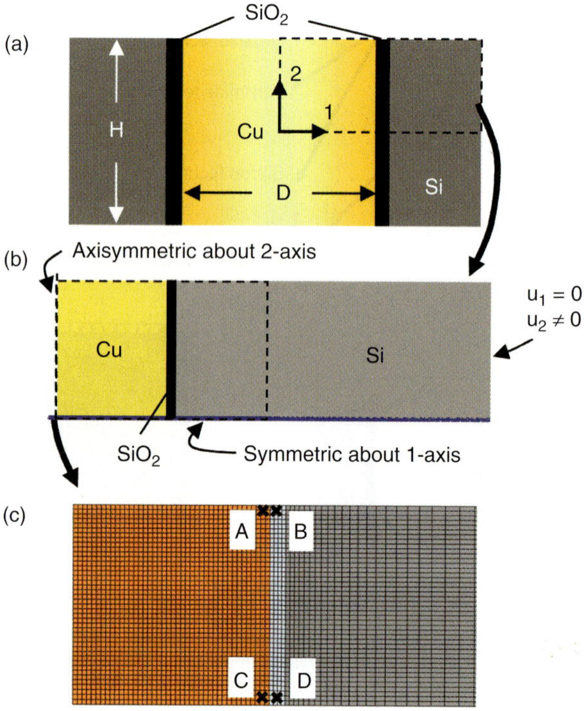

FIGURE 12.7 (a) Schematic of simplified TSV (b) Quarter model of TSV with applied boundary conditions (c) Mesh of critical regions.

temperature ramp-down analysis from 125°C to –40°C is carried out to simulate the maximum stress state. The modeling matrix and material properties used are shown in Tables 12.3 and 12.4, respectively.

12.3.2 DFR Results

During the temperature cycling of the TSV, each temperature ramp-up results in copper expanding more than 5 times as much as silicon and more than 10 times as much as silicon dioxide. The deformation resulting from heating from –40°C to 125°C and cooling from 125°C

Diameter (*D*)	**25 μm, 50 μm, 75 μm**
Aspect ratio (*H*/*D*)	1, 2, 3, 5, 7, 10, 12 for *D* = 25 μm 1, 2, 3, 5, 7 for *D* = 50 μm 1, 2, 3, 5 for *D* = 75 μm

TABLE 12.3 Simulation Matrix

Material	Silicon	Silicon Oxide	Electroplated Copper
Young's Modulus (GPa)	129.617@ 25°C 128.425@ 150°C	70	70
Poisson Ratio	0.28	0.16	0.34
CTE (ppm/°C)	2.813@ 25°C 3.107@ 150°C	0.6	18
Plastic properties (MPa)			240 @ 0 ε 250 @ 0.003 ε 255 @ 0.007 ε 255 @ 0.009 ε 250 @ 0.017 ε

TABLE 12.4 Material Properties for Simulations

to −40°C (exaggerated 100 times) is shown in Figs. 12.8*a* and 12.8*b*. It can be seen in Fig. 12.8*a* that the silicon dioxide layer is highly strained as it is dragged along by the axial expansion of the copper and compressed by the radial expansion of the copper. This deformation, however, is largely not transferred to the bulk of the silicon, as silicon is much stiffer than copper and silicon dioxide. Conversely, during temperature ramping-down a high strain is induced on the material near the interface indicated by points A and B, which are the potential failure locations.

Failure may occur at two critical points. First, failure could occur due to the tearing action during contraction at the interface between copper and silicon dioxide (points A and B in Fig. 12.7*c*). Second, cracking of copper or silicon dioxide could occur at the mid-plane of the TSV (points C and D in Fig. 12.7*c*). For the simulations carried out, the radial stress at points A and B, and the axial strains at points C and D were monitored to give an indication of the stress and strains causing delamination and cracking in those regions.

The radial stresses in the copper and silicon dioxide at the critical corner plotted against the diameter of the TSV for different aspect ratios are shown in Fig. 12.9. The radial stress increases linearly with the diameter of the TSV, with a steeper increase in TSV with larger aspect ratios. This is indicated by the slopes of the graphs marked on Fig. 12.9. Though a larger diameter of copper TSV induces a similar thermal strain, the deformation in copper is increased, resulting in a larger strain and therefore larger stresses. The stress in silicon dioxide is greater than that in copper for all cases due to its higher elastic modulus. In addition, the change in stress due to varying of the aspect ratio effect is most evident in TSV with smaller

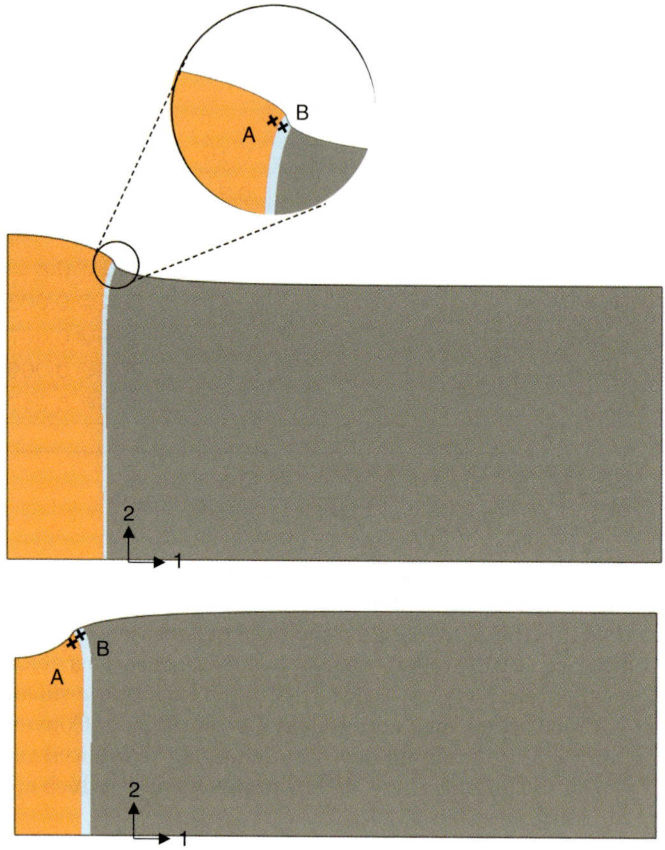

FIGURE 12.8 Deformation of TSV at critical region exaggerated 100 times heat-up (top), cool-down (bottom).

diameters—approximately 15% increase in radial stress is observed in a TSV of $D = 25$ μm compared to 2% increase for $D = 75$ μm.

The axial strains at points C and D in copper and silicon dioxide are plotted in Fig. 12.10. As expected, due to their close proximity, the strains in copper and silicon dioxide are identical for all aspect ratios and diameters. At an aspect ratio of 1, the axial strains are largest. As the aspect ratio increases, the strain decays. Beyond an aspect ratio of 5, axial strain is no longer dependant on aspect ratio. As the axial strain values are below the elongation values of copper (2%) and silicon dioxide (30%), cracking on the mid-plane is not expected for the copper-filled TSV.

The widened range of radial stress in the case of aspect ratio of 1 (Fig. 12.9) and the convergence of axial strain beyond an aspect ratio of 5 (Fig. 12.10) are caused by the proximity of the free edge on top of

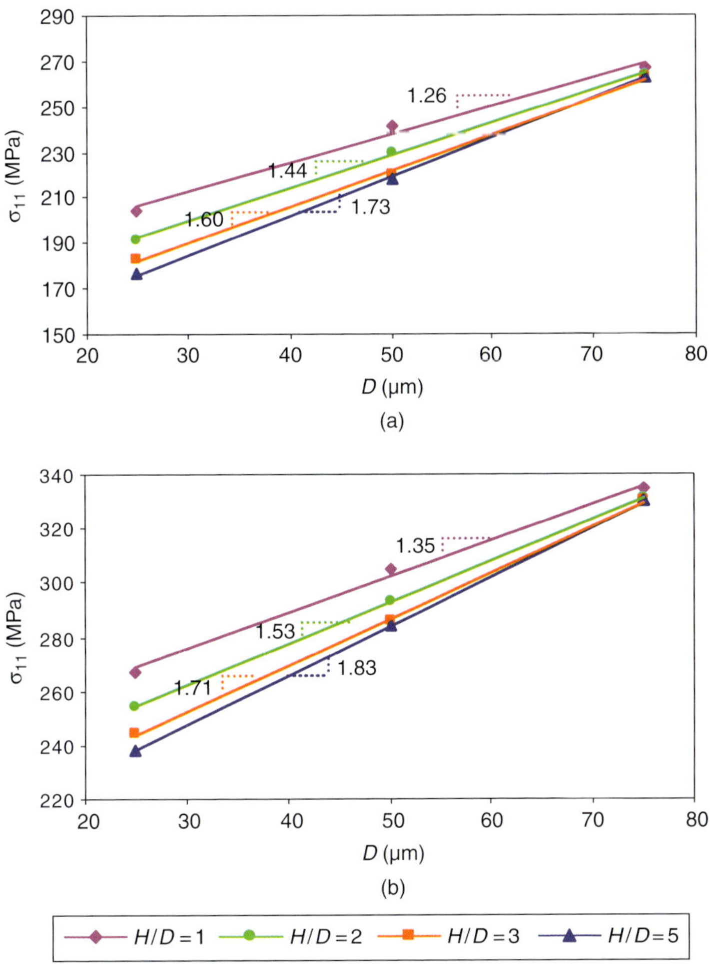

FIGURE 12.9 Radial stress (σ_{11}) at critical corner for (a) copper (point A) and (b) silicon dioxide (point B).

the TSV to the fixed edges (the axis and the mid-plane), which constraints the deformation. Figure 12.11 shows the axial strain contours of vias of various aspect ratios. Vias with an aspect ratio larger than 7 experience large strains at the top, instead of more distributed strains seen in vias with smaller aspect ratios.

12.3.3 TSV with Redistribution Layer

In TSV fabrication (Fig. 1.20), once the via is made by DRIE, a thin passivation layer of silicon oxide is deposited on the silicon. This is followed by via filling through electroplating. Next the wafer undergoes

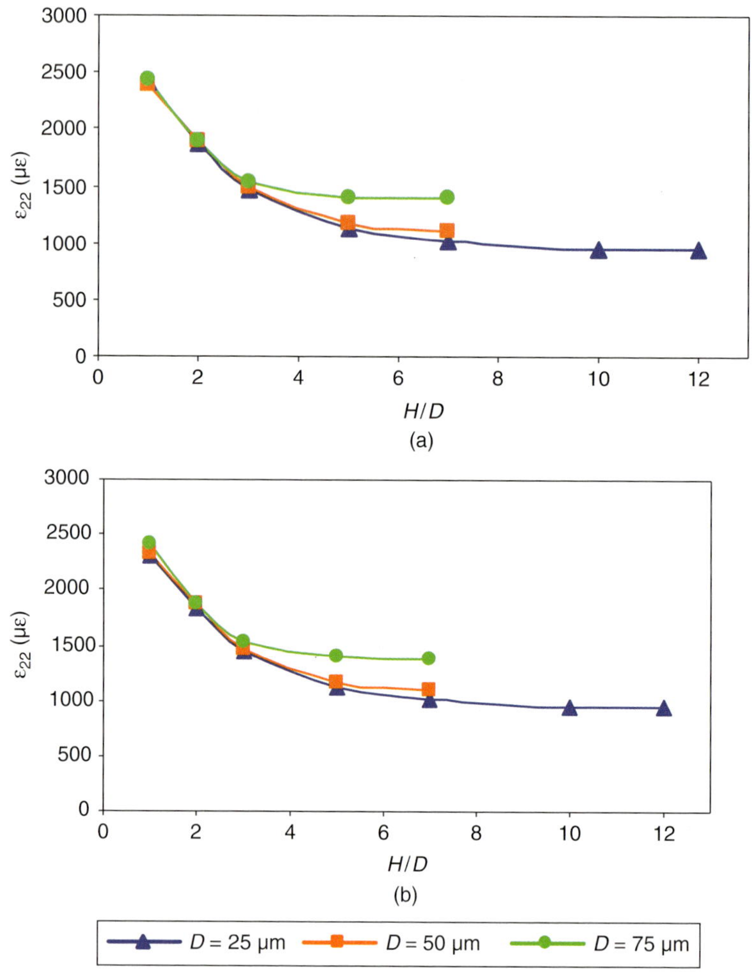

FIGURE 12.10 Axial strain (ε_{22}) at mid-plane for (a) copper and (b) silicon dioxide.

CMP and another thin passivation layer of silicon oxide is then deposited. The copper in the redistribution layer (RDL) is then sputtered at room temperature. Finally, the entire structure undergoes annealing to remove any residual stress induced by the deposition processes. As a result, residual stress is assumed to be negligible in the modeling work.

The redistribution layer in the model would simulate the real situation to a larger extent, as it would include the possibility of cracking of the RDL. A schematic drawing of the cross-section of the filled via, the one-quarter finite-element model, and the mesh at the

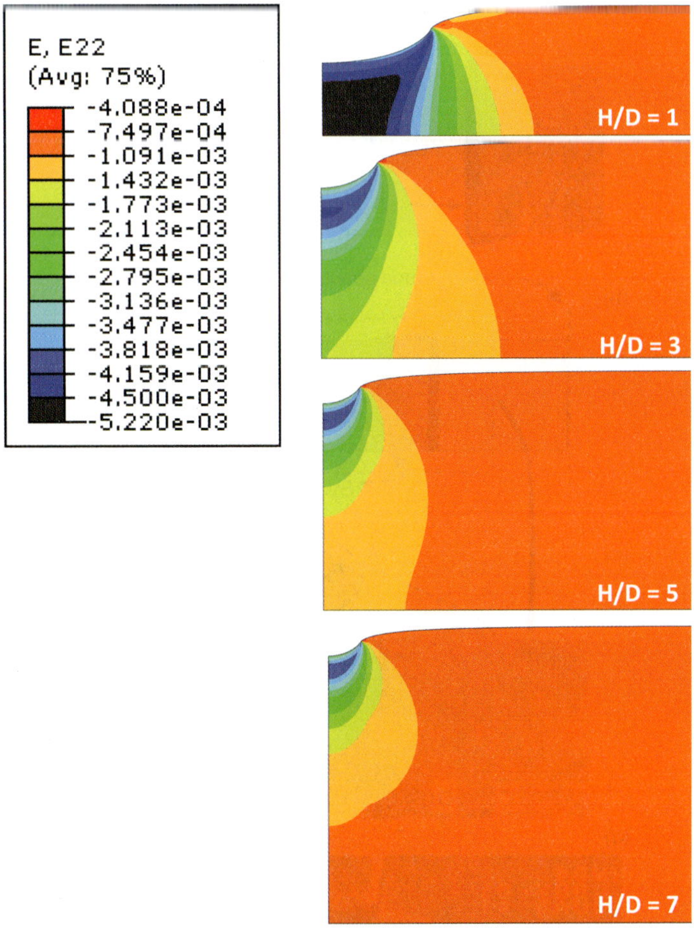

FIGURE 12.11 Axial strain (ε_{22}) contours of filled TSV of $D = 25$ μm at various aspect ratios.

critical location is shown in Fig. 12.12. For the model including the RDL layer, failure is expected to occur in the RDL layer at 45 degrees from the surface of the copper via. Hence, the strains along directions 1 and 2 have been resolved (transformed) to determine the strain in the diagonal-direction, $\varepsilon_D = \varepsilon_{1'}$, (indicated in Fig. 12.12).

The diagonal strain at points A and B are plotted in Fig. 12.13. It can be seen that there is a linear relationship between the diagonal strain and the diameter of the via. As expected, the strain in the silicon oxide is larger than that in the copper, and the silicon oxide could fail, as shown in Fig. 12.12*d*. This is due to the local thermal expansion mismatch between the Cu (17.5×10^{-6}/°C) and Si (2.5×10^{-6}/°C) in the z-direction reported in May at IEEE/ECTC2008 [28],

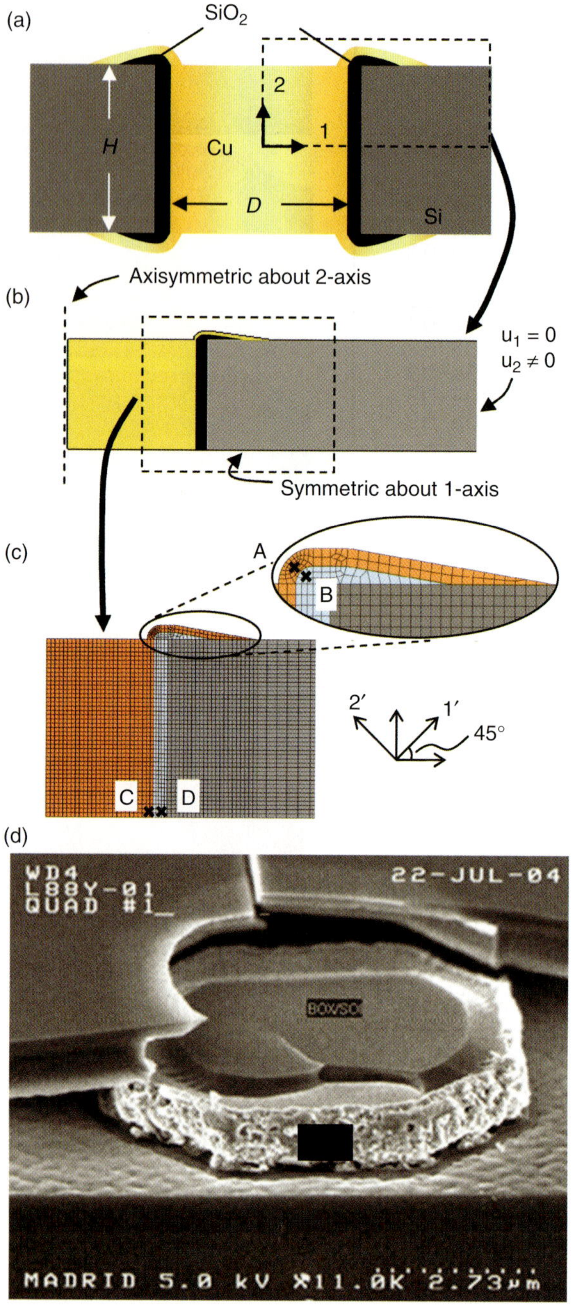

Figure 12.12 (a) Diagram of simplified TSV with RDL (b) Quarter model of TSV with applied boundary conditions (c) Mesh of critical region (d) Failure mode (*The SEM image is from Tezzaron*).

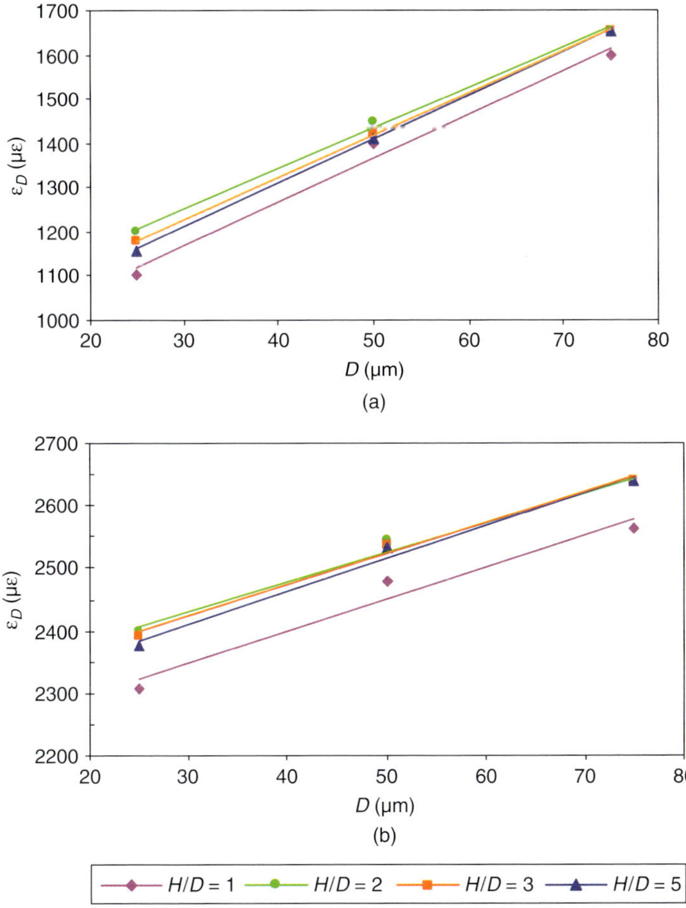

FIGURE 12.13 Diagonal strain ($\varepsilon_D = \varepsilon_I'$), at critical location for (a) copper (point A) and (b) silicon oxide (point B) in filled vias with RDL.

and the "mismatch" Cu pushes the silicon oxide to crack (see also Fig. 12.8). This phenomenon, called "copper pumping" (Fig. 12.12*d*), was reported in the fall of 2008 by Patti of Tezzaron (according to Philip Garrou, contributing editor of *Semiconductor International* on December 3, 2009).

The axial strain at point C is shown in Fig. 12.14 for the model with RDL. The axial strain curves at point D are not shown, as they had identical axial strains as the copper. Note that these curves are similar to those in Fig. 12.10*a*, which show the axial strains in copper in the model without the RDL. Hence, for the purpose of determining axial strain, the RDL may be ignored.

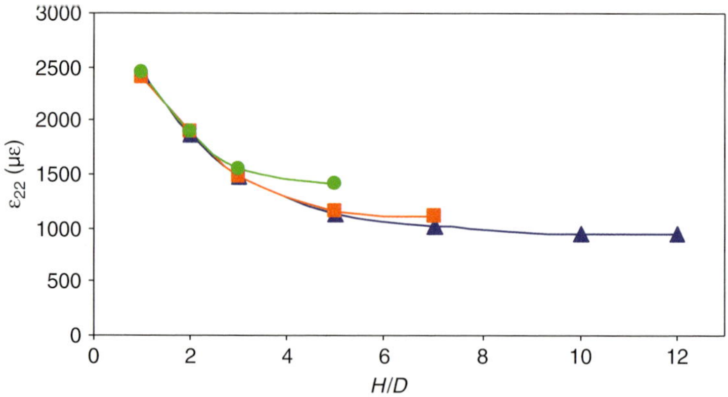

FIGURE 12.14 Axial strain (ε_{22}) at mid-plane for copper (point C).

12.3.4 Summary and Recommendations

Some important results are summarized in the follows:

(1) In general, above an aspect ratio (thickness/diameter) of 5, there is little dependence of stress and strain on the aspect ratio of the TSV.

(2) For the perfect TSV structures modeled, failures at the interfaces of the TSV are unlikely, as the strains in these elements are not large enough to cause failures.

(3) The imperfections that could arise during fabrication (such as the poor bonding between interfaces and the presence of asperities, which act as points of stress concentration) compounded with the stresses due to the local thermal expansion mismatch could result in failure. Thus, the TSV structure should be fabricated as perfectly as possible.

(4) The silicon dioxide dielectric layer is in a higher state of stress compared to copper as a results of its lower TCE compared to copper. Hence, extra care should be taken during the dielectric-layer fabrication to ensure a high-quality TSV.

12.4 TSV Reliability Due to Global Thermal Expansion Mismatch

This section investigates the TSV reliability due to the global thermal expansion mismatch between the Si chip and the Cu-filled TSV interposer. Also, the thermal cycling test results are provided.

12.4.1 2-D Boundary-Value Problem

Nonlinear stress/strain analyses of the global thermal expansion mismatches between the large Cu-low-k die and the TSV chip (interposer), and the TSV interposer and the BT substrate ($15 \times 10^{-6}/°C$) are performed to investigate the effect of the TSV interposer on the stress to cause failures of the Cu-low-k layers and the micro-solder bumps. The two-dimensional (2-D) plane-strain finite-element model of the package is shown in Fig. 12.15. Detailed structure of the micro-solder bump and the TSV is shown in Figs. 12.16 and 12.17, respectively.

All materials are modeled as elastic materials except solders. The solder materials are modeled as viscoplastic (creep) materials. The thermal loading for stress analysis of the Cu-low-k chip is from 260°C to 25°C. The thermal cycling (TC) profile for micro-solder bump reliability and failure analyses is 125°C to −40°C (60 min/cycle, 15 min ramp, and 15 min dwell).

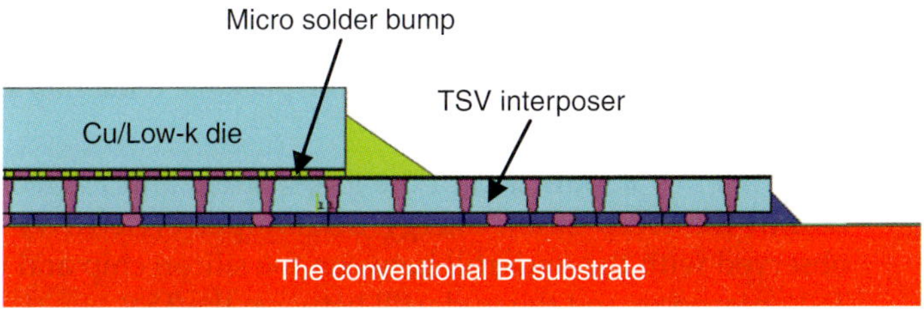

FIGURE 12.15 Finite-element 2-D model for global mismatch analysis.

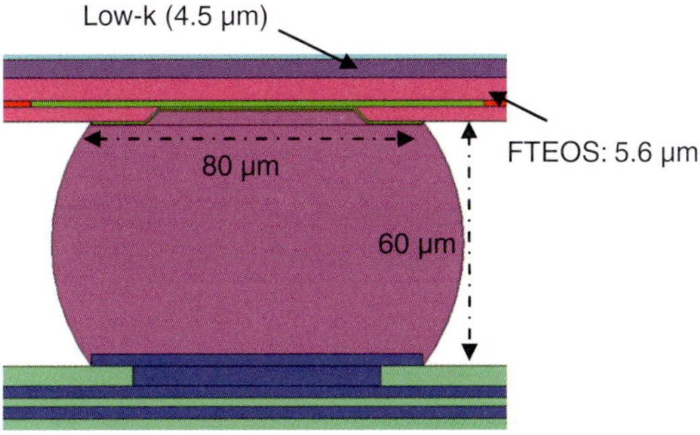

FIGURE 12.16 Detailed structure of the solder bumps.

FIGURE 12.17 Detailed structure of the tapered TSV.

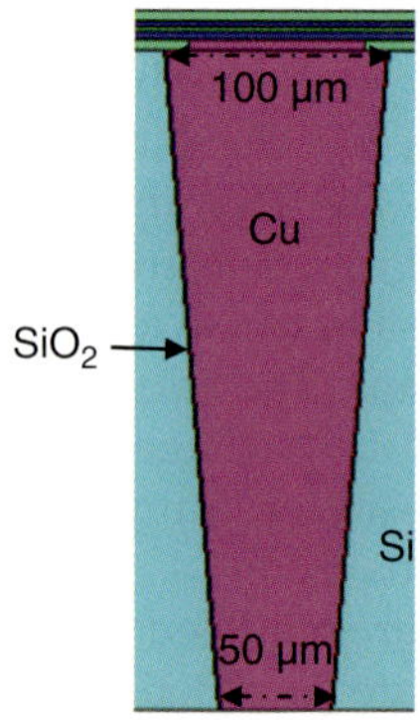

12.4.2 Summary and Recommendations

Some important results are shown in Tables 12.5 through 12.9, and recommendations are summarized as follows:

1) The impact of the TSV interposer is to reduce the low-k stress (by more than 50%) and reduce the creep strain energy density per cycle (ΔW) in the micro-solder bumps (by more than 28%). Thus, it can highly improve the package reliability and micro-solder joint reliability.
2) More than 60% reduction in the low-k stress and 7% reduction in the creep strain energy density per cycle (ΔW) are observed with the underfill between the silicon chip and the TSV interposer and the underfill between the TSV.
3) Based on various cases considered, two underfills that lead to lower stress/strain in the low-k layer and micro-solder bumps are recommended and shown in Sec. 12.5.1.
4) A 43% increase in the low-k stress is observed when TSV interposer thickness decreases from 300 μm to 100 μm. Thus, a thicker interposer is recommended. It should be pointed out that only the 2.5D IC integration SiP is modelled. For more accurate analyses, the PCB should also be included.

Effects of TSV Interposer without Underfill	The Max. Stress in Low-k Layer (MPa)	The Max. ΔW in the Micro Solder Bump (MPa)
Without TSV interposer	250.01	2.83
With TSV interposer	125.18	2.03

TABLE 12.5 Effects of TSV Interposer on Cu-Low-k Stress and Micro Solder Bump Reliability

Effects of Underfill (with TSV Interposer)	The Max. Stress in Low-k Layer (MPa)	The Max. ΔW in the Micro Solder Bump (MPa)
Without underfill	125.01	2.03
With underfill	41.98	1.89

TABLE 12.6 Effects of Underfills on Cu-Low-k Stress and Micro Solder Bump Reliability

TSV Interposer Thickness (μm)	The Max. Stress in Low-k Layer (MPa)	The Max. ΔW in the Micro Solder Bump (MPa)
300	41.98	1.89
200	50.15	2.00
100	59.97	2.09

TABLE 12.7 Effects of TSV Interposer Thickness on Cu-Low-k Stress and Micro Solder Bump Reliability

Via Top Diameter/via Bottom Diameter (μm)	The Max. Stress in Low-k Layer (MPa)	The Max. ΔW in the Micro Solder Bump (MPa)
100/50	41.98	1.89
80/35	34.79	1.83
70/30	30.73	1.80
60/20	25.49	1.75
50/10	19.82	1.70

TABLE 12.8 Effect of via Top Diameter/via Bottom Diameter on Cu-Low-k Stress and Micro Solder Bump Reliability

Low-k Die Thickness (μm)	The Max. Stress in Low-k Layer (MPa)	The Max. ΔW in the Micro Solder Bump (MPa)
750	41.98	1.89
300	17.12	1.22

TABLE 12.9 Effect of Cu-Low-k Chip Thickness on Cu-Low-k Stress and Micro Solder Bump Reliability

5) More than 50% reduction in the low-k stress and 10% reduction in the creep strain energy density per cycle (ΔW) in micro-solder bumps are observed when via top diameter/via bottom diameter decrease from 100 μm/50 μm to 50 μm/10 μm. Thus, a smaller via is recommended as, mentioned in Sec. 1.5.6 of Chap. 1.

6) A 59% reduction in the low-k stress and 35% reduction in the creep strain energy density per cycle (ΔW) in micro-solder bumps have been observed when low-k die thickness decreases from 750 μm to 300 μm. Thus, a thinner low-k chip is recommended.

12.4.3 3-D Boundary-Value Problem

Figure 12.18 shows the 3-D model (with the applied boundary conditions and the meshes of the critical region) for the analysis of the lead-free micro solder joints between the chip and interposer. (The BT-substrate is not modelled.) The creep material property of the structure was given in Chap. 2.

12.4.4 3-D Analysis Results

Results of the analysis indicate that as the thickness of the interposer increases, the creep strain energy density in the solder increases, as shown in Fig. 12.19 (H = interposer thickness). For interposer thicknesses of 50 to 200 μm, the maximum strain energy density occurs at the top of the solder joint. On the other hand, for interposer thicknesses of 300 to 500 μm, the maximum strain energy occurs at the bottom surface of the solder joint. The maximum creep strain energy location is transferred to the bottom when the expansion of the interposer begins to exert a shearing load on the solder. This effect can be seen in Fig. 12.20 where the solder joint on the thinnest interposer has greatly reduced creep strain energy density at its bottom surface. The use of underfill reduces the inelastic strain energy density by four times.

Increasing the pitch results in a decreasing of the creep strain energy density, as shown in Fig. 12.21 (P = TSV pitch). This is a result of the decreased copper content and hence decreased thermal expansion mismatch between the silicon chip and the silicon interposer with TSVs. Figure 12.22 shows the creep strain energy density for two different pitches decreased to almost the similar values when used with underfill.

12.4.5 Summary and Recommendations

Some important results and recommendations are summarized in the following.

1) The increased content of copper in the TSV interposer increased its effective CTE. Hence, the global thermal

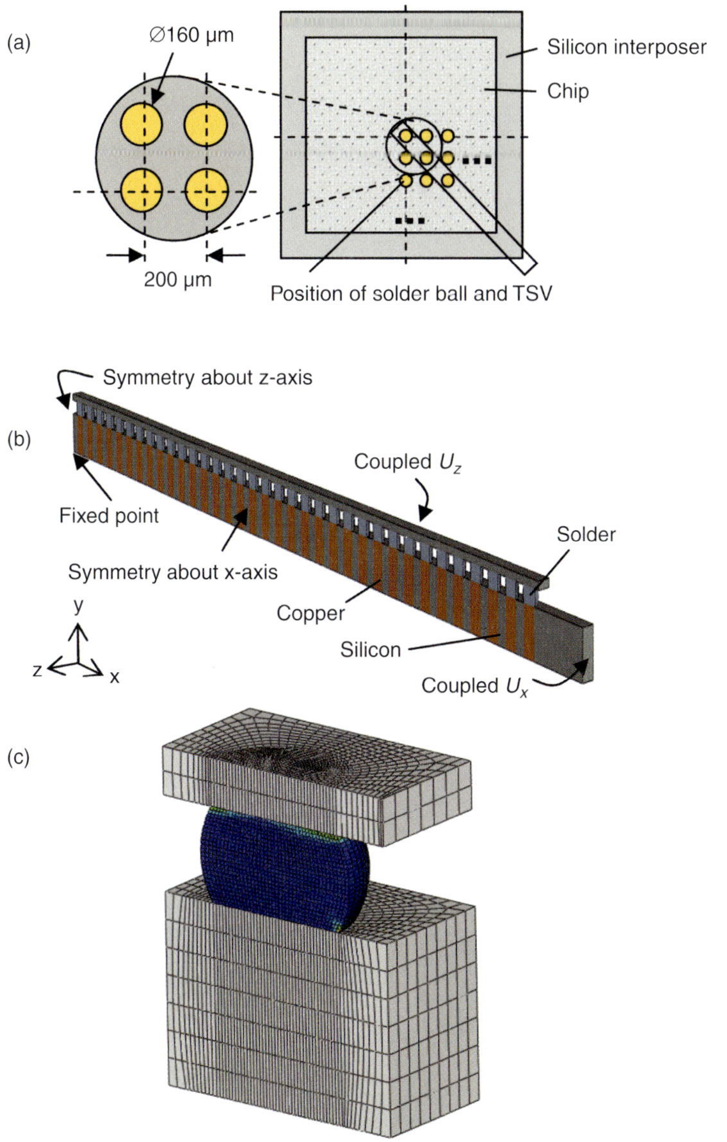

FIGURE 12.18 3-D models: (a) Diagram of global model (b) Diagonal slice model with applied boundary conditions (c) Mesh of critical region.

mismatch between the Si chip and the Cu-filled TSV interposer could be very large and the micro-solder joints connecting them are subjected to very large stresses and strains, especially the corner ones.

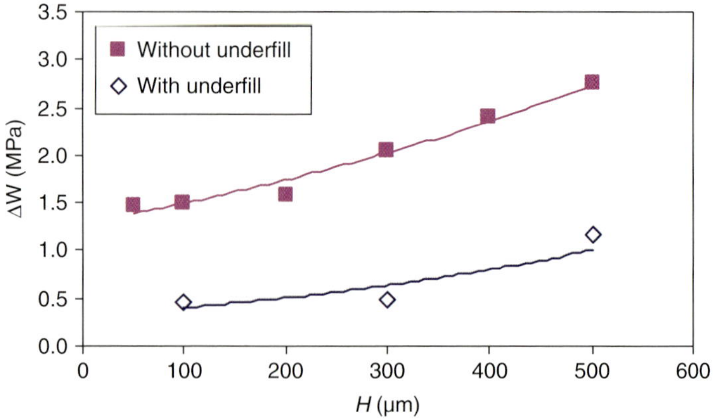

FIGURE 12.19 Creep strain energy density per cycle variation with interposer thickness (*H*).

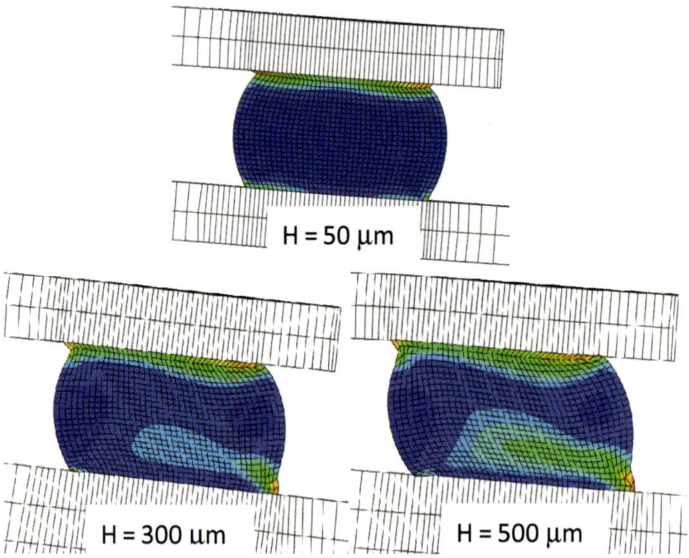

FIGURE 12.20 Creep strain energy density contours for three interposer thicknesses. (Not to scale.)

2) For a given TSV pitch and diameter, the thickness of the Cu-filled Si interposer plays a very important role. The larger the thickness, the larger the creep strain energy density per cycle, and thus the shorter the thermal-fatigue life of the micro-solder joints.

3) For a given TSV thickness and diameter, the pitch of the Cu-filled Si interposer also plays an important role. The larger

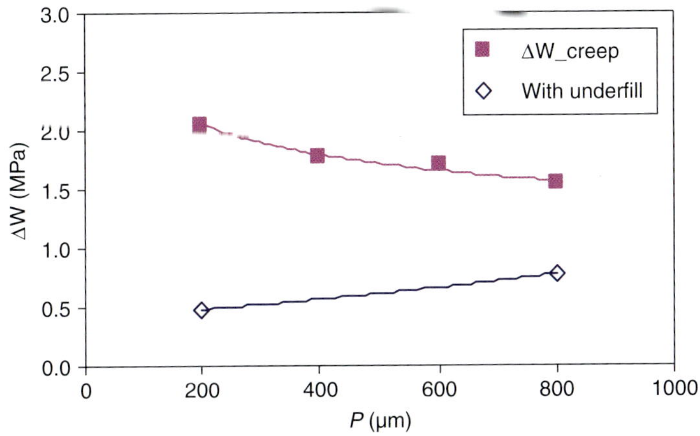

Figure 12.21 Creep strain energy density per cycle variation with via pitch (P).

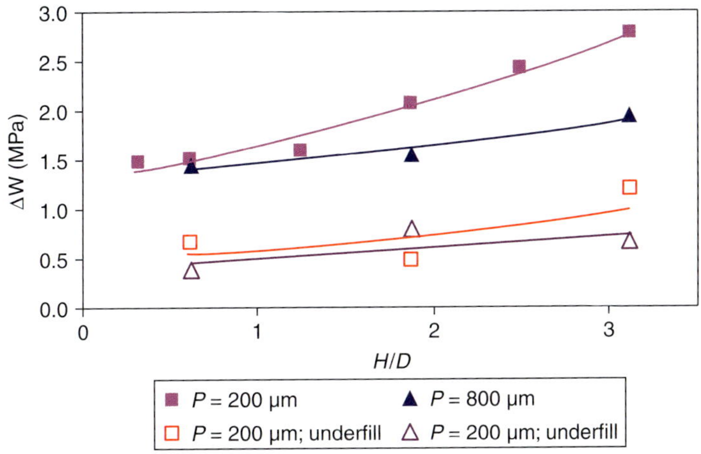

Figure 12.22 Creep strain energy density per cycle variation with aspect ratio (H/D).

the pitch, the smaller the creep strain energy density per cycle, and thus the longer the thermal-fatigue life of the micro-solder joints.

4) Underfill prolongs the life of the micro-solder joints by cementing the Si chip to the TSV interposer. This reduces the global thermal expansion mismatch and decreases the stresses and strains in the micro-solder joints. It has been found that with the presence of underfill, the maximum creep strain energy density per cycle acting at the corner micro-solder joint can be reduced by as much as four times.

5) Based on the cases considered, underfill is recommended, especially for thick and fine-pitch Cu-filled Si interposers.

6) It should be pointed out that, in order to have more accurate results, not only the BT-substrate but also the PCB should be included in the simulation.

12.5 Reliability Test and Failure Analysis of a 2.5D IC Integration SiP with a TSV Interposer

As mentioned earlier, the effective CTE of the Cu-filled silicon interposer is higher than that of the silicon chip. As a result, during temperature cycling, the interposer will expand and contract more than the chip, resulting in a fatigue loading action on the solder joints. Underfill material is used to cement the chip on the interposer so the relative displacements acting on the solder joints, especially the corner ones, are at a minimum.

12.5.1 Underfill Materials

Based on the simulation results reported in Sec. 12.4, two types of capillary flow underfill materials are considered, and their material properties and characteristics are shown in Table 12.10. The filler sizes of these two underfills are very different: The filler size of Underfill A is more than three times that of Underfill B.

12.5.2 Thermal Cycling Test and Results

The assembled package samples as shown in Fig. 12.23 are subjected to moisture sensitivity test (MST) level 3 (L3) and TC test between –40°C and 125°C according to JEDEC standards for components. Both C-mode scanning acoustic microscope (C-SAM) and ThruScan

Item	Unit	Underfill A	Underfill B
Filler content	Wt%	65	65
Filler size ave.	μm	2	0.6
Filler size max.	μm	10	3
T_g (TMA)	°C	100	100
CTE < T_g	ppm/C	28	26
CTE > T_g	ppm/C	90	90
Modulus	GPa	11	12
Viscosity	Pa.s	80	45
Cure profile	°C/mins	150/120	165/90

TABLE 12.10 Material Properties and Characteristics of Underfills

21 × 21 mm chip on the 25 × 25 mm TSV interposer on the BT substrate

A typical Thruscan image showing no delamination after 1000 thermal cycles

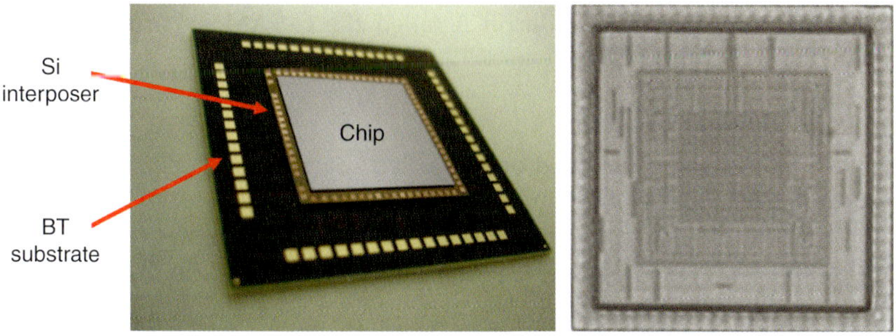

	MST level 3	1000 TC cycles (−40°C–125°C)
CSAM + Thruscan	Passed (0/8)	Passed (0/8)
Electrical test measurement	Passed (0/8)	Passed (0/8)

Figure 12.23 The assembled high-performance chip on TSV Si interposer, which is supported by a BT-substrate.

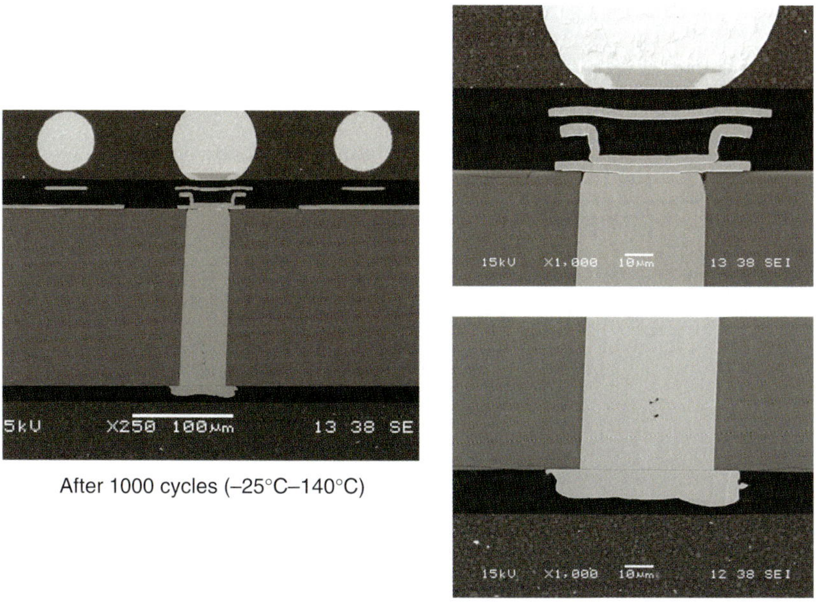

After 1000 cycles (−25°C–140°C)

Figure 12.24 Cross-section of the package after 1000 (−40 to 125°C) cycles test, which shows two RDLs on top of the TSV interposer.

are performed to examine the occurrence of delamination. No delamination was observed on all samples with Underfill B after MSL L3 and TC for 1000 cycles (−40°C to 125°C) (see Fig. 12.24). However, two samples with Underfill A failed the C-SAM + ThruScan after 1000 TC cycles.

12.5.3 Failure Analysis

Failure analysis is performed on the failed samples, and cracks are found at the top side of the micro-solder joint near the intermetallic compound (IMC) region, as shown in Fig. 12.25. Electric test is also performed to monitor the electrical resistance changes along the daisy-chain during MST L3 and TC tests. No electrical failure was encountered on any samples with underfill B after MSL L3 and 1000 TC cycles. However, three samples with underfill A have shown open circuits in the daisy chain. Failure analysis shows insufficient underfill coverage at the TSV interposer corner, as shown in Fig. 12.26. This corner delamination also leads to crack initiation and propagation of the underfill. Why? Because the filler size of underfill A is too big and it cannot fill the entire gap between the chip and the interposer. Thus, for 3-D IC integration underfills with a smaller filler size should be used, for example, average filler size < 0.6 μm and maximum filler size < 3 μm.

U underfill B

	MST level 3	1000 TC cycles
CSAM + thru-scan	Passed (0/8)	Passed (0/8)
Electrical test measurement	Passed (0/8)	Passed (0/8)

U underfill A

	MST level 3	1000 TC cycles
CSAM + thru-scan	Passed (0/8)	Passed (0/8)
Electrical test measurement	Passed (3/8)	Passed (0/8)

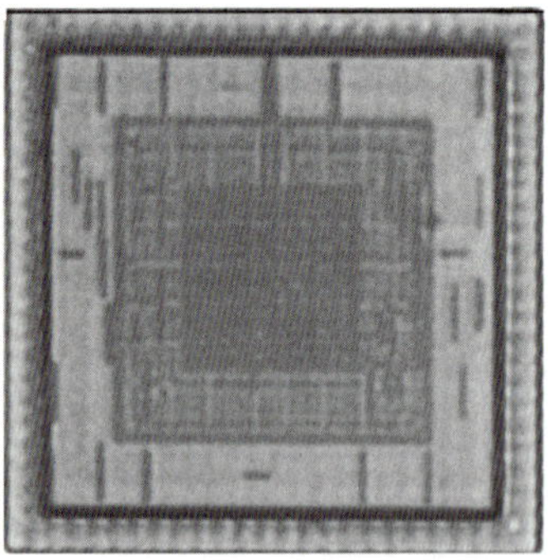

A typical Thruscan image showing no delamination after 1000 TC cycles (−44°C–125°C) (underfill B)

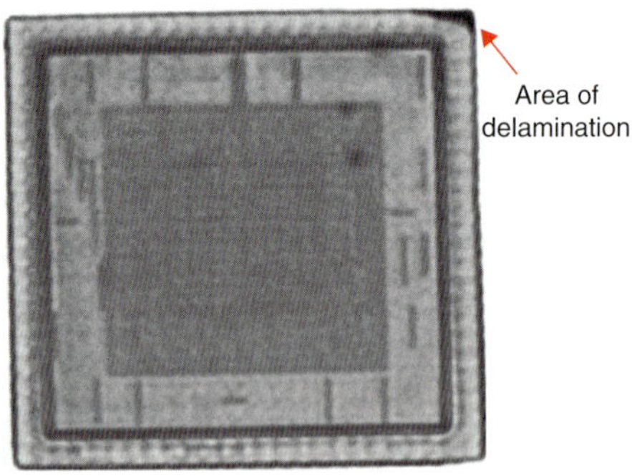

A delamination after 1000 TC cycles (−44°C–125°C) (underfill A)

Figure 12.25 Thermal cycling test results with underfill A and underfill B.

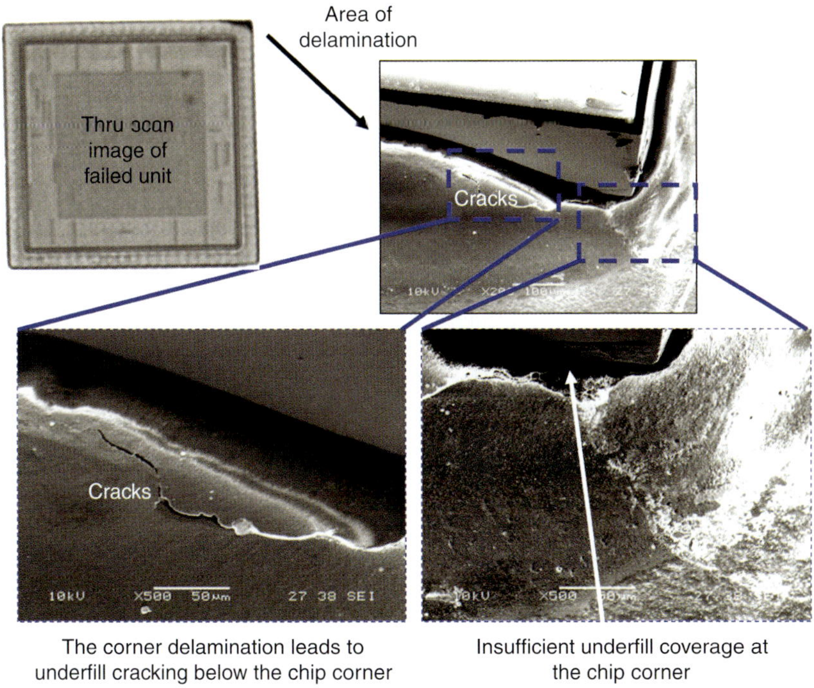

Figure 12.26 Failure modes of the failed sample showing the imperfect underfill structure (gap at the corner). Cracking occurs on the underfill as well as at the corner solder joint.

12.5.4 Summary and Recommendations

Some important results and recommendations are summarized in the following:

(1) Due the thermal expansion mismatch between the chip and the copper-filled Si interposer, underfill is needed for solder-joint reliability.

(2) For 3-D IC integration applications, the solder joint between the chip and the TSV interposer (either passive or active) is very small; thus, underfills with smaller filler sizes should be used in order to totally fill up the gap and enhance the solder-joint reliability.

12.6 Thermal-Enhanced and Cost-Effective 3-D IC Integration SiP with a TSV Interposer

A low-cost (with bare chips) and high (electrical, thermal, and mechanical) performance 3-D IC integration SiP is designed and described. This system consists of a silicon interposer with TSV, redistribution layers (RDL), and integrated passive devices (IPD), which carries the high-power flip chips with microbumps on its top surface and the low-power chips at its bottom surface. TSVs in the high- and low-power chips are optional but should be avoided. The backside of the high-power chips is attached to a heat spreader with or without a heat sink.

This 3-D IC integration system is supported (packaged) by a simple conventional organic substrate. The heat spreader (with or without heat sink) and the substrate are connected by a ring stiffener, which provides adequate standoff for the 3-D IC integration system. This novel structural design offers potential solutions for high-power, high-performance, high pin-count, ultra fine-pitch, small real-estate, and low-cost applications. Thermal management and reliability of the proposed systems are demonstrated by simulations based on heat-transfer theory and time and temperature dependent creep theory.

12.6.1 Introduction

As Moore predicted in 1965, silicon chips are getting larger while incorporating a higher pin-count and finer pad-pitch. Unfortunately, conventional substrates with build-up layers made by organic materials, such as BT (Fig. 12.6), are facing great challenges in supporting these high pin-count and ultra fine-pitch silicon chips. Thus, to address these issues, silicon interposers with high-density TSV has emerged as a perfect solution to provide high wiring density redistribution and interconnection.

TSV plays a very important role in modern 3-D IC integration and packaging. It can be used for stacking up a series of memory chips and providing a signal/heat path between the chips (Chap. 1). (This will not be discussed in the present chapter.) Also, it can be used in a passive interposer to support ultra fine-pitch, high pin-count, high-power, and high-density IC chips and in an active interposer such as logics, microprocessors, and high-power memories (Chap. 1).

For example, Fig. 12.27*a* shows a conventional face-down PBGA package. It can be seen that the chip is supported by the high-density BT-substrate with build-up layers [43]. As the chip gets bigger with a higher pin count and finer pitch, the BT substrate can no longer support it. Hence there is a need for an intermediate substrate (e.g., the passive TSV interposer) to redistribute the large array of fine-pitch pads on the chip to fewer and relatively larger pitch pads on a simpler and thinner BT substrate even without any build-up layers, as shown in Fig. 12.27*b*. The foot-print of this TSV package is much smaller than that with high-density BT-substrate with build-up layers (Fig. 12.27*a*).

Figure 12.28 shows the example of an active TSV interposer. It can be seen that in order to address the memory bandwidth challenge [44], the ultimate goal is to have the very high-power, high pin-count, and fine-pitch CPU (central processing unit) and 3-D memory chips

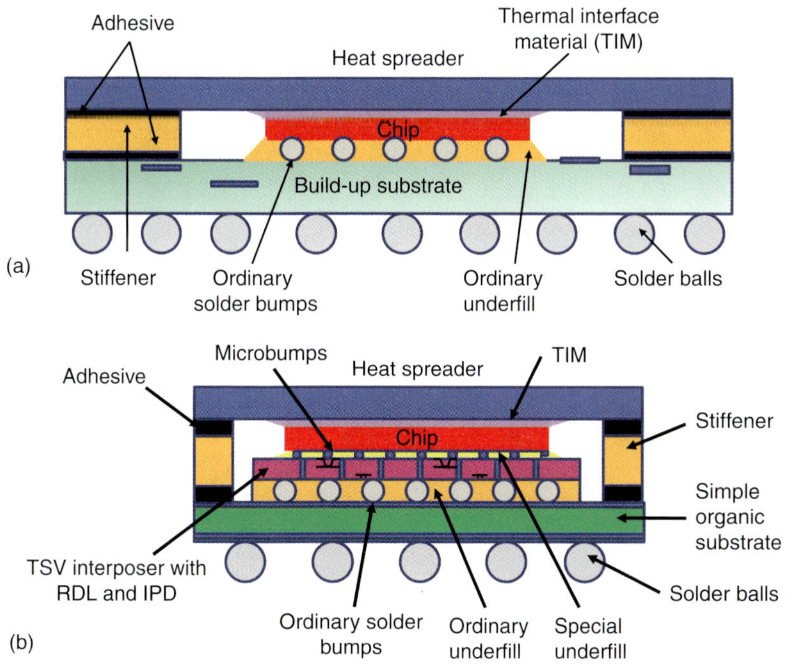

Figure 12.27 (a) Conventional face-down PBGA. (b) Face-down PBGA with a passive TSV interposer (2.5D IC integration SiP).

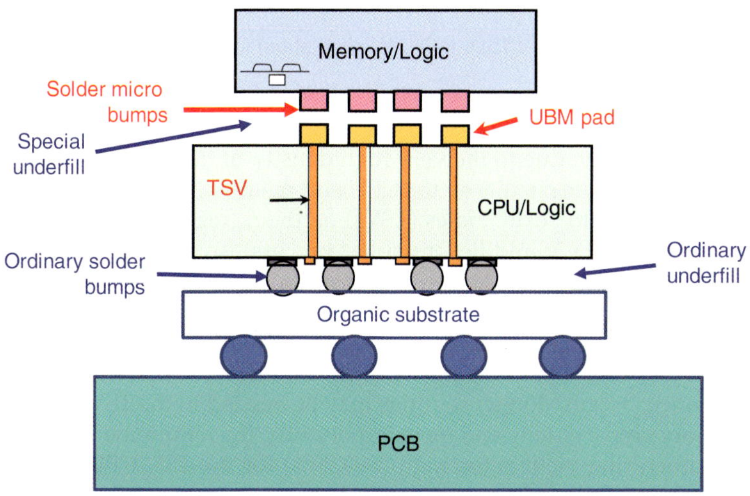

Figure 12.28 CPU/Logic as a TSV interposer to support a memory/Logic (3-D IC integration SiP).

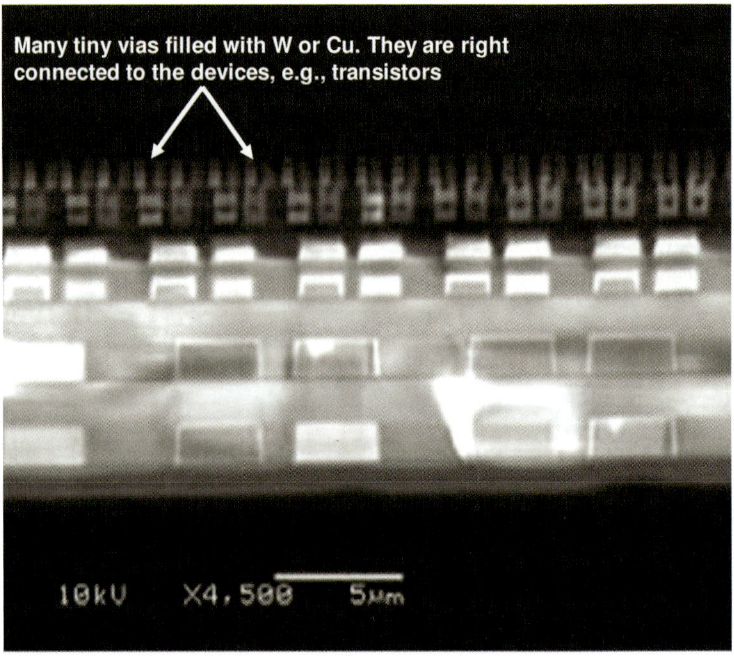

65nm Cu/low-k Chip by *GlobalFoundries (2006) to show the tiny vias.*

Besides the tiny devices such as transistors (which cannot be seen in this SEM image), there are many tiny vias (≤ 0.1 μm) on a chip. They are connected to devices (e.g., 4 tiny vias for each transistor) to build the first metal (M1) layer. Today, the number of these tiny vias, for many chips, already exceeds the world population of over 7 billion. One of the core competences and major businesses of Foundries are to make these tiny vias and they do best.

FIGURE 12.29 Vias (not TSVs in 3-D IC integration) in a chip.

attached together. In this case, the CPU acts like an active TSV interposer. However, due to the very high density of the devices and the complexity of the circuits on the CPU and memory, the chance to find places/spaces to "drill" holes (TSVs) on them (regardless it is made by the via-first, via-middle, or via-last process) is very slim (difficult) as shown in Fig. 12.29. Also, the chip size and number of pin outs of the CPU and memory chips, are most likely not the same. Forcing to attach them together will, therefore, reduce the design freedom and possibly compromise the functionality. Another issue is electrical performance, because longer routings may be needed in the TSV CPU or memory chips. Finally and most importantly, the requirement of TSV manufacturing yield is too high (>99.99%) for the TSV CPU to bear with the addition of costs due to TSV manufacturing yield loss [3]. Thus, a TSV interposer connecting the CPU and memory in certain 3-D configurations could be an effective solution to resolve these issues. That is the focus of this section.

One of the critical issues of 3-D IC integration is its thermal management. This is because: (1) 3-D circuits increase total power generated per unit surface area; (2) chips in the 3-D stack may overheat if proper and adequate cooling is not provided; (3) the space between the 3-D stack may be too small for cooling channels (i.e., no gap for fluid flow); and (4) thin chips may create extreme conditions for on-chip hot spots. Thus, low-cost and effective thermal management solutions are desperately needed for widespread use of 3-D IC integration and are therefore also the focus of this section.

12.6.2 Design Philosophy

The objective is to design a generic thermal-enhanced, electrical-sound, mechanical-integrity, and cost-effective 3-D IC integration system for SiP applications. In other words, this design addresses the electronic packaging of 3-D IC integration with a TSV/RDL/IPD interposer for high-power, high-performance, high pin-count, ultra fine-pitch, small real-estate, and low-cost applications. In order to achieve these goals, this design utilizes chip-to-chip interconnections through a TSV interposer in a 3-D SiP format with excellent thermal management. Avoid to "dig" holes (TSVs) on the active chips, that is, use whatever Moore's chips are made by the Foundries, and use a TSV interposer to perform the redistributions (to fan out the high pin-out and ultra fine-pitch circuitries of the chips) and most of the tasks, and then connect to the next level of interconnects (e.g., the package substrate) with much less pin-outs and coarse pitches.

Why not dig holes on the active chips now? Besides the technology issues mentioned earlier, the EDA (Electronic Design Automation) for 3-D IC is not ready! EDA is the soul of 3-D IC integration and is a category of software tools for designing electronic systems and the focuses of the software are design, simulation, analysis and verification, manufacturing preparation, and test. The key players of EDA are Synopsys, Cadence, Mentor Graphics, Zuken, Agilent, etc. Because the Ecosystem (e.g., standards and infrastructures) of 3-D IC does not exist now, thus, unfortunately they are not seriously, full-force, and systematically working on this area yet.

12.6.3 The New Designs

The proposed design is a 3-D IC integration system with high-power and low-power chips supported by a TSV/RDL/IPD interposer in a standard face-down PBGA package for high-performance, low-cost, effective thermal management, and high-reliability applications, as shown in Fig. 12.30. The entire system consists of a silicon interposer with high-density TSVs, redistribution layers to connect various chips with pads that have different pitches, sizes, locations, and IPD to enhance electrical performance. A simple organic substrate with standard (in size and pitch) solder balls for PCB assembly supports the TSV interposer. All the high-power chips such as the MPU (Micro-Processor

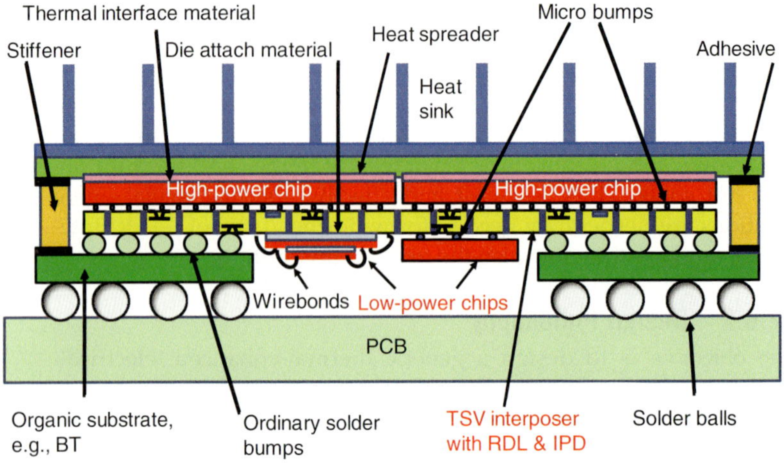

Special underful between the TSV interposer and the high-and low-power flip chips.
Ordinary underful between the TSV interposer and the organic substrate.
Encapsulant for wire bonding memory stack.

Figure 12.30 Thermal-enhanced and cost-effective 3-D IC integration SiP (without TSVs on active chips) supported by a passive TSV interposer with RDL and IPD (3-D IC integration SiP.)

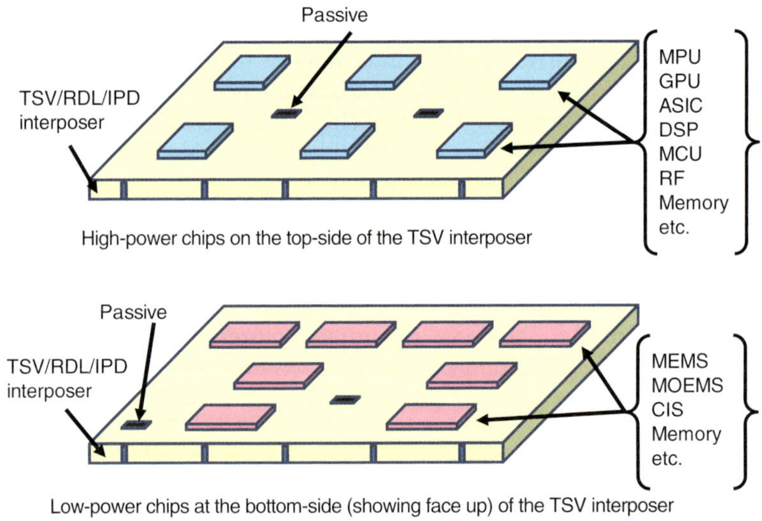

Figure 12.31 The passive TSV interposer with RDL and IPD supporting high-power chips on the top and low-power chips at its bottom.

Unit), GPU (Graph-Processor Unit), ASIC (Application Specific IC), DSP (Digital Signal Processor), MCU (Micro-Controller Unit), RF (Radio Frequency), and high-power memory chips, as shown schematically in Fig. 12.31, are on top of the TSV interposer in a flip-chip format so that the backside of these chips can be attached to a heat spreader via a thermal interface material (TIM). In this case, most of the heat from all the high-power chips can be dissipated through the heat spreader (with a heat sink if it is necessary).

All the low-power chips such as the MEMS (Micro-Electromechanical System), MOEMS (Micro-Optoelectromechanical System), CMOS image sensors, and memory chips, as shown schematically in Fig. 12.31, are at the bottom-side of the TSV interposer with either flip-chip or wire-bond formats. A ring-stiffener connecting the organic substrate and the heat spreader is used to provide the adequate standoff for the 3-D IC integration with the TSV interposer and to support the heat spreader with or w/o the heat sink. Underfill encapsulants are, most likely, needed between the TSV interposer and the high- and low-power flip chips, and between the TSV interposer and the BT substrate. However, underfill is not needed between the 3-D IC integration SiP and the PCB (this is a very important feature for system houses who don't like to have underfills on their PCBs). For wirebonding chips, encapsulants may be needed.

The appearance of the proposed 3-D IC integration SiP is very attractive to, for example, IDM (Integrated Device Manufacturers), OEM (Original Equipment Manufacturing), and EMS (Electronics Manufacturing Services). First of all, it is a standard face-down PBGA package and has been used by the electronic industry for more than 15 years [45]. This package is not only effective in thermal management [46–52] but its solder joints are very reliable [53–54]. Therefore, in conjunction with the proper design of the high-power and low-power chips above/below the TSV interposer inside the package, a SiP that is both cost-effective and displays high electrical and thermal performance can be achieved and manufactured.

Figure 12.32 is an example of a very high-performance application, consisting of MPUs, GPUs, and high-power memory chips. They are flip-chip bumped (with any geometric configuration and material make-up) on top of the TSV interposer. Their backside is attached to a heat spreader/sink with a TIM. Most of the heat from these chips is dissipated through this heat spreader/sink. The low-power memory chips are attached either by wire bond or flip-chip to the bottom-side of the TSV interposer. Underfills are needed between the TSV interposer (the effective thermal coefficient of expansion (TCE) $\sim 10 \times 10^{-6}/°C$) and the high-power chips (TCE $\sim 2.5 \times 10^{-6}/°C$), and between the low-power chips (TCE $\sim 2.5 \times 10^{-6}/°C$) and the organic substrate (TCE $\sim 18 \times 10^{-6}/°C$) to ensure the thermal-mechanical reliability of the bump interconnects. The thickness of the proposed TSV/RDL/IPD passive interposer should be ≤ 200 μm for electrical performance, the thinner the better. Thus thin-wafer handling methods are required for the passivation, metallization and/or wafer bumping processes. Fortunately, by using the passive interposer, thin-wafer handling is not necessary for the high- and low-power chips.

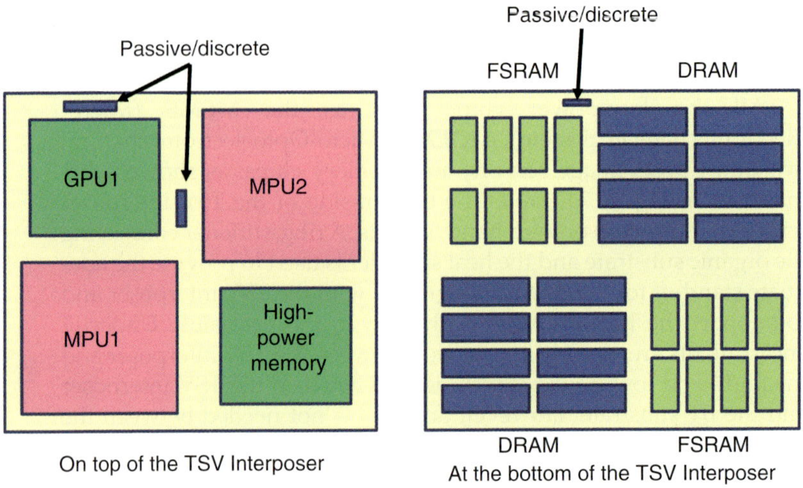

FIGURE 12.32 High-power chips (MPU, GPU, memory) on top of the TSV/RDL/IPD interposer and low-power chips (DRAM, FSRAM) at its bottom.

12.6.4 A 3-D IC Integration SiP Design Example for Demonstration

Figures 12.33 and 12.34 show a new design for detailed analysis of a 3-D IC integration system in a SiP. It consists of 4 identical high-power chips (e.g., microprocessors) uniformly distributed on the top-side of a TSV interposer and 16 identical low-power chips (e.g., memories) uniformly distributed at the bottom-side. The dimensions of the high-power chips are 10 mm × 10 mm × 200 μm, the low-power chips are 5 mm × 5 mm × 200 μm, and the TSV interposer are 35 mm × 35 mm × 200 μm. There are 1600 identical TSVs and their diameter is 20 μm and pitch is 850 μm (very large), since there are not any devices. The backsides of the high-power chips are attached to a heat spreader (44 mm × 44 mm × 1 mm) through a 100-μm-thick thermal interface material (TIM). An aluminum heat sink (44 mm × 44 mm × 20 mm) with 21 uniformly distributed fins is attached to the back of the heat spreader.

The 3-D IC integration system mentioned in Fig. 12.33 is attached (with ordinary solder bumps) to a BT-substrate, whose dimensions are 44 mm × 44 mm × 0.8 mm with a cavity of 33 mm × 33 mm × 0.8 mm. The substrate is also connected to the heat spreader with the heat sink through an aluminum ring-stiffener as shown in Fig. 12.34. Then, the substrate is lead-free (SnAgCu) soldered on a FR-4 PCB (50 mm × 50 mm × 2.5 mm).

12.6.5 Modeling of the 3-D IC Integration SiP Design Example

The finite-element model for the analyses is shown in Fig. 12.35 and the computer code is ANSYS. It can be seen that the 3-D IC integration

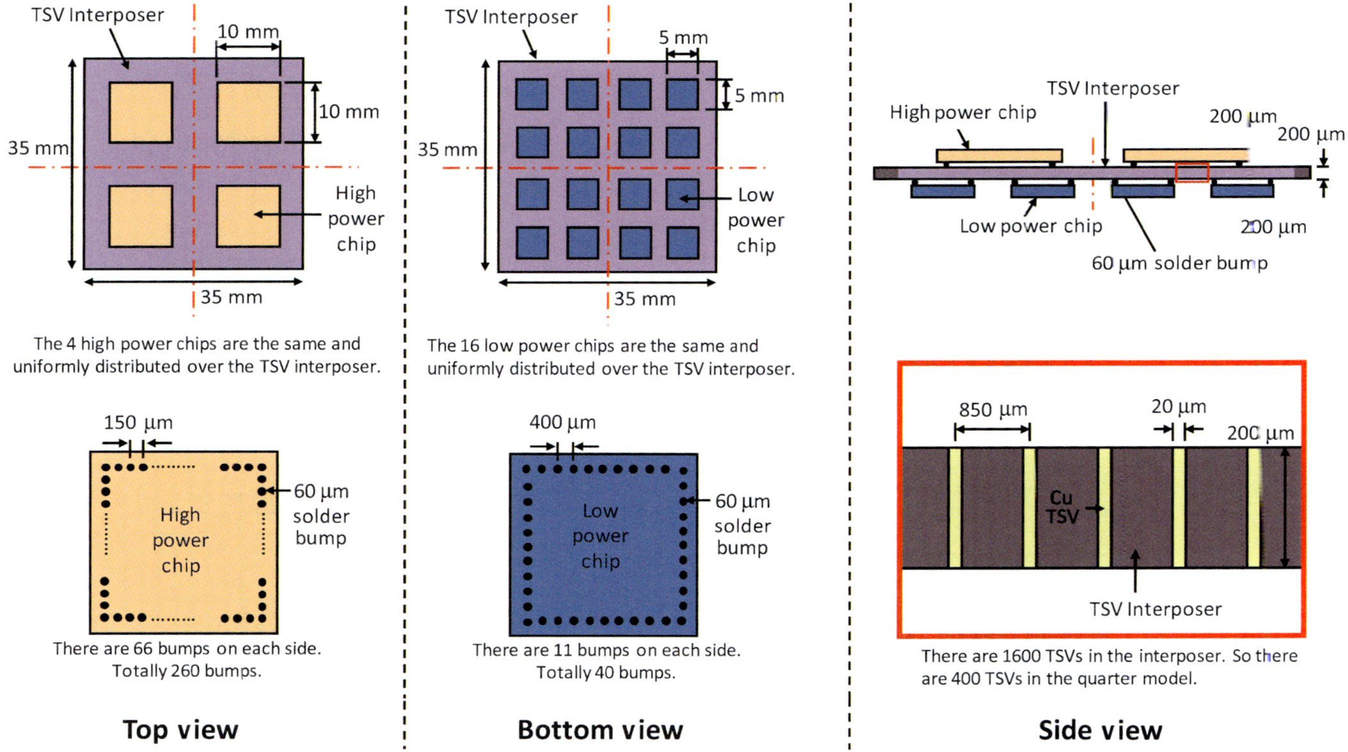

FIGURE 12.33 Dimensions of the TSV interposer and the high- and low-power flip chips and their top, bottom, and side views.

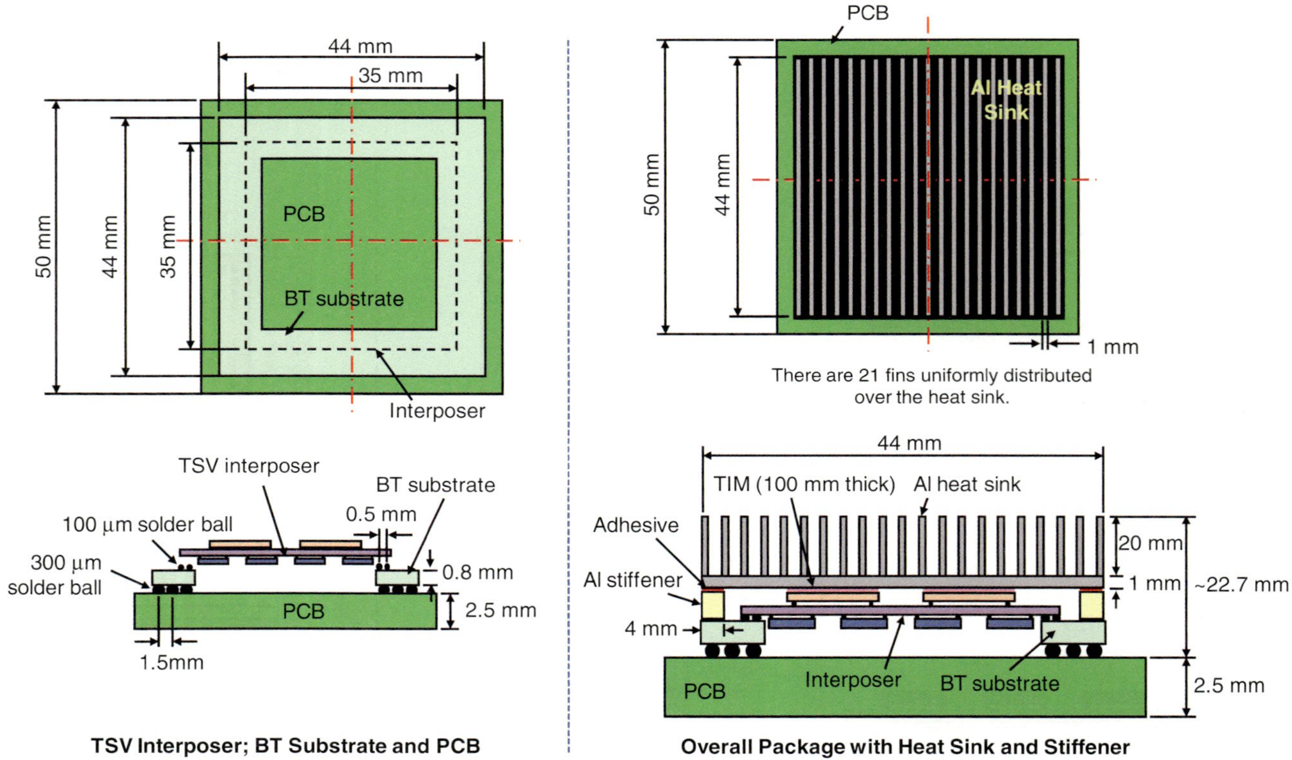

FIGURE 12.34 Dimensions of the overall 3-D IC integration SiP.

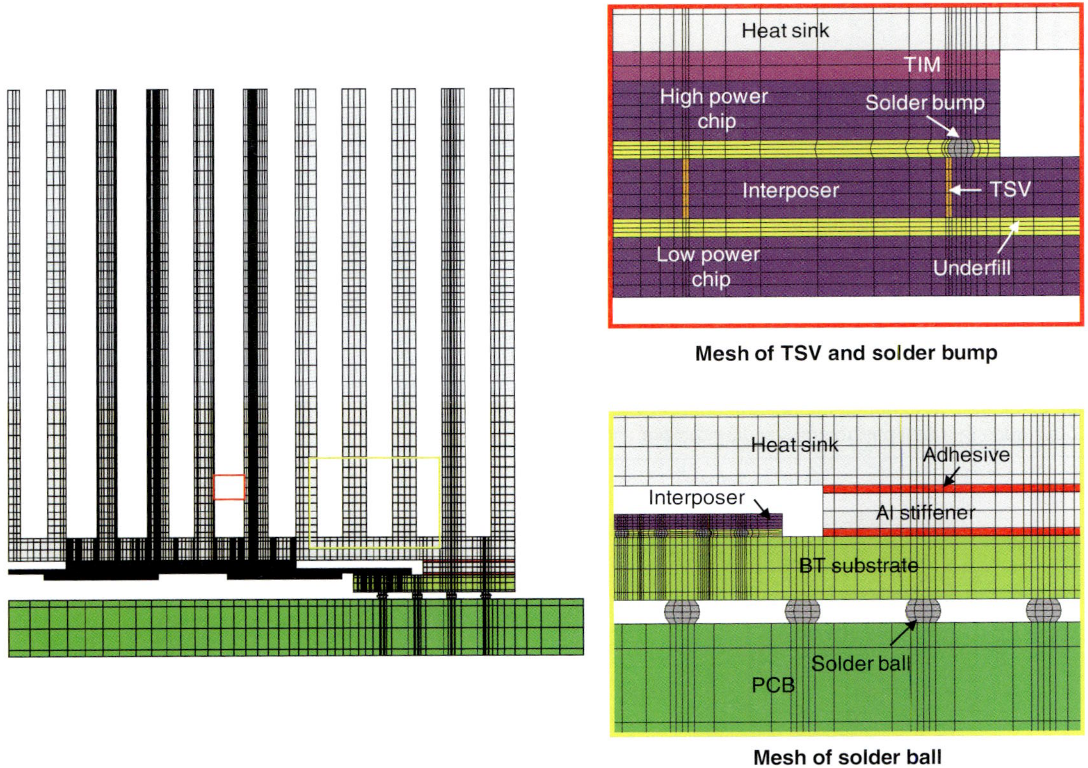

FIGURE 12.35 Finite-element model of the 3-D IC integration SiP.

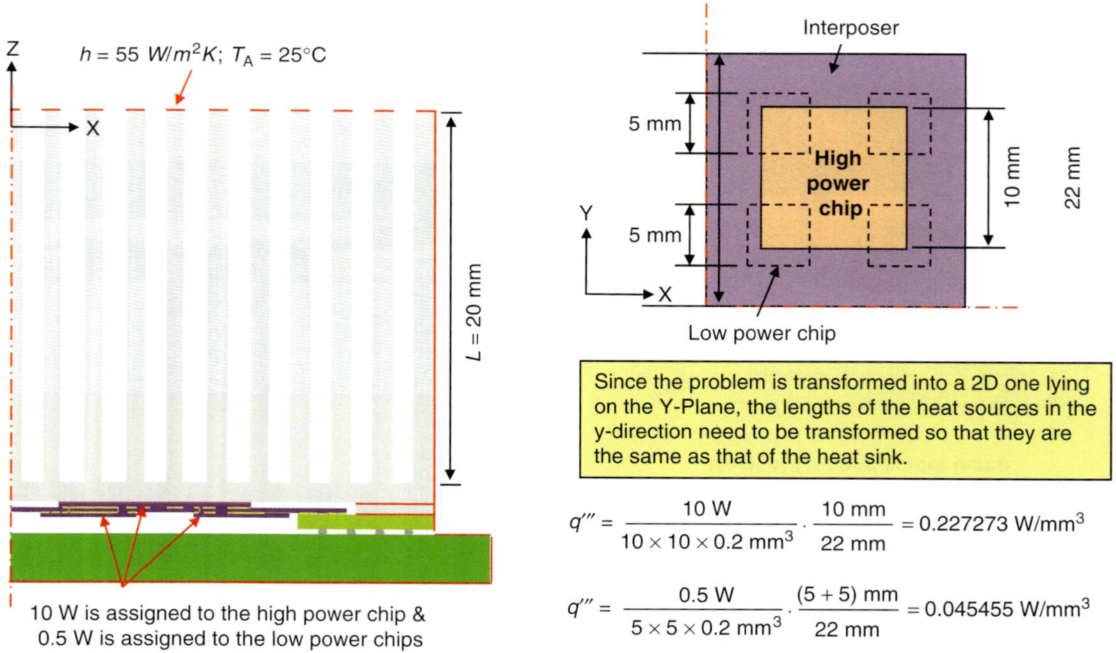

$$q''' = \frac{10\ \text{W}}{10 \times 10 \times 0.2\ \text{mm}^3} \cdot \frac{10\ \text{mm}}{22\ \text{mm}} = 0.227273\ \text{W/mm}^3$$

$$q''' = \frac{0.5\ \text{W}}{5 \times 5 \times 0.2\ \text{mm}^3} \cdot \frac{(5+5)\ \text{mm}}{22\ \text{mm}} = 0.045455\ \text{W/mm}^3$$

FIGURE 12.36 Boundary conditions for thermal simulation of the 3-D IC integration SiP.

Material	*k* (W/mK)
Silicon	150
Interposer	266
SnPb Solder	50.9
Aluminum	180
Copper	390
BT Substrate	0.19
PCB	0.3
TIM	1
Epoxy Adhesive	0.2
Underfill	0.2

TABLE 12.11 Material Properties for Thermal Simulation

SiP, including the PCB is modeled. Since the magnitude of the present boundary-value problem is very big, a 2-D analysis is a good start. The four-node PLANE55 is used for thermal analysis and the PLANE182 is used for plane strain stress analysis. Steady state thermal analysis is performed and the stress analysis is nonlinear time and temperature dependent.

12.6.6 Thermal Analyses and Results of the 3-D IC Integration SiP

The boundary conditions for thermal analyses are given in Fig. 12.36. The heat transfer coefficient (h) can be determined by $h = 3.886(V/L)$ with V (fan speed) = 4 m/s and L (fin height) = 20 mm. Thus, h = 55 W/m^2-K. For each high power chip, 10 W of power is assigned, while for each low power chip it is 0.5 W. Since a 2-D problem (lying on the y-plane) is on hands, the lengths of the heat sources in the y-direction need to be transformed so that they are the same as that of the heat sink. Thus, as shown in Fig. 12.36, for the high power chips, the corrected heat generation rate is 0.2273 W/mm^3, and for the low power chips it is 0.0455 W/mm^3. The material properties for thermal simulation are shown in Table 12.11.

Figure 12.37 shows the temperature distribution in the 3-D IC integration SiP for $V = 1$ m/s. It can be seen that: (1) the temperatures in the SiP are ranging between 27°C and 82°C; (2) the smallest temperature (27°C) occurs at the PCB; (3) the largest temperature (82°C) occurs at the high power chips; (4) the temperature distribution over the high power chips and the interposer is rather uniform (ranging from 79°C to 82°C); (5) the temperature variation across the heat sink

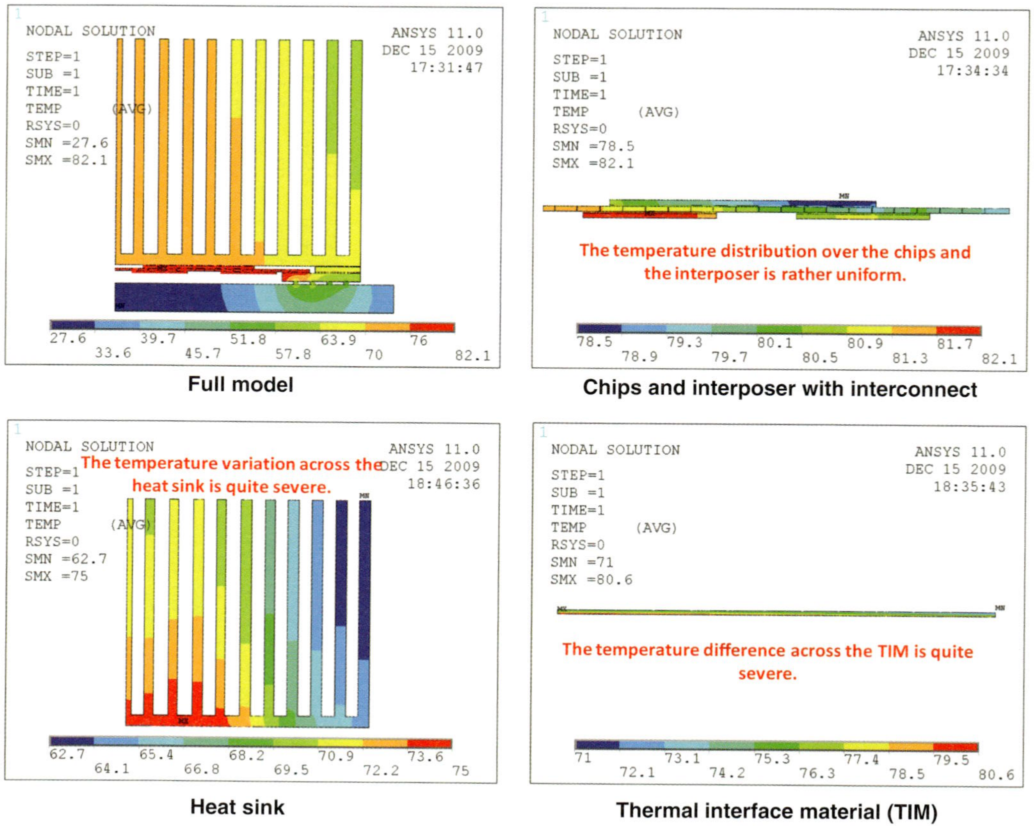

FIGURE 12.37 Temperature distributions in the 3-D IC integration SiP.

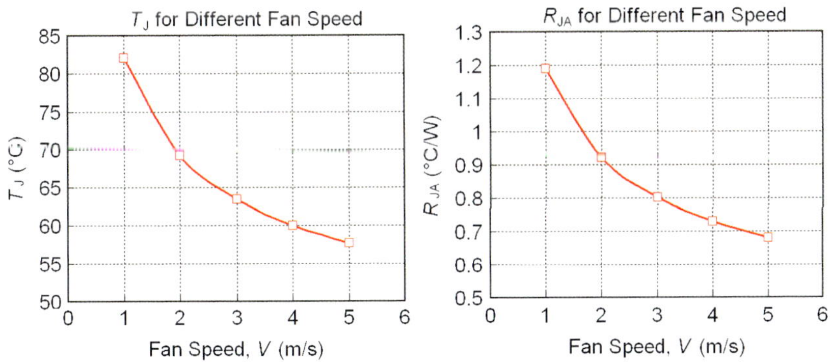

➤ The total power consumption P of the package is 48 W.

FIGURE 12.38 Junction temperature and thermal resistance vs. fan speed of the 3-D IC integration SiP.

is between 62°C and 75°C; and (6) the temperature difference across the TIM is quite severe (from 71°C to 81°C).

Figure 12.38 shows the junction temperature and thermal resistance of the 3-D IC integration SiP at various fan speeds (1 m/s to 5 m/s). The temperature distributions are very similar to those with $V = 1$ m/s except the magnitudes. The total power consumption (P) of the 3D IC integration SiP is 48 W and the thermal resistance (R_{JA}) can be calculated by $R_{JA} = (T_j - T_A)/P$. T_A is the ambient temperature. It can be seen and expected that the higher the fan speed the lower the junction temperature and thermal resistance. However, the relationship is not linear and it shows that the decrease-rate of the junction temperature and thermal resistance is reduced when the fan speed (V) is larger than 3 m/s.

12.6.7 Nonlinear Stress Analyses and Results of the 3-D IC Integration SiP

The boundary conditions for stress analysis are shown in Fig. 12.39. Due to symmetry, only half of the structure is modeled and along the symmetry line all the nodes cannot move in the x-direction ($u_x = 0$). In order to made the structure stable for analysis, one point (at the center) of the PCB is restricted to move in the z-direction ($u_z = 0$). The temperature loading is a function of time and the reference temperature for thermal strain calculation is 25°C.

The material properties of silicon chip, TSV interposer, copper, aluminum heat sink, BT-substrate, TIM, adhesive, FR-4 PCB, underfill, and SnAgCu solder for thermal stress analysis are shown in Tables 12.12 and 12.13, where the Young's modulus (E), Poisson's ratio (ν), thermal expansion of coefficient (α) and the creep constitutive coefficients (C_1, C_2, C_3, and C_4) are given.

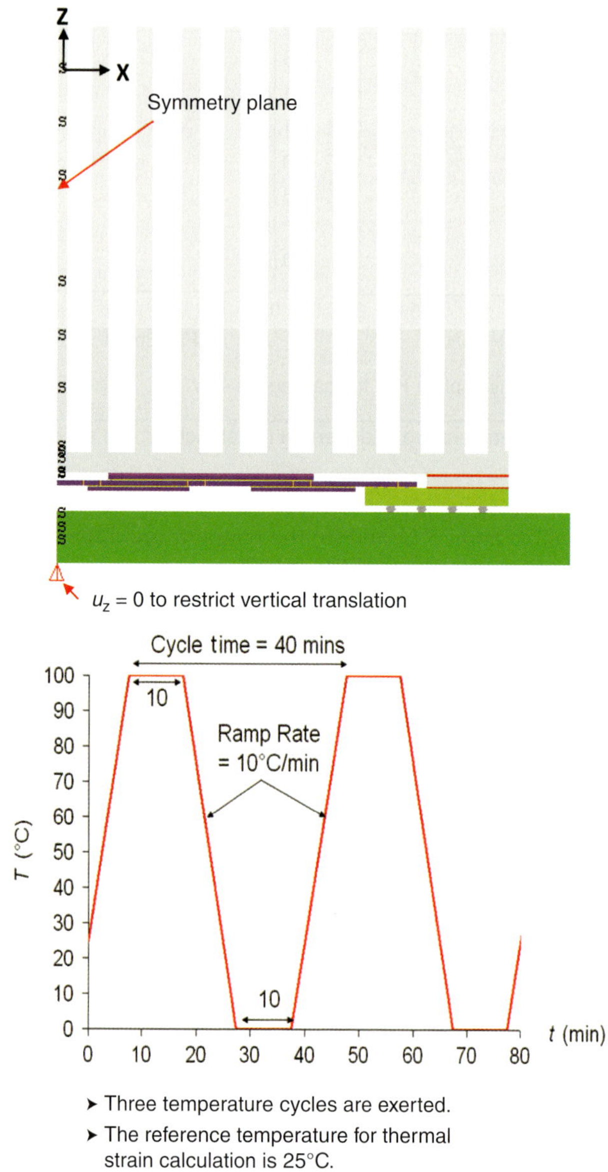

FIGURE 12.39 Boundary conditions for stress simulations of the 3-D IC integration SiP.

The global contour of displacement vector sum (scale @10) of the 3-D IC integration SiP after 3 temperature cycles (at T = 100°C) is shown in the upper half of Fig. 12.40 (the left-hand side is with underfill and the right-hand side is without underfill.) The local deformation of chips and interposer with interconnects are shown in the lower

Material	*E* (GPa)	ν	α (ppm/K)
Silicon	131	0.3	2.8
Interposer	124	0.32	10
Aluminum	69	0.33	23.6
Copper	117	0.35	17.7
BT Substrate	15.2	0.195	15
PCB	20	0.39	16
TIM	7.4	0.4	52
Epoxy Adhesive	7.4	0.4	52
Underfill	4	0.3	41

TABLE 12.12 Linear Material Properties for Stress Simulations

Hyperbolic Sine Law Model for SAC387	$\dot{\varepsilon}_{cr} = C_1[\sinh(C_2\sigma)]^{C_3}\exp\left(\frac{-C_4}{T}\right)$
E (MPa)	52770–76.4*T*(°C)
ν	0.36
α (ppm/K)	20
C_1 (1/s)	501.3
C_2 (1/MPa)	0.0316
C_3	5
C_4 (K)	5434

TABLE 12.13 Nonlinear Material Properties of SnAgCu Solder

half. It can be seen that: (1) the overall 3-D IC integration SiP displacement is almost the same for both with underfill and without, and (2) the local (chips and interposer) displacement of the 3-D IC integration SiP with underfill is 25% larger than that without underfill. This is because the one with underfill will make the chips and interposer deform like a unit with minimum relative displacement between the chips and the interposer.

Figure 12.41 shows the equivalent stress (σ_{eqv}) versus equivalent creep strain ($\varepsilon_{cr.eqv}$) hysteresis loops at the critical (corner) solder joints between the high power chips and TSV interposer of the 3-D IC integration SiP with and w/o underfill. It can be seen that, for both cases, the hysteresis loop is quite stabilized after the third temperature cycle.

Figure 12.42 shows the contour of accumulated equivalent creep strain of the whole 3-D IC integration SiP (upper half) and that at the

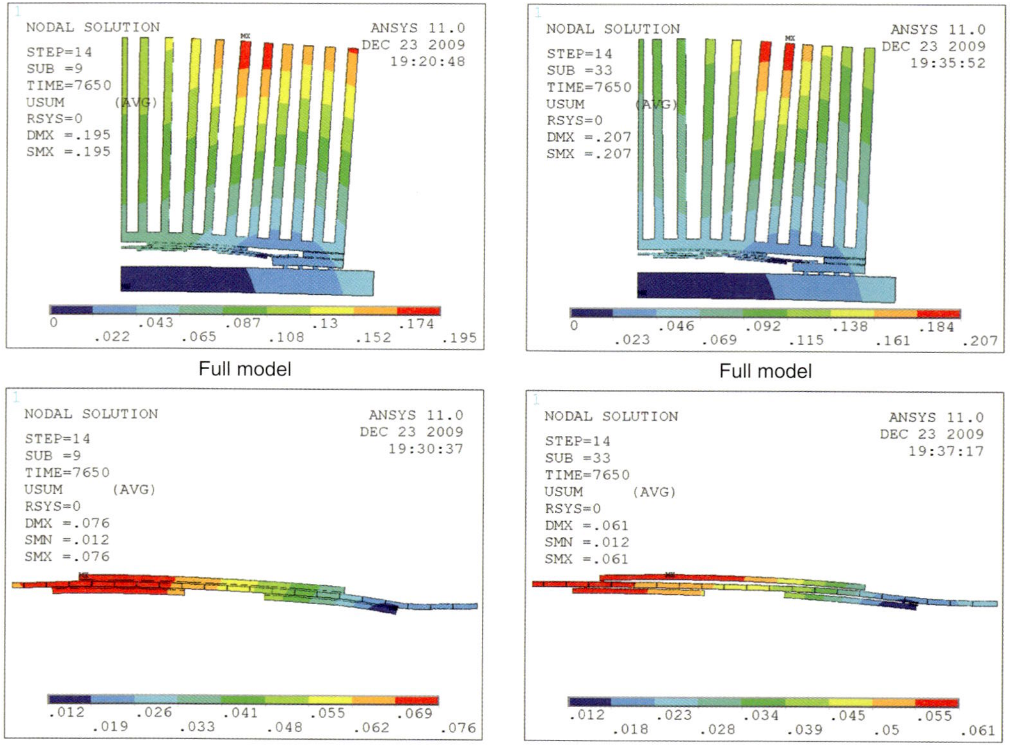

Full model | Full model

Chips and interposer with interconnect | Chips and interposer with interconnect

Model with underfill | **Model without underfill**

FIGURE 12.40 Global and local deformed shapes of the 3-D IC integration SiP with and without underfill.

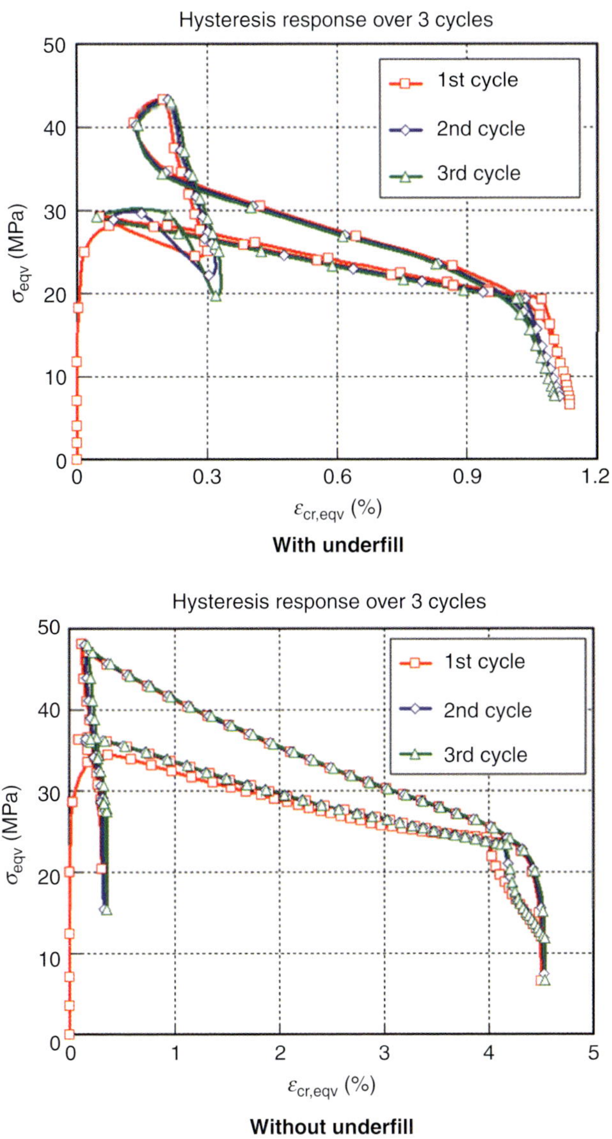

FIGURE 12.41 σ_{eqv} vs. $\varepsilon_{cr.eqv}$ hysteresis loops at the critical (corner) solder joint of the 3-D IC integration SiP with and without underfill.

critical (corner) solder joint (lower half) with underfill (left-hand side) and without (right-hand side). It can be seen that (1) for both cases (with and w/o underfill) the maximum strain occurs at the corner solder joints connecting the high power chips to the interposer, and (2) the maximum strain in the 3-D IC integration SiP without underfill

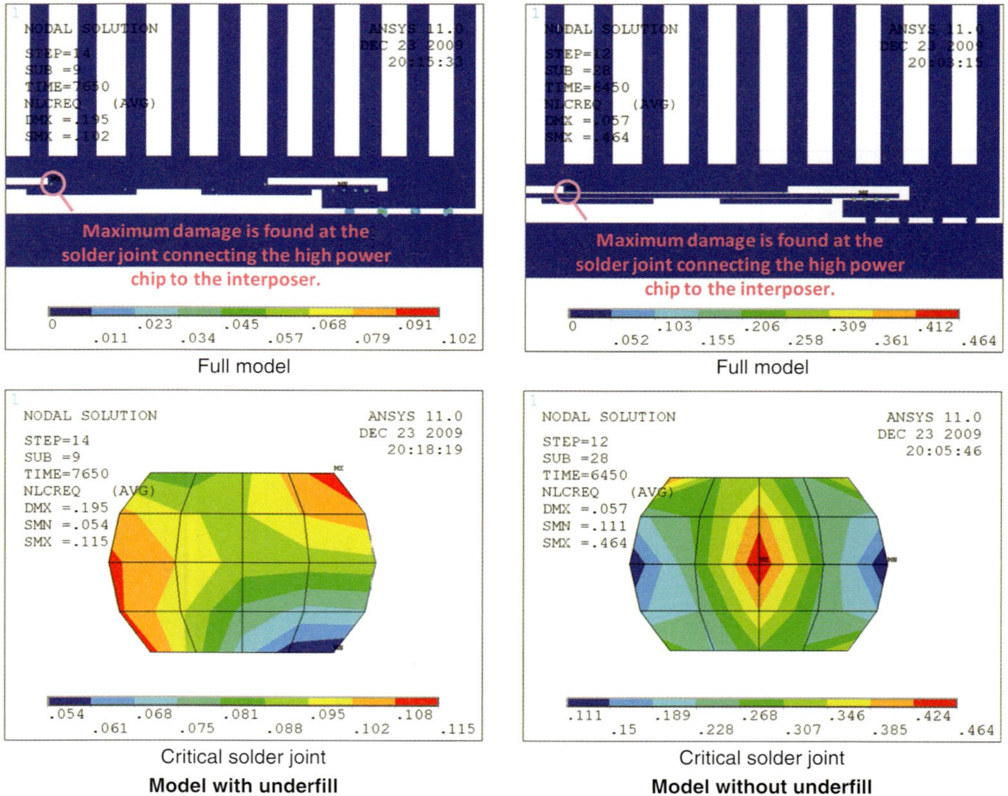

FIGURE 12.42 Equivalent creep strain in the 3-D IC integration SiP with and without underfill.

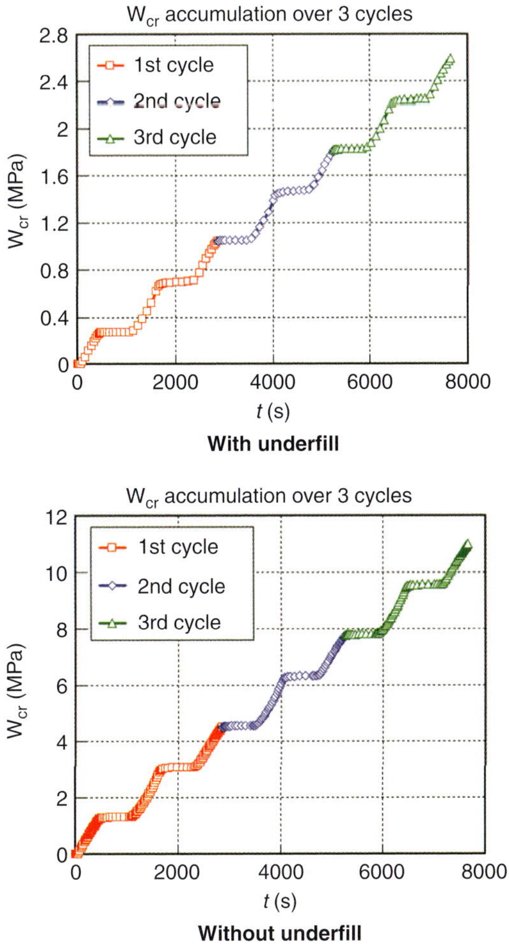

FIGURE 12.43 Creep strain energy density vs. time at the critical (corner) solder joint of the 3-D IC integration SiP with and without underfill.

is 3 times larger than that with underfill. Thus, underfill helps to reduce the strain in the solder joints.

Figure 12.43 shows the creep strain energy density (W_{cr}) as a function of time at the critical (corner) solder joints between the high power chips and the TSV interposer of the 3-D IC integration SiP. It can be seen that the effect of underfill is to reduce the creep strain energy density at the solder joint. The creep strain energy density per cycle (W_{crpc}) is 3.3 MPa (at the third cycle) for the 3-D IC integration SiP without underfill (the lower figure), which is four times larger than that (0.8 MPa) with underfill (the upper figure).

The solder joint thermal fatigue life (N_f) can be related to the creep strain energy density per cycle, see Sec. 2.6.8 of Chap. 2 for example, $N_f = \psi (W_{crpc})^{-\phi}$, where $\psi > 1$ and $\phi > 1$ are solder material constants. Thus, underfill prolongs the solder joint life.

12.6.8 Summary and Recommendations

Some important results and recommendations are summarized in the following:

(1) The TSV/RDL/IPD interposer, which supports the high-power chips on top- and low-power chips at its bottom, is the gut and workhorse of the current design.

(2) With the interposer, it is not necessary to "dig" holes on the active chips. As a matter of fact, try to avoid making TSVs in the active chips. Figure 12.44 shows an interposer roadmap.

(3) With the interposer, it provides a great flexibility of coupling whatever chips are available and/or necessary, and enhances the functionality and possibly the routings (shorter).

(4) With the interposer, the TSV manufacturing cost is lower because the requirement of TSV manufacturing yield is too high (>99.99%) for the active chips to bear with the addition of costs due to TSV manufacturing yield loss.

(5) With the interposer, the wafer thinning and thin-wafer handling costs (for the interposer) are lower because these are not needed for the active chips and thus no addition costs due to yield loss.

(6) With the current designs, all the chips are bare; the packaging cost for individual chips is eliminated.

(7) More than 90% of heats of the 3-D IC integration SiP are dissipated from the back-side of the high-power chips through the TIM and heat spreader/sink.

(8) The appearance and footprint of the current 3-D IC integration SiP designs are very attractive to the System Houses and EMS because they are standard packages.

(9) Underfills between the Cu-filled interposer and the high- and low-power chips are recommended in order to reduce the creep damage of the lead-free solder joints and prolong their lives.

(10) The industry should build an Ecosystem such as the standards and infrastructures for the 3-D IC integration as soon as possible; so the EDA vendors would seriously and systematically write the software for the design, simulation, analysis & verification, manufacturing preparation, and test of 3-D IC integration SiP.

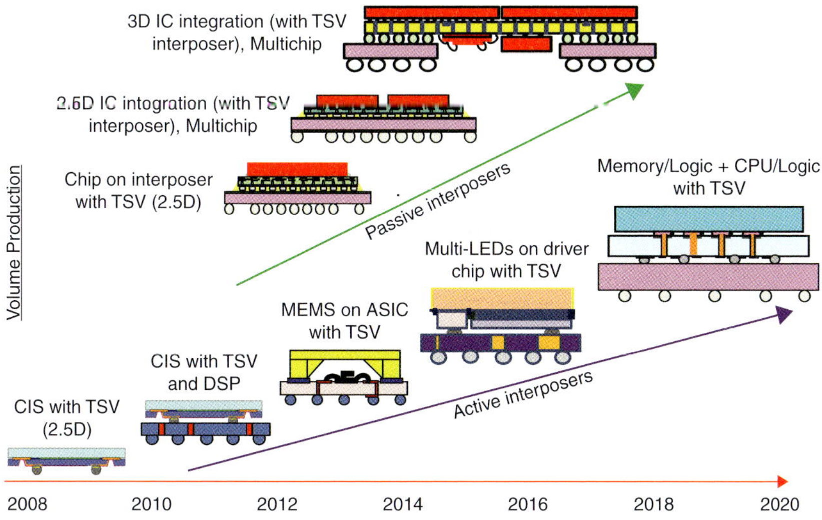

FIGURE 12.44 The road map for (active/passive) interposers.

12.7 References

1. Sekiguchi, M., Numata, H., Sato, N., Shirakawa, T., Matsuo, M., Yoshikawa, H., Yanagida, M., Nakayoshi, H., and Takahashi, K., "Novel Low Cost Integration of Through Chip Interconnection and Application to CMOS Image Sensor," *IEEE Proceedings of Electronic Components and Technology Conference*, San Diego, CA, May 2006, pp. 1367–1374.
2. Takahashi, K., and M. Sekiguchi, "Through Silicon Via and 3-D Wafer/Chip Stacking Technology," *IEEE Proceedings of Symposium on VLSI Circuits Digest of Technical Papers*, 2006, pp. 89–92.
3. Lau, J. H., "TSV Manufacturing Yield and Hidden Costs for 3D IC Integration," to be published in the IEEE Proceedings of *Electronic Components and Technology Conference*, Las Vegas, NV, June 2010, pp. 1031–1041.
4. Lau, J. H., C. K. Lee, C. S. Premachandran, A. Yu, *Advanced MEMS Packaging*, McGraw-Hill, New York, 2010.
5. Lau, J. H., "Critical Issues of 3D IC Integrations," *Proceedings of IMAPS International Symposium on Microelectronics*, San Jose, CA, November 2009, pp. 585–592.
6. Lau, J. H., "Key Enabling Technologies for 3D IC Integrations," *Professional Development Course given at IEEE Electronic & Components Technology Conference*, San Diego, CA May 2009.
7. Lau, J. H., "Design and Process of 3D MEMS Packaging," *Proceedings of IMAPS International Symposium on Microelectronics*, San Jose, CA, November 2009, pp. 1–9.
8. Lau, J. H., Lee, R., Yuen, M., and Chan, P., "3D LED and IC Wafer Level Packaging," *Journal of Microelectronics International*, Vol. 27, Issue 2, 2010, pp. 98–105.
9. Lau, J. H., and Tang, G., "Thermal Management of 3D IC Integration with TSV (Through Silicon Via)," *IEEE Proceedings of Electronic, Components & Technology Conference*, San Diego, CA, May, 2009, pp. 635–640.
10. Lau, J. H., X. Zheng, and C. Selvanayagam, "Failure Analyses of 3D SiP (System-in-Package) and WLP (Wafer-Level Package) by Finite Element Methods," *IEEE Proceedings of the 16th International Symposium on the Physical & Failure Analysis of Integrated Circuits*, Suzhou, China, Judy 2009, pp. 108–115.

11. Lau, J. H., G. Tang, G. Hoe, X. Zhang, C. Chong, P. Damaruganath, and K. Vaidyanathan, " Effects of TSV (Through Silicon Via) Interposer/Chip on the Thermal Performances of 3D IC Packaging," *ASME Paper No. IPACK2009-89380*, San Francisco, CA, July 2009, pp. 1–9.
12. Yu, A., J. H. Lau, Ho, S., Kumar, A., Wai, Y., Yu, D., Jong, M., Kripesh, V., Pinjala, D., Kwong, D., "Study of 15-µm-Pitch Solder Microbumps for 3D IC Integration." *IEEE Proceedings of Electronic Components and Technology Conference,* San Diego, CA, May 2009, pp. 6–10.
13. Yu, A., J. H. Lau, Ho, S., Kumar, A., Yin, H., Ching, J., Kripesh, V., Pinjala, D., Chen, S., Chan, C., Chao, C., Chiu, C., Huang, M., and Chen, C., "Three Dimensional Interconnects with High Aspect Ratio TSVs and Fine Pitch Solder Microbumps." *IEEE Proceedings of Electronic Components and Technology Conference,* San Diego, CA, May 2009, pp. 350–354.
14. Yu, A., A. Kumar, S. Ho, H. Yin, J. H. Lau, Ching, J., Kripesh, V., Pinjala, D., Chen, S., Chan, C., Chao, C., Chiu, C., Huang, M., and Chen, C., "Development of Fine Pitch Solder Microbumps for 3D Chip Stacking," *IEEE EPTC Proceedings*, Singapore, December 2008, pp. 387–392.
15. Yu, A., N. Khan, G. Archit, D. Pinjala1, K. Toh, V. Kripesh1, S. Yoon, and J. H. Lau, "Fabrication of Silicon Carriers with TSV electrical Interconnections and Embedded Thermal Solutions for High Power 3-D Package," *IEEE Proceedings of Electronic, Components & Technology Conference*, Orlando, FL, May 27-30, 2008, pp. 24–28.
16. Yu, A., N. Khan, G. Archit, D. Pinjalal, K. Toh, V. Kripesh, S. Yoon, and J. H. Lau, "Development of Silicon Carriers with Embedded Thermal Solutions for High Power 3-D Package." *IEEE Transactions on Components and Packaging Technology*, Vol. 32, No. 3, September 2009, pp. 566–571.
17. Tang, G., O. Navas, D. Pinjala, J. H. Lau, A. Yu, and V. Kripesh, "Integrated Liquid Cooling Systems for 3-D Stacked TSV Modules," accepted for publication in the *IEEE Transactions on Components and Packaging Technologies.*
18. Chen, K., C. Premachandran, K. Choi, C. Ong, X. Ling, A. Khairyanto, B. Ratmin, P. Myo, and J. H. Lau, "C2W Bonding Method for MEMS Applications," *IEEE Proceedings of Electronics Packaging Technology Conference*, December 2008, pp. 1283–1287.
19. Premachandran, C. S., J. H. Lau, X. Ling, A. Khairyanto, K. Chen, and Myo Ei Pa Pa,, "A Novel, Wafer-Level Stacking Method for Low-Chip Yield and Non-Uniform, Chip-Size Wafers for MEMS and 3D SIP Applications," *IEEE Proceedings of Electronic, Components & Technology Conference*, Orlando, FL, May 27-30, 2008, pp. 314–318.
20. Zhang, X., T. Chai, J. H. Lau, C. Selvanayagam, K. Biswas, S. Liu, D. Pinjala, G. Tang, Y. Ong, S. Vempati, E. Wai, H. Li, B. Liao, N. Ranganathan, V. Kripesh, J. Sun, J. Doricko, and C. Vath, "Development of Through Silicon Via (TSV) Interposer Technology for Large Die (21 × 21 mm) Fine-pitch Cu/low-k FCBGA Package," *IEEE Proceedings of Electronic, Components & Technology Conference*, San Diego, CA, May, 2009, pp. 305–312. (Also, accepted for publication in *IEEE Transactions in Advanced Packaging.*)
21. Hoe, G., G. Tang, P. Damaruganath, C. Chong, J. H. Lau, X. Zhang, and K. Vaidyanathan, "Effect of TSV Interposer on the Thermal Performance of FCBGA Package," *IEEE Proceedings of Electronics Packaging and Technology Conference,* Singapore, December 2009, pp. 778–786.
22. Choi, W. O., C. S. Premachandran, S. Ong, Ling, X., E. Liao, K. Ahmad, B. Ratmin, K. Chen, P. Thaw, and J. H. Lau, "Development of Novel Intermetallic Joints using Thin Film Indium Based Solder by Low Temperature Bonding Technology for 3D IC Stacking," *IEEE Proceedings of Electronic, Components & Technology Conference*, San Diego, CA, May, 2009, pp. 333–338.
23. Vempati1, S. R., S. Nandar, C. Khong, Y. Lim, K. Vaidyanathan, J. H. Lau, B. P. Liew, K. Y. Au, S. Tanary, A. Fenne, R. Erich, and J. Milla, "Development of 3-D Silicon Die Stacked Package Using Flip Chip Technology with Micro Bump Interconnects," *IEEE Proceedings of Electronic, Components & Technology Conference*, San Diego, CA, May, 2009, pp. 980–987. (Also, accepted for publication in *IEEE Transactions in CPMT.*)

24. Khan, N., L. Yu, P. Tan, S. Ho, N. Su, H. Wai, K. Vaidyanathan, D. Pinjala, J. H. Lau, T. Chuan, "3D Packaging with Through Silicon Via (TSV) for Electrical and Fluidic Interconnections," *IEEE Proceedings of Electronic, Components & Technology Conference*, San Diego, CA, May, 2009, pp. 1153–1158.
25. Sekhar, V. N., S. Lu, A. Kumar, T. C. Chai, V. Lee, S. Wang, X. Zhang, C. S. Premchandran, V. Kripesh, and J. H. Lau, "Effect of Wafer Back Grinding on the Mechanical Behavior of Multilayered Low-k for 3D-Stack Packaging Applications," *IEEE Proceedings of Electronic, Components & Technology Conference*, Orlando, FL, May 27-30, 2008, pp. 1517–1524.
26. Khan, N., V. Rao, S. Lim, S. Ho, V. Lee, X. Zhang, R. Yang, E. Liao, Ranganathan, T. Chai, V. Kripesh, and J. H. Lau, "Development of 3D Silicon Module with TSV for System in Packaging," *IEEE Proceedings of Electronic, Components & Technology Conference*, Orlando, FL, May 27–30, 2008, pp. 550–555.
27. Ho, S., S. Yoon, Q. Zhou, K. Pasad, V. Kripesh and J. H. Lau, "High RF Performance TSV for Silicon Carrier for High Frequency Application," *IEEE Proceedings of Electronic, Components & Technology Conference*, Orlando, FL, May 27–30, 2008, pp. 1956–1952.
28. Selvanayagam, C., J. H. Lau, X. Zhang, S. Seah, K. Vaidyanathan, and T. Chai, "Nonlinear Thermal Stress/Strain Analysis of Copper Filled TSV (Through Silicon Via) and Their Flip-Chip Microbumps," *IEEE Proceedings of Electronic Components and Technology* Conference, Orlando, FL, May 2008, pp. 1073–1081. Also, *IEEE Transactions in Advanced Packaging*, Vol. 32, No. 4, Nov. 2009, pp. 720–728.
29. Zhang, X., A. Kumar, Q. X. Zhang, Y. Y. Ong, S. W. Ho, C. H. Khong, V. Kripesh, J. H. Lau, D.L. Kwong, V. Sundaram, Rao R. Tummula, Georg Meyer-Berg, "Application of Piezoresistive Stress Sensors in Ultra Thin Device Handling and Characterization," *Journal of Sensors & Actuators: A. Physical*, Vol. 156, November 2009, pp. 2–7.
30. Garrou, P., C. Bower, and P. Ramm, *3D Integration: Technology and Applications*, Wiley, John & Sons, 2009.
31. Ramm, P., M. Wolf, A. Klumpp, R. Wieland, B. Wunderle, B. Michel, and H. Reichl, "Through Silicon Via Technology–Processes and Reliability for Wafer-Level 3D System Integration," *IEEE ECTC Proceedings*, Orlando, FL, May 2008, pp. 847–852.
32. Andry, P. S., Tsang, C. K., Webb, B. C., Sprogis E. J., Wright S. L., Bang, B., Manzer, D. G., "Fabrication and Characterization of Robust Through-Silicon Vias for Silicon-Carrier Applications," *IBM Journal of Research and Development*,, Vol. 52, No. 6, 2008, pp. 571–581.
33. Knickerbocker, J. U., P.S. Andry, B. Dang, R.R. Horton, C S. Patel, R.J. Polastre, K. Sakuma, E.S. Sprogis, C.K. Tsang, B.C. Webb, and S.L. Wright "3-D Silicon Integration," *IEEE Proceedings of Electronic Components and Technology Conf*, May 2008, pp. 538–543.
34. Kumagai, K., Yoneda, Y, Izumino, H., Shimojo, H., Sunohara, M., Kurihara, T., "A Silicon Interposer BGA Package with Cu-filled TSV and Multi-layer Cu-plating Interconnection," *IEEE Proceedings of Electronic Components and Technology Conf*, Orlando, FL, May 2008, pp. 571–576.
35. Sunohara, M., Tokunaga, T., Kurihara, T., Higashi, M., "Silicon Interposer with TSVs (Through Silicon Vias) and Fine Multilayer Wiring," *IEEE Proceedings of Electronic Components and Technology Conf*, Orlando, FL, May 2008, pp. 847–852.
36. Lee, H. S., Choi, Y-S., Song, E., Choi, K., Cho, T., Kang, S., "Power Delivery Network Design for 3D SIP Integrated over Silicon Interposer Platform," *IEEE Proceedings of Electronic Components and Technology Conf*, Reno, NV, May 2007, pp. 1193–1198.
37. Matsuo, M., Hayasaka, N., Okumura, K., "Silicon Interposer Technology for High-Density Package," *IEEE Proceedings of Electronic Components and Technology Conf*, Las Vegas, NV, May 2000, pp. 1455–1459.
38. Wong, E., Jacob Minz, and Sung Kyu Lim, "Effective Thermal Via and Decoupling Capacitor Insertion for 3D System-On-Package," *IEEE Proceedings of Electronic Components and Technology Conference*, San Diego, CA, May 2006, pp. 1795–1801.

39. Beica, R., Charles Sharbono, and Tom Ritzdorf, "Through Silicon Via Copper Electrodeposition for 3D Integration," *IEEE Proceedings of Electronic, Components & Technology Conference*, Orlando, FL, May 27–30, 2008, pp. 577–583.
40. Wolf, M., Bernhard Wunderle, Nils Jürgensen, Gunter Engelmann, Oswin Ehrmann, Bernd Michel, Thomas Dretschkow, Albrecht Uhlig, Herbert Reichl, "High Aspect Ratio TSV Copper Filling with Different Seed Layers," *IEEE Proceedings of Electronic, Components & Technology Conference*, Orlando, FL, May 27–30, 2008, pp. 563–570.
41. Henry, D., F. Jacquet, M. Neyret, X. Baillin, T. Enot, V. Lapras, C. Brunet-Manquat, J. Charbonnier, B. Aventurier, and N. Sillon, "Through Silicon Vias Technology for CMOS Image Sensors Packaging," *IEEE Proceedings of Electronic, Components & Technology Conference*, Orlando, FL, May 27–30, 2008, pp. 556–562.
42. Lau, J. H., "Who Invented Through Silicon Via (TSV) and When?" Posted in *3D InCites* http://www.semineedle.com/posting/31171 on April 24, 2010.
43. Lau, J. H., R. S. W. Lee, *Microvias for Low Cost High Density Interconnects*, McGraw-Hill, New York, 2001.
44. Hu, G., Kalyanam, H., Krishnamoorthy, S., and Polka, L., "Package technology to address the memory bandwidth challenge for tera-scale computing," *Intel Technology Journal*, Vol. 11, 2007, pp. 197–206.
45. Lau, J. H., *Ball Grid Array Technology*, McGraw-Hill, New York, 1995.
46. Lau, J. H., T. Tseng, and D. Cheng, "Heat Spreader with a Placement Recess and Bottom Saw-Teeth for Connection to Ground Planes on a Two-Side Single-Core BGA Substrate", *US Patent No. 6,057,601*, Granted: May 2, 2000, Filed: November 27, 1998.
47. Lau, J. H., K. L. Chen, "Thermal and Mechanical Evaluations of a Cost-Effective Plastic Ball Grid Array Package", *ASME Transactions, Journal of Electronic Packaging*, Vol. 119, September 1997, pp. 208–212.
48. Lau, J. H., T. Chen, "Cooling Assessment and Distribution of Heat Dissipation of A Cavity Down Plastic Ball Grid Array Package–NuBGA", *IMAPS Transactions, International Journal of Microelectronics & Electronic Packaging*, Vol. 21, No. 1, 1998, pp. 20–28.
49. Lau, J. H., T. Chou, "Electrical Design of a Cost-Effective Thermal Enhanced Plastic Ball Grid Array Package–NuBGA", *IEEE Transactions on CPMT, Part B*, Vol. 21, No. 1, Feb. 1998, pp. 35–42.
50. Lau, J. H., T. Chen, R. Lee, "Effect of Heat Spreader Sizes on the Thermal Performance of Large Cavity-Down Plastic Ball Grid Array Packages", *ASME Transactions, Journal of Electronic Packaging*, Vol. 121, No. 4, 1999, pp. 242–248.
51. Lau, J. H., "Design, Manufacturing, and Testing of a Novel Plastic Ball Grid Array Package", *Journal of Electronics Manufacturing*, Vol. 9, No. 4, December 1999, pp. 283–291.
52. Lau, J. H., T. Chen, "Low-Cost Thermal and Electrical Enhanced Plastic Ball Grid Array Package–NuBGA", *Microelectronics International*, 1999.
53. Lau, J. H., "Solder Joint Reliability of Flip Chip and Plastic Ball Grid Array Assemblies Under Thermal, Mechanical, and Vibration Conditions," *IEEE Transaction on CPMT, Part B*, Vol. 19, No. 4, pp. 728–735, November 1996.
54. Lau, J. H., W. Jung, Y. Pao, "Nonlinear Analysis of Full-Matrix and Perimeter Plastic Ball Grid Array Solder Joints", *ASME Transactions, Journal of Electronic Packaging*, Vol. 119, September 1997, pp. 163–170.

CHAPTER 13

Electromigration of Lead-Free Microbumps for 3-D IC Integration

13.1 Introduction

As mentioned earlier, microbumps are one of the important enabling technologies for three-dimensional (3-D) integrated circuit (IC) integrations, which require thin chips (could go down to 20 μm) in order to be useful and meaningful for stacking up the chips. The ordinary solder bumps (~50 to 100 μm) are too big for 3-D IC integrations, which require much smaller bumps (25 μm or less). The current crowding-induced electromigration failure of ordinary flip-chip solder bumps has been studied and published extensively in, for example, [1–39]. The current stress-induced failure of multiphase microbump interconnects [40] is presented in this chapter.

13.2 Test Vehicles and Methods

13.2.1 Solder Bumps

Figure 13.1 shows the solder bump used for the test vehicles, namely Sn2.5wt%Ag solder on the Cu-stud. Right after plating, the dimensions of the SnAg solder on the Cu-stud bump are 20 μm SnAg solder and 20 μm Cu-stud. The pad diameter is ~25 μm. The chip size is 10 × 8 mm with 13,413 bumps on either 50 μm or 100 μm pitches. The UBM on the substrate carrier chip is electroless Ni (5 μm) and immersion Au (0.1 μm), and is ~20 μm in diameter. For both chips, the Al trace is 1.5 μm thick and 25 μm wide.

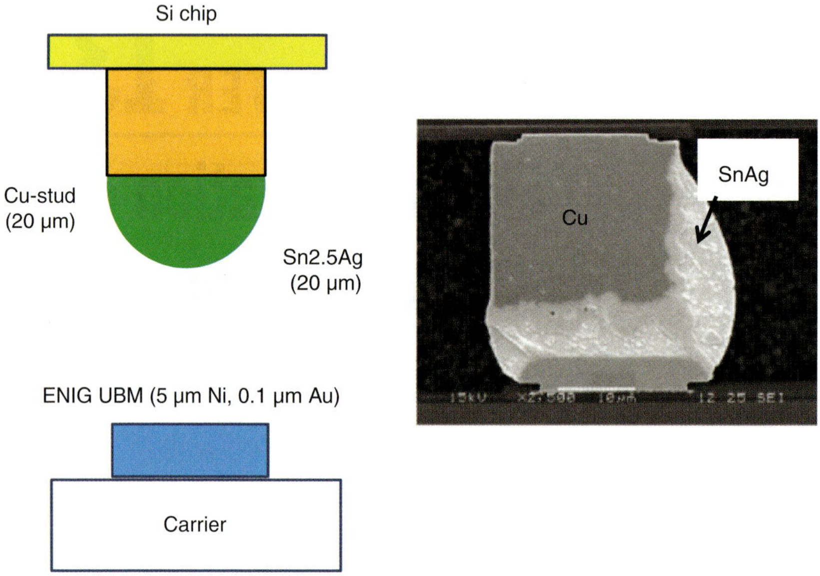

FIGURE 13.1 Test vehicle with Cu stud + SnAg + ENIG UBM. The UBM on the carrier is ENIG.

13.2.2 Test Methods

Figure 13.2 shows the assembled test vehicle, and Fig. 13.3 shows the schematic of a Kevin structure for the continuous resistance measurements of a single solder bump. It can be seen that: (1) there are four-point measurements of the bump through the Kevin circuit; (2) there are two abundant bumps adjacent to the bump under test; (3) the test vehicle is flipped and attached onto the (20 × 20 mm) carrier substrate; (4) the carrier is die-attached and Au wire-bonded onto a printed circuit board (PCB); and (5) the directions of current flow and voltage are measured, with two probes for applying current and the other two probes for sensing the voltage drop. In order to avoid the current-induced failure at the wire-bonding interface, four Au wires are used for one pair of pads. The current stress is at the solder joints near the edges of the chip.

13.3 Test Procedures

Electromigration (EM) tests are conducted by the modular integrated reliability analyzer (MIRA, QualiTau) system at the temperature of 140°C with two current densities: 2.04 and 4.08×10^4 A/cm^2 with respect to bump diameter. The direction of the electron flow in the bump is from ENIG (electronless nickel and immersion gold) to Cu post. For each condition, six samples are tested. Joule heating of the bumps is taken into account by measuring the temperature

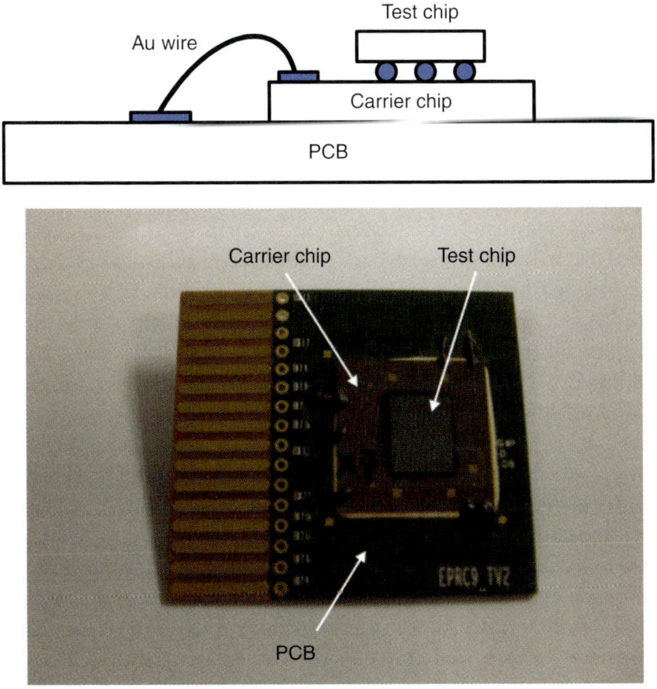

FIGURE 13.2 The assembled test vehicle.

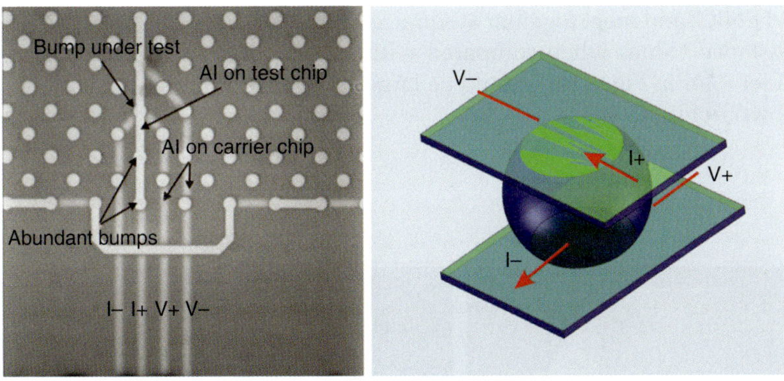

FIGURE 13.3 The four-point Kevin structure and the current flow direction and voltage measurement.

coefficient of resistance (TCR) of the bumps before the EM test. This method uses the bumps themselves as a thermometer. According to measurement results of the two different current densities, the real temperatures of the test chips are about 148°C and 165°C, respectively.

13.4 Microstructures of Samples Before Tests

Figure 13.4 shows two bump interconnects from the assembled chip with Cu-stud + SnAg + ENIG UBM before tests. According to energy-dispersive x-ray spectrometry (EDX) results, $(Cu,Ni)_6Sn_5$ compounds are formed on both the Cu side and ENIG side. From the lectures, it is known that when SnAg solder is wetted with Ni, NiSn intermetallic compounds (IMCs) will form [36]. Also, it is also reported that the change of Cu contents in SnCu and SnAgCu solders would lead to the IMC transformation, that is, from $(Cu,Ni)_3Sn_4$ to $(Cu,Ni)_6Sn_5$. Furthermore, a trace amount of Ni would decrease the Cu solubility in liquid Sn and could accelerate the growth of Cu_6Sn_5 on Cu UBM [36]. Considering the short distance between Cu UBM and ENIG pad (10~20 μm) in the present micro-bumps, Cu is expected to diffuse to the ENIG pad in a shorter time and the Cu contents can reach the threshold for the formation of $(Cu,Ni)_6Sn_5$ compounds on ENIG. The typical Ag_3Sn IMCs for SnAg solder joints can be seen in Figs. 13.4*a* and *b*.

The resistance of the Cu-stud and solder bump interconnect is in the range of 10 to 40 mΩ. The difference in resistance is mainly caused by the different initial microstructure after reflow. For the small bump interconnects, the initial interfacial microstructure may be slightly different after reflowing, even the neighbor bumps on the same chip. For example, in Fig. 13.4*a*, the IMC thickness on ENIG is thinner and the IMCs on UBM and ENIG are separated by the remaining solder. By way of contrast, in Fig. 13.4*b*, thick IMC layers formed on UBM and ENIG, and meet together at some regions. Since IMCs have larger resistance values when compared with solder, the resistance of the solder joint in Fig. 13.4*b* should be larger than that of the solder joint shown in Fig. 13.4*a*.

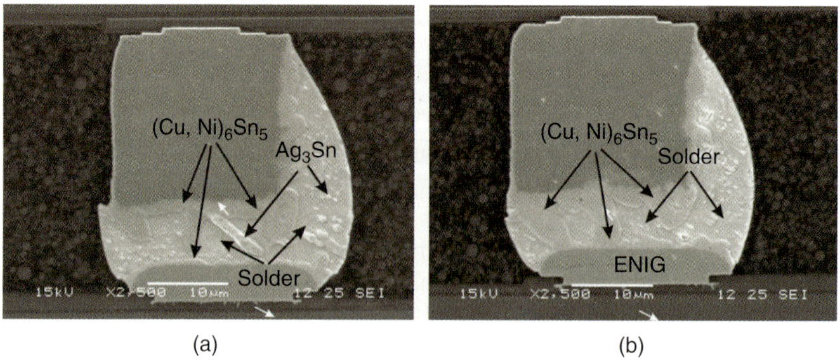

(a) (b)

FIGURE 13.4 Assembled solder joints with Cu stud + SnAg solder + ENIG UBM.

13.5 Samples Tested at 140°C with Low Current Density

13.5.1 Samples Tested at 140°C with Current Density of 2.04×10^4 A/cm²

The resistance evolution of three solder-joint interconnects subjected to a current density of 2.04×10^4 A/cm² is shown in Fig. 13.5*a*. It can be seen that there is no bump failure, and the resistance value gradually increases up to 1000 h. Figure 13.5*b* shows the plot of the relative

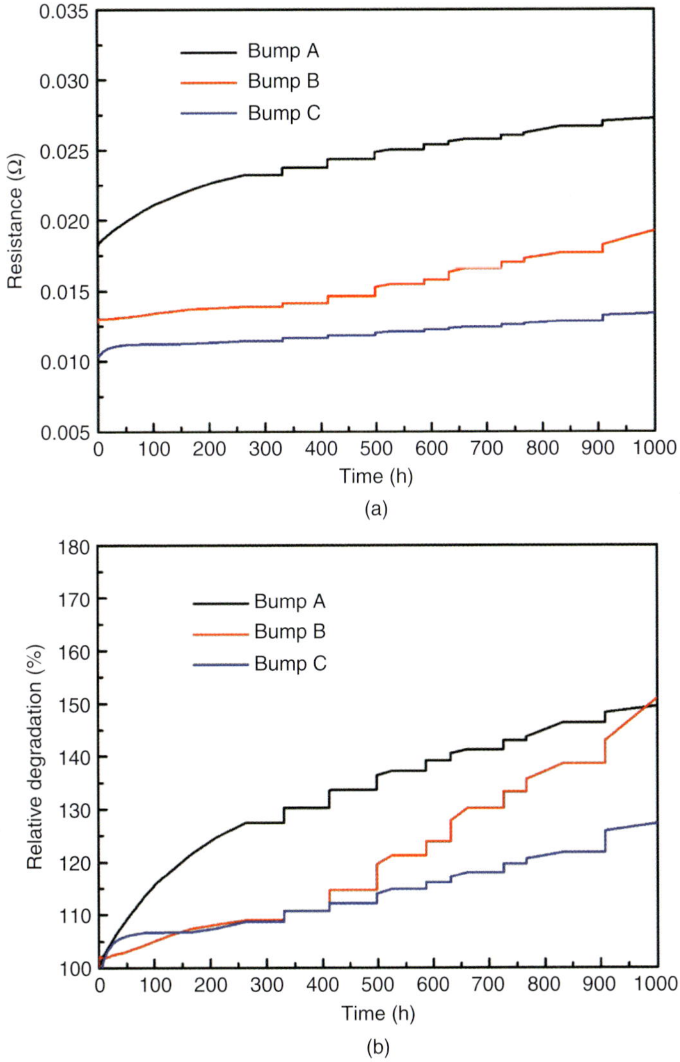

FIGURE 13.5 (a) Resistance vs. stress time and (b) relative degradation vs. stress time (under low current density, 2.04×10^4 A/cm²).

degradation versus stress time, which is easier to observe than the resistance changing rate. For the so-called relative degradation, the measured resistance is normalized to its initial value. The three curves in Fig. 13.5*b* show that the resistance of the solder-joint interconnects increases to 125%–150% of their original values at 140°C with the current density of 2.04×10^4 A/cm^2. Cross-sectional observation of the microbump interconnections reveals the microstructure evolution under current stressing.

13.5.2 Microstructures of Samples Tested at Low Current Density

Figure 13.6 shows the typical microstructure of the Cu-stud solder joint under the low current density of 2.04×10^4 A/cm^2 for 1000 h. Obviously, $(Cu,Ni)_6Sn_5$ becomes thicker and less solder material remains at the interface. Cu_3Sn compounds also form at the Cu side. In summary, the following reactions occur under present test conditions:

$$Cu + Sn + Ni \rightarrow (Cu,Ni)_6Sn_5 \tag{13.1}$$

$$Cu + Cu_6Sn_5 \rightarrow Cu_3Sn \tag{13.2}$$

For Eq. (13.1), solder is consumed and resistance increases since IMCs have a high resistivity. For Eq. (13.2), the formation and growth of Cu_3Sn will decrease "the increasing of resistance," since it has a low resistivity compared with Cu_6Sn_5. Therefore, the resistance change is closely related to the diffusion of materials and the growth of the

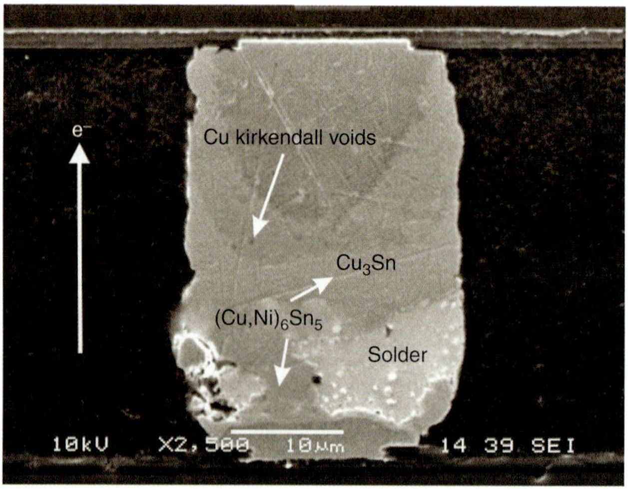

FIGURE 13.6 Typical microstructure of Cu-stud + SnAg + ENIG UBM solder joint under low current density (2.04×10^4 A/cm^2).

IMCs. The resistivity for Cu, SnAg solder, Cu_3Sn, and Cu_6Sn_5 are 1.7, 11.5, 8.3, and 17.5 μΩ-cm, respectively [41]. Thus, the consumption of Cu and solder, and the growth of IMCs would lead to increasing resistance of the solder joint.

Few Kirkendall voids are detected in the thin Cu_3Sn. Kirkendall voids are formed because the diffusion rate of Cu from Cu_3Sn to $(Cu,Ni)_6Sn_5$ is faster than that of Sn from $(Cu,Ni)_6Sn_5$ to Cu_3Sn [38]. With the growth of Cu_3Sn, more Kirkendall voids would form as a result. Xu et al. have studied the electromigration of 90 μm SnAgCu solder on a Cu column at 135°C with the current density of 1.6×10^4 A/cm^2 [26, 31]. Since the volume ratio between solder and Cu is larger, the main IMCs are $(Cu,Ni)_6Sn_5$ after 1290 h of current stressing. Kirkendall voids and crack formation cannot be found [31].

13.6 Samples Tested at 140°C with High Current Density

13.6.1 Samples Tested at 140°C with Current Density of 4.08×10^4 A/cm^2

When samples are tested with a higher current density (4.08×10^4 A/cm^2) there are failures. Figure 13.7*a* shows the resistance degradation curves of three samples from the chips with Cu-stud + SnAg + ENIG UBM solder joint. After the resistance reaches a certain value, it remains nearly constant until an abrupt increase, which means the solder-joint interconnect is opened at this time. In the three curves shown in Fig. 13.7*a*, two of them show a continuous increase in resistance before reaching the constant values. However, the resistance of one sample increases first and then drops after 100 h before it reaches the plateau stage. Apparently, it is hard to attribute the abnormal change of resistance to a restructuring effect of the solder matrix that is found in the larger solder bump interconnects because the spreading of resistance occurred at the beginning stage [5]. Also, after 100 h of testing under high temperature and high current density, the solder is nearly consumed, as can be seen in Fig. 13.8*a*.

13.6.2 Microstructures of Samples Tested at High Current Density

According to the discussion in Sec. 13.5, we know that the evolution of IMCs plays an important role in the resistance evolution. As shown in Fig. 13.7*a*, the initial resistance of bump D is about 0.01 Ω, which is smaller than those of the others. As mentioned in Sec. 13.4, it can be deduced that fewer IMCs are formed at the beginning in solder joint D. Thus, during the EM test, with the growth of $(Cu,Ni)_6Sn_5$ IMCs, the resistance would increase faster, which can be seen from Fig. 13.7*b*,

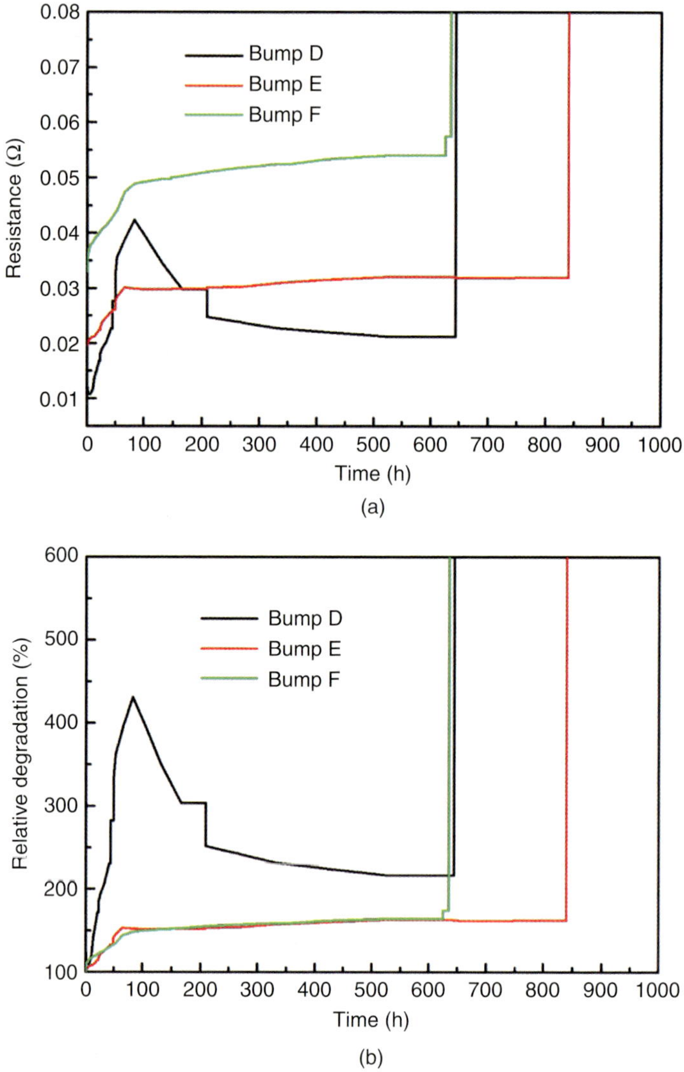

FIGURE 13.7 (a) Resistance vs. stress time and (b) relative degradation vs. stress time, under high current density (4.08×10^4 A/cm^2).

and the resistance is three times larger than its initial value. When the growth of Cu_3Sn becomes dominant, the resistance decreases until an equilibrium state is reached. In solder joints E and F, thicker $(Cu,Ni)_6Sn_5$ IMCs are formed after reflow. When under EM test, the growth rate of $(Cu,Ni)_6Sn_5$ is slower and Cu_3Sn would grow up earlier than in solder joint D. Therefore, there is no backward effect of Cu_3Sn formation. The diminishing of resistance due to Cu_3Sn formation is compensated by the growth of $(Cu,Ni)_6Sn_5$.

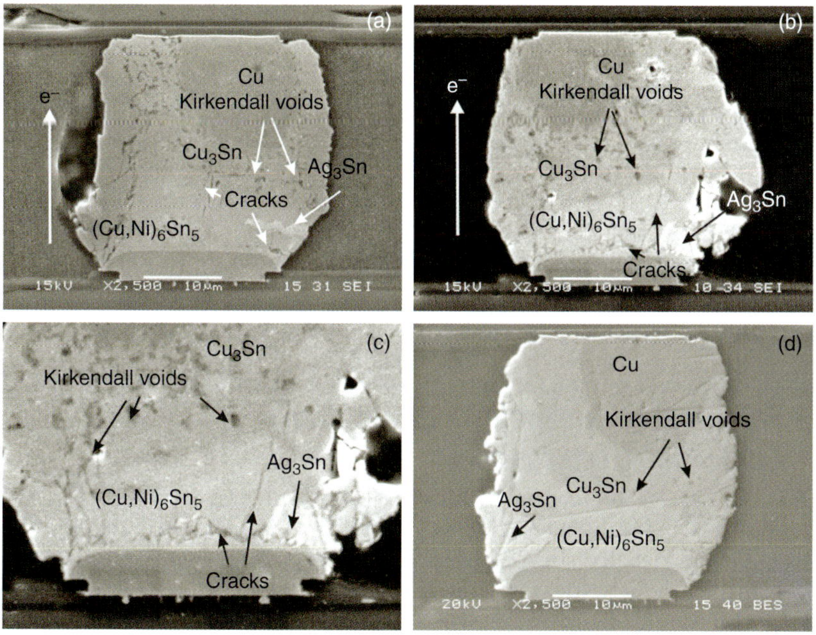

FIGURE 13.8 SnAg solder joint under high current density (4.08×10^4 A/cm^2), (a) 132 h, (b) 640 h, (c) 640 h close-up. In (d) the solder joint is not under any current density.

Figure 13.8*a* shows the solder-joint interconnect under high current density (4.08×10^4 A/cm^2) for 132 h. It can be seen that: (1) due to the effect of electron force and Joule heating, the diffusion rate of materials is accelerated; (2) there is no residual solder anymore; (3) $(Cu,Ni)_6Sn_5$ and Cu_3Sn compounds become thicker; and (4) a large number of Kirkendall voids are formed in the Cu_3Sn compounds. When the low temperature solder is consumed and interdiffusion between IMCs reaches equilibrium, the resistance would not change. This explains why a plateau of the resistance curve exists, as shown in Fig. 13.7. It can also be seen from Fig. 13.8*a* that vertical (Kirkendall voids) cracks are found and a small horizontal crack is initiated adjacent to the ENIG pad.

Figure 13.8*b* shows the failed solder joint subjected to the high current stressing for 640 h. Vertical cracks linking with Kirkendall voids penetrate into $(Cu,Ni)_6Sn_5$ compounds. Also, a continuous thin crack is formed in $(Cu,Ni)_6Sn_5$ compounds along the ENIG pad, which causes the solder joint to open. More detailed microstructure of the ENIG side is revealed in Fig. 13.8*c*. Such a thin crack is quite different from that (wider) formed at the normal flip-chip solder joint under current stressing [26, 27]. The reason is that in present case, the crack is formed by mechanical stress and it propagates along the grain or phase boundary of high-temperature components. It is interesting to note that after testing, some of the failed solder joints can resist slightly again, but when they are heated up to the test temperature, the openings reappear.

Figure 13.8*d* shows the microstructure of the neighboring solder joints without current stressing. It can be seen that: (1) there are fewer Kendendall voids and cracks; (2) no apparent cracks are found at the interface; and (3) the thickness of Cu_3Sn compounds is thinner and the size of Kirkendall voids is also smaller. These show that current stressing could enhance/accelerate the formation of Kirkendall voids [5].

13.7 Failure Mechanism of the Multiphase Solder-Joint Interconnect

It is known that electromigration triggers failures at zero-flux boundaries, that is, a heterogeneous boundary or an interface under high current density stressing. For microbump solder joints under high current density stressing, it is found that: (1) when all the solders convert into high-temperature IMCs, the potential zero-flux boundaries are Cu/Cu_3Sn, $Cu_3Sn/(Cu,Ni)_6Sn_5$, $(Cu,Ni)_6Sn_5/Ag_3Sn$, and $(Cu,Ni)_6Sn_5$/ENIG; (2) in such a multiple IMC joint, current stressing enhances the formation of Kirkendall voids at Cu/Cu_3Sn and their growth at $(Cu,Ni)_6Sn_5$/ENIG; and (3) vertical and horizontal cracks are formed and related to the Kirkendall voids.

It is well known that fractures along interfacial boundaries are accelerated by interfacial stresses generated by the difference in physical properties. The difference in coefficients of thermal expansion (CTE) of Cu_3Sn and Cu_6Sn_5 (16.3 and $19 \times 10^{-6}/°C$, respectively) [5] would produce stress in the solder joint. In addition, since the solder joint under current stressing is near the edge of the chip, the CTE mismatch between the chips and the carrier would also generate stress in the bilateral solder joints at the test temperature. Moreover, the fracture toughness of Cu_3Sn and Cu_6Sn_5 are 5.72 and 2.80 $MPa/m^{1/2}$, respectively [39]. Therefore, we can extrapolate that cracks propagate more easily in the $(Cu,Ni)_6Sn_5$ when the solder joints are under high current density stressing and mechanical stress, compared with Cu_3Sn. That explains why the opening of the solder joint interconnect is near the $(Cu,Ni)_6Sn_5$/ENIG interface instead of Cu_3Sn where a huge number of Kirkendall voids are formed. On the other hand, the fact that the microbump joints survive being subjected to the lower current density stressing is due to the following: (1) the electron wind force is smaller; (2) the joule heating effect is not significant, which also results in a smaller mechanical stress in the joint; and (3) the residual solder could release the mechanical stress in the solder joint.

13.8 Summary and Recommendations

The electromigration test for microjoints of the Cu-stud + SnAg + ENIG UBM was studied. It was found when the microjoint was subjected to a current density of 2.04×10^4 A/cm^2 at 140°C, no failure was found up to 1000 h. When the joints were tested under a current density of 4.08×10^4 A/cm^2 at the same temperature, due to the effect of joule heating and electron force, the initial low-temperature solder was quickly consumed and multiple IMCs, that is, Cu_3Sn and $(Cu,Ni)_6Sn_5$, were more prevalent within the joint. At the bump interconnect, besides the huge number of Kirkendall voids formed inside the Cu_3Sn compound, voids were also formed at the interface between $(Cu,Ni)_6Sn_5$ and ENIG, where crack formation and propagation triggered the sudden failure of the joint. The present study revealed that the formation of multiple IMCs within a microjoint would be a great concern for electromigration failure when subjected to high temperature and high current density stressing. Some of the important results and recommendations are summarized in the following:

1. The increase of electrical resistance of the microbump solder joints under high temperature (140°C) and current stressing (2.04×104 and 4.08×10^4 A/cm^2) is mainly due to the formation of high-temperature IMCs.
2. For the Cu-stud with SnAg + ENIG UBM solder joints tested at 140°C with a higher current density of 4.08×10^4 A/cm^2, the resistance would reach the plateau region after a certain time (50–200 h), where the diffusion between different materials, that is, Cu, Ni, and Sn, reaches equilibrium and IMCs become quite stable. A large number of Kirkendall voids in Cu_3Sn compounds are found under high-temperature current densities. Due to the brittle nature, large cracks along $(Cu,Ni)_6Sn_5$ grain boundaries are also found, which would cause sudden failure of the solder joints.
3. For the Cu-stud with SnAg + ENIG UBM solder joints tested at 140°C with a lower current density of 2.04×10^4 A/cm^2, the resistances do not reach the plateau region after 1000 h of testing due to the decreased joule heating effect. The electron flow direction has an effect on the diffusion of materials. The resistance increases faster when electrons flow from Cu UBM to ENIG.
4. Test results indicate that the EM failure mode and mechanism of microbump solder joints are totally different from ordinary solder joints, since the resistance increase of microbump solder joints is not due to the traditional pancake crack formation, but the IMC growth and phase transformation.

13.9 References

1. Tu, K. U., "Recent Advances on Electromigration in Very Large-Scale-Integration of Interconnects," *J. Appl. Phys.*, Vol. 94, No. 9 (2003), pp. 5451–5473.
2. Lin, K. L., and Lin, G. P. "The Electromigration Investigation on the Newly Developed Pb-Free Sn-Zn-Ag-Al-Ga Solder Ball Interconnect." In *IEEE Proceedings of Electronic & Components Technology Conference*, Reno, NV, May 29–June 1, 2007, pp. 1467–1472.
3. Lin, K. L., and Kuo, S. M. "The Electromigration and Thermomigration Behaviors of Pb-Free, Flip-Chip Sn-3Ag-0.5Cu Solder Bumps." In *IEEE Proceedings of Electronic & Components Technology Conference*, San Diego, CA, May 30–June 2, 2006, pp. 667–672.
4. Nah, J. W., Suh, J. O., Tu, K. N., Yoon, S. W., Chong, C. T., Kripesh, V., Su, B. R., and Chen, C. "Electromigration in Pb-Free Solder Bumps with Cu Column as Flip-Chip Joints." In *IEEE Proceedings of Electronic & Components Technology Conference*, San Diego, CA, May 30–June 2, 2006, pp. 657–662.
5. Ebersberger, B., Bauer, R., and Alexa, L. "Reliability of Lead-Free SnAg Solder Bumps: Influence of Electromigration and Temperature." In *IEEE Proceedings of Electronic & Components Technology Conference*, Lake Buena Vista, FL, May 31–June 3, 2005, pp. 1407–1415.
6. Rinne, G., "Electromigration in SnPb and Pb-Free Solder Bumps." In *IEEE Proceedings of Electronic & Components Technology Conference*, Las Vegas, June 1–4, 2004, pp. 974–978.
7. Jang, S. Y., Wolf, J., Kwon, W. S., and Paik, K. W. "UBM (Under Bump Metallization) Study for Pb-Free Electroplating Bumping: Interface Reaction and Electromigration." In *IEEE Proceedings of Electronic & Components Technology Conference*, San Diego, May 28–31, 2002, pp. 1213–1220.
8. Gan, H., and Tu, K. "Effect of Electromigration on Intermetallic Compound Formation in Pb-free Solder-Cu Interfaces." In *IEEE Proceedings of Electronic & Components Technology Conference*, San Diego, May 28–31, 2002, pp. 1206–1212.
9. Huang, Z., Chatterjee, R., Justison, P., Hernandez, R., Pozder, S., Jain, A., Acosta, E., Gajewski, D. A., Mathew, V., and Jones, R. E. "Electromigration of Cu-Sn-Cu Micropads in 3D Interconnect." In *IEEE Proceedings of Electronic Components & Technology Conference*, Lake Buena Vista, FL, May 27–30, 2008, pp. 12–17.
10. Rinne, G., "Emerging Issues in the Physics and Control of Electromigration in SnPb and Pb-Free Solder Bumps." In *IEEE Proceedings of Electronics Packaging Technology Conference*, Singapore, December 10–12, 2003, pp. 72–76.
11. Su, P., Uehling, T., Wontor, D., Ding, M., and Ho, P. S. "An Evaluation of Electromigration Performance of SnPb and Pb-free Flip Chip Solder Joints." In *IEEE Proceedings of Electronic & Components Technology Conference*, Lake Buena Vista, FL, May 31–June 3, 2005, pp. 1431–1436.
12. Lu, M., Shih, D. Y., Lauro, P., Kang, S., Goldsmith, C., and Seo, S. K. "The Effects of Ag, Cu Compositions and Zn Doping on the Electromigration Performance of Pb-Free Solders." In *IEEE Proceedings of Electronic & Components Technology Conference*, San Diego, CA, May 25–29, 2009, pp. 922–929.
13. Nicholls, L., Darveaux, R., Syed, A., Loo, S., Tee, T. Y., Wassick, T. A., and Batchelor, B. "Comparative Electromigration Performance of Pb-Free Flip Chip Joints with Varying Board Surface Condition." In *IEEE Proceedings of Electronic & Components Technology Conference*, San Diego, CA, May 25–29, 2009, pp. 914–921.
14. Su, P., Li, L., Lai, Y. S., Chiu, Y. T., and Kao, C. L. "A Comparison Study of Electromigration Performance of Pb-Free, Flip-Chip Solder Bumps." In *IEEE Proceedings of Electronic & Components Technology Conference*, San Diego, CA, May 25–29, 2009, pp. 903–908.
15. Lee, J. H., Lim, G. T., Park, Y. B., Yang, S. T., Suh, M. S., Chung, Q. H., and Byun, K. Y. "Size Effect on Electromigration Reliability of Pb-Free, Flip-Chip Solder Bump." In *IEEE Proceedings of Electronic & Components Technology Conference*, Lake Buena Vista, FL, May 27–30, 2008, pp. 2030–2034.

16. Lu, M., Lauro, P., Shih, D. Y., Polastre, R., Goldsmith, C., Henderson, D. W., Zhang, H., and Cho, M. G. "Comparison of Electromigration Performance for Pb-Free Solders and Surface Finishes with Ni UBM." In *IEEE Proceedings of Electronic & Components Technology Conference*, Lake Buena Vista, FL, May 27–30, 2008, pp. 360 365.
17. Chae, S. H., Chao, B., Zhang, X., Im, J., and Ho, P. S. "Investigation of Intermetallic Compound Growth Enhanced by Electromigration in Pb-Free Solder Joints." In *IEEE Proceedings of Electronic & Components Technology Conference*, Reno, NV, May 29–June 1, 2007, pp. 1442–1449.
18. Chae, S. H., Im, J., Ho, P. S., and Uehling, T. "Effects of UBM Thickness, Contact Trace Structure and Solder Joint Scaling on Electromigration Reliability of Pb-Free Solder Joints." In *IEEE Proceedings of Electronic & Components Technology Conference*, Lake Buena Vista, FL, May 27–30, 2008, pp. 354–359.
19. Chae, S. H., Zhang, X., Chao, H. L., Lu, K. H., Ho, P. S., Ding, M., Su, P., Uehling, T., and Ramanathan, L. N. "Electromigration Lifetime Statistics for Pb-Free Solder Joints with Cu and Ni UBM in Plastic Flip-Chip Packages." In *IEEE Proceedings of Electronic & Components Technology Conference*, San Diego, CA, May 30–June 2, 2006, pp. 650–656.
20. Kwon, Y. M., and Paik, K. W. "Electromigration of Pb-Free Solder Flip-Chip Using Electroless Ni-P/Au UBM." In *IEEE Proceedings of Electronic & Components Technology Conference*, Reno, NV, May 29–June 1, 2007, pp. 1472–1477.
21. Ebersberger, B., Bauer, R., and Alexa, L. "Reliability of Lead-Free SnAg Solder Bumps: Influence of Electromigration and Temperature." In *IEEE Proceedings of Electronic & Components Technology Conference*, Lake Buena Vista, FL, May 31–June 3, 2005, pp. 1407–1415.
22. Iwasaki, T., M. Watanabe, S. Baba, Y. Hatanaka, S. Idaka, Y. Yoloyama, M, Kimura, "Development of 30 micron Pitch Bump Interconnections for COC-FCBGA," *Proc56th Electronic Components and Technology Conf*, 2006, pp. 1216–1222.
23. Gan, H., S. L. Wright, R. Polastre, L. P. Buchwalter, R. Horton, R. Horton, P. S. Andry, C. Patel, C. Tsang, J. Knickerbocker, E. Sprogis, A. Pavlova, S. K. Kang, K. W. Lee, "Pb-free Micro-Joints (50 μm pitch) for the Next Generation Micro-Systems: The Fabrication, Assembly and Characterization," *Proc 56th Electronic Components and Technology Conf*, 2006, pp. 1210–1215.
24. Sunohara, M., T. Tokayunaga, T. Kurihara, and M. Higashi, "Slicon Interposer with TSVs (Through Silicon Vias) and Fine Multilayer Wiring," In *IEEE Proceedings of Electronic & Components Technology Conference*, Lake Buena Vista, FL, May 27–30, 2008, pp. 847–852.
25. Ren, F., Zhang, X., Nah, J. W., Tu, K. N., Xu, L., and Pang, J. H. L. "In-Situ Study of the Effect of Electromigration on Strain Evolution and Mechanical Property Change in Lead-Free Solder Joints." In *IEEE Proceedings of Electronic & Components Technology Conference*, San Diego, CA, May 30–June 2, 2006, pp. 1160–1163.
26. Yeh, E., W. J. Choi, K. N. Tu, P. Elenius, H. Balkan, "Current-Crowding-Induced Electromigation Failure in Flip Chip Solder Joints," *Appl. Phys. Lett.*, Vol. 80, (2002), pp. 580–582.
27. Zhang, L., S. Ou, J. Huang, K. N. Tu, S. Gee, L. Nguyen, "Effect of Cureent Crowding on Void Propagation at the Interface between Intermetallic Compound and Solder in Flip Chip Solder Joints," *Appl. Phys. Lett.*, Vol. 88, (2006), pp. 012106–1.
28. Ding, M., G. Wang, B. Chao, P. S. Ho, P. Su, T. Uehling, "Effect of Contact Metallization on Electromigration Reliability of Pb-Free Solder Joints," *J. Appl. Phys.*, Vol. 99, (2006), pp. 094906–094906-6.
29. Nah, J., J. O. Suh, K. N. Tu, S. W. Yoon, V. S. Rao, V. Kripesh, F. Hua, "Electromigration in Flip Chip Solder Joints Having a Thick Cu Column Bump *and a Shallow* Solder Interconnect," *J. Appl. Phys.*, Vol. 100, (2006), pp. 123513–123513-5.
30. Lai, Y., Y. T. Chia, C. W. Lee, Y. H. Shao, J. Chen, "Electromigration Reliability and Morphologies of Cu Pillar Flip-chip Solder Joints," *In IEEE Proceedings of Electronic & Components Technology Conference*, Lake Buena Vista, FL, May 27–30, 2008, pp. 330–335.

31. Xu, L., J. K. Han, J. J. Liang, K. N. Tu, and Y. S. Lai, "Electromigration Induced High Fraction of Compound Formation in SnAgCu Flip Chip Joints with Copper Column," *Appl. Phys. Lett.*, Vol. 92, (2008), pp. 262104–262104-3.
32. Chang, Y., T. H. Chiang, and C. Chen, "Effect of Voids Propagation on Bump Resistacne due to Electromigration in Flip-Chip Solder Joints Using Kelvin Structure," *Appl. Phys. Lett.*, Vol. 91, (2007), pp. 132113–132113-3.
33. Yeo, A., B. Ebersberger, C. Lee, " Consideration of Temperautre and Current Stress Testing on Flip Chip Solder Interconnects," *Miroelectron Reliab.*, Vol. 48, (2008), pp. 1847–1856.
34. Lu, M., D. Y. Shih, P. Lauro, C. Goldsmith, D. W. Henderson, "Effect of Sn Grain Orientation on Electromigration Mechanism in High Sn-Based Pb-Free Solders," *Appl. Phys. Lett.*, Vol. 92, (2008), pp. 211909–211909-3.
35. Yang, P., C. C. Kuo, C. Chen, "The Effect of Pre-Aging on the Electromigration of Flip-Chip Sn-Ag Solder Joints," *J of Materials*, (2008), pp. 77–80.
36. Laurila, T., V. Vuorinen, and J. K. Kivilahti, "Interfacial Reactions Between Lead-Free Solders and Common Base Materials," *Mater. Sci. Eng*. R 49, (2005), pp. 1–60.
37. Wei, C., C. F. Chen, P. C. Liu, C. Chen, "Electromigration in Sn-Cu Intermetallic Compounds," *J. Appl. Phys.*, Vol. 105, (2009), pp. 023715–023715-41.
38. Zeng, K., R. Stierman, T. C. Chiu, D. Edwards, K. Ano, K. N. Tu, "Kirkendall Voids Formation in Eutectic SnPb solder joints on Bare Cu and Its Effect on Joint Reliability," *J. Appl. Phys.*, Vol. 97, (2005), pp. 024508–024508-8.
39. Lee, C., P. Wand, J. Kim, "Are Intermetallics in Solder Joints Really Brittle?," In *IEEE Proceedings of Electronic & Components Technology Conference*, Reno, NV, May 29–June 1, 2007, pp. 648–652.
40. Yu, D., T. Chai, M. Thew, Y. Ong, V. Rao, L. Wai, and J. H. Lau, "Electromigration Study of 50 μm Pitch Micro Solder Bumps Using Four-Point Kelvin Structure," In *IEEE Proceedings of Electronic & Components Technology Conference*, San Diego, CA, May 25–29, 2009, pp. 930–935.
41. Frear, D., H. Morgan, S. Burchett, and J. H. Lau, *The Mechanics of Solder Alloy Interconnects*, Van Nostrand Reinhold, New York, 1994.

CHAPTER 14

Effects of Dwell Time and Ramp Rate on SAC Thermal Cycling Test Results

14.1 Introduction

For either reliability testing and data analysis or design for reliability (DFR) of lead-free interconnects, the most common loading condition is the temperature cycling [1–12]. Each temperature cycle consists of ramp-up time, dwell time at hot, ramp-down time, and dwell time at cold. The effects of temperature-cycling dwell time and ramp rate on the thermal-fatigue life of a 256-pin plastic ball grid array (PBGA) lead-free solder joints are investigated in this chapter.

14.2 Effects of Dwell Time on the Thermal-Fatigue Life of SnAgCu (SAC) Solder Joints

The effects of temperature-cycling dwell time on the thermal-fatigue life of lead-free solder joints are investigated in this section. Emphasis is placed on determining the creep responses and the creep strain energy density per cycle of this PBGA's solder joints subjected to various dwell times (namely, 15, 60, and 480 min) at peak cycling temperatures (0°C and 100°C).

14.2.1 The Structure, Material Properties, and Simulations

The geometry, material properties, and finite-element modeling of the 256-pin PBGA are exactly the same as those shown in Chap. 2 (Sec. 2.6.8)—the only difference is the temperature loading conditions.

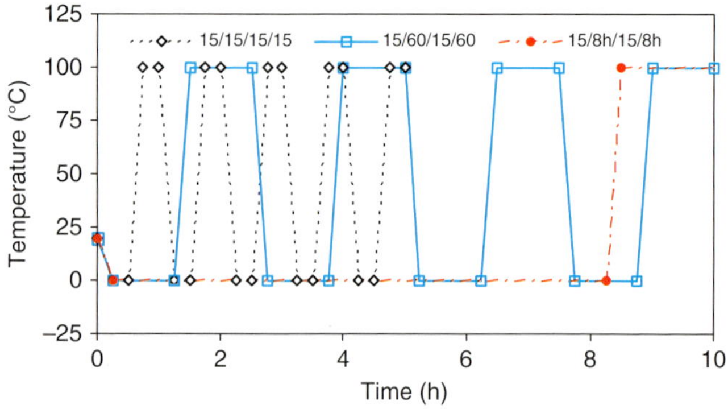

FIGURE 14.1 The thermal environments.

14.2.2 Temperature Loading Conditions

Figure 14.1 shows the temperature profiles to be imposed on the PBGA assembly. Their temperature range is 0°C to 100°C but with three different dwell times at peak temperatures: (1) 15 min; (2) 60 min; and (3) 480 min. For all these cases, the ramp-up and ramp-down times are 15 min. In each model simulation, the solder joints of the PBGA assembly are subjected to five cycles of one of the three aforementioned temperature regimes.

14.2.3 Creep Hysteresis Loops

It is important to study the creep responses for multiple cycles by observing when the hysteresis loops become stabilized. Figure 14.2 shows typical shear stress and creep shear strain hysteresis loops for multiple cycles at the upper corner of the most-likely-to-fail solder joint in the lead-free PBGA assembly under, respectively, various temperature profiles. The creep shear strain versus shear stress loop is quite stabilized after the fourth cycle.

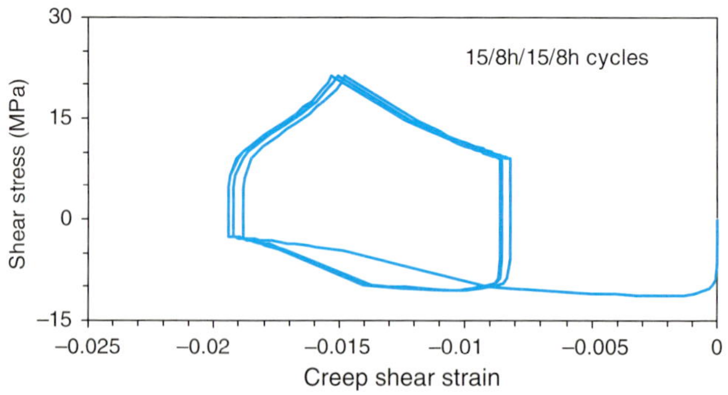

FIGURE 14.2 Hysteresis loops due to 15-min ramps and 8-h dwells.

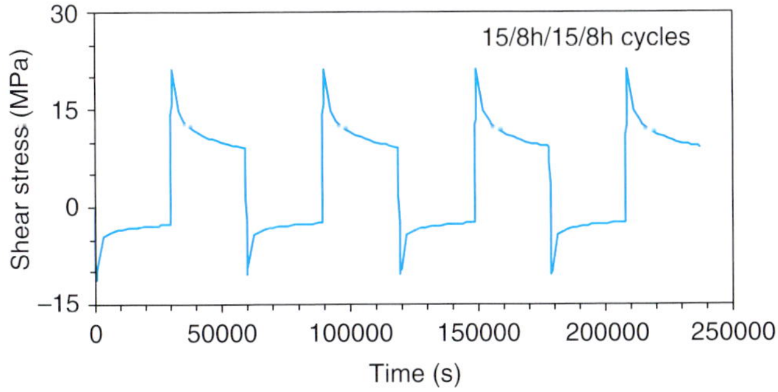

FIGURE 14.3 The shear stress time-history with 15-min ramps and 8-h dwells.

14.2.4 Shear Stress Time-History

Figure 14.3 shows a typical shear stress time-history near the upper corner of the solder joint most likely to fail. The responses of the lead-free PBGA assembly are described for the various temperature profiles [6]. It can be seen that: (1) for all the cases, the maximum shear stress range per cycle is about the same; (2) for all cases, the maximum shear stresses (at hot and at cold) per cycle are about the same; (3) for each cycle, the longer the dwell times at peak temperatures, the greater the stress relaxations within the dwell times; and (4) for each cycle, the longer the dwell times at peak temperatures, the faster the stress relaxations within the ramp-up and ramp-down times.

14.2.5 Shear Creep Strain Time-History

A typical creep shear strain time-history near the upper corner of the most-likely-to-fail solder joint is shown in Fig. 14.4 for the lead-free PBGA assembly exposed to the cited temperature profiles [6]. The longer the dwell time at peak temperatures, the larger the creep shear strain range per cycle.

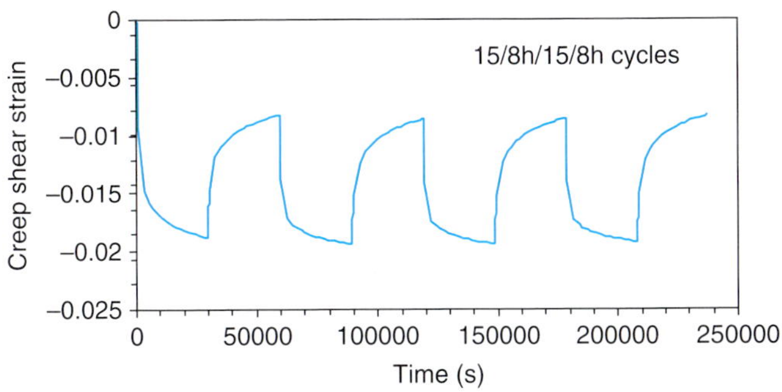

FIGURE 14.4 The creep shear strain time-history with 15-min ramps and 8-h dwells.

14.2.6 Creep Strain Energy Density Time-History

Figures 14.5 and 14.6 show the creep strain energy density time-history near the upper corner of the solder joint most likely to fail of the lead-free PBGA assembly for the specified temperature profiles. It can be seen that: (1) the longer the dwell time at peak temperatures, the larger the creep strain energy density ranges per cycle (ΔW); (2) the longer the dwell time at peak temperatures, the larger the creep strain energy density at any number of cycles; and (3) the creep strain energy densities per cycle are (at the fifth cycle) 0.126, 0.173, and 0.251 MPa, respectively, for 15, 60, and 480 min dwell times at peak temperatures, as shown in Fig. 14.7. These values can also be determined from the area within the fifth hysteresis loop as shown in Fig. 14.2. A curve fitting of the data points on Fig. 14.7 leads to the equation shown in Fig. 14.8, where the creep strain energy density per cycle (ΔW) versus dwell time is plotted. The rate of ΔW is larger at smaller dwell times.

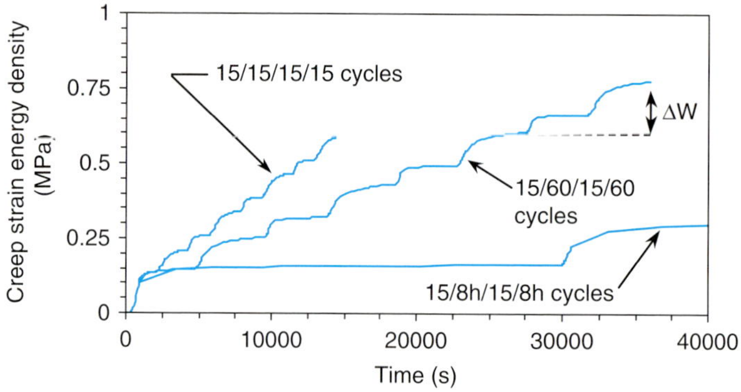

FIGURE 14.5 The critical creep strain energy density as a function of time.

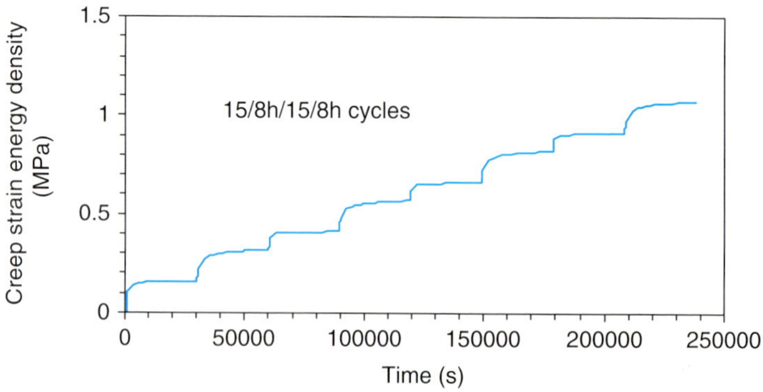

FIGURE 14.6 The critical creep strain energy density as a function of time for the 8-h dwell environment.

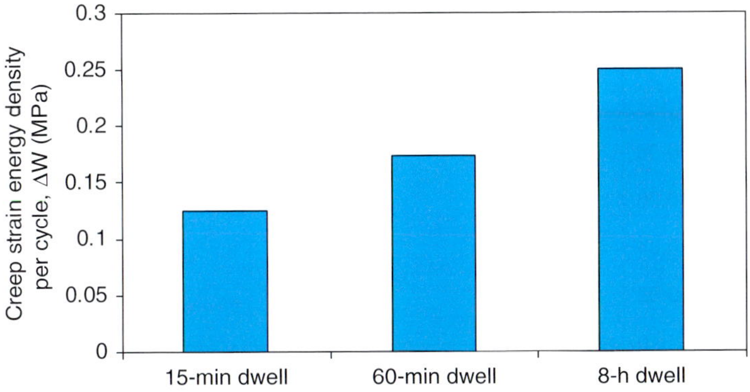

FIGURE 14.7 Creep strain energy density vs. dwell time.

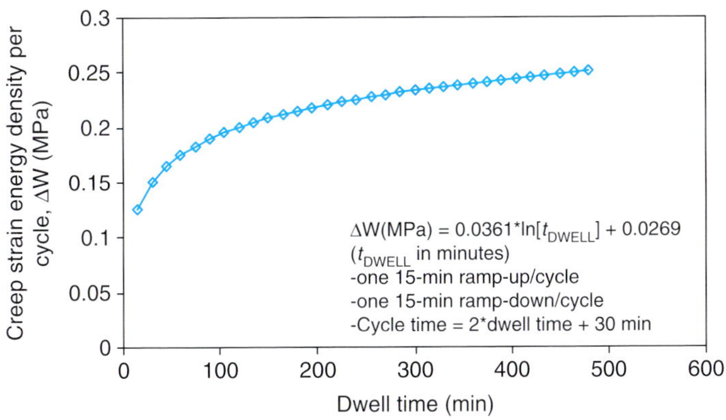

FIGURE 14.8 Imparted creep strain energy density per cycle, ΔW, as a function of dwell time.

When the dwell time is larger than 120 min, the change of ΔW with respect to the dwell time is almost constant.

Figure 14.8 presents the creep strain energy per cycle due to various dwell times. It is evident from the figure that longer dwell times correspond to greater accumulated hysteresis energies and, consequently, shorter fatigue-life expectations. When formulating a test, however, the engineer must consider pragmatisms such as the total time period involved in performing the test. From Fig. 14.8, creep strain energy density per cycle for an 8-h dwell is approximately twice that for a 15-min dwell. Testing with an 8-h dwell will, however, take approximately 16 times longer to obtain a given cycle count than will tests using the 15-min dwell. Figure 14.9 reconciles the accumulated strain energy densities and dwell times with the elapsed test time. The figure presents the average strain energy density accumulated per hour as obtained by normalizing the creep strain energy density by

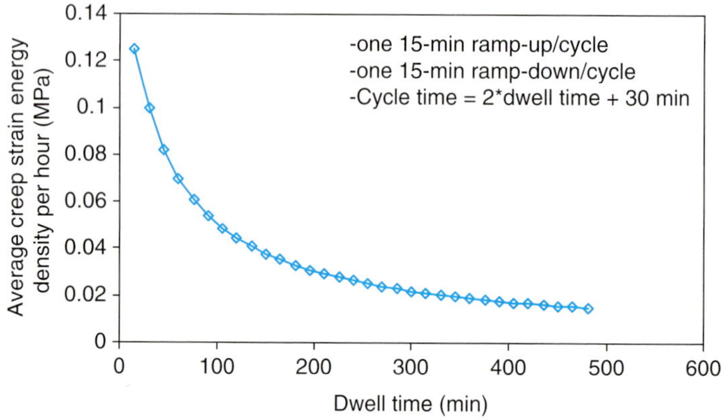

FIGURE 14.9 Imparted creep strain energy density as a function of test time.

the cycle time. Note that the faster cycles impart the greatest hysteresis energy for a given test (clock) time. This implies that tests using the faster, especially 15-min, dwell times may be most efficient at inciting failures in the least amount of test time.

14.2.7 Summary and Recommendations

Some important results and recommendations are summarized in the following:

1. For all cases, the creep hysteresis loops become stabilized before the fifth thermal cycle.
2. For all cases, the maximum shear stress range per cycle is about the same.
3. For all cases, the maximum shear stresses (at hot and at cold) per cycle are about the same.
4. For each cycle, the longer the dwell times at peak temperatures, the larger the stress relaxations within the dwell times.
5. For each cycle, the longer the dwell times at peak temperatures, the faster the stress relaxations within the ramp-up and ramp-down times.
6. For each cycle, the longer the dwell time at peak temperatures, the larger the creep shear strain ranges.
7. For each cycle, the longer the dwell time at peak temperatures, the larger the creep strain energy density ranges.
8. The change of creep strain energy density with respect to dwell time is larger at smaller dwell times. When the dwell time is larger than 120 min, the rate of creep strain energy density with respect to dwell time is almost constant.

9. Once the creep strain energy density per cycle is determined, the thermal-fatigue life of the lead-free solder joints can be determined by the fatigue crack growth equation; see, for example, Chap. 2.
10. Dwell time of 15 min is recommended at hot and cold during thermal cycling.

14.3 Effects of Ramp Rate on the Thermal-Fatigue Life of SAC Solder Joints

The effects of temperature-cycling ramp time on the thermal-fatigue life of lead-free solder joints are investigated in this section. Emphasis is placed on determining creep responses and the creep strain energy density per cycle of the PBGA package's lead-free solder joints on an FR(fiber-glass reinforced)-4 PCB subjected to various ramp times (0.5, 1, 5, 10, and 15 min) for two different ranges of cycling temperatures, namely, from 0°C to 100°C and from –25°C to 125°C. For all the cases, the dwell time at peak cycling temperatures (0°C, 100°C, –25°C, and 125°C) is 15 min.

14.3.1 Temperature Loading Conditions

Figures 14.10 and 14.11 show the temperature profiles to be imposed on the PBGA assembly. Their temperature range is 0°C to 100°C (Fig. 14.10) and –25°C to 125°C (Fig. 14.11). Also, each one has five different (ramp up = ramp down) times: (1) 30 s; (2) 60 s; (3) 300 s; (4) 600 s; and (5) 900 s. For all these cases, the dwell time at all the peak temperatures (0°C, 100°C, –25°C, and 125°C) is 15 min. In each model simulation, the solder joints of the PBGA assembly are subjected to five cycles of one of the ten aforementioned temperature regimes.

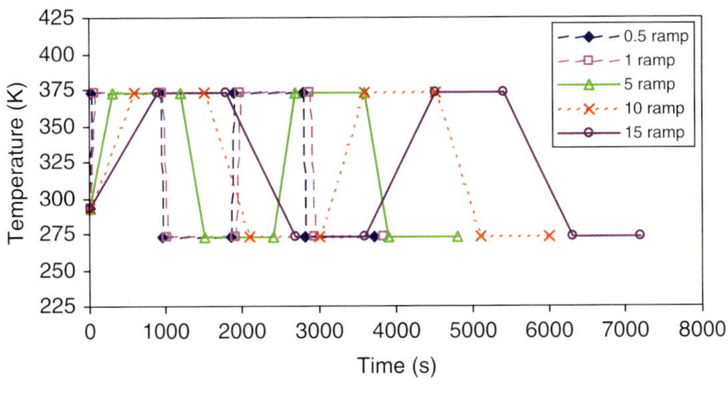

FIGURE 14.10 The first two cycles for the 0°C to 100°C profile with five different ramp times (min) thermal environments.

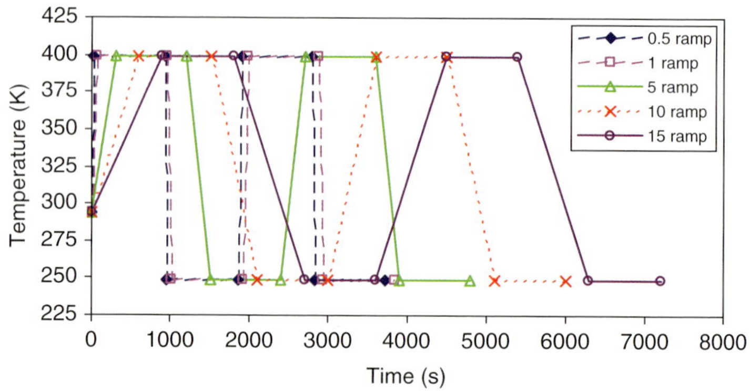

FIGURE 14.11 The first two cycles for the –25°C to 125°C profile with five different ramp times (min) thermal environments.

14.3.2 Maximum Stress/Strain Location

Figures 14.12 and 14.13 show, respectively, typical shear stress and creep shear strain contours of the most-likely-to-fail lead-free solder joint in the PBGA assembly at the end of the fifth cycle. (In this case, the temperature condition is from 0°C to 100°C with 5 min ramp

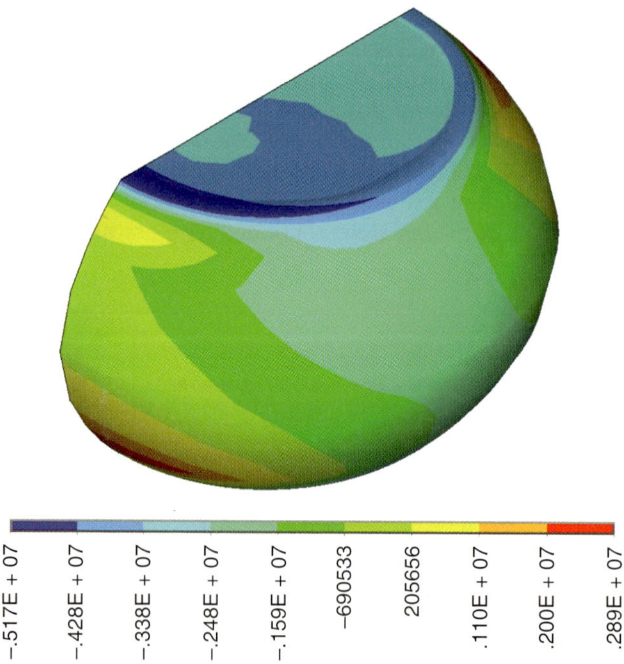

FIGURE 14.12 Shear stress contours at the end of the fifth cycle (0°C to 100°C with 5-min ramp).

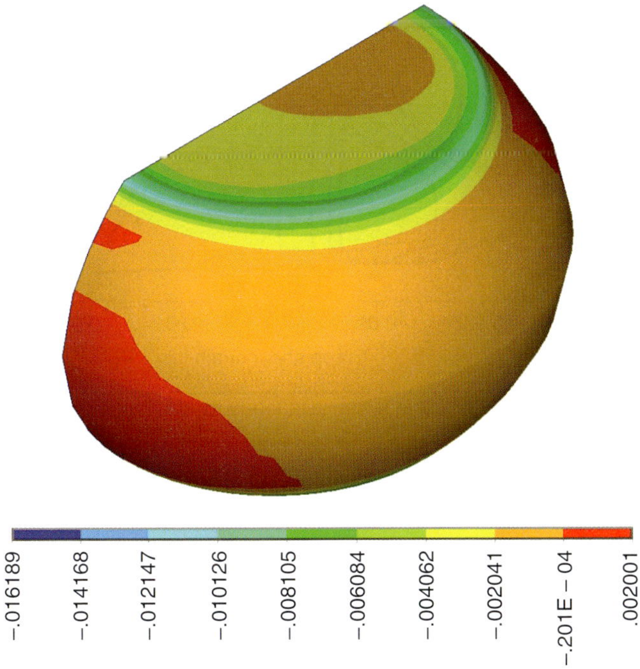

FIGURE 14.13 Creep shear strain contours at the end of the fifth cycle (–25°C to 125°C with 5-min ramp).

time.) The location for the maximum stress and strain is near the upper corner of the lead-free solder joint. This maximum stress and strain location happens for all other nine cases [7].

14.3.3 Creep Hysteresis Loops

Figure 14.14 shows a typical shear stress and creep shear strain hysteresis loop at the upper corner of the most-likely-to-fail solder joint in the lead-free PBGA assembly subjected to –25°C to 125°C (Fig. 14.11) environment with a 5-min ramp. The creep shear strain versus shear stress loop is quite stabilized after the fourth cycle. All the other cases have the similar behavior.

Figure 14.15 shows a typical fifth-cycle shear stress and creep shear strain hysteresis loop at the lower corner of the most-likely-to-fail solder joint in the lead-free PBGA assembly subjected to –25°C to 125°C (Fig. 14.11) temperature profiles with ramp time equal to 0.5 min (thermal shock) and 15 min (thermal cycling). For both temperature ranges, the area within the fifth shear stress and creep shear strain hysteresis loop (i.e., the creep shear strain energy density per cycle) for the shorter ramp time equals 0.5 min (i.e., faster ramp rate or thermal shock), which is larger than that for the longer ramp time: 15 min (i.e., slower ramp rate or thermal

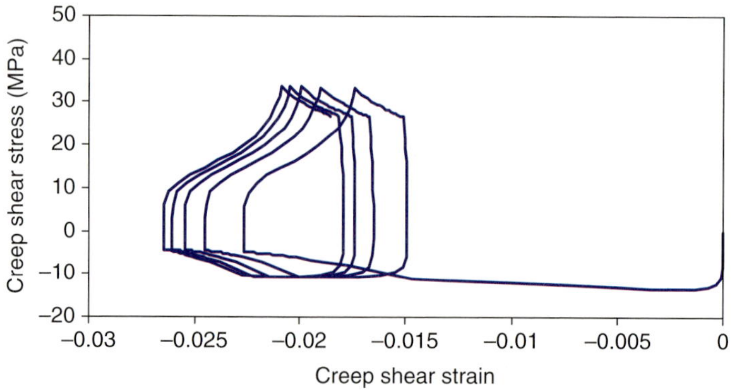

FIGURE 14.14 Evolution of the hysteresis loops for the 5-min ramp, −25°C to 125°C case.

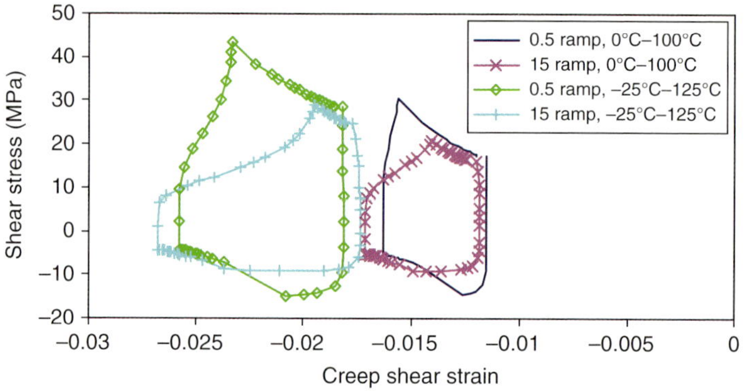

FIGURE 14.15 Shear stress vs. creep shear strain hysteresis loops for the fifth cycle.

cycling). Thus, the thermal-fatigue life (or number of cycles-to-failure) of the lead-free solder joints under the thermal shock condition is smaller than that under the thermal cycling condition. This is because the number of cycles-to-failure of lead-free solder joints is nonlinear and inversely proportional to the creep strain energy density per cycle; see, for example, Chap. 2. Similar results occur for other cases.

It can also be seen from Fig. 14.15 that for both ramp times (0.5 and 15 min i.e., thermal shock and thermal cycling) the creep shear strain energy density per cycle (i.e., the thermal-fatigue life) is strongly temperature-range–dependent. The area within the fifth shear stress and creep shear strain hysteresis loop at the critical location of the most-likely-to-fail solder joint in the lead-free PBGA assembly under the 0°C to 100°C temperature range is much smaller than that under −25°C to 125°C, as shown in Fig. 14.15 and Table 14.1.

Loadings (Min) (Ramp/ Dwell-Hot/Ramp/Dwell-Cold)	0°C to 100°C	–25°C to 125°C
0.5/15/0.5/15	0.158	0.330
1/15/1/15	0.153	0.310
5/15/5/15	0.136	0.263
10/15/10/15	0.129	0.250
15/15/15/15	0.125	0.243

TABLE 14.1 Change in Maximum Creep Strain Energy Density (MPa) per Cycle for the Thermal Environments

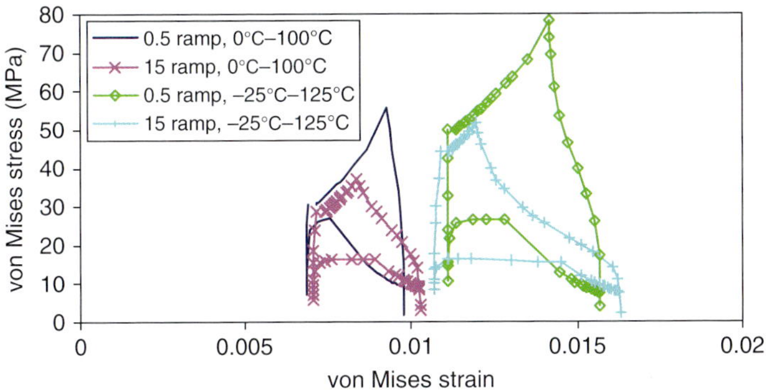

FIGURE 14.16 von Mises stress vs. creep strain hysteresis loops for the fifth cycle.

Figure 14.16 shows the fifth-cycle von Mises stress and creep strain hysteresis loop at the upper corner of the most-likely-to-fail solder joint in the lead-free PBGA assembly under 0°C to 100°C and –25°C to 125°C temperature profiles with thermal shock and thermal cycling conditions. Just like the cases under shear: (1) for both temperature ranges, the von Mises creep strain energy density per cycle in the lead-free solder joints under thermal shock condition is larger than that under the thermal cycling condition; and (2) for both ramp rates, the von Mises creep strain energy density per cycle in the lead-free solder joints is strongly temperature-dependent.

By comparing the hysteresis loops between Fig. 14.15 (shear stress vs. creep shear strain) and Fig. 14.16 (von Mises stress vs. creep strain), it is noted that the areas within the loops (i.e., von Mises creep strain energy density per cycle) in Fig. 14.16 are larger than those in Fig. 14.15. Thus, it is not safe to estimate the thermal-fatigue life of lead-free solder joints considering only the shear deformation. The expansion/compression of the solder joints due to bending/twisting of the whole assembly and the elongation/contraction of the solder joints are important, especially for area array packages like PBGA and flip chips.

14.3.4 Shear Stress Time-History

Figure 14.17 shows a typical shear stress time history for five complete cycles at the critical location of the most-likely-to-fail solder joint in the lead-free PBGA assembly under the –25°C to 125°C temperature profile with five different ramp times: 30, 60, 300, 600, and 900 s. It can be seen, and is expected, that for all ramp times, the larger temperature ranges produce larger shear stress ranges.

Figure 14.18 shows a typical shear stress time-history for five complete cycles at the critical location of the most-likely-to-fail solder joint in the lead-free PBGA assembly under the –25°C to 125°C temperature profile with the thermal shock (0.5 min ramp time) and thermal cycling (15 min ramp time) conditions. For both temperature ranges: (1) the shear stress range in the lead-free solder joint under thermal shock is larger than that under thermal cycling; and (2) the stress relaxation in the lead-free solder joint at both hot and cold

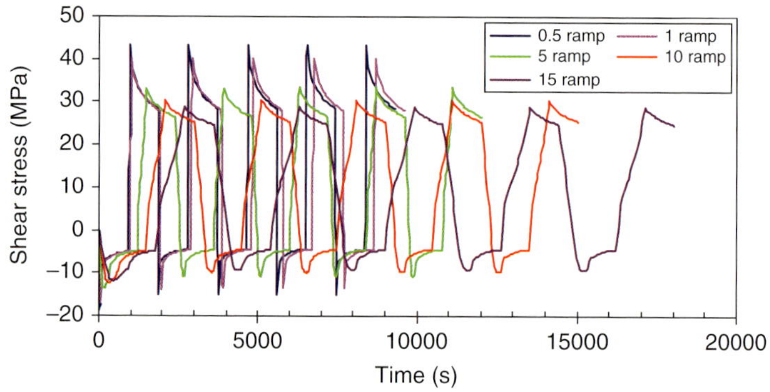

FIGURE 14.17 Shear stresses for the –25°C to 125°C thermal environments with five different ramp times (min).

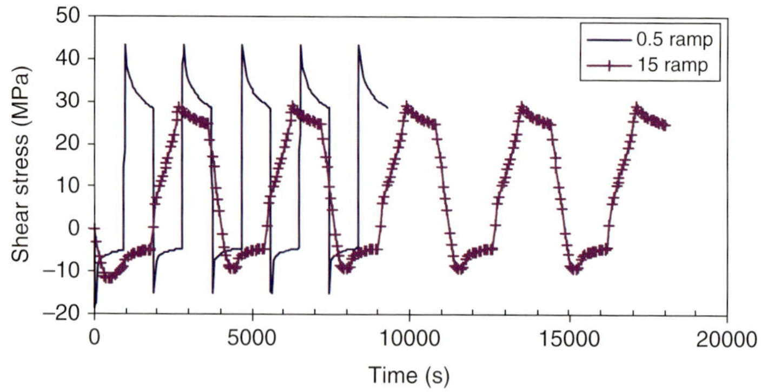

FIGURE 14.18 Shear stresses for the –25°C to 125°C thermal shock (0.5-min ramp) and thermal cycling (15-min ramp) environments.

dwells under thermal shock is faster and larger than that under thermal cycling. Similar behaviors apply to the other cases.

14.3.5 Shear Creep Strain Time-History

Figure 14.19 shows a typical creep shear strain time history for five complete cycles at the critical location of the most-likely-to-fail solder joint in the lead-free PBGA assembly under the –25°C to 125°C temperature profile with five different ramp times: 30, 60, 300, 600, and 900 s. It can be seen, and is expected, that for all ramp times, the larger the temperature ranges correspond with the larger creep shear strain ranges.

Figure 14.20 shows a typical creep shear strain time history for five complete cycles at the critical location of the most-likely-to-fail solder joint in the lead-free PBGA assembly under the –25°C to 125°C temperature profile with the thermal shock and thermal cycling

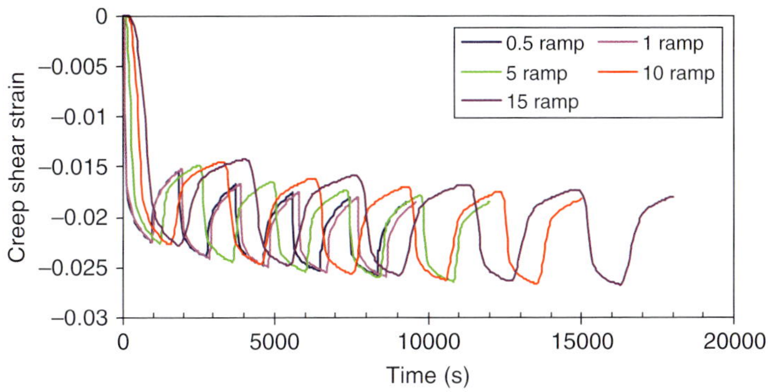

FIGURE 14.19 Creep shear strains for the –25°C to 125°C thermal environment.

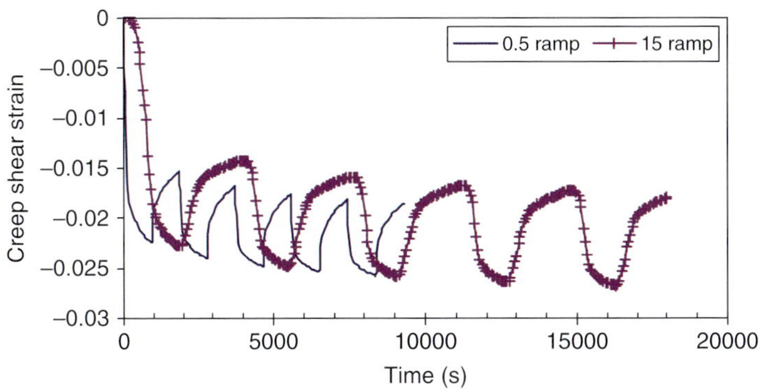

FIGURE 14.20 Creep shear strain for the –25°C to 125°C thermal shock (0.5-min ramp) and thermal cycling (15-min ramp) environments.

conditions. For both temperature ranges, the creep shear strain range in the lead-free solder joint under thermal shock is smaller than that under thermal cycling. Similar results occur for the other cases.

14.3.6 Creep Strain Energy Density Time-History

Figures 14.21 and 14.22 show the von Mises creep strain energy density time-history for five complete cycles near the upper corner of the solder joint most likely to fail of the lead-free PBGA assembly under, respectively, 0°C to 100°C and –25°C to 125°C temperature profiles, each with five different ramp times: 0.5, 1, 5, 10, and 15 min. For both temperature ranges: (1) the shorter the ramp times, the larger the von Mises creep strain energy density; and (2) the shorter the ramp times, the faster the rate of change of the von Mises creep strain energy density. Also, it can be seen, and is expected, that the larger the temperature ranges, the larger the von Mises creep strain energy density.

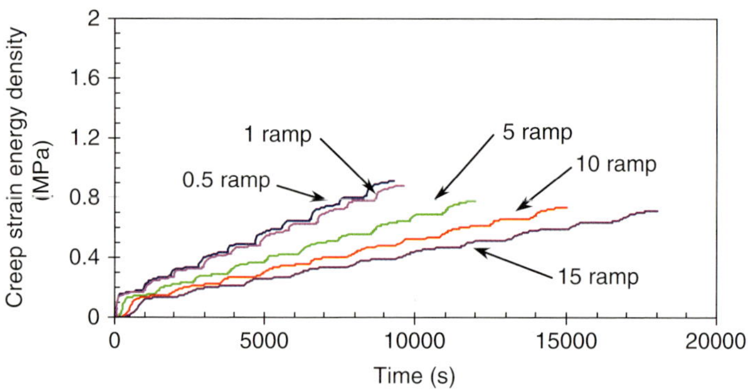

FIGURE 14.21 von Mises creep strain energy density for the 0°C to 100°C thermal environments.

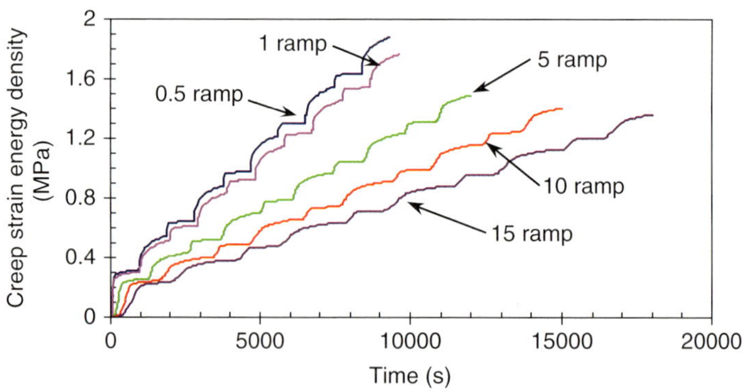

FIGURE 14.22 von Mises creep strain energy density for the –25°C to 125°C thermal environments.

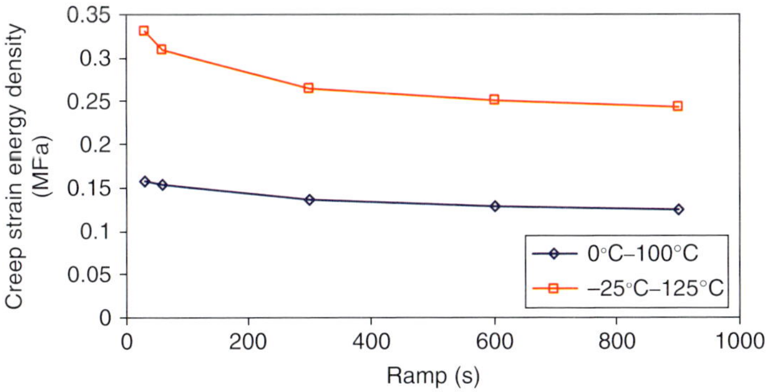

FIGURE 14-23 von Mises creep strain energy density per cycle vs. ramp time for two different temperature ranges.

Figure 14.23 shows the von Mises creep strain energy density per cycle (the fifth one), ΔW, near the upper corner of the solder joint most likely to fail the lead-free PBGA assembly under 0°C to 100°C and –25°C to 125°C temperature profiles, each with five different ramp times: 30, 60, 300, 600, and 900 s. The effects of temperature range and ramp time on the ΔW are obvious. After curve fitting, we have

$$\Delta W = 5.8 \times 10^{-5} (\Delta T)^{1.71} R^{-0.08} \tag{14.1}$$

where ΔW (von Mises creep strain energy density per cycle) is in terms of MPa and R (ramp time) is in terms of minutes. This equation could be useful in predicting the thermal-fatigue life of SnAuCu lead-free solder joints of PBGA packages. Also, this equation could be useful in preparing the thermal shock/thermal cycling tests.

14.3.7 Summary and Recommendations

Some important results and recommendations are summarized in the following:

1. For all cases, the maximum stress and strain occur near the upper corner of the most-likely-to-fail lead-free solder joint.
2. For all cases, the stress vs. creep strain hysteresis loops become stabilized after the fourth cycle.
3. For all cases, the stress, creep strain, creep strain energy density, and creep strain energy density per cycle are temperature dependent; their values are larger for larger temperature ranges.
4. For all cases, it has been shown that it is dangerous to estimate the thermal-fatigue life of lead-free solder joints based

solely on the shear deformation. The expansion/compression of the solder joints due to bending/twisting of the whole assembly and the elongation/contraction of the solder joints have to be considered.

5. For both temperature ranges considered, the shorter (or faster) the ramp time (rate), the larger the stress range in the lead-free solder joints.
6. For both temperature ranges considered, the shorter (or faster) the ramp time (rate), the smaller the creep strain range in the lead-free solder joints.
7. For both temperature ranges considered, the shorter (or faster) the ramp time (rate), the larger and faster the stress relaxation in the lead-free solder joints.
8. For both temperature ranges considered, the creep strain energy density per cycle of the lead-free solder joints under thermal shock is larger than that under thermal cycling. Thus, the thermal-fatigue life (or number of cycles-to-failure) of the lead-free solder joints under thermal shock is smaller than that under thermal cycling.
9. A simple yet useful equation that relates the creep strain energy density per cycle (ΔW) with the applied temperature range (ΔT) and the ramp time (R) has been recommended. With this equation, thermal shock/thermal cycling tests of PBGA solder joints could be designed.
10. Ramp time of 15 min is recommended at ramp-up and ramp-down during thermal cycling.

14.4 References

1. Fan, X., Rasier, G., and Vasudevan, V. S. "Effects of Dwell Time and Ramp Rate on Lead-Free Solder Joints in FCBGA Packages." In *IEEE Proceedings of Electronic & Components Technology Conference*, Lake Buena Vista, FL, May 31–June 3, 2005, pp. 901–906.
2. Lau, J. H., and Y. Pao, *Solder Joint Reliability of BGA, CSP, Flip Chip, and Fine Pitch SMT Assemblies*, McGraw-Hill, New York, 1997.
3. Lau, J. H., *Low Cost Flip Chip Technologies for DCA, WLCSP, and PBGA Assemblies*, McGraw-Hill, New York, 2000.
4. Lau, J. H., C. P. Wong, N. C. Lee, and R. Lee, *Electronics Manufacturing with Lead-Free, Halogen-Free, and Adhesive Materials*, McGraw-Hill, New York, 2003.
5. Lau, J. H., Lee, C., Premachandran, C., and Yu, A., *Advanced MEMS Packaging*, McGraw-Hill, New York, 2010.
6. Lau, J. H., and W. Dauksher, "Effects of Dwell-Time on the Thermal-Fatigue Life of SnAgCu Lead-Free Solder Joints," *ASME Paper No. IMECE2004-62492*, Anaheim, CA, November 2004.
7. Lau, J. H., and W. Dauksher, "Effects of Ramp-Time on The Thermal-Fatigue Life of SnAgCu Lead-Free Solder Joints," *IEEE Proceedings of Electronic Components Technology Conference*, Orlando, FL, June, 2005, pp. 1292–1298.

8. Quintero, P., Oberc, T., and McCluskey, P. "Reliability Assessment of High Temperature Lead-Free Device Attach Technologies." In *IEEE Proceedings of Electronic & Components Technology Conference*, Lake Buena Vista, FL, May 27–30, 2008, pp. 2131–2138.
9. Xie, D., Geiger, D., and Gektin, V. "Reliability Study of High-end Pb-Free CBGA Solder Joint Under Various Thermal Cycling Test Conditions." In *IEEE Proceedings of Electronic & Components Technology Conference*, San Diego, CA, May 25–29, 2009, pp. 109–116.
10. Yoon, S., Chen, Z., Osterman, M., Han, B., and Dasgupta, A. "Effect of Stress Relaxation on Board Level Reliability of Sn Based Pb-Free Solders." In *IEEE Proceedings of Electronic & Components Technology Conference*, Lake Buena Vista, FL, May 31–June 3, 2005, pp. 1210–1214.
11. Kim, I., and Lee, S. B. "Fatigue Life Evaluation of Lead-Free Solder Under Thermal and Mechanical Loads." In *IEEE Proceedings of Electronic & Components Technology Conference*, Reno, NV, May 29–June 1, 2007, pp. 95–104.
12. Ma, H., Suhling, J. C., Zhang, Y., Lall, P., and Bozack, M. J. "The Influence of Elevated Temperature Aging on Reliability of Lead-Free Solder Joints." In *IEEE Proceedings of Electronic & Components Technology Conference*, Reno, NV, May 29–June 1, 2007, pp. 653–668.

CHAPTER 15

Effects of High Strain Rate (Impact) on SAC Solder Balls/Bumps

15.1 Introduction

Due to the susceptibility of drop (impact) damage on lead-free solder interconnects, the popularity of portable products has led to a greater concern with interconnect performance under high-strain-rate loading conditions. Currently, high-strain-rate behaviors of interconnects have been evaluated using the high-speed shear test on package level or printed circuit board (PCB)-level drop testing [1–21]. While the board testing is time-consuming and laborious, the correlation of component-level high-speed shear testing on plastic ball grid array (PBGA) solder balls and flip-chip solder bumps is still under investigation by different researchers. In this chapter, a novel high-speed shear test that is incorporated with adjustable shear angle fitting for high-strain-rate testing of lead-free solder balls/bumps on a package level at different shear angles is presented.

15.2 Angled High-Speed Ball Shear Test of SnAgCu Solder Balled/Bumped Package

The responses of SnAgCu (SAC) solder balls under high-speed loading are presented in this section. Emphasis is placed on the development of an angled shear tester for the investigation of the ductile, brittle, and ductile + brittle failures of the SnAgCu solder balls/bumps [21].

15.2.1 Apparatus Setup with High-Speed Shear Tester

Figures 15.1 and 15.2 show the schematic of the angled ball shear setup and apparatus on a high-speed tester to investigate the responses of SnAgCu solder balls under high-speed loadings. It can

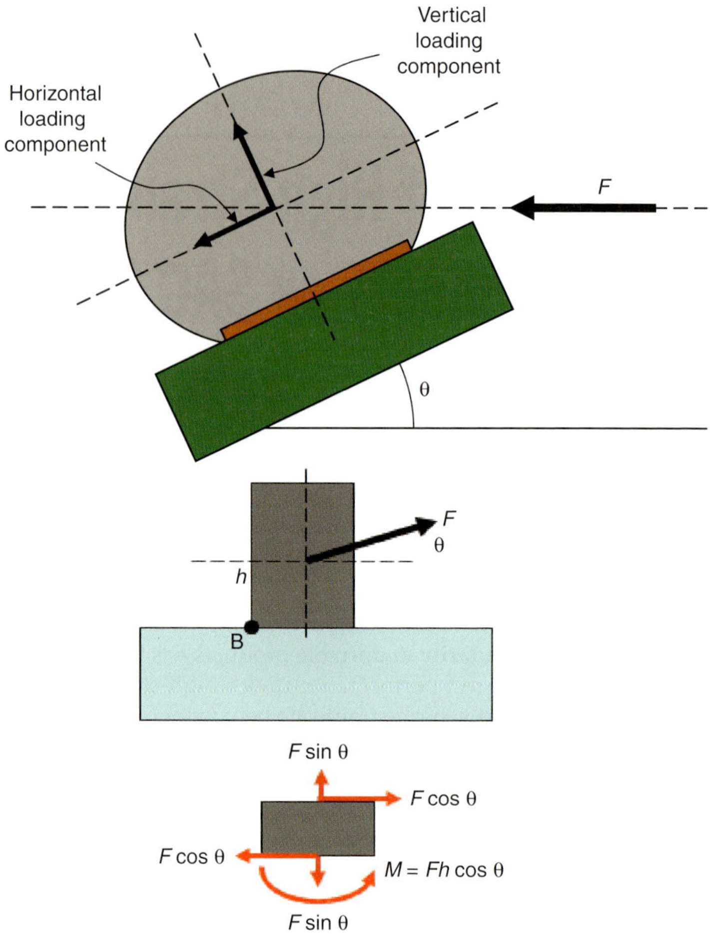

FIGURE 15.1 Schematic of angled ball shear test components.

be seen from Fig. 15.1 that a commercially available basic ball shear tester has a shear angle of attack (θ degrees) with regard to the solder ball orientation on the package. The objective is to introduce simultaneously vertical and horizontal loadings (stresses) to the ball under test in a single ball shear test. Such a mixed mode of loadings is useful for solder ball integrity assessment, as many realistic applications are under mixed mode conditions, as shown in Fig. 15.1.

The stress at point B for shear angle from 0 degrees increase to 10 degrees, for example, is to increase the peeling stress. The increase in peeling stress is due to the axial or normal stress component, while the shear stress component decreases with the shear angle. It will be shown that the improved sensitivity in revealing brittle failure in the angle shear test is believed to be related to the increase in peeling and axial stress component as the shear angle is increased. It is also noted that the peeling stress is strongly linked to the shear height. Its value can drastically change as the shear height changes. To ensure a balance of axial and peeling stress loading, it is necessary to keep the shear height low.

For demonstration and proof of this concept, sample-mount fitting that allows shear test at three angles of attack, for example, 3, 5, and 10 degrees from the horizontal direction, has been designed, as shown in Fig. 15.2. An ordinary shear tester allowing high-speed shear tool movement in a horizontal direction of up to 4000 mm per second will capture the load-displacement response of the shear tool.

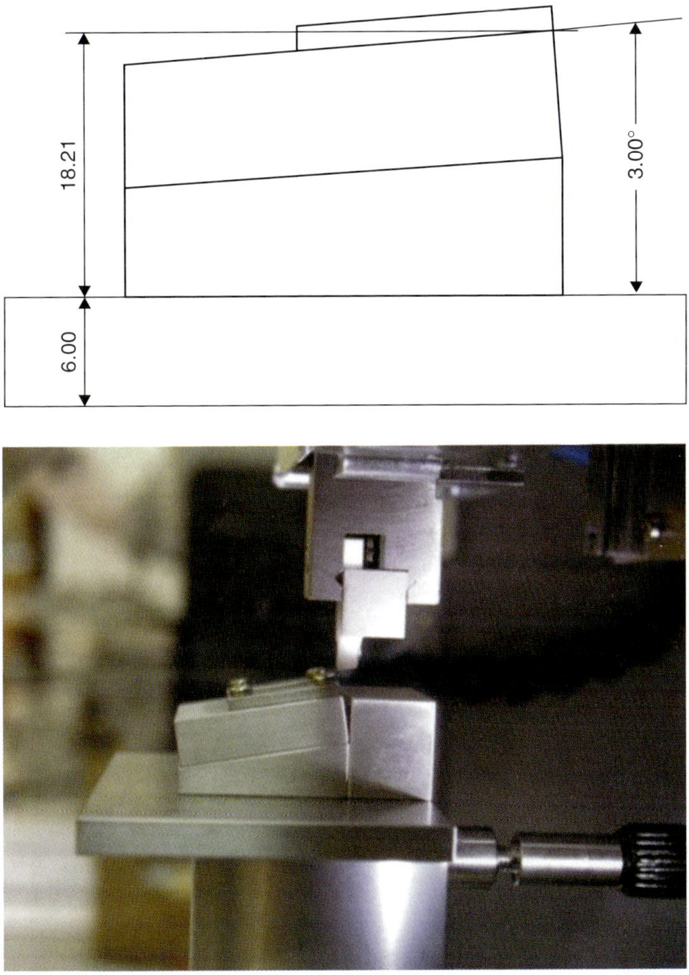

FIGURE 15.2 Test setup: Sample mount schematic (top); Apparatus setup on high-speed shear tester (bottom).

The fitting design is made to ensure that the point of contact between the shear tool and solder ball for all three shear angles will be at the same height position of the solder ball; hence, this ensures a consistent loading point on the solder ball. It should be noted that a firm sample clamping is important to avoid unnecessary secondary effects due to sample bending or vibration.

15.2.2 Test Specimens and Sample Preparation

Using the mentioned setup, high-speed shear characterization is performed on two types of lead-free solders—namely, Sn3Ag0.5Cu and Sn1Ag0.5Cu on the μBGA package with the Ni/Au pad finish, as shown in Fig. 15.3. It can be seen that: (a) the total I/O number is 54; (b) the solder ball size and pitch are at 0.4 mm and 0.8 mm, respectively;

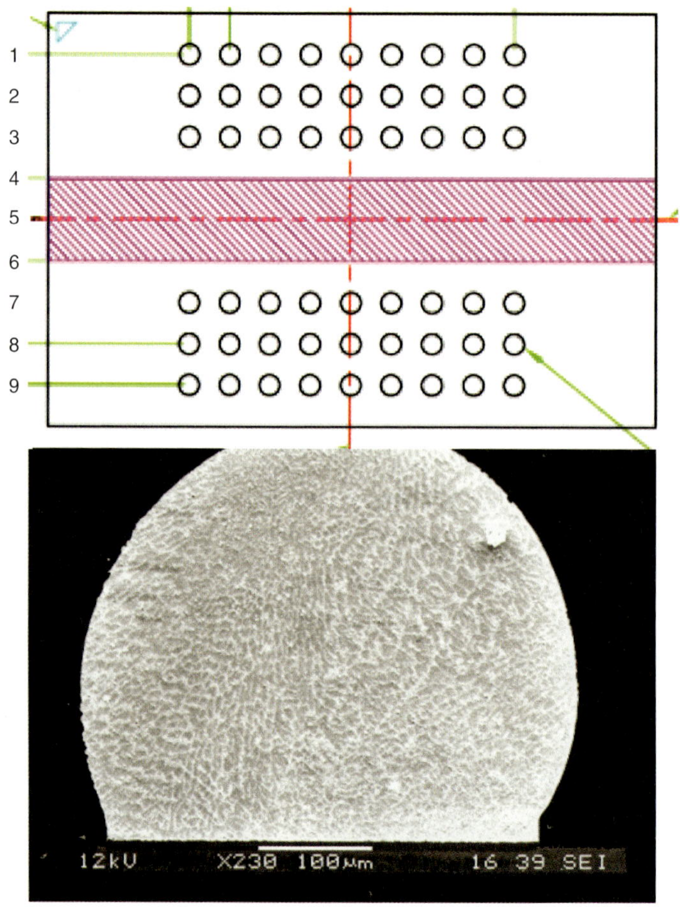

FIGURE 15.3 Samples for ball shear test: (top) μBGA package; (bottom) Solder ball cross-sectional profile.

(c) the solder ball consists of Sn3Ag0.5Cu and Sn1Ag0.5Cu; and (d) the pad material is Cu with solder-mask-defined (SMD) configuration and coated with electrolytic NiAu finishing.

Prior to testing, some sample preparation is required to avoid shearing multiple balls by removing the extra solders around the target one. The sample is clamped on the stainless steel frame with the solder ball facing up. The sample holder fitting includes a datum block for shear tool touchdown reference and ensures sufficient clearance is maintained to avoid collision between the shear tool and the specimen.

During shear tests, the load-displacement responses of the shear tool are captured for different shear speeds and shear angles. After the testing, the failed samples are subjected to microscope and scanning electron microscopy (SEM) inspection for failure mode identification. To determine a meaningful shear speed range for the experiments, some initial shear tests have been performed to ensure the ductile-brittle transitions are covered. In this case, tool speeds ranging from 50 mm/s to 2500 mm/s have been selected.

15.2.3 Test Results

Figure 15.4 shows the shear force (g) versus displacement (μm) curves from experiments at various shear speeds for the Sn3Ag0.5Cu samples. The curves plotted in Fig. 15.4 (from top to bottom) are, respectively, for 0, 3, 5, and 10 degrees. It can be seen that: (a) as the shearing speed increases, the displacement decreases steadily, while the maximum shear force increases (this is the strain hardening effect of solder material); (b) for low shear speeds below 100 mm/s, the changes of displacement are insignificant; (c) as the shear speed increases, there is an increase in the ramp rate of the shear force, with drastic changes in displacement (the sharp decrease of shear displacement is an indication that the solder ball is going from ductile failure to brittle failure); and (d) in between the two failure modes, where some irregular displacements have occurred, that is, the transition of ductile shear to brittle shear is considered to have happened.

Based on these observations of the load-displacement curves, the shear speeds for which different failure modes have occurred are summarized in Table 15.1, including the shear speed in which the ductile-brittle transition is considered to have occurred. This result provides a clear picture of the changes taking place for different shear angles. Generally, brittle failures tend to initiate at lower shear speeds and as shear angle increases. This includes the down-shift of the ductile-brittle transition to lower shear speeds. Despite a small increase in shear angle, the result indicates an improvement in the sensitivity of the angled shear test in revealing brittle failure than that of the horizontal shear test.

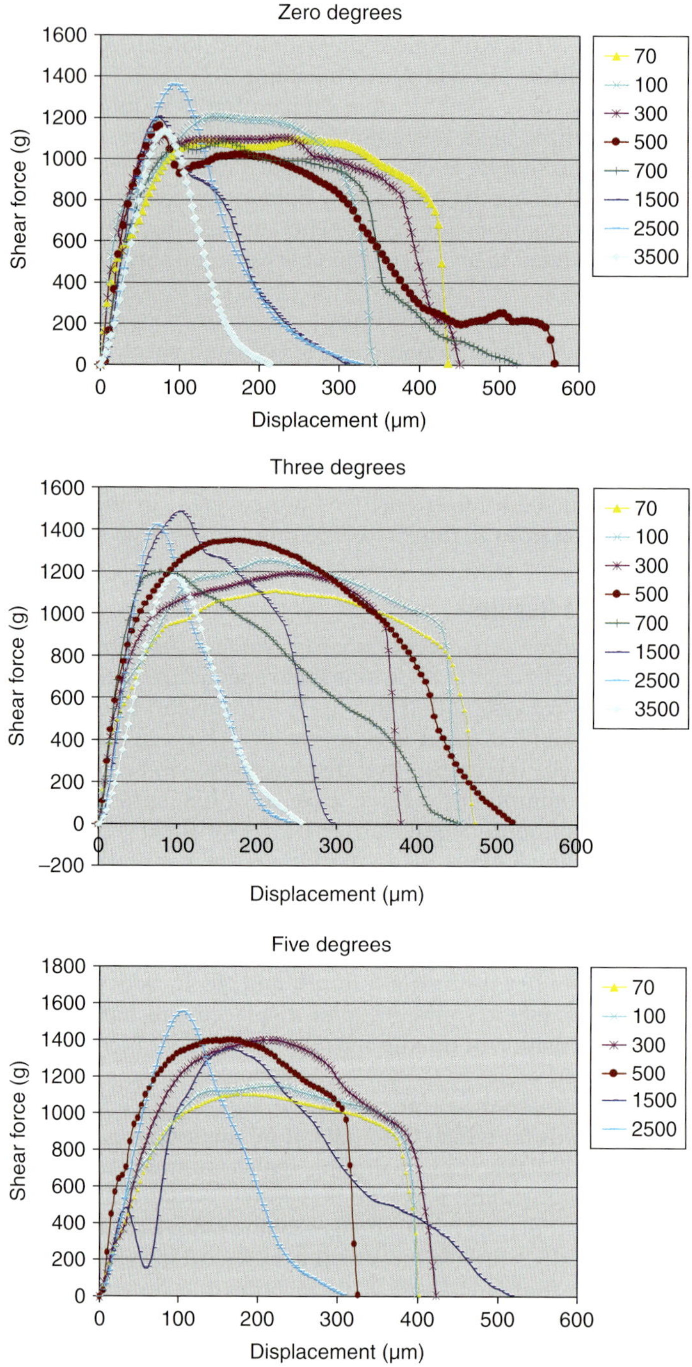

FIGURE 15.4 Typical force-displacement curves under variable shear speeds (mm/s) using different shear angles for Sn3Ag0.5Cu solder balls.

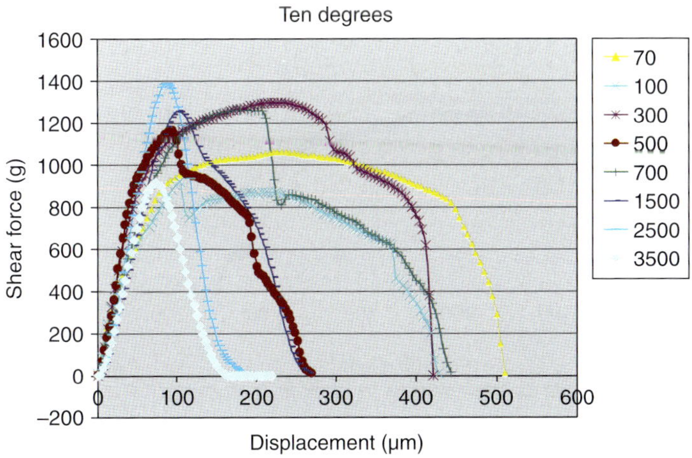

FIGURE 15.4 (*Continued*)

Angle	Shear Speed (mm/s)		
	Ductile Failure	**Transition**	**Brittle Failure**
Zero degrees	10, 50, 70, 100, 300, 500, 700	1500	2500, 3500
Three degrees	10, 50, 70, 100, 300, 500	700, 1500	2500, 3500
Five degrees	10, 50, 70, 100, 300	500, 700	1500, 2500, 3500
Ten degrees	10, 50, 70	100, 300	700, 1500, 2500, 3500

TABLE 15.1 Ductile-Brittle Transition as a Function of Shear Angle and Shear Speed

Figure 15.5 shows a direct comparison of load-displacement curves extracted from different shear angles and shear speeds around the ductile-brittle failure transition regions. The effect of shear angle on the load-displacement response described earlier can be clearly observed here.

15.2.4 Characterizations of Failure Samples

The failed samples are inspected under optical microscope and SEM for the failure mode after shear testing. For the sake of convenience, the failure modes are classified into five groups: (a) bulk failure, (b) bulk-intermetallic compound (IMC) partial failure, (c) IMC failure,

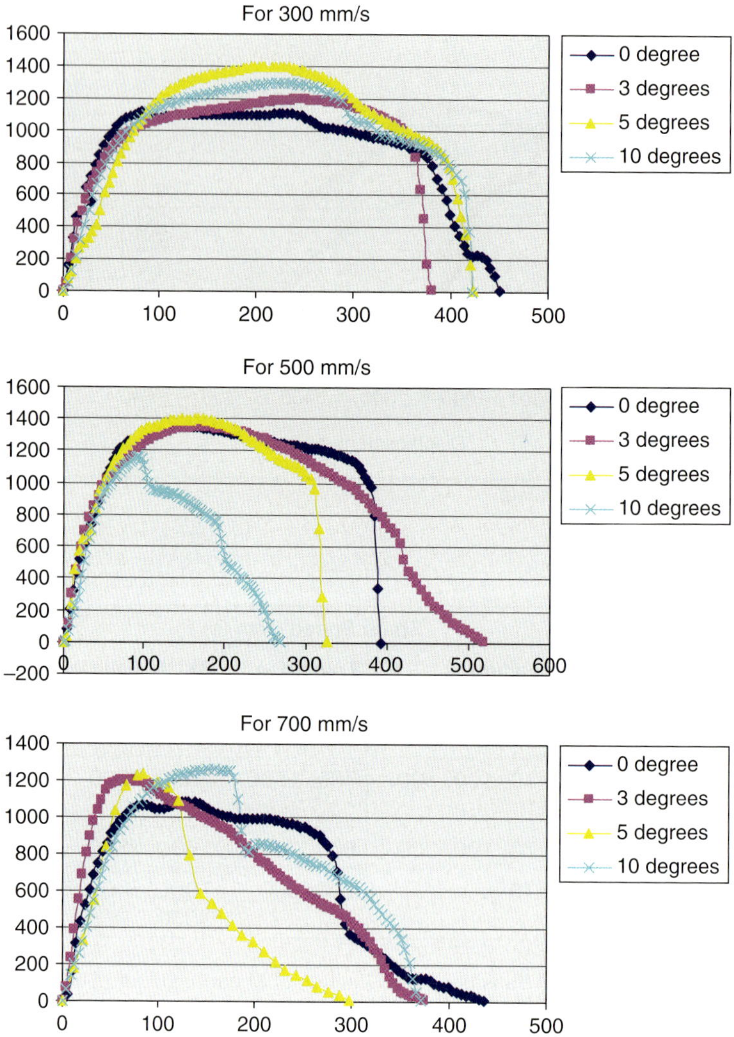

FIGURE 15.5 Comparison of loading force (g)-displacement (μm) curve as function of shear angle (degree).

(d) IMC + partial pad peel failure, and (e) pad peeling. Typical images of these failure modes are shown in Fig. 15.6.

Figure 15.7 shows the compilation of failure modes as a function of shear speed and shear angle for the Sn3Ag0.5Cu solder on Ni/Au finish. The results match well with the load-displacement curves, and once again the ductile-brittle transition is showing a down-shift in shear speed as shear angle increases. The result from a higher shear angle clearly shows improved sensitivity in revealing IMC brittle

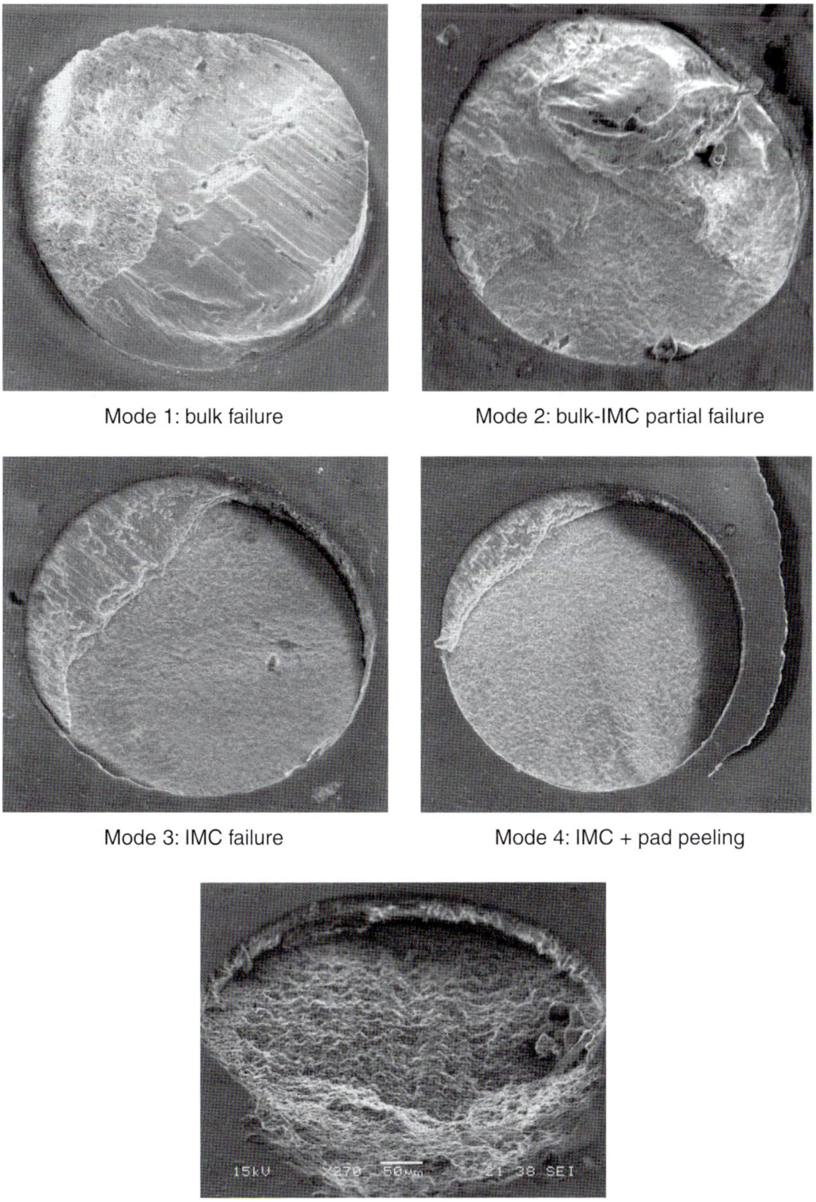

Mode 1: bulk failure

Mode 2: bulk-IMC partial failure

Mode 3: IMC failure

Mode 4: IMC + pad peeling

Mode 5: pad peeling

Figure 15.6 Failure modes observed from shear test, Mode 1: Bulk failure; Mode 2: Bulk-IMC partial failure; Mode 3: IMC failure; Mode 4: IMC + pad peeling; Mode 5: Pad peeling.

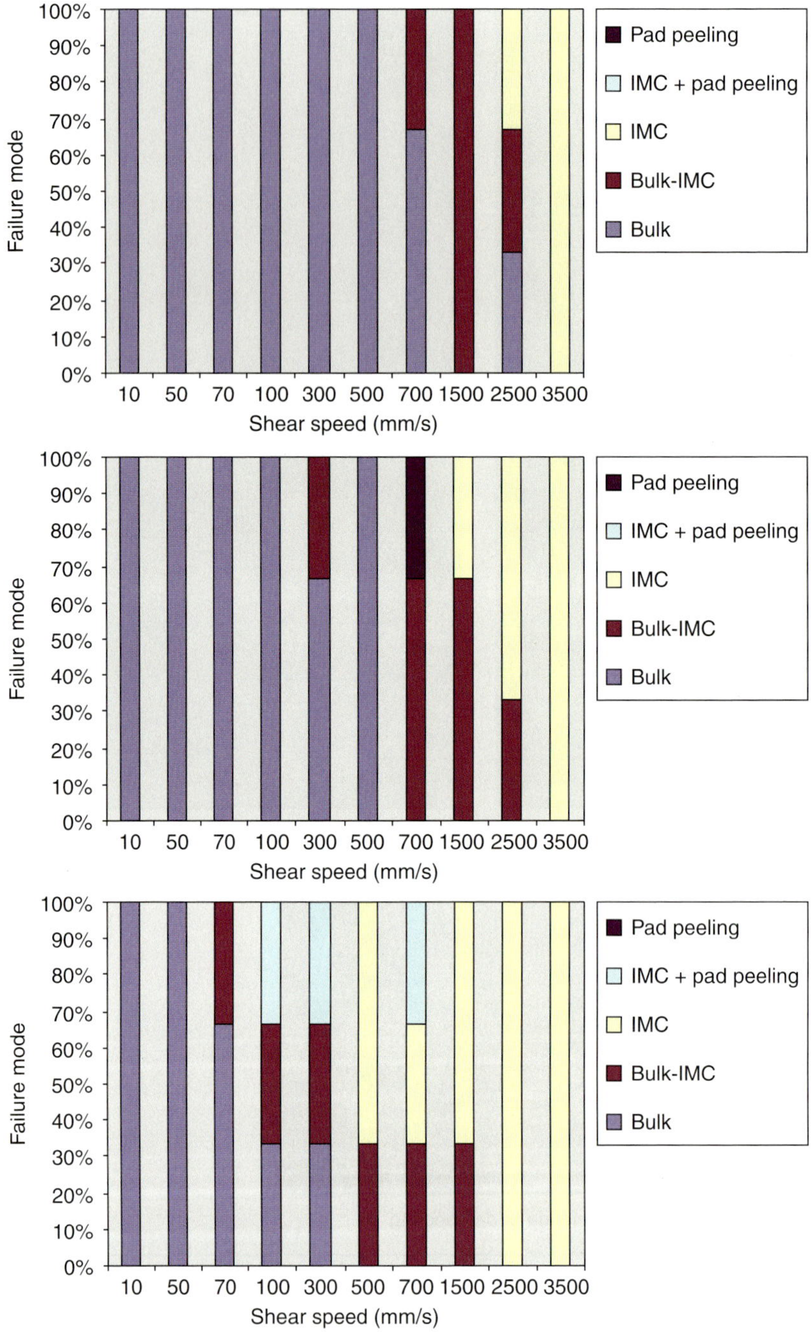

FIGURE 15.7 Compilation of failure modes from angled ball shear test, three degrees (top); five degrees (middle); and ten degrees (bottom).

failure. However, at a higher shear angle, for example, 10 degrees, the occurrence of IMC + partial pad peeling increases, compared to the lower shear angle samples. Since the pad peeling is an indication of weaker pad-to-core substrate interface integrity, the increase in angle shear reveals such weakness in the sample, subjected to the shear speed. At the same time, the pad peeling is competing with the bulk-IMC mixed failure; the result depends on which one is weaker. As the strain rate further increases to the higher shear speed region, the brittle IMC failure becomes more apparent.

15.2.5 Comparison of SAC305 and SAC105

When compared to Sn3Ag0.5Cu, the similar angled high-speed shear test is also performed on the Sn1Ag0.5Cu-soldered μBGA package and pad finish. Figure 15.8 shows the compilation of failure modes for the two solder types. The results show that SAC305 has a lower resistance to brittle IMC failures compared to SAC105. This result is consistent for all three angled shear tests. It is interesting to note that a large shear angle, for example, 10 degrees, shows better sensitivity in revealing the transition to brittle failure for the SAC305 as compared to SAC105. This could be because the higher angled shear tests bring out more subtle aspects of the solder ball integrity compared to the normal horizontal shear test.

15.3 Finite-Element Analyses of Angled High-Speed Ball Shear of the SnAgCu Solder-Balled Package

A further verification is also performed by using a three-dimensional (3-D), finite-element modeling of the ball shear test to investigate the effect of different shearing angles. This approach can provide more detailed information on the stress distribution in the solder ball and ball/Cu pad interface during the test.

15.3.1 Finite-Element Modeling

Figure 15.9 shows the finite-element model with a five-degree angle of attack. The solder and Cu pad are considered elastoplastic materials, while the solder mask and substrate of the package are elastic. (Creep is not considered because it is secondary in this case.) ANSYS/Mechanical and ANSYS/LS-Dyna are used for the static and dynamic simulation, respectively. Only static analysis results are presented here.

15.3.2 Finite-Element Analysis and Results

When the shear tool moves at a constant speed over a solder ball, a reaction force is activated from the solder ball when both come into

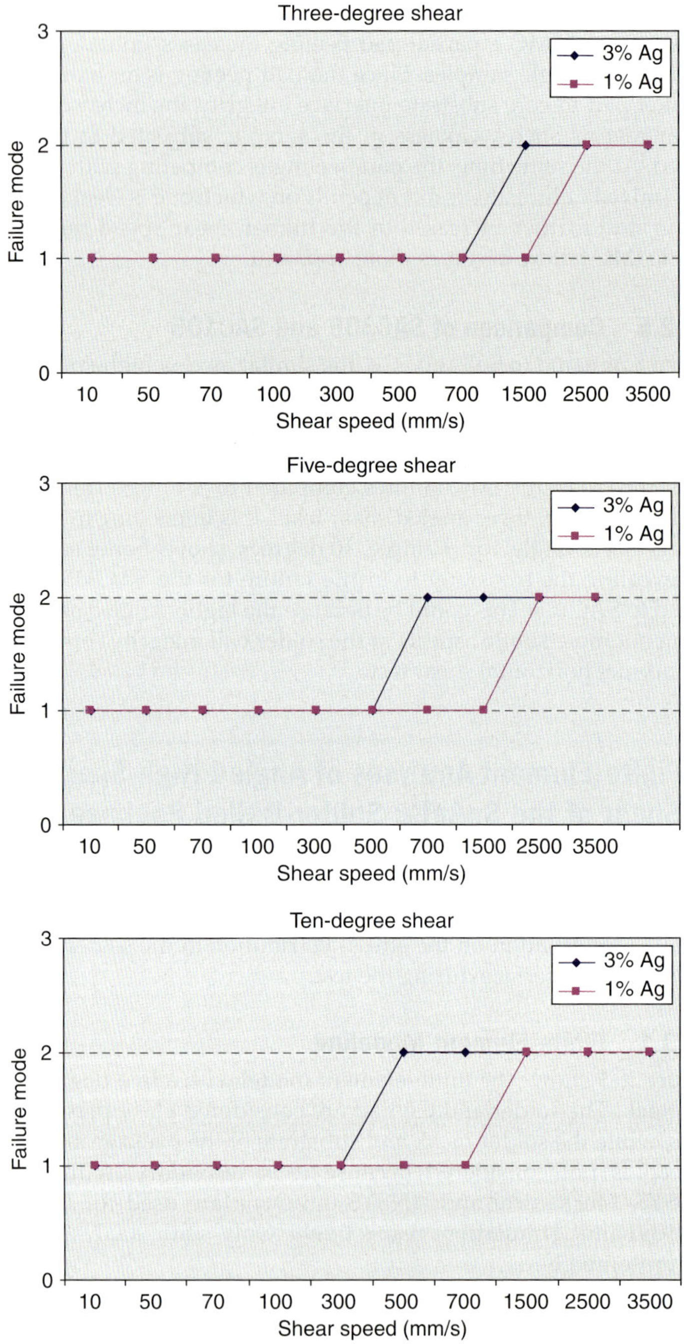

Figure 15.8 Comparison of ductile-to-brittle transition under variable shear angles for SAC305 and SAC105 samples, three degrees (top); five degrees; and ten degrees.

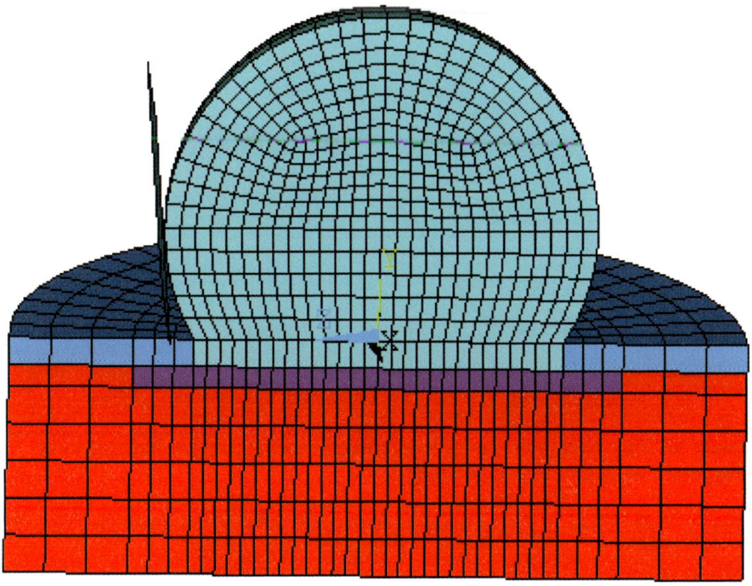

FIGURE 15.9 FEM (Finite element method) model for ball shear (angle 5°).

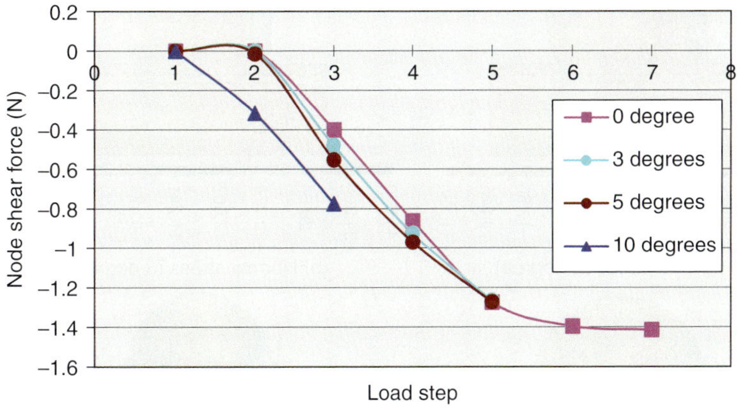

FIGURE 15.10 Node shear force in static shearing for zero degrees, three degrees, five degrees, and ten degrees.

contact. Figures 15.10 and 15.11 show the shearing and normal forces of the central contact node on the shearing ram. The normal force from the shearing ram increases as the shear angle increases, while the effect of shear angle on the change of the shearing force is insignificant.

Figures 15.12 and 15.13 shows the comparison of normal stress and shearing stress on the middle cross-section plane of the solder

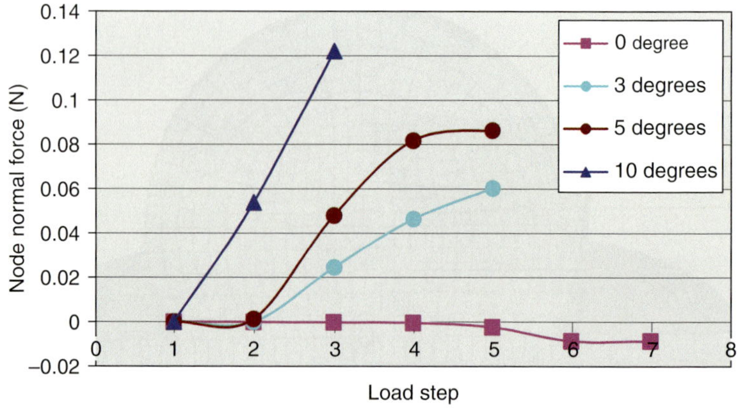

FIGURE 15.11 Node normal force in static shearing for zero degree, three degrees, five degrees, and ten degrees.

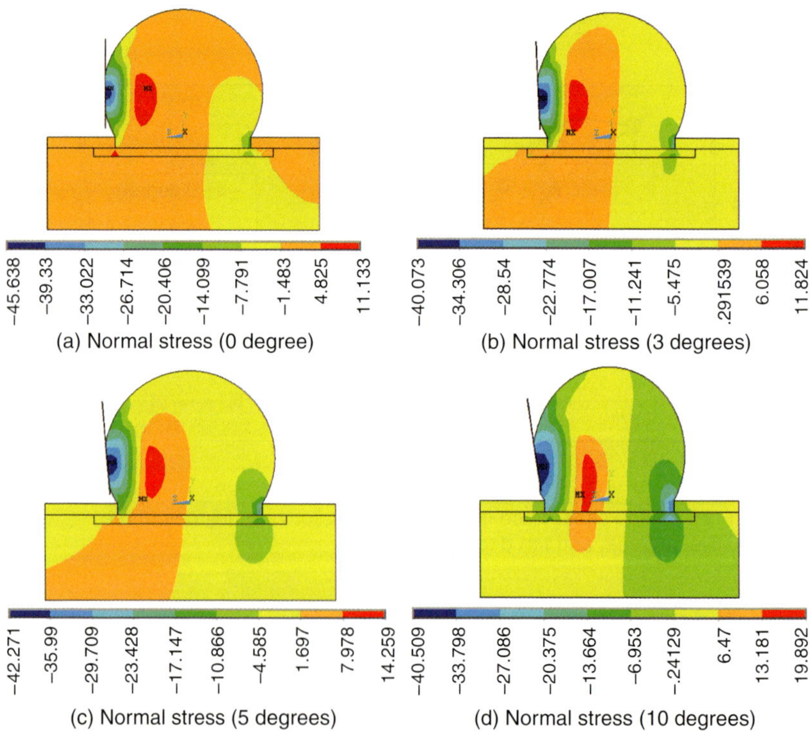

FIGURE 15.12 Normal stress contours in cross-sectional views for zero degree, three degrees, five degrees, and ten degrees.

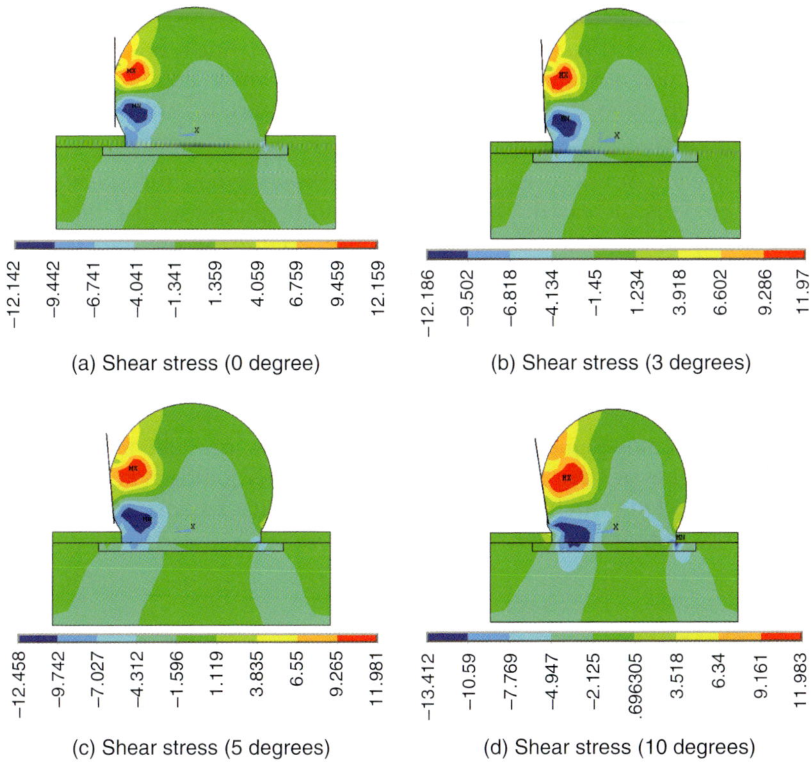

FIGURE 15-13 Shear stress contours in cross-sectional views for zero degree, three degrees, five degrees, and ten degrees.

ball for different shear angles. This provides a convenient two-dimensional (2-D) view of the stress distribution in the solder, although it is not necessarily a true reflection of the original 3-D stress distribution. Nevertheless, it sufficiently illustrates the normal and shear stress components as the shear angle changes.

15.4 Summary and Recommendations

Some important results and recommendation are summarized in the following:

1) Ball/bump shear test at a higher shear angle improves the sensitivity of revealing brittle IMC failure of solder balls.
2) The improvement of the angled shear test has been shown in both load-displacement responses of solder balls/bumps as well the failure modes of the samples.

3) The ductile-brittle transition of Sn3Ag0.5Cu and Sn1Ag0.5Cu solder on μBGA package with Ni/Au pad finish has been obtained for different shear speeds and shear angles. The results show that SAC305 has lower resistance to brittle IMC failures compared to SAC105.

4) Peeling stress in the solder ball increases when the shear angle is increased. This is believed to be the reason for the improved sensitivity in revealing brittle failure in the angled shear test.

5) An angled shear test that provides simultaneous shear and peeling stress to the solder ball/bump has been developed and is recommended for use in solder ball/bump integrity and reliability assessment purposes.

6) The finite-element modeling and analysis provide useful insights of stresses and strains in the solder joint when it is subjected to high-speed shearing test; thus, it is recommended not just for this case, but for every project.

15.5 References

1. Meier, K., Wolter, K. J., Wiese, S., and Roellig, M. "Mechanical Characterization of Lead-Free Solder Alloys under High Strain Rate Loads." In *IEEE Proceedings of Electronic & Components Technology Conference*, San Diego, CA, May 25–29, 2009, pp. 1781–1787.
2. Peterson, D., Chen, C. F., and Karulkar, P. C. "Characterization of Drop Impact Survivability of a 3D CSP Stack Module." In *IEEE Proceedings of Electronic Components & Technology Conference*, Lake Buena Vista, FL, May 27–30, 2008, pp. 1648–1653.
3. Amagai, M., Toyoda, Y., Ohnishi, T., and Akita, S., "High Drop Test Reliability: Lead-free Solders." In *IEEE Proceedings of Electronic & Components Technology Conference*, Las Vegas, June 1–4, 2004, pp. 1304–1309.
4. Kaulfersch, E., S. Rzepka, V. Ganeshan, A. Muller and B. Michle, "Thermal, Mechanical and Multi-Physics Simulation Experiments in Microelectronics and Micro-Systems," *EuroSime 2007*, April 2007, pp. 1–4.
5. Chong, D. Y. R., Ng, K., Tan, J. Y. N., Low, P. T. H., Pang, J. H. L., Che, F. X., Xiong, B. S., and Xu, L. H. "Drop Impact Reliability Testing for Lead-Free and Leaded Soldered IC Packages." In *IEEE Proceedings of Electronic & Components Technology Conference*, Lake Buena Vista, FL, May 31–June 3, 2005, pp. 622–629.
6. Xie, D., Rooney, D. T., Geiger, D., and Shangguan, D. "A Comparative Study on Drop Test Performance of Fine Pitch BGA Assemblies Using Pb-Free and Tin-Lead Solders." In *IEEE Proceedings of Electronic & Components Technology Conference*, San Diego, CA, May 30–June 2, 2006, pp. 865–874.
7. Song, F. B, Lee, S. W. R., Newman, K., Reynolds, H., Clark, S., and Sykes, Bob., "Comparison of Joint Strength and Fracture Energy of Lead-Free Solder Balls in High Speed Ball Shear/Pull Tests and Their Correlation with Board Level Drop Test." In *IEEE Proceedings of Electronics Packaging Technology Conference*, Singapore, December 10–12, 2007, pp. 450–458.
8. Wong, E., Y. Mai, R Rajoo, K. Tsai, F Liu, S. Seah, C. Yeh, "Micro Impact Characterization of Solder Joint for Drop Impact Application," Electronic Components and Technology Conference (56th ECTC), May 2006, pp. 64–71.

9. Syed, A., Scanlon, J., Kim, T. S., Cha, S. W., and Ryu, C. G. "Effect of Pb-Free Alloy Composition on Drop/Impact Reliability of 0.4, 0.5 & 0.8mm Pitch Chip Scale Packages with NiAu Pad Finish." In *IEEE Proceedings of Electronic & Components Technology Conference*, Reno, NV, May 29–June 1, 2007, pp. 951–956.
10. Kim, D., Suh, D., Millard, T., Kim, H., Kumar, C., Zhu, M., and Xu, Y. "Evaluation of High Compliant Low Ag Solder Alloys on OSP as a Drop Solution for the 2nd Level Pb-Free Interconnection." In *IEEE Proceedings of Electronic & Components Technology Conference*, Reno, NV, May 29–June 1, 2007, pp. 1614–1619.
11. Che, F. X., Zhu, W. H., Sun, W., Sun, A., Wang, C. K., Tan, H. B., and Pang, J. H. L. "Comprehensive Modeling of Stress-Strain Behavior for Lead-Free Solder Joints Under Board-Level Drop Impact Loading Condition." In *IEEE Proceedings of Electronic & Components Technology Conference*, Reno, NV, May 29–June 1, 2007, pp. 528–535.
12. Johnson, M. E., Lu, H. Y., Lawhead, D., Tessier, T., Scott, D., and Curtis, A. "Using High Speed Shear and Cold Ball Pull to Characterize Lead-Free Solder Alloys and Predict Board Level Drop Test Performance." In *IEEE Proceedings of Electronic & Components Technology Conference*, Reno, NV, May 29–June 1, 2007, pp. 536–542.
13. Pandher, R. S., Lewis, B. G., Vangaveti, R., and Singh, B. "Drop Shock Reliability of Lead-Free Alloys—Effect of Micro-Additives." In *IEEE Proceedings of Electronic & Components Technology Conference*, Reno, NV, May 29–June 1, 2007, pp. 669–676.
14. Kim, H., Zhang, M., Kumar, C. M., Suh, D., Liu, P., Kim, D., Xie, M., and Wang, Z. "Improved Drop Reliability Performance with Lead-Free Solders of Low Ag Content and Their Failure Modes." In *IEEE Proceedings of Electronic & Components Technology Conference*, Reno, NV, May 29–June 1, 2007, pp. 962–967.
15. Caers, J. F. J. M., Zhao, X. J., Wong, E. H., Seah, S. K. W., Selvanayagam, C. S., van Driel, W. D., Owens, N., Leoni, M., Tan, L. C., Eu, P. L., Lai, Y. S., and Yeh, C. L. "A Study of Crack Propagation in Pb-Free Solder Joints under Drop Impact." In *IEEE Proceedings of Electronic & Components Technology Conference*, Lake Buena Vista, FL, May 27–30, 2008, pp. 1166–1172.
16. Ramkumar, S. M., and Srihari, K. "Influence of Process Parameters on Component Assembly and Drop Test Performance Using a Novel Anisotropic Conductive Adhesive for Lead-Free Surface Mount Assembly." In *IEEE Proceedings of Electronic & Components Technology Conference*, Lake Buena Vista, FL, May 27–30, 2008, pp. 225–233.
17. Farris, A., Pan, J., Liddicoat, A., Toleno, B. J., Maslyk, D., Shangguan, D., Bath, J., Willie, D., and Geiger, D. A. "Drop Test Reliability of Lead-Free Chip Scale Packages." In *IEEE Proceedings of Electronic & Components Technology Conference*, Lake Buena Vista, FL, May 27–30, 2008, pp. 1173–1180.
18. K. Newman, "BGA Brittle Fracture – Alternative Solder Joint Integrity Test Method," In *IEEE Proceedings of Electronic Components & Technology Conference*, Lake Buena Vista, FL, May 31–June 3, 2005, pp. 1194–1201.
19. Newman, M. W., Muthukumar, S., Schuelein, M., Dambrauskas, T., Dunaway, P., Jordan, J. M., Kulkami, S., Linde, C., Opheim, T., Stingel, R., and Topic, L. A. "Fabrication and Electrical Characterization of 3D Vertical Interconnects." In *IEEE Proceedings of Electronic Components & Technology Conference*, San Diego, CA, May 30–June 2, 2006, pp. 394–398.
20. Zhao, X. J., Caers, J. F. J. M., Wong, E. H., Seah, S. K. W., van Driel, W. D., Owen, N., Lai, Y. S. "Frequency Dependent S-N Curves for Predicting Drop Impact Robustness of Pb-Free Solder Interconnects." In *IEEE Proceedings of Electronic & Components Technology Conference*, San Diego, CA, May 25–29, 2009, pp. 93–99.
21. Chai, T., D. Yu, J. H. Lau, W. Zhu, and X. Zhang, "Angled High Strain Rate Shear Testing for SnAgCu Solder Balls," In *IEEE Proceedings of Electronic & Components Technology Conference*, San Diego, CA, May 25–29, 2009, pp. 623–628.

CHAPTER 16

Effects of Voids on Solder Joint Reliability

16.1 Introduction

Voids in solder joints are unavoidable [1–29]! There are at least two classes of voids in electronics and optoelectronics packaging interconnects. One is due to the different diffusion rates of two different elements (e.g., Cu pad and Sn-rich solder), and usually occurs near the interface of these two elements. The element with a faster diffusion rate will dissolve into the other and leave some very tiny voids behind, which are called the Kirkendall voids. Usually, this kind of void cannot be detected by x-ray machines and is observed by cross-sections and scanning electron microscopy (SEM). The other kind of void is much larger, and usually is due to: (1) solder shrinkage during solidification (e.g., 2% for Sn); (2) entrapped flux during solder reflow (especially with solder pastes); (3) entrapped air during solder joining (not necessarily from solder pastes); and (4) laminate (e.g., printed circuit board [PCB] or organic substrate) outgassing during soldering. This kind of void is easily detected by x-ray machines undestructively, and is called a bulk void. In this chapter, the effects of bulk voids on solder joint reliability are presented. The effects of Kirkendall voids will not be discussed in this chapter, but were mentioned in Chap. 13.

In general, the presence of voids will affect the physical and mechanical properties of solder joints. For example, (1) voids in solder joints could produce spot overheating (like heat sinks), thus lessening the reliability of the solder joints; and (2) these voids could increase stresses and strains and thus reduce the structural integrity of the solder joints. In this chapter, the effects of void size, void location, and void percentage on the thermal-mechanical solder-joint reliability are presented. However, the heat-sinking effect due to voids in solder joints is not considered.

On the other hand, there is a school of thought that, due to the presence of voids, the thermal-fatigue life of solder joints increases. (The voids stop the cracks, and the stress concentrations around the voids are not large enough to initiate and propagate new cracks.) In this chapter, a method (through fracture mechanics) is presented to demonstrate this phenomenon for the solder joints of two different packages: bump chip carrier (BCC++) and wafer level chip scale package (WLCSP).

16.2 Effects of Voids on BCC++ Solder Joint Reliability

In this section, the effects of void size, void location, and void percentage on the solder-joint reliability of the bump chip carrier (BCC++) package on a PCB are presented. In addition, the effects of voids on the crack growth in the BCC++ solder joint are studied by the fracture mechanics method. Emphasis is placed on the demonstration that a crack in the solder joint may be stopped by a void in front of it.

16.2.1 The BCC++ Package and Structure

The BCC package family is designed for low-cost integrated circuits (ICs) with a low pin count (< 50) and as a low-profile (< 0.8 mm) leadless plastic package. The cross-section of a standard BCC package is shown in the top drawing of Fig. 16.1. The major feature of the BCC is the resin bumps at the bottom of the package. These bumps are covered with a thin metal film and serve as interconnects from the package to the PCB. The interconnects between the chip pads and the bump terminals are gold wire bonds as shown in

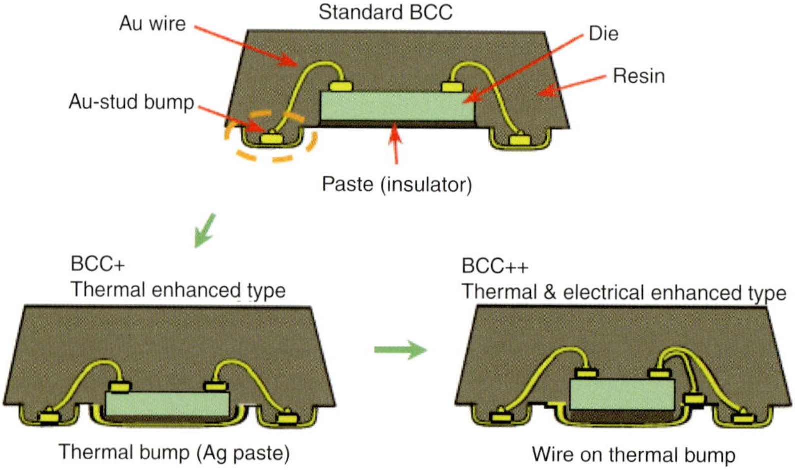

FIGURE 16.1 BCC package families (Shinko).

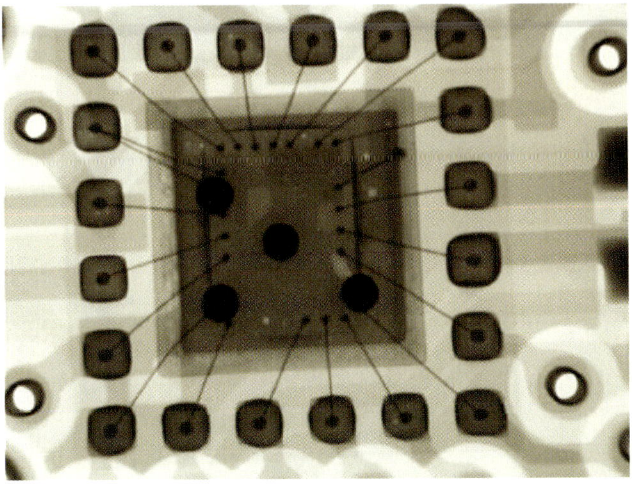

FIGURE 16.2 X-ray photo of the BCC++ PCB assembly.

Fig. 16.2. Like other plastic packages, the BCC is encapsulated by an epoxy molding compound. In order to enhance the thermal and electrical performance of the standard BCC, BCC+ (thermal enhanced type) and BCC++ (thermal and electrical enhanced type) expose the chip paddle, as shown in the bottom drawings of Fig. 16.1. This chip paddle becomes the thermal bump for the BCC+ package, and the thermal bump and ground for the BCC++ package. Furthermore, since the chip paddle can be directly soldered to the PCB, as a result, the solder-joint reliability of BCC+ and BCC++ has been improved dramatically. Figure 16.2 shows the x-ray image of a BCC++ assembly in a fiber-optic small form factor (SFF) transceiver, and Fig. 16.3 shows the typical cross-sections. Figure 16.4 shows the eye diagrams of the SFF transceiver after 1500 h of 85%RH-85%°C (with 3.47 V) test and –40°C to 100°C (with 5-min ramps and 15-min dwells) thermal cycling test.

16.2.2 Finite-Element Analyses

Figure 16.5 shows one-quarter of the PCB pad layout of the 20-pin BCC++ package used in this study. The pad layout for the chip paddle is 1.25 × 1.25 mm (a very large copper pad), and for the bumps it is about 0.43 × 0.43 mm. The corner solder joint of the BCC++ package will be subjected to the largest stresses and strains, since it has the largest distance to neutral point (DNP). Figure 16.5 shows the cross-section for nonlinear stress analysis, and Fig. 16.5 (bottom) shows one of the finite-element models (Case 1, the void-free case) of the 20-pin BCC++ PCB assembly.

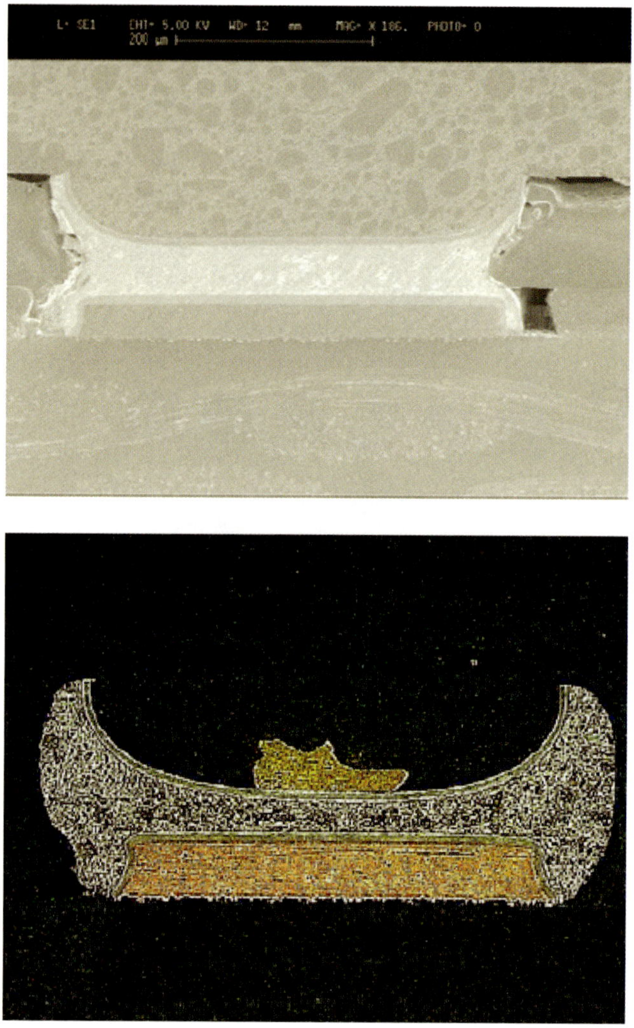

Figure 16.3 Cross-section of the BCC++ solder joint on PCB before (top) and after (bottom) thermal cycling test.

Due to symmetry, only one-half of the structure is modeled, and it is a two-dimensional (2-D) plane-strain analysis. The solder joints with different void sizes, void locations, and void percentages are shown in Fig. 16.6 and Table 16.1. There are 12 different cases in this study, and the percentage of voids ranges from 0 to 17.32%. (All the voids are assumed to be circular in two dimensions.) The physical and mechanical properties of the silicon chip, molding epoxy, Ni thin film, Cu pad, and FR-4 epoxy are tabulated in Table 16.2. The creep material properties of the solder have been given in Chap. 2.

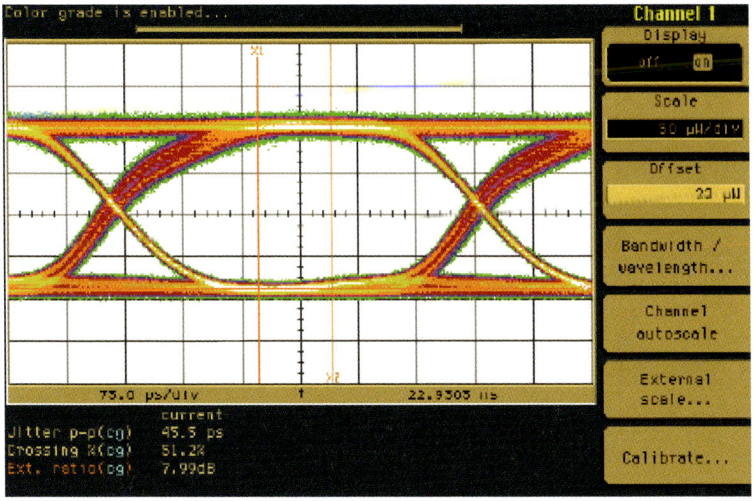

FIGURE 16.4 A typical eye diagram of SFF transceivers (using the BCC++ package) after 1000 temperature cycles from –40°C to 100°C with 15-min dwell and 10-min ramp.

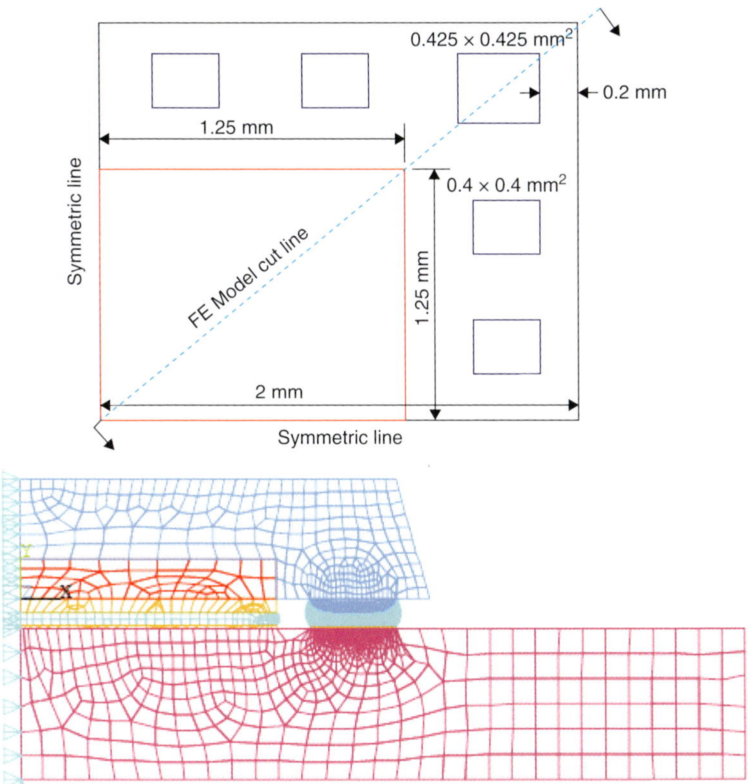

FIGURE 16.5 Footprints on PCB for quarter of the 20-pin BCC++ package (not to scale) and its finite-element model.

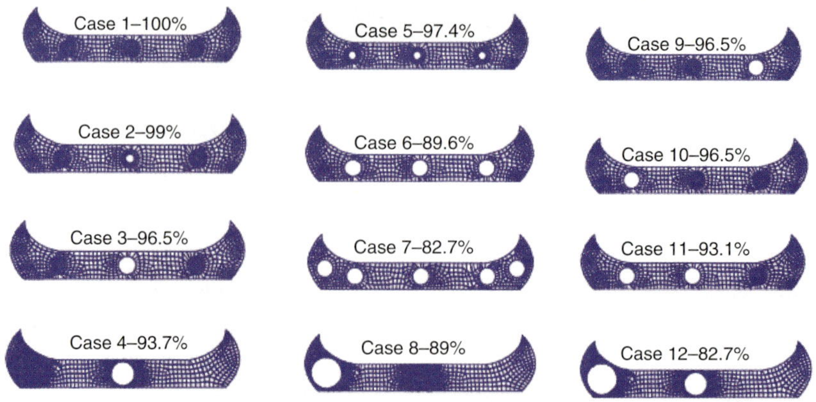

FIGURE 16.6 BCC solder joints with various void sizes, void locations, and void percentages.

Case	Voids (%)	Equivalent Creep Strain Range (%)	Equivalent Stress Range (MPa)
1	0.00	0.11	41
2	0.87	0.11	41
3	3.47	0.15	41
4	6.34	0.23	42
5	2.60	0.12	42
6	10.40	0.21	41
7	17.34	0.37	42
8	10.98	0.19	42
9	3.47	0.12	41.3
10	3.47	0.12	41
11	6.93	0.20	42.5
12	17.32	0.44	44

TABLE 16.1 Equivalent Creep Strain and Stress and Void Percentages for Different Cases Studied

Material	E (MPa)	Poisson Ratio	CTE (ppm/°C)
Solder	32,000	0.35	24.7
Die	131,000	0.3	2.8
Epoxy	28,420	0.38	8
FR4	22,000	0.28	18
Ni	205,000	0.3	12.3
Cu	76,000	0.35	17

TABLE 16.2 Material Properties for the BCC++ Assembly

16.2.3 Analysis Results

The temperature loading imposed on the BCC++ PCB assembly is shown in Fig. 16.7. For each cycle (60 min), the temperature is between –20°C and +110°C, with 15 min ramp, 20 min hold at hot, and 10 min hold at cold. Five full cycles are executed.

For creep analysis, it is important to study the responses for multiple cycles until the hysteresis loops become stabilized. Figure 16.8 shows the shear stress and shear creep strain hysteresis loops for multiple cycles at the center of the corner solder joint. It can be seen that: (1) the creep shear strain is quite stabilized after the third cycle;

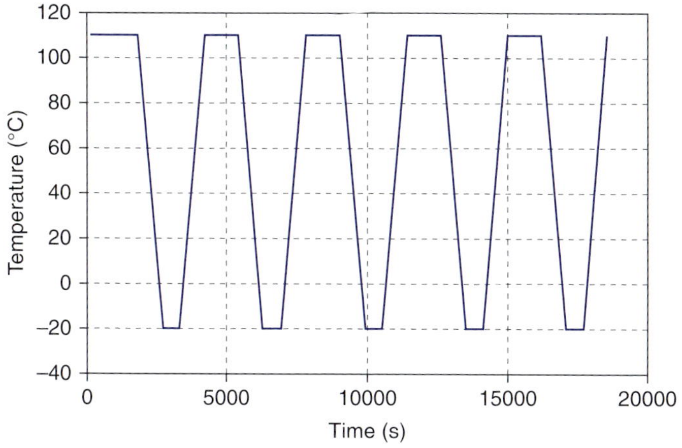

FIGURE 16.7 Temperature loading for finite-element analysis.

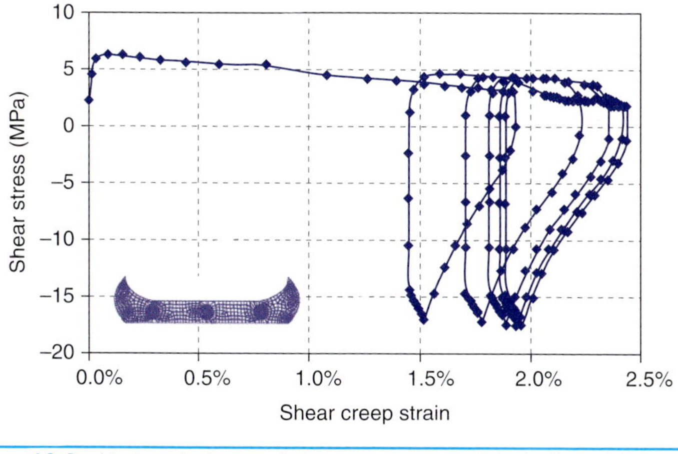

FIGURE 16.8 Hysteresis loops of stress and creep strain in Case 1 (no void) solder joint.

(2) the creep responses converge at the fourth thermal cycle; (3) the stress range (Fig. 16.8) in the corner solder joint is 20 MPa; and (4) the shear creep strain range in the corner solder joint is 0.006 (Fig. 16.8).

Table 16.1 tabulates the simulation results of the 12 different cases under consideration. It can be seen that: (1) in general, the larger the void percentage, the larger the creep strain, and Cases 12 and 7 are the worst cases, as shown in Fig. 16.9; (2) the center void right underneath the bump of the BCC++ plays a very important role (damage), and this can be seen by comparing Cases 4, 8, and 10, as shown in Fig. 16.10; (3) in general, the larger the void size, the larger the creep

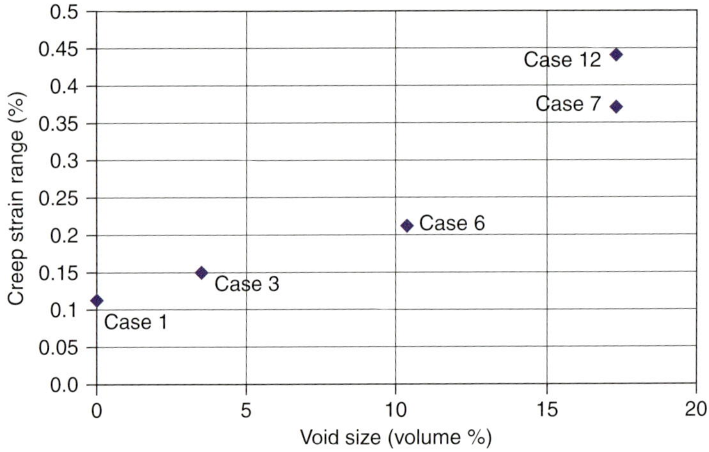

Figure 16.9 Void percentage effect.

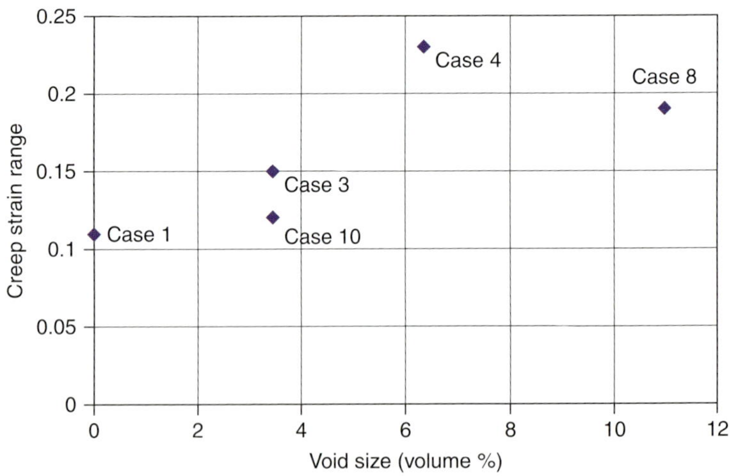

Figure 16.10 Void location effect.

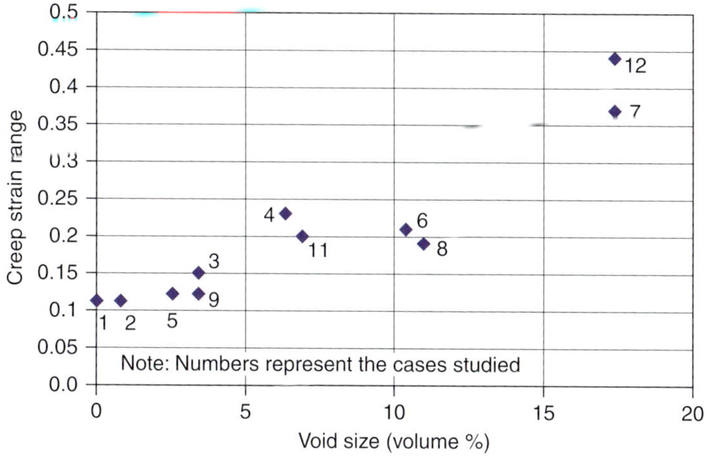

FIGURE 16.11 Void size effect.

strain, as shown in Fig. 16.11; however, sometimes it depends on the void location, and this can be seen from Case 4 and Case 8; and (4) for all cases, the creep strain ranges are too small to pose any solder-joint reliability problem under normal operating conditions. One of the major reasons is because of the huge solder mat connecting the chip paddle of the BCC++ package and the copper pad on the PCB.

16.2.4 Effects of Voids on Crack Growth

The objective of this section is to show that a crack in the BCC++ solder joint could be stopped by a void in front of the crack. In order to show this, the J-integrals around the crack tip, with and without a void, are determined and compared. Since in linear elastic fracture mechanics (LEMF) the path-independent J-integral is equal to the crack driving force, the one with the smaller J-integral should have a longer life.

16.2.5 J-Integral Evaluation around a Crack Tip

The J-integral is defined as a path-independent line integral that measures the strength of the singular stresses and strains near a crack tip. The following equation shows an expression for J in its two-dimensional form. It assumes that the crack lies in the global cartesian X-Y plane, with X parallel to the crack.

$$J = \int_{\Gamma} W\,dy - \int_{\Gamma} \left(t_x \frac{\partial u_x}{\partial x} + t_y \frac{\partial u_y}{\partial y} \right) ds$$

where Γ = any path surrounding the crack tip (Fig. 16.12)
W = strain energy density

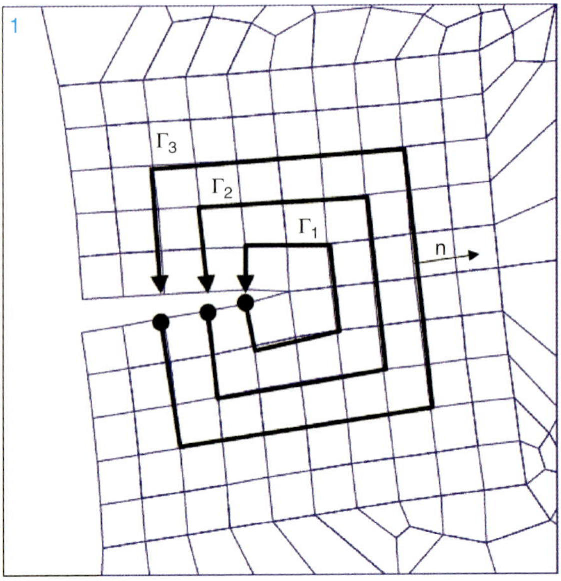

FIGURE 16.12 Finite-element meshes around crack tip and paths for J-integral calculations.

t_x = traction along x axis = $\sigma_x n_x + \sigma_{xy} n_y$
t_y = traction along y axis = $\sigma_y n_y + \sigma_{xy} n_x$
σ = stress component
n = unit outward normal to path
u = displacement
s = distance along the path

To evaluate the J-integral, singular elements around the crack tip are not necessary for accurately determining the stress responses. Regular (2-D plane strain) quad element meshes, which are more convenient to construct for different crack size configurations, can be used near the crack tip. In this section, the solder is assumed to be an elastic material and the cracked structures are assumed to be subjected to a 100°C temperature rise (stress-free at room temperature).

16.2.6 J-Integral for Various Crack Lengths without Voids

Figure 16.13 shows the deformed shape of the BCC++ solder joint with a crack length equal to 0.033 mm (no void). In this case, the J-integral is equal to 7.8×10^{-4} MPa-mm. For other crack lengths, 0.0102 mm, 0.0205 mm, 0.0455 mm, and 0.0588 mm, their J-integrals are shown in Fig. 16.14. It can be seen that for this particular structure and loading condition, the J-integral increases as the crack

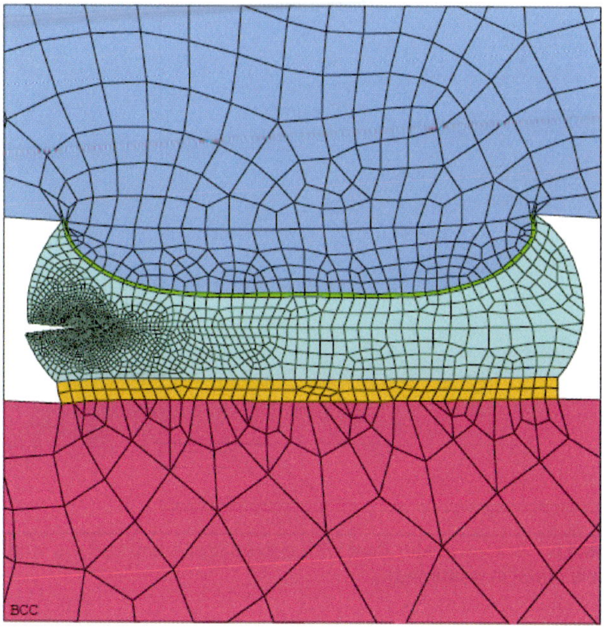

FIGURE 16.13 Deformed shape of a cracked (0.033 mm) BCC++ solder joint (no void).

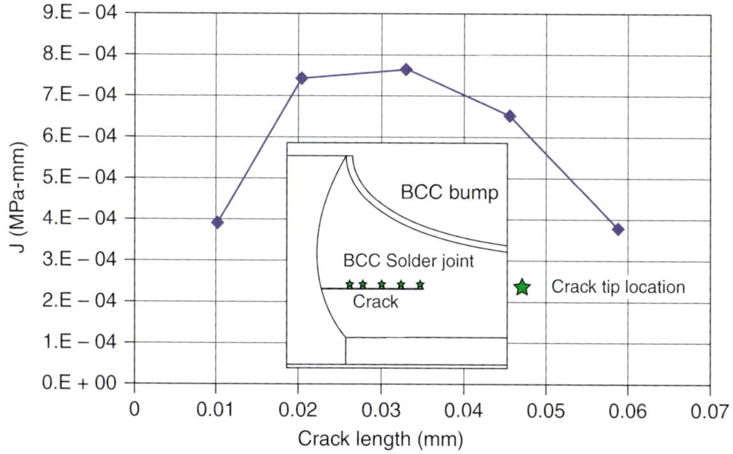

FIGURE 16.14 J-integral vs. crack length in the BCC++ solder joint (no void).

length increases and decreases when the crack length is larger than 0.033 mm. This convex shape of the J-integral indicates that under this fixed loading, the crack will be arrested after propagating to a certain length [18]. In other words, the crack will not grow further unless the loading is increased or there are new sources of loadings.

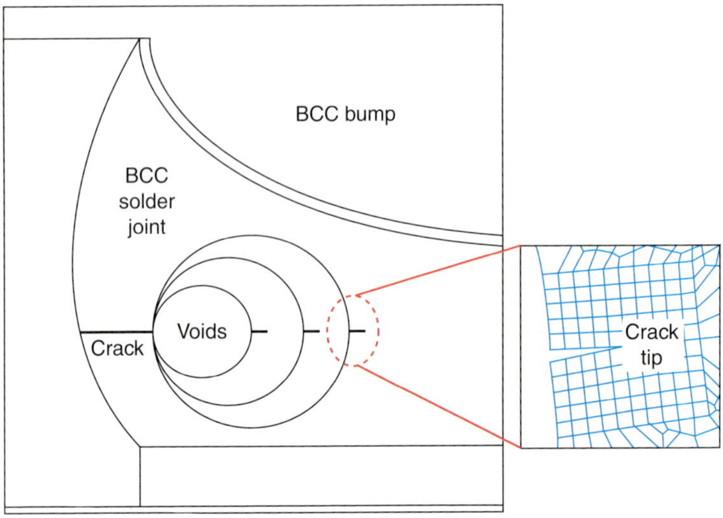

FIGURE 16.15 BCC++ solder joint with a fixed crack and various void sizes (a tiny new crack is assumed).

16.2.7 J-Integral for a Fixed Crack Length with Various Void Sizes

Figure 16.15 shows the BCC++ solder joint with a fixed crack length equal to 0.033 mm and with various void sizes (0.025 mm, 0.05 mm, and 0.075 mm) in front of the crack. Since the stress concentration factor for circular voids is 3 or less, in most of the cases, the stresses acting on the other side (180 degrees away) from the crack are not large enough to initiate a new crack. However, in order to determine the effect of the presence of a void on the cracked solder joint, a new tiny crack (0.0033 mm) is assumed for determining the J-integral around this new crack tip, as shown in Fig. 16.15.

Figure 16.16 shows the J-integrals for the BCC++ solder joints with a fixed crack length (0.033 mm) and various void sizes (0 mm, 0.0205 mm, 0.033 mm, and 0.0455 mm). The driving force (J-integral) around the crack tip with voids (B, C, and D) is a few times less than that without void (A). This means that the solder joint with voids in front of the crack is less likely to fracture than that without voids. Figure 16.17 shows the deformed shape of the BCC++ solder joint with a crack length of 0.033 mm and a 0.0455-mm void in front of the crack.

16.2.8 J-Integral for Various Crack Lengths and Void Sizes

The J-integrals for various pairs of crack lengths and void sizes are also determined. Figure 16.18 shows the J-integrals for crack length and void size (in front of the crack), respectively, equal to 0.0205 mm

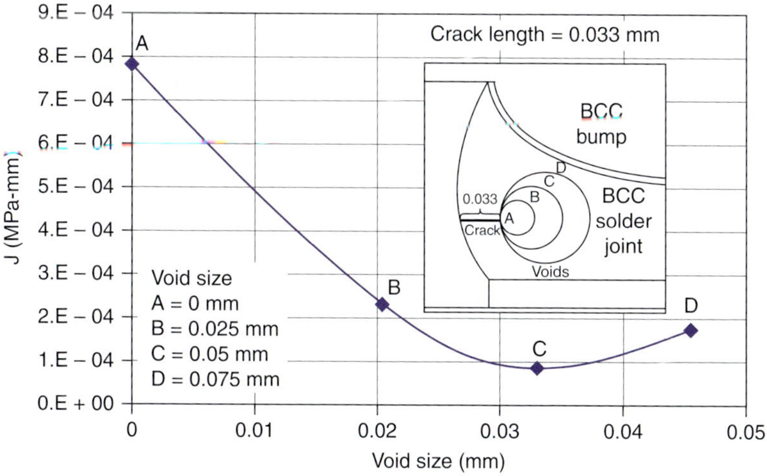

FIGURE 16.16 J-integral vs. void size with a fixed crack length.

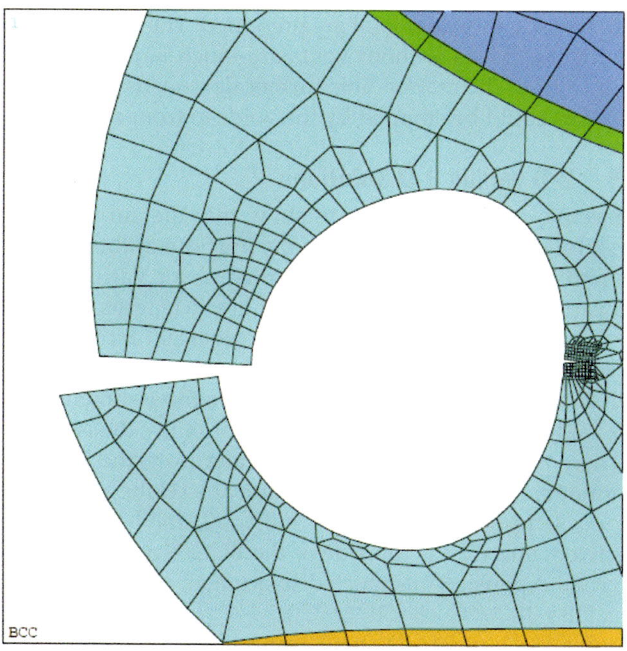

FIGURE 16.17 Deformed shape of a cracked (0.033 mm) solder joint with a void (0.0455 mm) in front of it. There is another tiny crack (0.0033 mm) on the opposite side of the void for J-integral calculation.

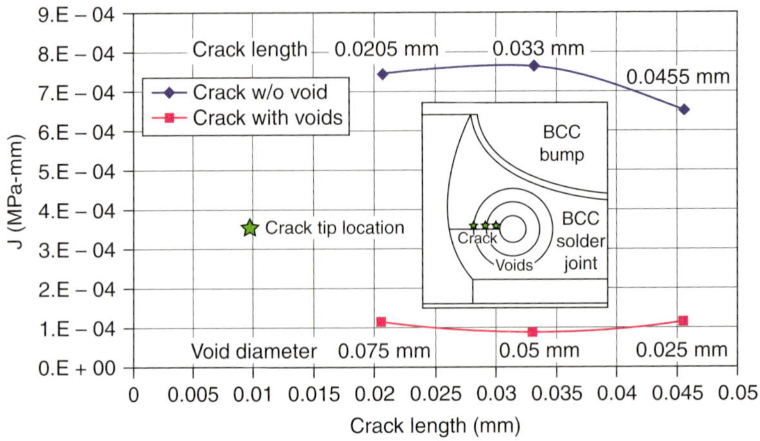

FIGURE 16.18 J-integral as a function of different pairs of crack length and void size.

and 0.075 mm, 0.033 mm and 0.05 mm, and 0.0455 mm and 0.025 mm. Also in Fig. 16.18, the J-integrals for various crack lengths without voids in the BCC++ solder joint are presented. The J-integrals (driving forces) for the cracks without voids are much larger than that with voids in front of the cracks. This further demonstrates that cracks in solder joints could be stopped by the voids in front of them.

16.2.9 2-D versus 3-D Solutions

The objective of this section is to show that the current 2-D plane-strain simulations are adequate for the purpose of this study, that is, to compare the stress, strain, and J-integral for various void sizes, void locations, void percentages, and crack lengths in the BCC++ solder joints and not to determine the thermal-fatigue life of the solder joints.

Figure 16.19 shows one-half of a single three-dimensional (3-D) BCC++ solder joint assembly, and Fig. 16.20 shows the isolated solder joint. There is a void (0.05 mm) right underneath the BCC++ bump. The boundary conditions of the structure are: (1) the assembly is subjected to a temperature rise, ΔT = 130°C; (2) the bottom side of the PCB is fixed; and (3) the top side of the BCC++ package is subjected to a thermal expansion mismatch displacement = 3.029 μm (2.33 mm × 130°C × [18×10^{-6}/°C – 8×10^{-6}/°C]).

Figures 16.21 and 16.22 show, respectively, the deformed shapes from the 3-D and 2-D simulations. They are very similar. Figures 16.23 and 16.24 show, respectively, the shear stress and shear strain of the 3-D and 2-D solutions through the bottom path of the BCC++ solder joint. For both the shear stress and strain, the 2-D solutions compare very well with the 3-D solutions.

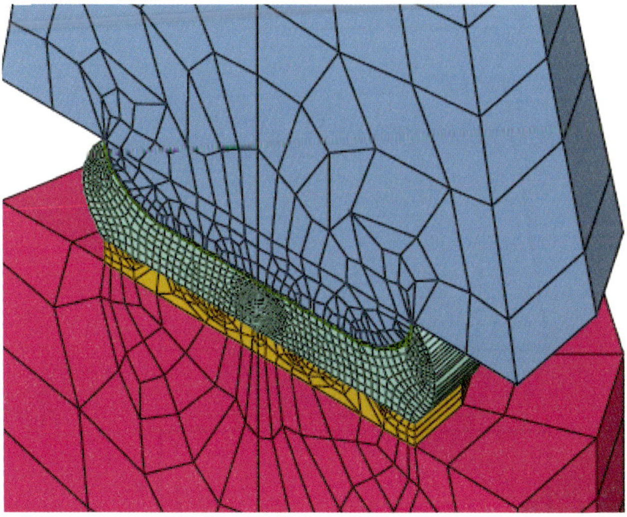

FIGURE 16.19 One-half of a single BCC++ solder joint assembly.

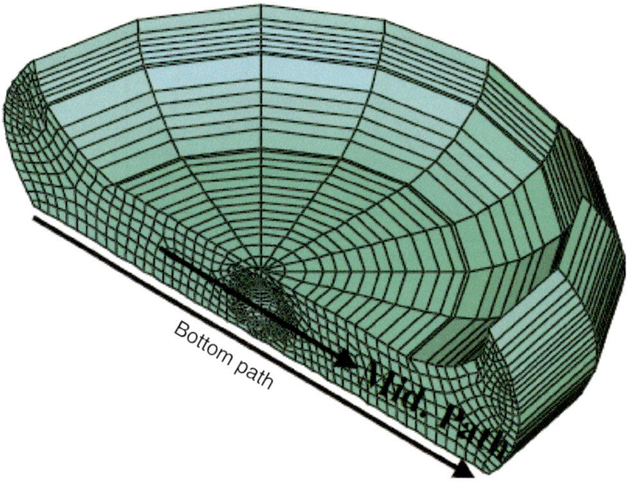

FIGURE 16.20 One-half of the isolated BCC++ solder joint.

16.2.10 Summary and Recommendations

Some important results are summarized in the following:

1. In general, the larger the void percentage of the solder joint, the larger the creep strains.
2. The center void right underneath the bump of the BCC++ package plays the most important role in damaging the solder joint.

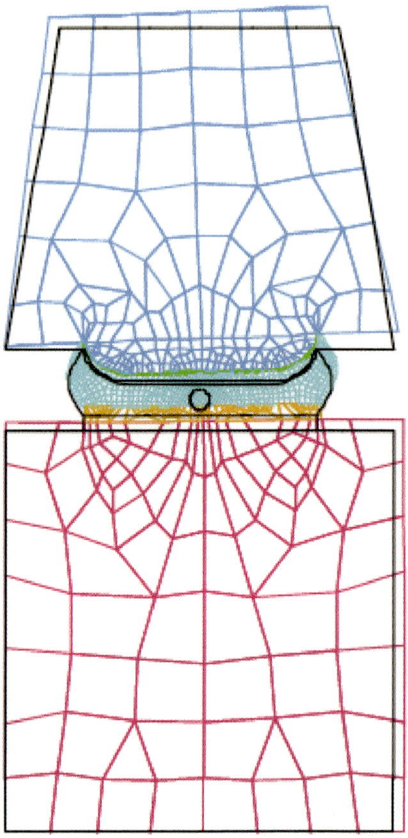

FIGURE 16.21 Deformed shape (3-D analysis).

3. In general, the larger the void sizes, the larger the creep strains. However, sometimes it depends on the void location.
4. For all the cases studied, the creep strains are very small (<1%). Thus, the BCC++ solder joints should be reliable for most of the operating conditions.
5. For inspection purpose, if the void percentage is less than 20% of the BCC++ solder joint, it should be accepted without solder-joint reliability concerns.
6. The J-integral (driving force) for the crack with a void in front of it is a few times smaller than that without a void in the BCC++ solder joint. This means that the crack could be stopped by the void. At least, the crack with a void is less likely to fracture than the crack without a void.

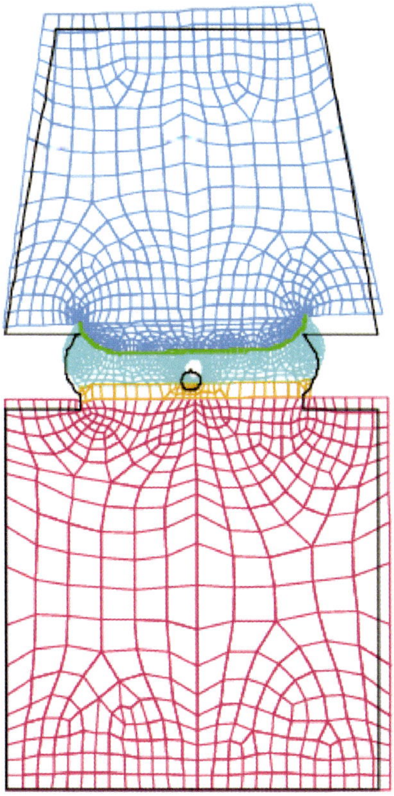

FIGURE 16.22 Deformed shape (2-D analysis).

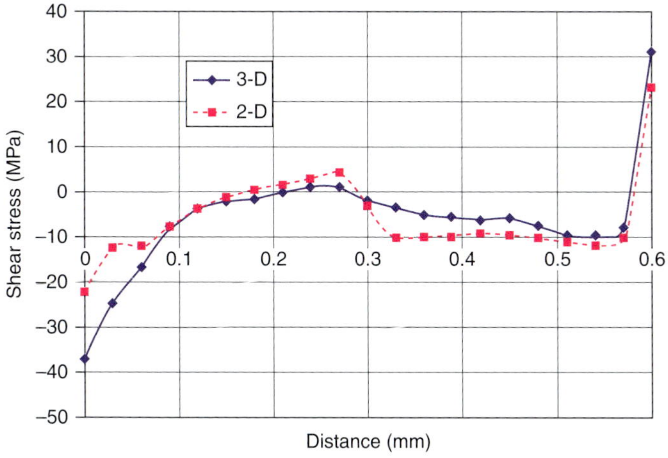

FIGURE 16.23 Shear stress distribution along the bottom path of the solder joint.

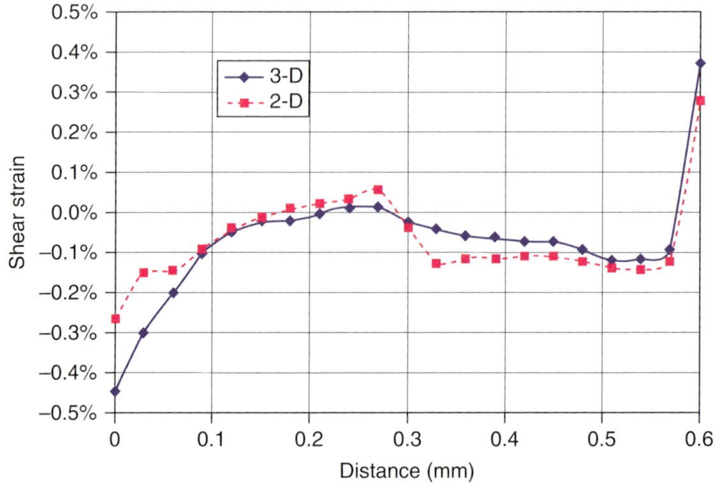

FIGURE 16.24 Shear strain distribution along the bottom path of the solder joint.

16.3 Effects of Voids on WLCSP Solder Joint Reliability

Because of the thermal expansion mismatch between the silicon WLCSP and the epoxy PCB, the solder joint is subjected to thermal stresses and strains, while the assembly is under temperature loadings. These stresses and strains are the driving force to initiate/propagate cracks in the solder joint. Almost all of the solder joints with crack(s) initiated are still functioning. These solder joints stop functioning when the crack is propagated through the cross-section (from a 2-D point of view) or plane (from a 3-D point of view) of the solder joints. (Very often, a crack is initiated from the other side of the flip-chip solder joint and propagated to meet with the other one.) In this section, the effects of crack length on the crack tip characteristics such as the J-integral in the WLCSP solder joint are presented. Also, the effects of voids on the crack growth in the WLCSP solder joint are discussed.

16.3.1 The Structure and Simulations

Figure 16.25 shows a schematic diagram of one-quarter of the silicon WLCSP under consideration. The dimensions of the WLCSP are 5 × 5 × 0.5 mm. It has 16 (0.3 mm in diameter)) peripheral pads with a pitch of 1 mm.

Figure 16.26 shows a typical finite-element model to simulate the solder-bumped WLCSP on the 1.57-mm-thick PCB. All computational analyses are two-dimensional. Due to symmetry, only one-half of the diagonal cross-section is modeled. The solder joint height is about 0.4 mm. The dimension of the copper pad on the PCB is 0.3 mm in diameter and the copper pads are non-solder mask defined. The solder mask on the PCB is 25 μm thick.

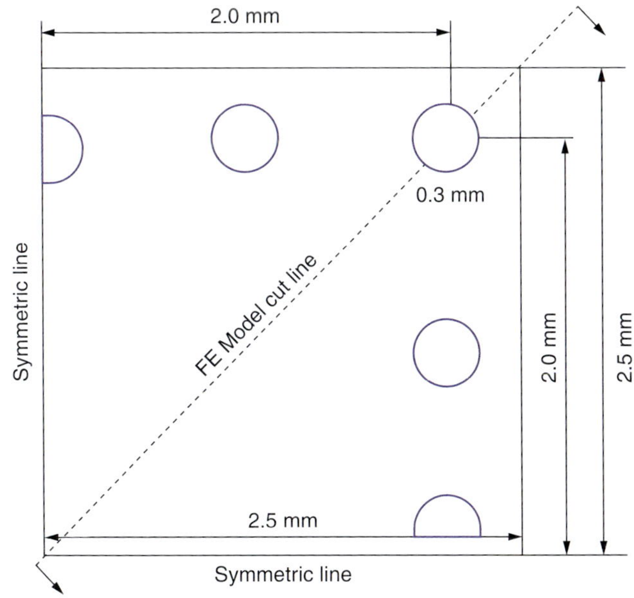

FIGURE 16.25 Schematic diagram of one-quarter of the silicon WLCSP.

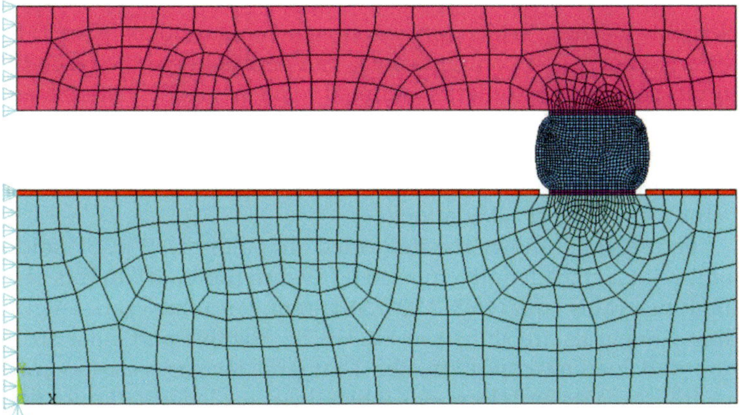

FIGURE 16.26 Typical finite-element model.

16.3.2 Effects of Cracks on the Reliability of WLCSP Solder Joints

In order to determine the effects of the crack length on the WLCSP corner solder-joint reliability, various crack lengths are considered and the J-integrals around the crack tips are calculated by the finite-element method. Since in LEFM the path-independent J-integral is

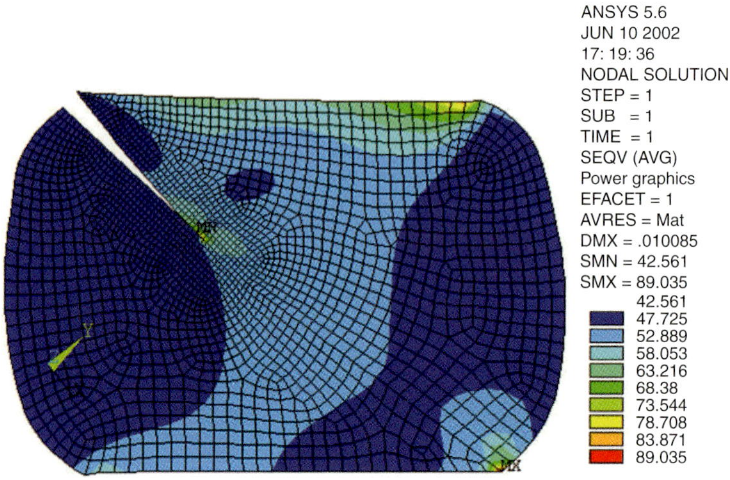

FIGURE 16.27 Deformed WLCSP solder joint (0.189-mm crack length).

equal to the crack driving force, the one with the smaller J-integral should have a longer life. To evaluate the J-integral, singular elements around the crack tip are not necessary for accurately determining the stress responses. Regular (2-D plane strain) quad element meshes, which are more convenient to construct for different crack size configurations, can be used near the crack tip. In this section, the solder is assumed to be an elastic material, and the cracked structures are assumed to be subjected to a 100°C temperature rise (stress-free at room temperature). Figure 16.12 shows the same finite-element mesh around the crack tip and the paths for J-integral calculations.

Figure 16.27 shows the deformed shape of the WLCSP solder joint with a crack length equal to 0.189 mm (no void). In this case, the J-integral is equal to 8×10^{-4} MPa-mm. For other crack lengths, their J-integrals are shown in Fig. 16.28. For this particular structure and loading condition, the J-integral increases as the crack length increases and decreases when the crack length is larger than 0.02 mm. This convex shape of the J-integral indicates that under this fixed loading, the crack will be arrested after propagating to a certain length [1, 3, 4, 15]. In other words, the crack will not grow further unless the loading is increased or other sources of loadings such as shock and vibration are introduced. However, it should be pointed out that cracks (both outer crack and inner crack toward the center of the chip, Figs. 12.46 and 12.47 of [17]), in most flip-chip solder joints under pure thermal cycling loadings, initiate from outer surface and inner surface, propagate toward the inside of the chip, then meet with each other to fail the solder joint.

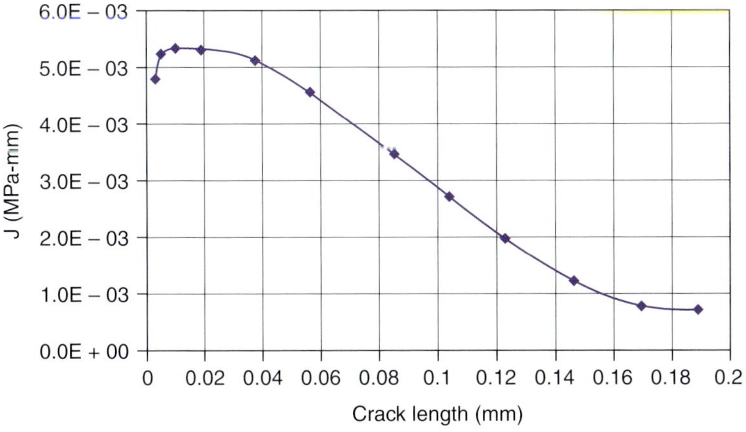

FIGURE 16.28 J-integrals for various crack lengths.

16.3.3 Effects of Voids on the Reliability of WLCSP Solder Joints

Figure 16.29 shows the WLCSP solder joint with a fixed crack length equal to 0.038 mm and with a void (0.05 mm) in front of the crack. Since the stress concentration factor for circular voids is 3 or less, in most of the cases, the stresses acting on the other side (180 degrees away) from the crack are not large enough to initiate a new crack. However, in order to determine the effect of the presence of a void on the cracked solder joint, a new tiny crack (0.0038 mm) is assumed for determining the J-integral around this new crack tip, as shown in Fig. 16.30.

Figure 16.31 shows the J-integrals for the WLCSP solder joints with a fixed crack length (0.038 mm) and various void sizes (0.025 mm, 0.05 mm, and 0.075 mm). The driving force (J-integral) around the crack tip with voids is less than that without voids. This means that the

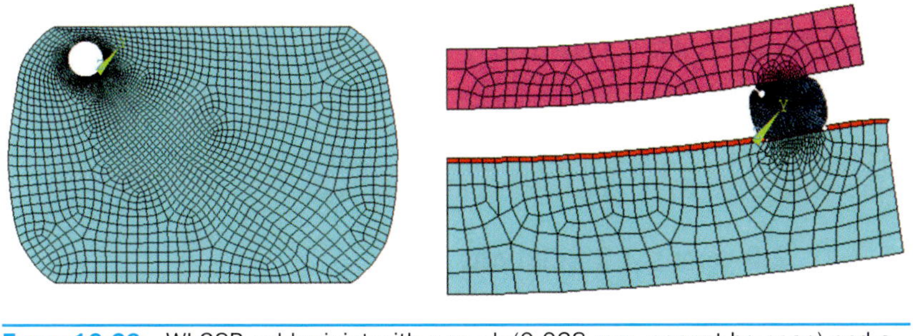

FIGURE 16.29 WLCSP solder joint with a crack (0.038 mm, cannot be seen) and a void (0.05 mm in front of the crack) (left) and with its deformed shape (right).

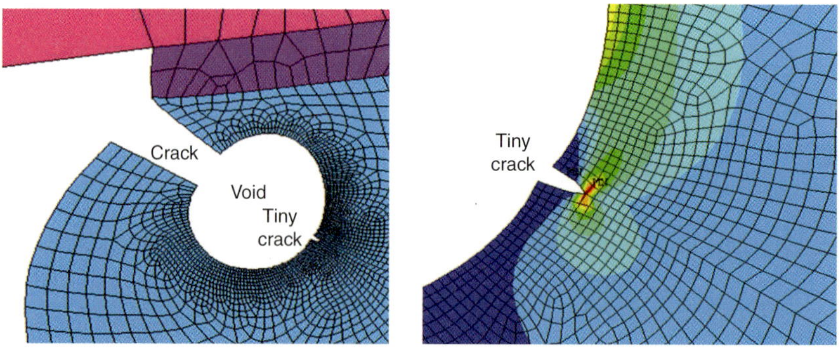

FIGURE 16.30 The deformed shape of WLCSP solder joint with a crack (0.038 mm), a void (0.05 mm) in front of the crack, and a tiny crack (0.0038 mm) on the void right opposite to the larger crack (left). The tiny crack and its deformed shape in a close-up look (right).

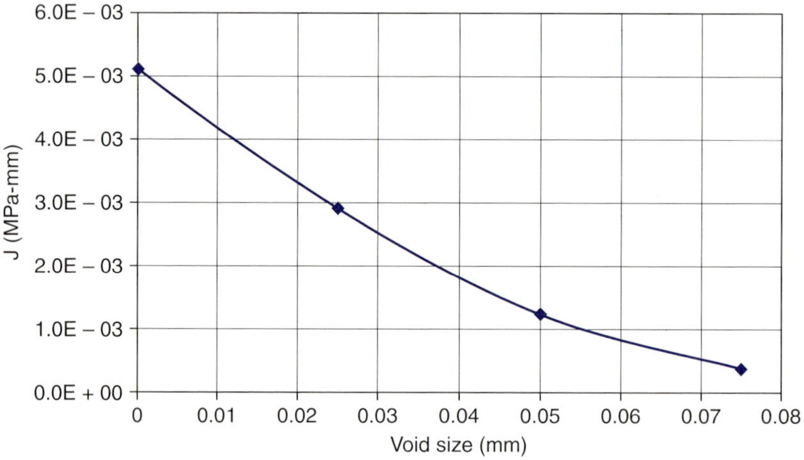

FIGURE 16.31 J-integrals with fixed crack length (0.038 mm) and various void sizes.

solder joint with voids in front of the crack is less likely to fracture than that without voids. Figure 16.30 shows the deformed shape of the WLCSP solder joint with a crack length of 0.038 mm and a 0.05-mm void in front of the crack. This further demonstrates that a crack in a solder joint could be stopped by a void in front of it.

16.3.4 2-D versus 3-D Solutions

The objective of this section is to show that the current 2-D plane-strain simulations are adequate for the purpose of this study, that is, to compare the stress, strain, and J-integral for various void sizes and

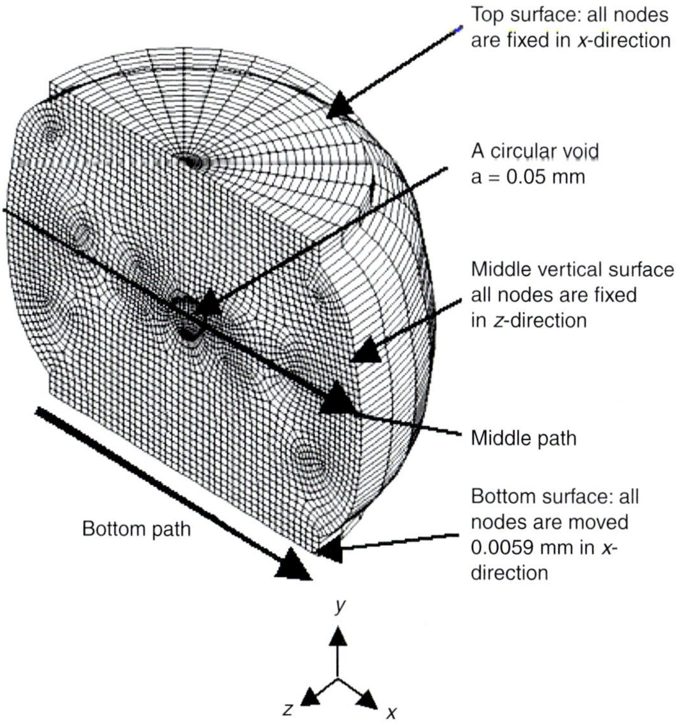

FIGURE 16.32 3-D finite-element model.

crack lengths in the WLCSP solder joints and not to determine the thermal-fatigue life of the solder joints.

Figure 16.32 shows one-half of the corner WLCSP solder joint. There is a void (0.05 mm) right at the center of the corner solder joint. The boundary conditions of the structure, as shown in Fig. 16.32 are: (1) the whole corner solder joint is subjected to a temperature rise, $\Delta T = 130°C$; (2) the middle (vertical) face is fixed in the z-direction; (3) the top face of the corner solder joint is fixed in the x-direction; and (4) the bottom surface of the corner solder joint is subjected to a thermal expansion mismatch displacement = 0.0059 mm = (2.828 mm × 130°C × [18.5×10^{-6}/°C − 2.5×10^{-6}/°C]).

Figures 16.33 and 16.34 show, respectively, the shear stress and shear strain of the 3-D and 2-D solutions through the bottom path of the WLCSP corner solder joint. For both the shear stress and strain, the 2-D solutions compare very well with the 3-D solutions. Figures 16.35 and 16.36 show, respectively, the shear stress and shear strain of the 3-D and 2-D simulations through the middle path with the void of the WLCSP corner solder joint. Again, it can be seen that the 2-D and 3-D solutions agree with each other very well.

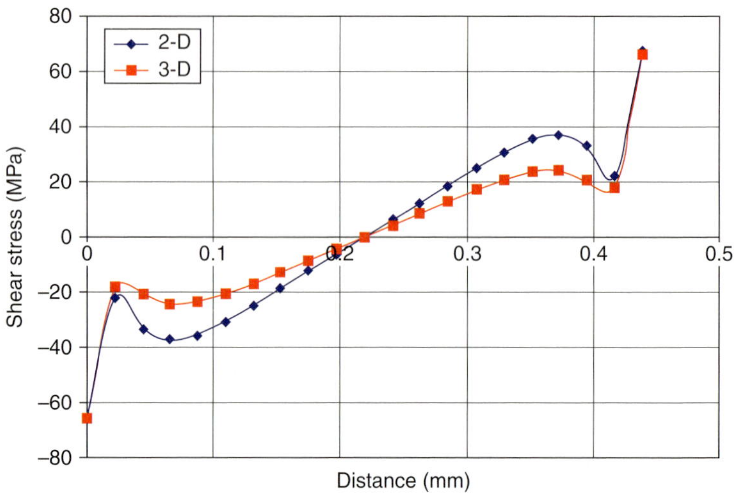

FIGURE 16.33 Shear stress distribution along the bottom path of the solder joint (2-D vs. 3-D).

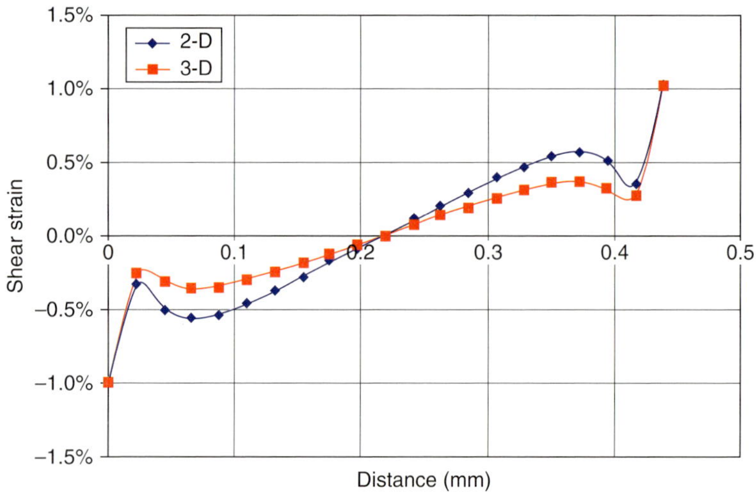

FIGURE 16.34 Shear strain distribution along the bottom path of the solder joint (2-D vs. 3-D).

16.3.5 Summary and Recommendations

Some important results and recommendation are summarized in the following:

1. For the crack lengths under consideration, the J-integral (driving force) increases as the crack length increases and decreases when the crack length is larger than 0.02 mm. Thus,

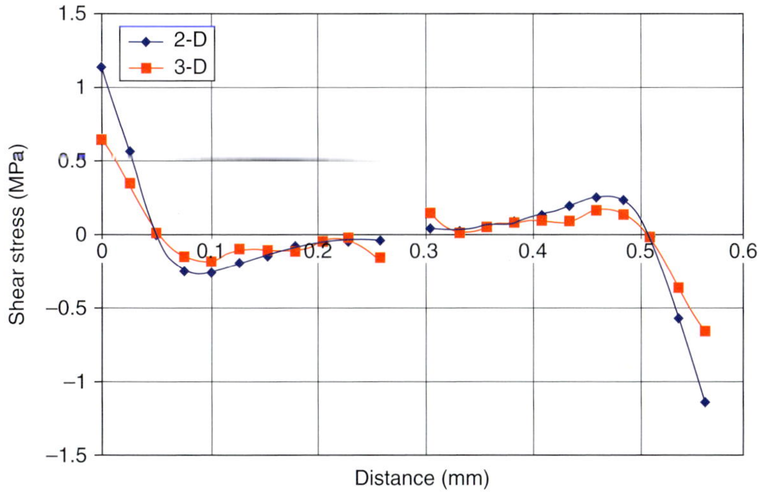

Figure 16.35 Shear stress distribution (2-D vs. 3-D) along the middle path (through the void) of the solder joint.

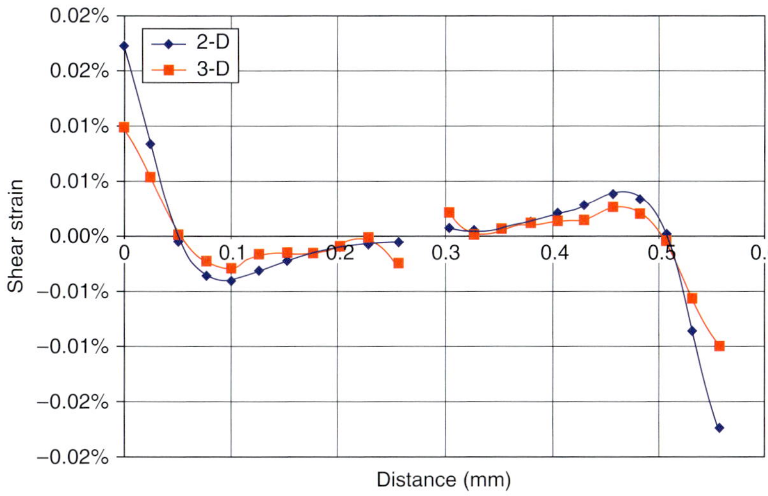

Figure 16.36 Shear strain distribution along the middle path (through the void) of the solder joint (2-D vs. 3-D).

under the fixed temperature rise condition (from room temperature to 100°C), the cracks will be arrested. However, for thermal cycling loadings, a crack most likely would be initiated from the other side of the solder joint, and propagated to meet with the other crack.

2. For the void sizes under consideration, the driving force for the crack with a void in front of it is smaller than that without a void in the corner WLCSP solder joint. This means that the crack with a void is less likely to fracture than the crack without a void. Thus, the cracks in solder joints could be stopped by the voids in front of them.
3. The 2-D and 3-D results (e.g., shear stress and shear strain) agree quite well with each other at various locations in the corner WLCSP solder joint.

16.4 References

1. Lau, J. H., and A. Jean, "Effects of Voids on SMT Solder Joint Reliability." *Proceedings of ASM International Symposium on Microelectronic Packaging Technology: Materials and Processes*, 1989, pp. 177–187.
2. Lau, J. H., S. Goh, G. Tay, C. Jiaa, S. Erasmus, F. Sporon-Fiedler, and S. Pan, "SMT Assembly and Solder Joint Reliability of Bump Chip Carrier (BCC++) Packages," *Proceedings of International Conference on Advances in Packaging*, Singapore, December 2001, pp. 338–344.
3. Lau, J. H., S. Erasmus, and S. Pan, "Effects of Voids on Bump Chip Carrier (BCC++) Solder Joint Reliability," *IEEE Proceedings of Electronic Components & Technology Conference*, San Diego, CA, May 28–31, 2002, pp. 992–1000. Also, accepted for publication in *IEEE Transactions on Components, Packaging, and Technology*.
4. Lau, J. H., and Y. Zou, "Effects of Void, Crack, and PCB Thickness on the Solder Joint Reliability of WLCSP Assemblies," *ASME Paper No. IMECE 2002/EPP-24720*, ASME IMEC, November 2002.
5. Tvergaard, V., "Material Failure by Void Growth to Coalescence," *Advances in Applied Mechanics*, Vol. 27, Pergamon Press, 1989, pp. 83–149.
6. Klein Wassink, R. J., *Soldering in Electronics*, Electrochemical Publications Limited, Ayr, Scotland, 1989, pp. 472–473.
7. Chiu, C., N. Lee, K. Randle, and C. Parrish, "Voiding in BGA at Solder Bumping Stage," *ISHM Proceedings of Microelectronics and Packaging Conference*, Philadelphia, PA, October 1997, pp. 462–471.
8. Hance, W., and N. Lee, "Voiding Mechanisms in SMT," *The 17th Annual Electronics Manufacturing Seminar*, China Lake, 1993.
9. Jo, H., B. Nieman, and N. Lee, "Voiding in Microvia," *IMAPS Proceedings of Microelectronics and Packaging Conference*, Denver, CO, September 2002.
10. O'Hara, W., and N. Lee, "Voiding Mechanism in BGA Assembly," *ISHM Proceedings of Microelectronics and Packaging Conference*, Los Angeles, CA, October 1995, pp. 24–30.
11. Hwang, J. S., *Modern Solder Technology for Competitive Electronics Manufacturing*, McGraw-Hill, New York, 1996.
12. Hwang, J. S., *Environmentally Friendly Electronic Lead-Free Technology*, Electrochemical Publications Limited, Ayr, Scotland, 2001.
13. Tummala, R. R., *Fundamentals of Microsystems Packaging*, McGraw-Hill New York, 2001.
14. Tummala, R. R., E. Rymaszewski, and A. Klopfenstein, *Microelectronics Packaging Handbook*, Chapman & Hall, New York, 1997.
15. Frear, D., H. Morgan, S. Burchett, and J. H. Lau, *The Mechanics of Solder Alloy Interconnects*, Van Nostrand Reinhold, New York, 1994.
16. Lau, J. H., C. P. Wong, N. C. Lee, and S.W.R. Lee, *Electronic Manufacturing with Lead-Free, Halogen-Free, and Conductive Adhesive Materials*, McGraw-Hill, New York, 2003.

17. Lau, J. H., and S.W.R. Lee, *Microvias for Low Cost High Density Interconnects*, McGraw-Hill, New York, 2001.
18. Lau, J. H., *Low Cost Flip Chip Technologies for DCA, WLCSP, and PBGA Assemblies*, McGraw-Hill, New York, 2000.
19. Lau, J. H., and S. W. R. Lee, *Chip Scale Package: Design, Materials, Process, Reliability, and Applications*, McGraw-Hill, New York, 1999.
20. Lau, J. H., C. P. Wong, J. L. Prince, and W. Nakayama, *Electronic Packaging, Design, Materials, Process, and Reliability*, McGraw-Hill, New York, 1998.
21. Lau, J. H., and Y.-H. Pao, *Solder Joint Reliability of BGA, CSP, Flip Chip, and Fine Pitch SMT Assemblies*, McGraw-Hill, New York, 1997.
22. Lau, J. H., *Flip Chip Technologies*, McGraw-Hill, New York, 1996.
23. Lau, J. H., *Ball Grid Array Technology*, McGraw-Hill, New York, 1995.
24. Lau, J. H., *Chip On Board Technologies for Multichip Modules*, Van Nostrand Reinhold, New York, 1994.
25. Lau, J. H., *Handbook of Fine Pitch Surface Mount Technology*, Van Nostrand Reinhold, New York, 1994.
26. Lau, J. H., *Thermal Stress and Strain in Microelectronics Packaging*, Van Nostrand Reinhold, New York, 1993.
27. Lau, J. H., *Handbook of Tape Automated Bonding*, Van Nostrand Reinhold, New York, 1992.
28. Lau, J. H., *Solder Joint Reliability, Theory and Applications*, Van Nostrand Reinhold, New York, 1991.
29. Lau, J. H., C. K Lee, C. Premachandran, A. Yu, *Advanced MEMS Packaging*, McGraw-Hill, New York, 2009.

Index

D

E

G

H

J

K

L

Q

R

S